Real and Abstract Analysis

A modern treatment of the theory
of functions of a real variable

by

Edwin Hewitt
Professor of Mathematics
The University of Washington

and

Karl Stromberg
Associate Professor of Mathematics
The University of Oregon

With 8 Figures

Springer-Verlag New York, Inc. 1965
175 Fifth Avenue, New York, N.Y. 10010

Library of Congress Catalog Card Number 65-26609
Printed in the United States of America

Title No. 1320

Table of Contents

This book is dedicated to

MARSHALL H. STONE

whose precept and example have taught us both.

Preface

This book is first of all designed as a text for the course usually called "theory of functions of a real variable". This course is at present customarily offered as a first or second year graduate course in United States universities, although there are signs that this sort of analysis will soon penetrate upper division undergraduate curricula. We have included every topic that we think essential for the training of analysts, and we have also gone down a number of interesting bypaths. We hope too that the book will be useful as a reference for mature mathematicians and other scientific workers. Hence we have presented very general and complete versions of a number of important theorems and constructions. Since these sophisticated versions may be difficult for the beginner, we have given elementary avatars of all important theorems, with appropriate suggestions for skipping. We have given complete definitions, explanations, and proofs throughout, so that the book should be usable for individual study as well as for a course text.

Prerequisites for reading the book are the following. The reader is assumed to know elementary analysis as the subject is set forth, for example, in TOM M. APOSTOL'S *Mathematical Analysis* [Addison-Wesley Publ. Co., Reading, Mass., 1957], or WALTER RUDIN'S *Principles of Mathematical Analysis* [2nd Ed., McGraw-Hill Book Co., New York, 1964]. There are no other prerequisites for reading the book: we define practically everything else that we use. Some prior acquaintance with abstract algebra may be helpful. The text *A Survey of Modern Algebra*, by GARRETT BIRKHOFF and SAUNDERS MAC LANE [3rd Ed., MacMillan Co., New York, 1965] contains far more than the reader of this book needs from the field of algebra.

Modern analysis draws on at least five disciplines. First, to explore measure theory, and even the structure of the real number system, one must use powerful machinery from the abstract theory of sets. Second, as hinted above, algebraic ideas and techniques are illuminating and sometimes essential in studying problems in analysis. Third, set-theoretic topology is needed in constructing and studying measures. Fourth, the theory of topological linear spaces ["functional analysis"] can often be applied to obtain fundamental results in analysis, with surprisingly little effort. Finally, analysis really is *analysis*. We think that handling inequalities, computing with actual functions, and obtaining actual numbers, is indispensable to the training of every mathematician. All five of these subjects thus find a place in our book. To make the book useful to probabilists, statisticians, physicists, chemists, and engineers, we have included many "applied" topics: Hermite functions; Fourier series and integrals, including PLANCHEREL's theorem and pointwise summability; the strong law of large numbers; a thorough discussion of complex-valued measures on the line. Such applications of the abstract theory are also vital to the pure mathematician who wants to know where his subject came from and also where it may be going.

With only a few exceptions, everything in the book has been taught by at least one of us at least once in our real variables courses, at the Universities of Oregon and Washington. As it stands, however, the book is undoubtedly too long to be covered *in toto* in a one-year course. We offer the following road map for the instructor or individual reader who wants to get to the center of the subject without pursuing byways, even interesting ones.

Chapter One. Sections 1 and 2 should be read to establish our notation. Sections 3, 4, and 5 can be omitted or assigned as outside reading. What is essential is that the reader should have facility in the use of cardinal numbers, well ordering, and the real and complex number fields.

Chapter Two. Section 6 is of course important, but a lecturer should not succumb to the temptation of spending too much time over it. Many students using this text will have already learned, or will be in the process of learning, the elements of topology elsewhere. Readers who are genuinely pressed for time may omit § 6 and throughout the rest of the book replace "locally compact Hausdorff space" by "real line", and "compact Hausdorff space" by "closed bounded subset of the real line". We do not recommend this, but it should at least shorten the reading. We urge everyone to cover § 7 in detail, except possibly for the exercises.

Chapter Three. This chapter is the heart of the book and must be studied carefully. Few, if any, omissions appear possible. Chapter Three is essential for all that follows, barring § 14 and most of § 16.

After Chapter Three has been completed, several options are open. One can go directly to § 21 for a study of product measures and FUBINI's theorem. [The applications of FUBINI's theorem in (21.32) *et seq.* require parts of §§ 13–18, however.] Also §§ 17–18 can be studied immediately after Chapter Three. Finally, of course, one can read §§ 13–22 in order.

Chapter Four. Section 13 should be studied by all readers. Subheads (13.40)–(13.51) are not used in the sequel, and can be omitted if necessary. Section 14 can also be omitted. [While it is called upon later in the text, it is not essential for our main theorems.] We believe nevertheless that § 14 is valuable for its own sake as a basic part of functional analysis. Section 15, which is an exercise in classical analysis, should be read by everyone who can possibly find the time. We use Theorem (15.11) in our proof of the LEBESGUE-RADON-NIKODÝM theorem [§ 19], but as the reader will see, one can get by with much less. Readers who skip § 15 must read § 16 in order to understand § 19.

Chapter Five. Sections 17 and 18 should be studied in detail. They are parts of classical analysis that every student should learn. Of § 19, only subheads (19.1)–(19.24) and (19.35)–(19.44) are really essential. Of § 20, (20.1)–(20.8) should be studied by all readers. The remainder of § 20, while interesting, is peripheral. Note, however, that subheads (20.55)–(20.59) are needed in the refined study of infinite product measures presented in § 22.

Chapter Six. Everyone should read (21.1)–(21.27) at the very least. We hope that most readers will find time to read our presentation of PLANCHEREL's theorem (21.31)–(21.53) and of the HARDY-LITTLEWOOD maximal theorems (21.74)–(21.83). Section 22 is optional. It is essential for all students of probability and in our opinion, its results are extremely elegant. However, it can be sacrificed if necessary.

Occasionally we use phrases like "obvious on a little thought", or "a moment's reflection shows...". Such phrases mean really that the proof is not hard but is clumsy to write out, and we think that more writing would only confuse the matter. We offer a very large number of exercises, ranging in difficulty from trivial to all but impossible. The harder exercises are supplied with hints. Heroic readers may of course ignore the hints, although we think that every reader will be grateful for some of them. Diligent work on a fairly large number of exercises is vital for a genuine mastery of the book: exercises are to a mathematician what CZERNY is to a pianist.

We owe a great debt to many friends. Prof. KENNETH A. ROSS has read the entire manuscript, pruned many a prolix proof, and uncovered myriad mistakes. Mr. LEE W. ERLEBACH has read most of the text and has given us useful suggestions from the student's point of view. Prof. KEITH L. PHILLIPS compiled the class notes that are the skeleton of the book, has generously assisted in preparing the typescript for the printer, and has written the present version of (21.74)–(21.83). Valuable conversations and suggestions have been offered by Professors ROBERT M. BLUMENTHAL, IRVING GLICKSBERG, WILLIAM H. SILLS, DONALD R. TRUAX, BERTRAM YOOD, and HERBERT S. ZUCKERMAN. Miss BERTHA THOMPSON has checked the references. The Computing Center of the University of Oregon and in particular Mr. JAMES H. BJERRING have generously aided in preparing the index. We are indebted to the several hundred students who have attended our courses on this subject and who have suffered, not always in silence, through awkward presentations. We are deeply grateful to Mrs. SHANTI THAYIL, who has typed the entire manuscript with real artistry.

Our thanks are also due to the Universities of Oregon and Washington for exemption from other duties and for financial assistance in the preparation of the manuscript. It is a pleasure to acknowledge the great help given us by Springer-Verlag, in their rapid and meticulous publication of the work.

Seattle, Washington EDWIN HEWITT

Eugene, Oregon KARL R. STROMBERG

July 1965

CHAPTER ONE

Set Theory and Algebra

From the logician's point of view, mathematics is the theory of sets and its consequences. For the analyst, sets and concepts immediately definable from sets are essential tools, and manipulation of sets is an operation he must carry out continually. Accordingly we begin with two sections on sets and functions, containing few proofs, and intended largely to fix notation and terminology and to form a review for the reader in need of one. Sections 3 and 4, on the axiom of choice and infinite arithmetic, are more serious: they contain detailed proofs and are recommended for close study by readers unfamiliar with their contents.

Plainly one cannot study real- and complex-valued functions seriously without knowing what the real and complex number fields are. Therefore, in § 5, we give a short but complete construction of these objects. This section may be read, recalled from previous work, or taken on faith.

This text is *not* rigorous in the sense of proceeding from the axioms of set theory. We believe in sets, and we believe in the rational numbers. Beyond that, we have tried to prove all we say.

§ 1. The algebra of sets

(1.1) The concept of a set. As remarked above, we take the notion of set as being already known. Roughly speaking, a set [collection, assemblage, aggregate, class, family] is any identifiable collection of objects of any sort. We identify a set by stating what its members [elements, points] are. The theory of sets has been described axiomatically in terms of the notion "member of". To build the complete theory of sets from these axioms is a long, difficult process, and it is remote from classical analysis, which is the main subject of the present text. Therefore we shall make no effort to be rigorous in dealing with the concept of sets, but will appeal throughout to intuition and elementary logic. Rigorous treatments of the theory of sets can be found in *Naive Set Theory* by P. Halmos [Princeton, N. J.: D. Van Nostrand Co. 1960] and in *Axiomatic Set Theory* by P. Suppes [Princeton, N. J.: D. Van Nostrand Co. 1960].

(1.2) Notation. We will usually adhere to the following notational conventions. Elements of sets will be denoted by small letters: $a, b, c, \ldots, x, y, z; \alpha, \beta, \gamma, \ldots$ Sets will be denoted by capital Roman letters: $A, B, C, \ldots$ Families of sets will be denoted by capital script letters: $\mathscr{A}, \mathscr{B}, \mathscr{C}, \ldots$ Occasionally we need to consider collections of families of sets. These entities will be denoted by capital Cyrillic letters: Ж, Ч, . . .

A set is often defined by some property of its elements. We will write $\{x : P(x)\}$ [where $P(x)$ is some proposition about x] to denote the set of all x such that $P(x)$ is true. We have done nothing here to sharpen the definition of a set, since "property" and "set" are from one point of view synonymous.

If the object x is an element of the set A, we will write $x \in A$; while $x \notin A$ will mean that x is not in A.

We write $\varnothing$ for the void [empty, vacuous] set; it has no members at all. Thus $\varnothing = \{x : x \text{ is a real number and } x^2 < 0\} = \{x : x \text{ is a unicorn in the Bronx Zoo}\}$, and so on.

For any object x, $\{x\}$ will denote the set whose only member is x. Similarly, $\{x_1, x_2, \ldots, x_n\}$ will denote the set whose members are precisely $x_1, x_2, \ldots, x_n$.

Throughout this text we will adhere to the following notations: N will denote the set $\{1, 2, 3, \ldots\}$ of all positive integers; Z will denote the set of all integers; Q will denote the set of all rational numbers; R will denote the set of all real numbers; and K will denote the set of all complex numbers. We assume a knowledge on the part of the reader of the sets N, Z, and Q. The sets R and K are constructed in § 5.

(1.3) Definitions. Let A and B be sets such that for all x, $x \in A$ implies $x \in B$. Then A is called a *subset of* B and we write $A \subset B$ or $B \supset A$. If $A \subset B$ and $B \subset A$, then we write $A = B$; $A \neq B$ denies $A = B$. If $A \subset B$ and $A \neq B$, we say that A is a *proper subset of* B and we write $A \subsetneqq B$. We note that under this idea of equality of sets, the void set is unique, for if $\varnothing_1$ and $\varnothing_2$ are any two void sets we have $\varnothing_1 \subset \varnothing_2$ and $\varnothing_2 \subset \varnothing_1$.

(1.4) Definitions. If A and B are sets, then we define $A \cup B$ as the set $\{x : x \in A \text{ or } x \in B\}$, and we call $A \cup B$ the *union of A and B*. Let $\mathscr{A}$ be a family of sets; then we define $\bigcup \mathscr{A} = \{x : x \in A \text{ for some } A \in \mathscr{A}\}$. Similarly if $\{A_\iota\}_{\iota \in I}$ is a family of sets indexed by iota, we write $\bigcup_{\iota \in I} A_\iota = \{x : x \in A_\iota \text{ for some } \iota \in I\}$. If $I = N$, the positive integers, $\bigcup_{n \in N} A_n$ will usually be written as $\bigcup_{n=1}^{\infty} A_n$. Other notations, such as $\bigcup_{n=-\infty}^{\infty} A_n$, are self-explanatory.

For given sets A and B, we define $A \cap B$ as the set $\{x : x \in A \text{ and } x \in B\}$, and we call $A \cap B$ the *intersection of A and B*. If $\mathscr{A}$ is any family of sets, we define $\bigcap \mathscr{A} = \{x : x \in A \text{ for all } A \in \mathscr{A}\}$; if $\{A_\iota\}_{\iota \in I}$ is a family of sets indexed by iota, then we write $\bigcap_{\iota \in I} A_\iota = \{x : x \in A_\iota \text{ for all } \iota \in I\}$. The notation $\bigcap_{n=1}^{\infty} A_n$ [and similar notations] have obvious meanings.

Example. If $A_n = \left\{x : x \text{ is a real number, } |x| < \frac{1}{n}\right\}$, $n = 1, 2, 3, \ldots$, then $\bigcap_{n=1}^{\infty} A_n = \{0\}$.

For a set A, the family of all subsets of A is a well-defined family of sets which is known as *the power set of* A and is denoted by $\mathscr{P}(A)$. For example, if $A = \{1, 2\}$, then $\mathscr{P}(A) = \{\varnothing, \{1\}, \{2\}, \{1, 2\}\}$.

(1.5) Theorem. *Let A, B, C be any sets. Then we have:*

(i) $A \cup B = B \cup A$; (i′) $A \cap B = B \cap A$;

(ii) $A \cup A = A$; (ii′) $A \cap A = A$;

(iii) $A \cup \varnothing = A$; (iii′) $A \cap \varnothing = \varnothing$;

(iv) $A \cup (B \cup C) = (A \cup B) \cup C$; (iv′) $A \cap (B \cap C) = (A \cap B) \cap C$;

(v) $A \subset A \cup B$; (v′) $A \cap B \subset A$;

(vi) $A \subset B$ *if and only if* $A \cup B = B$; (vi′) $A \subset B$ *if and only if* $A \cap B = A$.

The proof of this theorem is very simple and is left to the reader.

(1.6) Theorem.

(i) $A \cap (B \cup C) = (A \cap B) \cup (A \cap C)$;

(ii) $A \cup (B \cap C) = (A \cup B) \cap (A \cup C)$.

Proof. These and similar identities may be verified schematically; the verification of (i) follows:

Venn diagrams

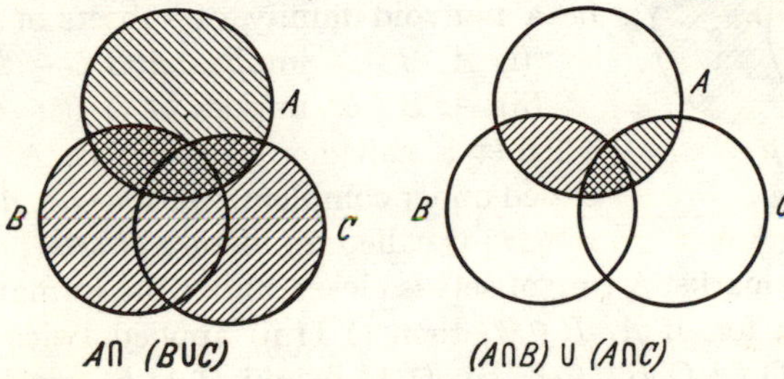

Fig. 1

A similar schematic procedure could be applied to (ii). However, we may use (i) and the previous laws as follows:

$$(A \cup B) \cap (A \cup C) = ((A \cup B) \cap A) \cup ((A \cup B) \cap C)$$
$$= (A \cap A) \cup (B \cap A) \cup (A \cap C) \cup (B \cap C) = A \cup (B \cap C)$$
$$\cup (B \cap A) \cup (A \cap C) = A \cup (B \cap C);$$

the last equality holds because $B \cap A \subset A$ and $A \cap C \subset A$. □[1]

(1.7) Definition. If $A \cap B = \varnothing$, then A and B are said to be *disjoint*. If $\mathscr{A}$ is a family of sets such that each pair of distinct members of $\mathscr{A}$ are disjoint, then $\mathscr{A}$ is said to be *pairwise disjoint*. Thus an indexed family $\{A_\iota\}_{\iota \in I}$ is pairwise disjoint if $A_\iota \cap A_\eta = \varnothing$ whenever $\iota \neq \eta$.

(1.8) Definition. In most of our ensuing discussions the sets in question will be subsets of some fixed "universal" set X. Thus if $A \subset X$, we define

[1] The symbol □ will be used throughout the text to indicate the end of a proof.

the *complement of* A [relative to X] to be the set $\{x : x \in X, x \notin A\}$. This set is denoted by the symbol A'. If there is any possible ambiguity as to which set is the universal set, we will write $X \cap A'$ for A'. Other common notations for what we call A' are $X - A$, $X \setminus A$, $X \sim A$, CA, and A^c; we will use A' exclusively.

(1.9) Theorem [DE MORGAN's laws].

(i) $(A \cup B)' = A' \cap B'$;

(ii) $(A \cap B)' = A' \cup B'$;

(iii) $(\bigcup_{\iota \in I} A_\iota)' = \bigcap_{\iota \in I} A'_\iota$;

(iv) $(\bigcap_{\iota \in I} A_\iota)' = \bigcup_{\iota \in I} A'_\iota$.

The proofs of these identities are easy and are left to the reader.

(1.10) Definition. For sets A and B, the *symmetric difference of* A *and* B is the set $\{x : x \in A$ or $x \in B$ and $x \notin A \cap B\}$, and we write $A \triangle B$ for this set. Note that $A \triangle B$ is the set consisting of those points which are in exactly one of A and B, and that it may also be defined by $A \triangle B = (A \cap B') \cup (A' \cap B)$. The symmetric difference is sketched in Fig. 2

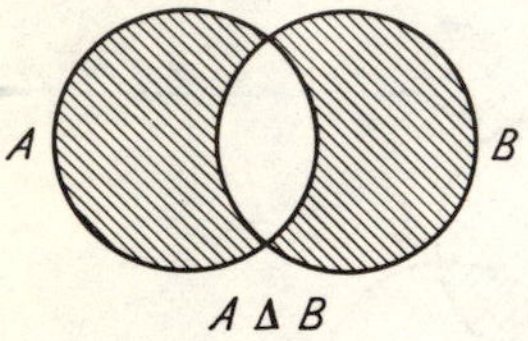

Fig. 2

(1.11) Definition. Let X be a set and let $\mathcal{R}$ be a nonvoid family of subsets of X such that

(i) $A, B \in \mathcal{R}$ implies $A \cup B \in \mathcal{R}$;

(ii) $A, B \in \mathcal{R}$ implies $A \cap B' \in \mathcal{R}$.

Then $\mathcal{R}$ is called a *ring of sets*. A ring of sets closed under complementation [*i.e.* $A \in \mathcal{R}$ implies $A' \in \mathcal{R}$] is called an *algebra of sets*.

(1.12) Remarks. A ring of sets is closed under the formation of finite intersections; for, if $A, B \in \mathcal{R}$, then (1.11.ii) applied twice shows that $A \cap B = A \cap (A \cap B')' \in \mathcal{R}$. By (1.11.i) and (1.11.ii), we have $A \triangle B = (A \cup B) \cap (A \cap B)' \in \mathcal{R}$. Note also that $\varnothing \in \mathcal{R}$ since $\mathcal{R}$ is nonvoid. Also $\mathcal{R}$ is an algebra if and only if $X \in \mathcal{R}$. There are rings of sets which are not algebras of sets; *e.g.*, the family of all finite subsets of N is a ring of sets but not an algebra of sets.

(1.13) Definition. A *σ-ring* [*σ-algebra*] *of sets* is a ring [algebra] of sets $\mathcal{R}$ such that if $\{A_n : n \in N\} \subset \mathcal{R}$, then $\bigcup_{n=1}^{\infty} A_n \in \mathcal{R}$.

Much of measure theory deals with families of sets which form σ-rings or σ-algebras. There are σ-rings which are not σ-algebras, *e.g.*, the family of all countable subsets of an uncountable set. [For the definitions of countable and uncountable, see § 4.]

(1.14) Remarks.[1] There are many axiomatic treatments of rings and algebras of sets, and in fact some very curious entities can be interpreted

[1] This subhead is included only for its cultural interest and may be omitted by anyone who is in a hurry.

as rings or algebras of sets [see (1.25)]. Let B be any set. Suppose that to each $a \in B$ there is assigned a unique element $a^* \in B$ and that to each pair of elements $a, b \in B$ there is assigned a unique element $a \vee b \in B$ such that these operations satisfy

(i) $a \vee b = b \vee a$,

(ii) $a \vee (b \vee c) = (a \vee b) \vee c$,

(iii) $(a^* \vee b^*)^* \vee (a^* \vee b)^* = a$.

Sets B with operations $\vee$ and $*$ [or similar operations] and satisfying axioms equivalent to (i)—(iii) were studied by many writers in the period 1890—1930. They bear the generic name *Boolean algebras*, after the English mathematician GEORGE BOOLE [1815—1864]. The axioms (i)—(iii) were given by the U.S. mathematician E. V. HUNTINGTON [1874—1952] [Trans. Amer. Math. Soc. 5, 288—309 (1904)].

The reader will observe that if a, b are interpreted as sets and $\vee$ and $*$ as union and complementation, then (i)—(iii) are simple identities. Other operations can be defined in a Boolean algebra, *e.g.*, $\wedge$ [the analogue of $\cap$ for sets], which is defined by $a \wedge b = (a^* \vee b^*)^*$. A great deal of effort has been devoted to investigating abstract Boolean algebras. In the 1930's, the contemporary U.S. mathematician M. H. STONE showed that any Boolean algebra can be interpreted as an algebra of sets in the following very precise way [Trans. Amer. Math. Soc. **40**, 37—111 (1936)]. Given any Boolean algebra B, there is a set X, an algebra $\mathscr{R}$ of subsets of X, and a one-to-one mapping τ of B onto $\mathscr{R}$ such that $\tau(a^*) = (\tau(a))'$ [$*$ becomes $'$] and $\tau(a \vee b) = \tau(a) \cup \tau(b)$ [$\vee$ becomes $\cup$]. Thus from the point of view of studying the operations in a Boolean algebra, one may as well study only algebras of sets.

STONE's treatment of the representation of Boolean algebras was based on a slightly different entity, namely, a *Boolean ring*. A Boolean ring is any ring S such that $x^2 = x$ for each $x \in S$. [For the definition of *ring*, see (5.3).]

STONE showed that Boolean algebras and Boolean rings having a multiplicative unit can be identified, and then based his treatment on Boolean rings. More precisely: for every Boolean ring S, there is a ring of sets $\mathscr{R}$ and a one-to-one mapping τ of S onto $\mathscr{R}$ such that

$$\tau(a + b) = \tau(a) \triangle \tau(b)$$

and

$$\tau(ab) = \tau(a) \cap \tau(b) .$$

That is, addition in a Boolean ring corresponds to the symmetric difference, and multiplication to intersection.

Proofs of the above results and a lengthy treatment of Boolean algebras and rings and of algebras and rings of sets can be found in G. BIRKHOFF, *Lattice Theory* [Amer. Math. Soc. Colloquium Publications, Vol. XXV, 2nd edition; Amer. Math. Soc., New York, N. Y., 1948].

(1.15) Exercise. Simplify as much as possible:

(a) $(A \cup (B \cap (C \cup W')))'$;

(b) $((X' \cup Y) \cap (X \cup Y'))'$;

(c) $(A \cap B \cap C) \cup (A' \cap B \cap C) \cup (A \cap B' \cap C) \cup (A \cap B \cap C')$
$\cup (A \cap B' \cap C') \cup (A' \cap B \cap C') \cup (A' \cap B' \cap C)$.

(1.16) Exercise [Poretsky]. Given two sets X and Y, prove that $X = \varnothing$ if and only if $Y = X \triangle Y$.

(1.17) Exercise. Describe in words the sets $\bigcup_{n=1}^{\infty}\left(\bigcap_{k=n}^{\infty} A_k\right)$ and $\bigcap_{n=1}^{\infty}\left(\bigcup_{k=n}^{\infty} A_k\right)$ where $\{A_1, A_2, \ldots, A_k, \ldots\}$ is any family of sets indexed by N. Also prove that the first set is a subset of the second.

(1.18) Exercise. Prove:

(a) $A \triangle (B \triangle C) = (A \triangle B) \triangle C$;

(b) $A \cap (B \triangle C) = (A \cap B) \triangle (A \cap C)$;

(c) $A \triangle A = \varnothing$;

(d) $\varnothing \triangle A = A$.

(1.19) Exercise. Let $\{A_\iota\}_{\iota \in I}$ and $\{B_\iota\}_{\iota \in I}$ be nonvoid families of sets. Prove that

$$\text{(i)} \qquad (\bigcup_\iota A_\iota) \triangle (\bigcup_\iota B_\iota) \subset \bigcup_\iota (A_\iota \triangle B_\iota).$$

Prove by an example that the inclusion may be proper. Can you assert anything about (i) if the $\cup$'s are changed to $\cap$'s?

(1.20) Exercise. For any sets A, B, and C, prove that

$$A \triangle B \subset (A \triangle C) \cup (B \triangle C),$$

and show by an example that the inclusion may be proper.

(1.21) Exercise. Let $\{M_n\}_{n=1}^{\infty}$ and $\{N_n\}_{n=1}^{\infty}$ be families of sets such that the sets N_n are pairwise disjoint. Define $Q_1 = M_1$ and $Q_n = M_n \cap (M_1 \cup \cdots \cup M_{n-1})'$ for $n = 2, 3, \ldots$ Prove that $N_n \triangle Q_n \subset \bigcup_{k=1}^{n} (N_k \triangle M_k)$ $(n = 1, 2, \ldots)$.

(1.22) Exercise. Consider an alphabet with a finite number of letters, say a, where $a > 1$. A *word* in this alphabet is a finite sequence of letters, not necessarily distinct. Two words are equal if and only if they have the same number of letters and if the letters are the same and in the same order. Consider all words of length l, where $l > 1$. How many words of length l have at least two repetitions of a fixed letter? How many have three such repetitions? In how many words of length l do there occur two specified distinct letters?

(1.23) Exercise.

(a) Let A be a finite set, and let $\nu(A)$ denote the number of elements of A: thus $\nu(A)$ is a nonnegative integer. Prove that

$$\nu(A \cup B) = \nu(A) + \nu(B) - \nu(A \cap B).$$

(b) Generalize this identity to $\nu(A \cup B \cup C)$ and to $\nu(A \cup B \cup C \cup D)$.

(c) A university registrar reported that the total enrollment in his university was 10,000 students. Of these, he stated, 2521 were married, 6471 were men, 3115 were over 21 years of age, 1915 were married men, 1873 were married persons over 21 years of age, and 1302 were married men over 21 years of age. Could this have been the case?

(d) Help the registrar. For a student body of 10,000 members, find positive integers for the categories listed in (c) that are consistent with the identity you found in (b).

(1.24) Exercise. Prove that in any Boolean ring we have the identities

(a) $x + x = 0$;

(b) $xy = yx$.

(1.25) Exercise. (a) Let B be the set of all positive integers that divide 30. For $x, y \in B$, let $x \vee y$ be the least common multiple of x and y, and let $x^* = \frac{30}{x}$. Prove that B is a Boolean algebra. Find an algebra of sets that represents B as in (1.14).

(b) Generalize (a), replacing 30 by any square-free positive integer.

(c) Generalize (b) by considering the set B of *all* square-free positive integers, defining $x \vee y$ as the least common multiple of x and y, $x \wedge y$ as the greatest common divisor of x and y, and $x \mathbin{\triangle} y$ as $\frac{x \vee y}{x \wedge y}$. Show that B can be represented as a certain ring of sets but not as an algebra of sets.

§ 2. Relations and functions

In this section we take up the concepts of relation and function, familiar in several forms from elementary analysis. We adopt the currently popular point of view that relations and functions are indistinguishable from their graphs, *i.e.*, they are sets of ordered pairs. As in the case of sets, we content ourselves with a highly informal discussion of the subject.

(2.1) Definition. Let X and Y be sets. The *Cartesian product of X and Y* is the set $X \times Y$ of all ordered pairs (x, y) such that $x \in X$ and $y \in Y$.

We write $(x, y) = (u, v)$ if and only if $x = u$ and $y = v$. Thus $(1, 2) \neq (2, 1)$ while $\{1, 2\} = \{2, 1\}$.

(2.2) Definition. A *relation* is any set of ordered pairs. Thus a relation is any set which is a subset of the Cartesian product of two sets. Observe that $\varnothing$ is a relation.

(2.3) Definitions. Let f be any relation. We define the *domain of f* to be the set $\operatorname{dom} f = \{x : (x, y) \in f \text{ for some } y\}$ and we define the *range of f* to be the set $\operatorname{rng} f = \{y : (x, y) \in f \text{ for some } x\}$. The symbol f^{-1} denotes the *inverse* of f: $f^{-1} = \{(y, x) : (x, y) \in f\}$.

(2.4) Definition. Let f and g be relations. We define the *composition* [*product, iterate* are also used] *of f and g* to be the relation $g \circ f = \{(x, z) :$ for some y, $(x, y) \in f$ and $(y, z) \in g\}$.

The composition of f and g may be void. In fact, $g \circ f \neq \varnothing$ if and only if $(\text{rng} f) \cap (\text{dom} g) \neq \varnothing$.

(2.5) Definition. Let f and g be relations such that $f \subset g$. Then we say that g is an *extension of f* and that f is a *restriction of g*.

We now discuss some special kinds of relations that are needed in the sequel. Wherever convenient, we will use the conventional notation $x f y$ to mean that $(x, y) \in f$.

(2.6) Definition. Let X be a set. An *equivalence relation on X* is any relation $\sim \subset X \times X$ such that, for all x, y, z in X we have:

(i) $x \sim x$ [*reflexive*];

(ii) $x \sim y$ implies $y \sim x$ [*symmetric*];

(iii) $x \sim y$ and $y \sim z$ imply $x \sim z$ [*transitive*].

(2.7) Definitions. Let P be a set. A *partial ordering on P* is any relation $\leqq \subset P \times P$ satisfying

(i) $x \leqq x$ [*reflexive*];

(ii) $x \leqq y$ and $y \leqq x$ imply $x = y$ [*antisymmetric*];

(iii) $x \leqq y$ and $y \leqq z$ imply $x \leqq z$ [*transitive*].

If $\leqq$ also satisfies

(iv) $x, y \in P$ implies $x \leqq y$ or $y \leqq x$ [*trichotomy*],

then $\leqq$ is called a *linear* [also called *simple, complete,* or *total*] *ordering on P*. If $x \leqq y$ and $x \neq y$, we write $x < y$. The expression $x \geqq y$ means $y \leqq x$ and $x > y$ means $y < x$.

If $\leqq$ is a linear ordering such that

(v) $\varnothing \neq A \subset P$ implies there exists an element $a \in A$ such that $a \leqq x$ for each $x \in A$ [a is the *smallest element of A*],

then $\leqq$ is called a *well ordering on P*.

A *partially ordered set* is an ordered pair $(P, \leqq)$ where P is a set and $\leqq$ is a partial ordering on P. If $\leqq$ is a linear ordering, $(P, \leqq)$ is called a *linearly* [*simply, completely, totally*] *ordered set*. If $\leqq$ is a well ordering, then $(P, \leqq)$ is called a *well-ordered set*.

Let P be a linearly ordered set. For $x, y \in P$, we define $\max\{x, y\} = y$ if $x \leqq y$, and $\max\{x, y\} = x$ if $y \leqq x$. For a finite subset $\{x_1, x_2, \ldots, x_n\}$ of P [not all x_j's necessarily distinct], we define $\max\{x_1, x_2, \ldots, x_n\}$ as $\max\{x_n, \max\{x_1, x_2, \ldots, x_{n-1}\}\}$. The expressions $\min\{x, y\}$ and $\min\{x_1, x_2, \ldots, x_n\}$ are defined analogously.

(2.8) Examples. (a) Let $\mathscr{F}$ be any family of sets. Then set inclusion $\subset$ is a partial ordering on $\mathscr{F}$ and $(\mathscr{F}, \subset)$ is a partially ordered set. For short we say that $\mathscr{F}$ is partially ordered by $\subset$. The reader should note that, depending on $\mathscr{F}$, this relation may fail to be a linear ordering; for example, take $\mathscr{F} = \mathscr{P}(\{0, 1\})$.

(b) Let P be the set of all nonnegative rational numbers:

$$P = \{x : x \in Q, x \geqq 0\},$$

and let $\leqq$ be the usual ordering on P. Then $\leqq$ is a linear ordering on P and P has a smallest element 0, but P, with this ordering, is not a well-ordered set, since there are nonvoid subsets of P containing no smallest element. For example, let $A = \{x \in P : x \neq 0\}$. Then $\frac{x}{2} \in A$ whenever $x \in A$, so A contains no smallest element.

(c) The set N of all positive integers with its usual ordering is a linearly ordered set. It is also a well-ordered set. This last assertion is equivalent to PEANO's axiom of mathematical induction.

(2.9) Definition. Let f be a relation and let A be a set. We define the *image of A under f* to be the set

$$f(A) = \{y : (x, y) \in f \text{ for some } x \in A\}.$$

Observe that $f(A) \neq \varnothing$ if and only if $A \cap \operatorname{dom} f \neq \varnothing$. The *inverse image* of A under f is the set $f^{-1}(A)$.

(2.10) Definition. A relation f is said to be *single-valued* if $(x, y) \in f$ and $(x, z) \in f$ imply $y = z$. If f and f^{-1} are both single-valued, then f is called a *one-to-one* relation. The definitions of *many-to-one, one-to-many*, and *many-to-many* relations are analogous.

Single-valued relations play such an important rôle in analysis that we make the following definition.

(2.11) Definition. A single-valued relation is called a *function* [*mapping, transformation, operation, correspondence, application*].

(2.12) Examples. The sine function, $\{(x, \sin x) : x \in R\}$ is many-to-one. The inverse of this function, $\{(\sin x, x) : x \in R\}$, is a one-to-many relation. The relation $\{(x, y) : x, y \in R,\ x^2 + y^2 = 1\}$ is a many-to-many relation. The function $\left\{(x, \tan x) : x \in R,\ -\frac{\pi}{2} < x < \frac{\pi}{2}\right\}$ is a one-to-one function.

(2.13) Definition. Let f be a function and let X and Y denote the domain and range of f, respectively. For $x \in X$, let $f(x)$ denote that unique element of Y such that $(x, f(x)) \in f$. The element $f(x)$ is called the *value of f at x* or the *image of x under f*.

Note that in order to specify a function completely, it is sufficient to specify the domain of the function and the value of the function at each point of its domain.

(2.14) Remark. Referring to (2.9), we observe that if f is a function and A is a set, then $f(A) = \{f(x) : x \in A \cap \operatorname{dom} f\}$ and $f^{-1}(A) = \{x : x \in \operatorname{dom} f,\ f(x) \in A\}$. The reader should verify these statements.

(2.15) Theorem. *Let X and Y be sets and let $f \subset X \times Y$ be a relation. Suppose that $\{A_\iota\}_{\iota \in I}$ is a family of subsets of X and that $\{B_\iota\}_{\iota \in I}$ is a family* *of subsets of Y. For $A \subset X$ we write A' for the complement of A relative to X and for $B \subset Y$ we write B' for the complement of B relative to Y. Then*

(i) $f(\bigcup_{\iota \in I} A_\iota) = \bigcup_{\iota \in I} f(A_\iota)$;

(ii) $f(\bigcap_{\iota \in I} A_\iota) \subset \bigcap_{\iota \in I} f(A_\iota)$.

The followings results are true if f is a function, but may fail for arbitrary relations:

(iii) $f^{-1}(\bigcap_{\iota \in I} B_\iota) = \bigcap_{\iota \in I} f^{-1}(B_\iota)$;

(iv) $f^{-1}(B') = (f^{-1}(B))'$;

(v) $f(f^{-1}(B) \cap A) = B \cap f(A)$.

The proof of this theorem is left to the reader.

(2.16) Remark. From Theorem (2.15) it follows that the domain and range of a one-to-one function cannot be distinguished from each other by any purely set-theoretic properties. If X and Y are sets for which there is a one-to-one function f with domain X and range Y, then for any subset A of X we have $f(A') = f(A)'$. For any family $\{A_\iota\}_{\iota \in I}$ of subsets of X, we have $f(\bigcup_{\iota \in I} A_\iota) = \bigcup_{\iota \in I} f(A_\iota)$ and $f(\bigcap_{\iota \in I} A_\iota) = \bigcap_{\iota \in I} f(A_\iota)$. Similar statements hold for subsets of Y and f^{-1}. Thus, all Boolean operations ($\cup$, $\cap$, $\triangle$, $'$) are preserved under f and f^{-1}.

(2.17) Definition. Let f be a function such that $\operatorname{dom} f = X$ and $\operatorname{rng} f \subset Y$. Then f is said to be a function *from* [*on*] X *into* [*to*] Y and we write $f: X \to Y$. If $\operatorname{rng} f = Y$, we say that f is *onto* Y.

(2.18) Definition. A *sequence* is a function having N, the set of all positive integers, as its domain. If x is a sequence, we will frequently write x_n instead of $x(n)$ for the value of x at n. The value x_n is called the nth *term* of the sequence. The sequence x whose n^{th} term is x_n will be denoted by $(x_n)_{n=1}^{\infty}$ or simply (x_n). A sequence (x_n) is said to be *in* X if $x_n \in X$ for each $n \in N$; we abuse our notation to write $(x_n) \subset X$.

The following theorem will be used several times in the sequel.

(2.19) Theorem. *Let $\mathfrak{F}$ be any family of functions such that $f, g \in \mathfrak{F}$ implies either $f \subset g$ or $g \subset f$,* i.e., *$\mathfrak{F}$ is linearly ordered relative to $\subset$. Let $h = \bigcup \mathfrak{F}$. Then:*

(i) *h is a function;*

(ii) $\operatorname{dom} h = \bigcup\{\operatorname{dom} f : f \in \mathfrak{F}\}$;

(iii) *$x \in \operatorname{dom} h$ implies $h(x) = f(x)$ for each $f \in \mathfrak{F}$ such that $x \in \operatorname{dom} f$;*

(iv) $\operatorname{rng} h = \bigcup\{\operatorname{rng} f : f \in \mathfrak{F}\}$.

Proof. (i) Obviously h is a relation since it is a union of sets of ordered pairs. We need only show that h is single-valued. Let $(x, y) \in h$ and $(x, z) \in h$. Then there exist f and g in $\mathfrak{F}$ such that $(x, y) \in f$ and $(x, z) \in g$. We know that $f \subset g$ or $g \subset f$; say $f \subset g$. Then $(x, y) \in g$ and $(x, z) \in g$. Since g is a function we have $y = z$. Thus h is a function.

The equality (ii) is true because the following statements are pairwise equivalent: $x \in \operatorname{dom} h$; $(x, y) \in h$ for some y; $(x, y) \in f$ for some $f \in \mathfrak{F}$; $x \in \operatorname{dom} f$ for some $f \in \mathfrak{F}$.

Let $x \in \operatorname{dom} h \cap \operatorname{dom} f = \operatorname{dom} f$ where $f \in \mathfrak{F}$. Then $(x, f(x)) \in f \subset h$ and h is single-valued so $h(x) = f(x)$. This proves (iii).

The equality (iv) follows from the previous conclusions and (2.15.i) since $\operatorname{rng} h = h(\operatorname{dom} h) = h(\bigcup\{\operatorname{dom} f : f \in \mathfrak{F}\}) = \bigcup\{h(\operatorname{dom} f) : f \in \mathfrak{F}\}$ $= \bigcup\{f(\operatorname{dom} f) : f \in \mathfrak{F}\} = \bigcup\{\operatorname{rng} f : f \in \mathfrak{F}\}$. □

(2.20) Definition. Let X be any set and E any subset of X. The function ξ_E with domain X and range contained in $\{0, 1\}$ such that

$$\xi_E(x) = \begin{cases} 1 \text{ if } x \in E, \\ 0 \text{ if } x \in X \cap E', \end{cases}$$

is called the *characteristic function of* E. It will always be clear from the context what the domain of ξ_E is. Characteristic functions are very useful in analysis, and will be encountered frequently throughout this text. One particular characteristic function is used so much that it has a special symbol. The *diagonal* D *of* $X \times X$ is defined as $D = \{(x, x) : x \in X\}$. The value of the characteristic function of D at (x, y) is written δ_{xy} and is called KRONECKER's *δ-symbol*. Thus $\delta_{xy} = 1$ if $x = y$ and $\delta_{xy} = 0$ if $x \neq y$; here x and y are arbitrary points in X.

(2.21) Exercise. Prove that $f \circ (g \circ h) = (f \circ g) \circ h$ for all relations f, g, and h.

(2.22) Exercise. Show that the equality $f(f^{-1}(B) \cap A) = B \cap f(A)$ fails for every relation f that is not a function.

(2.23) Exercise. For (a, b) and (c, d) in $N \times N$, define $(a, b) \leqq (c, d)$ if either: $a < c$, or $a = c$ and $b \leqq d$. Prove that, with this relation, $N \times N$ is a well-ordered set.

(2.24) Exercise. Let n be a positive integer and let $P_n = \{k \in N :$ k is a divisor of $n\}$. For $a, b \in P_n$ define $a \precsim b$ to mean that a is a divisor of b, *i.e.*, $a | b$.

(a) Prove that P_n, with $\precsim$, is a partially ordered set.

(b) Find necessary and sufficient conditions on n that P_n be a linearly ordered set.

(2.25) Exercise. Let X be a set with a binary operation p defined on it, *i.e.*, p is a function from $X \times X$ into X. Write $p(x, y) = xy$. Suppose that this operation satisfies

(i) $x(yz) = (xy)z$,

(ii) $xy = yx$,

(iii) $xx = x$,

for all x, y, z in X. Define $\leqq$ on X by $x \leqq y$ if and only if $xy = y$. Prove that: (a) X is a partially ordered set; (b) each pair of elements of X has a least upper bound, *i.e.*, if $x, y \in X$, then there exists $z \in X$ such that $x \leq z$, $y \leq z$, and if $x \leq w$, $y \leq w$, then $z \leq w$.

(2.26) Exercise. Let f be a function from X to Y. Suppose that there is a function g from Y to X such that $f \circ g(y) = y$ for all $y \in Y$ and $g \circ f(x) = x$ for all $x \in X$. Prove that f is a one-to-one function from X onto Y and that $g = f^{-1}$.

§ 3. The axiom of choice and some equivalents

In the study of algebra, analysis, and topology one frequently encounters situations in which the tools of elementary set theory [as they have been informally presented in §§ 1 and 2] are too weak to permit constructions, proofs, or even definitions that one may need. In the early 1900's the German mathematician ERNST ZERMELO propounded an innocent-appearing but actually very strong axiom, called the axiom of choice [*Auswahlpostulat*], which has many important consequences, and which has also excited vigorous controversy. In this section we take up the axiom of choice, establish the equivalence of four other assertions with it, and point out two important applications. Other applications of the axiom of choice will appear throughout the book.

(3.1) Definition. Let $\{A_\iota\}_{\iota \in I}$ be any family of sets. The *Cartesian product* of this family, written $\underset{\iota \in I}{\times} A_\iota$, is the set of all functions x having domain I such that $x_\iota = x(\iota) \in A_\iota$ for each $\iota \in I$. Each such function x is called a *choice function* for the family $\{A_\iota\}_{\iota \in I}$. For $x \in \underset{\iota \in I}{\times} A_\iota$ and $\iota \in I$, the value $x_\iota \in A_\iota$ is known as the ιth *coordinate of* x.

One may ask if there are any choice functions for a given family of sets. Of course if $I = \varnothing$, then the void function $\varnothing$ is a choice function for any family indexed by I. If $I \neq \varnothing$ and $A_\iota = \varnothing$ for some $\iota \in I$, then $\underset{\iota \in I}{\times} A_\iota = \varnothing$. These two special cases are of little interest. In general the question cannot be answered on the basis of the usual axioms of set theory. We will use the following axiom.

(3.2) Axiom of choice. The Cartesian product of any nonvoid family of nonvoid sets is a nonvoid set, *i. e.*, if $\{A_\iota\}_{\iota \in I}$ is a family of sets such that $I \neq \varnothing$ and $A_\iota \neq \varnothing$ for each $\iota \in I$, then there exists at least one choice function for the family $\{A_\iota\}_{\iota \in I}$.

P. J. COHEN has recently proved that this axiom is independent of the other axioms of set theory [Proc. Nat. Acad. Sci. U.S.A. **50**, 1143 — 1148 (1963); **51**, 105—110 (1964)].

(3.3) Definition. Let A and I be sets. We define A^I to be the Cartesian product $\underset{\iota \in I}{\times} A_\iota$, where $A_\iota = A$ for each $\iota \in I$. Thus A^I is the set of all functions f such that $\operatorname{dom} f = I$ and $\operatorname{rng} f \subset A$. If, for some $n \in N$, I is the set $\{1, \ldots, n\}$, then we write $A^I = A^n$. Some authors write A^∞ for A^N.

A typical member of A^n is, to be sure, a function and as such is a set of n ordered pairs. We follow conventional notation, however, and list the values of such a function as an ordered n-tuple. Thus $A^n = \{(a_1, \ldots, a_n) : a_k \in A$ for $k = 1, \ldots, n\}$. Similarly $A^N = \{(a_1, a_2, \ldots) : a_k \in A$ for $k \in N\}$. The set R^n is called *Euclidean n-space* and K^n is called *unitary n-space*.

(3.4) Example. Let $A = \{0, 1\}$. Then A^N is the set of all sequences $a = (a_1, a_2, \ldots, a_n, \ldots)$ where each a_n is 0 or 1. In many ways this set resembles CANTOR's ternary set $P = [0, 1] \cap \left(\left]\frac{1}{3}, \frac{2}{3}\right[\cup \left]\frac{1}{9}, \frac{2}{9}\right[\cup \left]\frac{7}{9}, \frac{8}{9}\right[\cup \cdots\right)'$ [see (6.62) *infra*]. The mapping φ defined by $\varphi(a) = 2 \sum_{n=1}^{\infty} \frac{a_n}{3^n}$ is a one-to-one mapping from A^N onto P. Anticipating future developments, we remark that A^N can be made a metric space by introducing the metric ϱ, where $\varrho(a, b) = \frac{1}{n}$ if $a_1 = b_1$, $a_2 = b_2, \ldots, a_{n-1} = b_{n-1}$, and $a_n \neq b_n$; and $\varrho(a, b) = 0$ if $a = b$. Under this metric on A^N, φ and φ^{-1} are both continuous. The set A^N becomes an Abelian group under the operation $+$ defined by $(a + b)_n = a_n + b_n \pmod 2$ for $n \in N$. [There are many other ways to make A^N into an Abelian group.]

(3.5) Definitions. Let $(P, \leqq)$ be any partially ordered set and let $A \subset P$. An element $u \in P$ is called an *upper bound for A* if $x \leqq u$ for each $x \in A$. An element $m \in P$ is called a *maximal element of P* if $x \in P$ and $m \leqq x$ implies $m = x$. Similarly we define *lower bound* and *minimal element*.[1] A *chain in P* is any subset C of P such that C is linearly ordered under the given order relation $\leqq$ on P.

This terminology of partially ordered sets will often be applied to an arbitrary family of sets. When this is done, it should be understood that the family is being regarded as a partially ordered set under the relation $\subset$ of set inclusion. Thus a maximal member of $\mathscr{A}$ is a set $M \in \mathscr{A}$ such that M is a proper subset of no other member of $\mathscr{A}$ and a chain of sets is a family $\mathscr{B}$ of sets such that $A \subset B$ or $B \subset A$ whenever $A, B \in \mathscr{B}$.

(3.6) Definition. Let $\mathscr{F}$ be a family of sets. Then $\mathscr{F}$ is said to be a *family of finite character* if for each set A we have $A \in \mathscr{F}$ if and only if each finite[2] subset of A is in $\mathscr{F}$.

We shall need the following technical fact.

---

[1] We agree that every element of P is both an upper bound and a lower bound for the void set $\varnothing$; but naturally $\varnothing$ *contains* neither a maximal nor a minimal element.

[2] A set F is said to be *finite* if either $F = \varnothing$ or there exist $n \in N$ and a one-to-one function from $\{1, 2, \ldots, n\}$ onto F. See (4.12).

(3.7) Lemma. *Let $\mathscr{F}$ be a family of finite character and let $\mathscr{B}$ be a chain in $\mathscr{F}$. Then $\cup\mathscr{B} \in \mathscr{F}$.*

Proof. It suffices to show that each finite subset of $\cup\mathscr{B}$ is in $\mathscr{F}$. Let $F = \{x_1, \ldots, x_n\} \subset \cup\mathscr{B}$. Then there exist sets $B_1, \ldots, B_n$ in $\mathscr{B}$ such that $x_j \in B_j$ $(j = 1, \ldots, n)$. Since $\mathscr{B}$ is a chain there is a $j_0 \in \{1, \ldots, n\}$ such that $B_j \subset B_{j_0}$ for each $j = 1, \ldots, n$. Then $F \subset B_{j_0} \in \mathscr{F}$. But $\mathscr{F}$ is of finite character, and so $F \in \mathscr{F}$. □

There are many problems in set theory, algebra, and analysis to which the axiom of choice in the form (3.2) is not immediately applicable, but to which one or another equivalent axiom is applicable at once. We next list four such statements. The names "lemma" and "theorem" are attached to them only for historical reasons, as they are all equivalent to Axiom (3.2).

(3.8) Tukey's Lemma. *Every nonvoid family of finite character has a maximal member.*

(3.9) Hausdorff Maximality Principle. *Every nonvoid partially ordered set contains a maximal chain.*

(3.10) Zorn's Lemma. *Every nonvoid partially ordered set in which each chain has an upper bound has a maximal element.*

(3.11) Well-ordering Theorem [Zermelo]. *Every set can be well ordered*; i.e., *if S is a set, then there exists some well-ordering $\leqq$ on S.*

(3.12) Theorem. *The following five propositions are pairwise equivalent:*

(i) *The axiom of choice;*
(ii) *Tukey's lemma;*
(iii) *The Hausdorff maximality principle;*
(iv) *Zorn's lemma;*
(v) *The well-ordering theorem.*

Proof. We will prove this theorem by showing successively that (i) implies (ii), (ii) implies (iii), (iii) implies (iv), (iv) implies (v), and finally that (v) implies (i). The most difficult of these five proofs is the first.

Suppose that (i) is true and assume that (ii) is false. Then there exists a nonvoid family $\mathscr{F}$ of finite character having no maximal member. For each $F \in \mathscr{F}$, let $\mathscr{A}_F = \{E \in \mathscr{F} : F \subsetneqq E\}$. Then $\{\mathscr{A}_F : F \in \mathscr{F}\}$ is a nonvoid family of nonvoid sets, so by (i) there is a function f defined on $\mathscr{F}$ such that $f(F) \in \mathscr{A}_F$ for each $F \in \mathscr{F}$. Thus we have $F \subsetneqq f(F) \in \mathscr{F}$ for each $F \in \mathscr{F}$.

A subfamily $\mathscr{I}$ of $\mathscr{F}$ will be called *f-inductive* if it has the following three properties:

(1) $\varnothing \in \mathscr{I}$;
(2) $A \in \mathscr{I}$ implies $f(A) \in \mathscr{I}$;
(3) $\mathscr{B}$ a chain $\subset \mathscr{I}$ implies $\cup\mathscr{B} \in \mathscr{I}$.

Since $\mathscr{F}$ is nonvoid, since $\varnothing$ is finite, and since (3.7) holds, the family $\mathscr{F}$ is f-inductive. Let $\mathscr{I}_0 = \cap\{\mathscr{I} : \mathscr{I}$ is f-inductive$\} = \{A \in \mathscr{F} : A \in \mathscr{I}$ for every f-inductive family $\mathcal{I}$}. It is easy to see that $\mathcal{I}_0$ is f-inductive. Thus $\mathcal{I}_0$ is the smallest f-inductive family, so any f-inductive family contained in $\mathcal{I}_0$ must be $\mathcal{I}_0$. We will make heavy use of this fact in proving that $\mathcal{I}_0$ is a chain.

Let $\mathcal{H} = \{A \in \mathcal{I}_0 \colon B \in \mathcal{I}_0$ and $B \subsetneqq A$ imply $f(B) \subset A\}$. We assert that if $A \in \mathcal{H}$ and $C \in \mathcal{I}_0$, then either $C \subset A$ or $f(A) \subset C$. To prove this assertion, let $A \in \mathcal{H}$ and define $\mathcal{G}_A = \{C \in \mathcal{I}_0 \colon C \subset A$ or $f(A) \subset C\}$. It suffices to show that $\mathcal{G}_A$ is f-inductive. Since $\varnothing \in \mathcal{I}_0$ and $\varnothing \subset A$, (1) is satisfied. Let $C \in \mathcal{G}_A$. Then we have either $C \subsetneqq A$, $C = A$, or $f(A) \subset C$. If $C \subsetneqq A$, then $f(C) \subset A$ because $A \in \mathcal{H}$. If $C = A$, then $f(A) \subset f(C)$. If $f(A) \subset C$, then $f(A) \subset f(C)$ because $C \subset f(C)$. Thus in every case $f(C) \in \mathcal{G}_A$ and (2) is satisfied. Next let $\mathcal{B}$ be a chain in $\mathcal{G}_A$. Then either $C \subset A$ for each $C \in \mathcal{B}$, in which case $\bigcup \mathcal{B} \subset A$, or there exists a $C \in \mathcal{B}$ such that $f(A) \subset C \subset \bigcup \mathcal{B}$. Thus $\bigcup \mathcal{B} \in \mathcal{G}_A$ and (3) is satisfied. We conclude that $\mathcal{G}_A$ is f-inductive and so $\mathcal{G}_A = \mathcal{I}_0$.

We next assert that $\mathcal{H} = \mathcal{I}_0$. We prove this by showing that $\mathcal{H}$ is f-inductive. Since $\varnothing$ has no proper subset, $\mathcal{H}$ satisfies (1) vacuously. Next let $A \in \mathcal{H}$ and $B \in \mathcal{I}_0$ be such that $B \subsetneqq f(A)$. Since $B \in \mathcal{I}_0 = \mathcal{G}_A$, we have $B \subset A$ [the inclusion $f(A) \subset B$ being impossible]. If $B \subsetneqq A$, the definition of $\mathcal{H}$ yields $f(B) \subset A \subset f(A)$. If $B = A$, then $f(B) \subset f(A)$. In either case, the inclusion $f(B) \subset f(A)$ obtains, so $f(A) \in \mathcal{H}$ and (2) holds for $\mathcal{H}$. Next, let $\mathcal{B}$ be a chain in $\mathcal{H}$ and let $B \in \mathcal{I}_0$ have the property that $B \subsetneqq \bigcup \mathcal{B}$. Since $B \in \mathcal{I}_0 = \mathcal{G}_A$ for each $A \in \mathcal{B}$, we have either $B \subset A$ for some $A \in \mathcal{B}$ or $f(A) \subset B$ for every $A \in \mathcal{B}$. If the latter alternative were true, we would have

$$B \subsetneqq \bigcup \mathcal{B} \subset \bigcup \{f(A) : A \in \mathcal{B}\} \subset B\,,$$

which is impossible. Thus there is some $A \in \mathcal{B}$ such that $B \subset A$. If $B \subsetneqq A$, then, since $A \in \mathcal{H}$, we have $f(B) \subset A \subset \bigcup \mathcal{B}$. If $B = A$, then $B \in \mathcal{H}$ and $\bigcup \mathcal{B} \in \mathcal{I}_0 = \mathcal{G}_B$. This implies that $f(B) \subset \bigcup \mathcal{B}$ [$\bigcup \mathcal{B} \subset B$ being impossible]. Thus in either case, we have $f(B) \subset \bigcup \mathcal{B}$ and so $\bigcup \mathcal{B} \in \mathcal{H}$. This proves that $\mathcal{H}$ satisfies (3). Therefore $\mathcal{H}$ is f-inductive and $\mathcal{H} = \mathcal{I}_0$.

We conclude from the above arguments that if $A \in \mathcal{I}_0 = \mathcal{H}$ and $B \in \mathcal{I}_0 = \mathcal{G}_A$, then either $B \subset A$ or $A \subset f(A) \subset B$. Accordingly $\mathcal{I}_0$ is a chain. Let $M = \bigcup \mathcal{I}_0$. Since $\mathcal{I}_0$ is f-inductive, (3) implies that $M \in \mathcal{I}_0$. Applying (2), we have $\bigcup \mathcal{I}_0 = M \subsetneqq f(M) \in \mathcal{I}_0$. This contradiction establishes the fact that (i) implies (ii).

We next show that (ii) implies (iii). Let $(P, \leqq)$ be any nonvoid partially ordered set. We want to show that P contains a maximal chain. This follows at once from Tukey's lemma since the family $\mathcal{C}$ of all chains in P is a nonvoid family of finite character [$\varnothing \in \mathcal{C}$ and $\{x\} \in \mathcal{C}$ for each $x \in P$].

To show that (iii) implies (iv), let $(P, \leqq)$ be any nonvoid partially ordered set in which each chain has an upper bound. By (iii) there is a maximal chain $M \subset P$. Let m be an upper bound for M. Then m is a maximal element of P, for if there is an $x \in P$ such that $m \leqq x$ and $m \neq x$, then $M \cup \{x\}$ is a chain which properly includes M, contradicting the maximality of M.

To prove that (iv) implies (v), let S be any nonvoid set and let $\mathscr{Z}$ denote the family of all well-ordered sets $(W, \leqq)$ such that $W \subset S$. For example, $(\{x\}, \{(x, x)\}) \in \mathscr{Z}$ for each $x \in S$. We next introduce an ordering on $\mathscr{Z}$ by defining $(W_1, \leqq_1) \lesssim (W_2, \leqq_2)$ to mean that either $W_1 = W_2$ and $\leqq_1 = \leqq_2$ or there exists $a \in W_2$ such that $W_1 = \{x \in W_2 : x \leqq_2 a, x \neq a\}$ and $\leqq_2$ agrees with $\leqq_1$ on W_1, *i.e.*, $\leqq_1 \subset \leqq_2$. We say that $(W_2, \leqq_2)$ is a *continuation of* $(W_1, \leqq_1)$. The reader should see without difficulty that $\lesssim$ is a partial ordering on $\mathscr{Z}$.

Let us show that ZORN's lemma can be applied to the partially ordered set $(\mathscr{Z}, \lesssim)$. Let $\mathscr{C} = \{(W_\iota, \leqq_\iota)\}_{\iota \in I}$ be any nonvoid chain (relative to $\lesssim$) in $\mathscr{Z}$. Set $W = \bigcup_{\iota \in I} W_\iota$ and $\leqq \; = \bigcup_{\iota \in I} \leqq_\iota$ [recall that each $\leqq_\iota$ is a set of ordered pairs]. We leave it to the reader to prove that $\leqq$ is a linear ordering on W. Let A be a nonvoid subset of W. There exists $\iota \in I$ such that $A \cap W_\iota \neq \varnothing$. Since $(W_\iota, \leqq_\iota)$ is a well-ordered set, there is an element $a \in A \cap W_\iota$ such that $a \leqq_\iota x$ for each $x \in A \cap W_\iota$. Assume that there is an element $b \in A$ such that $b \leqq a$. Then $b \in W_\iota$ and $b \leqq_\iota a$, so $b = a$. Thus A has a smallest element a in $(W, \leqq)$. We conclude that $(W, \leqq) \in \mathscr{Z}$ and is an upper bound for $\mathscr{C}$.

By ZORN's lemma, $\mathscr{Z}$ has a maximal element $(W_0, \leqq_0)$. If $W_0 = S$, then $\leqq_0$ is a well-ordering for S and we are through. Assume that $W_0 \neq S$. Let $z \in S \cap W_0'$. Define $\leqq \; = \; \leqq_0 \cup \{(x, z) : x \in W_0 \cup \{z\}\}$ on $W_0 \cup \{z\}$, *i.e.*, we place z after everything in W_0. Then $(W_0 \cup \{z\}, \leqq) \in \mathscr{Z}$. This contradicts the maximality of $(W_0, \leqq_0)$, and so we have proved that $W_0 = S$.

It remains only to show that (v) implies (i). Let $\{A_\iota\}_{\iota \in I}$ be any nonvoid family of nonvoid sets. Let $S = \bigcup_{\iota \in I} A_\iota$. Let $\leqq$ be a well-ordering for S. For each $\iota \in I$, let $f(\iota)$ be the smallest member of A_ι relative to the well-ordering $\leqq$. Then f is a choice function for the family $\{A_\iota\}_{\iota \in I}$. □

It is frequently useful to make definitions or carry out constructions by well ordering a certain set W and making the definition or construction at $a \in W$ depend upon what has been defined or done at all of the predecessors of a in the well-ordering. The general form of this process is described in (3.13) and (3.14) below.

(3.13) Definition. Let $(W, \leqq)$ be a well-ordered set and let $a \in W$. The set $I(a) = \{x \in W : x \leqq a, x \neq a\}$ is called the *initial segment of* W *determined by* a.

(3.14) Theorem [Principle of Transfinite Induction]. *Let* $(W, \leqq)$ *be a well-ordered set and let* $A \subset W$ *be such that* $a \in A$ *whenever* $I(a) \subset A$. *Then* $A = W$.

Proof. Assume that $W \cap A' \neq \varnothing$ and let a be the smallest member of $W \cap A'$. Then we have $I(a) \subset A$, so $a \in A$. But $a \in W \cap A'$. □

The axiom of choice does not perhaps play a central rôle in analysis, but when it is needed, it is needed most urgently. We shall encounter several such situations in our subsequent study of measure theory and linear functionals. To give an immediate and important application of the axiom of choice, we will prove from TUKEY's lemma that every vector space contains a basis. Exact definitions follow.

(3.15) Definition. A *vector space* [*linear space*] is an ordered triple $(X, \cdot, F)$ where X is an additive Abelian group[1], F is a field, and $\cdot$ is a function from $F \times X$ into X, whose value at (α, x) is denoted αx, such that for $\alpha, \beta \in F$ and $x, y \in X$ we have

(i) $\alpha(x + y) = \alpha x + \alpha y$;

(ii) $(\alpha + \beta)x = \alpha x + \beta x$;

(iii) $\alpha(\beta x) = (\alpha\beta)x$;

(iv) $1x = x$, where 1 is the multiplicative identity of F.

The members of X are called *vectors* and the members of F are called *scalars*. The operation $\cdot$ is called *scalar multiplication*. For short we say that X is a *vector space over the field* F.

(3.16) Remarks. In a vector space we have $0x = \alpha 0 = 0$ because $0x = (0 + 0)x = 0x + 0x$ and $\alpha 0 = \alpha(0 + 0) = \alpha 0 + \alpha 0$. Also $\alpha \neq 0$ and $x \neq 0$ imply $\alpha x \neq 0$, since otherwise we would have $x = 1x = (\alpha^{-1}\alpha)x = \alpha^{-1}(\alpha x) = \alpha^{-1}0 = 0$.

(3.17) Examples. (a) Let F be any field, let $n \in N$, and let $X = F^n$. For $\boldsymbol{x} = (x_1, \ldots, x_n)$ and $\boldsymbol{y} = (y_1, \ldots, y_n)$ in X and $\alpha \in F$ define $\boldsymbol{x} + \boldsymbol{y} = (x_1 + y_1, \ldots, x_n + y_n)$ and $\alpha\boldsymbol{x} = (\alpha x_1, \ldots, \alpha x_n)$. Then X is a vector space over F.

(b) Let F be any field, let A be any nonvoid set, and let $X = F^A$. For $f, g \in X$ and $\alpha \in F$ define $(f + g)(x) = f(x) + g(x)$ and $(\alpha f)(x) = \alpha f(x)$ for all $x \in A$. Then X is a vector space over F. Note that (a) is the special case of (b) in which $A = \{1, \ldots, n\}$.

(c) Let $X = R$ with its usual addition and let $F = Q$. For $x \in R$ and $\alpha \in Q$ let αx be the usual product in R. Then R is a vector space over Q.

(3.18) Definition. Let X be a vector space over F. A subset A of X is said to be *linearly independent* [*over* F] if for every finite subset $\{x_1, x_2, \ldots, x_n\}$ of distinct elements of A and every sequence $(\alpha_1, \alpha_2, \ldots, \alpha_n)$ of elements of F, the equality $\sum_{k=1}^{n} \alpha_k x_k = 0$ implies the equalities

[1] The reader will find a discussion of groups, rings, and fields in § 5.

$\alpha_1 = \alpha_2 = \cdots = \alpha_n = 0$.[1] A nonvoid linearly independent set B such that $B \subsetneqq E \subset X$ implies that E is not linearly independent is called a *Hamel basis* [or merely *basis*] for X over F. Thus a Hamel basis is a maximal linearly independent set.

(3.19) Theorem. *Every vector space with at least two elements contains a Hamel basis.*

Proof. Let X be a vector space with at least two elements. Let $x \neq 0$ in X. Then (3.16) shows that $\{x\}$ is a linearly independent set. Thus the family $\mathscr{F}$ of all linearly independent subsets of X is nonvoid. The definition of linear independence shows at once that $\mathscr{F}$ is of finite character. TUKEY's lemma proves that $\mathscr{F}$ contains a maximal member, *i.e.*, X contains a basis. □

(3.20) Theorem. *Let X be a vector space over a field F and let B be a Hamel basis for X over F. Then for each $x \in X$ there exists a unique function α from B into F such that $\alpha(b) = 0$ except for finitely many $b \in B$ and $x = \sum_{b \in B} \alpha(b)\, b$,* i.e., *$x$ can be expressed in just one way as a finite linear combination of members of B.*

Proof. Let $x \in X$. If $x \in B$, define $\alpha(x) = 1$ and $\alpha(b) = 0$ for $b \in B$, $b \neq x$. Then $\sum_{b \in B} \alpha(b)\, b = 1x = x$. Suppose $x \notin B$. Then $B \cup \{x\}$ is not linearly independent, so there is a finite set $\{x, x_1, x_2, \ldots, x_n\} \subset B \cup \{x\}$ and a finite sequence $(\beta, \beta_1, \ldots, \beta_n) \subset F$ not all 0, such that $\beta x + \beta_1 x_1 + \cdots + \beta_n x_n = 0$. Since B is independent, we see at once that $\beta \neq 0$. Therefore $x = -\beta^{-1}\beta_1 x_1 - \cdots - \beta^{-1}\beta_n x_n$. Now define $\alpha(x_j) = -\beta^{-1}\beta_j$ $(j = 1, \ldots, n)$ and $\alpha(b) = 0$ for $b \in B \cap \{x_1, \ldots, x_n\}'$. Then $x = \sum_{b \in B} \alpha(b)\, b$. This proves the existence statement.

To prove uniqueness, suppose that $\sum_{b \in B} \alpha_1(b)\, b = \sum_{b \in B} \alpha_2(b)\, b$. Then $\sum_{b \in B} (\alpha_1(b) - \alpha_2(b))\, b = 0$, and this is a finite linear combination of elements of B. By independence, $\alpha_1(b) - \alpha_2(b) = 0$ for each $b \in B$ and therefore the two functions α_1 and α_2 are the same. □

(3.21) Exercise. Given a nonvoid set A and a field F, let $\mathfrak{L}$ be the subset of F^A consisting of those functions f for which the set $\{a \in A : f(a) \neq 0\}$ is finite. Let the linear operations in $\mathfrak{L}$ be as in (3.17.b). Prove that $\mathfrak{L}$ is a vector space over F. Prove that every vector space is isomorphic *qua* vector space with some vector space $\mathfrak{L}$.[2]

(3.22) Exercise. Prove that if P is a set and $\leqq$ is a partial ordering on P, then there exists a linear ordering $\leqq_0$ on P such that $\leqq \subset \leqq_0$.

---

[1] Note that $\varnothing$ is linearly independent.

[2] Let X_1 and X_2 be linear spaces over F. An isomorphism τ of X_1 onto X_2 is a one-to-one mapping of X_1 onto X_2 such that $\tau(x + y) = \tau(x) + \tau(y)$ and $\tau(\alpha x) = \alpha\tau(x)$ for all $x, y \in X_1$ and all $\alpha \in F$. Isomorphic linear spaces cannot be told apart by any linear space properties.

(3.23) Exercise. Let $(L, \leqq)$ be a linearly ordered set. Prove that there exists a set $W \subset L$ such that $\leqq$ well orders W and such that for each $x \in L$ there is a $y \in W$ for which $x \leqq y$.

(3.24) Exercise. Let G be a group and let H be an Abelian subgroup of G. Prove that there exists a maximal Abelian subgroup J of G such that $H \subset J$; *i.e.*, J is Abelian, but no subgroup J^* such that $J \subsetneqq J^*$ is Abelian.

(3.25) Exercise. Prove that the following assertion is equivalent to the axiom of choice: If A and B are nonvoid sets and f is a function from A onto B, then there exists a function g from B into A such that $g(y) \in f^{-1}(y)$ for each $y \in B$.

(3.26) Exercise. Let X be a vector space over a field F. Let A be a nonvoid linearly independent subset of X and let S be a subset of X such that each element of X is a finite linear combination of elements of S. [The set S is said to *span* X.] Suppose that $A \subset S$. Prove that X has a Hamel basis B such that $A \subset B \subset S$.

§ 4. Cardinal numbers and ordinal numbers

As noted in (2.16), two sets which can be placed in a one-to-one correspondence cannot be told apart by any purely set-theoretic properties although, of course, they may be quite different entities. This observation leads us to the following definition.

(4.1) Definition. With every set A we associate a symbol, called the *cardinal number of* A, such that two sets A and B have the same symbol attached to them if and only if there exists a one-to-one function f with $\operatorname{dom} f = A$ and $\operatorname{rng} f = B$. We will write $A \sim B$ to mean that such a one-to-one function exists. If $A \sim B$, we say that A and B are *equivalent* [*equipollent*, *equipotent*, have the *same cardinality*, have the *same power*]. We write $\overline{\overline{A}}$ to denote the cardinal number of A. Thus $\overline{\overline{A}} = \overline{\overline{B}}$ if and only if $A \sim B$.

(4.2) Examples. Some sets are so commonly encountered that we name their cardinal numbers by special symbols. Thus $\overline{\overline{\varnothing}} = 0$, $\overline{\overline{\{1, 2, \ldots, n\}}} = n$ for each $n \in N$, $\overline{\overline{N}} = \aleph_0$ [read "aleph nought"], and $\overline{\overline{R}} = \mathfrak{c}$ [for *continuum*].

(4.3) Remark. The reader will easily verify, by considering the identity, inverse, and composite functions, that set equivalence, as defined in (4.1), is reflexive, symmetric, and transitive. This fact makes Definition (4.1) reasonable and also extremely useful.

(4.4) Remark. Our definition of cardinal number is somewhat vague since, among other things, it is not made clear what these "symbols" are to be. Some such vagueness is inevitable because of our intuitive approach to set theory. However our definition is adequate for our purposes. In one version of axiomatic set theory, the cardinal number of a set is taken to be a very specific well-ordered set, *viz.*, the smallest ordinal number that is equivalent to the given set.

We next define an order relation for cardinal numbers.

(4.5) Definition. Let $\mathfrak{u}$ and $\mathfrak{v}$ be cardinal numbers and let U and V be sets such that $\bar{\bar{U}} = \mathfrak{u}$ and $\bar{\bar{V}} = \mathfrak{v}$. We write $\mathfrak{u} \leqq \mathfrak{v}$ or $\mathfrak{v} \geqq \mathfrak{u}$ to mean that U is equivalent to some subset of V. One sees by considering composite functions that this definition is unambiguous. We write $\mathfrak{u} < \mathfrak{v}$ or $\mathfrak{v} > \mathfrak{u}$ to mean that $\mathfrak{u} \leqq \mathfrak{v}$ and $\mathfrak{u} \neq \mathfrak{v}$.

(4.6) Theorem. *Let* $\mathfrak{u}$, $\mathfrak{v}$, *and* $\mathfrak{w}$ *be cardinal numbers. Then:*

(i) $\mathfrak{u} \leqq \mathfrak{u}$;

(ii) $\mathfrak{u} \leqq \mathfrak{v}$ *and* $\mathfrak{v} \leqq \mathfrak{w}$ *imply* $\mathfrak{u} \leqq \mathfrak{w}$.

Proof. Exercise.

(4.7) Theorem [SCHRÖDER-BERNSTEIN]. *If* $\mathfrak{u}$ *and* $\mathfrak{v}$ *are cardinal numbers such that* $\mathfrak{u} \leqq \mathfrak{v}$ *and* $\mathfrak{v} \leqq \mathfrak{u}$, *then* $\mathfrak{u} = \mathfrak{v}$.

Proof. Let U and V be sets such that $\bar{\bar{U}} = \mathfrak{u}$ and $\bar{\bar{V}} = \mathfrak{v}$. By hypothesis there exist one-to-one functions f and g such that $\operatorname{dom} f = U$, $\operatorname{rng} f \subset V$, $\operatorname{dom} g = V$, and $\operatorname{rng} g \subset U$. Define a function φ on $\mathscr{P}(U)$ into $\mathscr{P}(U)$ by the following rule:

$$\varphi(E) = U \cap [g(V \cap (f(E))')]' . \tag{1}$$

It is easy to see that

$$E \subset F \subset U \quad \text{implies} \quad \varphi(E) \subset \varphi(F) . \tag{2}$$

Define $\mathscr{D} = \{E \in \mathscr{P}(U) : E \subset \varphi(E)\}$. Notice that $\varnothing \in \mathscr{D}$. Next let $D = \bigcup \mathscr{D}$. Since $E \subset D$ for each $E \in \mathscr{D}$, (2) implies that $E \subset \varphi(E) \subset \varphi(D)$ for each $E \in \mathscr{D}$. Therefore $D \subset \varphi(D)$. Applying (2) again, we have $\varphi(D) \subset \varphi(\varphi(D))$ so $\varphi(D) \in \mathscr{D}$. Thus we have the reversed inclusion $D = \bigcup \mathscr{D} \supset \varphi(D)$, so that $\varphi(D) = D$. According to (1), this means that

$$D = U \cap [g(V \cap (f(D))')]' .$$

Thus $U \cap D' = g(V \cap (f(D))')$. It follows that the function h defined on U by

$$h(x) = \begin{cases} f(x) & \text{for} \quad x \in D , \\ g^{-1}(x) & \text{for} \quad x \in U \cap D' , \end{cases}$$

is one-to-one and onto V. □

The proof of the Schröder-Bernstein theorem does not require the axiom of choice. Also it does not tell us all that we would like to know about comparing cardinal numbers: it merely asserts that $\mathfrak{u} < \mathfrak{v}$ and $\mathfrak{v} < \mathfrak{u}$ cannot occur. To prove that all pairs of cardinals are actually comparable, as we do in (4.8), the axiom of choice is needed.

(4.8) Theorem. *Let* $\mathfrak{u}$ *and* $\mathfrak{v}$ *be cardinal numbers. Then either* $\mathfrak{u} \leqq \mathfrak{v}$ *or* $\mathfrak{v} \leqq \mathfrak{u}$.

Proof. Let U and V be sets such that $\overline{\overline{U}} = \mathfrak{u}$ and $\overline{\overline{V}} = \mathfrak{v}$. Let $\mathfrak{F}$ denote the family of all one-to-one functions f such that $\operatorname{dom} f \subset U$ and $\operatorname{rng} f \subset V$. It is easily seen that $\mathfrak{F}$ is a family of finite character so, by TUKEY's lemma (3.8), $\mathfrak{F}$ contains a maximal member h. We assert that either $\operatorname{dom} h = U$ or $\operatorname{rng} h = V$. Assume that this is false. Then there exist $x \in U \cap (\operatorname{dom} h)'$ and $y \in V \cap (\operatorname{rng} h)'$. But then $h \cup \{(x,y)\} \in \mathfrak{F}$, contradicting the maximality of h. Thus our assertion is true. If $\operatorname{dom} h = U$, then h shows that $\mathfrak{u} \leqq \mathfrak{v}$. If $\operatorname{rng} h = V$, then h^{-1} shows that $\mathfrak{v} \leqq \mathfrak{u}$. □

(4.9) Theorem. *The ordering $\leqq$ for cardinal numbers makes any set of cardinal numbers a linearly ordered set.*

Proof. This theorem is just a summary of Theorems (4.6), (4.7), and (4.8). □

Our next theorem shows that there is no largest cardinal number.

(4.10) Theorem [CANTOR]. *Let U be any set. Then $\overline{\overline{U}} < \overline{\overline{\mathscr{P}(U)}}$.*

Proof. We suppose that $U \neq \varnothing$, since $\overline{\overline{\mathscr{P}(\varnothing)}} = 1 > 0 = \overline{\overline{\varnothing}}$. Let $\mathfrak{u} = \overline{\overline{U}}$ and $\mathfrak{v} = \overline{\overline{\mathscr{P}(U)}}$. The function f defined on U by $f(x) = \{x\} \in \mathscr{P}(U)$ is one-to-one, so $\mathfrak{u} \leqq \mathfrak{v}$. Assume that $\mathfrak{u} = \mathfrak{v}$. Then there exists a [one-to-one] function h such that $\operatorname{dom} h = U$ and $\operatorname{rng} h = \mathscr{P}(U)$. Define

$$S = \{x \in U : x \notin h(x)\}.$$

Since $S \subset U$ [perhaps $S = \varnothing$], we have $S \in \mathscr{P}(U)$. Thus, because h is onto $\mathscr{P}(U)$, there exists an element $a \in U$ such that $h(a) = S$. There are only two alternatives: either $a \in S$ or $a \notin S$. If $a \in S$, then, by the definition of S, we have $a \notin h(a) = S$. Therefore $a \notin S$. But S is the set $h(a)$, so $a \notin h(a)$, which implies that $a \in S$. This contradiction shows that $\mathfrak{u} \neq \mathfrak{v}$, and so we have proved that $\mathfrak{u} < \mathfrak{v}$. □

(4.11) Remark. Intuitive set theory suffers from the presence of several well-known paradoxes. These known paradoxes are avoided in axiomatic set theory by the elimination of "sets" that are "too large". For example, let C be the "set" of all cardinal numbers. For each $\mathfrak{a} \in C$, let $A_{\mathfrak{a}}$ be a set such that $\overline{\overline{A}}_{\mathfrak{a}} = \mathfrak{a}$. Define $B = \bigcup \{A_{\mathfrak{a}} : \mathfrak{a} \in C\}$. Let $\mathfrak{b} = \overline{\overline{B}}$. Since $A_{\mathfrak{a}} \subset B$ we have $\mathfrak{a} \leqq \mathfrak{b}$ for every cardinal number $\mathfrak{a}$. This conclusion is incompatible with Theorem (4.10). The trouble is that the "set" C is "too large". It is indeed *very* large. We shall have no occasion in this book to consider such large sets.

(4.12) Definition. A set S is said to be *finite* if either $S = \varnothing$ or $\overline{\overline{S}} = n = \overline{\overline{\{1, 2, \ldots, n\}}}$ for some $n \in N$. Any set that is not finite is said to be *infinite*.

Definitions of "finite" and "infinite" that make no mention of the natural numbers have been given by TARSKI and DEDEKIND. We state them in the form of a theorem.

(4.13) Theorem. *Let S be a set. Then:*

(i) [TARSKI] *the set S is finite if and only if each nonvoid family of subsets of S has a minimal member;*

(ii) [DEDEKIND] *the set S is infinite if and only if S is equivalent to some proper subset of itself.*

Proof. Exercise. [Use (4.15).]

(4.14) Definition. A set S is called *countable* if either S is finite or $\overline{\overline{S}} = \overline{\overline{N}} = \aleph_0$. Any set that is not countable is *uncountable*. A set S is *countably infinite* [or *denumerable*] if $\overline{\overline{S}} = \aleph_0$. If S is countably infinite and f is a one-to-one function from N onto S, then the sequence (x_n) where $x_n = f(n)$ is called an *enumeration* of S. Note that $x_n \neq x_m$ if $n \neq m$.

(4.15) Theorem. *Every infinite set has a countably infinite subset.*

Proof. Let A be any infinite set. We show by induction that for each $n \in N$ there exists a set $A_n \subset A$ such that $\overline{\overline{A}}_n = n$. Indeed $A \neq \varnothing$ so there exists an $A_1 \subset A$. If $A_n \subset A$ and $\overline{\overline{A}}_n = n$, then, since A is infinite, there exists an element $x \in A \cap A_n'$. Letting $A_{n+1} = A_n \cup \{x\}$, we have $A_{n+1} \subset A$ and $\overline{\overline{A}}_{n+1} = n + 1$.

Next let $\{A_n\}_{n \in N}$ be any family of subsets of A as described above. [Notice the use of the axiom of choice in selecting this family.] For each $n \in N$, define

$$B_n = A_{2^n} \cap \left(\bigcup_{k=0}^{n-1} A_{2^k}\right)' .$$

Then the family $\{B_n\}_{n \in N}$ is a pairwise disjoint family of subsets of A, and for each $n \in N$ we have

$$\overline{\overline{B}}_n \geqq \overline{\overline{A}}_{2^n} - \sum_{k=0}^{n-1} \overline{\overline{A}}_{2^k} = 2^n - \sum_{k=0}^{n-1} 2^k = 2^n - (2^n - 1) = 1 ,$$

so each B_n is nonvoid. Apply the axiom of choice to $\{B_n\}_{n \in N}$ to get a choice function f. Then f is a one-to-one mapping of N into A, so rng$f$ is a countably infinite subset of A. □

(4.16) Corollary. *If $\mathfrak{a}$ is any infinite cardinal number*, i.e., *the cardinal number of an infinite set, then $\aleph_0 \leqq \mathfrak{a}$.*

(4.17) Theorem. *Any subset of a countable set is countable.*

Proof. Let A be any countable set and let $B \subset A$. If B is finite, there is nothing to prove. Thus suppose that B is infinite. Then A is countably infinite. Let (a_n) be an enumeration of A. Define a one-to-one function f from N onto B recursively as follows:

$f(1) = a_{n_1}$ where n_1 is the smallest $n \in N$ such that $a_n \in B$; $f(k + 1) = a_{n_{k+1}}$ where n_{k+1} is the smallest $n \in N$ such that $a_n \in B \cap \{a_1, a_2, \ldots, a_{n_k}\}'$. □

(4.18) Theorem. *The Cartesian product $N \times N$ is a countable set.*

Proof. We must show that $N \sim N \times N$. One way to do this is to define the mapping f from $N \times N$ onto N by $f(m, n) = 2^{m-1}(2n - 1)$.

Since each positive integer is a power of 2 [possibly the 0th power] times an odd integer, f is onto N. We see that f is one-to-one, for otherwise there would be an integer which is both even and odd. □

(4.19) Lemma. *If A is any nonvoid countable set, then there exists a mapping from N onto A.*

Proof. Since A is countable, there exists a one-to-one mapping g from A into N. Let $a \in A$. Define f on N by

$$f(n) = \begin{cases} g^{-1}(n) & \text{for} \quad n \in \operatorname{rng} g\,, \\ a & \text{for} \quad n \notin \operatorname{rng} g\,. \end{cases} \quad \square$$

(4.20) Lemma. *If A and B are two nonvoid sets and if there is a mapping f from A onto B, then $\overline{\overline{A}} \geqq \overline{\overline{B}}$.*

Proof. Let g be a choice function for the family $\{f^{-1}(b)\}_{b \in B}$. Then g is a one-to-one mapping from B into A. □

(4.21) Theorem. *The union of any countable family of countable sets is a countable set*, i.e., *if $\{A_i\}_{i \in I}$ is a family of sets such that I is countable and each A_i is countable, then $A = \bigcup_{i \in I} A_i$ is countable.*

Proof. Let $\{A_i\}_{i \in I}$ be as in the theorem. We obviously may, and do, suppose that I and each A_i are nonvoid. Apply Lemma (4.19) to obtain mappings f_i and g such that $\operatorname{dom} f_i = \operatorname{dom} g = N$, $\operatorname{rng} g = I$, and $\operatorname{rng} f_i = A_i$ for all $i \in I$. Now define h on $N \times N$ by $h(m, n) = f_{g(m)}(n)$. Then h is onto A. It follows from (4.20) and (4.18) that

$$\overline{\overline{A}} \leqq \overline{\overline{N \times N}} = \aleph_0\,.$$

By (4.16), A is countable. □

(4.22) Corollary. *Each of the following sets is countable:*

(i) *Z, the set of all integers;*

(ii) *Q, the set of all rational numbers.*

Proof. We have

$$Z = N \cup \{0\} \cup \{-n : n \in N\}$$

and

$$Q = \bigcup_{n=1}^{\infty} \left\{\frac{m}{n} : m \in Z\right\}. \quad \square$$

We next introduce arithmetical operations for cardinal numbers. We will show that the arithmetic of infinite cardinals is quite simple.

(4.23) Definition. Let $\mathfrak{a}$ and $\mathfrak{b}$ be cardinal numbers and let A and B be sets for which $\overline{\overline{A}} = \mathfrak{a}$ and $\overline{\overline{B}} = \mathfrak{b}$. If $A \cap B = \varnothing$, we define $\mathfrak{a} + \mathfrak{b} = \overline{\overline{A \cup B}}$. We define $\mathfrak{a}\mathfrak{b} = \overline{\overline{A \times B}}$ and $\mathfrak{a}^{\mathfrak{b}} = \overline{\overline{(A^B)}}$.

It is easy to show that these are unambiguous definitions. Also we hasten to point out that $\mathfrak{a} + \mathfrak{b}$ is always defined since it is always possible to find appropriate sets A and B that are disjoint. In fact if $A \cap B \neq \varnothing$, then define $A_0 = \{(a, 0) : a \in A\}$ and $B_0 = \{(b, 1) : b \in B\}$ to obtain $A \sim A_0$, $B \sim B_0$, and $A_0 \cap B_0 = \varnothing$.

(4.24) Theorem. *Let* $\mathfrak{u}, \mathfrak{v}$, *and* $\mathfrak{w}$ *be any three cardinal numbers. Then:*

(i) $\mathfrak{u} + (\mathfrak{v} + \mathfrak{w}) = (\mathfrak{u} + \mathfrak{v}) + \mathfrak{w}$;

(ii) $\mathfrak{u} + \mathfrak{v} = \mathfrak{v} + \mathfrak{u}$;

(iii) $\mathfrak{u}(\mathfrak{v} + \mathfrak{w}) = \mathfrak{u}\mathfrak{v} + \mathfrak{u}\mathfrak{w}$;

(iv) $\mathfrak{u}(\mathfrak{v}\mathfrak{w}) = (\mathfrak{u}\mathfrak{v})\mathfrak{w}$;

(v) $\mathfrak{u}\mathfrak{v} = \mathfrak{v}\mathfrak{u}$;

(vi) $\mathfrak{u}^{\mathfrak{v}}\mathfrak{u}^{\mathfrak{w}} = \mathfrak{u}^{\mathfrak{v}+\mathfrak{w}}$;

(vii) $\mathfrak{u}^{\mathfrak{w}}\mathfrak{v}^{\mathfrak{w}} = (\mathfrak{u}\mathfrak{v})^{\mathfrak{w}}$;

(viii) $(\mathfrak{u}^{\mathfrak{v}})^{\mathfrak{w}} = \mathfrak{u}^{\mathfrak{v}\mathfrak{w}}$;

(ix) $\mathfrak{u} \leqq \mathfrak{v}$ *implies* $\mathfrak{u} + \mathfrak{w} \leqq \mathfrak{v} + \mathfrak{w}$;

(x) $\mathfrak{u} \leqq \mathfrak{v}$ *implies* $\mathfrak{u}\mathfrak{w} \leqq \mathfrak{v}\mathfrak{w}$;

(xi) $\mathfrak{u} \leqq \mathfrak{v}$ *implies* $\mathfrak{u}^{\mathfrak{w}} \leqq \mathfrak{v}^{\mathfrak{w}}$;

(xii) $\mathfrak{u} \leqq \mathfrak{v}$ *implies* $\mathfrak{w}^{\mathfrak{u}} \leqq \mathfrak{w}^{\mathfrak{v}}$.

Proof. All twelve of these conclusions are proved by defining appropriate one-to-one mappings. As a sample we prove (viii) while the remaining eleven are left as exercises.

Let U, V, and W be sets such that $\overline{\overline{U}} = \mathfrak{u}$, $\overline{\overline{V}} = \mathfrak{v}$, and $\overline{\overline{W}} = \mathfrak{w}$. We must show that $(U^V)^W \sim U^{V \times W}$. To do this we define a mapping φ on $(U^V)^W$ by the rule

$$\varphi(f) = g \in U^{V \times W}$$

where

$$g(y, z) = (f(z))(y) \in U \quad \text{for} \quad (y, z) \in V \times W .$$

Now φ is onto $U^{V \times W}$ since if $g \in U^{V \times W}$, we define f to be that function on W whose value at $z \in W$ is that function on V which assigns to each $y \in V$ the value $g(y, z) \in U$. Then $\varphi(f) = g$. To see that φ is one-to-one, suppose that $f_1 \neq f_2$ in $(U^V)^W$. Then there is a $z_0 \in W$ such that $f_1(z_0) \neq f_2(z_0)$. Since these two functions on V are different, there must be a $y_0 \in V$ such that $f_1(z_0)(y_0) \neq f_2(z_0)(y_0)$. Thus $[\varphi(f_1)](y_0, z_0) \neq [\varphi(f_2)](y_0, z_0)$ so $\varphi(f_1)$ and $\varphi(f_2)$ are different functions. □

(4.25) Theorem. *If* $\mathfrak{a}$ *is any cardinal number, then* $\mathfrak{a} < 2^{\mathfrak{a}}$.

Proof. Let A be a set such that $\overline{\overline{A}} = \mathfrak{a}$. We know (4.10) that $\mathfrak{a} < \overline{\overline{\mathscr{P}(A)}}$ and that $2^{\mathfrak{a}} = \overline{\overline{(\{0, 1\}^A)}}$. It suffices to show that $\mathscr{P}(A) \sim \{0, 1\}^A$. Define φ on $\mathscr{P}(A)$ by

$$\varphi(E) = \xi_E \in \{0, 1\}^A \quad \text{for} \quad E \subset A ,$$

where as in (2.20)

$$\xi_E(x) = \begin{cases} 1 & \text{for} \quad x \in E , \\ 0 & \text{for} \quad x \in A \cap E' . \end{cases} \quad \square$$

We next consider the cardinal number $\mathfrak{c} = \overline{\overline{R}}$. The reader is invited to look ahead to § 5 for a detailed construction of R and for the relevant properties of R that we use here.

(4.26) Theorem. *Let*

$$]0, 1[= \{x \in R : 0 < x < 1\},$$
$$[0, 1[= \{x \in R : 0 \leqq x < 1\},$$
$$[0, 1] = \{x \in R : 0 \leqq x \leqq 1\}.$$

Then $\overline{\overline{]0, 1[}} = \overline{\overline{[0, 1[}} = \overline{\overline{[0, 1]}} = \mathfrak{c}$.

Proof. The function f defined by $f(x) = \frac{1 - 2x}{x(1 - x)}$ is a one-to-one mapping of $]0, 1[$ onto R. Therefore $\overline{\overline{]0, 1[}} = \overline{\overline{R}} = \mathfrak{c}$. The rest follows from the inequalities $\mathfrak{c} = \overline{\overline{]0, 1[}} \leqq \overline{\overline{[0, 1[}} \leqq \overline{\overline{[0, 1]}} \leqq \overline{\overline{R}} = \mathfrak{c}$ and the Schröder-Bernstein theorem. □

(4.27) Theorem. $2^{\aleph_0} = \mathfrak{c}$.

Proof. Let $A = \{0, 1\}^N$. Definition (4.23) shows that $\overline{\overline{A}} = 2^{\aleph_0}$. By (4.26), $\overline{\overline{[0, 1[}} = \mathfrak{c}$. Define f on A by $f(\varphi) = \sum_{n=1}^{\infty} \frac{\varphi(n)}{3^n}$. Then [see (5.40)] f is a one-to-one mapping of A into $[0, 1[$, and so $2^{\aleph_0} \leqq \mathfrak{c}$. For each $x \in [0, 1[$ there is a unique representation of x in the form

$$x = \sum_{n=1}^{\infty} \frac{x_n}{2^n}$$

where each x_n is 0 or 1 and $x_n = 0$ for infinitely many $n \in N$: see (5.40). Define g on $[0, 1[$ into A by $g(x) = \varphi$ where $\varphi(n) = x_n$ for each $n \in N$. Then g is a one-to-one mapping, so that $\mathfrak{c} \leqq 2^{\aleph_0}$. Now apply the Schröder-Bernstein theorem. □

We next point out a few curious arithmetical properties of infinite cardinal numbers. First we need a lemma.

(4.28) Lemma. *If D is any infinite set and F is any finite set such that $D \cap F = \varnothing$, then $\overline{\overline{D}} = \overline{\overline{D \cup F}}$.*

Proof. Let $F = \{y_1, y_2, \ldots, y_n\}$ where $y_i \neq y_j$ for $i \neq j$ and let $C = \{x_j : j \in N\}$ be a countably infinite subset of D where $x_i \neq x_j$ for $i \neq j$ (4.15). Define f from D onto $D \cup F$ by

$$f(x) = \begin{cases} y_j & \text{for } x = x_j,\ 1 \leqq j \leqq n, \\ x_{j-n} & \text{for } x = x_j,\ j > n, \\ x & \text{for } x \in D \cap C'. \end{cases}$$

Then f is one-to-one. □

(4.29) Theorem. *Let $\mathfrak{a}$ be any infinite cardinal number. Then $\mathfrak{a} + \mathfrak{a} = \mathfrak{a}$.*

Proof. Let A be any set such that $\overline{\overline{A}} = \mathfrak{a}$. Let $B = A \times \{0, 1\}$. Then $B = \{(a, 0) : a \in A\} \cup \{(a, 1) : a \in A\}$ so, by Definition (4.23), we have $\overline{\overline{B}} = \mathfrak{a} + \mathfrak{a}$. Let $\mathfrak{F}$ denote the set of all one-to-one functions f such that $\operatorname{dom} f \subset A$ and $\operatorname{rng} f = (\operatorname{dom} f) \times \{0, 1\}$. Since A is infinite, there exists a countably infinite set C such that $C \subset A$ (4.15). In view of (4.21), we see that $C \times \{0, 1\}$ is also countably infinite. Hence there is a one-to-one function f with $\operatorname{dom} f = C$ and $\operatorname{rng} f = C \times \{0, 1\}$. This proves that $\mathfrak{F} \neq \varnothing$. Partially order $\mathfrak{F}$ by $\subset$. According to the Hausdorff Maximality Principle (3.9), $\mathfrak{F}$ contains a maximal chain $\mathfrak{C}$. Let $g = \bigcup \mathfrak{C}$. It is easily checked that $g \in \mathfrak{F}$. Let $D = \operatorname{dom} g$. The existence of the function g shows that $\overline{\overline{D}} = \overline{\overline{D}} + \overline{\overline{D}}$. Thus, to complete the proof, it suffices to show that $\overline{\overline{D}} = \mathfrak{a}$. Let $E = A \cap D'$. If E is finite, our Lemma (4.28) shows that $\overline{\overline{D}} = \overline{\overline{D \cup E}} = \mathfrak{a}$. If E is infinite, let G be a countably infinite subset of E. Let f be any one-to-one mapping of G onto $G \times \{0, 1\}$. Then $h = f \cup g \in \mathfrak{F}$ and $g \subsetneqq h$. This contradicts the maximality of $\mathfrak{C}$. Therefore E is finite and $\overline{\overline{D}} = \mathfrak{a}$. □

(4.30) Corollary. *If* $\mathfrak{a}$ *is any infinite cardinal number and* $\mathfrak{b}$ *is any cardinal number such that* $\mathfrak{b} \leqq \mathfrak{a}$, *then* $\mathfrak{a} + \mathfrak{b} = \mathfrak{a}$.

Proof. Since $\mathfrak{b} \leqq \mathfrak{a}$, we have $\mathfrak{a} \leqq \mathfrak{a} + \mathfrak{b} \leqq \mathfrak{a} + \mathfrak{a} = \mathfrak{a}$. □

(4.31) Theorem. *If* $\mathfrak{a}$ *is any infinite cardinal number, then* $\mathfrak{a}^2 = \mathfrak{a}\mathfrak{a} = \mathfrak{a}$.

Proof. Let A be any set such that $\overline{\overline{A}} = \mathfrak{a}$. Let $\mathfrak{F}$ denote the set of all one-to-one functions f such that $\operatorname{dom} f \subset A$ and $\operatorname{rng} f = (\operatorname{dom} f) \times (\operatorname{dom} f)$. Since A contains a countably infinite subset (4.15) and since $\aleph_0 \aleph_0 = \aleph_0$ (4.18), we see that $\mathfrak{F} \neq \varnothing$. As in (4.29), we use the Hausdorff Maximality Principle to prove that $\mathfrak{F}$ contains a maximal member g. Let $D = \operatorname{dom} g$. Then the existence of g shows that $D \sim D \times D$. To finish the proof we need only show that $\overline{\overline{D}} = \mathfrak{a}$. Let $E = A \cap D'$ and let $\mathfrak{d} = \overline{\overline{D}}$. If $\overline{\overline{E}} \leqq \mathfrak{d}$, then (4.30) shows that $\mathfrak{d} = \mathfrak{d} + \overline{\overline{E}} = \overline{\overline{D \cup E}} = \overline{\overline{A}} = \mathfrak{a}$. The only other possibility is that $\mathfrak{d} < \overline{\overline{E}}$ (4.8). Assume that this is the case. Then there is a set $G \subset E$ such that $\overline{\overline{G}} = \mathfrak{d}$. Since $D \sim D \times D$, we know that $\mathfrak{d}^2 = \mathfrak{d}$. Thus $\overline{\overline{D \times G}} = \overline{\overline{G \times D}} = \overline{\overline{G \times G}} = \mathfrak{d}$. We appeal to (4.29) to see that $\mathfrak{d} = \mathfrak{d} + \mathfrak{d} + \mathfrak{d}$. It follows that $\overline{\overline{(D \times G) \cup (G \times D) \cup (G \times G)}} = \mathfrak{d} = \overline{\overline{G}}$. Consequently there exists a one-to-one function f from G onto $(D \times G) \cup (G \times D) \cup (G \times G)$. Define $h = f \cup g$. Then h is a one-to-one correspondence between $D \cup G$ and $(D \cup G) \times (D \cup G)$. Thus we have $h \in \mathfrak{F}$. Since $g \subsetneqq h$, we have contradicted the maximality of g. Consequently $\overline{\overline{E}} \leqq \mathfrak{d}$ and $\mathfrak{d} = \mathfrak{a}$. The accompanying figure may be helpful.

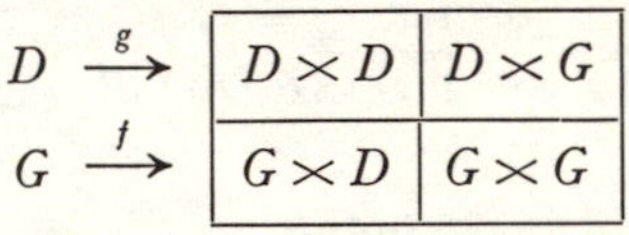

□

(4.32) Corollary. *If* $\mathfrak{a}$ *is an infinite cardinal number and* $\mathfrak{b}$ *is a cardinal number such that* $0 < \mathfrak{b} \leqq \mathfrak{a}$, *then* $\mathfrak{a}\mathfrak{b} = \mathfrak{a}$.

Proof. We have $\mathfrak{a} \leqq \mathfrak{a}\mathfrak{b} \leqq \mathfrak{a}\mathfrak{a} = \mathfrak{a}$. □

(4.33) Exercise. Prove that our ordering and our arithmetical operations for cardinal numbers agree on the set N with the usual ordering and arithmetical operations for positive integers.

(4.34) Exercise. Let $\mathfrak{a}$ be any cardinal number such that $2 \leqq \mathfrak{a} \leqq \mathfrak{c}$. Prove that $\mathfrak{a}^{\aleph_0} = \mathfrak{c}$ and that $\mathfrak{a}^{\mathfrak{c}} = 2^{\mathfrak{c}}$.

(4.35) Exercise. Let A be any infinite set and let $\mathscr{F}$ be the family of all finite subsets of A. Prove that $\overline{\overline{\mathscr{F}}} = \overline{\overline{A}}$.

(4.36) Exercise. Let A be any infinite set such that $\overline{\overline{A}} \leqq \mathfrak{c}$, and let $\mathscr{C}$ denote the family of all countable subsets of A. Prove that $\overline{\overline{\mathscr{C}}} = \mathfrak{c}$.

(4.37) Exercise. Let W be a set and suppose that $\varphi \subset W \times W$ is a relation such that (W, φ) and (W, φ^{-1}) are both well-ordered sets. Prove that W is finite.

(4.38) Exercise. Without appealing to the axiom of choice or its equivalents prove that $\mathfrak{c}^2 = \mathfrak{c}$. [Hint. Use decimal representations of real numbers.]

(4.39) Exercise [König]. Let I be a nonvoid set and let $\{A_\iota\}_{\iota \in I}$ and $\{B_\iota\}_{\iota \in I}$ be families of sets such that $\overline{\overline{A}}_\iota < \overline{\overline{B}}_\iota$ for each $\iota \in I$. Let $A = \bigcup_{\iota \in I} A_\iota$ and $B = \underset{\iota \in I}{\times} B_\iota$. Prove that $\overline{\overline{A}} < \overline{\overline{B}}$. [For each $\iota \in I$, let π_ι denote the projection mapping of B onto B_ι, *i.e.*, $\pi_\iota(b) = b_\iota$ for each $b \in B$. Let f be any mapping from A into B. Then $\pi_\iota \circ f(A_\iota) \neq B_\iota$ (4.20), so that there exists $c_\iota \in B_\iota \cap [\pi_\iota \circ f(A_\iota)]'$ for each $\iota \in I$. It follows that the element $c \in B$ having c_ι as ι^{th} coordinate is not in rng f. Thus there is no mapping of A onto B.]

We next present a brief introduction to the theory of ordinal numbers. The chief distinction between cardinal numbers and ordinal numbers is that each set has a cardinal number while only well-ordered sets have ordinal numbers. There may be many essentially different ways to well order a given set. Each of these ways has its own ordinal number even though the set has only one cardinal number. For example, we can well order N in the following two ways:

$$1 < 2 < 3 < \cdots,$$

$$2 < 3 < 4 < \cdots < 1.$$

The first of these is the usual ordering and the second of these is the same except that 1 has been removed from the beginning and placed at the end. These well orderings are different since the second has a last element while the first has no last element. We need some precise definitions.

(4.40) Definition. Let A and B be linearly ordered sets. An *order isomorphism from A onto B* is a one-to-one function f from A onto B such that $x \leqq y$ in A implies $f(x) \leqq f(y)$ in B. We write $A \approx B$ to mean that such an order isomorphism exists. It is easy to see that the relation $\approx$ is reflexive, symmetric, and transitive. With every linearly ordered set A we associate a symbol, called the *order type of A*, such that two linearly ordered sets A and B have the same symbol attached to them if and only if $A \approx B$. If $A \approx B$, we say that A and B are *order isomorphic* or *have the same order type*. We write $\operatorname{ord} A$ to denote the order type of A. If, in particular, A is well ordered, we call $\operatorname{ord} A$ an *ordinal number*.

(4.41) Examples. Let $\{1, 2, \ldots, n\}$ $(n \in N)$, N, and Q have their usual orderings. We write $\operatorname{ord} \varnothing = 0$, $\operatorname{ord}\{1, 2, \ldots, n\} = n$, $\operatorname{ord} Q = \eta$, and $\operatorname{ord} N = \omega$. Thus 0, n, and ω are ordinal numbers, but η is *not* an ordinal number, since Q is far from being well ordered.

(4.42) Definition. Let A be a linearly ordered set and let $x \in A$. The *initial segment of A determined by x* is the set $A_x = \{y \in A : y < x\}$. If α and β are ordinal numbers and A and B are well-ordered sets such that $\operatorname{ord} A = \alpha$ and $\operatorname{ord} B = \beta$, we write $\alpha < \beta$ to mean that there is an $x \in B$ such that $A \approx B_x$. We write $\alpha \leqq \beta$ to mean that either $\alpha < \beta$ or $\alpha = \beta$. It is easy to verify that this definition of ordering for ordinal numbers does not depend on the particular sets A and B that are used, but only upon their order types.

Let us investigate the properties of this ordering.

(4.43) Theorem. *If A is a well-ordered set and f is an order isomorphism from A into A, then $x \leqq f(x)$ for each $x \in A$.*

Proof. Assume that there is an x in A such that $f(x) < x$ and let a be the smallest such x. Then $f(a) < a$ so $f(f(a)) < f(a)$. But this contradicts the minimality of a. □

(4.44) Theorem. *Let A and B be well-ordered sets. Then:*

(i) *A is order isomorphic to no initial segment of A;*

(ii) *$A_x \approx A_y$ for some $x, y \in A$ implies $x = y$;*

(iii) *if $A \approx B$, there exists exactly one order isomorphism from A onto B.*

Proof. Assume that there are an $x \in A$ and an order isomorphism f from A onto A_x. Then by (4.43), we have $x \leqq f(x)$. But $f(x) \in A_x$, so that $f(x) < x$. This contradiction proves (i).

Now suppose that $A_x \approx A_y$ for some $x, y \in A$. Assume that $x \neq y$. We may suppose that $x < y$. Then A_x is an initial segment of the well-ordered set A_y. As shown in (i), this is impossible, and so (ii) is established.

Suppose that f and g are order isomorphisms from A onto B. Then $h = f^{-1} \circ g$ is an order isomorphism of A onto A. Applying (4.43), we have $x \leqq h(x)$ for each $x \in A$, *i. e.*, $f(x) \leqq g(x)$ for each $x \in A$. Interchanging the rôles of f and g in this argument, we obtain $g(x) \leqq f(x)$ for each $x \in A$. It follows that $f(x) = g(x)$ for each $x \in A$, *i. e.*, $f = g$. □

(4.45) Theorem. *Let α and β be ordinal numbers. Then exactly one of the following three alternatives obtains: $\alpha < \beta$, $\alpha = \beta$, $\beta < \alpha$.*

Proof. Theorem (4.44) shows that at most one of these alternatives can prevail. We now show that at least one of them must.

Let A and B be well-ordered sets such that $\operatorname{ord} A = \alpha$ and $\operatorname{ord} B = \beta$. Let $\mathfrak{F}$ denote the family of all mappings f such that f is an order isomorphism from either an initial segment of A or A itself onto either an initial segment of B or B itself [we may obviously suppose that $A \neq \varnothing \neq B$]. If a is the least element of A and b is the least element of B, then $\{(a, b)\} \in \mathfrak{F}$, and so $\mathfrak{F} \neq \varnothing$. By the Hausdorff Maximality Principle, there is a maximal chain $\mathfrak{C} \subset \mathfrak{F}$. [Actually $\mathfrak{C} = \mathfrak{F}$, but we do not need this fact.] Let $h = \bigcup \mathfrak{C}$. It is easy to check that h belongs to $\mathfrak{F}$. If $\operatorname{dom} h$ and $\operatorname{rng} h$ are initial segments A_x and B_y of A and B, respectively, then $h \cup \{(x, y)\}$ can be adjoined to $\mathfrak{C}$, and this violates the maximality of $\mathfrak{C}$. Thus we have either $\operatorname{dom} h = A$ or $\operatorname{rng} h = B$. If $\operatorname{dom} h = A$, then either $\operatorname{rng} h = B$ [*i. e.*, $\alpha = \beta$], or $\operatorname{rng} h$ is an initial segment of B [*i. e.*, $\alpha < \beta$]. If $\operatorname{dom} h \neq A$, then $\operatorname{dom} h$ is an initial segment of A and $\operatorname{rng} h = B$, and so the existence of h^{-1} shows in this case that $\beta < \alpha$. □

(4.46) Corollary. *With the ordering defined in* (4.42), *any set of ordinal numbers is linearly ordered.*

(4.47) Theorem. *Let α be any ordinal number >0 and let P_α denote the set of all ordinal numbers $<\alpha$. Then P_α, with the order relation of* (4.42), *is a well-ordered set and* $\operatorname{ord} P_\alpha = \alpha$.

Proof. Let $\beta \in P_\alpha$ and let A and B be well-ordered sets such that $\operatorname{ord} A = \alpha$ and $\operatorname{ord} B = \beta$. Since $\beta < \alpha$, there is an $x \in A$ such that $A_x \approx B$. In view of (4.44.ii), this x is uniquely determined by β. Thus define $\varphi(\beta) = x$. The reader should have no difficulty in verifying that this defines an order isomorphism φ from P_α onto A. Thus P_α is well ordered and $\operatorname{ord} P_\alpha = \operatorname{ord} A = \alpha$. □

(4.48) Theorem. *Let $\mathfrak{a}$ be a cardinal number. Then there exists an ordinal number α such that $\overline{\overline{P}}_\alpha = \mathfrak{a}$.*

Proof. Let A be any set such that $\overline{\overline{A}} = \mathfrak{a}$. According to the Well-ordering Theorem (3.11), there is a well ordering on A, making A a well-ordered set. Let $\alpha = \operatorname{ord} A$. Then (4.47) $A \approx P_\alpha$. Consequently $A \sim P_\alpha$ and $\overline{\overline{P}}_\alpha = \overline{\overline{A}} = \mathfrak{a}$. □

(4.49) Theorem. *There is a smallest ordinal number Ω such that P_Ω is uncountable. The set P_Ω has the following properties:*

(i) *P_Ω is well ordered;*

(ii) *$\alpha \in P_\Omega$ implies P_α is countable;*

(iii) *P_Ω is uncountable;*

(iv) *$C \subset P_\Omega$ and C countable imply there is a $\beta \in P_\Omega$ such that $\alpha \leqq \beta$ for each $\alpha \in C$.*

Proof. Choose an ordinal number γ such that $\overline{\overline{P}}_\gamma = \mathfrak{c}$. If each member of P_γ has only countably many predecessors, set $\Omega = \gamma$. Otherwise some members of P_γ have uncountably many predecessors and we let Ω be the smallest of these (4.47). Conclusion (i) follows at once from (4.47). Conclusions (ii) and (iii) follow from the definition of Ω. Suppose that C is a countable subset of P_Ω. Let $D = \bigcup \{P_\alpha : \alpha \in C\}$. Then D is a countable union of countable sets so D is countable (4.21). Let $\beta \in P_\Omega \cap D'$.

Clearly $\alpha \leqq \beta$ for each $\alpha \in C$ for otherwise $\beta \in D$. This proves (iv). □

(4.50) Remark. The cardinal number of the set P_Ω is denoted $\aleph_1$. The *continuum hypothesis* is the assertion that $\aleph_1 = \mathfrak{c}$. This is equivalent to the assertion that each infinite subset of R is either countable or of cardinal number $\mathfrak{c}$. Many interesting and important theorems have been proved with the aid of this hypothesis. It was recently shown by PAUL J. COHEN [*loc. cit.* (3.2)] that the continuum hypothesis is independent of the Zermelo-Fraenkel axioms of set theory.

(4.51) Exercise. Prove that every nonvoid set of cardinal numbers has a smallest member.

(4.52) Exercise. Let A be an infinite linearly ordered set. Suppose that no infinite subset of A has a largest element. Prove that $\operatorname{ord} A = \omega$. [Recall that $\omega = \operatorname{ord} N$.]

(4.53) Exercise. Let A be a linearly ordered set such that:

(i) A is countably infinite;

(ii) A has no first or last element;

(iii) $x, y \in A$, $x < y$ imply that there is a $z \in A$ such that $x < z < y$.

Prove that $\operatorname{ord} A = \eta$. [Recall that $\eta = \operatorname{ord} Q$.]

(4.54) Exercise. Prove that there are uncountably many different ways to well order the set N such that no two of these different well-ordered sets are order isomorphic.

(4.55) Exercise. Let A be any infinite set. Prove that A can be well ordered in such a way that it has no last element. Also show that there is a well-ordering of A in which there is a last element.

(4.56) Exercise. A *permutation* of a set A is any one-to-one mapping of A onto A. Let A be a set such that $\overline{\overline{A}} > 1$.

(a) Prove that there exists a permutation f of A such that $f(x) \neq x$ for all $x \in A$.

(b) Show that if $\overline{\overline{A}}$ is an even integer or is infinite, then the permutation f in (a) can be chosen so that $f \circ f(x) = x$ for all $x \in A$. What happens if $\overline{\overline{A}}$ is an *odd* integer?

(c) Show that the permutation f in (a) can always be chosen so that $f \circ f \circ f \circ f \circ f \circ f(x) = x$ for all $x \in A$.

(4.57) Exercise. Let B be a set, let $\mathfrak{b} = \overline{\overline{B}}$, and let

$$\mathfrak{b}! = \overline{\overline{\{f : f \text{ is a permutation of } B\}}}.$$

Prove that if B is infinite, then $\mathfrak{b}! = 2^{\mathfrak{b}}$.

We now prove a theorem which allows us to define the algebraic dimension of any vector space.

(4.58) Theorem. *Let X be a vector space over a field F and let A and B be any two Hamel bases for X over F. Then $\overline{\overline{A}} = \overline{\overline{B}}$.*

Proof. We will first use ZORN's lemma to produce a one-to-one function from A into B. To this end let $\mathfrak{Z}$ denote the set of all one-to-one functions f such that:

(1) $\operatorname{dom} f \subset A$;

(2) $\operatorname{rng} f \subset B$;

(3) $(\operatorname{rng} f) \cup [A \cap (\operatorname{dom} f)']$ is linearly independent over F.

The fact that A is linearly independent shows that the empty function $\varnothing$ is an element of $\mathfrak{Z}$. Thus $\mathfrak{Z} \neq \varnothing$. Partially order $\mathfrak{Z}$ by inclusion. To show that ZORN's lemma applies to $\mathfrak{Z}$, let $\mathfrak{C}$ be any nonvoid chain contained in $\mathfrak{Z}$ and let $g = \bigcup \mathfrak{C}$. An application of (2.19) shows that g is a function and that (1) and (2) hold for the function g. One easily sees that g is one-to-one. We have

$$(\operatorname{rng} g) \cup \big[A \cap (\operatorname{dom} g)'\big] = \big(\bigcup_{f \in \mathfrak{C}} \operatorname{rng} f\big) \cup \big[A \cap \big(\bigcup_{f \in \mathfrak{C}} \operatorname{dom} f\big)'\big]. \tag{4}$$

Now let F be any finite subset of the set in (4). Since $\{\operatorname{rng} f : f \in \mathfrak{C}\}$ is a chain under inclusion, there is a function $f_0 \in \mathfrak{C}$ such that

$$\begin{aligned} F &\subset (\operatorname{rng} f_0) \cup \big[A \cap \big(\bigcup_{f \in \mathfrak{C}} \operatorname{dom} f\big)'\big] \\ &\subset (\operatorname{rng} f_0) \cup \big[A \cap (\operatorname{dom} f_0)'\big]. \end{aligned}$$

Therefore F is linearly independent, and so g satisfies condition (3). Thus g is in $\mathfrak{Z}$, and g is an upper bound for $\mathfrak{C}$. By ZORN's lemma, $\mathfrak{Z}$ has a maximal member, say h.

We assert that $\operatorname{dom} h = A$. Assume that $\operatorname{dom} h \neq A$ and let $a_0 \in A \cap (\operatorname{dom} h)'$. According to (3), a_0 is not a linear combination of elements of $\operatorname{rng} h$. Since a_0 is a linear combination of elements of B, it follows that $\operatorname{rng} h \neq B$. Let b_0 be any element of $B \cap (\operatorname{rng} h)'$. If the set

$$\{b_0\} \cup (\operatorname{rng} h) \cup \big[A \cap (\operatorname{dom} h)'\big]$$

is linearly independent, then, as is easily seen, the function $h \cup \{(a_0, b_0)\}$ is in $\mathfrak{Z}$, contrary to the maximality of h. We infer that b_0 is a linear combination of elements of the set $(\operatorname{rng} h) \cup [A \cap (\operatorname{dom} h)']$; we write

$$b_0 = \sum_{k=1}^{n} \alpha_k x_k .$$

Since B is linearly independent, b_0 is not a linear combination of elements of $\operatorname{rng} h$. Hence there exists a k such that $x_k \in A \cap (\operatorname{dom} h)'$ and $\alpha_k \neq 0$. Thus b_0 is not a linear combination of elements of the linearly independent set $(\operatorname{rng} h) \cup [A \cap (\{x_k\} \cup \operatorname{dom} h)']$ and therefore the function $h \cup \{(x_k, b_0)\}$ is an element of $\mathfrak{Z}$. This contradicts the maximality of h. Consequently $\operatorname{dom} h = A$ and $\overline{\overline{A}} \leqq \overline{\overline{B}}$.

Interchanging the rôles of A and B in the above argument, we see also that $\overline{\overline{B}} \leqq \overline{\overline{A}}$. The proof is completed by invoking the Schröder-Bernstein theorem (4.7). □

(4.59) Definition. Let X be a vector space over a field F. We define the *algebraic* [*linear*] *dimension of* X to be 0 if $X = \{0\}$ and to be the cardinal number of an arbitrary Hamel basis for X over F if $X \neq \{0\}$.

(4.60) Exercise. Let X be a vector space over a field F and let B be a Hamel basis for this space. Prove that:

(a) $\overline{\overline{X}} = \max\{\overline{\overline{B}}, \overline{\overline{F}}\}$ if B is infinite;

(b) $\overline{\overline{X}} = \overline{\overline{F}}^{\overline{\overline{B}}}$ if B is finite.

(4.61) Exercise. Without using the continuum hypothesis, find the algebraic dimension of the vector space R over the field Q.

(4.62) Exercise [proposed by M. Hewitt]. Let A be a nonvoid set. Suppose that there is a family $\mathcal{S}$ of subsets of A with the following properties:

(i) $\overline{\overline{B}} = 3$ for all $B \in \mathcal{S}$;

(ii) $\bigcup \mathcal{S} = A$;

(iii) $\overline{\overline{B_1 \cap B_2}} = 1$ for distinct $B_1, B_2 \in \mathcal{S}$;

(iv) if $x, y \in A$ and $x \neq y$, then there is exactly one $B \in \mathcal{S}$ containing $\{x, y\}$.

Prove that such an $\mathcal{S}$ exists if and only if $\overline{\overline{A}} = 3$, $\overline{\overline{A}} = 7$, or $\overline{\overline{A}} \geqq \aleph_0$.

§ 5. Construction of the real and complex number fields

We give in this section a short and reasonably sophisticated construction of the real and complex numbers, assuming the rational numbers as known. It seems appropriate to do this, since completeness of the real number field is the rock on which elementary analysis rests. Also there is a strong interplay between algebra and contemporary analysis, which demands the use of the ideas and methods of algebra in analysis. We begin with a few facts about groups and other algebraic structures.

(5.1) Definition. A set G together with a binary operation $(x, y) \rightarrow xy$ mapping $G \times G$ into G is called a *group* provided that:

(i) $x(yz) = (xy)z$ for all $x, y, z \in G$ [*associative law*];

(ii) there is an element $e \in G$ such that $ex = x$ for all $x \in G$ [e is a *left identity*];

(iii) for all e as in (ii) and all $a \in G$ there exists $a^{-1} \in G$ such that $a^{-1}a = e$ [a^{-1} is a *left inverse for* a].

If also we have

(iv) $ab = ba$ for all $a, b \in G$,

then G is called an *Abelian group* [after the Norwegian mathematician N. H. Abel (1802–1829)].

(5.2) Remarks. (a) Every left inverse is a right inverse. In fact, for any e as in (ii) we have

$$(a^{-1}a)\, a^{-1} = e a^{-1} = a^{-1}.$$

Now let b be a left inverse of a^{-1}, *i. e.*, $ba^{-1} = e$. Then

$$b(a^{-1}aa^{-1}) = ba^{-1} = e,$$

$$(ba^{-1})\,(aa^{-1}) = e,$$

$$e(aa^{-1}) = e,$$

and so by (ii)

$$aa^{-1} = e .$$

Note that the last equality also implies that a is a left inverse of a^{-1}.

(b) For any e as in (ii) and any $a \in G$, we have

$$ae = a(a^{-1}a) = (aa^{-1})\, a = ea = a ,$$

i. e., e is also a right identity. If e_1 and e_2 satisfy (ii), then they are both right identities also, and so

$$e_1 e_2 = e_2 \quad [e_1 \text{ is a left identity}] ,$$
$$e_1 e_2 = e_1 \quad [e_2 \text{ is a right identity}] ,$$

so that $e_1 = e_2$, *i. e.*, there is a unique left and right identity in G.

(c) Similarly one sees that a^{-1} is unique.

(d) For Abelian groups, we often use additive notation [+ denotes the binary operation]; in this case we denote the identity by 0, the inverse of a by $-a$, and $a + (-b)$ by $a - b$.

(5.3) Definition. Consider a set A with two binary operations $+$ and $\cdot$ [called *addition* and *multiplication*, respectively], which is an Abelian group under $+$, with identity 0, and in which the equalities

$$a \cdot (b + c) = (a \cdot b) + (a \cdot c) \quad [\textit{left distributive law}] ,$$

$$(a + b) \cdot c = (a \cdot c) + (b \cdot c) \quad [\textit{right distributive law}] ,$$

and

$$(a \cdot b) \cdot c = a \cdot (b \cdot c) \quad [\textit{associative law for multiplication}]$$

hold for all $a, b, c \in A$. Then A is called a *ring*. If $a \cdot b = b \cdot a$ for all a, b in a ring A, A is called *commutative*. An element $1 \in A$ such that $1 \cdot a = a \cdot 1 = a$ for all $a \in A$ is called a [*two-sided*] *unit for* A. A nonvoid subset I of a ring A is called a *left* [*right*] *ideal* if $a - b \in I$ for all $a, b \in I$ and $x \cdot a \in I$ [$a \cdot x \in I$] for all $a \in I$ and $x \in A$. A subset I of A that is a left and a right ideal is called a *two-sided ideal*.

(5.4) Remarks. (a) The notation in the statements of the distributive laws is correct but clumsy. From now on we will follow the universal algebraic convention that ab means $a \cdot b$ and that $ab + cd$ means $(a \cdot b) + (c \cdot d)$.

(b) For all x in a ring A, we have $xx = x(x + 0) = xx + x0$, so that $x0 = 0$. Similarly $0x = 0$.

(c) Evidently two groups or rings can be distinct objects and still be indistinguishable as groups or rings. Formally, we say that rings A and A' are *isomorphic* if there is a one-to-one mapping τ carrying A onto A' such that $\tau(a + b) = \tau(a) + \tau(b)$ and $\tau(ab) = \tau(a)\,\tau(b)$ for all $a, b \in A$. The mapping τ is called an *isomorphism* or an *isomorphic mapping*. An analogous definition is made for groups. An isomorphism of a ring or group onto itself is called an *automorphism*.

We now define an important special type of ring.

(5.5) Definition. A ring F such that $F \cap \{0\}'$ is an Abelian group under multiplication is called a *field*.

(5.6) Remarks.

(a) Since a group contains an element, our definition of a field shows that $1 \neq 0$ and that a field contains at least two elements.

(b) The identities (5.4.b) show that 0 has to be excluded from F in order to obtain a group under multiplication.

(c) The simplest field is $\{0, 1\}$, with operations addition and multiplication modulo 2. The addition and multiplication tables for this field are

+	0	1
0	0	1
1	1	0

,

·	0	1
0	0	0
1	0	1

.

(d) If p is a prime, the set $\{0, 1, 2, \ldots, p-1\}$ is a field under addition and multiplication modulo p. All necessary verifications are easy; we will give the least easy one, namely that each nonzero element has a multiplicative inverse. If $a \in \{1, 2, \ldots, p-1\}$, then we must show that there exists $x \in \{1, 2, \ldots, p-1\}$ such that

$$ax \equiv 1 \pmod{p}.$$

Since p is prime, the greatest common divisor of a and p is 1. It follows that there are integers x and y, $x \neq 0$, such that

$$1 = ax + py .$$

In particular, there are integers x' and y' such that

$$1 = ax' + py' , \quad 1 \leq x' < p;$$

hence $ax' \equiv 1 \pmod{p}$ and $x' \in \{1, 2, \ldots, p-1\}$.

(e) If F is any field, then the elements $0, 1, 1+1, \ldots, n1, \ldots$ [where n is any positive integer and $n1$ has an obvious meaning] are all members of F. If $n1 = 0$ for some positive integer n, then the smallest positive integer p such that $p1 = 0$ is obviously a prime. In this case F is said to have *characteristic* p; otherwise F is said to have *characteristic* 0. Fields of characteristic p are of no interest at present in elementary analysis. If F has characteristic 0, then F has a subfield [the definition of a subfield is obvious] which is isomorphic to the rational number field. We will always denote the rational number field by the symbol Q.

To see that F contains a [unique] isomorph of Q, consider a mapping τ of part of F onto Q, and see what properties it must have in order to be an isomorphism. For notational convenience, we ignore the distinction between the zeros of F and of Q and between the units of F and of Q. It is clear that $\tau(0) = 0$ and $\tau(1) = 1$. A trivial induction shows that $\tau(n1) = n$ for all $n \in N$. If τ is an isomorphism, we must also have $\tau((n1)^{-1}) = \frac{1}{n}$ and $\tau(-(n1)) = -n$ for all $n \in N$. It follows that we must have

$$\tau((m1)(n1)^{-1}) = \frac{m}{n}$$

and

$$\tau((-(m1))(n1)^{-1}) = -\frac{m}{n}$$

for all $m, n \in N$. It is a routine matter to show that the mapping τ so constructed is an isomorphism of a subfield of F onto the field Q.

(f) Let G be any group. For $A, B \subset G$, we define

$$AB = \{xy : x \in A, y \in B\},$$
$$A^{-1} = \{x^{-1} : x \in A\},$$
$$A^2 = AA.$$

In additive notation, these sets are written as $A + B$, $-A$, and $2A$. The set A^n [nA in additive notation], where n is a positive integer, has the obvious definition.

(g) For a nonzero element b of a field, we frequently write the multiplicative inverse b^{-1} as $\frac{1}{b}$, and for an element a of the field, we frequently write ab^{-1} as $\frac{a}{b}$.

(5.7) Definition. A field F is said to be *ordered* if there is a subset P of F such that:

(i) $P \cap (-P) = \varnothing$;

(ii) $P \cup \{0\} \cup (-P) = F$;

(iii) $a, b \in P$ imply $a + b \in P$ and $ab \in P$.

If one thinks of F as the rational or real numbers, then P is just the set of positive rational or real numbers. Accordingly, in the general case elements of P are called *positive*; elements of $-P$ are called *negative*. Since $0 = -0$, 0 cannot be an element of P.

(5.8) Theorem. *Let F be an ordered field, and let P be as in* (5.7). *If $a \in F$ and $a \neq 0$, then $a^2 \in P$. In particular, $1 \in P$. If $a, b \in F$, $ab \in P$, and $a \in P$, then also $b \in P$.*

Proof. If $a \in P$, then $a^2 \in P$ by (5.7.iii). If $a \notin P$ and $a \neq 0$, then $a \in -P$ by (5.7.ii), *i. e.*, $-a \in P$. Again by (5.7.iii), we have $(-a)^2 \in P$. In any ring, the identity $(-a)(-b) = ab$ holds, and so $a^2 = (-a)^2 \in P$. Since $1^2 = 1$, the second assertion holds. To prove the third assertion, assume that $b \notin P$. If $b = 0$, then $ab = 0 \notin P$, which contradicts the hypothesis. If $b \in -P$, then $-b \in P$ and

$$a(-b) = -(ab) \in P;$$

therefore we have $ab \in -P$, which is a contradiction. Hence $b \in P$. □

(5.9) Theorem. *Every ordered field F contains an isomorph of Q, and the isomorphism can be taken as order-preserving.*

Proof. Since $1 \in P$, (5.7.iii) implies that $n1 \in P$ for all $n \in N$. This implies that F has characteristic 0, and so the isomorphism τ of (5.6.e) can be constructed. It is easy to check that τ preserves order. □

(5.10) Definition. Let F be an ordered field. We write $a < b$ and $b > a$ if $b - a \in P$; the expressions $a \leqq b$ and $b \geqq a$ have obvious meanings.

(5.11) Theorem. *Let F be an ordered field. For all $a, b \in F$, we have $a < b$ or $a = b$ or $a > b$, and only one of these relations holds.*

Proof. The proof is immediate from the definition of P and the fact that

$$b - a = 0 \quad \text{if and only if} \quad b = a\,,$$
$$b - a \in P \quad \text{if and only if} \quad b > a\,,$$
$$a - b \in P \quad \text{if and only if} \quad a > b\,. \;\square$$

Many elementary facts about inequalities are consequences of the axioms of order (5.7). We now list a few of them.

(5.12) Theorem. *If F is an ordered field, if $a, b, c, d \in F$, and $a < b$ and $c \leqq d$, then $a + c < b + d$.*

The proof is left to the reader.

(5.13) Theorem. *If F is an ordered field, if $a, b, c \in F$, $a < b$, and $c > 0$ $[c < 0]$, then $ac < bc$ $[ac > bc]$.*

Proof. If $a < b$, then $b - a \in P$, and therefore if $c > 0$, we have $c(b - a) \in P$ and $cb > ca$. If $c < 0$, then $-c \in P$, and so $(-c)(b - a) \in P$; hence $ac - bc \in P$, *i. e.*, $ac > bc$. □

(5.14) Theorem. *In an ordered field F, the inequalities $0 < a < b$ imply that $0 < \frac{1}{b} < \frac{1}{a}$.*

Proof. Since $b\frac{1}{b} = 1 \in P$ and $b \in P$, (5.8) implies that $\frac{1}{b} \in P$. Hence $\frac{1}{b}$, $\frac{1}{a}$, and $b - a$ are in P, and it follows that

$$\frac{1}{a} - \frac{1}{b} = \frac{1}{a}\,\frac{1}{b}\,(b - a) \in P\,,$$

so that

$$0 < \frac{1}{b} < \frac{1}{a}. \;\square$$

(5.15) Definition. For an element a of an ordered field F, we define

$$|a| = \begin{cases} a & \text{if} \quad a \geqq 0\,, \\ -a & \text{if} \quad a < 0\,. \end{cases}$$

(5.16) Theorem. *For all elements a, b in an ordered field F, we have:*

(i) $|a| = |-a|$;

(ii) $|ab| = |a|\,|b|$;

(iii) $|a + b| \leqq |a| + |b|$;

(iv) $|\,|a| - |b|\,| \leqq |a - b|$.

Proof. Statements (i) and (ii) are obvious consequences of the definition of $|a|$, and (iv) follows from (iii). We will prove (iii). If $a \geqq 0$, we have $|a| = a$; if $a < 0$, then we have $|a| = -a > 0$. Hence we always have

$$a \leqq |a|\,,$$
$$b \leqq |b|\,,$$

and so (5.10) implies that

$$a + b \leqq |a| + |b|\,.$$

Since $-a \leqq |-a| = |a|$, we also have

$$-(a + b) \leqq |a| + |b|\,,$$

and (iii) follows from these two inequalities. □

(5.17) Definition. An ordered field F is said to be *Archimedean ordered* if for all $a \in F$ and all $b \in P$ there exists a positive integer n such that $nb > a$. In intuitive language, this definition means that no matter how large a is and how small b is, successive repetitions of b will eventually exceed a. There are ordered fields which are not Archimedean ordered; see (5.39).

(5.18) Theorem. *Let F be an Archimedean ordered field, and let $a, b \in F$ be such that $a < b$. Then there exists $\frac{m}{n} \in F$, where m and n are integers, such that $a < \frac{m}{n} < b$.*[1]

Proof. Since $b - a > 0$, we have $(b - a)^{-1} > 0$; and so, since F is Archimedean ordered and $1 > 0$, there exists an integer n such that

$$n1 > (b - a)^{-1} > 0\,.$$

Using (5.14), we have

$$0 < (n1)^{-1} < b - a;$$

or, using an obvious notation,

$$\frac{1}{n} 1 < b - a\,.$$

Let n be any integer satisfying the last inequality, and let S be the set

$$S = \left\{k : k \text{ is an integer and } k \cdot \frac{1}{n} > a\right\}.$$

---

[1] The expression $\frac{m}{n}$ really means $\frac{m1}{n1} \in F$; in view of (5.3.e), it does no harm to suppose that $F \supset Q$.

Since $\frac{1}{n} > 0$ and F is Archimedean ordered, we have $S \neq \varnothing$. Also, again by the Archimedean order property of F, there is a positive integer p such that

$$p\frac{1}{n} > -a,$$

$$-\left(p\frac{1}{n}\right) < a,$$

$$(-p)\frac{1}{n} < a.$$

It follows that $-p$, $-(p+1)$, $-(p+2)$, ... lie outside of S. Hence S is a nonvoid set of integers which is bounded below, and so it contains a least element m. Since $m \in S$, we have $a < \frac{m}{n}$.

Since $m - 1 \notin S$, we have $\frac{m-1}{n} \leqq a$; and so

$$\frac{m}{n} = \frac{m-1}{n} + \frac{1}{n} \leqq a + \frac{1}{n} < a + (b - a) = b;$$

i. e.,

$$a < \frac{m}{n} < b. \quad \square$$

(5.19) Definition. Let F be an ordered field. A sequence (a_n) $[= (a_1, a_2, \ldots, a_n, \ldots)]$ of elements of F is called *bounded* if there is an element $b \in F$ such that $|a_n| \leqq b$ for each positive integer n. A sequence (a_n) is called *Cauchy* if for every $e \in F$ such that $e > 0$, there is a positive integer $N(e)$ such that $|a_p - a_q| < e$ for all $p, q \geqq N(e)$. [The French mathematician AUGUSTIN LOUIS CAUCHY (1789–1857) first considered this class of sequences, for the case in which F is the real number field.] A sequence (a_n) is called *null* if for every $e \in F$ such that $e > 0$, there is a positive integer $N(e)$ such that $|a_p| < e$ for all $p \geqq N(e)$. The families of sequences satisfying these conditions will be denoted by $\mathfrak{B}$, $\mathfrak{C}$, and $\mathfrak{N}$, respectively.

(5.20) Theorem. *The inclusions* $\mathfrak{N} \subset \mathfrak{C} \subset \mathfrak{B}$ *obtain.*

Proof. If $(a_n) \in \mathfrak{C}$, then $p, q \geqq N(1)$ implies $|a_p - a_q| < 1$. In particular,

$$|a_{N(1)+k} - a_{N(1)}| < 1 \quad \text{for} \quad k = 0, 1, 2, \ldots.$$

Let $b = \max\{|a_1|, |a_2|, \ldots, |a_{N(1)}|, |a_{N(1)}| + 1\}$ [see (2.7) for the definition of max]; then $|a_p| \leqq b$ for $p = 1, 2, \ldots$, and so $(a_n) \in \mathfrak{B}$.

If $(a_n) \in \mathfrak{N}$, then for any given positive $e \in F$, we have

$$|a_p - a_q| \leqq |a_p| + |a_q| < \frac{1}{2}e + \frac{1}{2}e = e$$

provided that $p, q \geqq N\left(\frac{1}{2}e\right)$. It follows that $(a_n) \in \mathfrak{C}$, and so $\mathfrak{N} \subset \mathfrak{C}$. $\square$

(5.21) Theorem. *For* (a_n), $(b_n) \in \mathfrak{C}$, *let* $(a_n) + (b_n) = (a_n + b_n)$ *and* $(a_n)(b_n) = (a_n b_n)$. *With these definitions of sum and product,* $\mathfrak{C}$ *is a commutative ring with unit, and* $\mathfrak{N}$ *is an ideal in* $\mathfrak{C}$ *such that* $\mathfrak{N} \subsetneqq \mathfrak{C}$.

Proof. Let us first show that sums and products of Cauchy sequences are again Cauchy sequences. Let (a_n), $(b_n) \in \mathfrak{C}$ be given. For a positive $e \in F$, let $N(e)$ and $M(e)$ be the positive integers associated with the Cauchy sequences (a_n) and (b_n), respectively. If $p, q \geqq \max\left\{N\left(\frac{e}{2}\right), M\left(\frac{e}{2}\right)\right\}$, then we have $|a_p + b_p - (a_q + b_q)| \leqq |a_p - a_q| + |b_p - b_q| < \frac{1}{2}e + \frac{1}{2}e = e$. It follows that $(a_n) + (b_n)$ is a Cauchy sequence.

Since $\mathfrak{C} \subset \mathfrak{B}$, there are positive elements c and d of F such that:

$$|a_p| \leqq c \quad \text{for} \quad p = 1, 2, 3, \ldots ;$$
$$|b_p| \leqq d \quad \text{for} \quad p = 1, 2, 3, \ldots .$$

If $p, q \geqq \max\left\{N\left(\frac{e}{2d}\right), M\left(\frac{e}{2c}\right)\right\}$, then we have

$$\begin{aligned} |a_p b_p - a_q b_q| &= |a_p b_p - a_p b_q + a_p b_q - a_q b_q| \\ &\leqq |a_p|\,|b_p - b_q| + |b_q|\,|a_p - a_q| < \frac{ce}{2c} + \frac{de}{2d} = e . \end{aligned}$$

It follows that $(a_n)(b_n)$ is a Cauchy sequence.

It is now obvious that $\mathfrak{C}$ is an additive Abelian group in which $(0, 0, \ldots)$ is the zero. Also multiplication in $\mathfrak{C}$ is clearly commutative and $\mathfrak{C}$ has the multiplicative unit $(1, 1, \ldots)$. The distributive law for $\mathfrak{C}$ follows immediately from the distributive law for F. We have thus shown that $\mathfrak{C}$ is a commutative ring with unit.

We will now show that $\mathfrak{N}$ is a proper ideal of $\mathfrak{C}$. If (a_n), $(b_n) \in \mathfrak{N}$, then it is clear that $(a_n) - (b_n) \in \mathfrak{N}$; thus $\mathfrak{N}$ is an additive subgroup of $\mathfrak{C}$. If $(a_n) \in \mathfrak{N}$ and $(b_n) \in \mathfrak{C}$, then we must show that $(a_n b_n) \in \mathfrak{N}$. Let c be a positive element of F such that

$$|b_n| \leqq c\,, \qquad n = 1, 2, 3, \ldots .$$

For a given positive $e \in F$, there is a positive integer $N\left(\frac{e}{c}\right)$ such that $p \geqq N\left(\frac{e}{c}\right)$ implies that

$$|a_p| < \frac{e}{c}.$$

Hence $p \geqq N\left(\frac{e}{c}\right)$ implies that

$$|a_p b_p| \leqq |a_p|\, c < \frac{e}{c}\, c = e\,,$$

and so $(a_n b_n) \in \mathfrak{N}$. Thus we have shown that $\mathfrak{N}$ is an ideal. Since $(1, 1, 1, \ldots) \in \mathfrak{C} \cap \mathfrak{N}'$, $\mathfrak{N}$ is a proper ideal. $\square$

Note that $\mathfrak{C}$ is not a field; *e. g.*, $(0, 1, 0, 0, \ldots)$ has no multiplicative inverse in $\mathfrak{C}$.

(5.22) Theorem. *Let $\mathfrak{C}/\mathfrak{N}$ denote the set whose elements are the sets $(a_n) + \mathfrak{N}$ [called* cosets *of $\mathfrak{N}$], where $(a_n) \in \mathfrak{C}$. Addition and multiplication* *in $\mathfrak{C}/\mathfrak{N}$ are defined by*

$$((a_n) + \mathfrak{N}) + ((b_n) + \mathfrak{N}) = (a_n) + (b_n) + \mathfrak{N} = (a_n + b_n) + \mathfrak{N},$$

and

$$((a_n) + \mathfrak{N})\,((b_n) + \mathfrak{N}) = (a_n)\,(b_n) + \mathfrak{N} = (a_n b_n) + \mathfrak{N}.$$

These definitions are unambiguous, and with addition and multiplication so defined, $\mathfrak{C}/\mathfrak{N}$ is a field.

Proof. Since $\mathfrak{N}$ is an additive subgroup of $\mathfrak{C}$, any two cosets $(a_n) + \mathfrak{N}$, $(b_n) + \mathfrak{N}$ are either disjoint or identical. Two sequences (a_n), $(a'_n) \in \mathfrak{C}$ belong to the same coset if and only if $(a'_n) = (a_n) + (c_n)$, for some $(c_n) \in \mathfrak{N}$. If (c_n), $(d_n) \in \mathfrak{N}$, then

$$\begin{aligned}\big(((a_n) + (c_n)) + \mathfrak{N}\big) &+ \big(((b_n) + (d_n)) + \mathfrak{N}\big) \\ &= (a_n) + (b_n) + (c_n) + (d_n) + \mathfrak{N} = (a_n) + (b_n) + \mathfrak{N},\end{aligned}$$

and

$$\begin{aligned}\big(((a_n) + (c_n)) + \mathfrak{N}\big)&\big(((b_n) + (d_n)) + \mathfrak{N}\big) \\ &= (a_n)\,(b_n) + (a_n)\,(d_n) + (c_n)\,(b_n) + (c_n)\,(d_n) + \mathfrak{N} \\ &= (a_n)\,(b_n) + \mathfrak{N}.\end{aligned}$$

Thus addition and multiplication of cosets are defined unambiguously. It is a simple matter to verify that $\mathfrak{C}/\mathfrak{N}$ is a commutative ring with unit. For example, $(1, 1, 1, \ldots) + \mathfrak{N}$ is the unit of $\mathfrak{C}/\mathfrak{N}$.

It remains to show that every nonzero element of $\mathfrak{C}/\mathfrak{N}$ has a multiplicative inverse. Let $(a_n) + \mathfrak{N} \neq \mathfrak{N}$; *i. e.*, let $(a_n) \in \mathfrak{C} \cap \mathfrak{N}'$. We are required to find $(x_n) \in \mathfrak{C}$ such that

$$(a_n x_n) + \mathfrak{N} = (1_{(n)}) + \mathfrak{N};$$

i. e., such that

$$(a_n x_n - 1_{(n)}) \in \mathfrak{N}.$$

Since (a_n) is not a null sequence, there is a positive $e \in F$ such that for every positive integer r there is some integer $s > r$ for which $|a_s| \geqq e$. For all $p, q \geqq N\left(\frac{1}{2}e\right)$ we have

$$|a_p - a_q| < \frac{1}{2}e.$$

Let $s > N\left(\frac{1}{2}e\right)$ be such that $|a_s| \geqq e$; then for arbitrary $p \geqq N\left(\frac{1}{2}e\right)$ we have

$$e \leqq |a_s| = |a_s - a_p + a_p| \leqq |a_s - a_p| + |a_p| < \frac{1}{2}e + |a_p|.$$

Hence

$$|a_p| > \frac{1}{2}e \quad \text{if} \quad p \geqq N\left(\frac{1}{2}e\right).$$

We now define (x_n). Write $N\left(\frac{1}{2}e\right)$ as m, and let

$$x_1 = x_2 = \cdots = x_{m-1} = 1,$$

and

$$x_p = \frac{1}{a_p} \quad \text{if} \quad p \geqq m\,.$$

We have

$$(x_n a_n) = (a_1, a_2, \ldots, a_{m-1}, 1, 1, \ldots)\,,$$

and hence

$$(x_n a_n - 1_{(n)}) = (a_1 - 1, a_2 - 1, \ldots, a_{m-1} - 1, 0, 0, \ldots)\,.$$

Thus it is obvious that $(x_n a_n - 1_{(n)}) \in \mathfrak{N}$. To complete the proof, we need only show that $(x_n) \in \mathfrak{C}$. If $p, q \geqq N\left(\frac{1}{2} e\right)$, then we have

$$|x_p - x_q| = \left|\frac{1}{a_p} - \frac{1}{a_q}\right| = \left|\frac{1}{a_p}\right| \cdot \left|\frac{1}{a_q}\right| |a_p - a_q| < \frac{2}{e} \cdot \frac{2}{e} |a_p - a_q|\,.$$

For arbitrary positive $d \in F$, it is now obvious that

$$p, q \geqq \max\left\{N\left(\frac{1}{2} e\right), N\left(\frac{e^2 d}{4}\right)\right\}$$

implies that

$$|x_p - x_q| < d\,.$$

Hence $(x_n) \in \mathfrak{C}$. □

(5.23) Notation. The field $\mathfrak{C}/\mathfrak{N}$ will be written as $\overline{F}$. Throughout (5.23–5.30), elements $(a_n) + \mathfrak{N}$ of $\mathfrak{C}/\mathfrak{N}$ will be denoted by small Greek letters: $\alpha, \beta, \ldots$. If $a \in F$, then the element $(a_{(n)}) + \mathfrak{N}$ of $\overline{F}$ will be written as $\bar{a}$; it is the coset of $\mathfrak{N}$ containing the constant sequence all of whose terms are a.

(5.24) Theorem. *In $\overline{F}$, let $\overline{P} = \{\alpha \in \overline{F} : \alpha \neq \bar{0}$ and there exists $(a_n) \in \alpha$ such that $a_n > 0$ for $n = 1, 2, 3, \ldots\}$. With this set $\overline{P}$, $\overline{F}$ is an ordered field in the sense of* (5.7). *The mapping $\tau : \tau(a) = \bar{a}$ is an order-preserving algebraic isomorphism of F into $\overline{F}$.*

The proof of this theorem is left to the reader.

(5.25) Definition. Given a sequence (a_n) in an ordered field F and $b \in F$, we say that the *limit of (a_n) is b*, and we write

$$\lim_{n \to \infty} a_n = b \quad \text{or} \quad a_n \to b\,,$$

if for every positive e in F there exists a positive integer $L(e)$ such that $|a_n - b| < e$ for all $n \geqq L(e)$. An ordered field is said to be *complete* if every Cauchy sequence in F has a limit in F.

(5.26) Lemma. *A sequence with a limit is a Cauchy sequence. If (a_n) is a Cauchy sequence and $(a_{n_k})_{k=1}^{\infty}$ $(1 \leqq n_1 < n_2 < \cdots < n_k < \cdots)$ is a subsequence with limit b, then (a_n) has limit b.*

Proof. The first assertion is trivial. To prove the second, choose $e > 0$ in F. If $L\left(\frac{1}{2} e\right) \leqq k$, then we have $|a_{n_k} - b| < \frac{1}{2} e$. Since (a_n) is a Cauchy sequence, we have

$$|a_p - a_q| < \frac{1}{2} e$$

if $p, q \geqq N\left(\frac{1}{2}e\right)$. Choose any fixed k such that $k \geqq L\left(\frac{1}{2}e\right)$ and $n_k \geqq N\left(\frac{1}{2}e\right)$. Then for $q \geqq N\left(\frac{1}{2}e\right)$, we have

$$|a_q - b| \leqq |a_q - a_{n_k}| + |a_{n_k} - b| < e. \quad \square$$

(5.27) Lemma. *For $\alpha > \bar{0}$, $\alpha \in \bar{F}$, there exists $e \in F$ such that $\bar{0} < \bar{e} < \alpha$. If F is Archimedean ordered, then $\bar{F}$ is also Archimedean ordered.*

Proof. Since $\alpha > \bar{0}$, there exists $(a_n) \in \alpha$ such that $a_n > 0$ for $n = 1, 2, 3, \ldots$ and $(a_n) \notin \mathfrak{N}$. Hence there is a $d \in F$ such that $a_s = |a_s| \geqq d$ for arbitrarily large s. We have

$$|a_p - a_q| < \frac{1}{2}d$$

if $p, q \geqq N\left(\frac{1}{2}d\right)$. Choose an s as above such that $s \geqq N\left(\frac{1}{2}d\right)$; then we have

$$d \leqq a_s = a_s - a_p + a_p \leqq |a_s - a_p| + |a_p| < \frac{1}{2}d + a_p;$$

i. e., $\frac{1}{2}d < a_p$ whenever $p \geqq N\left(\frac{1}{2}d\right)$. It follows that $\left(a_n - \frac{1}{2}d_{(n)}\right) + \mathfrak{N} = \alpha - \overline{\frac{1}{2}d} \geqq \bar{0}$. We have

$$\alpha - \overline{\frac{1}{3}d} > \alpha - \overline{\frac{1}{2}d} \geqq \bar{0},$$

and so

$$\alpha > \overline{\frac{1}{3}d}.$$

Hence $e = \frac{1}{3}d$ satisfies the first assertion of the theorem.

Suppose that F is Archimedean ordered, and let $\alpha, \beta \in \bar{F}$ be such that

$$\bar{0} < \alpha \leqq \beta.$$

By the first assertion of the theorem, there is a positive $e \in F$ such that $\bar{e} < \alpha$. Since every Cauchy sequence is bounded, there is a $d \in F$ such that $\beta < \bar{d}$. If m is any positive integer such that $me > d$, then we have

$$m\alpha > m\bar{e} > \bar{d} > \beta;$$

hence $m\alpha > \beta$, and so $\bar{F}$ is Archimedean ordered. $\square$

(5.28) Lemma. *Let $\alpha \in \bar{F}$ and $(a_n) \in \alpha$. Then we have*

$$\lim_{n \to \infty} \bar{a}_n = \alpha.$$

Proof. Choose any $\varepsilon > \bar{0}$ in $\bar{F}$ and any $e > 0$ in F such that $0 < \bar{e} < \varepsilon$; this is possible by (5.27). We have

$$|a_p - a_q| < e \quad \text{if} \quad p, q \geqq N(e).$$

Now fix $p \geqq N(e)$. For $n \geqq N(e)$ we have

$$a_p - a_n < e\,,$$
$$a_n - a_p < e\,.$$

It follows that

$$\bar{a}_p - \alpha \leqq \bar{e} < \varepsilon\,,$$
$$\alpha - \bar{a}_p \leqq \bar{e} < \varepsilon\,;$$

i. e.,

$$|\alpha - \bar{a}_p| < \varepsilon \quad \text{if} \quad p \geqq N(e)\,. \quad \square$$

We now state and prove our main result about $\bar{F}$.

(5.29) Theorem. *The field $\bar{F}$ is complete.*

Proof. Let (α_p) be any Cauchy sequence in $\bar{F}$. If (α_p) is ultimately constant, there is nothing to prove; if not, there exists a subsequence $(\alpha_{p_l})_{l=1}^{\infty}$ such that $\alpha_{p_l} \neq \alpha_{p_{l+1}}$ for $l = 1, 2, \ldots$. By (5.26), it suffices to prove that $(\alpha_{p_l})_{l=1}^{\infty}$ has a limit. Hence we suppose with no loss of generality that $(\alpha_p)_{p=1}^{\infty}$ is such that $\alpha_p \neq \alpha_{p+1}$ $(p = 1, 2, 3, \ldots)$. Write $\bar{0} < |\alpha_p - \alpha_{p+1}| = \mu_p$.

For all $\varepsilon > \bar{0}$ in $\bar{F}$, there exists $N(\varepsilon)$ such that

$$|\alpha_p - \alpha_q| < \varepsilon \quad \text{if} \quad p, q \geqq N(\varepsilon)\,;$$

in particular,

$$\mu_p < \varepsilon \quad \text{if} \quad p \geqq N(\varepsilon)\,.$$

Using (5.28), we choose $a_p \in F$ such that $|\bar{a}_p - \alpha_p| < \mu_p$ $(p = 1, 2, 3, \ldots)$. Now choose any $e > 0$ in F. For $p, q \geqq N\left(\frac{1}{3}\bar{e}\right)$, we have

$$\begin{aligned} |\bar{a}_p - \bar{a}_q| &\leqq |\bar{a}_p - \alpha_p| + |\alpha_p - \alpha_q| + |\alpha_q - \bar{a}_q| \\ &< \mu_p + \frac{1}{3}\bar{e} + \mu_q < \frac{1}{3}\bar{e} + \frac{1}{3}\bar{e} + \frac{1}{3}\bar{e} = \bar{e}\,. \end{aligned}$$

Since the mapping τ of (5.24) is an order-preserving isomorphism, it follows that $|a_p - a_q| < e$ if $p, q \geqq N\left(\frac{1}{3}\bar{e}\right)$, *i. e.*, $(a_p) \in \mathfrak{C}$. Define β as $(a_p) + \mathfrak{N}$.

We claim that $\lim_{p\to\infty} \alpha_p = \beta$. To prove this, choose any positive ε in $\bar{F}$. For $p \geqq N\left(\frac{1}{2}\varepsilon\right)$, we have $|\bar{a}_p - \alpha_p| < \mu_p < \frac{1}{2}\varepsilon$. Also (5.28) shows that there is a positive integer $M\left(\frac{1}{2}\varepsilon\right)$ such that $|\bar{a}_p - \beta| < \frac{1}{2}\varepsilon$ for $p \geqq M\left(\frac{1}{2}\varepsilon\right)$. Hence

$$\begin{aligned} |\alpha_p - \beta| &\leqq |\alpha_p - \bar{a}_p| + |\bar{a}_p - \beta| \\ &< \frac{1}{2}\varepsilon + \frac{1}{2}\varepsilon = \varepsilon \end{aligned}$$

if $p \geqq \max\left\{N\left(\frac{1}{2}\varepsilon\right), M\left(\frac{1}{2}\varepsilon\right)\right\}$. $\square$

(5.30) Theorem. *For any ordered field F, $\overline{F}$ is isomorphic with $\overline{F}$; every Cauchy sequence of $\overline{F}$ differs from a constant sequence by a null sequence.*

Proof. For a Cauchy sequence (α_p) of elements of $\overline{F}$, let $\beta = \lim_{p\to\infty} \alpha_p$ (5.29). Then $(\alpha_p - \beta_{(p)})$ is a null sequence, and the theorem follows. □

(5.31) Theorem. *In any Archimedean ordered field, the sequence $(2^{-p})_{p=1}^{\infty}$ is null.*

Proof. We have $2^p = \sum_{k=0}^{p} \binom{p}{k} > p$, and so $2^{-p} < \frac{1}{p}$. Since $\left(\frac{1}{p}\right)_{p=1}^{\infty}$ is null, (2^{-p}) is also null. □

(5.32) Definition. Let F be an ordered field and $\varnothing \subsetneqq A \subset F$. An element $b \in F$ is said to be an *upper* [*lower*] *bound for A* if $x \leqq b$ [$x \geqq b$] for all $x \in A$. An upper bound b is called the *least upper bound* or *supremum* of A, and we write $b = \sup A$, if b is less than all other upper bounds for A. The *greatest lower bound* or *infimum of A*, written $\inf A$, is defined analogously[1]. The notations l.u.b.A and g.l.b.A. are sometimes used for what we call $\sup A$ and $\inf A$.

(5.33) Theorem. *Let F be a complete Archimedean ordered field, and let A be a nonvoid subset of F that is bounded above* [*below*]. *Then* $\sup A$ [$\inf A$] *exists.*

Proof. Let b be any upper bound for A, and let $a \in A$. There exist positive integers M and $-m$ such that $M > b$ and $-m > -a$, *i.e.*, $m < a \leqq b < M$. For each positive integer p, let

$$S_p = \left\{k : k \text{ is an integer and } \frac{k}{2^p} \text{ is an upper bound for } A\right\}.$$

If $k \leqq 2^p m$, then k is not in S_p. Thus S_p is bounded below. Since we have $2^p M \in S_p$, S_p is nonvoid. It follows that S_p has a least element, say k_p. We define $a_p = \frac{k_p}{2^p}$ ($p = 1, 2, 3, \ldots$). By the definition of k_p, $\frac{2k_p}{2^{p+1}} = \frac{k_p}{2^p}$ is an upper bound for A and $\frac{2k_p - 2}{2^{p+1}} = \frac{k_p - 1}{2^p}$ is not. Therefore we have either

$$k_{p+1} = 2k_p \quad \text{or} \quad k_{p+1} = 2k_p - 1\,,$$

so that

$$a_{p+1} = \frac{2k_p}{2^{p+1}} = a_p \quad \text{or} \quad a_{p+1} = \frac{2k_p - 1}{2^{p+1}} = a_p - \frac{1}{2^{p+1}}\,,$$

and hence

$$a_{p+1} \leqq a_p \quad \text{and} \quad a_p - a_{p+1} \leqq \frac{1}{2^{p+1}} \qquad (p = 1, 2, 3, \ldots).$$

---

[1] Clearly we can also define suprema and infima in arbitrary partially ordered sets.

If $q > p \geqq 1$, then

$$0 \leqq a_p - a_q = (a_p - a_{p+1}) + (a_{p+1} - a_{p+2}) + \cdots + (a_{q-1} - a_q)$$
$$\leqq \frac{1}{2^{p+1}} + \frac{1}{2^{p+2}} + \cdots + \frac{1}{2^q} = \frac{1}{2^{p+1}}\left(1 + \frac{1}{2} + \cdots + \frac{1}{2^{q-p-1}}\right)$$
$$= \frac{1}{2^{p+1}}\left(2 - \frac{1}{2^{q-p-1}}\right) < \frac{1}{2^p}.$$

We thus have $|a_p - a_q| = a_p - a_q < \frac{1}{2^p}$ whenever $q > p \geqq 1$. From (5.31) we infer that (a_p) is a Cauchy sequence, and so $\lim\limits_{p \to \infty} a_p$ exists; call it c. It is plain that $a_p \geqq c$.

We claim that $\sup A = c$. To prove it, assume first that c is not an upper bound for A. Then there is an $x \in A$ such that $x > c$, and hence there is a positive integer p such that $a_p - c = |a_p - c| < x - c$; *i. e.*, $a_p < x$. Since a_p is an upper bound for A, the last inequality cannot obtain. Therefore c is an upper bound for A. Assume next that there exists an upper bound c' for A such that $c' < c$, and choose a positive integer p such that $\frac{1}{2^p} < c - c'$. We then have $a_p - \frac{1}{2^p} \geqq c - \frac{1}{2^p} > c + c' - c = c'$, and so $a_p - \frac{1}{2^p}$ is an upper bound for A. However, $a_p - \frac{1}{2^p}$ is by definition $\frac{k_p - 1}{2^p}$, and $\frac{k_p - 1}{2^p}$ is not an upper bound for A. It follows that $c = \sup A$.

A similar proof can be given that $\inf A$ exists if A is bounded below; or it can be shown that

$$\inf A = -\sup(-A). \quad \square$$

(5.34) Theorem. *Any two complete Archimedean ordered fields F_1 and F_2, with sets of positive elements P_1 and P_2, respectively, are algebraically and order isomorphic,* i. e., *there exists a one-to-one mapping τ of F_1 onto F_2 such that*

$$\tau(x + y) = \tau(x) + \tau(y),$$
$$\tau(xy) = \tau(x)\,\tau(y),$$
$$\tau(x) \in P_2 \quad \textit{if and only if} \quad x \in P_1.$$

Proof. Let 1_1 and 1_2 be the units of F_1 and F_2 and 0_1 and 0_2 the zeros. The mapping τ [cf. (5.6.e)] is first defined on the rational elements of F_1; thus:

$$\tau(1_1) = 1_2;$$
$$\tau(0_1) = 0_2;$$
$$\tau(m\,1_1) = m\,1_2, \quad \text{where } m \text{ is an integer};$$
$$\tau\left(\frac{1}{n}\,1_1\right) = \frac{1}{n}\,1_2, \quad \text{where } n \text{ is a nonzero integer};$$
$$\tau\left(\frac{m}{n}\,1_1\right) = \frac{m}{n}\,1_2.$$

If $x \in F_1$ and x is not of the form $\frac{m}{n} 1_1$, then we define

$$\tau(x) = \sup\left\{\frac{m}{n} 1_2 : \frac{m}{n} 1_1 < x\right\}.$$

It is left to the reader to prove that τ has the desired properties. □

(5.35) Definition. The *real number field* is any complete Archimedean ordered field; *e. g.*, $\overline{Q}$. We will always denote this field by R.

(5.36) Exercise. Let F be any ordered field. For $a, b \in F$, prove that

$$\max\{a, b\} = \frac{1}{2}(|a - b| + a + b),$$

$$\min\{a, b\} = \frac{1}{2}(-|a - b| + a + b).$$

(5.37) Exercise. Let F be any ordered field, and let a, b, c be any elements of F. Define

$$\text{mode}\{a, b, c\} \quad \text{as} \quad \min\{\max\{a, b\}, \max\{b, c\}, \max\{a, c\}\}.$$

Describe the mode in words, and write it in terms of absolute values and the field operations.

(5.38) Exercise. Let F be any ordered field. A subset D of F is called a *Dedekind cut in F* if:

(i) $\varnothing \subsetneqq D \subsetneqq F$;

(ii) the relations $x \in D$ and $y < x$ imply $y \in D$.

(a) Let D be a Dedekind cut in R. Prove that $D = \{x \in R : x < a\}$ for some $a \in R$ or $D = \{x \in R : x \leqq a\}$ for some $a \in R$.

(b) If F is an ordered field not order isomorphic to R, prove that F contains a Dedekind cut that is of neither of these two forms.

(c) Using (a) for the field R, prove that every positive real number has a unique positive k^{th} root ($k = 2, 3, 4, \ldots$).

(5.39) Exercise. Consider the field of all rational functions with coefficients in Q in a single indeterminate t, and denote this field by the symbol $Q(t)$. Thus a generic nonzero element of $Q(t)$ has the form $\frac{A(t)}{B(t)}$, where $A(t) = \sum_{k=0}^{n} a_k t^k$ and $B(t) = \sum_{j=0}^{m} b_j t^j$. The numbers a_k and b_j are in Q, and $a_n \neq 0$ and $b_m \neq 0$. Addition and multiplication are defined as usual. We order $Q(t)$ by the rule that $\frac{A(t)}{B(t)}$ is in P if and only if $a_n b_m$ is a positive rational number. Prove that $Q(t)$ is an ordered field and that the order is non-Archimedean. Prove also that *every* non-Archimedean ordered field contains a subfield algebraically and order isomorphic with $Q(t)$. Find the completion of $Q(t)$.

(5.40) Exercise. Let $(a_n)_{n=1}^{\infty}$ be any sequence of integers all greater than 1. Prove that every real number x such that $0 \leqq x < 1$ has an expansion of the form

$$\sum_{k=1}^{\infty} \frac{x_k}{a_1 a_2 \cdots a_k},$$

where each $x_k \in \{0, 1, \ldots, a_k - 1\}$. Find a necessary and sufficient condition for two distinct expansions to be the same real number.

We next construct the complex number field, a much simpler process than our construction of R.

(5.41) Theorem. *Consider the ring $R[t]$ of all polynomials in the indeterminate t with coefficients in R, and with addition and multiplication defined as usual. Let $J = \{(t^2+1)\,p(t) : p(t) \in R[t]\}$. Then J is an ideal in $R[t]$. Let $R[t]/J$ be the set of cosets $p(t) + J$. Addition and multiplication in $R[t]/J$ are defined by*

$$(p(t) + J) + (q(t) + J) = (p(t) + q(t)) + J$$

and

$$(p(t) + J)\,(q(t) + J) = (p(t)\,q(t)) + J\,.$$

These definitions are unambiguous, and with addition and multiplication so defined, $R[t]/J$ is a field.

Proof. It is obvious that J is an ideal in $R[t]$. Exactly as in the proof of (5.22), we see that the definitions of addition and multiplication in $R[t]/J$ are unambiguous, and that $R[t]/J$ is a commutative ring with zero J and unit $1 + J$.

To describe $R[t]/J$ more closely, suppose that $(a + bt) + J = (a' + b't) + J$. Then $[(a + bt) - (a' + b't)] \in J$, and so $(a - a') + (b - b')t = (t^2 + 1)\,p(t)$, for some $p(t) \in R[t]$. Comparing the degrees of these two polynomials, we see that $p(t) = 0$ and that $a = a'$, $b = b'$. In other words, each element of the set $\{(a + bt) + J : (a, b) \in R \times R\}$ is a distinct element of $R[t]/J$. It is an elementary algebraic fact, whose proof we omit, that every $p(t) \in R[t]$ can be written

$$p(t) = (t^2 + 1)\,q(t) + r(t)\,,$$

where $q(t) \in R[t]$ and $r(t) = a + bt$. Thus the coset $p(t) + J$ is equal to $(t^2 + 1)\,q(t) + (a + bt) + J = (a + bt) + J$. This proves that $R[t]/J = \{a + bt + J : (a, b) \in R \times R\}$, where distinct pairs (a, b) yield distinct elements of $R[t]/J$.

Routine computations show that

$$((a + bt) + J) + ((a' + b't) + J) = ((a + a') + (b + b')\,t) + J$$

and that

$$((a + bt) + J)((a' + b't) + J) = ((aa' - bb') + (ab' + a'b)t) + J\,.$$

If $(a + bt) + J \neq J$, then $a \neq 0$ or $b \neq 0$. Since R is an ordered field, we have $a^2+b^2>0$, and so $\frac{1}{a^2+b^2}$ exists in R. It is clear that

$$[(a+bt)+J]\left[\left(\frac{a}{a^2+b^2}-\frac{b}{a^2+b^2}t\right)+J\right]=1+J\,.$$

This shows that every nonzero element of $R[t]/J$ has a multiplicative inverse, and so $R[t]/J$ is a field. □

(5.42) Definitions. The field $R[t]/J$ is called the *complex number field* or the *field of complex numbers* and is denoted by the symbol K. We write the coset $(a+bt)+J$ as $a+bi$; $a+bi$ is called a *complex number*. The number a is called the *real part of* $a+bi$ and is written $\mathrm{Re}(a+bi)$. The number b is called the *imaginary part of* $a+bi$ and is written $\mathrm{Im}(a+bi)$. The symbols $z=x+iy$, $w=u+iv$, $\sigma+i\tau$, $\alpha+\beta i$, etc., will be used to denote complex numbers. The complex number $a+0i$ will be written as a alone and $0+bi$ as bi alone. For $z=x+iy\in K$, the *absolute value of* z is defined as $(x^2+y^2)^{\frac{1}{2}}$ [the nonnegative square root!] and is written $|z|$. The *complex conjugate of* z [or simply *conjugate*] is defined as $x-iy$ and is written $\bar{z}$.

(5.43) Theorem. *The field K cannot be ordered.*

Proof. Assuming the existence in K of a subset P as in (5.7), we have $i\in P$ or $-i\in P$. If $i\in P$, then $i^2=-1\in P$, which contradicts (5.8). If $-i\in P$, then $(-i)^2=-1\in P$, also a contradiction. □

(5.44) Theorem. *For all $z, z_1, z_2\in K$ we have:*

(i) $\bar{\bar{z}}=z$;

(ii) $\overline{z_1+z_2}=\bar{z}_1+\bar{z}_2$;

(iii) $\overline{z_1z_2}=\bar{z}_1\bar{z}_2$.

Proof. Routine calculation.

(5.45) Remark.[1] The foregoing theorem shows that conjugation is an automorphism of K. The field R has no automorphisms save the identity. In fact let φ be a function with domain R, range contained in R, $\varphi(R)\neq\{0\}$, and such that $\varphi(x+y)=\varphi(x)+\varphi(y)$, $\varphi(xy)=\varphi(x)\varphi(y)$. It is easy to show that $\varphi(1)=1$, $\varphi(0)=0$, and in general that $\varphi(r)=r$ for all $r\in Q$. If $x\neq 0$ and $\varphi(x)=0$, then

$$1=\varphi(1)=\varphi\left(x\frac{1}{x}\right)=\varphi(x)\;\varphi\left(\frac{1}{x}\right)=0\,.$$

Hence $\varphi(x)\neq 0$ if $x\neq 0$. If $a<b$, then

$$\varphi(b)-\varphi(a)=\varphi(b-a)=\varphi\left(\left((b-a)^{\frac{1}{2}}\right)^2\right)=\left(\varphi\left((b-a)^{\frac{1}{2}}\right)\right)^2>0\,.$$

Hence $\varphi(a)<\varphi(b)$ if $a<b$. For an arbitrary real number x, choose $r_1, r_2\in Q$ such that $r_1<x<r_2$. Then

$$r_1=\varphi(r_1)<\varphi(x)<\varphi(r_2)=r_2\,.$$

---

[1] Subheads (5.45) and (5.46) are included only for cultural interest and are not referred to in the sequel.

Since $r_2 - r_1$ can be made arbitrarily small, it follows that $\varphi(x) = x$.

(5.46) The functional equation $\varphi(x+y) = \varphi(x) + \varphi(y)$ has $2^{\mathfrak{c}}$ discontinuous solutions on R. In fact, regard R as a vector space over Q (3.17.c) and let B be a Hamel basis for R over Q (3.19). For each $x \in R$ let α_x denote that unique function from B into Q as in (3.20) such that $x = \sum_{b \in B} \alpha_x(b)\, b$. Now for each $f \in R^B$ define $\varphi_f : R \to R$ by the rule

$$\varphi_f(x) = \sum_{b \in B} \alpha_x(b)\, f(b) .$$

The reader can easily verify that each such φ_f satisfies the desired functional equation and that $\varphi_f(rx) = r\varphi_f(x)$ for $r \in Q$, $x \in R$. Thus $\varphi_f(r) = r\varphi_f(1)$ $(r \in Q)$, so that if φ_f is continuous, then $\varphi_f(x) = x\varphi_f(1)$ for all $x \in R$. Since $\varphi_f(1)$ has just $\mathfrak{c}$ possible values we see that there are just $\mathfrak{c}$ continuous φ_f's. But $\overline{\overline{R^B}} = \mathfrak{c}^{\mathfrak{c}} = 2^{\mathfrak{c}}$ [see (4.34)] and $f \neq g$ in R^B implies $\varphi_f \neq \varphi_g$, so there exist $2^{\mathfrak{c}}$ discontinuous φ_f's. The preceding paragraph shows that the additional requirement that $\varphi(xy) = \varphi(x)\,\varphi(y)$ forces φ to be continuous.

To illustrate the bizarre nature of some of these additive functions, define $\psi(x) = \sum_{b \in B} \alpha_x(b)$ for $x \in R$, *i. e.*, $\psi = \varphi_f$ where $f(b) = 1$ for each $b \in B$. Now consider $b_1 \neq b_2$ in B, $c < d$ in R, and $r \in Q$.
Next choose $s \in Q$ such that $c < rb_1 + s(b_1 - b_2) < d$. Let

$$u = rb_1 + s(b_1 - b_2) = (r+s)\, b_1 - sb_2.$$

Then $c < u < d$ and $\psi(u) = (r+s) - s = r$. Therefore $c < d$ in R implies that $\psi(\{x : c < x < d\}) = Q$. This function is wildly discontinuous.

The field K has $2^{\mathfrak{c}}$ automorphisms. This fact depends on the fact that K is algebraically closed. Only the identity $z \to z$ and conjugation $z \to \bar{z}$ are continuous in the usual topology on K (6.17).

(5.47) Theorem. *For* $z, w \in K$ *we have* $|zw| = |z| \cdot |w|$, $|z| = |\bar{z}|$, $|z|^2 = z\bar{z}$, $z + \bar{z} = 2\,\mathrm{Re}(z)$, *and* $z - \bar{z} = 2i\,\mathrm{Im}(z)$.

Proof. Computation.

(5.48) Lemma. *Let* $z = x + yi$ *be a complex number. Then* $|\mathrm{Re}(z)| \leqq |z|$, *and* $\mathrm{Re}(z) = |z|$ *if and only if* $x \geqq 0$ *and* $y = 0$. *Also* $|\mathrm{Im}(z)| \leqq |z|$, *and* $\mathrm{Im}(z) = |z|$ *if and only if* $x = 0$ *and* $y \geqq 0$.

Proof. The following relations are evident:

$$\begin{aligned} -|z| = -(x^2+y^2)^{\frac{1}{2}} \leqq -(x^2)^{\frac{1}{2}} = -|x| \leqq x \\ \leqq |x| = (x^2)^{\frac{1}{2}} \leqq (x^2+y^2)^{\frac{1}{2}} = |z| . \end{aligned}$$

Clearly $x = (x^2+y^2)^{\frac{1}{2}}$ if and only if $y = 0$ and $x \geqq 0$. The proof for $\mathrm{Im}(z)$ is the same. □

(5.49) Theorem. *For* $z, w \in K$, *we have* $|z+w| \leqq |z| + |w|$, *and equality holds if and only if* $\alpha z = \beta w$, *where* α *and* β *are nonnegative real numbers not both zero.*

Proof. Applying (5.47) and (5.48), we write

$$\begin{aligned} |z+w|^2 &= (z+w)(\bar{z}+\bar{w}) = z\bar{z} + w\bar{w} + z\bar{w} + \bar{z}w \\ &= |z|^2 + |w|^2 + 2\,\mathrm{Re}(z\bar{w}) \\ &\leqq |z|^2 + |w|^2 + 2|z\bar{w}| \\ &= |z|^2 + |w|^2 + 2|z|\,|w| \\ &= (|z| + |w|)^2 . \end{aligned}$$

This shows that $|z+w| \leqq |z| + |w|$. Equality holds if and only if $\mathrm{Re}(z\bar{w}) = |z\bar{w}|$, and so by (5.48) if and only if $z\bar{w}$ is a nonnegative real number. If $z = 0$, take $\alpha = 1$ and $\beta = 0$. If $w = 0$, take $\alpha = 0$ and $\beta = 1$ If $z \neq 0$ and $w \neq 0$ and $z\bar{w}$ is a positive real number β, then $z|w|^2 = z\bar{w}w = \beta w$, and we can take $\alpha = |w|^2 > 0$. □

(5.50) Geometric interpretation. As the reader will already know, the field K can be very usefully regarded as the Euclidean plane $R \times R$, in which the point (a, b) corresponds to the complex number $a + bi$. Thus the Euclidean distance between (a, b) and $(0, 0)$ is the absolute value of $a + bi$. Conjugation is simply reflection in the X-axis.

(5.51) Definition. Let $z = x + iy$ be a complex number different from 0. Then $\arg(z)$ is the set of all real numbers θ such that

$$\cos(\theta) = \frac{x}{|z|} \quad \text{and} \quad \sin(\theta) = \frac{y}{|z|} .$$

Any element θ of $\arg(z)$ such that $-\pi < \theta \leqq \pi$ will be denoted by $\mathrm{Arg}(z)$. We define $\arg(0) = R$ and do not define $\mathrm{Arg}(0)$.

(5.52) Theorem. *For every nonzero complex number z, $\arg(z)$ is a countably infinite set, and $\mathrm{Arg}(z)$ contains exactly one real number. If $\theta \in \arg(z)$, then $\arg(z) = \{\theta + 2\pi n : n \in Z\}$.*

Proof. We only sketch the proof; details may be found, for example, in Saks and Zygmund, *Analytic Functions*, pp. 62–64 [Monografie Matematyczne, Warszawa, Vol. 28 (1952)]. The real-valued functions

$$\sin(x) = \sum_{n=0}^{\infty} \frac{(-1)^n x^{2n+1}}{(2n+1)!} \quad \text{and} \quad \cos(x) = \sum_{n=0}^{\infty} \frac{(-1)^n x^{2n}}{(2n)!}$$

are defined, continuous, and in fact infinitely differentiable for all $x \in R$. In particular, $\sin(0) = 0$ and $\cos(0) = 1$. The number $\frac{\pi}{2}$ is defined as the least positive zero of cos. One then proves that for every pair (c, d) of real numbers such that $c^2 + d^2 = 1$, there is a unique real number θ such that $-\pi < \theta \leqq \pi$, $\cos(\theta) = c$, and $\sin(\theta) = d$. This number is $\mathrm{Arg}(z)$. One also shows that $\cos(\theta) = \cos(\theta + 2\pi n)$ and $\sin(\theta) = \sin(\theta + 2\pi n)$ for all $\theta \in R$ and $n \in Z$, and that 2π is the smallest period of cos and of sin. These facts imply the last two statements of the present theorem. □

(5.53) Exercise. For a nonzero complex number z, prove that $\arg\left(\frac{1}{z}\right) = -\arg(z)$. If z is not a negative real number, prove that

$\text{Arg}\left(\frac{1}{z}\right) = -\text{Arg}(z)$. If z is a negative real number, prove that $\text{Arg}(z) = \text{Arg}\left(\frac{1}{z}\right) = \pi$.

(5.54) Exercise. For z and w nonzero complex numbers, prove that $\arg(zw) = \arg(z) + \arg(w)$. For every positive integer k, $\arg(z^k) = k \cdot \arg(z)$.

(5.55) Recapitulation.

In the accompanying Figure 3 we illustrate addition and multiplication of complex numbers. Addition is componentwise; graphically, one applies the parallelogram law for addition of vectors. Multiplication is a little more complicated. We have

$$\text{Arg}(zw) = \text{Arg}(z) + \text{Arg}(w) \text{ modulo } 2\pi, \quad \text{and} \quad |zw| = |z| \cdot |w| .$$

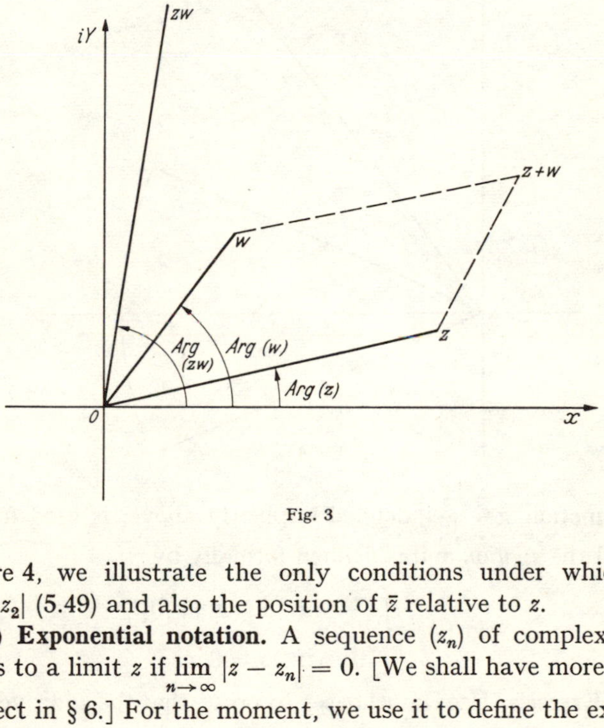

Fig. 3

In Figure 4, we illustrate the only conditions under which $|z_1 + z_2| = |z_1| + |z_2|$ (5.49) and also the position of $\bar{z}$ relative to z.

(5.56) Exponential notation. A sequence (z_n) of complex numbers converges to a limit z if $\lim_{n\to\infty} |z - z_n| = 0$. [We shall have more to say on this subject in § 6.] For the moment, we use it to define the exponential function exp by

$$\exp(z) = \sum_{n=0}^{\infty} \frac{z^n}{n!} = \lim_{k\to\infty} \sum_{n=0}^{k} \frac{z^n}{n!} .$$

Just as with real power series, one proves that $\exp(z)$ exists [*i. e.*, the limit exists] for all $z \in K$. The identity

$$\exp(z + w) = \exp(z)\exp(w)$$

holds and is proved by multiplying out $\left(\sum_{n=0}^{k} \frac{z^n}{n!}\right)\left(\sum_{m=0}^{k} \frac{w^m}{m!}\right)$ and taking the limit as $k \to \infty$. It is easy to show that

$$\exp(i\theta) = \cos(\theta) + i\sin(\theta)$$

for all $\theta \in R$ and so $|\exp(i\theta)| = 1$. Every nonzero complex number z can thus be written as

$$z = |z|\left(\frac{z}{|z|}\right) = |z| \exp(i\theta) = |z|\,(\cos(\theta) + i\sin(\theta))\,.$$

Here θ is any number in $\arg(z)$.

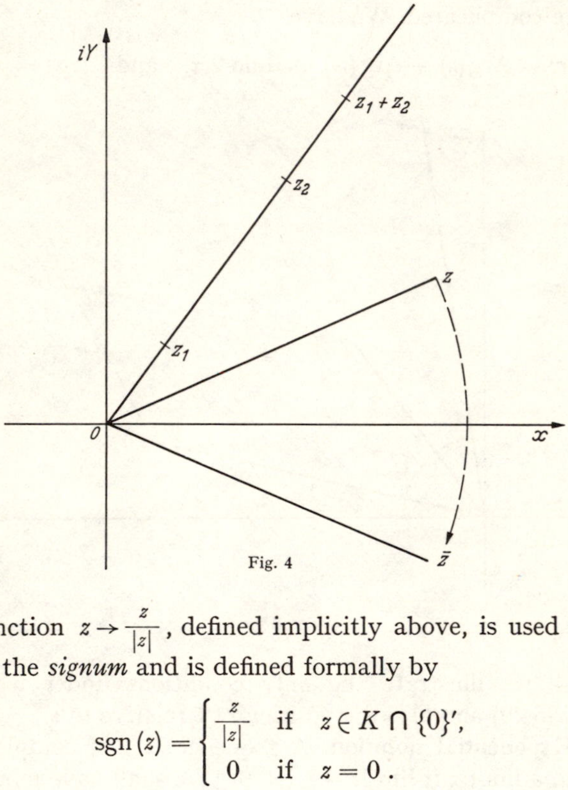

Fig. 4

The function $z \to \frac{z}{|z|}$, defined implicitly above, is used frequently. It is called the *signum* and is defined formally by

$$\operatorname{sgn}(z) = \begin{cases} \frac{z}{|z|} & \text{if} \quad z \in K \cap \{0\}', \\ 0 & \text{if} \quad z = 0\,. \end{cases}$$

(5.57) Exercise. Use Hamel bases to prove that the additive groups R and K are isomorphic.

(5.58) Exercise. Define addition in K as usual and define multiplication "coordinatewise":

$$(x + iy)\,(u + iv) = xu + iyv\,.$$

Prove that, with these operations, K is a commutative ring with unit. Prove also that K is not a field.

CHAPTER TWO

Topology and Continuous Functions

The main goal of this text is to give a complete presentation of integration and differentiation. Plainly a detailed study of set-theoretic topology would be out of place here. Similarly, a detailed treatment of continuous functions is outside our purview. Nevertheless, topology and continuity can be ignored in no study of integration and differentiation having a serious claim to completeness.

First of all, there is an intimate connection between measure theory [which is almost coextensive with the theory of integration] and the topological notion of compactness. Second, many important facts in the theories of integrals and derivatives rest in the end on properties of continuous functions. Third, purely topological notions play a vital part both in constructing the objects studied in abstract analysis and in carrying out proofs. Fourth, a great many proofs are just as simple for arbitrary topological spaces as they are for the real line.

Therefore, in asking the reader to consider constructions involving topological spaces far more general than the line, we ask for a not inconsiderable preliminary effort, as the length of § 6 will show. In return, we promise a much more thorough presentation of contemporary analysis.

Section 6 is a self-contained if rather terse treatment of those parts of set-theoretic topology that have proved important for analysis. With some reluctance we have omitted the topics of paracompactness and compactifications of completely regular spaces. But a line had to be drawn somewhere. In § 7, we embark on a study of continuous functions and of functions closely related to continuous functions. We are particularly concerned with spaces of such functions and properties that they may have. The section culminates with the STONE-WEIERSTRASS theorem, surely an indispensable tool for every analyst.

§ 6. Topological preliminaries

Set-theoretic topology is the study of abstract forms of the notions of *nearness*, *limit point*, and *convergence*. Consider as a special but extremely important case the line R. For $x, y \in R$, we can define the distance between x and y as the absolute value of $x - y$:

$$\varrho(x, y) = |x - y| .$$

We can also associate with a point $x \in R$ all of the points $y \in R$ such that $\varrho(x, y)$ is less than a specified positive number α, *i. e.*, the open interval $]x - \alpha, x + \alpha[$. This gives us a systematic notion of nearness in R based on the distance-function ϱ. Even a more general idea is needed. A subset A of R is called *open* if it contains all points "sufficiently close" to each of its members. The abstract and axiomatized notion of open set is one common and convenient way to approach the study of set-theoretic topology.

To make all of this precise, we begin with some definitions. We first extend the real number system, and describe certain important subsets of the extended real numbers.

(6.1) Definition. (a) Let ∞ and $-\infty$ be two distinct objects, neither of which is a real number[1]. The set $R^{\#} = R \cup \{-\infty\} \cup \{\infty\}$ is known as the set of *extended real numbers*. We make $R^{\#}$ a linearly ordered set by taking the usual ordering in R and defining $-\infty < \infty$ and $-\infty < x < \infty$ for each $x \in R$. For a and b in $R^{\#}$ such that $a < b$, the sets

$$]a, b[= \{x \in R^{\#} : a < x < b\},$$
$$[a, b] = \{x \in R^{\#} : a \leqq x \leqq b\},$$
$$[a, b[= \{x \in R^{\#} : a \leqq x < b\},$$

and

$$]a, b] = \{x \in R^{\#} : a < x \leqq b\}$$

are called *intervals*, with *endpoints* a and b. The interval $]a, b[$ is an *open interval*, $[a, b]$ is a *closed interval*, and $[a, b[$ and $]a, b]$ are *half-open intervals*. If a and b are in R, these intervals are said to be *bounded*. Otherwise they are said to be *unbounded*. Note that R itself is the open interval $]-\infty, \infty[$ and that all intervals under our definition have cardinal number $\mathfrak{c}$.

(b) For future use, we define sums and products in $R^{\#}$, with a few restrictions, by the following rules. For $x, y \in R \subset R^{\#}$, $x + y$ and xy are defined as usual. For $x \in R$, we define:

$$\infty + x = x + \infty = \infty;$$
$$(-\infty) + x = x + (-\infty) = x - \infty = -\infty;$$

we also define

$$\infty + \infty = \infty;$$
$$(-\infty) + (-\infty) = -\infty - \infty = -\infty;$$
$$-(\infty) = -\infty \quad \text{and} \quad -(-\infty) = \infty;$$

the expressions $\infty + (-\infty)$ and $(-\infty) + \infty$ are not defined. For $x \in R$

[1] Many writers use the symbol $+\infty$ for what we write as ∞. The $+$ sign is a mere nuisance and so we omit it.

and $x > 0$, we define:

$$\infty \cdot x = x \cdot \infty = \infty;$$
$$(-\infty) \cdot x = x \cdot (-\infty) = -\infty.$$

We define

$$\infty \cdot 0 = 0 \cdot \infty = (-\infty) \cdot 0 = 0 \cdot (-\infty) = 0,$$

and

$$\infty \cdot \infty = \infty.$$

For $x \in R$ and $x < 0$, we define

$$\infty \cdot x = x \cdot \infty = -\infty;$$
$$(-\infty) \cdot x = x \cdot (-\infty) = \infty;$$

the expressions $\infty \cdot (-\infty)$, $(-\infty) \cdot \infty$, and $(-\infty) \cdot (-\infty)$ are not defined.

Open subsets of R are defined in terms of open intervals, as follows.

(6.2) Definition. A set $U \subset R$ is said to be *open* if for each $x \in U$ there is a positive real number ε such that $]x - \varepsilon, x + \varepsilon[\subset U$.

Thus we may say informally that a subset U of R is open if for each of its points x, it contains all points y such that y is sufficiently close to x. [The "sufficiently close" depends of course on x.] Plainly every open interval is an open set. Some important properties of open subsets of R are listed in the following theorem.

(6.3) Theorem. *Let $\mathcal{O}$ denote the family of all open subsets of R. Then:*

(i) *$\varnothing \in \mathcal{O}$ and $R \in \mathcal{O}$;*

(ii) *if $\mathcal{U}$ is a subfamily of $\mathcal{O}$, then $\bigcup \mathcal{U} \in \mathcal{O}$* [1]*;*

(iii) *if $\{U_1, U_2, \ldots, U_n\} \subset \mathcal{O}$, then $U_1 \cap U_2 \cap \cdots \cap U_n \in \mathcal{O}$.*

The reader can supply the proof of this very simple result.

The properties of open sets in R given in Theorem (6.3) form the basis for the concept of a topological space, which we now define.

(6.4) Definition. Let X be a set and $\mathcal{O}$ a family of subsets of X with the following properties:

(i) $\varnothing \in \mathcal{O}$ and $X \in \mathcal{O}$;

(ii) if $\mathcal{U}$ is a subfamily of $\mathcal{O}$, then $\bigcup \mathcal{U} \in \mathcal{O}$;

(iii) if $\{U_1, U_2, \ldots, U_n\} \subset \mathcal{O}$, then $U_1 \cap U_2 \cap \cdots \cap U_n \in \mathcal{O}$.

[That is, $\mathcal{O}$ is closed under the formation of arbitrary unions and of finite intersections.] Then $\mathcal{O}$ is called a *topology for* X and the pair $(X, \mathcal{O})$ is called a *topological space*. When confusion appears impossible, we will call X itself a topological space. The members of $\mathcal{O}$ are called *open sets in* X.

Definition (6.4) by itself is rather barren. No great number of exciting theorems can be proved about arbitrary topological spaces. However, certain entities definable in terms of open sets are of considerable interest in showing the connections among various topological concepts and in

[1] Recall that the union of a void family is $\varnothing$.

showing the simple nature of such ideas as continuity. Also by the addition of only two more axioms, we obtain a class of topological spaces in which many of the processes of analysis can be carried out with great profit. We first give a few examples of topological spaces.

(6.5) Examples. (a) Naturally, the real line R with the topology described in (6.2) and (6.3) is a topological space. This topology for R is known as the *usual topology for R*. We will always suppose that R is equipped with its usual topology unless the contrary is specifically stated.

(b) Consider $R^{\#}$ and the family $\mathcal{O}^{\#}$ of all sets having any of the following four forms: U, $U \cup]t, \infty]$, $U \cup [-\infty, s[$, $U \cup [-\infty, s[\cup]t, \infty]$, where U is an open subset of R and $s, t \in R$. Then $(R^{\#}, \mathcal{O}^{\#})$ is a topological space; $\mathcal{O}^{\#}$ is called the *usual topology for* $R^{\#}$.

(c) Let X be any set. The pair $(X, \mathcal{P}(X))$ is a topological space, obviously. The family $\mathcal{P}(X)$ is called the *discrete topology for X*. A set X with the discrete topology will frequently be denoted by X_d.

(d) Let X be any set and let $\mathcal{O} = \{X, \varnothing\}$. Then $\mathcal{O}$ is called the *indiscrete* [or *concrete*] *topology for X*. This topology is of little interest to us.

We proceed to the definition of various topological concepts from our basic notion of open sets.

(6.6) Definitions. Let X be a topological space. A *neighborhood of a point* $x \in X$ is any open subset U of X such that $x \in U$. The space X is known as a *Hausdorff space* if each pair of distinct points of X have disjoint neighborhoods. A set $A \subset X$ is said to be *closed* if $X \cap A'$ is open. For $x \in X$ and $A \subset X$, we say that x is a *limit point of A* if $(U \cap \{x\}') \cap A \neq \varnothing$ for each neighborhood U of x. For $A \subset X$, the *closure of A* is the set $A^- = \bigcap \{F : F \text{ is closed}, A \subset F \subset X\}$; the *interior of A* is the set $A^\circ = \bigcup \{U : U \text{ is open}, U \subset A\}$; and the *boundary of A* is the set $\partial A = A^- \cap (A')^-$.

(6.7) Theorem. *Let X be a topological space.*

(i) *The union of any finite collection of closed subsets of X is a closed set.*

(ii) *The intersection of any nonvoid family of closed subsets of X is a closed set.*

(iii) *The closure A^- of a subset A of X is the smallest closed set containing A, and A is closed if and only if $A = A^-$.*

(iv) *The interior A° of A is the largest open set contained in A, and A is open if and only if $A = A^\circ$.*

(v) *A subset A of X is closed if and only if it contains all of its limit points.*

For subsets A and B of X, we have:

(vi) $A^\circ = A'^{-\prime}$;

(vii) $\partial A = A^- \cap A^{\circ\prime}$;

(viii) $(A \cup B)^- = A^- \cup B^-$;

(ix) $(A \cap B)^\circ = A^\circ \cap B^\circ$.

For an arbitrary family $\{A_\iota\}$ *of subsets of* X, *we have:*

(x) $\bigcup A_\iota^- \subset (\bigcup A_\iota)^-$;

(xi) $\bigcap A_\iota^\circ \supset (\bigcap A_\iota)^\circ$.

Finally,

(xii) $\varnothing$ *and* X *are closed.*

Proof. Assertions (i) and (ii) follow at once from de MORGAN's laws (1.9.iii) and (1.9.iv) applied to axioms (6.4.iii) and (6.4.ii) for open sets.

Since A^- is the intersection of all closed supersets of A, assertion (ii) proves (iii). Assertion (iv) is all but obvious.

We next prove (v). Suppose first that A is closed. Then A' is a neighborhood of each point in A' and $A' \cap A = \varnothing$, so that no point of A' is a limit point of A, *i.e.*, A contains all of its limit points. Conversely, if no point of A' is a limit point of A, then for each $x \in A'$, there is a neighborhood U_x of x such that $U_x \cap A = \varnothing$, and therefore $A' = \bigcup \{U_x : x \in A'\}$ is open, *i.e.*, A is closed.

To prove (vi), we compute as follows:

$$\begin{aligned} A'^{-\prime} &= [\bigcap \{F : F \text{ is closed and } F \supset A'\}]' \\ &= \bigcup \{F' : F \text{ is closed and } F \supset A'\} \\ &= \bigcup \{F' : F' \text{ is open and } F' \subset A\} \\ &= A^\circ. \end{aligned}$$

Assertion (vii) is immediate from (vi) and the definition of ∂A.

To prove (viii), notice that $(A \cup B)^-$ is a closed set containing both A and B, so it must contain both A^- and B^-. Thus we have

$$(A \cup B)^- \supset A^- \cup B^-.$$

But $A^- \cup B^-$ is a closed set containing $A \cup B$, so that

$$(A \cup B)^- \subset A^- \cup B^-$$

and hence

$$(A \cup B)^- = A^- \cup B^-.$$

To prove (ix), we write

$$\begin{aligned} (A \cap B)^\circ &= (A \cap B)'^{-\prime} = (A' \cup B')^{-\prime} \\ &= (A'^- \cup B'^-)' = (A'^{-\prime} \cap B'^{-\prime}) \\ &= A^\circ \cap B^\circ. \end{aligned}$$

Assertion (x) follows from the inclusions $A_{\iota_0} \subset \bigcup A_\iota$ and $A_{\iota_0}^- \subset (\bigcup A_\iota)^-$, both of which are obvious for all indices ι_0. Assertion (xi) is obvious from (x), and (xii) from (6.4.i) and the definition of a closed set. □

(6.8) Definition. A topological space X is said to be *connected* if $\varnothing$ and X are the only subsets of X that are both open and closed.

(6.9) Theorem. *The space R with its usual topology is connected.*

Proof. Let A be a nonvoid subset of R which is both open and closed. Assume that $A \neq R$ and let $c \in R \cap A'$. Since $A \neq \varnothing$, we have either $A \cap]-\infty, c[\neq \varnothing$ or $A \cap]c, \infty[\neq \varnothing$. Suppose that $B = A \cap]-\infty, c[\neq \varnothing$ and let a be the supremum of this set (5.33). It is clear that $a \leqq c$. If $\varepsilon > 0$, then $a - \varepsilon$ is not an upper bound for B, and so there is some $x \in B$ such that $a - \varepsilon < x \leqq a$. This proves that every neighborhood of a meets A so, since A is closed, a is in A. Since A is open, there is a $\delta > 0$ such that $]a - \delta, a + \delta[\subset A$. Choose any $b \in R$ such that $a < b < \min\{a + \delta, c\}$. [Note that $a \neq c$ since $c \in A'$.] It follows that $b \in A$ and $b < c$, so that $b \in B$. The inequality $b > a$ contradicts the choice of a. A similar contradiction is obtained if $A \cap]c, \infty[\neq \varnothing$. We are thus forced to the conclusion that $A = R$. □

It is often convenient to define a topology not by specifying all of the open sets but only some of them.

(6.10) Definition. Let $(X, \mathcal{O})$ be a topological space. A family $\mathcal{B} \subset \mathcal{O}$ is called a *base for the topology* $\mathcal{O}$ if for each $U \in \mathcal{O}$ there exists some subfamily $\mathcal{A} \subset \mathcal{B}$ such that $U = \bigcup \mathcal{A}$. That is, every open set is a union of sets in $\mathcal{B}$. A subfamily $\mathcal{S}$ of $\mathcal{O}$ is called a *subbase for the topology* $\mathcal{O}$ if the family of all finite intersections of sets in $\mathcal{S}$ is a base for the topology $\mathcal{O}$.

(6.11) Theorem. *Let X be a set and let $\mathcal{B} \subset \mathcal{P}(X)$. Define $\mathcal{O} = \{\bigcup \mathcal{A} : \mathcal{A} \subset \mathcal{B}\}$. Then $(X, \mathcal{O})$ is a topological space, and $\mathcal{B}$ is a base for $\mathcal{O}$, if and only if*

(i) $\bigcup \mathcal{B} = X$

and

(ii) *$U, V \in \mathcal{B}$ and $x \in U \cap V$ imply that there exists $W \in \mathcal{B}$ such that $x \in W \subset U \cap V$.*

Proof. Suppose that $\mathcal{O}$ is a topology for X. Then $X \in \mathcal{O}$, so there exists $\mathcal{A} \subset \mathcal{B}$ such that $X = \bigcup \mathcal{A} \subset \bigcup \mathcal{B} \subset X$. That is, (i) is true. Next let U, V be sets in $\mathcal{B}$ and let $x \in U \cap V$. Then $U \cap V$ is in $\mathcal{O}$, so there is some $\mathcal{W} \subset \mathcal{B}$ such that $U \cap V = \bigcup \mathcal{W}$. Thus we have $x \in W \subset U \cap V$ for some $W \in \mathcal{W}$. This proves (ii).

Conversely, suppose that (i) and (ii) hold. We must show that $\mathcal{O}$ is a topology. Let $\{U_\iota\}_{\iota \in I}$ be any subfamily of $\mathcal{O}$. Then, by the definition of $\mathcal{O}$, for each ι there exists $\mathcal{A}_\iota \subset \mathcal{B}$ such that $U_\iota = \bigcup \mathcal{A}_\iota$. [Here we use the axiom of choice to choose just one $\mathcal{A}_\iota$ for each $\iota \in I$.] Let $\mathcal{A} = \bigcup_{\iota \in I} \mathcal{A}_\iota$. It is clear that $\mathcal{A} \subset \mathcal{B}$ and that $\bigcup \mathcal{A} = \bigcup_{\iota \in I} U_\iota$; thus $\mathcal{O}$ is closed under the formation of arbitrary unions. Next let U, V be in $\mathcal{O}$. Then there exist subfamilies $\{U_\iota\}_{\iota \in I}$ and $\{V_\eta\}_{\eta \in H}$ of $\mathcal{B}$ such that $U = \bigcup_{\iota \in I} U_\iota$ and $V = \bigcup_{\eta \in H} V_\eta$. Thus for each $x \in U \cap V$, there exist $\iota \in I$ and $\eta \in H$ such that $x \in U_\iota \cap V_\eta$ and therefore, by (ii), there is a W_x

in $\mathscr{B}$ such that $x \in W_x \subset U_\iota \cap V_\eta \subset U \cap V$. Let $\mathscr{A} = \{W_x : x \in U \cap V\}$. Then $\mathscr{A} \subset \mathscr{B}$ [$\mathscr{A}$ may be void!] and $U \cap V = \bigcup \mathscr{A} \in \mathcal{O}$. Thus $\mathcal{O}$ is closed under the formation of finite intersections. According to (i) X is in $\mathcal{O}$, and, since $\varnothing \subset \mathscr{B}$, we have $\varnothing = \bigcup \varnothing \in \mathcal{O}$. This proves that $\mathcal{O}$ is a topology for X. Clearly $\mathscr{B}$ is a base for $\mathcal{O}$. □

The function $(x, y) \to |x - y|$ defined on $R \times R$ is an obvious distance-function. An important although special class of topological spaces are those in which the topology can be defined from a reasonable distance-function. The axiomatic definition follows.

(6.12) Definition. Let X be a set and let ϱ be a function from $X \times X$ into R such that for all $x, y, z \in X$ we have:

(i) $\varrho(x, y) \geqq 0$;

(ii) $\varrho(x, y) = 0$ if and only if $x = y$;

(iii) $\varrho(x, y) = \varrho(y, x)$;

(iv) $\varrho(x, z) \leqq \varrho(x, y) + \varrho(y, z)$ [the *triangle inequality*].

Then ϱ is called a *metric* [or *distance-function*] *for* X; $\varrho(x, y)$ is called the *distance from* x *to* y, and the pair (X, ϱ) is called a *metric space*. When no confusion seems possible, we will refer to X as a metric space.

(6.13) Examples. (a) Let n be a positive integer, let $X = R^n$ or K^n, and let p be a real number such that $p \geqq 1$. For $\boldsymbol{x} = (x_1, \ldots, x_n)$ and $\boldsymbol{y} = (y_1, \ldots, y_n)$ in X, define

$$\varrho_p(\boldsymbol{x}, \boldsymbol{y}) = \left(\sum_{j=1}^{n} |x_j - y_j|^p\right)^{\frac{1}{p}} .$$

Properties (6.12.i)–(6.12.iii) are obvious for ϱ_p. The triangle inequality (6.12.iv) is a special case of MINKOWSKI's inequality, which we will prove in (13.7) *infra*.

The metric ϱ_2 is known as the *Euclidean metric on* R^n or K^n.

(b) For $\boldsymbol{x}, \boldsymbol{y} \in R^n$ or K^n define $\varrho(\boldsymbol{x}, \boldsymbol{y}) = \max\{|x_j - y_j| : 1 \leqq j \leqq n\}$. It is easy to verify that (K^n, ϱ) and (R^n, ϱ) are metric spaces.

(c) Let X be any set. For $x, y \in X$ define $\varrho(x, y) = \delta_{xy}$ [δ is KRONECKER's δ-symbol as in (2.20)]. Plainly ϱ is a metric. It is known as the *discrete metric for* X.

(d) Consider the set N^{N_0}, which we realize in concrete form as the set of all sequences $(a_k)_{k=1}^{\infty}$ of positive integers. For $\boldsymbol{a} = (a_k)$ and $\boldsymbol{b} = (b_k)$ in N^{N_0}, define:

$$\varrho(\boldsymbol{a}, \boldsymbol{b}) = 0 \quad \text{if} \quad \boldsymbol{a} = \boldsymbol{b};$$

$$\varrho(\boldsymbol{a}, \boldsymbol{b}) = \frac{1}{n} \quad \text{if} \quad a_1 = b_1,\, a_2 = b_2, \ldots, a_{n-1} = b_{n-1},$$

and $a_n \neq b_n$.

Then (N^{N_0}, ϱ) is a metric space.

(e) Let $D = \{z \in K : |z| \leqq 1\}$ be the closed unit disk in the complex plane. For $z, w \in D$ define

$$\varrho(z, w) = \begin{cases} |z - w| & \text{if } \arg(z) = \arg(w) \text{ or one of } z \text{ and } w \text{ is zero,} \\ |z| + |w| & \text{otherwise.} \end{cases}$$

Then (D, ϱ) is a metric space. This space is called the "French railroad space" or the "Washington D. C. space". A picture should be sketched to appreciate the reasons for these names. Actually this rather artificial-looking space is [essentially] a certain closed subset of the closed unit ball in a Hilbert space of dimension $\mathfrak{c}$. See (16.54) *infra*.

(6.14) Definition. Let (X, ϱ) be any metric space. For $\varepsilon > 0$ and $x \in X$, let

$$B_\varepsilon(x) = \{y \in X : \varrho(x, y) < \varepsilon\}.$$

This set is called the *ε-neighborhood of x* or the *open ball of radius ε centered at x.*

(6.15) Theorem. *Let (X, ϱ) be a metric space. Let*

$$\mathscr{B}_\varrho = \{B_\varepsilon(x) : \varepsilon > 0, x \in X\}.$$

Then $\mathscr{B}_\varrho$ is a base for a topology $\mathscr{O}_\varrho$ for X. We call $\mathscr{O}_\varrho$ the topology generated *by ϱ. The members of $\mathscr{O}_\varrho$ are called* ϱ-open sets.

Proof. We need only show that $\mathscr{B}_\varrho$ satisfies (6.11.i) and (6.11.ii). Property (6.11.i) is obvious. Let $B_\varepsilon(x)$ and $B_\delta(y)$ be in $\mathscr{B}_\varrho$ and let $z \in B_\varepsilon(x) \cap B_\delta(y)$. Then we have $\varrho(x, z) < \varepsilon$ and $\varrho(y, z) < \delta$. Define

$$\gamma = \min\{\varepsilon - \varrho(x, z), \delta - \varrho(y, z)\}.$$

Thus γ is positive, and for $u \in B_\gamma(z)$, we have $\varrho(x, u) \leqq \varrho(x, z) + \varrho(z, u) < (\varepsilon - \gamma) + \gamma = \varepsilon$ and $\varrho(y, u) \leqq \varrho(y, z) + \varrho(z, u) < (\delta - \gamma) + \gamma = \delta$. This proves that $B_\gamma(z) \subset B_\varepsilon(x) \cap B_\delta(y)$ and so (6.11.ii) is satisfied. □

(6.16) Remark. Restated slightly, (6.15) says that a set $U \subset X$ is ϱ-open if and only if for each $x \in U$ there is an $\varepsilon > 0$ such that $y \in U$ whenever $\varrho(x, y) < \varepsilon$. When we make statements of a topological nature about a metric space X, we will always mean the topology generated by the given metric, unless we make some explicit statement to the contrary.

(6.17) Exercise. Let n be a positive integer and let X denote either R^n or K^n. Prove that all of the metrics defined in (6.13.a) and (6.13.b) for X generate exactly the same topology for X, *i.e.*, any two of those metrics yield the same open sets. This topology is known as the *usual topology for R^n [K^n]*.

Every subset of a topological space can be made into a topological space in a natural way.

(6.18) Definition. Let $(X, \mathscr{O})$ be a topological space and let S be a subset of X. The *relative topology on S induced by $\mathscr{O}$* is the family $\{U \cap S : U \in \mathscr{O}\}$ and the set S with this topology is called a *subspace of X.*

Thus a set $V \subset S$ is *relatively open* if and only if $V = U \cap S$ for some set U that is open in X.

(6.19) Examples. (a) Let $X = R$ [with its usual topology] and let $S = [0, 1]$. Then the set $\left]\frac{1}{2}, 1\right]$ is open relative to $[0, 1]$ since $\left]\frac{1}{2}, 1\right] = \left]\frac{1}{2}, 2\right[\cap [0, 1]$ and $\left]\frac{1}{2}, 2\right[$ is open in R. However $\left]\frac{1}{2}, 1\right]$ is obviously not open in R.

(b) Let $X = R$ and $S = Q$. Then $[\sqrt{2}, \sqrt{3}\,] \cap Q$ is open relative to Q since $[\sqrt{2}, \sqrt{3}\,] \cap Q =]\sqrt{2}, \sqrt{3}\,[\cap Q$.

(c) Consider the set $L = \{(x, x) : x \in R\} \subset \{(x, y) : x \in R, y \in R\} = R^2$. With the usual topology for R^2, L has the usual topology for R.

(d) Consider the set $C = \left\{\left(\frac{1}{x} \cos(x), \frac{1}{x} \sin(x)\right) : x \in R, 0 < x < \infty\right\} \subset R^2$. The relative topology of C in R^2 [which has its usual topology] is the usual topology of $]0, \infty[$. [Identify C with $]0, \infty[$ in the natural way.]

We pass on to some additional important notions.

(6.20) Definition. A subset D of a topological space X is said to be *dense in* X if $D^- = X$. A space X is said to be *separable* if X contains a countable dense subset. A space X is said to have a *countable base* if there is a base for the topology of X which is a countable family.

(6.21) Example. The space R^n with its usual topology is separable since the set $D = \{(x_1, \ldots, x_n) : x_j \in Q, 1 \leq j \leq n\}$ is countable and dense. However R_d is not separable since each countable subset [like all subsets] of R_d is closed and R is uncountable. Also the French railroad space (6.13.e) is not separable.

(6.22) Theorem. *Any space with a countable base is separable.*

Proof. Let X be a space with a countable base $\mathscr{B}$. For each nonvoid $B \in \mathscr{B}$ let $x_B \in B$. Then the set $D = \{x_B : B \in \mathscr{B}\}$ is countable and dense. □

(6.23) Theorem. *Any separable metric space has a countable base.*

Proof. Let X be a metric space containing a countable dense subset D. Let $\mathscr{B} = \{B_r(x) : x \in D, r \in Q, r > 0\}$. Then $\mathscr{B}$ is countable. To see that $\mathscr{B}$ is a base, let U be open and let $z \in U$. Then there exists $\varepsilon > 0$ such that $B_\varepsilon(z) \subset U$. Since D is dense in X, there is an $x \in B_{\varepsilon/3}(z) \cap D$. Now choose a rational number r such that $\frac{1}{3}\varepsilon < r < \frac{2}{3}\varepsilon$. Then if $y \in B_r(x)$, we have

$$\varrho(y, z) \leq \varrho(x, y) + \varrho(x, z) < r + \frac{1}{3}\varepsilon < \varepsilon,$$

so that $B_r(x) \subset B_\varepsilon(z) \subset U$. Also

$$\varrho(x, z) < \frac{1}{3}\varepsilon < r,$$

so that $z \in B_r(x)$. Thus U is a union of members of $\mathscr{B}$. □

(6.24) Definition. A sequence $(x_n)_{n=1}^{\infty}$ in a topological space X is said to *converge to an element* $x \in X$, or *to have limit* x, if for each neighborhood U of x there exists a positive integer n_0 such that $x_n \in U$ whenever $n \geqq n_0$. We write $\lim_{n \to \infty} x_n = x$ and also $x_n \to x$ if $(x_n)_{n=1}^{\infty}$ converges to x.

(6.25) Theorem. *A subset A of a metric space X is closed if and only if whenever (x_n) is a sequence with values in A and (x_n) has limit x in X, we have $x \in A$.*

Proof. Suppose that A is closed and let (x_n) be a sequence with values in A for which a limit x in X exists. If x were in A', then A' would be a neighborhood of x, and so all but a finite number of the values x_n would lie in A' — a contradiction.

Conversely, suppose that A is not closed. Then by (6.7.v), A has a limit point x such that $x \notin A$. For each $n \in N$, choose $x_n \in A \cap B_{\frac{1}{n}}(x)$. Then $(x_n) \subset A$, $x_n \to x$, and $x \notin A$. □

(6.26) Theorem. *Let X be a Hausdorff space. Suppose that $A \subset X$ and that x is a limit point of A. Then each neighborhood of x contains infinitely many points of A.*

Proof. Exercise.

(6.27) Theorem. *Every metric space is a Hausdorff space.*

Proof. Exercise.

One of the most important concepts in topology is compactness. There are several versions of this concept, which we next discuss.

(6.28) Definition. If (x_n) is a sequence and $\{n_1 < n_2 < \cdots < n_k < \cdots\}$ is an infinite set of positive integers, then the sequence (x_{n_k}), defined by $k \to x_{n_k}$ for $k \in N$, is said to be a *subsequence of* (x_n).

(6.29) Definition. A topological space X is said to be *sequentially compact* if every sequence in X admits a subsequence converging to some point of X.

(6.30) Definition. A topological space X is said to be *Fréchet compact* [or to have the *Bolzano-Weierstrass property*] if every infinite subset of X admits a limit point in X.

Sequential compactness and Fréchet compactness are useful enough, but the most useful notion of this sort is compactness alone, which we now define.

(6.31) Definition. Let X be a topological space. A *cover of* X is any family $\mathscr{A}$ of subsets of X such that $\cup \mathscr{A} = X$. A cover in which each member is an open set is called an *open cover*. A subfamily of a cover which is also a cover is called a *subcover*.

(6.32) Definition. A topological space X is said to be *compact* if each open cover of X admits a finite subcover.

(6.33) Definition. A family of sets is said to have the *finite intersection property* if each finite subfamily has nonvoid intersection.

(6.34) Theorem. *A topological space X is compact if and only if each family of closed subsets of X having the finite intersection property has nonvoid intersection.*

Proof. This is nothing but an application of de MORGAN's laws (1.9). In fact, $\mathscr{U}$ is an open cover of X if and only if $\mathscr{F} = \{U' : U \in \mathscr{U}\}$ is a family of closed sets with void intersection. Thus every open cover has a finite subcover if and only if every family of closed sets having void intersection has a finite subfamily with void intersection. □

(6.35) Theorem. *Every compact topological space is Fréchet compact.*

Proof. Let X be a compact space. Assume that X has an infinite subset A with no limit points in X. Then A is a closed set (6.7.v). Moreover each $a \in A$ has a neighborhood U_a containing no point of $A \cap \{a\}'$. Then $\{U_a : a \in A\} \cup \{A'\}$ is an open cover of X with no finite subcover. This contradiction completes the proof. □

(6.36) Theorem. *Every sequentially compact metric space is separable.*

Proof. Let X be a sequentially compact metric space. TUKEY's lemma (3.8) shows that for each positive integer n there is a maximal subset A_n of X having the property that $\varrho(x, y) \geqq \frac{1}{n}$ for each pair of distinct points $x, y \in A_n$. Each A_n is a finite set since otherwise, for some n, A_n would have an infinite sequence of distinct points with no convergent subsequence. Thus the set $A = \bigcup_{n=1}^{\infty} A_n$ is countable. We assert that A is dense in X. If this is not the case, then there exists an $x \in X \cap A^{-\prime}$. Since $A^{-\prime}$ is open, there is an $\varepsilon > 0$ such that $B_\varepsilon(x) \subset A^{-\prime}$. Choose $n \in N$ such that $\frac{1}{n} < \varepsilon$. Then we have $\varrho(x, y) \geqq \varepsilon > \frac{1}{n}$ for each $y \in A_n$, and the existence of the set $A_n \cup \{x\}$ contradicts the maximality of A_n. It follows that $A^- = X$. □

(6.37) Theorem. *Let X be a metric space. Then the following three assertions are pairwise equivalent:*

(i) *X is compact;*

(ii) *X is Fréchet compact;*

(iii) *X is sequentially compact.*

Proof. The fact that (i) implies (ii) follows from (6.35). Suppose that (ii) holds and let (x_n) be a sequence with values in X. If (x_n) has only finitely many distinct terms, it is clear that there exists an infinite set $\{n_k : k \in N\} \subset N$ such that $n_1 < n_2 < \cdots$ and $x_{n_k} = x_{n_1}$ for each $k \in N$. In this case the subsequence (x_{n_k}) converges to x_{n_1}. Therefore we suppose that (x_n) has infinitely many distinct values. Then the set $\{x_n : n \in N\}$ has a limit point $x \in X$. Let $x_{n_1} = x_1$. Suppose that $x_{n_1}, \ldots, x_{n_k}$ have been chosen. Since each neighborhood of x contains infinitely many distinct x_n's, we choose $x_{n_{k+1}} \in B_{\frac{1}{k+1}}(x)$ such that $n_{k+1} > n_j$ $(1 \leqq j \leqq k)$.

Then the subsequence (x_{n_k}) converges to x. Thus (ii) implies (iii). Next suppose that (iii) holds. According to (6.36) X is separable so, by (6.23), X has a countable base $\mathscr{B}$. Now let $\mathscr{U}$ be any open cover of X. Let $\mathscr{A} = \{B \in \mathscr{B} : B \subset U \text{ for some } U \in \mathscr{U}\}$. For each $B \in \mathscr{A}$, choose $U_B \in \mathscr{U}$ such that $B \subset U_B$ and let $\mathscr{V} = \{U_B : B \in \mathscr{A}\}$. Clearly $\mathscr{V}$ is a countable family. If $x \in X$, then $x \in U$ for some $U \in \mathscr{U}$, and since $\mathscr{B}$ is a base, there is a $B \in \mathscr{B}$ such that $x \in B \subset U$. Then $B \in \mathscr{A}$ and $x \in B \subset U_B$. We conclude that $\mathscr{V}$ is a countable subcover of $\mathscr{U}$. Enumerate $\mathscr{V}$ in a sequence $\mathscr{V} = (V_n)$. For each $k \in N$ let $W_k = \bigcup_{n=1}^{k} V_n$. To prove (i), we need only show that $W_k = X$ for some $k \in N$. Assume that this is false. For each k choose $x_k \in X \cap W_k'$. Then (x_k) has a subsequence (x_{k_j}) converging to some $x \in X$. Since $\mathscr{V}$ is a cover there exists a $k_0 \in N$ such that $x \in V_{k_0} \subset W_{k_0}$. Thus W_{k_0} is a neighborhood of x which contains x_k for only finitely many k. This contradiction establishes the fact that (iii) implies (i). □

(6.38) Theorem. *Let X be a Hausdorff space and let A be a subspace of X that is compact in its relative topology. Then A is a closed subset of X.*

Proof. We will show that A' is open. Let $z \in A'$. For each $x \in A$ choose disjoint open sets U_x and V_x such that $x \in U_x$, $z \in V_x$. Then $\{U_x \cap A : x \in A\}$ is an open cover of A, in its relative topology, so there exists a finite set $\{x_1, \ldots, x_n\} \subset A$ such that $A \subset \bigcup_{j=1}^{n} U_{x_j}$. Let $V = \bigcap_{j=1}^{n} V_{x_j}$. Then V is a neighborhood of z and $V \cap A = \varnothing$, *i.e.*, $V \subset A'$. □

(6.39) Theorem. *Let X be a compact space and let A be a closed subset of X. Then A is a compact subspace of X.*

Proof. Let $\mathscr{F}$ be any family of closed [in the relative topology] subsets of A having the finite intersection property. Then each member of $\mathscr{F}$ is closed in X, so (6.34) implies that $\bigcap \mathscr{F} \neq \varnothing$. Thus A is compact, by (6.34). □

We next present a striking characterization of compactness which shows that we may restrict our attention to very special open covers in proving that a space is compact.

(6.40) Theorem [ALEXANDER]. *Let X be a topological space and let $\mathscr{S}$ be any subbase for the topology of X* [*see* (6.10)]. *Then the following two assertions are equivalent.*

(i) *The space X is compact.*

(ii) *Every cover of X by a subfamily of $\mathscr{S}$ admits a finite subcover.*

Proof. Obviously (i) implies (ii). To prove the converse, assume that (ii) holds and (i) fails. Consider the family Ж of all open covers of X without finite subcovers. The family Ж is partially ordered by inclusion, and plainly the union of a nonvoid chain in Ж is a cover in Ж. ZORN'S

Lemma (3.10) implies that $\mathfrak{K}$ contains a maximal cover $\mathscr{V}$. That is, $\mathscr{V}$ is an open cover of X, $\mathscr{V}$ has no finite subcover, and if U is any open set not in $\mathscr{V}$, then $\mathscr{V} \cup \{U\}$ admits a finite subcover. Let $\mathscr{W} = \mathscr{V} \cap \mathscr{S}$. Then no finite subfamily of $\mathscr{W}$ covers X, and so (ii) implies that $\mathscr{W}$ is not a cover of X. Let x be a point in $X \cap (\cup \mathscr{W})'$, and select a set V in the cover $\mathscr{V}$ that contains x. Since $\mathscr{S}$ is a subbase, there are sets $S_1, \ldots, S_n$ in $\mathscr{S}$ such that $x \in \bigcap_{j=1}^{n} S_j \subset V$. Since $x \notin (\cup \mathscr{W})$, no S_j is in $\mathscr{V}$. Since $\mathscr{V}$ is maximal, there exists for each j a set A_j which is the union of a finite number of sets in $\mathscr{V}$ such that $S_j \cup A_j = X$. Hence

$$V \cup \bigcup_{j=1}^{n} A_j \supset \left(\bigcap_{j=1}^{n} S_j\right) \cup \left(\bigcup_{j=1}^{n} A_j\right) = X,$$

and therefore X is a union of finitely many sets from $\mathscr{V}$. This contradicts our choice of $\mathscr{V}$. □

Another important class of topological spaces are those obtained by taking the Cartesian product of a given family of topological spaces. We need a definition.

(6.41) Definition. Let $\{X_\iota\}_{\iota \in I}$ be a nonvoid family of topological spaces and let $X = \mathop{\times}_{\iota \in I} X_\iota$ [see (3.1)]. For each $\iota \in I$, define π_ι on X by $\pi_\iota(\boldsymbol{x}) = x_\iota$. The function π_ι is known as the *projection of* X *onto* X_ι. We define the *product topology on the set* X by using as a subbase the family of all sets of the form $\pi_\iota^{-1}(U_\iota)$, where ι runs through I and U_ι runs through the open sets of X_ι. Thus a base for the product topology is the family of all finite intersections of inverse projections of open sets. A base for the product topology is the family of all sets of the form $\mathop{\times}_{\iota \in I} U_\iota$, where U_ι is open in X_ι for each $\iota \in I$ and $U_\iota = X_\iota$ for all but a finite number of the ι's. Whenever we discuss the Cartesian product of a family of topological spaces, it is to be understood that the product is endowed with the product topology unless the contrary is specified.

(6.42) Exercise. Let $I = \{1, 2, \ldots, n\}$ for some $n \in N$, and for each $\iota \in I$ let $X_\iota = R$ [or K] with its usual topology. Clearly $X = \mathop{\times}_{\iota \in I} X_\iota = R^n$ [or K^n]. Prove that the product topology on X is the usual topology on X.

(6.43) TIHONOV's Theorem[1]. *Let* $\{X_\iota\}_{\iota \in I}$ *be a nonvoid family of compact topological spaces. Then the Cartesian product* X *of these spaces is compact* [*in the product topology*].

Proof. According to ALEXANDER's theorem (6.40) it suffices to consider open covers of X by subbasic open sets as described in (6.41). Let $\mathscr{U}$ be any cover of X by subbasic open sets. For each $\iota \in I$, let $\mathscr{U}_\iota$ denote the family of all open sets $U \subset X_\iota$ such that $\pi_\iota^{-1}(U) \in \mathscr{U}$. We

[1] This theorem was proved by A. TIHONOV for the case in which each X_ι is the closed unit interval [0,1] [Math. Annalen **102**, 544–561 (1930)]. The general case was first proved by E. ČECH [Ann. of Math. (2) 38, 823–844 (1937)].

assert that $\bigcup \mathscr{U}_\iota = X_\iota$ for some $\iota \in I$. If this were not the case, there would be a point $\boldsymbol{x} \in X$ such that for every $\iota \in I$, $\pi_\iota(\boldsymbol{x}) = x_\iota \in X_\iota \cap (\bigcup \mathscr{U}_\iota)'$; hence $\boldsymbol{x} \notin \pi_\iota^{-1}(U)$ for all $\pi_\iota^{-1}(U) \in \mathscr{U}$. That is, $\mathscr{U}$ would not be a cover of X. Hence we can [and do] choose an $\eta \in I$ such that $\bigcup \mathscr{U}_\eta = X_\eta$. Since X_η is compact, there is a finite family $\{U_1, \ldots, U_n\} \subset \mathscr{U}_\eta$ such that $X_\eta = U_1 \cup U_2 \cup \cdots \cup U_n$. Plainly $\{\pi_\eta^{-1}(U_j) : j = 1, \ldots, n\}$ is a finite subcover of $\mathscr{U}$ for X. □

We next characterize the compact subspaces of R^n and K^n.

(6.44) Theorem [HEINE-BOREL-BOLZANO-WEIERSTRASS]. *Let $n \in N$ and let $A \subset R^n$ [or K^n]. Then A is compact [in the relativized usual topology] if and only if A is closed and bounded*[1].

Proof. The mapping $x + iy \to (x, y)$ of K onto R^2 preserves distance: $|(x + iy) - (u + iv)| = ((x - u)^2 + (y - v)^2)^{\frac{1}{2}} = \varrho((x, y), (u, v))$. Thus K^n and R^{2n} are indistinguishable as topological spaces. We therefore restrict our attention to the case in which $A \subset R^n$.

We first take the case that $n = 1$ and $A = [a, b]$, a bounded closed interval in R. A subbase for the topology of $[a, b]$ is the family $\mathscr{S}$ of all intervals of the form $[a, d[$ or $]c, b]$ where $c, d \in [a, b]$. Let $\mathscr{U}$ be any cover of $[a, b]$ by sets in $\mathscr{S}$. Since b is covered by $\mathscr{U}$, there is a set of the form $]c, b]$ in $\mathscr{U}$. Let $c_0 = \inf\{c :]c, b] \in \mathscr{U}\}$. Since c_0 is covered by $\mathscr{U}$, there is an interval $[a, d_1[\in \mathscr{U}$ such that $c_0 < d_1$. By the definition of infimum, there is an interval $]c_1, b] \in \mathscr{U}$ such that $c_1 < d_1$. Thus $\{[a, d_1[,]c_1, b]\} \subset \mathscr{U}$ and $[a, b] = [a, d_1[\cup]c_1, b]$. It follows from this and ALEXANDER's theorem (6.40) that $[a, b]$ is compact.

Now let n be arbitrary and suppose that A is closed and bounded. Since A is bounded, there exists a *cube* $C = \mathop{\times}\limits_{j=1}^{n} [a_j, b_j]$ such that $A \subset C$. The preceding paragraph and TIHONOV's theorem (6.43) show that C is compact. Using the fact that A is closed and citing (6.39), we see that A is compact.

Conversely, suppose that A is compact. By (6.38), A is closed. It is clear that $A \subset \bigcup\limits_{k=1}^{\infty} B_k(\mathbf{0}) = R^n$, where $B_k(\mathbf{0})$ is the *open* ball of radius k centered at $\mathbf{0} = (0, 0, \ldots, 0)$ in R^n. Since A is compact, there exists $k_0 \in N$ such that $A \subset B_{k_0}(\mathbf{0})$, *i.e.*, A is bounded. □

(6.45) Exercise. Prove the following.

(a) Any compact subset of a metric space is bounded.

(b) Theorem (6.44) is not true for arbitrary metric spaces.

(c) Every bounded sequence in R^n [or K^n] admits a convergent subsequence.

---

[1] A subset A of a metric space X is said to be *bounded* if there exist $p \in X$ and $\beta \in R$ such that $\varrho(p, x) \leqq \beta$ for all $x \in A$.

We next take up the study of completeness for metric spaces.

(6.46) Definition. A sequence (x_n) in a metric space X is said to be a *Cauchy sequence* if for each $\varepsilon > 0$ there exists $n_0 \in N$ such that $\varrho(x_m, x_n) < \varepsilon$ whenever $m, n \geqq n_0$. A metric space X is said to be *complete* if each Cauchy sequence in X converges to a point of X.

(6.47) Example. The real line R is complete (5.25), (5.35). Also it is easy to see that a subset of a complete metric space is complete if and only if it is closed.

(6.48) Theorem. *Any compact metric space is complete.*

Proof. Let X be a compact metric space and let (x_n) be a Cauchy sequence in X. Then (x_n) has a subsequence (x_{n_k}) converging to some $x \in X$. Let $\varepsilon > 0$ be given. Choose $n_0, k_0 \in N$ such that $m, n \geqq n_0$ implies $\varrho(x_m, x_n) < \varepsilon/2$ and $k \geqq k_0$ implies $\varrho(x_{n_k}, x) < \varepsilon/2$. Choose $k_1 \geqq k_0$ such that $n_{k_1} \geqq n_0$. Then $n \geqq n_{k_1}$ implies $\varrho(x_n, x) \leqq \varrho(x_n, x_{n_{k_1}}) + \varrho(x_{n_{k_1}}, x) < \varepsilon/2 + \varepsilon/2 = \varepsilon$. Thus $\lim\limits_{n\to\infty} x_n = x$, and so X is complete. □

(6.49) Theorem. *Any Cauchy sequence in a metric space is bounded.*

Proof. Let (x_n) be a Cauchy sequence in a metric space X. Choose $n_0 \in N$ such that $n \geqq n_0$ implies $\varrho(x_n, x_{n_0}) < 1$. Let $\alpha = \max\{1, \varrho(x_1, x_{n_0}), \ldots, \varrho(x_{n_0-1}, x_{n_0})\}$. Then $\varrho(x_n, x_{n_0}) \leqq \alpha$ for each $n \in N$. □

(6.50) Theorem. *Let $n \in N$. Then R^n and K^n are complete in the Euclidean metric.*

Proof. Let $X = R^n$ or K^n and let $(\boldsymbol{x}_k)$ be a Cauchy sequence in X. Since $(\boldsymbol{x}_k)$ is bounded (6.49), there exists a real number β such that $\varrho(\boldsymbol{0}, \boldsymbol{x}_k) \leqq \beta$ for each $k \in N$. Then $(\boldsymbol{x}_k)$ is a Cauchy sequence in the compact metric space $(B_\beta(\boldsymbol{0}))^-$ (6.44), so $(\boldsymbol{x}_k)$ converges (6.48). □

(6.51) Definition. Let A be a nonvoid bounded set in a metric space X. The *diameter of* A is the number

$$\operatorname{diam}(A) = \sup\{\varrho(x, y) : x, y \in A\}.$$

(6.52) Theorem [CANTOR]. *Let X be a metric space. Then X is complete if and only if whenever (A_n) is a decreasing sequence of nonvoid closed subsets of X,* i.e. $A_1 \supset A_2 \supset \cdots$, *such that* $\lim\limits_{n\to\infty} \operatorname{diam}(A_n) = 0$, *we have* $\bigcap\limits_{n=1}^{\infty} A_n = \{x\}$ *for some* $x \in X$.

Proof. Suppose that (A_n) is a decreasing sequence of nonvoid closed subsets of X such that $\operatorname{diam}(A_n) \to 0$. For each $n \in N$ let $x_n \in A_n$. Then $m \geqq n$ implies that $\varrho(x_m, x_n) \leqq \operatorname{diam}(A_n) \to 0$ so (x_n) is a Cauchy sequence. Let $x = \lim\limits_{n\to\infty} x_n$. For each m, $x_n \in A_m$ for all large n, and A_m is closed, so $x \in A_m$. Thus $x \in \bigcap\limits_{n=1}^{\infty} A_n$. If $x' \in \bigcap\limits_{n=1}^{\infty} A_n$, then $\varrho(x, x') \leqq \operatorname{diam}(A_n)$ for every n. Therefore $\varrho(x, x') = 0$. Hence $\bigcap\limits_{n=1}^{\infty} A_n = \{x\}$.

Conversely, suppose that X has the decreasing closed sets property. Let (x_n) be a Cauchy sequence in X. For each $n \in N$, let $A_n = \{x_m : m \geqq n\}^-$. Then (A_n) is a decreasing sequence of closed sets and, since (x_n) is a Cauchy sequence, $\operatorname{diam}(A_n) \to 0$. Let $\bigcap_{n=1}^{\infty} A_n = \{x\}$. If $\varepsilon > 0$, then there is an $n_0 \in N$ such that $\operatorname{diam}(A_{n_0}) < \varepsilon$. But $x \in A_{n_0}$, so $n \geqq n_0$ implies that $\varrho(x_n, x) < \varepsilon$. □

(6.53) Definition. Let X be a topological space. A set $A \subset X$ is said to be *nowhere dense* if $A^{-\circ} = \varnothing$. A set $F \subset X$ is said to be of *first category* if F is a countable union of nowhere dense sets. All other subsets of X are said to be of *second category*.

(6.54) Baire Category Theorem. *Let X be a complete metric space. Suppose that $A \subset X$ and that A is of first category in X. Then $X \cap A'$ is dense in X. Thus X is of second category* [*as a subset of itself*].

Proof. Let $A = \bigcup_{n=1}^{\infty} A_n$, where each A_n is nowhere dense in X. We suppose that each A_n is closed [at worst this makes $X \cap A'$ smaller]. Let V be any nonvoid open subset of X. We will show that $V \cap A' \neq \varnothing$. Choose a nonvoid open set $U_1 \subset V$ such that $\operatorname{diam}(U_1^-) < 1$. For example we may take U_1 to be an open ball of radius $< \frac{1}{2}$. Then U_1 is not a subset of A_1, so $U_1 \cap A_1'$ is a nonvoid open set. Let U_2 be a nonvoid open set such that $U_2^- \subset U_1 \cap A_1'$ and $\operatorname{diam}(U_2^-) < \frac{1}{2}$. Suppose that $U_1, \ldots, U_n$ have been chosen such that U_{j+1} is a nonvoid open set, $U_{j+1}^- \subset U_j \cap A_j'$, and $\operatorname{diam}(U_{j+1}^-) < \frac{1}{j+1}$ for $1 \leqq j \leqq n-1$. Then $U_n \cap A_n' \neq \varnothing$, so there exists a nonvoid open set U_{n+1} such that $U_{n+1}^- \subset U_n \cap A_n'$ and $\operatorname{diam}(U_{n+1}^-) < \frac{1}{n+1}$. We thus obtain a decreasing sequence (U_n^-) of nonvoid closed sets such that $\operatorname{diam}(U_n^-) \to 0$. Since X is complete, there exists an $x \in X$ such that $\bigcap_{n=1}^{\infty} U_n^- = \{x\}$. Then $x \in \bigcap_{n=1}^{\infty} U_{n+1}^- \subset U_1 \cap \bigcap_{n=1}^{\infty} A_n'$ $\subset V \cap \left(\bigcup_{n=1}^{\infty} A_n\right)' = V \cap A'$. Since V was arbitrary, it follows that A' is dense in X. □

The Baire category theorem has many interesting and important applications throughout analysis, as we shall see several times in the sequel. For the moment, we content ourselves with an unimportant though interesting application.

(6.55) Definition. Let X be a topological space and let $A \subset X$. The set A is called a G_δ *set* if A is a countable intersection of open sets, and A is called an F_σ *set* if it is a countable union of closed sets.

(6.56) Theorem. *The set Q of rational numbers is not a G_δ set in R.*

Proof. Assume that $Q = \bigcap_{n=1}^{\infty} U_n$, where each U_n is open in R. Then each U_n' is nowhere dense since it is closed and contains no rational numbers. Let $Q = (x_n)_{n=1}^{\infty}$ be an enumeration of Q (4.22). Then $R = \bigcup_{n=1}^{\infty} (U_n' \cup \{x_n\})$. But $U_n' \cup \{x_n\}$ is nowhere dense for each $n \in N$ and R is a complete metric space. This contradicts (6.54). □

We next examine the structure of open subsets and closed subsets of R.

(6.57) Definition. Let A be a nonvoid subset of $R^{\#}$. If A has no upper [lower] bound in R, we say that the supremum [infimum] of A is ∞ [$-\infty$] and write $\sup A = \infty$ [$\inf A = -\infty$].

(6.58) Remark. In view of (5.33) and (6.57), every nonvoid subset of R has both a supremum and an infimum in $R^{\#}$.

(6.59) Theorem. *Let U be a nonvoid open subset of R. Then there exists one and only one pairwise disjoint family $\mathscr{I}$ of open intervals of R such that $U = \bigcup \mathscr{I}$. The family $\mathscr{I}$ is countable and the members of $\mathscr{I}$ are called* component intervals of U. *For each $I \in \mathscr{I}$, the endpoints of I are not in U.*

Proof. Let $x \in U$ and define $a_x = \inf\{t :]t, x] \subset U\}$ and $b_x = \sup\{t : [x, t[\subset U\}$. Since U is open, it is clear that a_x and b_x exist in $R^{\#}$. We first assert that $]a_x, b_x[\subset U$ and begin by proving that $]a_x, x] \subset U$. If $a_x \in R$, let $x_n = a_x + \frac{1}{n}$ and if $a_x = -\infty$, let $x_n = -n$. In either case $a_x = \inf\{x_n : n \in N\}$. By the definition of a_x it follows that for each sufficiently large $n \in N$ there exists a real number t_n such that $a_x \leqq t_n < x_n$ and $]t_n, x] \subset U$. Then $]a_x, x] = \bigcup_{n=n_0}^{\infty}]x_n, x] \subset \bigcup_{n=n_0}^{\infty}]t_n, x] \subset U$. Likewise, we have $[x, b_x[\subset U$, and hence $]a_x, b_x[\subset U$.

We next show that $a_x \notin U$, $b_x \notin U$. Assume that $b_x \in U$. Since U is open, there is a $\delta > 0$ such that $]b_x - \delta, b_x + \delta[\subset U$. But then $[x, b_x + \delta[= [x, b_x[\cup [b_x, b_x + \delta[\subset U$ and $b_x + \delta > b_x$. This contradicts the definition of b_x. Thus $b_x \notin U$. Likewise $a_x \notin U$.

Let $\mathscr{I} = \{]a_x, b_x[: x \in U\}$. Since $x \in U$ implies $x \in]a_x, b_x[$, we have $U = \bigcup \mathscr{I}$. We next show that $\mathscr{I}$ is a pairwise disjoint family. Let $x, y \in U$ and suppose that there exists $u \in]a_x, b_x[\cap]a_y, b_y[$. If $a_x < a_y < u$, then $a_y \in U$ and if $a_y < a_x < u$, then $a_x \in U$. But neither a_x nor a_y is in U. Therefore $a_x = a_y$. Likewise $b_x = b_y$. Accordingly any two intervals in $\mathscr{I}$ are either disjoint or identical, *i.e.*, $\mathscr{I}$ is pairwise disjoint.

For each $I \in \mathscr{I}$ there is a rational number $r_I \in I$. Since $\mathscr{I}$ is pairwise disjoint, the mapping $f : I \to r_I$ of $\mathscr{I}$ into Q is one-to-one, and so $\overline{\overline{\mathscr{I}}} \leqq \overline{\overline{Q}} = \aleph_0$. Thus $\mathscr{I}$ is countable.

It remains only to prove that $\mathscr{I}$ is unique. Thus suppose that $U = \bigcup \mathscr{J}$ where $\mathscr{J}$ is a pairwise disjoint family of open intervals. Let $]a, b[\in \mathscr{J}$. Assume that $a \in U$. Then there exists an interval $]c, d[\in \mathscr{J}$ such that

$a \in]c, d[$. Thus $]a, b[\neq]c, d[$, but $]a, b[\cap]c, d[=]a, \min\{b, d\}[\neq \varnothing$. This contradiction shows that $a \notin U$. Likewise $b \notin U$. Let $x \in]a, b[$. Then $]a, x] \subset U$ and $[x, b[\subset U$ so $]a, b[\subset]a_x, b_x[\subset U$. Since $a \notin U$ and $b \notin U$, we have $]a, b[=]a_x, b_x[\in \mathscr{I}$. Therefore $\mathscr{J} \subset \mathscr{I}$. If there exists $]a_x, b_x[\in \mathscr{I} \cap \mathscr{J}'$, then $x \in U$ while $x \notin \bigcup \mathscr{J} = U$, a contradiction. Therefore $\mathscr{I} = \mathscr{J}$. □

(6.60) Remark. The simple structure of open sets in R has no analogue in Euclidean spaces of dimension >1. For example, in the plane R^2 open disks play the rôle that open intervals play on the line as the building blocks for open sets, *i.e.*, the base for the topology. But it is plain that the open square $\{(x, y) : 0 < x < 1, 0 < y < 1\}$ is not a union of disjoint open disks, for if it were, the diagonal $\{(x, x) : 0 < x < 1\}$ would be a union of [more than one] disjoint open intervals, contrary to the uniqueness statement of (6.59).

Neither do the closed subsets of R have such a simple structure as the open ones do. The next few paragraphs show this rather complicated structure. We begin with a definition.

(6.61) Definition. Let X be a topological space and let $A \subset X$. A point $a \in A$ is called an *isolated point of* A if it is not a limit point of A, *i.e.*, if there exists a neighborhood U of a such that $U \cap A = \{a\}$. The set A is said to be *perfect* if it is closed and has no isolated points, *i.e.*, if A is equal to the set of its own limit points.

We will now construct a large class of nowhere dense perfect subsets of $[0, 1]$.

(6.62) Definition. Remove any open interval $I_{1,1}$ of length < 1 from the center of $[0, 1]$. This leaves two disjoint closed intervals $J_{1,1}$ and $J_{1,2}$ each having length $< \frac{1}{2}$. This completes the first stage of our construction. If the nth step of the construction has been completed, leaving 2^n disjoint closed intervals $J_{n,1}, J_{n,2}, \ldots, J_{n,2^n}$ [numbered from left to right], each of length $< \frac{1}{2^n}$, we perform the $(n+1)^{st}$ step by removing any open interval $I_{n+1,k}$ from the center of $J_{n,k}$ such that the length of $I_{n+1,k}$ is less than the length of $J_{n,k}$ $(1 \leqq k \leqq 2^n)$. This leaves 2^{n+1} closed intervals $J_{n+1,1}, \ldots, J_{n+1,2^{n+1}}$ each of length $< \frac{1}{2^{n+1}}$. Let $V_n = \bigcup_{k=1}^{2^{n-1}} I_{n,k}$ and $P_n = \bigcup_{k=1}^{2^n} J_{n,k}$ $(n \in N)$. Let $P = \bigcap_{n=1}^{\infty} P_n = [0, 1] \cap \left(\bigcup_{n=1}^{\infty} V_n\right)'$. Any set P constructed in the above manner is known as a *Cantor-like set*. In the case that $I_{1,1} = \left]\frac{1}{3}, \frac{2}{3}\right[$ and the length of $I_{n+1,k}$ is exactly $\frac{1}{3}$ of the length of $J_{n,k}$ for all $k, n \in N$, $1 \leqq k \leqq 2^n$, the resulting set P is known as the *Cantor ternary set* [or simply the *Cantor* set]. In this latter case $J_{1,1} = \left[0, \frac{1}{3}\right]$, $J_{1,2} = \left[\frac{2}{3}, 1\right]$, $I_{2,1} = \left]\frac{1}{9}, \frac{2}{9}\right[$, $I_{2,2} = \left]\frac{7}{9}, \frac{8}{9}\right[$, $J_{2,1} = \left[0, \frac{1}{9}\right]$, etc.

(6.63) Theorem. *Let P be any Cantor-like set. Then P is compact, nowhere dense in R, and perfect.*

Proof. We use the notation of (6.62). Obviously each P_n is closed, so that P is closed and bounded and hence compact (6.44). Since no P_n contains an interval of length $\geqq \frac{1}{2^n}$ and $P \subset P_n$ for each $n \in N$, it follows that P contains no interval. Thus $P^{-\circ} = P^\circ = \varnothing$; that is, P is nowhere dense in R. Next let $x \in P$. For each $n \in N$ we have $x \in P_n$, so that there exists k_n such that $x \in J_{n,k_n}$. Thus, given $\varepsilon > 0$, there is an $n \in N$ such that $\frac{1}{2^n} < \varepsilon$, and therefore the endpoints of J_{n,k_n} are both in $]x - \varepsilon, x + \varepsilon[$. But these endpoints are in P. Hence x is a limit point of P. We conclude that P is perfect. □

(6.64) Theorem. *Let P be the Cantor ternary set. Then* $P = \left\{ \sum_{n=1}^{\infty} \frac{x_n}{3^n} : x_n \in \{0, 2\} \text{ for each } n \in N \right\}$, *and therefore* $\overline{\overline{P}} = \mathfrak{c}$.

Proof. Each number $x \in [0, 1]$ has a ternary [base three] expansion in the form $x = \sum_{n=1}^{\infty} \frac{x_n}{3^n}$, where each x_n is 0, 1, or 2. This expansion is unique except for the case that $x = \frac{a}{3^m}$ for some $a, m \in N$ where $0 < a < 3^m$ and 3 does not divide a. In this case x has a finite expansion of the form $x = \frac{x_1}{3} + \cdots + \frac{x_m}{3^m}$ where $x_m = 1$ [if $a \equiv 1 \pmod 3$] or $x_m = 2$ [if $a \equiv 2 \pmod 3$]. If $x_m = 2$ we use this finite expansion for x, but if $x_m = 1$, we prefer the expansion $x = \frac{x_1}{3} + \cdots + \frac{x_{m-1}}{3^{m-1}} + \frac{0}{3^m} + \sum_{n=m+1}^{\infty} \frac{2}{3^n}$. We leave it to the reader to verify these assertions [cf. (5.40)]. Thus we have assigned a unique ternary expansion to each $x \in [0, 1]$. One sees by induction that $P_n = \{x : 0 \leqq x \leqq 1, \{x_1, \ldots, x_n\} \subset \{0, 2\}\}$. For example $P_1 = \left[0, \frac{1}{3}\right] \cup \left[\frac{2}{3}, 1\right]$ and $P_2 = \left[0, \frac{1}{9}\right] \cup \left[\frac{2}{9}, \frac{1}{3}\right] \cup \left[\frac{2}{3}, \frac{7}{9}\right] \cup \left[\frac{8}{9}, 1\right]$ $\left[\text{we write } \frac{1}{3} = \sum_{n=2}^{\infty} \frac{2}{3^n}\right]$. Thus $x \in P = \bigcap_{n=1}^{\infty} P_n$ if and only if $x_n \in \{0, 2\}$ for each $n \in N$. Clearly the mapping $\sum_{n=1}^{\infty} \frac{x_n}{3^n} \to (x_n)$ is a one-to-one correspondence between P and $\{0, 2\}^N$. Therefore $\overline{\overline{P}} = 2^{\aleph_0} = \mathfrak{c}$. □

In view of the following theorem, it is no accident that the Cantor set has cardinal number $\mathfrak{c}$.

(6.65) Theorem. *Let X be a complete metric space and let A be a nonvoid perfect subset of X. Then $\bar{\bar{A}} \geqq \mathfrak{c}$.*

Proof. We will construct a one-to-one mapping of $\{0, 1\}^N$ into A. Since A is nonvoid, it has a limit point and therefore A is infinite (6.26). Let $x_0 \neq x_1$ in A. Let $\varepsilon_1 = \min\left\{\frac{1}{2}, \frac{1}{3}\varrho(x_1, x_2)\right\}$ and define $A(0) = \{x \in A : \varrho(x_0, x) \leqq \varepsilon_1\}$ and $A(1) = \{x \in A : \varrho(x_1, x) \leqq \varepsilon_1\}$. Then $A(0)$ and $A(1)$ are disjoint infinite closed sets each of diameter $\leqq 1$. Suppose that n is a positive integer and for each n-tuple $(a_1, \ldots, a_n) \in \{0, 1\}^n$ we have an infinite closed subset $A(a_1, \ldots, a_n)$ of A having diameter $\leqq \frac{1}{n}$ and such that no two of these sets have a common point. For $(a_1, \ldots, a_n) \in \{0, 1\}^n$, choose $x(a_1, \ldots, a_n, 0) \neq x(a_1, \ldots, a_n, 1)$ in $A(a_1, \ldots, a_n)$ and let $\varepsilon_{n+1} = \min\left\{\frac{1}{2(n+1)}, \frac{1}{3}\varrho(x(a_1, \ldots, a_n, 0), x(a_1, \ldots, a_n, 1))\right\}$. Define $A(a_1, \ldots, a_n, j) = \{x \in A(a_1, \ldots, a_n) : \varrho(x(a_1, \ldots, a_n, j), x) \leqq \varepsilon_{n+1}\}$ $(j = 0, 1)$. Then $\{A(a_1, \ldots, a_{n+1}) : (a_1, \ldots, a_{n+1}) \in \{0, 1\}^{n+1}\}$ is a pairwise disjoint family of closed infinite sets each having diameter $\leqq \frac{1}{n+1}$. Thus for each $\boldsymbol{a} = (a_n) \in \{0, 1\}^N$ we have a decreasing sequence $(A(a_1, \ldots, a_n))_{n=1}^{\infty}$ of infinite closed subsets of A with diameters tending to 0. Hence by CANTOR'S theorem (6.52), there exists a point $x(\boldsymbol{a}) \in A$ such that $\bigcap_{n=1}^{\infty} A(a_1, \ldots, a_n) = \{x(\boldsymbol{a})\}$. Suppose $\boldsymbol{a} \neq \boldsymbol{b}$ in $\{0, 1\}^N$. Then, for some n_0, $a_{n_0} \neq b_{n_0}$ so $x(\boldsymbol{a}) \in A(a_1, \ldots, a_{n_0})$ while $x(\boldsymbol{b}) \notin A(a_1, \ldots, a_{n_0})$ and therefore $x(\boldsymbol{a}) \neq x(\boldsymbol{b})$. It follows that the mapping $\boldsymbol{a} \to x(\boldsymbol{a})$ is one-to-one. Thus $\bar{\bar{A}} \geqq \overline{\overline{\{0, 1\}^N}} = \mathfrak{c}$. □

We next present a structure theorem for closed sets.

(6.66) Theorem [CANTOR-BENDIXSON]. *Let X be a topological space with a countable base $\mathscr{B}$ for its topology and let A be any closed subset of X. Then X contains a perfect subset P and a countable subset C such that $A = P \cup C$.*

Proof. A point $x \in X$ will be called a *condensation point of A* if $U \cap A$ is uncountable for each neighborhood U of x. Let $P = \{x \in X : x$ is a condensation point of $A\}$ and let $C = A \cap P'$. Since each condensation point is a limit point, it follows that $P \subset A$. Clearly $A = P \cup C$. Since no point of C is a condensation point of A, each $x \in C$ has a neighborhood $V_x \in \mathscr{B}$ such that $A \cap V_x$ is countable. But $\mathscr{B}$ is countable so $C \subset \bigcup\{A \cap V_x : x \in C\}$, and C is countable.

Next let $x \in P$ and let U be a neighborhood of x. Then $U \cap A$ is uncountable and $U \cap C$ is countable, so $U \cap P = (U \cap A) \cap (U \cap C)'$ is uncountable, and hence x is a limit point of P. Thus P has no isolated points. To show that P is closed, let $x \in P'$. Then x has a neighborhood V such that $V \cap A$ is countable. If there is a $y \in V \cap P$, then V is a neighborhood of y and y is a condensation point of A, so $V \cap A$ is uncountable. It follows that $V \cap P = \varnothing$, so that x is not a limit point of P. Therefore P contains all of its limit points, *i.e.*, P is closed. We conclude that P is perfect. □

(6.67) Remark. In view of (6.21) and (6.23), every Euclidean space satisfies the hypothesis of (6.66).

We now make a brief study of continuity.

(6.68) Definition. Let X and Y be topological spaces and let f be a function from X into Y. Then f is said to be *continuous at a point* $x \in X$ if for each neighborhood V of $f(x)$ there exists a neighborhood U of x such that $f(U) \subset V$. The function f is said to be *continuous on* X if f is continuous at each point of X.

(6.69) Theorem. *Let X, Y, and f be as in* (6.68). *Then f is continuous on X if and only if $f^{-1}(V)$ is open in X whenever V is open in Y.*

Proof. Suppose that f is continuous on X and let V be open in Y. We must show that $f^{-1}(V)$ is open in X. For $x \in f^{-1}(V)$, we know that f is continuous at x, so there exists a neighborhood U_x of x such that $f(U_x) \subset V$, *i.e.*, $U_x \subset f^{-1}(V)$. It follows that $f^{-1}(V) = \bigcup \{U_x : x \in f^{-1}(V)\}$ which is a union of open sets, so that $f^{-1}(V)$ is open.

Conversely, suppose that $f^{-1}(V)$ is open in X whenever V is open in Y. Let $x \in X$ and let V be a neighborhood of $f(x)$. Then $f^{-1}(V)$ is a neighborhood of x and $f(f^{-1}(V)) \subset V$. Thus f is continuous at x. Since x is arbitrary, f is continuous on X. □

(6.70) Theorem. *Let X, Y, and f be as in* (6.68). *Suppose that $\mathcal{S}$ is a subbase for the topology of Y and that $f^{-1}(S)$ is open in X for every $S \in \mathcal{S}$. Then f is continuous on X.*

Proof. Let $\mathcal{B}$ be the family of all sets of the form $B = \bigcap_{j=1}^{n} S_j$, where $\{S_1, \ldots, S_n\}$ is a finite subfamily of $\mathcal{S}$. Then $\mathcal{B}$ is a base for the topology of Y (6.10), and the set $f^{-1}(B) = \bigcap_{j=1}^{n} f^{-1}(S_j)$, being a finite intersection of open sets, is open for every $B \in \mathcal{B}$. Next, let V be open in Y. Then $V = \bigcup_{\iota \in I} B_\iota$ for some family $\{B_\iota\}_{\iota \in I} \subset \mathcal{B}$. Therefore $f^{-1}(V) = f^{-1}\left(\bigcup_{\iota \in I} B_\iota\right)$ $= \bigcup_{\iota \in I} f^{-1}(B_\iota)$ which, being a union of open sets, is open in X. □

(6.71) Theorem. *Let X, Y, and f be as in* (6.68). *Suppose that X is a metric space and $x \in X$. Then f is continuous at x if and only if $f(x_n) \to f(x)$ whenever (x_n) is a sequence in X such that $x_n \to x$.*

Proof. Suppose that $f(x_n) \to f(x)$ whenever $x_n \to x$ and assume that f is not continuous at x. Then there is a neighborhood V of $f(x)$ such that $f(U) \subset V$ for no neighborhood U of x. For each $n \in N$, choose $x_n \in B_{\frac{1}{n}}(x)$ such that $f(x_n) \notin V$. Then $x_n \to x$ but $f(x_n) \not\to f(x)$. This contradiction shows that f is continuous at x.

Conversely, suppose that f is continuous at x and let (x_n) be any sequence in X such that $x_n \to x$. Let V be any neighborhood of $f(x)$. Then there is a neighborhood U of x such that $f(U) \subset V$. Since $x_n \to x$, there exists $n_0 \in N$ such that $n \geqq n_0$ implies $x_n \in U$. Then $n \geqq n_0$ implies $f(x_n) \in f(U) \subset V$. Thus $f(x_n) \to f(x)$. □

(6.72) Theorem. *Let X, Y, and f be as in* (6.68). *Suppose that X is compact and that f is continuous on X. Then $f(X)$ is a compact subspace of Y.*

Proof. Let $\mathscr{V}$ be any open cover of $f(X)$. Then $\{f^{-1}(V) : V \in \mathscr{V}\}$ is an open cover of X, so there exist $V_1, \ldots, V_n \in \mathscr{V}$ such that $X = \bigcup_{k=1}^{n} f^{-1}(V_k) = f^{-1}\left(\bigcup_{k=1}^{n} V_k\right)$. It follows that $f(X) \subset \bigcup_{k=1}^{n} V_k$. □

(6.73) Corollary. *Let X be a compact space and let f be a continuous real-valued function on X. Then f is bounded* [i.e., $f(X)$ *is a bounded set*] *and there exist points a and b in X such that* $f(a) = \sup\{f(x) : x \in X\}$, $f(b) = \inf\{f(x) : x \in X\}$.

Proof. According to (6.72), $f(X)$ is a compact subspace of R. Thus $f(X)$ is closed and bounded (6.44). Let $\alpha = \sup f(X)$ and $\beta = \inf f(X)$. Since $f(X)$ is bounded, we have $\alpha, \beta \in R$. Since $f(X)$ is closed, we have $\alpha, \beta \in f(X)$. Choose $a \in f^{-1}(\{\alpha\})$, $b \in f^{-1}(\{\beta\})$. □

(6.74) Theorem. *Let A, B, and C be topological spaces. Let f be a function from A into B and let g be a function from B into C. Let $x \in A$ and suppose that f is continuous at x and g is continuous at $f(x)$. Then $g \circ f$ is continuous at x.*

Proof. Let W be any neighborhood of $g \circ f(x) = g(f(x))$. Then there is a neighborhood V of $f(x)$ such that $g(V) \subset W$. Since f is continuous at x, there is a neighborhood U of x such that $f(U) \subset V$. Thus we have found a neighborhood U of x such that $g \circ f(U) = g(f(U)) \subset g(V) \subset W$. □

(6.75) Corollary. *Let A, B, C, f, and g be as in* (6.74). *Suppose that f is continuous on A and g is continuous on B. Then $g \circ f$ is continuous on A.*

(6.76) Theorem. *Let X and Y be topological spaces and let f be a continuous function from X into Y. Let $S \subset X$. Then the function f [with its domain restricted to S] is a continuous function from S [with its relative topology] into Y.*

Proof. Let $x \in S$ and let V be a neighborhood of $f(x)$. Then there is a neighborhood U [open in X] of x such that $f(U) \subset V$. But then $U \cap S$ is a neighborhood of x in the relative topology on S and $f(U \cap S) \subset f(U) \subset V$. □

We next discuss locally compact spaces. These spaces are of great importance in our treatment of measure theory.

(6.77) Definition. A topological space X is said to be *locally compact* if each point $x \in X$ has a neighborhood U such that U^- is compact.

(6.78) Theorem. *Let X be a locally compact Hausdorff space. Let $x \in X$ and let U be a neighborhood of x. Then there exists a neighborhood V of x such that V^- is compact and $V^- \subset U$.*

Proof. Let W be any neighborhood of x such that W^- is compact. Let $G = U \cap W$. Then G is a neighborhood of x; since G^- is a closed subset of W^-, it follows from (6.39) that G^- is compact. We have $G \subset U$, but we do not know that $G^- \subset U$. Recall (6.7.vii) that $\partial G = G^- \cap G^{\circ\prime} = G^- \cap G'$. Thus ∂G is compact (6.39). If $\partial G = \varnothing$, we may take $V = G$. Thus suppose $\partial G \neq \varnothing$. For each $y \in \partial G$, choose neighborhoods V_y and H_y of x and y respectively such that $V_y \cap H_y = \varnothing$. We may suppose that $V_y \subset G$, for otherwise intersect it with G. Then $\{H_y : y \in \partial G\}$ is an open cover of ∂G, and by compactness there exist $y_1, \ldots, y_n \in \partial G$ such that $\partial G \subset H_{y_1} \cup \cdots \cup H_{y_n} = H$. Let $V = V_{y_1} \cap \cdots \cap V_{y_n}$. Then V is a neighborhood of x and $V \cap H = \varnothing$. Clearly $V \subset G$, so $V^- \subset G^-$ and V^- is compact. Moreover $V \subset H'$ and H' is closed so $V^- \subset H'$. Thus $V^- \subset G^- \cap H' \subset G^- \cap (\partial G)' = G$. □

(6.79) Theorem. *Let X be a locally compact Hausdorff space and let A be a compact subspace of X. Suppose that U is an open subset of X such that $A \subset U$. Then there exists an open $V \subset X$ such that $A \subset V \subset V^- \subset U$ and V^- is compact.*

Proof. Apply (6.78) to each $x \in A$. Thus for each $x \in A$, there exists a neighborhood V_x of x such that V_x^- is compact and $V_x^- \subset U$. The family $\{V_x : x \in A\}$ is an open cover of A, so there exist $x_1, \ldots, x_n \in A$ such that $A \subset \bigcup_{k=1}^{n} V_{x_k} = V$. Then $V^- = \bigcup_{k=1}^{n} V_{x_k}^- \subset U$ (6.7.viii) und V^-, being a finite union of compact sets, is plainly compact. □

The following locally compact version of URYSOHN'S lemma will be adequate for our purposes.

(6.80) Theorem [URYSOHN]. *Let X be a locally compact Hausdorff space, let A be a compact subspace of X, and let U be an open set such that $A \subset U$. Then there exists a continuous function f from X into* [0, 1] *such that $f(x) = 1$ for all $x \in A$ and $f(x) = 0$ for all $x \in U'$.*

Proof. Let $D_0 = \{0, 1\}$ and for each $n \in N$ define $D_n = \left\{\frac{a}{2^n} : a \in N, a \text{ is odd}, 0 < a < 2^n\right\}$. Let $D = \bigcup_{n=0}^{\infty} D_n$. Thus D is the set of all dyadic rational numbers in [0, 1]. We shall define by induction on n a chain $\{U_t\}_{t \in D}$ of subsets of X. First let $U_1 = A$ and $U_0 = U$. For $n = 1$ we have $D_1 = \left\{\frac{1}{2}\right\}$ and we apply (6.79) to obtain an open set $U_{\frac{1}{2}}$ such that $U_1^- \subset U_{\frac{1}{2}} \subset U_{\frac{1}{2}}^- \subset U_0$. Next let $n \geqq 2$ and suppose that open sets U_t have been defined for all $t \in \bigcup_{k=1}^{n-1} D_k$ so that $s < t$ in $\bigcup_{k=0}^{n-1} D_k$ implies $U_t^- \subset U_s$. For $t = \frac{a}{2^n} \in D_n$, we set

$t' = \frac{a-1}{2^n}$ and $t'' = \frac{a+1}{2^n}$ and notice that $U_{t'}$ and $U_{t''}$ are already defined [$a - 1$ and $a + 1$ are even]. We again use (6.79) to obtain an open set U_t such that $U_{t''}^- \subset U_t \subset U_t^- \subset U_{t'}$. Thus we obtain the desired family $\{U_t\}_{t \in D}$, and we have $U_t^- \subset U_s$ whenever $s < t$ in D.

Now define f on X by $f(x) = 0$ for $x \in U'$ and $f(x) = \sup\{t \in D : x \in U_t\}$ for $x \in U$. Clearly $f(x) = 1$ for all $x \in A = U_1$. It remains to show that f is continuous. To this end, let $0 \leqq \alpha < 1$ and $0 < \beta \leqq 1$. Clearly $f(x) > \alpha$ if and only if $x \in U_t$ for some $t > \alpha$ and therefore $f^{-1}(]\alpha, 1]) = \bigcup \{U_t : t \in D, t > \alpha\}$, which is open. In like manner $f(x) \geqq \beta$ if and only if $x \in U_s$ for every $s < \beta$. Therefore $f^{-1}([\beta, 1]) = \bigcap \{U_s : s \in D, s < \beta\} = \bigcap \{U_t^- : t \in D, t < \beta\}$, which is closed. Taking complements we see that $f^{-1}([0, \beta[)$ is open. These facts together with (6.70) show that f is continuous. □

We now take up the notions of limit superior and limit inferior for sequences of real numbers.

(6.81) Definition. A *nondecreasing* [*nonincreasing*] *sequence* in $R^\#$ is a sequence $(x_n) \subset R^\#$ such that $m \leqq n$ implies $x_m \leqq x_n$ [$x_m \geqq x_n$]. A sequence $(x_n) \subset R^\#$ is said to have limit ∞ [$-\infty$] if to each $\alpha \in R$ there corresponds an $n_\alpha \in N$ such that $n \geqq n_\alpha$ implies $x_n \geqq \alpha$ [$x_n \leqq \alpha$], and we write $\lim_{n\to\infty} x_n = \infty$ [$\lim_{n\to\infty} x_n = -\infty$] or $x_n \to \infty$ [$x_n \to -\infty$]. A sequence that is either nondecreasing or nonincreasing is called *monotone*.

(6.82) Theorem. *Every monotone sequence in $R^\#$ has a limit in $R^\#$.*

Proof. Let (x_n) be nondecreasing and let $x = \sup\{x_n : n \in N\}$. Then $\lim_{n\to\infty} x_n = x$. □

(6.83) Definition. Let (x_n) be any sequence in $R^\#$. We define the *limit superior of* (x_n) to be the extended real number

$$\overline{\lim_{n\to\infty}}\, x_n = \inf_{k \in N} \left(\sup_{n \geqq k} x_n\right)$$

and the *limit inferior of* (x_n) to be the extended real number

$$\underline{\lim_{n\to\infty}}\, x_n = \sup_{k \in N} \left(\inf_{n \geqq k} x_n\right).$$

Obviously the sequences $\left(\sup_{n \geqq k} x_n\right)_{k=1}^\infty$ and $\left(\inf_{n \geqq k} x_n\right)_{k=1}^\infty$ are monotone sequences, so that $\overline{\lim_{n\to\infty}}\, x_n$ and $\underline{\lim_{n\to\infty}}\, x_n$ are just their respective limits. The alternative notations $\lim_{n\to\infty} \sup x_n = \overline{\lim_{n\to\infty}}\, x_n$ and $\lim_{n\to\infty} \inf x_n = \underline{\lim_{n\to\infty}}\, x_n$ are often used.

(6.84) Theorem. *Let (x_n) be a sequence in $R^\#$ and let $L = \{x \in R^\# : x$ is the limit of some subsequence of $(x_n)\}$. Then $\underline{\lim_{n\to\infty}}\, x_n$ and $\overline{\lim_{n\to\infty}}\, x_n$ are in L and $\underline{\lim_{n\to\infty}}\, x_n = \inf L$, $\overline{\lim_{n\to\infty}}\, x_n = \sup L$.*

Proof. We prove only the assertions about the limit superior, the others being obvious duals. Let $x = \overline{\lim\limits_{n\to\infty}}\, x_n$, and for each $k \in N$, let $y_k = \sup\{x_n : n \geqq k\}$. Then $x = \inf\{y_k : k \in N\}$.

Case I: $x = \infty$. Then $y_k = \infty$ for each $k \in N$, so that for each $m \in N$ there are infinitely many $n \in N$ such that $x_n > m$. Choose n_1 so that $x_{n_1} > 1$. When $n_1, \ldots, n_m$ have been chosen, choose $n_{m+1} > n_m$ such that $x_{n_{m+1}} > m + 1$. Then $(x_{n_m})_{m=1}^{\infty}$ is a subsequence of (x_n) and $\lim\limits_{m\to\infty} x_{n_m} = \infty$. Thus $x = \infty \in L$ and clearly $x = \infty = \sup L$.

Case II: $x \in R$. We have $x = \inf\{y_k : k \in N\}$. Thus for each $\beta > x$ there is a $y_k < \beta$ and therefore $x_n > \beta$ for only finitely many n $[n < k]$. This proves that there is no element of L greater than x. On the other hand, $y_k \geqq x$ for all k so for each $m \in N$ there exist arbitrarily large n's such that $x_n > x - \frac{1}{m}$. We conclude that $\left\{n \in N : x - \frac{1}{m} < x_n < x + \frac{1}{m}\right\}$ is an infinite set for each $m \in N$. Consequently, as in Case I, we can choose a subsequence (x_{n_m}) of (x_n) such that $\lim\limits_{m\to\infty} x_{n_m} = x$. Therefore $x \in L$.

Case III: $x = -\infty$. The argument given in Case II proves that there is no element of L greater than x. But for each $m \in N$ there is a y_k such that $y_k < -m$. Thus $x_n \leqq -m$ for all but finitely many $n \in N$; and so $\lim\limits_{n\to\infty} x_n = -\infty = x$. □

(6.85) Exercise. Let (X, ϱ) be a metric space. Prove that:

(a) there exists a complete metric space $(\overline{X}, \bar{\varrho})$ and a function f from X into $\overline{X}$ such that $f(X)$ is dense in $\overline{X}$ and $\bar{\varrho}(f(x), f(y)) = \varrho(x, y)$ for all $x, y \in X$ [$(\overline{X}, \bar{\varrho})$ is called the *completion of* (X, ϱ)];

(b) $(\overline{X}, \bar{\varrho})$ is unique in the sense that if (Y, σ) is a complete metric space and g is a function from X into Y such that $g(X)$ is dense in Y and $\sigma(g(x), g(y)) = \varrho(x, y)$ for all $x, y \in X$, then there is a function h from $\overline{X}$ onto Y such that $\sigma(h(\alpha), h(\beta)) = \bar{\varrho}(\alpha, \beta)$ for all $\alpha, \beta \in \overline{X}$. [Functions such as f, g, and h which preserve distance are called *isometries*.]

[Hints. Let $\mathfrak{C}$ be the set of all Cauchy sequences in X. Define $(x_n) \sim (y_n)$ if $\varrho(x_n, y_n) \to 0$. Let $\overline{X}$ be the set of equivalence classes. Define $\bar{\varrho}(\alpha, \beta) = \lim\limits_{n\to\infty} \varrho(x_n, y_n)$ where $(x_n) \in \alpha$, $(y_n) \in \beta$. [cf. the completion of an ordered field in § 5].]

(6.86) Exercise. Let (X, ϱ) be a metric space. For each nonvoid subset A of X and each $x \in X$, define

$$\varrho(x, A) = \inf\{\varrho(x, a) : a \in A\}.$$

The number $\varrho(x, A)$ is called the *distance from* x *to* A. Prove each of the following statements.

(a) If $\varnothing \neq A \subset X$, then $A^{-} = \{x \in X : \varrho(x, A) = 0\}$.

(b) If $\varnothing \neq A \subset X$ and $x, y \in X$, then $|\varrho(x, A) - \varrho(y, A)| \leqq \varrho(x, y)$.

Thus the function f defined on X by $f(x) = \varrho(x, A)$ is continuous on X.

(c) If A and B are two nonvoid disjoint closed subsets of X, then the function h defined on X by

$$h(x) = \frac{\varrho(x, B)}{\varrho(x, A) + \varrho(x, B)}$$

is continuous on X. Also $h(A) = \{1\}$ and $h(B) = \{0\}$. Notice that this gives a simple proof of (6.80) in the case that X is a metric space.

(6.87) Exercise. Let (X, ϱ) be a metric space and let A and B be nonvoid subsets of X. Define the *distance from A to B* to be the number

$$\varrho(A, B) = \inf\{\varrho(a, b) : a \in A, b \in B\}.$$

Prove the following assertions.

(a) If A is compact, then there exists a point $a \in A$ such that $\varrho(a, B) = \varrho(A, B)$.

(b) If A and B are both compact, then there exist points $a \in A$ and $b \in B$ such that $\varrho(a, b) = \varrho(A, B)$.

(c) If A is compact and B is closed, then $\varrho(A, B) = 0$ if and only if $A \cap B \neq \varnothing$.

(d) If X is a noncompact metric space with no isolated points, then X contains nonvoid, closed, disjoint sets A and B such that $\varrho(A, B) = 0$.

(6.88) Exercise. Let X be a nonvoid complete metric space. Suppose that f is a function from X into X such that for some constant $c \in]0, 1[$ we have

$$\varrho(f(x), f(y)) \leq c\varrho(x, y)$$

for all $x, y \in X$. Prove that there exists a unique point $u \in X$ such that $f(u) = u$. [Let $x \in X$ and consider the sequence $x, f(x), f(f(x)), \ldots$]. This result is known as BANACH'S *fixed-point theorem.* It implies several existence theorems in the theory of differential and integral equations.

(6.89) Exercise. Prove that the closed interval $[0,1]$ cannot be expressed as the union of a pairwise disjoint family of closed [nondegenerate] intervals each of length less than 1.

(6.90) Exercise. Suppose that: X is a topological space; Y is a metric space; and f is a function from X into Y. For each $x \in X$, define

$$\omega(x) = \inf\{\operatorname{diam}(f(U)) : U \text{ is a neighborhood of } x\}.$$

The function ω is called the *oscillation function for f.* Prove the following statements.

(a) The function f is continuous at x if and only if $\omega(x) = 0$.

(b) For each real number α, the set $\{x \in X : \omega(x) < \alpha\}$ is open in X.

(c) The set $\{x \in X : f \text{ is continuous at } x\}$ is a G_δ set.

(d) There is no real-valued function f defined on R such that $\{x \in R : f \text{ is continuous at } x\} = Q$.

(e) There exists a real-valued function f on R such that $\{x \in R : f \text{ is discontinuous at } x\} = Q$.

(6.91) Exercise. Prove that every locally compact Hausdorff space is of second category [as a subset of itself]. [Mimic the proof of the Baire category theorem (6.54) by constructing an appropriate decreasing sequence of compact sets.]

(6.92) Exercise. Let X be a topological space and let Y be a metric space. Suppose that f and $(f_n)_{n=1}^{\infty}$ are functions from X into Y such that each f_n is continuous and $\lim_{n\to\infty} f_n(x) = f(x)$ for each $x \in X$. Let

$$A = \bigcap_{k=1}^{\infty} \bigcup_{m=1}^{\infty} \left[\bigcap_{n=m}^{\infty}\left\{x \in X : \varrho(f_m(x), f_n(x)) \leqq \frac{1}{k}\right\}\right]^{\circ}.$$

Prove that:

(a) f is continuous at each point of A;

(b) $X \cap A'$ is of first category in X;[1]

(c) if X is of second category [in itself], then $\{x \in X : f \text{ is continuous at } x\}$ is dense in X;

(d) $f^{-1}(V)$ is an F_σ set for each open set $V \subset Y$ $\Big[$prove that $f^{-1}(V)$

$$= \bigcup_{k=1}^{\infty} \bigcup_{m=1}^{\infty} \bigcap_{n=m}^{\infty} \left\{x \in X : \varrho(f_n(x), V') \geqq \frac{1}{k}\right\}\Big];$$

(e) the function ξ_Q is the pointwise limit of no sequence of continuous real-valued functions on R.[2]

(f) Prove that $\xi_Q(x) = \lim_{m\to\infty} [\lim_{n\to\infty} \{\cos(m!\pi x)\}^{2n}]$ for all $x \in R$.

(g) Prove that $\operatorname{sgn}(x) = \lim_{n\to\infty} \frac{2}{\pi} \arctan(nx)$ for all $x \in R$.

(h) Prove that $1 - \xi_Q(x) = \lim_{m\to\infty} \operatorname{sgn}\{\sin^2(m!\pi x)\}$ for all $x \in R$.

(6.93) Exercise. Let $l_\infty^r(N)$ denote the set of all bounded sequences $x = (x_n)$ of real numbers. For $x, y \in l_\infty^r(N)$, define $d(x, y) = \sup\{|x_n - y_n| : n \in N\}$. Prove the following.

(a) The function d is a metric for $l_\infty^r(N)$.

(b) The metric space $l_\infty^r(N)$ is not separable.

(c) If (X, ϱ) is any separable metric space, then there exists an isometry f from X into $l_\infty^r(N)$, *i.e.*, $d(f(x), f(y)) = \varrho(x, y)$ for all $x, y \in X$. [Let $(p_n)_{n=1}^{\infty}$ be dense in X and define $f(x) = (\varrho(x, p_n) - \varrho(p_n, p_1))_{n=1}^{\infty}$].

(6.94) Exercise. Prove that if X is a compact metric space and f is an isometry from X into X, then f is onto X.

(6.95) Exercise. Let X be a locally compact Hausdorff space and let D be a dense subset of X such that D is locally compact in its relative topology. Prove that D is open in X.

(6.96) Exercise. Let X be a linearly ordered set. The *order topology for* X is the topology on X obtained by taking as a subbase the family

---

[1] Show first that if $(E_m)_{m=1}^{\infty}$ is any sequence of subsets of X, then $\bigcap_{m=1}^{\infty} E_m^- \subset \left(\bigcap_{m=1}^{\infty} E_m\right) \cup \left(\bigcup_{m=1}^{\infty} E_m^- \cap E_m'\right)$.

[2] Recall that ξ_Q is the characteristic function of Q (2.20).

of all sets of the form $\{x \in X : c < x\}$ and $\{x \in X : x < d\}$ for $c, d \in X$. Prove that X, with its order topology, is compact if and only if every nonvoid subset of X has both a supremum and an infimum in X. [Use (6.40) as in the proof of (6.44).]

(6.97) Exercise. (a) Use (6.96) to prove that a well-ordered set is compact [in its order topology] if and only if it contains a greatest element.

(b) Use (6.96) to show that $R^{\#}$ with the usual topology (6.5.b) is compact.

(6.98) Exercise. Prove that the set P_{Ω} of all countable ordinal numbers [see (4.49)] with its order topology (6.96) is sequentially compact but not compact.

(6.99) Exercise. Let P be CANTOR'S ternary set and let $X = \{0, 1\}^N$ have the product topology, where $\{0, 1\}$ has the discrete topology. For $\boldsymbol{x} = (x_1, x_2, \ldots) \in X$, define $\varphi(\boldsymbol{x}) = \sum_{n=1}^{\infty} \frac{2x_n}{3^n}$. According to (6.64), φ is a one-to-one mapping from X onto P. Prove that both φ and φ^{-1} are continuous.

(6.100) Exercise. Prove that if Y is a compact metric space and P is CANTOR'S ternary set, then there exists a continuous function f from P onto Y. [Let $\{V_n\}_{n=1}^{\infty}$ be a countable base for the topology of Y. For each $n \in N$, set $A_{n,0} = V_n^-$ and $A_{n,1} = Y \cap V_n'$. For a point $x = \sum_{n=1}^{\infty} \frac{2x_n}{3^n}$ in P, the set $\bigcap_{n=1}^{\infty} A_{n,x_n}$ is either void or contains just one point. Let $B = \left\{x \in P : \bigcap_{n=1}^{\infty} A_{n,x_n} \neq \varnothing\right\}$ and for each $x \in B$, let $g(x) \in \bigcap_{n=1}^{\infty} A_{n,x_n}$. Prove that g is continuous from B onto Y. Show that B is closed in P and that there exists a continuous function h from P onto B. Finally set $f = g \circ h$.]

(6.101) Exercise [BANACH]. Let f be a continuous real-valued function on $[a, b] \subset R$.

(a) For each positive integer n, let $F_n = \left\{x : x \in [a, b], \text{ and } f(x') = f(x) \text{ for some } x' \geqq x + \frac{1}{n}\right\}$. Prove that F_n is a closed set.

(b) Let $E = [a, b] \cap \bigcap_{n=1}^{\infty} F_n'$. Prove that f is one-to-one on E and that $f(E) = f([a, b])$. In fact, each $x \in E$ is equal to $\sup\{y : y \in [a, b], f(y) = f(x)\}$. Note that E is a G_δ set.

(6.102) Exercise. Let f be a real-valued function with domain R having a relative minimum at each point of R, *i.e.*, for each $a \in R$, there is a number $\delta(a) > 0$ such that $f(t) \geqq f(a)$ if $|t - a| < \delta(a)$.

(a) Prove that $f(R)$ is a countable set.

(b) Find a function as above that is unbounded and also monotone on no interval containing 0.

(6.103) Exercise. Consider a function f with domain R and range contained in R such that $f \circ f = f$. Describe f completely. If f is continuous, what more can you say? [Recall that R and $f(R)$ are connected (6.9) and so $f(R)$ is an interval.] If f is differentiable, what more can you say?

(6.104) Exercise. Prove the following.

(a) A continuous image of a connected space is connected.

(b) A Cartesian product $\underset{\iota \in I}{\times} X_\iota$ is connected if and only if every X_ι is connected.

§ 7. Spaces of continuous functions

Functions — both real- and complex-valued — are a major object of study in this text. Given a set X and a set $\mathfrak{F}$ of functions defined on X, we are frequently interested not only in individual functions f in $\mathfrak{F}$, but also in $\mathfrak{F}$ as an entity, or space, in its own right. Often $\mathfrak{F}$ admits a natural topology [or several natural topologies] of interest by themselves and also for proving facts about $\mathfrak{F}$. Often too $\mathfrak{F}$ is a vector space over K or R, and vector space notions can be most helpful in studying analytic questions regarding $\mathfrak{F}$. In the present section we take up a simple class of function spaces — spaces of continuous functions — and a simple topology for these spaces. Many other function spaces will be studied in the sequel.

We begin with a few definitions and some notation.

(7.1) Definition. Let X be any nonvoid set [no topology as yet], and consider the set K^X of all complex-valued functions defined on X. For $f, g \in K^X$, let $f + g$ be the function in K^X defined by

(i) $(f+g)(x) = f(x) + g(x)$ for all $x \in X$;

let fg be defined by

(ii) $(fg)(x) = f(x)\,g(x)$ for all $x \in X$;

for $f \in K^X$ and $\alpha \in K$, let αf be defined by

(iii) $(\alpha f)(x) = \alpha(f(x))$ for all $x \in X$.

For $f \in K^X$, let $|f|$ be the function such that

(iv) $|f|(x) = |f(x)|$ for all $x \in X$

and $\bar{f}$ the function such that

(v) $\bar{f}(x) = \overline{f(x)}$ for all $x \in X$.

That is, sums, products, scalar multiples, absolute values, and complex conjugates of functions on X are defined *pointwise*. The set R^X of all real-valued functions on X can be considered in an obvious way as a subset of K^X, and so definitions (i), (ii), (iii) [for real α], (iv) and (v) [$f = \bar{f}$ if and only if $f \in R^X$] hold for R^X as well as K^X. In addition, R^X admits a natural partial order. For $f, g \in R^X$, we write $f \leqq g$ [or $g \geqq f$] if

(vi) $f(x) \leqq g(x)$ for all $x \in X$.

We define $\max\{f, g\}$ and $\min\{f, g\}$ by

(vii) $\max\{f, g\}(x) = \max\{f(x), g(x)\}$ for all $x \in X$

and

(viii) $\min\{f, g\}(x) = \min\{f(x), g(x)\}$ for all $x \in X$.

For some purposes, we also need extended real-valued functions on X. For $\mathfrak{D} \subset (R^{\#})^X$, we define $\sup\{f : f \in \mathfrak{D}\}$ by

(ix) $\sup\{f : f \in \mathfrak{D}\}(x) = \sup\{f(x) : f \in \mathfrak{D}\}$,

which can be any element of $R^{\#}$, and $\inf\{f : f \in \mathfrak{D}\}$ by

(x) $\inf\{f : f \in \mathfrak{D}\}(x) = \inf\{f(x) : f \in \mathfrak{D}\}$.

Thus all of our operations on and relations between functions are defined *pointwise*.

Finally, for a subset $\mathfrak{F}$ of K^X, we define $\mathfrak{F}^r$ by

(xi) $\mathfrak{F}^r = \{f \in \mathfrak{F} : f(x) \in R$ for all $x \in X\} = \mathfrak{F} \cap R^X$

and $\mathfrak{F}^+$ by

(xii) $\mathfrak{F}^+ = \{f \in \mathfrak{F}^r : f(x) \geqq 0$ for all $x \in X\}$.

The set $\mathfrak{F}^+$ also can be defined for $\mathfrak{F} \subset (R^{\#})^X$.

(7.2) Remarks. (a) For $\alpha \in K$, the function ψ in K^X such that $\psi(x) = \alpha$ for all $x \in X$ is called the *constant function with value* α or the *function identically* α. This function is a quite different entity from the number α. It would be unwieldy to use a distinct symbol [*e.g.* $C_{\alpha, X}$] for this function whenever we need to write it. We will therefore write the function identically α simply as α, trusting to the reader's good sense to avoid confusion.

(b) It is easy to check that K^X is a vector space over K and that R^X is a vector space over R. Also these spaces are commutative rings, with [multiplicative] unit the constant function 1. It is further obvious that

(i) $\alpha(fg) = (\alpha f)g = f(\alpha g)$

for all functions f, g and scalars α. That is, K^X and R^X are *algebras over* K and R, respectively. [A vector space over a field F that is also a ring in which (i) holds is called an *algebra over* F.]

(c) It is also clear that the relation $\leqq$ in R^X satisfies (2.7.i)–(2.7.iii), *i.e.*, $\leqq$ is a genuine partial ordering. If $\overline{\overline{X}} > 1$, then $\leqq$ is not a linear order. It is also easy to see that $(R^X, \leqq)$ is a *lattice*: for $f, g \in R^X$, there is a unique $h \in R^X$ such that $h \geqq f$, $h \geqq g$, and $h \leqq h'$ if $h' \geqq f$ and $h' \geqq g$; that is, h is the smallest *majorant* of f and g. Similarly there is a largest *minorant* k of f and g. It is obvious that $h = \max\{f, g\}$ and that $k = \min\{f, g\}$.

(d) The partially ordered set R^X enjoys a much stronger property than (c). Let $\mathfrak{F}$ be any nonvoid subset of R^X bounded above by a function $\varphi \in R^X$, *i.e.*, $f \leqq \varphi$ for all $f \in \mathfrak{F}$. Then $\mathfrak{F}$ admits a smallest majorant. Its value at $x \in X$ is of course $\sup\{f(x) : f \in \mathfrak{F}\}$. Similar statements hold for sets $\mathfrak{F} \subset R^X$ that admit minorants.

For infinite sets X, the algebras K^X and R^X are too large to be of much use in analysis, although their algebraic structure is of great interest to specialists. By a first restriction we obtain a metrizable space.

(7.3) Definition. Let X be a nonvoid set. Let $\mathfrak{B}(X)$ denote the set of all functions $f \in K^X$ such that

(i) $\sup\{|f(x)| : x \in X\}$

is finite. Such functions are said to be *bounded*. The number (i), written as $\|f\|_u$, is called the *uniform norm of* f.

(7.4) Theorem. *Let X be a nonvoid set, and consider $f, g \in \mathfrak{B}(X)$ and $\alpha \in K$. Then the following relations hold:*

(i) $\|0\|_u = 0$, $\|f\|_u > 0$ *if* $f \neq 0$;

(ii) $\|\alpha f\|_u = |\alpha| \, \|f\|_u$;

(iii) $\|f + g\|_u \leqq \|f\|_u + \|g\|_u$;

(iv) $\|fg\|_u \leqq \|f\|_u \, \|g\|_u$.

Similar assertions hold for $f, g \in \mathfrak{B}^r(X)$ and $\alpha \in R$.

Proof. Simple exercise.

The linear space $\mathfrak{B}(X)$ with its norm $\| \ \|_u$ is an important example of a class of analytico-algebraic objects which we shall encounter repeatedly.

(7.5) Definition. Let E be a linear space over K [or R]. Suppose that there is a function $x \to \|x\|$ with domain E and range contained in R such that:

(i) $\|0\| = 0$ and $\|x\| > 0$ if $x \neq 0$;

(ii) $\|\alpha x\| = |\alpha| \, \|x\|$ for all $x \in E$ and $\alpha \in K$ [or R];

(iii) $\|x + y\| \leqq \|x\| + \|y\|$ for all $x, y \in E$.

The pair $(E, \| \ \|)$ is called a *complex* [or *real*] *normed linear space*, and $\| \ \|$ is called a *norm*.[1]

If E is a normed linear space and also an algebra over K [or R], and if

(iv) $\|xy\| \leqq \|x\| \, \|y\|$ for all $x, y \in E$,

then E is called a *complex* [or *real*] *normed algebra*. If a normed algebra has a multiplicative unit u, then we will postulate that

(v) $\|u\| = 1$.[2]

(7.6) Theorem. *Let E be a complex or real normed linear space. Let ϱ be the function on $E \times E$ defined by*

(i) $\varrho(x, y) = \|x - y\|$.

Then ϱ is a metric on E.

Proof. Trivial.

---

[1] As usual, where confusion seems unlikely we will call E itself a normed linear space.

[2] Since $\|x\| = \|ux\| \leqq \|u\| \, \|x\|$, we have $\|u\| \geqq 1$ without (v). Also, a normed algebra with unit can be renormed so that the unit has norm 1 and nothing essential is changed. See Exercise (7.42) *infra*.

(7.7) Definition. A complex [real] normed linear space that is complete in the metric $\|x - y\|$ is called a *complex [real] Banach space.* A complex [real] Banach space that is also a normed algebra is called a *Banach algebra.*

Banach spaces are very important in contemporary analysis; many basic theorems can be couched in abstract terms as assertions about Banach spaces of one kind or another. We will give many examples throughout the text of this technique [see in particular § 14]. We turn next to the principal object of study in the present section, and one of the important objects of study in the entire text.

(7.8) Definition. Let X be a nonvoid topological space. Let $\mathfrak{C}(X)$ denote the set of all functions in $\mathfrak{B}(X)$ that are continuous complex-valued functions on X.

(7.9) Theorem. *With the algebraic operations* (7.1.i)–(7.1.iii) *and the norm* $\| \ \|_u$ *of* (7.3), $\mathfrak{C}(X)$ *is a commutative complex Banach algebra with unit.*

Proof. The only nonobvious point is the completeness of $\mathfrak{C}(X)$ in the uniform metric. Let $(f_n)_{n=1}^{\infty}$ be a sequence of functions in $\mathfrak{C}(X)$ such that

$$\lim_{m,n\to\infty} \|f_n - f_m\|_u = 0 . \tag{1}$$

That is,

$$\lim_{m,n\to\infty} \left[\sup\{|f_n(x) - f_m(x)| : x \in X\}\right] = 0 . \tag{2}$$

For every fixed $x \in X$, (2) implies that

$$\lim_{m,n\to\infty} |f_n(x) - f_m(x)| = 0 ,$$

and so $(f_n(x))_{n=1}^{\infty}$ is a Cauchy sequence in K. Since K is complete [use (6.50) with $n = 1$], the sequence $(f_n(x))$ has a limit in K, which we denote by $f(x)$. The mapping $x \to f(x)$ is thus an element of K^X. We claim that $f \in \mathfrak{C}(X)$ and that

$$\lim_{n\to\infty} \|f - f_n\|_u = 0 . \tag{3}$$

Actually, it is easiest to prove (3) first. Let ε be an arbitrary positive real number and let the integer p [depending only on ε] be so large that

$$\|f_n - f_m\|_u < \frac{\varepsilon}{3} \tag{4}$$

for all $m, n \geqq p$. Now consider a fixed but arbitrary $x \in X$, and choose m [depending on both x and ε] so large that $m \geqq p$ and also

$$|f_m(x) - f(x)| < \frac{\varepsilon}{3} . \tag{5}$$

Combining (4) and (5), we see that

$$\begin{aligned}|f_n(x) - f(x)| &\leqq |f_m(x) - f_n(x)| + |f_m(x) - f(x)| \\ &< \frac{\varepsilon}{3} + \frac{\varepsilon}{3} \\ &= \frac{2}{3}\varepsilon \end{aligned} \tag{6}$$

for all $n \geqq p$. The integer p is independent of x, and as x is arbitrary in (6), we may take the supremum in (6) to write

$$\begin{aligned}\|f_n - f\|_u &= \sup\{|f_n(x) - f(x)| : x \in X\} \\ &\leqq \frac{2}{3}\varepsilon \\ &< \varepsilon\end{aligned}$$

if $n \geqq p$. That is, (3) holds.

It remains to prove that $f \in \mathfrak{C}(X)$. Choose n so that $\|f - f_n\|_u < 1$. Then it is evident that $|f(x)| < |f_n(x)| + 1$ for all $x \in X$, and so $\|f\|_u$ exists and does not exceed $\|f_n\|_u + 1$. To prove that f is continuous, let x be any point in X, let ε be a positive number, and let n be so large that $\|f_n - f\|_u < \frac{\varepsilon}{3}$. Let U be a neighborhood of x such that

$$|f_n(y) - f_n(x)| < \frac{\varepsilon}{3} \quad \text{for all} \quad y \in U.$$

For $y \in U$, we thus see that

$$\begin{aligned}|f(y) - f(x)| &\leqq |f(y) - f_n(y)| + |f_n(y) - f_n(x)| + |f_n(x) - f(x)| \\ &\leqq \|f - f_n\|_u + |f_n(y) - f_n(x)| + \|f_n - f\|_u \\ &< \frac{\varepsilon}{3} + \frac{\varepsilon}{3} + \frac{\varepsilon}{3} \\ &= \varepsilon.\end{aligned}$$

This is just the defining property (6.68) of continuity of f at x. □

(7.10) Remarks. Some topological spaces X admit no nonconstant continuous real- or complex-valued functions. A simple but trivial example is $(X, \mathcal{O})$ where X is infinite and $\mathcal{O}$ consists of $\varnothing$ plus all subsets of X with finite complements. Such spaces are of no interest for the present text. However, if $\overline{\overline{X}} > 1$ and X is a locally compact Hausdorff space, then $\mathfrak{C}(X)$ contains an abundance of nonconstant functions, as Theorem (6.80) shows. At the present time, locally compact Hausdorff spaces seem to be an ideal vehicle for integration theory, as well as for a number of classical theorems of analysis. Much of the present section will accordingly be devoted to these spaces.

(7.11) Exercise. Many, although not all, noncompact topological spaces admit unbounded continuous complex-valued functions. We will make no detailed study of the space of *all* continuous functions on a topological space, but in this exercise the reader is invited to consider some of the possibilities, and to prove the following assertions.

(a) Every noncompact metric space admits an unbounded continuous real-valued function.

(b) Let P_Ω be the well-ordered set defined in (4.49). Let $\mathscr{B}$ be the family of all sets of the form $\{0\}$ or $\{\gamma \in P_\Omega : \alpha < \gamma \leqq \beta\}$ where $\alpha < \beta < \Omega$. Then $\mathscr{B}$ is a base for the order topology of P_Ω [cf. (6.96) and (6.98)]. The topological space P_Ω is a locally compact Hausdorff space; in fact every subspace $\{\gamma \in P_\Omega : \alpha \leqq \gamma \leqq \beta\}$ is compact. Also P_Ω is noncompact, although it is both sequentially compact and Fréchet compact. [Again see (6.96) and (6.98).]

(c) Every continuous complex-valued function f on P_Ω is ultimately constant: there are an ordinal $\alpha \in P_\Omega$ and a complex number t such that $f(\gamma) = t$ for all $\gamma \geqq \alpha$. Consequently every continuous complex-valued function on P_Ω is bounded.

(d) Let X be a topological space, and let $\mathfrak{C}_a(X)$ denote the set of *all* continuous complex-valued functions on X, bounded or unbounded. If $\mathfrak{C}_a(X)$ contains unbounded functions, we cannot impose the uniform norm $\| \ \|_u$ on $\mathfrak{C}_a(X)$. However, an analogue of (7.9) does hold. If $(f_n)_{n=1}^\infty$ is a sequence of functions in $\mathfrak{C}_a(X)$ for which all differences $f_n - f_m$ are bounded, and if

$$\lim_{m,n\to\infty} \|f_n - f_m\|_u = 0,$$

then there is an $f \in \mathfrak{C}_a(X)$ such that all $f - f_n$ are bounded and

$$\lim_{n\to\infty} \|f - f_n\|_u = 0 .$$

We return to a study of function spaces on locally compact Hausdorff spaces.

(7.12) Definition. Let X be a nonvoid locally compact Hausdorff space. Let $\mathfrak{C}_{00}(X)$ be the subset of $\mathfrak{C}(X)$ consisting of all $f \in \mathfrak{C}(X)$ such that for some compact subset F of X [depending on f], $f(x) = 0$ for all $x \in F' \cap X$. Let $\mathfrak{C}_0(X)$ be the subset of $\mathfrak{C}(X)$ consisting of all $f \in \mathfrak{C}(X)$ such that for every positive number ε, there is a compact subset F of X [depending on f and ε] such that $|f(x)| < \varepsilon$ for all $x \in F' \cap X$. Functions in $\mathfrak{C}_{00}(X)$ are said to *vanish in a neighborhood of infinity* and functions in $\mathfrak{C}_0(X)$ to *vanish at infinity*: both phrases are loose but expressive.

(7.13) Exercise. Prove the following.

(a) The inclusions $\mathfrak{C}_{00}(X) \subset \mathfrak{C}_0(X) \subset \mathfrak{C}(X)$ obtain. If X is noncompact, then $\mathfrak{C}_{00}(X) \subsetneqq \mathfrak{C}(X)$ (6.80). For the space P_Ω of (7.11.b), we have $\mathfrak{C}_{00}(X) = \mathfrak{C}_0(X)$. If X is compact, then $\mathfrak{C}_{00}(X) = \mathfrak{C}_0(X) = \mathfrak{C}(X)$.

(b) Let X be an arbitrary locally compact Hausdorff space, and consider $\mathfrak{C}(X)$ and $\mathfrak{B}(X)$ as metric spaces under the uniform metric $\|f - g\|_u$. The space $\mathfrak{C}_0(X)$ is closed in $\mathfrak{B}(X)$ [and hence also in $\mathfrak{C}(X)$]: if $f_n \in \mathfrak{C}_0(X)$, $f \in \mathfrak{B}(X)$, and $\lim\limits_{n\to\infty} \|f_n - f\|_u = 0$, then $f \in \mathfrak{C}_0(X)$. Also $\mathfrak{C}_0(X)$ is complete as a metric space. [Recall (6.47).] The space $\mathfrak{C}_{00}(X)$

is not in general closed: its closure is $\mathfrak{C}_0(X)$. [A proof of the last statement is indicated in (7.41) below.]

Several concepts closely related to, but not identical with, continuity are needed in our treatment of integration theory. We now take them up.

(7.14) Definition. Let X and Y be metric spaces with metrics ϱ and σ respectively. A mapping φ with domain X and range contained in Y is said to be *uniformly continuous* if for every $\varepsilon > 0$ there is a $\delta > 0$ such that $\varrho(x, x') < \delta$ implies that $\sigma(\varphi(x), \varphi(x')) < \varepsilon$.

(7.15) Remarks. (a) It is easy to see from (6.71) that a uniformly continuous mapping is indeed continuous.

(b) There is a notion of uniform continuity more general than that in (7.14), based on what are called *uniform spaces*. We do not need this concept and hence omit it.

(7.16) Exercise. (a) Consider the function exp (5.56) defined on K. Prove that exp is continuous but not uniformly continuous.

(b) Let X be a nonvoid set and let ϱ be the discrete metric on X (6.13.c). Prove that every mapping of (X, ϱ) into a metric space is uniformly continuous.

(c) Find reasonable necessary and sufficient conditions for a metric space to admit a continuous real-valued function that is not uniformly continuous.

(d) Let $(f_n)_{n=1}^{\infty}$ be a sequence of complex-valued uniformly continuous functions on a metric space X such that all differences $f_n - f_m$ are bounded and

$$\lim_{m, n \to \infty} \|f_n - f_m\|_u = 0 .$$

Prove that the limit function f (7.11.d) is uniformly continuous.

(7.17) Theorem. *Let X and Y be as in* (7.14) *and let φ be a continuous mapping of X into Y with the following property. For every $\varepsilon > 0$, there is a compact subset A_ε of X such that $\sigma(\varphi(x), \varphi(x')) < \varepsilon$ for all x, x' in $A'_\varepsilon \cap X$. Then φ is uniformly continuous.*

Proof. Let ε be an arbitrary positive number, and let A_ε be as in the statement of the theorem: if x, x' are in $A'_\varepsilon \cap X$, then

$$\sigma(\varphi(x), \varphi(x')) < \varepsilon . \tag{1}$$

Now look at an arbitrary point $y \in A_\varepsilon$. Since φ is continuous, there is a positive number η_y [depending on y] such that

$$z \in B_{2\eta_y}(y) \quad \text{implies} \quad \sigma(\varphi(z), \varphi(y)) < \frac{\varepsilon}{2} \cdot \tag{2}$$

[Notation is as in (6.14).] Now consider the family of sets $\{B_{\eta_y}(y) : y \in A_\varepsilon\}$. This is an open covering of A_ε, and so by (6.32) and (6.18) there is a finite subfamily $\{B_{\eta_{y_1}}(y_1), \ldots, B_{\eta_{y_m}}(y_m)\}$ that covers A_ε. Let δ be the number $\min\{\eta_{y_1}, \eta_{y_2}, \ldots, \eta_{y_m}\}$. We claim that this δ will satisfy (7.14) for the preassigned ε. If x and x' are in $A'_\varepsilon \cap X$, than (1) may be applied. If at least one of x and x' is in A_ε, we may suppose that $x \in A_\varepsilon$. Then x is in some $B_{\eta_{y_k}}(y_k)$. If $\varrho(x, x') < \delta$, then we have

$$\begin{aligned}\varrho(y_k, x') &\leqq \varrho(y_k, x) + \varrho(x, x')\\ &< \eta_{y_k} + \delta\\ &\leqq 2\eta_{y_k}.\end{aligned}$$

That is, x' is in $B_{2\eta_{y_k}}(y_k)$ and so (2) shows that $\sigma(\varphi(x'), \varphi(y_k)) < \frac{\varepsilon}{2}$. Since x is also in $B_{2\eta_{y_k}}(y_k)$, (2) implies that $\sigma(\varphi(x), \varphi(y_k)) < \frac{\varepsilon}{2}$, and hence

$$\begin{aligned}\sigma(\varphi(x), \varphi(x')) &\leqq \sigma(\varphi(x), \varphi(y_k)) + \sigma(\varphi(y_k), \varphi(x'))\\ &< \frac{\varepsilon}{2} + \frac{\varepsilon}{2}\\ &= \varepsilon.\end{aligned}$$

Thus φ is uniformly continuous. □

(7.18) Corollary. *If X is a locally compact metric space, then all functions in $\mathfrak{C}_0(X)$ are uniformly continuous. If X is a compact metric space, then all functions in $\mathfrak{C}(X)$ are uniformly continuous.*

Proof. See (7.17). □

(7.19) Our next notion is restricted to real-valued functions f on a topological space X. The definition of continuity for f at $x_0 \in X$ may be [somewhat artificially] broken up into two parts: for every $\varepsilon > 0$, there is a neighborhood U of x_0 such that

(i) $f(x) > f(x_0) - \varepsilon$ for all $x \in U$,

and

(ii) $f(x) < f(x_0) + \varepsilon$ for all $x \in U$.

Taken separately, (i) and (ii) define useful classes of functions. It is advisable also to consider *extended* real-valued functions.

(7.20) Definition. Let X be a topological space and f an extended real-valued function defined on X that does not assume the value $-\infty$, that is, $f(x)$ is real or ∞ for all $x \in X$. The function f is said to be *lower semicontinuous at* $x_0 \in X$ if the following conditions hold. If $f(x_0) < \infty$, then for every $\varepsilon > 0$, there is a neighborhood U of x_0 such that $f(x) > f(x_0) - \varepsilon$ for all $x \in U$. If $f(x_0) = \infty$, then for every positive number α there is a neighborhood U of x_0 such that $f(x) > \alpha$ for all $x \in U$. The function f is called *lower semicontinuous* if it is lower semicontinuous at every point of X. The set of all lower semicontinuous functions on X is denoted by the symbol $\mathfrak{M}(X)$. A function on X with values in $[-\infty, \infty[$ is called *upper semicontinuous* if analogous conditions $f(x) < f(x_0) + \varepsilon$ [$f(x_0)$ finite] or $f(x) < -\alpha$ [$f(x_0) = -\infty$] hold near x_0. The set of all upper semicontinuous functions on X is denoted by $\mathfrak{N}(X)$.

(7.21) Exercise. Prove the following.

(a) Nonconstant semicontinuous functions exist on every topological space X having an open set U such that $\varnothing \subsetneqq U \subsetneqq X$. The characteristic function ξ_U is lower semicontinuous and $\xi_{U'}$ is upper semicontinuous.

(b) A function f is lower semicontinuous if and only if $-f$ is upper semicontinuous. [Recall that $-(\infty) = -\infty$ and that $-(-\infty) = \infty$ (6.1).] Thus for any fact about lower semicontinuous functions there is a dual fact about upper semicontinuous functions. [We will ordinarily state explicitly only assertions about lower semicontinuous functions.]

(c) The definition of lower semicontinuity at x_0 can be recast: for every real number $\alpha < f(x_0)$, there is a neighborhood U of x_0 such that $f(x) > \alpha$ for all $x \in U$. This deals with the cases $f(x_0) < \infty$ and $f(x_0) = \infty$ simultaneously.

(d) A function f from X into $]-\infty, \infty]$ is lower semicontinuous if and only if $f^{-1}(]t, \infty])$ is open in X for every $t \in R$. [This characterization is frequently useful.]

(7.22) Theorem. *Let X be a topological space.*

(i) *If $f \in \mathfrak{M}(X)$ and α is a nonnegative real number, then $\alpha f \in \mathfrak{M}(X)$.*

(ii) *If $f, g \in \mathfrak{M}(X)$, then* $\min\{f, g\} \in \mathfrak{M}(X)$.

(iii) *If $\mathfrak{D}$ is a nonvoid subset of $\mathfrak{M}(X)$, then* $\sup\{f : f \in \mathfrak{D}\}$ *is a function in $\mathfrak{M}(X)$.*

(iv) *If $f, g \in \mathfrak{M}(X)$, then $f + g \in \mathfrak{M}(X)$.*

(v) *Suppose that X is a locally compact Hausdorff space. Then if $f \in \mathfrak{M}^+(X)$, we have* $f = \sup\{\varphi : \varphi \in \mathfrak{C}_{00}^+(X),\ \varphi \leq f\}$.

Proof. Assertions (i), (ii), and (iv) are all but obvious, and we leave their proof to the reader. To prove (iii), write g for the function $\sup\{f : f \in \mathfrak{D}\}$. Consider a fixed but arbitrary point $x_0 \in X$. For every real number $\alpha < g(x_0)$, the definition of supremum (5.32) shows that there is an $f \in \mathfrak{D}$ such that $\alpha < f(x_0) \leq g(x_0)$. [This holds both for $g(x_0) < \infty$ and $g(x_0) = \infty$; note that $f(x_0)$ may be finite or infinite if $g(x_0) = \infty$]. Since f is in $\mathfrak{M}(X)$, there is a neighborhood U of x_0 such that

$$f(x) > \alpha \quad \text{for all} \quad x \in U$$

[this holds for $f(x_0) < \infty$ as well as for $f(x_0) = \infty$]. For all $x \in U$, we clearly have

$$g(x) \geq f(x) > \alpha .$$

Thus g satisfies (7.21.c) and so (iii) is proved.

To prove (v), we use URYSOHN's theorem (6.80). If $f = 0$, there is nothing to prove. If $f(x_0) > 0$ for some $x_0 \in X$, consider any real number α such that $0 < \alpha < f(x_0)$. There is a neighborhood U of x_0 such that $f(x) > \alpha$ for all $x \in U$. By (6.80) there is a function $\varphi \in \mathfrak{C}_{00}^+(X)$ such that $\varphi(X) \subset [0, \alpha]$, $\varphi(x_0) = \alpha$, and $\varphi(U') \subset \{0\}$. Plainly we have $\varphi \leq f$, and as α can be arbitrarily close to $f(x_0)$ [$f(x_0) < \infty$] or arbitrarily large

$[f(x_0) = \infty]$, we infer that $f(x_0) = \sup\{\varphi(x_0) : \varphi \in \mathfrak{C}_{00}^{+}(X),\ \varphi \leqq f\}$. Since x_0 is arbitrary, (v) is established. □

(7.23) Exercise. (a) State and prove the analogue of (7.22) for upper semicontinuous functions.

(b) Let X be a topological space admitting a nonclosed open set. Prove that $\mathfrak{M}^r(X)$, the real-valued lower semicontinuous functions on X, do not form a linear space. If every open subset of X is closed, prove that $\mathfrak{M}^r(X)$ is a linear space.

(c) Prove that uniform limits of sequences in $\mathfrak{M}^r(X)$ are in $\mathfrak{M}^r(X)$.

(7.24) We now take up a famous and vitally important approximation theorem. The German mathematician KARL WEIERSTRASS [1815—1897] published in 1885 a proof that polynomials with real coefficients are dense in the space $\mathfrak{C}^r([0, 1])$ in the topology induced by the uniform metric. We will prove a far-reaching generalization of this theorem due to the contemporary U.S. mathematician M. H. STONE. Every proof of the STONE-WEIERSTRASS theorem requires *some* "hard" analysis. Our proof uses only a little such analysis, which is presented in the next theorem. We prove somewhat more than we need.

(7.25) Theorem. *For any real number* α, *let* $\binom{\alpha}{0} = 1$ *and*

$$\binom{\alpha}{n} = \frac{\alpha(\alpha-1)(\alpha-2)\cdots(\alpha-n+1)}{n!}$$

for $n = 1, 2, \ldots$. *The infinite series*

$$\text{(i)} \qquad \sum_{n=0}^{\infty} \binom{\alpha}{n} x^n$$

converges for all $x \in\ ]-1, 1[$. *For* $\alpha > 0$, *the series*

$$\text{(ii)} \qquad \sum_{n=0}^{\infty} \left|\binom{\alpha}{n}\right|$$

converges, and the series (i) *converges uniformly and absolutely in* $[-1, 1]$. *Finally, we have*

$$\text{(iii)} \qquad \sum_{n=0}^{\infty} \binom{\alpha}{n} x^n = (1+x)^\alpha$$

for $x \in\ ]-1, 1[$ *for all real* α. *For* $\alpha > 0$, (iii) *holds for all* $x \in [-1, 1]$.

Proof. To prove (i), we use the ratio test. If α is a nonnegative integer, then all but finitely many of the numbers $\binom{\alpha}{n}$ are 0, and so (i) trivially converges. Otherwise, for $|x| < 1$ and $n = 0, 1, 2, \ldots$, we have

$$\left|\frac{\binom{\alpha}{n+1} x^{n+1}}{\binom{\alpha}{n} x^n}\right| = \frac{|\alpha - n|}{n+1}\,|x| \to |x| < 1.$$

Hence the series (i) converges absolutely for $|x| < 1$. Next we prove (ii). The case in which α is a nonnegative integer is again trivial. For α not a nonnegative integer, let $a_n = \left|\binom{\alpha}{n}\right|$. Then we have

$$\frac{a_{n+1}}{a_n} = \frac{|\alpha - n|}{n+1} = \frac{n-\alpha}{n+1},$$

the last equality holding for $n \geqq [\alpha] + 1$ $\big[[\alpha]$ is the largest integer not exceeding $\alpha\big]$. Hence for $n \geqq [\alpha] + 1$, we have $(n+1)\,a_{n+1} = n\,a_n - \alpha\,a_n$, and so

$$n\,a_n - (n+1)\,a_{n+1} = \alpha a_n > 0\,. \tag{1}$$

Thus for $n \geqq [\alpha] + 1$, $(n\,a_n)$ is a decreasing sequence, and so has a limit; let $\lim\limits_{n\to\infty} n\,a_n = \gamma \geqq 0$. Now consider the series $\sum\limits_{n=0}^{\infty} (n\,a_n - (n+1)\,a_{n+1})$. The pth partial sum of this series is $-(p+1)\,a_{p+1}$, which converges to $-\gamma$. Hence the series $\sum\limits_{n=0}^{\infty} (n\,a_n - (n+1)\,a_{n+1})$ converges, and since by (1)

$$a_n = \frac{1}{\alpha}(n\,a_n - (n+1)\,a_{n+1})$$

for sufficiently large n, the series $\sum\limits_{n=0}^{\infty} a_n$ converges; *i.e.*, the series (ii) converges. Since $\left|\binom{\alpha}{n} x^n\right| \leqq \left|\binom{\alpha}{n}\right|$ for $|x| \leqq 1$, the series (i) converges absolutely and uniformly in $[-1, 1]$ and so defines a continuous function on $[-1, 1]$.

We now prove (iii). For $x \in\,]-1, 1[$ and $\alpha \in R$, let $f_\alpha(x) = \sum\limits_{n=0}^{\infty} \binom{\alpha}{n} x^n$. A power series may be differentiated term by term in its *open* interval of convergence, and so we have

$$f'_\alpha(x) = \sum_{n=1}^{\infty} n\binom{\alpha}{n} x^{n-1} = \sum_{n=0}^{\infty} (n+1)\binom{\alpha}{n+1} x^n.$$

Using the identity $(n+1)\binom{\alpha}{n+1} = \alpha\binom{\alpha-1}{n}$,

we see that

$$f'_\alpha(x) = \alpha \sum_{n=0}^{\infty} \binom{\alpha-1}{n} x^n = \alpha f_{\alpha-1}(x)\,. \tag{2}$$

Next we have

$$(1+x)\, f_{\alpha-1}(x) = (1+x) \sum_{n=0}^{\infty} \binom{\alpha-1}{n} x^n$$

$$= 1 + \sum_{n=1}^{\infty} \left[\binom{\alpha-1}{n} + \binom{\alpha-1}{n-1}\right] x^n = \sum_{n=0}^{\infty} \binom{\alpha}{n} x^n = f_\alpha(x)\,,$$

i.e.,

$$(1 + x) f_{\alpha-1}(x) = f_\alpha(x) . \tag{3}$$

Combining (2) and (3), we see that

$$(1 + x) f'_\alpha(x) - \alpha f_\alpha(x) = 0$$

for all $x \in]-1, 1[$. Also we have

$$\frac{d}{dx} [f_\alpha(x) (1 + x)^{-\alpha}] = (1 + x)^{-\alpha-1}((1 + x) f'_\alpha(x) - \alpha f_\alpha(x)) = 0 .$$

Hence $\frac{f_\alpha(x)}{(1+x)^\alpha}$ is a constant function in $]-1, 1[$. Setting $x = 0$, we determine that the constant value of this function is 1; *i.e.*, $f_\alpha(x) = (1 + x)^\alpha$ for $|x| < 1$. If $\alpha > 0$, then the series also converges for $x = \pm 1$, and hence the identity (iii) also holds for these values of x. [Two continuous complex-valued functions that are equal on a dense subset of their common domain are equal everywhere.] □

(7.26) Remarks. (a) In proving the important Theorem (7.27), we need only the special case of (7.25.iii) in which $\alpha = \frac{1}{2}$. That is, we need the identity

(i) $$(1 + x)^{\frac{1}{2}} = \sum_{n=0}^{\infty} \binom{\frac{1}{2}}{n} x^n ,$$

and the fact that the series converges absolutely and uniformly for $|x| \leqq 1$.

(b) The STONE-WEIERSTRASS theorem depends ultimately upon order properties of the space $\mathfrak{C}^r(X)$ for a compact Hausdorff space X. It is evident that $\max\{f, g\}$ and $\min\{f, g\}$ are continuous if f, g are in $\mathfrak{C}^r(X)$, and this simple fact will be very useful. The following theorem connects polynomials with maxima and minima.

(7.27) Theorem. *Let X be any nonvoid set [no topology] and let ψ and φ be functions in $\mathfrak{B}^r(X)$. For every $\varepsilon > 0$, there is a polynomial $P = \sum_{j=0}^{m} \sum_{k=0}^{n} \alpha_{jk} \psi^j \varphi^k$ with real coefficients such that*

(i) $$\|\max\{\psi, \varphi\} - P\|_u < \varepsilon .$$

A like assertion holds for $\min\{\psi, \varphi\}$.

Proof. For every real number t, the identity

$$|t| = (1 + t^2 - 1)^{\frac{1}{2}}$$

is obvious. For $|t^2 - 1| \leqq 1$, that is, for $|t| \leqq 2^{\frac{1}{2}}$, we have

$$(1 + t^2 - 1)^{\frac{1}{2}} = \sum_{n=0}^{\infty} \binom{\frac{1}{2}}{n} (t^2 - 1)^n ,$$

and the series converges absolutely and uniformly for all t such that

$|t| \leqq 2^{\frac{1}{2}}$ (7.26). For every positive integer p and for all real t such that $|t| \leqq 2^{\frac{1}{2}}$, we thus have

$$\left| |t| - \sum_{n=0}^{p} \binom{\frac{1}{2}}{n} (t^2 - 1)^n \right| \leqq \sum_{n=p+1}^{\infty} \left| \binom{\frac{1}{2}}{n} \right| ; \tag{1}$$

and the right side of (1) is arbitrarily small for p sufficiently large. Now consider any nonzero function ψ in $\mathfrak{B}^r(X)$, and for brevity write the number $\|\psi\|_u$ as β. For $x \in X$ and $p \in N$, we apply (1) to write

$$\left| |\psi(x)| - \sum_{n=0}^{p} \beta \binom{\frac{1}{2}}{n} \left[\frac{\psi^2(x)}{\beta^2} - 1 \right]^n \right|$$

$$= \beta \left| \beta^{-1} |\psi(x)| - \sum_{n=0}^{p} \binom{\frac{1}{2}}{n} \left[\frac{\psi^2(x)}{\beta^2} - 1 \right]^n \right|$$

$$\leqq \beta \sum_{n=p+1}^{\infty} \left| \binom{\frac{1}{2}}{n} \right| .$$

It follows that for $\psi \in \mathfrak{B}^r(X)$ and $\varepsilon > 0$, there is a real polynomial Q in the functions 1 and ψ^2 [*i.e.*, a linear combination with real coefficients of $1, \psi^2, \psi^4, \ldots$], the coefficients of which depend solely upon β and ε, for which

$$\| |\psi| - Q \|_u < 2\varepsilon . \tag{2}$$

If $\varphi = \psi$, the relation (i) is trivial. If $\varphi \neq \psi$, then $\psi - \varphi$ is not the zero function, and we may apply (2) with ψ replaced by $\psi - \varphi$:

$$\| |\psi - \varphi| - Q(\psi - \varphi) \|_u < 2\varepsilon .$$

As noted in (5.36), the identity

$$\max\{\psi, \varphi\} = \frac{1}{2}(|\psi - \varphi| + (\psi + \varphi))$$

obtains, and so we have

$$\left\| \max\{\psi, \varphi\} - \frac{1}{2}(Q(\psi - \varphi) + \psi + \varphi) \right\|_u < \varepsilon .$$

Setting $P(\psi, \varphi) = \frac{1}{2}(Q(\psi - \varphi) + \psi + \varphi)$, we obtain (i). To prove (i) with "max" replaced by "min", note that $\min\{\psi, \varphi\} = -\max\{-\psi, -\varphi\}$. □

(7.28) Definitions. Let X be a set and $\mathfrak{S}$ a family of functions on X with values in a set Y. Suppose that for all $x, y \in X$ such that $x \neq y$ there is an $f \in \mathfrak{S}$ such that $f(x) \neq f(y)$. Then we say that $\mathfrak{S}$ is a *separating family of functions on* X. Next suppose that $\mathfrak{S}$ is a family of real-valued functions on X. A *real polynomial in functions from* $\mathfrak{S}$ is any finite sum of functions $\alpha f_1^{n_1} f_2^{n_2} \cdots f_l^{n_l}$, where the coefficient α is a real number and the exponents n_j are positive integers. Equivalently, a real polynomial in functions from $\mathfrak{S}$ is an element of the smallest subalgebra of R^X that contains $\mathfrak{S}$. Complex polynomials are defined similarly in K^X.

We now prove one version of the STONE-WEIERSTRASS theorem.

(7.29) Theorem. *Let X be a nonvoid compact Hausdorff space, and let $\mathfrak{S}$ be a subset of $\mathfrak{C}^r(X)$ such that:*

(i) *$\mathfrak{S}$ is a separating family;*

(ii) *$\mathfrak{S}$ contains the function 1;*

(iii) *$f \in \mathfrak{S}$ and $\alpha \in R$ imply $\alpha f \in \mathfrak{S}$;*

(iv) *$f, g \in \mathfrak{S}$ implies $f + g \in \mathfrak{S}$;*

(v) *$f, g \in \mathfrak{S}$ implies $\max\{f, g\} \in \mathfrak{S}$.*

Then $\mathfrak{S}$ is dense in $\mathfrak{C}^r(X)$ in the topology induced by the uniform metric.[1]

Proof. Consider $f_0 \in \mathfrak{C}^r(X)$. If f_0 is constant, then the approximation is trivial. If not, we have

$$c = \inf\{f_0(x) : x \in X\} < d = \sup\{f_0(x) : x \in X\} .$$

Let $f = \frac{2}{d - c}(f_0 - d) + 1$, so that $f(X) \subset [-1, 1]$ and $\inf f = -1$, $\sup f = 1$. It obviously suffices to prove that f lies in the closure of $\mathfrak{S}$. Consider the nonvoid compact sets $E = \left\{x \in X : f(x) \leqq -\frac{1}{3}\right\}$ and $F = \left\{x \in X : f(x) \geqq \frac{1}{3}\right\}$. For every $x \in E$ and $y \in F$ there exists $g_{x,y} \in \mathfrak{S}$ such that $g_{x,y}(x) \neq g_{x,y}(y)$. Define

$$h_{x,y} = \frac{4}{3(g_{x,y}(y) - g_{x,y}(x))}(g_{x,y} - g_{x,y}(y)) + \frac{2}{3} .$$

We have $h_{x,y}(x) = -\frac{2}{3}$ and $h_{x,y}(y) = \frac{2}{3}$. Since $\mathfrak{S}$ is a linear space, it is clear that $h_{x,y} \in \mathfrak{S}$. Since $h_{x,y}$ is continuous, there exists for each $x \in E$ and $y \in F$ a neighborhood U_x of x such that $h_{x,y}(w) < -\frac{1}{3}$ for all $w \in U_x$. Since E is compact and $\bigcup_{x \in E} U_x \supset E$, there are points $x_1, x_2, \ldots, x_m \in E$ such that

$$U_{x_1} \cup U_{x_2} \cup \cdots \cup U_{x_m} \supset E .$$

Let $\varphi_y = \min\{h_{x_1,y}, h_{x_2,y}, \ldots, h_{x_m,y}\}$. Since $\min\{a, b\} = -\max\{-a, -b\}$, hypothesis (v) shows that $\varphi_y \in \mathfrak{S}$. It is clear that $\varphi_y(y) = \frac{2}{3}$ and that $\varphi_y(x) < -\frac{1}{3}$ for all $x \in E$. Note that the function φ_y is defined for every fixed $y \in F$.

We repeat the above technique to find points $y_1, y_2, \ldots, y_n \in F$ and functions $\varphi_{y_1}, \varphi_{y_2}, \ldots, \varphi_{y_n} \in \mathfrak{S}$ such that $\varphi_{y_j}(x) < -\frac{1}{3}$ for all $x \in E$ and such that for each $x \in F$, some $\varphi_{y_j}(x)$ is greater than $\frac{1}{3}$.

[1] Note that our hypotheses are slightly redundant: if X admits a separating family of continuous real-valued functions, then X has to be a Hausdorff space.

Hence the function $\psi = \max\{\varphi_{y_1}, \varphi_{y_2}, \ldots, \varphi_{y_n}\}$ is in $\mathfrak{S}$ and satisfies the following inequalities: $\psi(x) < -\frac{1}{3}$ for all $x \in E$ and $\psi(x) > \frac{1}{3}$ for all $x \in F$. Now define w_1 by

$$w_1 = \min\left\{\max\left\{\psi, -\frac{1}{3}\right\}, \frac{1}{3}\right\}.$$

It is clear that $w_1 \in \mathfrak{S}$, $w_1(E) = \left\{-\frac{1}{3}\right\}$, $w_1(F) = \left\{\frac{1}{3}\right\}$, and $w_1(X) \subset \left[-\frac{1}{3}, \frac{1}{3}\right]$. The definitions of E and F show that

$$\|f - w_1\|_u = \frac{2}{3}.$$

The function $\frac{3}{2}(f - w_1)$ is in $\mathfrak{C}^r(X)$ and has minimum -1 and maximum 1. The method used to construct w_1 can again be used to approximate $\frac{3}{2}(f - w_1)$. Thus there exists $w_2 \in \mathfrak{S}$ such that

$$\left\|\frac{3}{2}(f - w_1) - w_2\right\|_u = \frac{2}{3}.$$

Multiplying by $\frac{2}{3}$, we have

$$\left\|f - w_1 - \frac{2}{3}w_2\right\|_u = \left(\frac{2}{3}\right)^2.$$

Our scheme is now clear. In the next step we approximate

$$\left(\frac{3}{2}\right)^2\left(f - w_1 - \frac{2}{3}w_2\right)$$

by a suitable function w_3 in $\mathfrak{S}$, obtaining the equality

$$\left\|f - w_1 - \frac{2}{3}w_2 - \left(\frac{2}{3}\right)^2 w_3\right\|_u = \left(\frac{2}{3}\right)^3.$$

In general, if n is any positive integer, there are functions $w_1, \ldots, w_n$ such that

$$\left\|f - w_1 - \frac{2}{3}w_2 - \cdots - \left(\frac{2}{3}\right)^{n-1} w_n\right\|_u = \left(\frac{2}{3}\right)^n,$$

where each w_j is in $\mathfrak{S}$. Since $\mathfrak{S}$ is a linear space and $\lim_{n\to\infty}\left(\frac{2}{3}\right)^n = 0$, the proof is complete. □

The standard version of the approximation theorem is simple to prove from (7.27) and (7.29).

(7.30) Stone-Weierstrass Theorem. *Let X be a nonvoid compact Hausdorff space and $\mathfrak{S}$ a separating family of functions in $\mathfrak{C}^r(X)$ containing the function 1. Then polynomials with real coefficients in functions from $\mathfrak{S}$ are a dense subalgebra of $\mathfrak{C}^r(X)$ in the topology induced by the uniform metric.*

Proof. Let $\mathfrak{P}$ be the set of all polynomials with real coefficients in functions from $\mathfrak{S}$, and let $^-$ be the closure operation in $\mathfrak{C}^r(X)$. Clearly $\mathfrak{P}$

is a subalgebra of $\mathfrak{C}^r(X)$. Suppose that f and g are in $\mathfrak{P}^-$, and that

$$\lim_{n\to\infty} \|f - f_n\|_u = 0\,, \quad \lim_{n\to\infty} \|g - g_n\|_u = 0\,,$$

where f_n and g_n are in $\mathfrak{P}$. Theorem (7.27) proves that $\max\{f_n, g_n\}$ is in $\mathfrak{P}^-$. It is easy to see from (5.36) that

$$\|\max\{f_n, g_n\} - \max\{f, g\}\|_u \leqq \|f_n - f\|_u + \|g_n - g\|_u\,,$$

and so $\max\{f, g\}$ is in $(\mathfrak{P}^-)^- = \mathfrak{P}^-$. Thus $\mathfrak{P}^-$ satisfies all the hypotheses imposed on $\mathfrak{S}$ in (7.29), and (7.29) therefore implies that $(\mathfrak{P}^-)^- = \mathfrak{P}^-$ is $\mathfrak{C}^r(X)$. □

(7.31) Corollary [WEIERSTRASS]. *Let X be a compact subset of R and let $f \in \mathfrak{C}^r(X)$. Then there is a real polynomial $P = P(x)$ such that $\|f - P\|_u$ is arbitrarily small.*

Proof. The corollary is simply (7.30) with $\mathfrak{S} = \{\iota, 1\}$ where $\iota(x) = x$ for all $x \in X$. □

(7.32) Remarks. The hypothesis in (7.29) and (7.30) that $\mathfrak{S}$ be a separating family of functions is obviously necessary. Let X be a compact Hausdorff space containing at least two points. Suppose that $\mathfrak{F}$ is a nonvoid subset of $\mathfrak{C}^r(X)$ such that for some distinct $a, b \in X$, we have $f(a) = f(b)$ for all $f \in \mathfrak{F}$. Then $P(a) = P(b)$ for every polynomial P in functions from $\mathfrak{F}$. It follows that polynomials in functions from $\mathfrak{F}$ cannot be dense in $\mathfrak{C}^r(X)$. In fact, no function having different values at a and b can be arbitrarily uniformly approximated by polynomials in the functions of $\mathfrak{F}$, and Theorem (6.80) shows that there is a $\varphi \in \mathfrak{C}^r(X)$ such that $\varphi(a) = 1$ and $\varphi(b) = 0$. The beauty of the STONE-WEIERSTRASS theorem lies in the fact that its hypothesis, trivially necessary, is also *sufficient*. We will use the STONE-WEIERSTRASS theorem very frequently. It is an essential tool for the analyst.

(7.33) Examples. (a) Let X be CANTOR's ternary set (6.62), which we write as the set of all numbers $2\sum_{k=1}^{\infty} \frac{y_k}{3^k}$, where each y_k is 0 or 1. For $n \in N$, let φ_n be the function on X such that

$$\varphi_n\left(2\sum_{k=1}^{\infty} \frac{y_k}{3^k}\right) = (-1)^{y_n}\,.$$

For $n_1 < n_2 < \cdots < n_l$, let $\varphi_{n_1, n_2, \ldots, n_l} = \prod_{j=1}^{l} \varphi_{n_j}$.

Then each φ_n is continuous on X, the set $\{\varphi_1, \varphi_2, \ldots, \varphi_n, \ldots\}$ is a separating family on X, and $\varphi_n^2 = 1$. Hence every function in $\mathfrak{C}^r(X)$ is arbitrarily uniformly approximable by a linear combination of the functions 1 and $\varphi_{n_1, n_2, \ldots, n_l}$.

(b) The function exp defined as in (5.56) is real-valued on R and satisfies, as every schoolboy should know, the inequality $\exp(x_1) < \exp(x_2)$

if $x_1 < x_2$. Hence the one-element family $\{\exp\}$ is a separating family on R, and the family of polynomials in exp and 1 is dense in $\mathfrak{C}^r(X)$ for every compact subset X of R. These polynomials are precisely all functions P of the form $P(x) = \sum_{k=0}^{n} \alpha_k \exp(n_k x)$, where the α_k's are real numbers, the n_k's are nonnegative integers, and $n = 1, 2, 3. \ldots$

(c) Polynomials in the cosine function and 1 are dense in $\mathfrak{C}^r([0, \pi])$. For every $n \in N$, $\cos^n(x)$ can be written as a linear combination of terms of the form $\cos(kx)$ and 1 [this fact can be proved by induction on n]. Hence the family of all linear combinations of the functions $1, \cos(x)$, $\cos(2x), \ldots$ is dense in $\mathfrak{C}^r([0, \pi])$.

(d) Consider $\mathfrak{C}^r([0, 1])$ and the smallest subset $\mathfrak{S}$ of $\mathfrak{C}^r([0, 1])$ containing 1, the function ι $[\iota(x) = x]$, and satisfying (7.29.iii)–(7.29.v). It is not hard to see that $\mathfrak{S}$ is exactly the linear lattice consisting of all piecewise linear, continuous, real-valued functions f on $[0, 1]$. That is, there are a finite number of subintervals $[0, x_1], [x_1, x_2], \ldots, [x_{n-1}, 1]$ such that $0 < x_1 < \cdots < x_{n-1} < 1$ and such that f is linear on each $[x_{k-1}, x_k]$. Theorem (7.29) shows that $\mathfrak{S}$ is dense in $\mathfrak{C}^r([0, 1])$.

One would expect a complex version of the STONE-WEIERSTRASS theorem, and indeed there is one. However, in the complex case some additional hypothesis is required.

(7.34) Theorem. *Let X be a nonvoid compact Hausdorff space. Let $\mathfrak{S}$ be a separating family of functions in $\mathfrak{C}(X)$ containing the function 1 and such that $\bar{f} \in \mathfrak{S}$ whenever $f \in \mathfrak{S}$. Then polynomials with complex coefficients in functions from $\mathfrak{S}$ are dense in $\mathfrak{C}(X)$ in the topology induced by the uniform metric.*

Proof. Let g be in $\mathfrak{C}(X)$. We will show that the real-valued continuous functions $\mathrm{Re}\,g$ and $\mathrm{Im}\,g$ can be approximated by polynomials in functions belonging to $\mathfrak{S}$. For $f \in \mathfrak{S}$, we have $\mathrm{Re}\,f = \frac{f + \bar{f}}{2}$ and $\mathrm{Im}\,f = \frac{f - \bar{f}}{2i}$, and so $\mathrm{Re}\,f$ and $\mathrm{Im}\,f$ are polynomials in functions from $\mathfrak{S}$. The family of functions $\mathfrak{S}_0$, consisting of all $\mathrm{Re}\,f$ and $\mathrm{Im}\,f$ for $f \in \mathfrak{S}$, is a separating family on X, for if $x, y \in X$ and $f(x) \neq f(y)$, then either $\mathrm{Re}\,f(x) \neq \mathrm{Re}\,f(y)$ or $\mathrm{Im}\,f(x) \neq \mathrm{Im}\,f(y)$. Also $\mathfrak{S}_0$ contains the function 1. Theorem (7.30) shows that for every $\varepsilon > 0$, there are polynomials P and Q in functions from $\mathfrak{S}_0$ such that

$$\|P - \mathrm{Re}\,g\|_u < \frac{1}{2}\varepsilon,$$

$$\|Q - \mathrm{Im}\,g\|_u < \frac{1}{2}\varepsilon,$$

and so

$$\|P + iQ - g\|_u < \varepsilon. \quad \square$$

(7.35) Examples. (a) Let $X = \{z \in K : |z| \leqq 1\}$, and let $\mathfrak{S} = \{\iota\}$ where $\iota(z) = z$. Uniform limits of polynomials in ι and 1 are analytic on the open unit disk $\{z \in K : |z| < 1\}$ and continuous on the closed unit disk $\{z \in K : |z| \leqq 1\}$. There are certainly nonanalytic continuous functions on the unit disk, and so polynomials in functions from $\mathfrak{S}$ are *not* dense in $\mathfrak{C}(X)$. Thus some additional hypothesis is necessary in the complex STONE-WEIERSTRASS theorem. Conditions on X under which the theorem holds for $\mathfrak{C}(X)$ with no additional hypothesis have been found by W. RUDIN [Proc. Amer. Math. Soc. 7, 825–830 (1956)].

(b) Let $T = \{z \in K : |z| = 1\}$, and consider the function ι on T given by $\iota(z) = z$. For any $z \in T$, we have $z = \exp(ix) = \cos(x) + i\sin(x)$, for exactly one $x \in [0, 2\pi[$. Hence $\iota(\exp(ix)) = \exp(ix)$, and we see that $\{\iota\}$ is a separating family for T. We have $\bar{\iota}(z) = \overline{\iota(\exp(ix))} = \cos(x) - i\sin(x) = \cos(-x) + i\sin(-x) = \exp(-ix)$. Recall that $\exp(ix)^n = \exp(inx)$, $n = 1, 2, \ldots$. The family $\{\iota, \bar{\iota}, 1\}$ satisfies the conditions of (7.34). Now let $f \in \mathfrak{C}(T)$ and let $\varepsilon > 0$. By the above remarks there is a function

$$\exp(ix) \to \sum_{k=-n}^{n} \alpha_k \exp(ikx)$$

such that

$$\left| \sum_{k=-n}^{n} \alpha_k \exp(ikx) - f(\exp(ix)) \right| < \varepsilon$$

for all $x \in [0, 2\pi[$ [actually for all $x \in R$]. This result is important in the theory of trigonometric series [see (16.34)], and the proof we have given is the shortest one that we know of.

(7.36) Exercise. (a) Let X be any noncompact subset of R. Find a separating family $\mathfrak{S}$ in $\mathfrak{C}^r(X)$ such that polynomials in the family $\mathfrak{S} \cup \{1\}$ are not dense in $\mathfrak{C}^r(X)$. Be careful to consider all possible cases. How small can you make $\overline{\overline{\mathfrak{S}}}$?

(b) Let A be a nonvoid, finite, discrete space. Prove a sharpened form of the STONE-WEIERSTRASS theorem for this set, and do it without recourse to (7.30).

(7.37) Exercise. There are several versions of the STONE-WEIERSTRASS theorem besides (7.29) and (7.30), which are listed below. Prove them, using (7.30).

(a) For a compact Hausdorff space X, a *closed* subalgebra $\mathfrak{A}$ of $\mathfrak{C}^r(X)$ that separates points is either $\mathfrak{C}^r(X)$ or $\{f \in \mathfrak{C}^r(X) : f(x_0) = 0\}$ for a fixed point $x_0 \in X$. A subalgebra $\mathfrak{A}$ of $\mathfrak{C}^r(X)$ that separates points and vanishes identically at no point of X is dense in $\mathfrak{C}^r(X)$. [This statement differs from (7.30) in that $\mathfrak{A}$ need not contain the function 1.]

(b) For a locally compact Hausdorff space X, a closed subalgebra $\mathfrak{A}$ of $\mathfrak{C}_0^r(X)$ that separates points is either $\mathfrak{C}_0^r(X)$ or $\{f \in \mathfrak{C}_0^r(X) : f(x_0) = 0\}$

for a fixed point $x_0 \in X$. A separating subalgebra $\mathfrak{A}$ of $\mathfrak{C}_0^r(X)$ vanishing identically at no point of X is dense in $\mathfrak{C}_0^r(X)$.

(c) State and prove complex versions of (a) and (b).

(7.38) Exercise. Theorems (7.29) and (7.30) are incomparable, in the sense that subalgebras of $\mathfrak{C}^r(X)$ need not be sublattices and sublattices need not be subalgebras.

(a) Let P and Q be real polynomials on $[0, 1]$. Prove that $\max\{P, Q\}$ is a polynomial if and only if $P \geqq Q$ or $Q \geqq P$. Prove, however, that if P_1, Q_1, P_2, and Q_2 are polynomials such that $P_1 \leqq P_2$, $P_1 \leqq Q_2$, $Q_1 \leqq P_2$, and $Q_1 \leqq Q_2$, then there is a polynomial Φ such that $\max\{P_1, Q_1\} \leqq \Phi \leqq \min\{P_2, Q_2\}$.

(b) The sublattice $\mathfrak{S}$ of $\mathfrak{C}^r([0, 1])$ defined in (7.33.d) is not a subalgebra of $\mathfrak{C}^r([0, 1])$. Prove that if f and f^2 are both in $\mathfrak{S}$, then f is a constant. Find conditions on f and g in $\mathfrak{S}$ necessary and sufficient for fg to be in $\mathfrak{S}$.

(7.39) Exercise. Let X be a nonvoid compact subset of R^n, where $n \in N$. For $\boldsymbol{x} = (x_1, \ldots, x_n) \in X$ and $(k_1, \ldots, k_n)$ a finite sequence of nonnegative integers, consider the function

$$\boldsymbol{x} \to x_1^{k_1} x_2^{k_2} \cdots x_n^{k_n}.$$

Prove that linear combinations of these functions are dense in $\mathfrak{C}^r(X)$. State and prove an analogue for $X \subset K^n$ and $\mathfrak{C}(X)$.

We close this section with an important application of the STONE-WEIERSTRASS theorem.

(7.40) TIETZE's Extension Theorem. *Let X be a locally compact Hausdorff space, let Y be a nonvoid compact subspace of X, and let U be an open set such that $Y \subset U \subset X$. Then every function in $\mathfrak{C}(Y)$ can be extended to a function in $\mathfrak{C}_{00}(X)$ that vanishes on $X \cap U'$.*

Proof. In view of (6.79), we lose no generality in supposing that U^- is compact. It obviously suffices to show that every function f in $\mathfrak{C}^r(Y)$ admits an extension $f^\dagger$ in $\mathfrak{C}^r(X)$ such that $f^\dagger(x) = 0$ for all $x \in X \cap U'$. Let $\mathfrak{S}$ be the set of all functions f in $\mathfrak{C}^r(Y)$ admitting such extensions $f^\dagger$. It is trivial that $\mathfrak{S}$ is a subalgebra of $\mathfrak{C}^r(Y)$. Let us show that $\mathfrak{S}$ separates points of Y. For $a \neq b$ and $a, b \in Y$, there is a neighborhood W of a such that $b \notin W$. By (6.79), there is a neighborhood V of a such that $V^- \subset U \cap W$ and V^- is compact. By (6.80) there is a continuous function φ from X into $[0, 1]$ such that $\varphi(V^-) = \{1\}$ and $\varphi((U \cap W)') = \{0\}$. Thus $\varphi(a) = 1$, $\varphi(b) = 0$, $\varphi(X \cap U') \subset \{0\}$, and the restriction of φ to the domain Y is plainly in $\mathfrak{S}$. That is, $\mathfrak{S}$ separates points of Y. Theorem (6.80) shows too that all constant functions on Y are in $\mathfrak{S}$.

Next consider any $f \in \mathfrak{S}$, and write $\alpha = \max\{f(y) : y \in Y\}$, $\beta = \min\{f(y) : y \in Y\}$. It is obvious that $\|f\|_u = \max\{|\alpha|, |\beta|\}$. Let $f^\dagger$ be a continuous real-valued extension of f over X such that

$f^\dagger(X \cap U') \subset \{0\}$. Let $\alpha^\dagger = \max\{\alpha, 0\}$ and $\beta^\dagger = \min\{\beta, 0\}$. Then define the function φ on X as

$$\varphi = \min\{\max\{f^\dagger, \beta^\dagger\}, \alpha^\dagger\}.$$

It is clear that φ is in $\mathfrak{C}^r(X)$, and it is easy to see that: $\varphi(x) = f(x)$ for $x \in Y$; $\alpha^\dagger = \max\{\varphi(x) : x \in X\}$; $\beta^\dagger = \min\{\varphi(x) : x \in X\}$; $\varphi(x) = 0$ for $x \in X \cap U'$. Thus φ is an extension of f of the sort we want for which $\|\varphi\|_u = \|f\|_u$. That is, if $f \in \mathfrak{S}$, then f admits an extension $f^\dagger$ such that $\|f^\dagger\|_u = \|f\|_u$.

Now let g be a function in $\mathfrak{C}^r(Y)$ that is the uniform limit *on* Y of a sequence $(f_n)_{n=1}^{\infty}$ of functions in $\mathfrak{S}$. Choose a subsequence $(f_{n_k})_{k=1}^{\infty}$ such that $\|f_{n_k} - f_{n_{k+1}}\|_u < 2^{-k}$ and write $g_k = f_{n_k} - f_{n_{k+1}}$ $(k = 1, 2, 3, \ldots)$. Then we have

$$g - f_{n_1} = \sum_{k=1}^{\infty} g_k,$$

where the infinite series converges uniformly on Y, and $\|g_k\|_u < 2^{-k}$. Now let $g_k^\dagger$ be a continuous extension of g_k over X such that $g_k^\dagger$ vanishes on U' and $\|g_k^\dagger\|_u = \|g_k\|_u < 2^{-k}$. The infinite series $\sum_{k=1}^{\infty} g_k^\dagger$ converges uniformly on X to a function in $\mathfrak{C}^r(X)$ that vanishes on U'. Therefore $g - f_{n_1}$ is in $\mathfrak{S}$ and hence g itself is in $\mathfrak{S}$.

We have therefore shown that $\mathfrak{S}$ satisfies the hypotheses of (7.30) and is uniformly closed in $\mathfrak{C}^r(Y)$. Therefore $\mathfrak{S}$ is all of $\mathfrak{C}^r(Y)$. □

(7.41) Exercise. Let X be a locally compact Hausdorff space. Every function f in $\mathfrak{C}_0(X)$ can be arbitrarily uniformly approximated by functions in $\mathfrak{C}_{00}(X)$. [Hint: consider a compact set outside of which f is small and use (7.40).]

(7.42) Exercise. Let E be a real or complex normed algebra with multiplicative unit u. For $x \in E$, let

$$|||x||| = \sup\{\|yx\| : y \in E, \|y\| \leqq 1\}.$$

Prove the following:

(a) $|||u||| = 1$;

(b) $\frac{1}{\|u\|}\|x\| \leqq |||x||| \leqq \|x\|$ for all $x \in E$;

(c) the algebra E with the norm $|||\ \ |||$ is a normed algebra in the sense of (7.5);

(d) if E is complete in the metric $\varrho(x, y) = \|x - y\|$, it is also complete in the metric $|||x - y|||$.

(7.43) Exercise. Let X be a nonvoid compact Hausdorff space and let $\mathfrak{L}$ be a lattice of real-valued continuous functions on X, *i.e.*, $f, g \in \mathfrak{L}$ implies $\max\{f, g\} \in \mathfrak{L}$ and $\min\{f, g\} \in \mathfrak{L}$. Suppose that φ is a real-valued continuous function on X having the property that for each $\varepsilon > 0$ and each pair of points $x, y \in X$ there exists a function $f \in \mathfrak{L}$ such that

$|\varphi(x) - f(x)| < \varepsilon$ and $|\varphi(y) - f(y)| < \varepsilon$. Prove that for each $\varepsilon > 0$ there exists a function $h \in \mathfrak{L}$ such that $\|\varphi - h\|_u < \varepsilon$.

(7.44) Exercise [ROBERT I. JEWETT]. Let I denote the interval $[0, 1]$. A family $\mathfrak{F}$ of functions is said to have property V if

(i) $\mathfrak{F} \subset I^X$ for some set X,

(ii) $f \in \mathfrak{F}$ implies $(1 - f) \in \mathfrak{F}$,

and

(iii) $f, g \in \mathfrak{F}$ implies $fg \in \mathfrak{F}$.

Supply I^X with the topology of uniform convergence [the topology induced by the uniform metric]. Prove assertions (a)–(k). [This exercise is rather difficult.]

(a) If X is a set and $\mathfrak{A} \subset I^X$, then $\mathfrak{A}$ is contained in a smallest subfamily [closed subfamily] of I^X having property V. [Intersect a collection of families.]

Suppose that X is a topological space, and let $\mathfrak{D}(X) = I^X \cap \mathfrak{C}(X)$. For $n \in N$, let $\mathfrak{P}_n$ be the smallest subfamily of $\mathfrak{D}(I^n)$ that has property V and contains the n projections

$$(x_1, \ldots, x_n) \to x_k \quad (k = 1, \ldots, n).$$

(b) If $\mathfrak{F}$ has property V, if $\{f_1, \ldots, f_n\} \subset \mathfrak{F}$, and if $p \in \mathfrak{P}_n$, then the function f defined on X by

$$f(x) = p(f_1(x), \ldots, f_n(x))$$

is in $\mathfrak{F}$. [Consider the set of all $p \in \mathfrak{D}(I^n)$ for which the conclusion holds.]

(c) For every $\varepsilon > 0$, there exist functions p and q in $\mathfrak{P}_1$ such that $p < \varepsilon$ and $q > 1 - \varepsilon$ on I. [Consider $p(x) = x^m(1 - x)^m$.]

(d) If $\varepsilon > 0$ and $0 < a < b < 1$, then there is a function $p \in \mathfrak{P}_1$ such that

$$p > 1 - \varepsilon \quad \text{on} \quad [0, a],$$
$$p < \varepsilon \quad \text{on} \quad [b, 1].$$

[Take $p(x) = (1 - x^m)^n$ for appropriate $m, n \in N$.]

(e) If $\{a_1, \ldots, a_n\} \cup \{b_1, \ldots, b_n\} \subset I$, then

$$\left| \prod_{k=1}^{n} b_k - \prod_{k=1}^{n} a_k \right| \leqq \sum_{k=1}^{n} |b_k - a_k|.$$

[Use induction on n.]

(f) If $(a, b) \in I^2 = I \times I$ and ε and δ are positive real numbers, then there is a function $p \in \mathfrak{P}_2$ such that

$$p(x, y) > 1 - \varepsilon \quad \text{if} \quad (x - a)^2 + (y - b)^2 \leqq \delta^2$$

and

$$p(x, y) < \varepsilon \quad \text{if} \quad (x - a)^2 + (y - b)^2 \geqq (4\delta)^2$$

for $(x, y) \in I^2$.

[Take $p(x, y) = (1 - p_1(x))p_2(x)(1 - p_3(y))p_4(y)$, where the p_j's are obtained by using (d).]

(g) Let A and B be disjoint compact subsets of I^2, and consider arbitrary $\varepsilon > 0$ and $p \in \mathfrak{P}_2$. Then there exists $q \in \mathfrak{P}_2$ such that

$$q \geqq p \quad \text{on} \quad I^2,$$

$$q > 1 - \varepsilon \quad \text{on} \quad A,$$

and

$$q < p + \varepsilon \quad \text{on} \quad B.$$

[Let $4\delta = \text{dist}(A, B)$. Use compactness and (f) to piece together a $q_0 \in \mathfrak{P}_2$ such that q_0 is near 1 on B and near 0 on A. Then let $q = 1 - (1 - p)q_0$.]

(h) Define φ and ψ on I^2 by $\varphi(x, y) = \max\{x, y\}$ and $\psi(x, y) = \min\{x, y\}$. For each $\varepsilon > 0$ there exist functions $r, q \in \mathfrak{P}_2$ such that

$$\|\varphi - r\|_u < \varepsilon$$

and

$$\|\psi - q\|_u < \varepsilon.$$

[Let $8\delta = \varepsilon < 1$. Let $C = \{(x, y) \in I^2 : \delta \leqq \psi(x, y) \leqq 1 - \delta\}$ and choose $m \in N$ such that $(xy)^m < \delta$ on C. Let $p(x, y) = 1 - (xy)^m$. Then $p \in \mathfrak{P}_2$ and $1 - \delta < p < 1$ on C. For $k \geqq 0$ let $A_k = \{(x, y) \in C : p^k(x, y) \leqq \psi(x, y)\}$ and $B_k = \{(x, y) \in C : p^k(x, y) \geqq \psi(x, y)\}$. Choose $n \in N$ such that $B_n = \varnothing$. Note that $A_{k-1} \cap B_k = \varnothing$ for all k. For $k = 1, \ldots, n$, use (g) to find $q_k \in \mathfrak{P}_2$ such that $q_k \geqq p$ on I^2, $q_k > 1 - \frac{\delta}{n}$ on A_{k-1}, and $q_k < p + \frac{\delta}{n}$ on B_k. Put $q' = q_1 q_2 \cdots q_n$. Prove that $0 \leqq p^k - \psi < \delta$ on $B_k \cap B'_{k+1}$ and use (e) to show that $|p^k - q'| < 3\delta$ on C $(k = 1, \ldots, n-1)$. Thus conclude that $|\psi - q'| < 4\delta$ on C. Use (g) again to find $q'' \in \mathfrak{P}_2$ such that $q'' \geqq q'$ on I^2, $q'' > 1 - \delta$ if $\psi \geqq 1 - \delta$, and $q'' < q' + \delta$ if $\psi \leqq 1 - 2\delta$. Notice that $|\psi - q''| < 6\delta$ if $\psi \geqq \delta$. Use a slight variant of (g) to find $q \in \mathfrak{P}_2$ such that $q \leqq q''$ on I^2, $q < \delta$ if $\psi \leqq \delta$, and $q > q'' - \delta$ if $\psi \geqq 2\delta$. Then $|q - \psi| < 8\delta = \varepsilon$ on I^2.]

(i) Let X be a nonvoid set and let $\mathfrak{F}$ be a closed subfamily of I^X having property V. Then $\mathfrak{F}$ is a lattice. [Use (b) and (h).]

(j) Let X be a compact Hausdorff space and let $\mathfrak{F}$ be a closed separating subset of $\mathfrak{D}(X)$ having property V. Let $S = \{x \in X : f(x) \in \{0, 1\}$ for all $f \in \mathfrak{F}\}$. Then $\mathfrak{F} = \{f : f \in \mathfrak{D}(X), f(S) \subset \{0, 1\}\}$. [Use (7.43), (i), (b), and (d).]

(k) If $n \in N$ and $\mathfrak{F}$ is a closed subfamily of $\mathfrak{D}(I^n)$ which has property V and contains the n projections as well as some function which is never 0 or 1, then $\mathfrak{F} = \mathfrak{D}(I^n)$.

(7.45) Exercise. The *algebra of quaternions*[1] H is defined as the 4-dimensional vector space over R having a basis, traditionally written

[1] Discovered by the Irish mathematician Sir William Rowan Hamilton (1805–1865).

as $\{1, i, j, k\}$, with the following multiplication table:

	1	i	j	k
1	1	i	j	k
i	i	-1	k	$-j$
j	j	$-k$	-1	i
k	k	j	$-i$	-1

Products $(a1 + bi + cj + dk)(a'1 + b'i + c'j + d'k)$ [real coefficients!] are defined by supposing that H is an associative algebra over R. Prove the following.

(a) For every $a1 + bi + cj + dk \in H$,

$$(a1 + bi + cj + dk)(a1 - bi - cj - dk) = (a^2 + b^2 + c^2 + d^2)1.$$

(b) The set $H \cap \{01 + 0i + 0j + 0k\}'$ is a non-Abelian group under multiplication.

(c) For $x = a1 + bi + cj + dk \in H$, we have $x - ixi - jxj - kxk = 4a1$.

(d) For $x = a1 + bi + cj + dk \in H$, let $\|x\| = (a^2 + b^2 + c^2 + d^2)^{\frac{1}{2}}$. Prove that $\|\ \|$ is a norm on H (7.5) and that $\|xy\| = \|x\| \|y\|$ for all $x, y \in H$.

(e) Let X be a topological space and $\mathfrak{C}(X, H)$ the set of all continuous mappings f of X into H [make H into a metric and hence topological space *via* the norm] for which $\|f\|_u = \sup\{\|f(t)\| : t \in X\}$ is finite. Show that $\mathfrak{C}(X, H)$ is a noncommutative Banach algebra over R with a multiplicative unit, all operations being pointwise.

(f) [J. C. HOLLADAY]. Let X be a compact Hausdorff space and let $\mathfrak{A}$ be any subalgebra of $\mathfrak{C}(X, H)$ [closed in particular under multiplication by all constant functions] that separates points and contains 1. Prove that $\mathfrak{A}$ is dense in $\mathfrak{C}(X, H)$. [Use (c) and (7.30), regarding $\mathfrak{C}(X, R)$ as a subring of $\mathfrak{C}(X, H)$.]

CHAPTER THREE

The Lebesgue Integral

Integration from one point of view is an averaging process for functions, and it is in this spirit that we will introduce and discuss integration. In applying an averaging process to a class $\mathfrak{F}$ of real- or complex-valued functions, a number $I(f)$ is assigned to each $f \in \mathfrak{F}$. If $I(f)$ is to be an average, then it should certainly satisfy the conditions

$$I(f+g) = I(f) + I(g)\,,$$
$$I(\alpha f) = \alpha I(f)$$

for $f, g \in \mathfrak{F}$ and $\alpha \in R$. A less essential but often desirable property for I is that $I(f) \geqq 0$ if $f \geqq 0$. In some cases these three properties suffice to identify the averaging process completely.

Let us mention a few such averaging processes. Suppose for example that $\mathfrak{F}$ is all real-valued functions on the finite set $\{1, 2, 3, \ldots, n\}$, *i.e.*, $\mathfrak{F} = R^n$, and define

$$e_j(k) = \begin{cases} 1 & \text{if} \quad j = k, \\ 0 & \text{if} \quad j \neq k; \end{cases}$$

that is, $e_j(k) = \delta_{jk}$. For each $f \in \mathfrak{F}$, we have

$$f = \sum_{j=1}^{n} f(j)\, e_j\,.$$

For any "integral" I we must have

$$I(f) = \sum_{j=1}^{n} f(j)\, I(e_j)\,.$$

In fact if we choose the numbers $I(e_1), I(e_2), \ldots, I(e_n)$ arbitrarily, then an integral I satisfying the first two conditions is determined by the above sum for all $f \in \mathfrak{F}$. Hence in this case the integral is merely a finite sum in which certain weights have been assigned to the points in the domain of the function. To satisfy the third property, we need only require that $I(e_j) \geqq 0$.

If $\mathfrak{F} = \mathfrak{C}([0, 1])$ and $I(f) = \int_0^1 f(x)\, dx$ [the Riemann integral], then I is an averaging process for $\mathfrak{F}$. A simple-minded average defined for any class of functions defined at $x = \frac{1}{2}$ is given by $I(f) = f\left(\frac{1}{2}\right)$. Both of these averaging processes satisfy the third property: $I(f) \geqq 0$ if $f \geqq 0$.

Let $(x_n)_{n=1}^{\infty}$ be an enumeration of a countably infinite set C, let (α_n) be a sequence of complex numbers such that $\sum_{n=1}^{\infty} |\alpha_n| < \infty$, and let $\mathfrak{F}$ be the set of all bounded complex-valued functions defined on C. Then I defined on $\mathfrak{F}$ by $I(f) = \sum_{n=1}^{\infty} \alpha_n f(x_n)$ is an average. It satisfies the third property if and only if $\alpha_n \geqq 0$ for all n.

As a final example, let $\mathfrak{F}$ be all complex-valued functions f on the set C of our last example such that $\sum_{n=1}^{\infty} |f(x_n)|^2 < \infty$. For fixed $g \in \mathfrak{F}$ let $I_g(f) = \sum_{n=1}^{\infty} f(x_n)\, \overline{g(x_n)}$. This series converges absolutely by CAUCHY's inequality (13.13), *i. e.*,

$$|I_g(f)| \leqq \left(\sum_{n=1}^{\infty} |f(x_n)|^2\right)^{\frac{1}{2}} \left(\sum_{n=1}^{\infty} |g(x_n)|^2\right)^{\frac{1}{2}}.$$

In this chapter we will first discuss averages of continuous functions and then extend the process to wider classes of functions. In particular, we will extend the Riemann integral *qua* averaging process to obtain the Lebesgue integral.

We begin with a rapid review of the Riemann-Stieltjes integral, which is a classical device for averaging continuous functions on intervals $[a, b]$.

§ 8. The Riemann-Stieltjes integral

(8.1) Definition. Let A be a subset of R. An extended real-valued function α defined on A is said to be *nondecreasing* on A if $\alpha(x) \leqq \alpha(y)$ whenever $x < y$ in A. In case $\alpha(x) < \alpha(y)$ whenever $x < y$ in A, we say that α is *strictly increasing* on A. The terms *nonincreasing* and *strictly decreasing* are defined analogously. A function is said to be *monotone* or *monotonic* if it is either nondecreasing or nonincreasing[1].

(8.2) Definition. Let $[a, b]$ be any closed interval in R. A *subdivision of* $[a, b]$ is any finite [ordered] set $\Delta \subset [a, b]$ of the form

$$\Delta = \{a = t_0 < t_1 < \cdots < t_n = b\}.$$

The family of all subdivisions of $[a, b]$ is denoted by $\mathscr{D}([a, b])$.

(8.3) Definition. Let $[a, b]$ be a closed interval in R, let α be a real-valued nondecreasing function on $[a, b]$, and let f be a bounded real-valued function on $[a, b]$. For each $\Delta = \{a = t_0 < t_1 < \cdots < t_n = b\} \in$

[1] This definition plainly generalizes that given in (6.81) for sequences: simply take $A = N$ in (8.1) to obtain (6.81).

$\mathfrak{D}([a, b])$, define

$$L(f, \alpha, \Delta) = \sum_{j=1}^{n} \inf\{f(x) : x \in [t_{j-1}, t_j]\} \cdot (\alpha(t_j) - \alpha(t_{j-1})),$$

and

$$U(f, \alpha, \Delta) = \sum_{j=1}^{n} \sup\{f(x) : x \in [t_{j-1}, t_j]\} \cdot (\alpha(t_j) - \alpha(t_{j-1})).$$

These numbers are known as *lower Darboux sums* and *upper Darboux sums*, respectively. Suppose that for every $\varepsilon > 0$ there exists a $\Delta \in \mathfrak{D}([a, b])$ [depending on f, α, and ε] such that

$$U(f, \alpha, \Delta) - L(f, \alpha, \Delta) < \varepsilon.$$

Then f is said to be *Riemann-Stieltjes integrable with respect to* α *on* $[a, b]$.

(8.4) Lemma. *Let* $[a, b]$, f, *and* α *be as in* (8.3). *Suppose that* Δ, $\Delta^* \in \mathfrak{D}([a, b])$ *and that* $\Delta \subset \Delta^*$. *Then we have*

(i) $L(f, \alpha, \Delta) \leqq L(f, \alpha, \Delta^*) \leqq U(f, \alpha, \Delta^*) \leqq U(f, \alpha, \Delta)$.

Proof. The middle inequality is obvious. Let us show that $L(f, \alpha, \Delta) \leqq L(f, \alpha, \Delta^*)$. Suppose first that Δ^* contains just one more point than Δ. Thus if $\Delta = \{a = t_0 < t_1 < \cdots < t_n = b\}$, then

$$\Delta^* = \{a = t_0 < t_1 < \cdots < t_{k-1} < u < t_k < \cdots < t_n = b\}$$

for some k. We have

$$\begin{aligned} L(f, \alpha, \Delta^*) - L(f, \alpha, \Delta) &= \inf\{f(x) : x \in [t_{k-1}, u]\} \cdot (\alpha(u) - \alpha(t_{k-1})) \\ &\quad + \inf\{f(x) : x \in [u, t_k]\} \cdot (\alpha(t_k) - \alpha(u)) \\ &\quad - \inf\{f(x) : x \in [t_{k-1}, t_k]\} \cdot (\alpha(t_k) - \alpha(t_{k-1})) \\ &= (\inf\{f(x) : x \in [t_{k-1}, u]\} - \inf\{f(x) : x \in [t_{k-1}, t_k]\})(\alpha(u) - \alpha(t_{k-1})) \\ &\quad + (\inf\{f(x) : x \in [u, t_k]\} - \inf\{f(x) : x \in [t_{k-1}, t_k]\})(\alpha(t_k) - \alpha(u)) \\ &\geqq 0. \end{aligned}$$

Hence $L(f, \alpha, \Delta^*) \geqq L(f, \alpha, \Delta)$. The proof by induction on the number of points in Δ^* and not in Δ is now clear. Similarly, suprema decrease on subintervals of an interval, and so the last inequality in (i) also holds. □

(8.5) Lemma. *If* Δ, Δ^* *are in* $\mathfrak{D}([a, b])$, *then* $L(f, \alpha, \Delta) \leqq U(f, \alpha, \Delta^*)$.

Proof. By (8.4) we have $L(f, \alpha, \Delta) \leqq L(f, \alpha, \Delta \cup \Delta^*) \leqq U(f, \alpha, \Delta^*)$. □

(8.6) Theorem. *Let* f *be Riemann-Stieltjes integrable with respect to* α *on* $[a, b]$. *Then there exists a unique real number* γ *such that*

$$L(f, \alpha, \Delta) \leqq \gamma \leqq U(f, \alpha, \Delta)$$

for every $\Delta \in \mathfrak{D}([a, b])$. *In fact,*

$$\gamma = \sup\{L(f, \alpha, \Delta) : \Delta \in \mathfrak{D}([a, b])\} = \inf\{U(f, \alpha, \Delta) : \Delta \in \mathfrak{D}([a, b])\}.$$

This number γ is called the *Riemann-Stieltjes integral of* f *with respect to* α *over* $[a, b]$. Historically it is denoted by $\int_a^b f(x)\, d\alpha(x)$, but in this text we write it as $S_\alpha(f; [a, b])$. In case $\alpha(x) = x$ for each $x \in [a, b]$, we call γ the *Riemann integral of f over* $[a, b]$ and denote it by $S(f; [a, b])$.

Proof. Let $\gamma = \sup\{L(f, \alpha, \Delta) : \Delta \in \mathscr{D}([a, b])\}$ and $\delta = \inf\{U(f, \alpha, \Delta) : \Delta \in \mathscr{D}([a, b])\}$. It follows from (8.5) that γ and δ are real numbers and that $\gamma \leqq \delta$. We need only show that $\gamma = \delta$. Assume that $\gamma < \delta$. Since $\delta - \gamma > 0$, the definition of integrability (8.3) shows that there is a $\Delta \in \mathscr{D}([a, b])$ such that $U(f, \alpha, \Delta) - L(f, \alpha, \Delta) < \delta - \gamma$. Then

$$\delta \leqq U(f, \alpha, \Delta) = (U(f, \alpha, \Delta) - L(f, \alpha, \Delta)) + L(f, \alpha, \Delta) < (\delta - \gamma) + \gamma = \delta,$$

a clear contradiction. Thus $\gamma = \delta$. □

(8.7) Theorem. *Let* $[a, b]$, f *and* α *be as in* (8.3). *If* f *is continuous on* $[a, b]$, *then* f *is Riemann-Stieltjes integrable with respect to* α *over* $[a, b]$.

Proof. Let $\varepsilon > 0$ be given. Since $[a, b]$ is compact (6.44), the function f is uniformly continuous on $[a, b]$ (7.18). Thus there exists a $\delta > 0$ such that $|f(x) - f(y)| < \dfrac{\varepsilon}{\alpha(b) - \alpha(a) + 1}$ whenever $x, y \in [a, b]$ and $|x - y| < \delta$. Choose $\Delta = \{a = t_0 < t_1 < \cdots < t_n = b\}$ such that $t_j - t_{j-1} < \delta$ for $j = 1, \ldots, n$. Select $x_j, y_j \in [t_{j-1}, t_j]$ such that

$$f(x_j) = \sup\{f(x) : x \in [t_{j-1}, t_j]\}$$

and

$$f(y_j) = \inf\{f(x) : x \in [t_{j-1}, t_j]\}$$

for $j = 1, \ldots, n$. [Such selections are possible by (6.73).] Then we have $|x_j - y_j| \leqq t_j - t_{j-1} < \delta$, so that

$$0 \leqq f(x_j) - f(y_j) < \frac{\varepsilon}{\alpha(b) - \alpha(a) + 1}.$$

Thus we also have

$$\begin{aligned} U(f, \alpha, \Delta) - L(f, \alpha, \Delta) &= \sum_{j=1}^{n} (f(x_j) - f(y_j))\,(\alpha(t_j) - \alpha(t_{j-1})) \\ &< \sum_{j=1}^{n} \frac{\varepsilon}{\alpha(b) - \alpha(a) + 1} (\alpha(t_j) - \alpha(t_{j-1})) \\ &= \frac{\varepsilon}{\alpha(b) - \alpha(a) + 1} (\alpha(b) - \alpha(a)) < \varepsilon. \quad \square \end{aligned}$$

(8.8) Theorem. *If* f_1 *and* f_2 *are both Riemann-Stieltjes integrable with respect to* α *over* $[a, b]$, *then so is* $f_1 + f_2$ *and* $S_\alpha(f_1 + f_2; [a, b]) = S_\alpha(f_1; [a, b]) + S_\alpha(f_2; [a, b])$.

Proof. Let $\varepsilon > 0$ be given. Choose Δ_j such that $U(f_j, \alpha, \Delta_j) - L(f_j, \alpha, \Delta_j) < \frac{\varepsilon}{2}$ for $j = 1, 2$, and let $\Delta = \Delta_1 \cup \Delta_2$. Using (8.4) and the simple inequalities $\inf(f_1 + f_2) \geqq \inf f_1 + \inf f_2$ and $\sup(f_1 + f_2) \leqq \sup f_1 + \sup f_2$ [valid for any bounded real-valued functions on any nonvoid set], we have

$$\begin{aligned} L(f_1, \alpha, \Delta_1) + L(f_2, \alpha, \Delta_2) &\leqq L(f_1, \alpha, \Delta) + L(f_2, \alpha, \Delta) \\ &\leqq L(f_1 + f_2, \alpha, \Delta) \\ &\leqq U(f_1 + f_2, \alpha, \Delta) \\ &\leqq U(f_1, \alpha, \Delta) + U(f_2, \alpha, \Delta) \\ &\leqq U(f_1, \alpha, \Delta_1) + U(f_2, \alpha, \Delta_2) \\ &< L(f_1, \alpha, \Delta_1) + L(f_2, \alpha, \Delta_2) + \varepsilon . \end{aligned}$$

It follows that $U(f_1 + f_2, \alpha, \Delta) - L(f_1 + f_2, \alpha, \Delta) < \varepsilon$ and hence $f_1 + f_2$ is integrable.

Next let $S_\alpha(f_j; [a, b]) = \gamma_j$ and let Γ_j be such that

$$|L(f_j, \alpha, \Gamma_j) - \gamma_j| < \frac{\varepsilon}{6}, \tag{1}$$

and

$$|U(f_j, \alpha, \Gamma_j) - \gamma_j| < \frac{\varepsilon}{6}. \tag{2}$$

It follows that $0 \leqq U(f_j, \alpha, \Gamma_j) - L(f_j, \alpha, \Gamma_j) < \frac{\varepsilon}{3}$ for $j = 1, 2$. Setting $\Gamma = \Gamma_1 \cup \Gamma_2$, we have

$$\begin{aligned} L(f_1, \alpha, \Gamma_1) + L(f_2, \alpha, \Gamma_2) &\leqq L(f_1, \alpha, \Gamma) + L(f_2, \alpha, \Gamma) \\ &\leqq L(f_1 + f_2, \alpha, \Gamma) \\ &\leqq U(f_1 + f_2, \alpha, \Gamma) \\ &\leqq U(f_1, \alpha, \Gamma) + U(f_2, \alpha, \Gamma) \\ &\leqq U(f_1, \alpha, \Gamma_1) + U(f_2, \alpha, \Gamma_2) \\ &< L(f_1, \alpha, \Gamma_1) + L(f_2, \alpha, \Gamma_2) + \frac{2\varepsilon}{3}, \end{aligned}$$

from which we see that

$$L(f_1 + f_2, \alpha, \Gamma) - L(f_1, \alpha, \Gamma_1) - L(f_2, \alpha, \Gamma_2) < \frac{2\varepsilon}{3} \tag{3}$$

and

$$U(f_1, \alpha, \Gamma_1) + U(f_2, \alpha, \Gamma_2) - U(f_1 + f_2, \alpha, \Gamma) < \frac{2\varepsilon}{3}. \tag{4}$$

From (1) and (3) we infer that

$$|L(f_1 + f_2, \alpha, \Gamma) - (\gamma_1 + \gamma_2)| < \varepsilon ,$$

and from (2) and (4) that

$$|U(f_1 + f_2, \alpha, \Gamma) - (\gamma_1 + \gamma_2)| < \varepsilon . \quad \square$$

(8.9) Theorem. *If f is Riemann-Stieltjes integrable with respect to α over $[a, b]$ and if $c \in R$, then so also is cf and $S_\alpha(cf; [a, b]) = cS_\alpha(f; [a, b])$.*

Proof. Exercise.

(8.10) Theorem. *Let $[a, b]$, f, and α be as in (8.3). If f is Riemann-Stieltjes integrable with respect to α over $[a, b]$ and $f \geqq 0$, then $S_\alpha(f; [a, b])$ is nonnegative.*

Proof. Trivial.

(8.11) Theorem. *Let f be Riemann-Stieltjes integrable with respect to α over $[a, b]$ and let $a < c < b$. Then f is Riemann-Stieltjes integrable with respect to α over both $[a, c]$ and $[c, b]$ and $S_\alpha(f; [a, b]) = S_\alpha(f; [a, c]) + S_\alpha(f; [c, b])$.*

Proof. Let $\varepsilon > 0$ be given. Choose $\Delta \in \mathfrak{D}([a, b])$ such that

$$U(f, \alpha, \Delta) - L(f, \alpha, \Delta) < \varepsilon.$$

In view of (8.4) we may, and do, suppose that $c \in \Delta$; say

$$\Delta = \{a = t_0 < \cdots < t_m = c < t_{m+1} < \cdots < t_n = b\}.$$

Let

$$\Delta_1 = \{a = t_0 < \cdots < t_m = c\} \in \mathfrak{D}([a, c])$$

and

$$\Delta_2 = \{c = t_m < \cdots < t_n = b\} \in \mathfrak{D}([c, b]).$$

Then

$$(U(f, \alpha, \Delta_1) - L(f, \alpha, \Delta_1)) + (U(f, \alpha, \Delta_2) - L(f, \alpha, \Delta_2)) = U(f, \alpha, \Delta) - L(f, \alpha, \Delta) < \varepsilon.$$

It follows that f is integrable over both $[a, c]$ and $[c, b]$. Let $S_\alpha(f;[a,c]) = \gamma_1$ and $S_\alpha(f; [c, b]) = \gamma_2$. Clearly $0 \leqq U(f, \alpha, \Delta_1) - \gamma_1 < \varepsilon$ and $0 \leqq U(f, \alpha, \Delta_2) - \gamma_2 < \varepsilon$. Adding these two inequalities, we get $0 \leqq U(f, \alpha, \Delta) - (\gamma_1 + \gamma_2) < 2\varepsilon$. Since a similar statement is true for $L(f, \alpha, \Delta)$, we conclude that $S_\alpha(f; [a, b]) = \gamma_1 + \gamma_2$. □

(8.12) Theorem. *Let α be a real-valued nondecreasing function defined on R. For $f \in \mathfrak{C}_{00}^r(R)$, define $S_\alpha(f) = S_\alpha(f; [a, b])$ where $[a, b]$ is any closed interval in R such that f vanishes outside of $[a, b]$. Then $S_\alpha(f)$ is unambiguously defined for each $f \in \mathfrak{C}_{00}^r([a, b])$. Moreover the function S_α has the following properties:*

(i) $S_\alpha(f + g) = S_\alpha(f) + S_\alpha(g)$;

(ii) $S_\alpha(cf) = cS_\alpha(f)$;

(iii) $S_\alpha(f) \geqq 0$ *if* $f \in \mathfrak{C}_{00}^+(R)$.

Proof. Let $f \in \mathfrak{C}_{00}^r(R)$ and suppose that $f = 0$ outside of both $[a, b]$ and $[a', b']$. Let $a'' = \min\{a, a'\}$ and $b'' = \max\{b, b'\}$. By (8.11), we have

$$\begin{aligned} S_\alpha(f; [a'', b'']) &= S_\alpha(f; [a'', a]) + S_\alpha(f; [a, b]) + S_\alpha(f; [b, b'']) \\ &= 0 + S_\alpha(f; [a, b]) + 0 \\ &= S_\alpha(f; [a, b]). \end{aligned}$$

Similarly we see that $S_\alpha(f; [a'', b'']) = S_\alpha(f; [a', b'])$. Therefore $S_\alpha(f; [a, b]) = S_\alpha(f; [a', b'])$ and so $S_\alpha(f)$ is well defined. The remaining assertions of the theorem are now trivial consequences of previous results. □

(8.13) Theorem. *Let α be as in (8.12). For $f \in \mathfrak{C}_{00}(R)$, define $S_\alpha(f)$ as $S_\alpha(\mathrm{Re} f) + iS_\alpha(\mathrm{Im} f)$. Then for $f, g \in \mathfrak{C}_{00}(R)$ and $c \in K$, we have*

(i) $S_\alpha(f + g) = S_\alpha(f) + S_\alpha(g)$,

(ii) $S_\alpha(cf) = cS_\alpha(f)$,

(iii) $S_\alpha(f) \geqq 0$ *if* $f \in \mathfrak{C}_{00}^{+}(R)$.

This simple computation is left to the reader.

(8.14) Definition. For any function f defined on R and any $t \in R$, let f_t be defined on R by $f_t(x) = f(x+t)$. The function f_t is called the *translate of f by t*. Let f^* be defined on R by $f^*(x) = f(-x)$. The function f^* is called the *reflection of f*.

(8.15) Theorem. *Let S be the Riemann integral on* $\mathfrak{C}_{00}(R)$, i.e., $S = S_\alpha$ *where $\alpha(x) = x$ for all $x \in R$. Then S has the following properties:*

(i) $S(f+g) = S(f) + S(g)$;

(ii) $S(cf) = cS(f)$ *for all* $c \in K$;

(iii) $S(f) > 0$ *if* $f \in \mathfrak{C}_{00}^{+}(R)$ *and* $f \neq 0$;

(iv) $S(f_t) = S(f)$ *for all* $t \in R$;

(v) $S(f^*) = S(f)$.

Proof. Conclusions (i) and (ii) are contained in (8.13). To prove (iii), let $f \in \mathfrak{C}_{00}^{+}(R)$ where f is not the zero function. Then there exists $u \in R$ such that $f(u) > 0$. Since f is continuous at u, there is some neighborhood $]u-\delta, u+\delta[$ of u such that $f(x) > \frac{1}{2} f(u)$ whenever $u - \delta < x < u + \delta$. Suppose that f vanishes off of $[a, b]$, where $a < u - \delta < u + \delta < b$, and let $\Delta = \{a, u-\delta, u+\delta, b\}$. Then $S(f) = S(f; [a, b]) \geqq L(f, \alpha, \Delta)$ $\geqq \frac{1}{2} f(u)\,(2\delta) > 0$.

The proofs of (iv) and (v) are left to the reader. □

(8.16) Remark. The function S is [except for a positive multiple] the *only* complex-valued function on $\mathfrak{C}_{00}(R)$ satisfying (8.15.i)–(8.15.iv), i.e., if S' is another complex-valued function on $\mathfrak{C}_{00}(R)$ satisfying (8.15.i) to (8.15.iv), then $S' = \gamma S$ for some positive real number γ. We will prove this fact *infra* [see (19.50)] when more analytical machinery has been developed.

Theorem (8.13) shows that for each nondecreasing α defined on R, the Riemann-Stieltjes integral S_α is an averaging process on $\mathfrak{C}_{00}(R)$. We shall see later (19.50) that every averaging process on $\mathfrak{C}_{00}(R)$ is of the form S_α for some α. Our next theorem gives a sufficient condition that two different α's give rise to the same averaging process on $\mathfrak{C}_{00}(R)$.

(8.17) Theorem. *Let α and β be two real-valued nondecreasing functions defined on R and suppose that there exists a $c \in R$ such that the set $D = \{x \in R : \alpha(x) = \beta(x) + c\}$ is dense in R. Then $S_\alpha(f) = S_\beta(f)$ for every f in* $\mathfrak{C}_{00}(R)$.

Proof. In view of (8.13) it obviously suffices to consider real-valued f's. Thus let $f \in \mathfrak{C}_{00}^{r}(R)$. Choose $a < b$ in D such that f vanishes off of $[a, b]$. Note that $\alpha(x) - \alpha(y) = \beta(x) - \beta(y)$ whenever $x, y \in D$. Let $\varepsilon > 0$ be given. Since f is uniformly continuous on R, there is a $\delta > 0$

such that $|f(x) - f(y)| < \frac{\varepsilon}{\alpha(b) - \alpha(a) + 1}$ whenever $|x - y| < \delta$. Because D is dense in R, there exists a subdivision $\Delta = \{a = t_0 < \cdots < t_n = b\} \subset D$ such that $t_j - t_{j-1} < \delta$ for $j = 1, 2, \ldots, n$. Then, as in the proof of (8.7), it follows that

$$U(f, \alpha, \Delta) - L(f, \alpha, \Delta) < \varepsilon .$$

Since $\Delta \subset D$, we have $U(f, \beta, \Delta) = U(f, \alpha, \Delta)$ and $L(f, \beta, \Delta) = L(f, \alpha, \Delta)$. Therefore $|S_\beta(f) - S_\alpha(f)| < \varepsilon$ [see (8.6)]. Since ε is arbitrary, the theorem is proved. □

We next examine a few interesting properties of nondecreasing functions. First, a definition is in order.

(8.18) Definition. Let f be a complex-valued function defined on an open interval $]a, b[$ of the real line. We say that the *right-hand limit of f at* a is $f(a+)$ [or that the *left-hand limit of f at b* is $f(b-)$], and we write

$$\lim_{x \downarrow a} f(x) = f(a+) \quad \left[\text{or} \lim_{x \uparrow b} f(x) = f(b-)\right],$$

if there exists a complex number $f(a+)$ [or $f(b-)$] such that for each $\varepsilon > 0$ there is a $\delta > 0$ such that

$$|f(x) - f(a+)| < \varepsilon \ [|f(x) - f(b-)| < \varepsilon]$$

whenever $x \in]a, b[$ and $x - a < \delta$ [or $b - x < \delta$].

Next suppose that f is defined on $[a, b]$. We say that f is *right continuous at a [left continuous at b]* if $\lim_{x \downarrow a} f(x) = f(a)$ $\left[\lim_{x \uparrow b} f(x) = f(b)\right]$.

(8.19) Theorem. *Let α be a real-valued nondecreasing function defined on R. Then α has finite right- and left-hand limits at all points of R, and α is continuous except at a countable set of points of R.*

Proof. Let $x \in R$ and set $\alpha(x+) = \inf\{\alpha(t) : x < t\}$. This infimum exists in R since $\alpha(x) \leqq \alpha(t)$ whenever $x < t$. For $\varepsilon > 0$, $\alpha(x+) + \varepsilon$ is not a lower bound for $\{\alpha(t) : x < t\}$, so there exists a $\delta > 0$ such that $\alpha(x + \delta) < \alpha(x+) + \varepsilon$. It follows that $x < t < x + \delta$ implies $\alpha(x+) \leqq \alpha(t) \leqq \alpha(x+) + \varepsilon$, *i.e.*, $\lim_{t \downarrow x} \alpha(t) = \alpha(x+)$. Similarly we prove that

$$\lim_{t \uparrow x} \alpha(t) = \sup\{\alpha(t) : t < x\} = \alpha(x-) .$$

Let $D = \{x \in R : \alpha \text{ is discontinuous at } x\}$. Plainly $x \in D$ if and only if $\alpha(x-) < \alpha(x+)$. Also $x < y$ in D implies that $\alpha(x+) \leqq \alpha(y-)$. Thus the family $\mathscr{I} = \{]\alpha(x-), \alpha(x+)[: x \in D\}$ is a pairwise disjoint family of nonvoid open intervals of R. By (6.59), $\mathscr{I}$ is countable, and so D is countable as well. □

(8.20) Remarks. (a) Let α be a real-valued nondecreasing function on R. Define β by the rule: $\beta(x) = \alpha(x-) - \alpha(0-)$ for $x \in R$. Then β is nondecreasing, $\beta(0) = 0$, and β is left continuous at each point of R. Moreover, since α is continuous except on a countable set, we have

$\beta(x) = \alpha(x) - \alpha(0-)$ on a dense subset of R. It follows from (8.17) that $S_\beta(f) = S_\alpha(f)$ for all $f \in \mathfrak{C}_{00}(R)$. Because of these facts, we lose nothing, in constructing Riemann-Stieltjes integrals on $\mathfrak{C}_{00}(R)$, by *normalizing* our α's to be left continuous and to vanish at 0. It must be pointed out however that this normalization may affect the value of the integral of a discontinuous function. For example, let $\alpha(x) = 0$ for $x < 0$ and $\alpha(x) = 1$ for $x \geqq 0$; $\beta(x) = 0$ for $x \leqq 0$ and $\beta(x) = 1$ for $x > 0$; and take $f = \beta$. Integrating over $[-1, 1]$, we have $L(\beta, \alpha, \Delta) = L(\beta, \beta, \Delta) = 0$ and $U(\beta, \beta, \Delta) = 1$ for all Δ in $\mathscr{D}([-1, 1])$ while $U(\beta, \alpha, \Delta) = 0$ if $0 \in \Delta \in \mathscr{D}([-1, 1])$. Therefore $S_\alpha(\beta, [-1, 1])$ is zero and $S_\beta(\beta, [-1, 1])$ does not exist.

(b) The foregoing example of a non Stieltjes integrable function is undramatic, to say the least. A bit more interesting is the function on $[0, 1]$ such that $f(x) = 0$ if x is rational and $f(x) = 1$ if x is irrational. The Riemann-Stieltjes integral $S_\alpha(f; [0, 1])$ exists only if $\alpha(1) = \alpha(0)$, since all lower Darboux sums are 0 and all upper Darboux sums are $\alpha(1) - \alpha(0)$. A complete description of Riemann integrable functions appears in (12.51) *infra*.

(8.21) Exercise. Prove that if f is a bounded real-valued function on $[a, b]$ having only a finite number of discontinuities and if α is a real-valued nondecreasing function on $[a, b]$ having no discontinuities in common with f, then f is Riemann-Stieltjes integrable with respect to α over $[a, b]$.

(8.22) Exercise. Let α be a real-valued nondecreasing function on $[a, b]$ and let (f_n) be a sequence of bounded real-valued functions on $[a, b]$, each of which is Riemann-Stieltjes integrable with respect to α over $[a, b]$. Suppose that $\lim_{n\to\infty} \|f - f_n\|_u = 0$, where f is in $\mathfrak{B}^r([a, b])$. Prove that f is Riemann-Stieltjes integrable with respect to α over $[a, b]$ and that

$$\lim_{n\to\infty} S_\alpha(f_n; [a, b]) = S_\alpha(f; [a, b]) .$$

(8.23) Exercise. By way of contrast with (8.22), find a sequence (f_n) where each $f_n \in \mathfrak{C}_{00}(R)$, such that $f_n \to 0$ uniformly on R, but $\lim_{n\to\infty} S(f_n) \neq 0 = S(0)$.

(8.24) Exercise. Let g be defined on R by $g(x) = x - [x]$, where $[x]$ is the largest integer not exceeding x. Using (8.21) and (8.22), prove that the function f defined by $f(x) = \sum_{n=1}^{\infty} \frac{g(nx)}{2^n}$ is Riemann integrable over $[0, 1]$. Also evaluate $S(f; [0, 1])$.

(8.25) Exercise. Let α and β be continuous, real-valued, nondecreasing functions on R such that $\alpha(0) = \beta(0) = 0$ and $\alpha \neq \beta$. Find a function $f \in \mathfrak{C}_{00}^+(R)$ such that $S_\alpha(f) \neq S_\beta(f)$.

(8.26) Exercise. Let $(x_n)_{n=1}^{\infty}$ be an enumeration of Q. Define α on R by the rule: $\alpha(x) = \sum \frac{1}{2^n}$, where the summation is extended over all n such that $x_n < x$. Prove that:

(a) α is strictly increasing;
(b) α is left continuous;
(c) α is discontinuous at each rational number;
(d) α is continuous at each irrational number;
(e) $\lim\limits_{x \to \infty} \alpha(x) = 1$;
(f) $\lim\limits_{x \to -\infty} \alpha(x) = 0$.

[This example gives some indication of how bad a monotone function can be.]

(8.27) Exercise. Let (x_n) and α be as in (8.26). Prove that $S_\alpha(f) = \sum_{n=1}^{\infty} \frac{f(x_n)}{2^n}$ for every $f \in \mathfrak{C}_{00}(R)$. [First consider $f \in \mathfrak{C}_{00}^r(R)$, show that every L is $\leqq$ this sum and that every U is $\geqq$ it.]

(8.28) Exercise. Let P denote the Cantor ternary set. Notation is as in (6.62) and (6.64). Define a function ψ on $[0, 1]$ as follows:

$\psi(0) = 0$;

$\psi(x) = \frac{2k-1}{2^n}$ for $x \in I_{n,k}$ $(n = 1, 2, \ldots;\ k = 1, 2, \ldots, 2^{n-1})$;

$\psi(x) = \sup\{\psi(t) : t \in P',\ t < x\}$ for $x \in P \cap \{0\}'$.

The function ψ is called *Lebesgue's singular function*[1]. Prove that:

(a) $\psi(x)$ is defined for all $x \in [0, 1]$;
(b) ψ is nondecreasing;
(c) ψ is continuous on $[0, 1]$;
(d) $\operatorname{rng}\psi = [0, 1]$;
(e) if $x \in P$ and $x = \sum_{n=1}^{\infty} \frac{2x_n}{3^n}$, where each $x_n = 0$ or 1, then $\psi(x) = \sum_{n=1}^{\infty} \frac{x_n}{2^n}$.

The accompanying Fig. 5, showing part of the graph of ψ may be helpful. [Use the fact that the dyadic rationals in [0, 1] [on the y-axis] form a dense subset of [0, 1].]

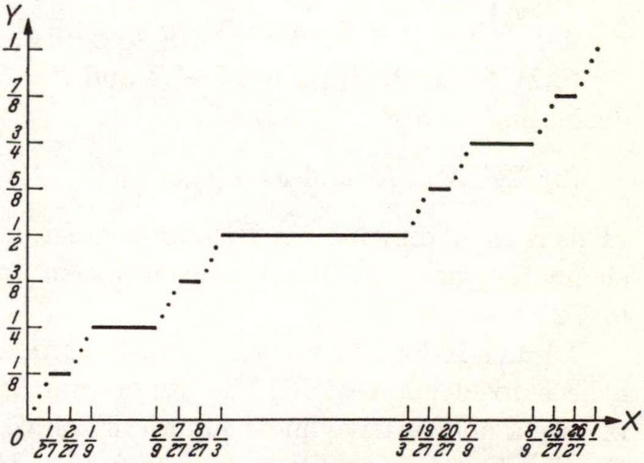

Fig. 5

[1] The French mathematician Henri Lebesgue (1875–1941) was a principal founder of the theories of measure, integration, and differentiation treated in this book. His name occurs frequently in the sequel.

(8.29) Exercise. Let ψ be as in (8.28) and let $\iota(x) = x$ for all $x \in [0, 1]$. Prove that $S_\psi(\iota; [0, 1]) = \frac{1}{2}$. [Consider the sequence of subdivisions (Δ_n) where Δ_n consists of the endpoints of the closed intervals $J_{n,k}$ making up P_n.] For the computation of $S_\psi(\iota^k; [0, 1])$ for $k = 2, 3, \ldots$, see (22.44) and (22.45) *infra*.

§ 9. Extending certain functionals

In § 8 we have presented Riemann-Stieltjes integrals as averaging processes on $\mathfrak{C}_{00}(R)$. We wish to extend these processes to much wider classes of functions. To be sure, there exist many discontinuous functions on R which are Riemann-Stieltjes integrable, but these integrals suffer from serious defects. For example, very simple functions are nonintegrable [see (8.20)]. Our method of extension is a very general one. It consists of starting with an averaging process on a space of continuous functions, extending it to nonnegative lower semicontinuous functions, extending a second time to all nonnegative, extended real-valued functions, and finally realizing our process as an integral [the *Lebesgue integral*] on a large class of functions. This procedure is often called the *Daniell approach* to integration theory, after the British mathematician P. J. Daniell (1889–1946).

(9.1) Definition. Let X be a nonvoid locally compact Hausdorff space. A complex-valued function I defined on $\mathfrak{C}_{00}(X)$ is called a *nonnegative linear functional* [sometimes *Radon measure*] if for all $f, g \in \mathfrak{C}_{00}(X)$ and $\alpha \in K$ we have

(i) $I(f + g) = I(f) + I(g)$ [*additive*],

(ii) $I(\alpha f) = \alpha I(f)$ [*homogeneous*],

(iii) $I(f) \geqq 0$ if $f \in \mathfrak{C}_{00}^+(X)$ [*nonnegative*].

(9.2) Examples. (a) Let $X = R$ and $I = S_\alpha$ for any real-valued nondecreasing α on R.

(b) Let $X = R^2$ and let $I(f) = \int\limits_{-\infty}^{\infty} \int\limits_{-\infty}^{\infty} f(x, y)\, dx\, dy$ for all $f \in \mathfrak{C}_{00}(R^2)$. [This is the 2-dimensional Riemann integral, familiar to the reader from elementary analysis. We will discuss such "multiple" integrals in detail in § 21.]

(c) Let X be any nonvoid locally compact Hausdorff space and let a be a fixed point of X. The functional E_a defined by $E_a(f) = f(a)$ is plainly a nonnegative linear functional on $\mathfrak{C}_{00}(X)$. It is called an *evaluation functional*. In addition, E_a satisfies the identity

$$E_a(fg) = E_a(f)\, E_a(g)$$

for all $f, g \in \mathfrak{C}_{00}(X)$. Such linear functionals are called *multiplicative*.

They are of great importance in the theory of Banach algebras, and we shall return to them later [see (20.52) and (21.65) *infra*].

(d) Let $X = R$ and let w be any nonnegative real-valued continuous function on R [perhaps unbounded]. Then $I(f) = S(fw)$ [S being the Riemann integral] defines a nonnegative linear functional I on $\mathfrak{C}_{00}(R)$.

Throughout the rest of § 9, except where the contrary is specified, X will denote a fixed but arbitrary nonvoid locally compact Hausdorff space, and I will denote a fixed but arbitrary nonnegative linear functional on $\mathfrak{C}_{00}(X)$. We will write $\mathfrak{C}_{00}$ instead of $\mathfrak{C}_{00}(X)$, for brevity's sake.

(9.3) Theorem.

(i) *If* $f \in \mathfrak{C}_{00}^r$, *then* $I(f)$ *is a real number.*

(ii) *If* $f \leqq g$ *in* $\mathfrak{C}_{00}^r$, *then* $I(f) \leqq I(g)$.

Proof. For $f \in \mathfrak{C}_{00}^r$, write $f^+ = \max\{f, 0\}$ and $f^- = -\min\{f, 0\}$. It is clear that $f = f^+ - f^-$, $f^+ \geqq 0$, and $f^- \geqq 0$ Thus $I(f) = I(f^+) - I(f^-)$, so $I(f)$ is the difference of two nonnegative real numbers. This proves (i).

If $f, g \in \mathfrak{C}_{00}^r$ and $f \leqq g$, then $g - f \geqq 0$ and so $I(g) - I(f) = I(g-f) \geqq 0$. This proves (ii). □

(9.4) Theorem. *Let* $f \in \mathfrak{C}_{00}$. *Then* $|I(f)| \leqq I(|f|)$.

Proof. Write $I(f) = \varrho \exp(i\theta)$ where $0 \leqq \varrho < \infty$ and $-\pi < \theta \leqq \pi$. Let $\exp(-i\theta)f = g_1 + ig_2$ where $g_1, g_2 \in \mathfrak{C}_{00}^r$. Then $\varrho = \exp(-i\theta)I(f) = I(\exp(-i\theta)f) = I(g_1 + ig_2) = I(g_1) + iI(g_2)$. But ϱ is real and $I(g_1)$ and $I(g_2)$ are real. Therefore $I(g_2) = 0$ and $\varrho = I(g_1)$. Clearly $g_1 \leqq |\exp(-i\theta)f| = |f|$. Thus $|I(f)| = \varrho = I(g_1) \leqq I(|f|)$. □

(9.5) Theorem. *Let A be any compact subset of X. Then there exists a real number β [depending only on A] such that* $|I(f)| \leqq \beta \|f\|_u$ *for all* $f \in \mathfrak{C}_{00}$ *such that* $f(A') \subset \{0\}$.

Proof. According to (6.79) there is an open set $U \supset A$ such that U^- is compact. By (6.80) there is a continuous function ω from X into $[0, 1]$ such that $\omega(x) = 1$ for all $x \in A$ and $\omega(x) = 0$ for all $x \in X \cap U'$. Let $f \in \mathfrak{C}_{00}$ be such that $f = 0$ on A'. Then $f(x) = f(x)\omega(x)$ for all $x \in X$. It follows that $|f| \leqq \|f\|_u \omega$ and hence $|I(f)| \leqq I(|f|) \leqq I(\|f\|_u \omega) = \|f\|_u I(\omega)$. Set $\beta = I(\omega)$. □

Our linear functional I is given to us *a priori* as only *finitely* additive: $I(f + g) = I(f) + I(g)$. For many purposes, it is useful to have *countable* additivity: $I\left(\sum_{n=1}^{\infty} f_n\right) = \sum_{n=1}^{\infty} I(f_n)$. This equality is seldom true for all convergent sequences of functions on X, and to prove it for some sequences of functions, we must extend the domain of definition of I. In (9.6)–(9.18), we carry out this extension and establish properties of the extended functional that lead up to countable additivity.

(9.6) Theorem. *Let $\mathfrak{D}$ be a nonvoid subset of* $\mathfrak{C}_{00}^+$ *such that for all* $f_1, f_2 \in \mathfrak{D}$ *there exists* $f_3 \in \mathfrak{D}$ *such that* $f_3 \leqq \min\{f_1, f_2\}$ *[we say that $\mathfrak{D}$ is* directed downward]. *Suppose also that* $\inf\{f(x) : f \in \mathfrak{D}\} = 0$ *for every* $x \in X$. *Then we have*

(i) $\inf\{I(f) : f \in \mathfrak{D}\} = 0$,

and for every $\varepsilon > 0$ *there is an* $f_\varepsilon \in \mathfrak{D}$ *such that* $\|f_\varepsilon\|_u < \varepsilon$.

Proof. Let $\varepsilon > 0$ be fixed. Choose $f_0 \in \mathfrak{D}$ and define $A_0 = \{x \in X : f_0(x) > 0\}^-$. Since $f_0 \in \mathfrak{C}_{00}$, A_0 is compact. For each $f \in \mathfrak{D}$, let $A_f = \{x \in A_0 : f(x) \geqq \varepsilon\}$. Then $\{A_f : f \in \mathfrak{D}\}$ is a family of closed subsets of the compact space A_0 and $\bigcap \{A_f : f \in \mathfrak{D}\} = \varnothing$. It follows (6.34) that this family does not have the finite intersection property, *i.e.*, there is a finite subset $\{f_1, \ldots, f_n\}$ of $\mathfrak{D}$ such that $A_{f_1} \cap A_{f_2} \cap \cdots \cap A_{f_n} = \varnothing$. Since $\mathfrak{D}$ is directed downward, there is a function $f_\varepsilon \in \mathfrak{D}$ such that $f_\varepsilon \leqq \min\{f_0, f_1, \ldots, f_n\}$; plainly $\|f_\varepsilon\|_u < \varepsilon$.

To prove (i), apply (9.5) to find a real number β [depending only on A_0 and not on ε] such that $|I(f)| \leqq \beta \|f\|_u$ for all $f \in \mathfrak{C}_{00}$ that vanish on A_0'. It is clear that $f_\varepsilon = 0$ on A_0'. Thus $0 \leqq I(f_\varepsilon) \leqq \beta \|f_\varepsilon\|_u < \beta\varepsilon$. Since ε is arbitrary, (i) is established. □

(9.7) Remarks. The reader will note that compactness is used in an essential way in the proof of Theorem (9.6). This is no accident. Our function space $\mathfrak{C}_{00}$ was chosen so that compactness could be applied. Many other plausible choices of function spaces and positive linear functionals fail to produce countably additive extensions. Theorem (9.6) is the key to obtaining countable additivity of our extended functional, and so countable additivity depends in the end upon compactness.

Our first extension of I is to the class $\mathfrak{M}^+$ of all lower semicontinuous functions on X with range contained in $[0, \infty]$. [See (7.20)–(7.22).]

(9.8) Definition. For each $g \in \mathfrak{M}^+$, define

$$\bar{I}(g) = \sup\{I(f) : f \in \mathfrak{C}_{00}^+, f \leqq g\}.$$

Notice that $\bar{I}(g)$ may be ∞. For example, if $X = R$ and $I = S$, then the function 1 is in $\mathfrak{M}^+$ and $\bar{S}(1) = \infty$. For arbitrary X and $I \neq 0$, the function ∞ is in $\mathfrak{M}^+$, and $\bar{I}(\infty) = \infty$.

(9.9) Exercise. (a) Make a careful computation of $\bar{S}(\xi_U)$, where U is any open subset of R and S is the Riemann integral. [Hint. Use (6.59).] The number $\bar{S}(\xi_U)$ is called the *Lebesgue measure of* U. It will be discussed in detail below.

(b) For the evaluation functional E_a (9.2.c) and for $g \in \mathfrak{M}^+$, compute $\bar{E}_a(g)$.

(9.10) Theorem.

(i) *If* $f \in \mathfrak{C}_{00}^+$, *then* $I(f) = \bar{I}(f)$.

(ii) *If* $g_1, g_2 \in \mathfrak{M}^+$ *and* $g_1 \leqq g_2$, *then* $\bar{I}(g_1) \leqq \bar{I}(g_2)$.

(iii) *If* $g \in \mathfrak{M}^+$ *and* $\alpha \geqq 0$, *then* $\bar{I}(\alpha g) = \alpha \bar{I}(g)$.

These assertions are all obvious upon a moment's reflection.

The next theorem is a counterpart of (9.6).

(9.11) Theorem. *Let $\mathfrak{D}$ be a nonvoid subset of $\mathfrak{M}^+$ such that for all $g_1, g_2 \in \mathfrak{D}$ there is a $g_3 \in \mathfrak{D}$ such that $g_3 \geqq \max\{g_1, g_2\}$ [we say that $\mathfrak{D}$ is directed upward]. Then*

(i) $\bar{I}(\sup \mathfrak{D}) = \sup\{\bar{I}(g) : g \in \mathfrak{D}\}$.

Proof. Let $g_0 = \sup\{g : g \in \mathfrak{D}\}$, *i.e.*, $g_0(x) = \sup\{g(x) : g \in \mathfrak{D}\}$ for all $x \in X$. Then (7.22.iii) shows that $g_0 \in \mathfrak{M}^+$. There are two cases to consider.

Case I: $\{g_0\} \cup \mathfrak{D} \subset \mathfrak{C}_{00}^+$. Then the family $\{g_0 - g : g \in \mathfrak{D}\}$ satisfies the hypotheses of (9.6), for if $g_3 \geqq \max\{g_1, g_2\}$, then

$$g_0 - g_3 \leqq \min\{g_0 - g_1, g_0 - g_2\},$$

and

$$\inf\{g_0(x) - g(x) : g \in \mathfrak{D}\} = g_0(x) - \sup\{g(x) : g \in \mathfrak{D}\} = g_0(x) - g_0(x) = 0$$

for every $x \in X$. Then (9.6) implies that

$$0 = \inf\{I(g_0 - g) : g \in \mathfrak{D}\} = \inf\{I(g_0) - I(g) : g \in \mathfrak{D}\} = I(g_0) - \sup\{I(g) : g \in \mathfrak{D}\}.$$

But I and $\bar{I}$ agree on $\mathfrak{C}_{00}^+$, and so $\bar{I}(g_0) = \sup\{\bar{I}(g) : g \in \mathfrak{D}\}$.

Case II: $\{g_0\} \cup \mathfrak{D} \subset \mathfrak{M}^+$, *i.e.*, the general case. Since $g \leqq g_0$ for all $g \in \mathfrak{D}$, we have $\sup\{\bar{I}(g) : g \in \mathfrak{D}\} \leqq \bar{I}(g_0)$. To prove the reversed inequality, we introduce the family $\mathfrak{E} = \{f \in \mathfrak{C}_{00}^+ : f \leqq g \text{ for some } g \in \mathfrak{D}\}$. Making use of (7.22.v), we have

$$g_0 = \sup\{g : g \in \mathfrak{D}\} = \sup\{\sup\{f \in \mathfrak{C}_{00}^+ : f \leqq g\} : g \in \mathfrak{D}\} = \sup \mathfrak{E}.$$

Let β be any real number less than $\bar{I}(g_0)$. By the definition of $\bar{I}$, there exists a $\varphi \in \mathfrak{C}_{00}^+$ such that $\varphi \leqq g_0$ and $I(\varphi) > \beta$. We have $\varphi = \min\{\varphi, g_0\} = \min\{\varphi, \sup \mathfrak{E}\} = \sup\{\min\{\varphi, f\} : f \in \mathfrak{E}\}$ [the reader should check this sleight of hand carefully]. The result for Case I now applies with the family $\{\min\{\varphi, f\} : f \in \mathfrak{E}\}$ taking the rôle of $\mathfrak{D}$ and φ the rôle of g_0. Thus

$$\beta < I(\varphi) = \sup\{I(\min\{\varphi, f\}) : f \in \mathfrak{E}\} \leqq \sup\{I(f) : f \in \mathfrak{E}\}$$
$$= \sup\{\bar{I}(g) : g \in \mathfrak{D}\}.$$

Since β is arbitrary, we have proved that

$$\bar{I}(g_0) \leqq \sup\{\bar{I}(g) : g \in \mathfrak{D}\}. \quad \square$$

(9.12) Corollary. *If $g_1 \leqq g_2 \leqq \cdots \leqq g_n \leqq \cdots$, where $g_n \in \mathfrak{M}^+$ for $n = 1, 2, \ldots$, then*

$$\lim_{n \to \infty} \bar{I}(g_n) = \bar{I}\left(\lim_{n \to \infty} g_n\right).$$

The proof is immediate.

(9.13) Corollary. *For $g_1, g_2 \in \mathfrak{M}^+$, we have $\bar{I}(g_1+g_2)=\bar{I}(g_1)+\bar{I}(g_2)$.*

Proof. We have $g_1+g_2=\sup\{f_1+f_2 : f_j \in \mathfrak{C}_{00}^+ \text{ and } f_j \leqq g_j, j=1,2\}$, and therefore

$$\begin{aligned}\bar{I}(g_1+g_2) &= \sup\{I(f_1)+I(f_2) : f_j \in \mathfrak{C}_{00}^+, f_j \leqq g_j\}\\ &= \sup\{I(f_1) : f_1 \in \mathfrak{C}_{00}^+, f_1 \leqq g_1\} + \sup\{I(f_2) : f_2 \in \mathfrak{C}_{00}^+, f_2 \leqq g_2\}\\ &= \bar{I}(g_1)+\bar{I}(g_2)\,. \quad \square\end{aligned}$$

(9.14) Corollary. *For any nonvoid $\mathfrak{D} \subset \mathfrak{M}^+$, we have*

$$\bar{I}\Big(\sum_{g\in\mathfrak{D}} g\Big) = \sum_{g\in\mathfrak{D}} \bar{I}(g)\,.$$[1]

Proof. For finite $\mathfrak{D}$, (9.13) and an easy induction yield the result. For infinite $\mathfrak{D}$, apply (9.13) and (9.11). □

We now make our second, and last, extension of I.

(9.15) Definition. Let $\mathfrak{F}^+$ be the set of all functions defined on X with values in $[0,\infty]$. For $h \in \mathfrak{F}^+$, let

$$\bar{\bar{I}}(h) = \inf\{\bar{I}(g) : g \in \mathfrak{M}^+, g \geqq h\}\,.$$

(9.16) Theorem.

(i) *For $g \in \mathfrak{M}^+$, we have $\bar{I}(g) = \bar{\bar{I}}(g)$.*

(ii) *If $h_1 \leqq h_2$, $h_j \in \mathfrak{F}^+$, then we have $\bar{\bar{I}}(h_1) \leqq \bar{\bar{I}}(h_2)$.*

(iii) *For $h \in \mathfrak{F}^+$ and $0 \leqq \alpha < \infty$, we have $\bar{\bar{I}}(\alpha h) = \alpha\bar{\bar{I}}(h)$.*

(iv) *For $h_1, h_2 \in \mathfrak{F}^+$, we have $\bar{\bar{I}}(h_1+h_2) \leqq \bar{\bar{I}}(h_1)+\bar{\bar{I}}(h_2)$.*

The proof is left to the reader.

Our next theorem is a generalization of B. Levi's monotone convergence theorem.

(9.17) Theorem [Generalized B. Levi theorem]. *If $h_n \in \mathfrak{F}^+$ for $n = 1, 2, 3, \ldots,$ and $h_1 \leqq h_2 \leqq \cdots \leqq h_n \leqq \cdots$, then $\bar{\bar{I}}(\lim_{n\to\infty} h_n) = \lim_{n\to\infty} \bar{\bar{I}}(h_n)$.*

Proof. Let $h = \lim_{n\to\infty} h_n$; we obviously have

$$\bar{\bar{I}}(h_1) \leqq \bar{\bar{I}}(h_2) \leqq \cdots \leqq \lim_{n\to\infty} \bar{\bar{I}}(h_n) \leqq \bar{\bar{I}}(h)\,.$$

We must now prove that $\lim_{n\to\infty} \bar{\bar{I}}(h_n) \geqq \bar{\bar{I}}(h)$; clearly we may suppose that $\lim_{n\to\infty} \bar{\bar{I}}(h_n) < \infty$. Choose $\varepsilon > 0$. For each positive integer n, select $g_n \in \mathfrak{M}^+$ such that $g_n \geqq h_n$ and $\bar{I}(g_n) - \frac{\varepsilon}{2^n} < \bar{\bar{I}}(h_n)$. We wish to apply (9.12). To do this, we must replace the functions g_n by a *nondecreasing* sequence. Thus define $g_n' = \max\{g_1, g_2, \ldots, g_n\}$ for $n = 1, 2, \ldots$; we have $g_n' \in \mathfrak{M}^+$

[1] The function $\sum_{g\in\mathfrak{D}} g$ is defined by $\big(\sum_{g\in\mathfrak{D}} g\big)(x) = \sup\{g_1(x)+g_2(x)+\cdots+g_n(x) : \{g_1, g_2, \ldots, g_n\} \subset \mathfrak{D}\}$; similarly, we have $\sum_{g\in\mathfrak{D}} \bar{I}(g) = \sup\{\bar{I}(g_1)+\bar{I}(g_2)+\cdots+\bar{I}(g_n) : \{g_1, g_2, \ldots, g_n\} \subset \mathfrak{D}\}$.

for $n = 1, 2, \ldots$. The reader can easily check the identity

$$g'_{n+1} + \min\{g'_n, g_{n+1}\} = g'_n + g_{n+1}.$$

All of the functions above are in $\mathfrak{M}^+$, and so we have

$$\bar{I}(g'_{n+1}) + \bar{I}(\min\{g'_n, g_{n+1}\}) = \bar{I}(g'_n) + \bar{I}(g_{n+1}).$$

From the inequalities $g_{n+1} \geqq h_{n+1} \geqq h_n$ and $g'_n = \max\{g_1, g_2, \ldots, g_n\} \geqq \max\{h_1, h_2, \ldots, h_n\} = h_n$, it follows that $\min\{g'_n, g_{n+1}\} \geqq h_n$. Hence $-\bar{I}(\min\{g'_n, g_{n+1}\}) \leqq -\bar{\bar{I}}(h_n)$, and we have

$$\begin{aligned} \bar{I}(g'_{n+1}) &= \bar{I}(g'_n) + \bar{I}(g_{n+1}) - \bar{I}(\min\{g'_n, g_{n+1}\}) \\ &\leqq \bar{I}(g'_n) + \bar{I}(g_{n+1}) - \bar{\bar{I}}(h_n) \\ &< \bar{I}(g'_n) + \bar{\bar{I}}(h_{n+1}) + \frac{\varepsilon}{2^{n+1}} - \bar{\bar{I}}(h_n). \end{aligned}$$

Summing this inequality over $n = 1, 2, \ldots, p$, we have

$$\sum_{n=1}^{p} \bar{I}(g'_{n+1}) < \sum_{n=1}^{p} \bar{I}(g'_n) + \sum_{n=1}^{p} \bar{\bar{I}}(h_{n+1}) - \sum_{n=1}^{p} \bar{\bar{I}}(h_n) + \frac{\varepsilon}{2} \sum_{n=1}^{p} 2^{-n},$$

and so

$$\begin{aligned} \bar{I}(g'_{p+1}) &< \bar{I}(g'_1) + \bar{\bar{I}}(h_{p+1}) - \bar{\bar{I}}(h_1) + \frac{\varepsilon}{2} \\ &= \bar{\bar{I}}(h_{p+1}) + \left(\bar{I}(g_1) - \bar{\bar{I}}(h_1)\right) + \frac{\varepsilon}{2} \\ &< \bar{\bar{I}}(h_{p+1}) + \varepsilon. \end{aligned}$$

Thus the inequality $\bar{I}(g'_{p+1}) < \bar{\bar{I}}(h_{p+1}) + \varepsilon$ is valid for $p = 1, 2, 3, \ldots$. The same inequality obtains with $p = 0$, and so we have

$$\bar{I}(g'_n) < \bar{\bar{I}}(h_n) + \varepsilon \tag{1}$$

for $n = 1, 2, 3, \ldots$. The sequence (g'_n) is nondecreasing and $g'_n \in \mathfrak{M}^+$ for $n = 1, 2, 3, \ldots$; (9.12) thus implies that

$$\lim_{n\to\infty} \bar{I}(g'_n) = \bar{I}\left(\lim_{n\to\infty} g'_n\right) = \bar{I}(\sup g'_n).$$

Since $g'_n \geqq g_n \geqq h_n$ for $n = 1, 2, 3, \ldots$, we have $\sup g'_n \geqq \sup h_n = h$, and so $\bar{I}(\sup g'_n) \geqq \bar{\bar{I}}(h)$. Using (1), we now find that $\lim\limits_{n\to\infty} \bar{\bar{I}}(h_n) \geqq \lim\limits_{n\to\infty} \bar{I}(g'_n) - \varepsilon \geqq \bar{\bar{I}}(h) - \varepsilon$. The inequality $\lim\limits_{n\to\infty} \bar{\bar{I}}(h_n) \geqq \bar{\bar{I}}(h)$ follows, since ε is arbitrary. □[1]

(9.18) Corollary. *Let $(h_n)_{n=1}^{\infty}$ be any sequence of functions in $\mathfrak{F}^+$; then* $\bar{\bar{I}}\left(\sum_{n=1}^{\infty} h_n\right) \leqq \sum_{n=1}^{\infty} \bar{\bar{I}}(h_n)$.

---

[1] Theorem (9.17) is of course (9.12) with $\bar{I}$ replaced by $\bar{\bar{I}}$. It is important to note that (9.13) and (9.11) fail in general if $\bar{I}$ is replaced by $\bar{\bar{I}}$: see (10.41). The truth of (9.17) is by itself fairly remarkable.

Proof. Write $\psi_n = h_1 + h_2 + \cdots + h_n$; then

$$\bar{I}\left(\sum_{n=1}^{\infty} h_n\right) = \bar{I}\left(\lim_{n\to\infty} \psi_n\right) = \lim_{n\to\infty} \bar{I}(\psi_n) = \lim_{n\to\infty} \bar{I}\left(\sum_{k=1}^{n} h_k\right)$$

$$\leq \lim_{n\to\infty} \sum_{k=1}^{n} \bar{I}(h_k) = \sum_{n=1}^{\infty} \bar{I}(h_n). \quad \square$$

We now define the measure [with respect to $\bar{I}$] of a subset of X.

(9.19) Definition. For $A \subset X$ let $\iota(A) = \bar{I}(\xi_A)$. We call $\iota(A)$ the [*outer*] *measure of* A. The function ι, defined on $\mathscr{P}(X)$, is known as the [*outer*] *measure induced by* I. In case $X = R$ and $I = S_\alpha$ for some real-valued nondecreasing function α on R, we write $\lambda_\alpha(A) = \bar{S}_\alpha(\xi_A)$ and call λ_α the *Lebesgue-Stieltjes* [*outer*] *measure on* R *induced by* α. If $X = R$ and $I = S$ [the Riemann integral], then we write $\lambda(A) = \bar{S}(\xi_A)$ and call λ *Lebesgue* [*outer*] *measure on* R. For arbitrary X and $I = E_a$ (9.2.c), we write $\varepsilon_a(A) = \bar{E}_a(\xi_A)$ and call ε_a the *unit point mass* [or *Dirac measure*] *concentrated at* a.

(9.20) Exercise. For arbitrary X, $a \in X$, and $A \subset X$, prove that $\varepsilon_a(A) = 1$ if $a \in A$ and $\varepsilon_a(A) = 0$ if $a \notin A$. That is, $\varepsilon_a(A) = \xi_A(a)$.

We propose now to investigate properties of the measure ι and to construct an integral [which is the classical Lebesgue integral for $\iota = \lambda$] from it. This integral turns out to be the functional $\bar{I}$ whenever the integral exists. The program is somewhat long; it will be completed in Theorem (12.35) *infra*. We begin by pointing out some properties of the set function ι.

(9.21) Theorem. *The set function ι has the following properties:*

(i) $0 \leq \iota(A) \leq \infty$ *for all* $A \subset X$;

(ii) $\iota(A) \leq \iota(B)$ *if* $A \subset B \subset X$;

(iii) $\iota(\varnothing) = 0$;

(iv) *if* $(A_n)_{n=1}^{\infty}$ *is any sequence of subsets of* X, *then* $\iota\left(\bigcup_{n=1}^{\infty} A_n\right) \leq \sum_{n=1}^{\infty} \iota(A_n)$

[countably subadditive].

Proof. Assertions (i)–(iii) are trivial consequences of the definition of ι. To prove (iv), we write

$$\iota\left(\bigcup_{n=1}^{\infty} A_n\right) = \bar{I}\left(\xi_{\bigcup_{n=1}^{\infty} A_n}\right) \leq \bar{I}\left(\sum_{n=1}^{\infty} \xi_{A_n}\right) \leq \sum_{n=1}^{\infty} \bar{I}(\xi_{A_n}) = \sum_{n=1}^{\infty} \iota(A_n);$$

here we have used the inequality $\xi_{\bigcup_{n=1}^{\infty} A_n} \leq \sum_{n=1}^{\infty} \xi_{A_n}$ and (9.18). $\square$

(9.22) Theorem. *Let $\{U_\theta : \theta \in \Theta\}$ be any pairwise disjoint family of open subsets of X. Then* $\iota\left(\bigcup_{\theta\in\Theta} U_\theta\right) = \sum_{\theta\in\Theta} \iota(U_\theta)$.

Proof. If $U = \bigcup_{\theta \in \Theta} U_\theta$, it is clear that $\xi_{U_\theta} \in \mathfrak{M}^+$ for each $\theta \in \Theta$ and that $\xi_U = \sum_{\theta \in \Theta} \xi_{U_\theta}$. Applying (9.14), we have $\iota(U) = \bar{I}(\xi_U) = \sum_{\theta \in \Theta} \bar{I}(\xi_{U_\theta}) = \sum_{\theta \in \Theta} \iota(U_\theta)$. □

(9.23) Corollary. *If* $(]a_n, b_n[)_{n=1}^\infty$ *is any pairwise disjoint sequence of open intervals of* R, *then* $\lambda\left(\bigcup_{n=1}^\infty]a_n, b_n[\right) = \sum_{n=1}^\infty (b_n - a_n)$.

Proof. This is an immediate consequence of (9.22) and the fact [which is a trivial exercise] that $\lambda(]a, b[) = b - a$. [Note that a or b may be infinite.] See also Exercise (9.9.a). □

(9.24) Theorem. *For every* $A \subset X$ *we have*

$$\iota(A) = \inf\{\iota(U) : U \text{ is open, } A \subset U\}.$$

Proof. If $\iota(A) = \infty$, then the result is trivial since $A \subset X$, X is open, and $\infty = \iota(A) \leqq \iota(X) = \infty$. Thus suppose that $\iota(A) < \infty$ and let $\varepsilon > 0$ be arbitrary. Choose a real number δ such that $0 < \delta < \dfrac{\varepsilon}{1 + \varepsilon + \iota(A)} < 1$. Next select $g \in \mathfrak{M}^+$ such that $g \geqq \xi_A$ and $\bar{I}(g) - \delta < \bar{I}(\xi_A) = \iota(A)$. Let $U = \{x \in X : g(x) > 1 - \delta\}$. Clearly U is open and $A \subset U$. For $x \in U$, we have $\dfrac{1}{1-\delta} g(x) > 1$, and so $\dfrac{1}{1-\delta} g \geqq \xi_U$. Thus

$$\iota(U) = \bar{I}(\xi_U) \leqq \bar{I}\left(\frac{1}{1-\delta} g\right) = \frac{1}{1-\delta} \bar{I}(g) < \frac{1}{1-\delta}(\iota(A) + \delta) < \iota(A) + \varepsilon. \ \square$$

(9.25) Remark. It follows from (9.24), (6.59), and (9.23) that for $A \subset R$ we have $\lambda(A) = \inf\left\{\sum_{n=1}^\infty (b_n - a_n) : \bigcup_{n=1}^\infty]a_n, b_n[\supset A \text{ and } \{]a_n, b_n[\}_{n=1}^\infty \text{ are pairwise disjoint}\right\}$. This is how λ was originally defined by LEBESGUE himself, in 1902. LEBESGUE's fundamental idea was to consider *countable* coverings of A by open intervals. Earlier attempts at defining a suitable notion of measure for subsets of R were similar to LEBESGUE's, but in each case only finite coverings of the set in question were considered. For example, the *content of* A was defined by C. JORDAN to be the number $\inf\left\{\sum_{n=1}^p (b_n - a_n) : A \subset \bigcup_{n=1}^p]a_n, b_n[, \ p = 1, 2, \ldots\right\}$. The Jordan content [*Inhalt* in German] is still studied by some mathematicians[1], but it has proved to be unsatisfactory for the purposes of modern analysis.

(9.26) Theorem. *Let* U *be any open set in* X; *then we have* $\iota(U) = \sup\{\iota(F) : F$ *is compact and* $F \subset U\} = \sup\{\iota(V) : V$ *is open in* X, V^- *is compact, and* $V^- \subset U\}$.

[1] See for example K. MAYRHOFER, *Inhalt und Maß*, Springer-Verlag, Wien, 1952.

Proof. Take any real number β such that $\beta < \iota(U)$. Since U is open, we have $\iota(U) = \bar{I}(\xi_U) = \sup\{I(f) : f \in \mathfrak{C}_{00}^+, f \leqq \xi_U\}$. Thus we can choose $f \in \mathfrak{C}_{00}^+$ such that $\beta < I(f) \leqq \iota(U)$. For $n = 1, 2, 3, \ldots$, define $F_n = \left\{x \in X : f(x) \geqq \frac{1}{n}\right\}$, and $W_n = \left\{x \in X : f(x) > \frac{1}{n}\right\}$; F_n is compact, W_n is open, and W_n^- is compact. Let $W = \{x \in X : f(x) > 0\}$. It is clear that $\lim_{n\to\infty} \xi_{F_n}(x) = \lim_{n\to\infty} \xi_{W_n}(x) = \xi_W(x)$ for each $x \in X$, and it is clear too that the sequences are nondecreasing. Applying (9.17), we have

$$\beta < I(f) \leqq \bar{I}(\xi_W) = \lim_{n\to\infty} \bar{I}(\xi_{F_n}) = \lim_{n\to\infty} \bar{I}(\xi_{W_n}) = \lim_{n\to\infty} \iota(F_n) = \lim_{n\to\infty} \iota(W_n)\ .$$

The theorem follows from these inequalities. □

(9.27) Theorem. *If $A \subset X$ and A^- is compact, then $0 \leqq \iota(A) < \infty$.*

Proof. According to (6.79), there is an open set U such that U^- is compact and $A^- \subset U$. Apply (6.80) to find a continuous $f: X \to [0, 1]$ such that $f(x) = 1$ for all $x \in A^-$ and $f(x) = 0$ for all $x \in U'$. Then $\xi_A \leqq f \in \mathfrak{C}_{00}^+$, and so $\iota(A) = \bar{\bar{I}}(\xi_A) \leqq I(f) < \infty$. □

(9.28) Theorem. *There exists a unique set $E \subset X$ having the following properties:*

(i) *E is closed in X;*

(ii) *$\iota(E \cap U) > 0$ if $E \cap U \neq \varnothing$ and U is open in X;*

(iii) *$\iota(X \cap E') = 0$.*

The set E is called the support [*or* carrier, *or* spectrum] *of ι.*

Proof. Let $\mathcal{U} = \{U : U$ is open in X, $\iota(U) = 0\}$ and let $V = \bigcup \mathcal{U}$, $E = V'$. Since $\xi_V \leqq \sum_{U\in\mathcal{U}} \xi_U$ and $\xi_U \in \mathfrak{M}^+$ for each $U \in \mathcal{U}$, (9.14) yields

$$\iota(E') = \iota(V) = \bar{I}(\xi_V) \leqq \bar{I}\Big(\sum_{U\in\mathcal{U}} \xi_U\Big) = \sum_{U\in\mathcal{U}} \bar{I}(\xi_U) = \sum_{U\in\mathcal{U}} \iota(U) = 0\ .$$

Thus (i) and (iii) are established. To prove (ii), let W be any open subset of X such that $E \cap W \neq \varnothing$. Then $W \notin \mathcal{U}$ and $V \cap W \in \mathcal{U}$. Thus

$$0 < \iota(W) \leqq \iota(E \cap W) + \iota(V \cap W) = \iota(E \cap W)\ .$$

Thus E satisfies (i), (ii), and (iii).

To prove that E is unique, assume that both E_1 and E_2 satisfy (i), (ii), and (iii) and $E_1 \neq E_2$. At least one of $E_1' \cap E_2$ and $E_1 \cap E_2'$ is nonvoid; say $E_1' \cap E_2 \neq \varnothing$. Since E_1' is open, (ii) implies that $\iota(E_1' \cap E_2) > 0$. But $E_1' \cap E_2 \subset E_1'$ and so $0 < \iota(E_1' \cap E_2) \leqq \iota(E_1')$. According to (iii), we have $\iota(E_1') = 0$. This is a contradiction. □

(9.29) Definition. A subset A of X for which $\iota(A) = 0$ is called an *ι-null set*. If $\iota(B \cap F) = 0$ for every compact set $F \subset X$, then B is called a *locally ι-null set*. A property which holds for all $x \in X$ except for those x in some ι-null set is said to hold *ι-almost everywhere* [abbreviated ι-a.e.]. If a property holds for all $x \in X$ except for those x in some locally ι-null set, then the property is said to hold *locally ι-almost everywhere* [locally ι-a.e.]. A complex or extended real-valued function f on X such that $f(x) = 0$ ι-a.e. [locally ι-a.e.] is called an *ι-null function* [*locally ι-null function*]. Where no confusion can result, we will drop the prefix "ι-".

(9.30) Theorem. *For $h \in \mathfrak{F}^+$, we have $\bar{I}(h) = 0$ if and only if h is an ι-null function. If $\bar{I}(h) < \infty$, then h is finite ι-a.e.*

Proof. Let $A = \{x \in X : h(x) > 0\}$. The functions nh, $n = 1, 2, \ldots$, are all in $\mathfrak{F}^+$, and it is obvious that $\lim\limits_{n\to\infty} nh \geqq \xi_A$. Thus if $\bar{I}(h) = 0$, (9.17) shows that $\iota(A) = \bar{I}(\xi_A) \leqq \bar{I}\left(\lim\limits_{n\to\infty} nh\right) = \lim\limits_{n\to\infty} \bar{I}(nh) = \lim\limits_{n\to\infty} n\bar{I}(h) = 0$. If h is an ι-null function, then $\iota(A) = 0$; using the inequality $h \leqq \lim\limits_{n\to\infty} n\xi_A$, we have $\bar{I}(h) \leqq \lim\limits_{n\to\infty} \bar{I}(n\xi_A) = \lim\limits_{n\to\infty} n\bar{I}(\xi_A) = \lim\limits_{n\to\infty} n\iota(A) = 0$.

Next suppose that $\bar{I}(h) < \infty$, and let $B = \{x \in X : h(x) = \infty\}$. For all $\varepsilon > 0$ we have $\xi_B \leqq \varepsilon h$, and so $\iota(B) = \bar{I}(\xi_B) \leqq \bar{I}(\varepsilon h) = \varepsilon\bar{I}(h)$. Since $\bar{I}(h)$ is finite, we infer that $\iota(B) = 0$. □

(9.31) Corollary. *Let $(h_n)_{n=1}^{\infty}$ be a sequence of functions in $\mathfrak{F}^+$ and suppose that $\lim\limits_{n\to\infty} \bar{I}(h_n) = 0$. Then there is a subsequence $(h_{n_k})_{k=1}^{\infty}$ such that $\sum\limits_{k=1}^{\infty} h_{n_k}(x) < \infty$ ι-a.e. on X and in particular, $\lim\limits_{k\to\infty} h_{n_k}(x) = 0$ ι-a.e.*

Proof. We first select a subsequence (h_{n_k}) of (h_n) such that $\sum\limits_{k=1}^{\infty} \bar{I}(h_{n_k}) < \infty$. Using (9.18), we see that $\bar{I}\left(\sum\limits_{k=1}^{\infty} h_{n_k}\right) < \infty$, and it follows from (9.30) that $\sum\limits_{k=1}^{\infty} h_{n_k}(x) < \infty$ ι-a.e. □

The next theorem is a technicality, but a very useful one for later purposes.

(9.32) Theorem. *Let U be an open subset of X. Then*

$$\iota(T) = \iota(T \cap U) + \iota(T \cap U')$$

for every set $T \subset X$.

Proof. Let $T \subset X$. It is an immediate consequence of (9.21) that $\iota(T) \leqq \iota(T \cap U) + \iota(T \cap U')$. The reversed inequality is obvious if $\iota(T) = \infty$. Thus suppose that $\iota(T) < \infty$, and let $\varepsilon > 0$ be arbitrary. By (9.24), there is an open set $V \supset T$ such that $\iota(V) < \iota(T) + \frac{\varepsilon}{2}$. Use (9.24) again to choose an open set $H \supset V \cap U'$ such that $\iota(H) < \iota(V \cap U') + \frac{\varepsilon}{4}$. Applying (9.26), choose an open set W such that $W^- \subset V \cap U$ and $\iota(W) + \frac{\varepsilon}{4} > \iota(V \cap U)$. Let $W_0 = V \cap H \cap (W^-)'$; then W and W_0 are disjoint open sets. Since $V \cap U'$ is a subset of each of the sets V, H, and $(W^-)'$, it follows that $V \cap U' \subset W_0 \subset H$, and so

$$0 \leqq \iota(W_0) - \iota(V \cap U') \leqq \iota(H) - \iota(V \cap U') < \frac{\varepsilon}{4} \,.$$

Therefore

$$|\iota(W) + \iota(W_0) - (\iota(V \cap U) + \iota(V \cap U'))| \leqq |\iota(W_0) - \iota(V \cap U')| + |\iota(W) - \iota(V \cap U)| < \frac{\varepsilon}{4} + \frac{\varepsilon}{4} = \frac{\varepsilon}{2} \,.$$

Combining this with the fact that $W \cup W_0 \subset V$ and using (9.22), we have

$$\iota(T) + \varepsilon > \iota(V) + \frac{\varepsilon}{2} \geqq \iota(W \cup W_0) + \frac{\varepsilon}{2} = \iota(W) + \iota(W_0) + \frac{\varepsilon}{2} > \iota(V \cap U) + \iota(V \cap U') \geqq \iota(T \cap U) + \iota(T \cap U') \,.$$

Since ε is arbitrary, we conclude that

$$\iota(T) \geqq \iota(T \cap U) + \iota(T \cap U'). \quad \square$$

(9.33) Exercise. Prove that:

(a) if $a < b$ in $R^\#$, then $\lambda(]a, b[) = b - a$; and

(b) if $a < b$ in R, then $\lambda(]a, b[) = \lambda([a, b[) = \lambda(]a, b]) = \lambda([a, b]) = b - a$.

(9.34) Exercise. Let A be a countable subset of R. Prove that $\lambda(A) = 0$.

(9.35) Exercise. Let P be the Cantor ternary set. Prove that $\lambda(P) = 0$.

(9.36) Exercise. Construct a nowhere dense perfect subset F of $[0, 1]$ such that $\lambda(F) = \alpha$, where α is any real number, $0 \leqq \alpha < 1$.

(9.37) Exercise. Let F be a nonvoid perfect subset of R. Prove that F contains a nonvoid perfect subset of Lebesgue measure zero.

(9.38) Exercise. Let $(a_n)_{n=1}^\infty$ be a sequence of positive real numbers such that $\sum_{n=1}^\infty a_n = 1$. Prove that there exists a pairwise disjoint sequence $(I_n)_{n=1}^\infty$ of open intervals such that $\bigcup_{n=1}^\infty I_n \subset [0, 1]$, $\lambda(I_n) = a_n$ for each $n \in N$, and $[0, 1] \cap \left(\bigcup_{n=1}^\infty I_n\right)'$ is nowhere dense and perfect in R. [See (8.26).]

(9.39) Exercise [FATOU's lemma]. Let X and I again be arbitrary. Suppose that (h_n) is a sequence of functions in $\mathfrak{F}^+$. Prove that

$$\bar{I}(\varliminf_{n\to\infty} h_n) \leqq \varliminf_{n\to\infty} \bar{I}(h_n) \,.$$

Also find a sequence $(h_n) \subset \mathfrak{C}_{00}^+$ for which the strict inequality holds, where $X = R$ and $I = S$, the Riemann integral.

(9.40) Exercise. Let X be a locally compact Hausdorff space. Let X^* be a nonvoid closed subset of X [with the relative topology]. Let I^*

be a nonnegative linear functional on $\mathfrak{C}_{00}(X^*)$ and let ι^* be the set function defined on the subsets of X^* constructed as in (9.19) from the functional I^*.

(a) For a function f on X, let f^* be f with its domain restricted to X^*. Prove that if $f \in \mathfrak{C}_{00}(X)$, then $f^* \in \mathfrak{C}_{00}(X^*)$.

(b) Let $g \in \mathfrak{C}_{00}(X^*)$. Prove that there is a function $f \in \mathfrak{C}_{00}(X)$ such that $g = f^*$. [Use TIETZE's extension theorem (7.40).]

(c) For $f \in \mathfrak{C}_{00}(X)$, let $I(f) = I^*(f^*)$. Prove that I is a nonnegative linear functional on $\mathfrak{C}_{00}(X)$.

(d) Let ι be the set function obtained from I as in (9.19). Prove that $\iota(X^{*\prime}) = 0$ and $\iota(A) = \iota(A \cap X^*) = \iota^*(A \cap X^*)$ for every $A \subset X$.

(9.41) Exercise. Let X be the product space $R_d \times R$, where the first factor is the real line with the discrete topology and the second factor is the real line with its usual topology. For $x, a, b \in R$ with $a < b$, let

$$U(x, a, b) = \{(x, y) \in X : a < y < b\} = \{x\} \times]a, b[\,.$$

(a) Show that $\{U(x, a, b) : x, a, b \in R \text{ and } a < b\}$ is a base for the product topology on X (6.41).

(b) Prove that X with this topology is a locally compact Hausdorff space.

For any function f defined on X and any $x \in R$, let $f_{[x]}$ be the function defined on R by the rule $f_{[x]}(y) = f(x, y)$.

(c) Prove that if $f \in \mathfrak{C}_{00}(X)$, then $f_{[x]}$ is identically zero except for finitely many $x \in R$.

Define I on $\mathfrak{C}_{00}(X)$ by the rule $I(f) = \sum\limits_{x \in R} S(f_{[x]})$, where S is the ordinary Riemann integral.

(d) Prove that I is a nonnegative linear functional on $\mathfrak{C}_{00}(X)$. Let ι be the measure obtained from I as in (9.19).

(e) Prove that the set $A = \{(x, 0) : x \in R\}$ is locally ι-null but is not ι-null.

(9.42) Exercise. Prove that if X is a countable union of compact sets [such a space is said to be *σ-compact*], then every locally ι-null set is ι-null.

§ 10. Measures and measurable sets

(10.1) Introduction. Section 9 was devoted to the construction of the functional $\overline{I}$ and the set function ι, defined on $\mathscr{P}(X)$. Our ultimate aim [to be realized in § 12] is to find a reasonably large class of functions on X for which the equality

(i) $$\sum_{n=1}^{\infty} \overline{I}(f_n) = \overline{I}\left(\sum_{n=1}^{\infty} f_n\right)$$

holds. [We have already proved (i) for functions in $\mathfrak{M}^+$.] The avenue we choose toward this goal is through abstract measures and integrals defined in terms of these abstract measures. The problem of finding a *largest* family of functions on X for which (i) holds is unsolved and apparently very difficult. Our approach to the problem is not the only one possible, but it has the advantages of simplicity and also of introducing abstract integrals, which every reader should know about anyway.

The present section is mainly concerned with properties of the set function ι and in particular with its good behavior on a certain well-defined family of subsets of X. The properties of ι that we need in defining this family are set down in Theorem (9.21). It turns out that set functions enjoying properties (9.21.i)–(9.21.iv) can be studied *in abstracto*, with no reference to a topological space or to positive functionals. We make a formal definition, as follows.

(10.2) Definition. Let X be a set [no topology]. A function μ defined on $\mathscr{P}(X)$ is called a *[CARATHÉODORY[1]] outer measure* if the following relations hold:

(i) $0 \leqq \mu(A) \leqq \infty$ for all $A \subset X$;

(ii) $\mu(\varnothing) = 0$;

(iii) $\mu(A) \leqq \mu(B)$ if $A \subset B \subset X$;

(iv) $\mu\left(\bigcup_{n=1}^{\infty} A_n\right) \leqq \sum_{n=1}^{\infty} \mu(A_n)$ for all sequences $(A_n)_{n=1}^{\infty}$ of subsets of X.

Outer measures are not by themselves of great use. Far more important for integration theory are measures, which we next define.

(10.3) Definition. Let X be a set and $\mathscr{A}$ an algebra of subsets of X. A set function μ defined only on $\mathscr{A}$ is called a *finitely additive measure* if:

(i) $0 \leqq \mu(A) \leqq \infty$ for all $A \in \mathscr{A}$;

(ii) $\mu(\varnothing) = 0$;

(iii) $\mu(A \cup B) = \mu(A) + \mu(B)$ if $A, B \in \mathscr{A}$ and $A \cap B = \varnothing$ [2].

A finitely additive measure μ such that

(iv) $\mu\left(\bigcup_{n=1}^{\infty} A_n\right) = \sum_{n=1}^{\infty} \mu(A_n)$ for all pairwise disjoint sequences $(A_n)_{n=1}^{\infty}$

such that $A_n \in \mathscr{A}$ and $\bigcup_{n=1}^{\infty} A_n \in \mathscr{A}$ is called a *countably additive measure* or simply a *measure*. If $\mathscr{A}$ is a σ-algebra of subsets of X and μ is a [countably additive] measure on $\mathscr{A}$, then the triple $(X, \mathscr{A}, \mu)$ is called a *measure space*. Let $(X, \mathscr{A}, \mu)$ be a measure space. If $\mu(X) < \infty$, then

[1] CONSTANTIN CARATHÉODORY (1873–1950), the inventor of outer measures, was a distinguished German mathematician [of Greek descent], who made many vital contributions to modern analysis.

[2] Finitely additive measures as such are only of peripheral interest for this text. We include their definition largely for the sake of completeness, but also with an eye to certain applications in § 20.

$(X, \mathscr{A}, \mu)$ is called a *finite measure space* and μ a *finite measure*. If X is the union of a countable family of sets in $\mathscr{A}$ each having finite measure, then $(X, \mathscr{A}, \mu)$ and μ are called *σ-finite*. If $\mu = 0$ or μ assumes only the values 0 and ∞, then $(X, \mathscr{A}, \mu)$ and μ are called *degenerate*.

Outer measures offer a convenient method of obtaining measures, if we suitably restrict the family of subsets on which the outer measure is defined. Before entering on the technicalities of this construction, we give a few examples.

(10.4) Examples. (a) Let X be any set. For $A \subset X$, let

$$\nu(A) = \begin{cases} \bar{A} & \text{if } A \text{ is finite}, \\ \infty & \text{if } A \text{ is infinite}. \end{cases}$$

Then ν is a measure and also an outer measure on $\mathfrak{P}(X)$. [It is easy to see that a measure on $\mathfrak{P}(X)$ is also an outer measure.] The measure ν is usually called *counting measure*.

(b) Let X be any set. For $A \subset X$, let

$$\mu(A) = \begin{cases} 0 & \text{if } A = \varnothing, \\ \infty & \text{if } A \neq \varnothing. \end{cases}$$

Then μ is a [plainly degenerate] measure and also outer measure on $\mathfrak{P}(X)$.

(c) The set function identically zero is a degenerate measure on $\mathfrak{P}(X)$, for any set X.

(d) The most important outer measures for our purposes are the set functions ι constructed as in § 9 for all subsets of a locally compact Hausdorff space X. See Theorem (9.21) for the proof that ι is an outer measure. The reader should be aware, however, that not all measures used in analysis are derived from set functions ι.

We now begin our construction of measures.

(10.5) Definition [Carathéodory]. Let X be a set [no topology] and μ an outer measure on $\mathfrak{P}(X)$. A subset A of X is said to be *μ-measurable* if

$$\mu(T) = \mu(T \cap A) + \mu(T \cap A')$$

for all $T \subset X$. Let $\mathscr{M}_\mu$ denote the family of all μ-measurable subsets of X and $\mathscr{N}_\mu$ the family of all subsets A of X such that $\mu(A) = 0$.

(10.6) Remarks. (a) In view of (10.2.iv) and (10.2.ii), a subset A of X is μ-measurable provided that $\mu(T) \geqq \mu(T \cap A) + \mu(T \cap A')$ for all $T \subset X$.

(b) Definition (10.5) has a somewhat artificial air. It singles out the subsets A of X for which A splits *all* subsets T of X into two pieces on which μ adds. How Carathéodory came to think of this definition seems mysterious, since it is not in the least intuitive. Carathéodory's definition has many useful implications. It gives us a σ-algebra, although not necessarily the *largest* possible σ-algebra, on which μ is a countably additive measure. [In (10.40), we point out conditions under which $\mathscr{M}_\mu$ is the largest σ-algebra on which μ is countably additive.]

(c) If X is a locally compact Hausdorff space and ι is an outer measure as in § 9, then the family $\mathscr{M}_\iota$ of ι-measurable sets contains all open sets. This fact, proved in (9.32), has important consequences as we shall see.

(d) Let λ be Lebesgue outer measure on R. The family $\mathscr{M}_\lambda$ of λ-measurable sets is often called the family of *Lebesgue measurable sets*. We will use this phrase when convenient.

We proceed to develop the properties of μ-measurable sets. Throughout (10.7)–(10.11), X is an arbitrary set and μ is an arbitrary outer measure on $\mathscr{P}(X)$.

(10.7) Theorem. *Every subset A of X such that $\mu(A) = 0$ is μ-measurable, and $\mu(T) = \mu(T \cap A')$ for all $T \subset X$.*

Proof. Let T be any subset of X. Then $\mu(T \cap A) = 0$, since $T \cap A \subset A$. Also, we have $\mu(T) \leqq \mu(T \cap A) + \mu(T \cap A') = \mu(T \cap A') \leqq \mu(T)$. It follows that $\mu(T) = \mu(T \cap A) + \mu(T \cap A') = \mu(T \cap A')$; the first equality shows that A is measurable. □

(10.8) Theorem. *If A is μ-measurable, then A' is also.*

Proof. Trivial.

(10.9) Theorem. *Let $(A_n)_{n=1}^\infty$ be a sequence of pairwise disjoint μ-measurable subsets of X. Then*

(i) $\mu(T) = \sum_{n=1}^{\infty} \mu(T \cap A_n) + \mu\left(T \cap \left(\bigcup_{n=1}^{\infty} A_n\right)'\right)$ *for all $T \subset X$.*

Proof. By countable subadditivity, we have $\mu(T) \leqq \sum_{n=1}^{\infty} \mu(T \cap A_n) + \mu\left(T \cap \left(\bigcup_{n=1}^{\infty} A_n\right)'\right)$. If $\mu(T) = \infty$, (i) follows immediately. Hence we may suppose that $\mu(T) < \infty$. We first prove

$$\mu(T) = \sum_{n=1}^{p} \mu(T \cap A_n) + \mu\left(T \cap \left(\bigcup_{n=1}^{p} A_n\right)'\right) \text{ for all } p \in N. \tag{1}$$

We prove this by induction on p. For $p = 1$, (1) becomes $\mu(T) = \mu(T \cap A_1) + \mu(T \cap A_1')$. This is true for all $T \subset X$ since A_1 is μ-measurable. Suppose that (1) is true for a positive integer p and all subsets T of X. Since A_{p+1} is μ-measurable, we have

$$\begin{aligned} \mu(T) &= \mu(T \cap A_{p+1}) + \mu(T \cap A_{p+1}') \\ &= \mu(T \cap A_{p+1}) + \sum_{n=1}^{p} \mu(T \cap A_{p+1}' \cap A_n) + \mu\left(T \cap A_{p+1}' \cap \left(\bigcup_{n=1}^{p} A_n\right)'\right). \end{aligned} \tag{2}$$

[We apply the inductive hypothesis to the set $T \cap A_{p+1}'$.] Since

$A_n \subset A'_{p+1}$ for $n \neq p + 1$, (2) can be written as

$$\mu(T) = \mu(T \cap A_{p+1}) + \sum_{n=1}^{p} \mu(T \cap A_n) + \mu\left(T \cap A'_{p+1} \cap \left(\bigcup_{n=1}^{p} A_n\right)'\right)$$
$$= \sum_{n=1}^{p+1} \mu(T \cap A_n) + \mu\left(T \cap \left(\bigcup_{n=1}^{p+1} A_n\right)'\right),$$

and this is (1) for $p + 1$.

The sequence of numbers $\left(\mu\left(T \cap \left(\bigcup_{n=1}^{p} A_n\right)'\right)\right)_{p=1}^{\infty}$ is a nonincreasing sequence bounded below by the number $\mu\left(T \cap \left(\bigcup_{n=1}^{\infty} A_n\right)'\right)$. It thus has a limit, which is greater than or equal to $\mu\left(T \cap \left(\bigcup_{n=1}^{\infty} A_n\right)'\right)$. Taking limits in the equality (1), we obtain

$$\mu(T) = \lim_{p\to\infty} \sum_{n=1}^{p} \mu(T \cap A_n) + \lim_{p\to\infty} \mu\left(T \cap \left(\bigcup_{n=1}^{p} A_n\right)'\right)$$
$$\geq \sum_{n=1}^{\infty} \mu(T \cap A_n) + \mu\left(T \cap \left(\bigcup_{n=1}^{\infty} A_n\right)'\right).$$

Since the reversed inequality has already been established, this completes the proof. □

(10.10) Theorem. *If A and B are μ-measurable, then $A \cap B'$ is μ-measurable.*

Proof. It suffices to prove that if $E \subset A \cap B'$ and $F \subset (A \cap B')'$, then $\mu(E \cup F) = \mu(E) + \mu(F)$. Since $F = (F \cap B) \cup (F \cap B')$ and B is μ-measurable, we have

$$\mu(E) + \mu(F) = \mu(E) + \mu((F \cap B) \cup (F \cap B'))$$
$$= \mu(E) + \mu(F \cap B) + \mu(F \cap B').$$

Now since $E \subset A$, $F \cap B' \subset A'$, and A is μ-measurable, we have

$$\mu(E) + \mu(F \cap B') + \mu(F \cap B) = \mu(E \cup (F \cap B')) + \mu(F \cap B).$$

Again $E \cup (F \cap B') \subset B'$ and $F \cap B \subset B$ so that

$$\mu(E \cup (F \cap B')) + \mu(F \cap B) = \mu(E \cup (F \cap B') \cup (F \cap B)) = \mu(E \cup F).$$

Combining these equalities, we have $\mu(E) + \mu(F) = \mu(E \cup F)$. □

(10.11) Theorem. *The family $\mathscr{M}_\mu$ of μ-measurable sets is a σ-algebra of subsets of X, and μ is a countably additive measure on the σ-algebra $\mathscr{M}_\mu$.*

Proof. Let $(A_n)_{n=1}^{\infty}$ be a sequence of μ-measurable subsets of X. Then $\bigcup_{n=1}^{\infty} A_n = A_1 \cup (A_2 \cap A'_1) \cup (A_3 \cap A'_2 \cap A'_1) \cup \cdots \cup (A_n \cap A'_{n-1} \cap \cdots \cap A'_1) \cup \cdots$. By (10.10), each set of the form $B_n = (A_n \cap A'_{n-1} \cap \cdots \cap A'_1)$ is μ-measurable. Furthermore, the sets B_n are pairwise disjoint.

Let $T \subset X$. By (10.9.i) and countable subadditivity, we have

$$\mu(T) = \sum_{n=1}^{\infty} \mu(T \cap B_n) + \mu\left(T \cap \left(\bigcup_{n=1}^{\infty} B_n\right)'\right) \geqq \mu\left(T \cap \left(\bigcup_{n=1}^{\infty} B_n\right)\right) + \mu\left(T \cap \left(\bigcup_{n=1}^{\infty} B_n\right)'\right).$$

By countable subadditivity, we have

$$\mu(T) \leqq \mu\left(T \cap \left(\bigcup_{n=1}^{\infty} B_n\right)\right) + \mu\left(T \cap \left(\bigcup_{n=1}^{\infty} B_n\right)'\right),$$

and so

$$\mu(T) = \mu\left(T \cap \left(\bigcup_{n=1}^{\infty} B_n\right)\right) + \mu\left(T \cap \left(\bigcup_{n=1}^{\infty} B_n\right)'\right).$$

This implies that $\bigcup_{n=1}^{\infty} B_n$ is μ-measurable. Thus $\bigcup_{n=1}^{\infty} A_n = \bigcup_{n=1}^{\infty} B_n$ is μ-measurable. This fact and (10.8) imply that the family of μ-measurable sets is a σ-algebra (1.13).

Upon setting $T = \bigcup_{k=1}^{\infty} B_k$ in (10.9.i), we obtain

$$\mu\left(\bigcup_{n=1}^{\infty} B_n\right) = \sum_{n=1}^{\infty} \mu\left(\left(\bigcup_{k=1}^{\infty} B_k\right) \cap B_n\right) + \mu\left(\left(\bigcup_{k=1}^{\infty} B_k\right) \cap \left(\bigcup_{n=1}^{\infty} B_n\right)'\right)$$
$$= \sum_{n=1}^{\infty} \mu(B_n) + \mu(\varnothing) = \sum_{n=1}^{\infty} \mu(B_n).$$

Thus μ is countably additive on the σ-algebra of all μ-measurable subsets of X. □

Having proved in (10.11) that nontrivial measure spaces exist, we digress to prove some useful facts about arbitrary measure spaces.

(10.12) Theorem. *Let $(X, \mathscr{A}, \mu)$ be a measure space. If $A, B \in \mathscr{A}$ and $A \subset B$,* then $\mu(A) \leqq \mu(B)$.

Proof. By (10.3.iii) we have

$$\mu(B) = \mu(A) + \mu(B \cap A'),$$

and by (10.3.i) $\mu(B \cap A') \geqq 0$. □

(10.13) Theorem. *Let $(X, \mathscr{A}, \mu)$ be a measure space. Let $(A_n)_{n=1}^{\infty}$ be a sequence of sets in $\mathscr{A}$ such that $A_1 \subset A_2 \subset \cdots \subset A_n \subset \cdots$. Then*

$$\mu\left(\bigcup_{n=1}^{\infty} A_n\right) = \lim_{n \to \infty} \mu(A_n).$$

Proof. Write $A_0 = \varnothing$. Then clearly $\bigcup_{n=1}^{\infty} A_n = \bigcup_{n=1}^{\infty} (A_n \cap A'_{n-1})$. By (10.3.iv), countable additivity, we have

$$\mu\left(\bigcup_{n=1}^{\infty} A_n\right) = \sum_{n=1}^{\infty} \mu(A_n \cap A'_{n-1}) = \lim_{p \to \infty} \sum_{n=1}^{p} \mu(A_n \cap A'_{n-1})$$
$$= \lim_{p \to \infty} \mu\left(\bigcup_{n=1}^{p} (A_n \cap A'_{n-1})\right) = \lim_{p \to \infty} \mu(A_p). \quad \square$$

(10.14) Remark. A result strictly analogous to Theorem (10.13) for intersections of measurable sets cannot be proved. To see this let $X = R$, $\mu = \lambda$, and $\mathscr{A} = \mathscr{M}_\lambda$, the Lebesgue measurable sets. Let $A_n = [n, \infty[$. Then $\lambda(A_n) = \infty$ for $n = 1, 2, 3, \ldots$, so that $\lim_{n\to\infty} \lambda(A_n) = \infty$. On the other hand, $\lambda\left(\bigcap_{n=1}^{\infty} A_n\right) = \lambda(\varnothing) = 0$. However, we do obtain the following result.

(10.15) Theorem. *Let $(X, \mathscr{A}, \mu)$ be a measure space. If $(A_n)_{n=1}^{\infty}$ is a sequence of sets in $\mathscr{A}$ such that $\mu(A_1) < \infty$ and $A_1 \supset A_2 \supset \cdots \supset A_n \supset \cdots$, then*

$$\mu\left(\bigcap_{n=1}^{\infty} A_n\right) = \lim_{n\to\infty} \mu(A_n)\,.$$

In particular, if $\bigcap_{n=1}^{\infty} A_n = \varnothing$, then $\lim_{n\to\infty} \mu(A_n) = 0$.

Proof. The sequence $(A_1 \cap A_n')_{n=1}^{\infty}$ is nondecreasing, and all $A_1 \cap A_n'$ are in $\mathscr{A}$. Applying (10.13), we get

$$\begin{aligned}\mu(A_1) - \lim_{n\to\infty} \mu(A_n) &= \lim_{n\to\infty} (\mu(A_1) - \mu(A_n)) = \lim_{n\to\infty} \mu(A_1 \cap A_n') \\ &= \mu\left(\bigcup_{n=1}^{\infty}\left(A_1 \cap A_n'\right)\right) = \mu\left(A_1 \cap \left(\bigcup_{n=1}^{\infty} A_n'\right)\right) \\ &= \mu\left(A_1 \cap \left(\bigcap_{n=1}^{\infty} A_n\right)'\right) = \mu(A_1) - \mu\left(\bigcap_{n=1}^{\infty} A_n\right).\end{aligned}$$

Subtracting $\mu(A_1)$ [which we may do since $\mu(A_1) < \infty$], we have

$$\lim_{n\to\infty} \mu(A_n) = \mu\left(\bigcap_{n=1}^{\infty} A_n\right). \quad \square$$

(10.16) Theorem. *Let $(A_n)_{n=1}^{\infty}$ be any sequence of sets in $\mathscr{A}$. Then*

(i) $\mu\left(\bigcup_{k=1}^{\infty}\left(\bigcap_{n=k}^{\infty} A_n\right)\right) \leq \underline{\lim}_{k\to\infty} \mu(A_k)$.

We also have

(ii) $\mu\left(\bigcap_{k=1}^{\infty}\left(\bigcup_{n=k}^{\infty} A_n\right)\right) \geq \overline{\lim}_{k\to\infty} \mu(A_k)$ *provided that* $\mu\left(\bigcup_{k=1}^{\infty} A_k\right) < \infty$.

Proof. It is clear that $\bigcap_{n=1}^{\infty} A_n \subset \bigcap_{n=2}^{\infty} A_n \subset \cdots \subset \bigcap_{n=k}^{\infty} A_n \subset \cdots$. Theorem (10.13) implies that $\mu\left(\bigcup_{k=1}^{\infty}\left(\bigcap_{n=k}^{\infty} A_n\right)\right) = \lim_{k\to\infty} \mu\left(\bigcap_{n=k}^{\infty} A_n\right)$. We also have $\mu(A_k) \geq \mu\left(\bigcap_{n=k}^{\infty} A_n\right)$ for all k. This implies that $\lim_{k\to\infty} \mu\left(\bigcap_{n=k}^{\infty} A_n\right) \leq \underline{\lim}_{k\to\infty} \mu(A_k)$, from which (i) follows. The inequality (ii) is proved in like manner. $\square$

(10.17) Corollary. *Hypotheses are as in* (10.16). *Suppose also that* $\bigcap_{k=1}^{\infty}\left(\bigcup_{n=k}^{\infty} A_n\right) = \bigcup_{k=1}^{\infty}\left(\bigcap_{n=k}^{\infty} A_n\right) = B$ *and* $\mu\left(\bigcup_{k=1}^{\infty} A_k\right) < \infty$. *Then* $\lim_{k\to\infty} \mu(A_k)$ *exists and is equal to* $\mu(B)$.

Proof. The assertion follows from the inequalities $\overline{\lim_{k\to\infty}}\, \mu(A_k) \leqq \mu(B) \leqq \underline{\lim_{k\to\infty}}\, \mu(A_k)$. □

(10.18) Note. In (10.3), as we trust the reader has noted, countably additive measures are defined on algebras $\mathscr{A}$ that need not be σ-algebras. This is a technicality, but it is occasionally quite useful. There is a technique for extending a countably additive measure to a σ-algebra containing $\mathscr{A}$. This is not an essential point in our development of measure theory, and we relegate it to an exercise (10.36). Note however that the term *measure space* is reserved for a triple $(X, \mathscr{A}, \mu)$ in which $\mathscr{A}$ is a σ-algebra and μ is countably additive on $\mathscr{A}$.

We return to locally compact Hausdorff spaces and outer measures ι as in § 9.

(10.19) Definition. For an arbitrary set X and an arbitrary family $\mathscr{E}$ of subsets of X, let $\mathscr{S}(\mathscr{E})$ denote the intersection of all σ-algebras of subsets of X that contain $\mathscr{E}$. Clearly $\mathscr{S}(\mathscr{E})$ is a σ-algebra. Thus $\mathscr{S}(\mathscr{E})$ is the smallest σ-algebra of subsets of X containing $\mathscr{E}$. If X is a topological space, let $\mathscr{B}(X)$ be the smallest σ-algebra of subsets of X that contains every open set, *i.e.*, $\mathscr{B}(X) = \mathscr{S}(\mathscr{O})$ where $\mathscr{O}$ is the family of all open sets. The members of $\mathscr{B}(X)$ are called the *Borel sets of* X.

(10.20) Theorem. *Let X be a locally compact Hausdorff space and let ι be as in* (9.19). *Then* $\mathscr{B}(X) \subset \mathscr{M}_\iota$, i.e., *every Borel set is ι-measurable, and $(X, \mathscr{M}_\iota, \iota)$ is a measure space.*

Proof. Theorem (9.32) is just the statement that all open subsets of X are ι-measurable [although this phrase is not used in (9.32)]. Thus $\mathscr{M}_\iota$ contains the family $\mathscr{O}$ of all open subsets of X, and so by (10.11),

$$\mathscr{M}_\iota \supset \mathscr{S}(\mathscr{O}) = \mathscr{B}(X). \quad \square$$

(10.21) Remarks. (a) For many choices of X and ι, there are ι-measurable sets that are not Borel sets. For example, let $X = R$, $\iota = \lambda$, and let P be Cantor's ternary set. Since $\lambda(P) = 0$ (9.35), we have $\lambda(A) = 0$ for all $A \subset P$, and so all subsets of P are λ-measurable. Thus $\mathscr{P}(P) \subset \mathscr{M}_\lambda \subset \mathscr{P}(R)$. Since $\overline{\overline{P}} = \overline{\overline{R}} = \mathfrak{c}$, we have $2^{\mathfrak{c}} = \overline{\overline{\mathscr{P}(P)}} \leqq \overline{\overline{\mathscr{M}_\lambda}} \leqq \overline{\overline{\mathscr{P}(R)}} = 2^{\mathfrak{c}}$, and so $\overline{\overline{\mathscr{M}_\lambda}} = 2^{\mathfrak{c}}$. There are exactly $\mathfrak{c}$ open subsets of R. This follows from the fact that each open subset of R is a union of open intervals with *rational* endpoints. It is therefore a corollary to the next theorem that $\overline{\overline{\mathscr{B}(R)}} = \mathfrak{c}$. This crude cardinal number argument shows that there are $2^{\mathfrak{c}}$ λ-measurable sets that are not Borel sets, but it gives no indication of how to construct such sets.

(b) It is possible actually to construct a very large class of λ-measurable sets, the so-called *analytic* sets, which includes all Borel sets as well as $\mathfrak{c}$ other sets. The interested reader is referred to the discussions of analytic sets in S. SAKS, *Theory of the Integral* [2nd Edition, Monografie Matematyczne, Warszawa-Lwów, 1937], Chapter III, and in W. SIERPIŃSKI, *General Topology* [2nd Edition, Univ. of Toronto Press, Toronto, 1956], Chapter VII.

(10.22) Exercise. (a) Let X, a, E_a, and ε_a be as in (9.19). Prove that $\mathscr{M}_{\varepsilon_a} = \mathscr{P}(X)$.

(b) Let X be a locally compact Hausdorff space, let $\{x_n\}_{n=1}^{\infty}$ be a countable [possibly finite] subset of X, and let $(\alpha_n)_{n=1}^{\infty}$ be a sequence of positive numbers such that $\sum_{n=1}^{\infty} \alpha_n < \infty$. Let $I(\varphi) = \sum_{n=1}^{\infty} \alpha_n \varphi(x_n)$ for all $\varphi \in \mathfrak{C}_{00}(X)$. Prove that: I is a nonnegative linear functional on $\mathfrak{C}_{00}(X)$; the corresponding measure ι is $\sum_{n=1}^{\infty} \alpha_n \varepsilon_{x_n}$; $\mathscr{M}_\iota = \mathscr{P}(X)$.

(c) Extend (b) to the case in which $\sum_{n=1}^{\infty} \alpha_n = \infty$. Find extra conditions on $\{x_n\}_{n=1}^{\infty}$ necessary and sufficient for I to be finite for all $\varphi \in \mathfrak{C}_{00}(X)$, and prove the second and third assertions of part (b) for the new I and ι.

The proof of the next theorem gives a method of "constructing" the σ-algebra generated by a given family of sets.

(10.23) Theorem[1]. *Let X be a set, let $\mathscr{E}$ be a family of subsets of X such that $\varnothing \in \mathscr{E}$, and let $\mathscr{S} = \mathscr{S}(\mathscr{E})$ be the smallest σ-algebra of subsets of X containing $\mathscr{E}$. If $\overline{\overline{\mathscr{E}}} = \mathfrak{e}$ and $\mathfrak{e} \geqq 2$, then $\overline{\overline{\mathscr{S}}} \leqq \mathfrak{e}^{\aleph_0}$.*

Proof. For each nonvoid family $\mathscr{F}$ of subsets of X, let $\mathscr{F}^*$ be the family of all sets of the form $\bigcup_{n=1}^{\infty} A_n$, where, for each $n \in N$, either A_n or A_n' is an element of $\mathscr{F}$. Let Ω denote the smallest uncountable ordinal number (4.49). We use transfinite induction to define a family $\mathscr{E}_\alpha$ for each ordinal number $\alpha < \Omega$. Define $\mathscr{E}_0$ as the family $\mathscr{E}$. Suppose that $0 < \alpha < \Omega$ and that $\mathscr{E}_\beta$ has been defined for each β such that $0 \leqq \beta < \alpha$. Define $\mathscr{E}_\alpha$ as $\left(\bigcup_{0 \leqq \beta < \alpha} \mathscr{E}_\beta\right)^*$, and write $\mathscr{A}$ for the family $\bigcup_{0 \leqq \alpha < \Omega} \mathscr{E}_\alpha$. We assert that $\mathscr{A} = \mathscr{S}$.

It is clear that $\mathscr{E}_0 = \mathscr{E} \subset \mathscr{S}$. Suppose that $\mathscr{E}_\beta \subset \mathscr{S}$ for every $\beta < \alpha$ and let $\bigcup_{n=1}^{\infty} A_n \in \mathscr{E}_\alpha$. For each $n \in N$, either A_n or A_n' is an element of $\bigcup_{\beta < \alpha} \mathscr{E}_\beta \subset \mathscr{S}$ so that $A_n, A_n' \in \mathscr{S}$. Thus $\bigcup_{n=1}^{\infty} A_n$ is in $\mathscr{S}$; *i.e.*, $\mathscr{E}_\alpha$ is contained

[1] This theorem is not needed in the sequel and may be omitted by any reader who is pressed for time; similarly for (10.24) and (10.25).

in $\mathscr{S}$. Since $\mathscr{A} = \bigcup_{0 \leqq \alpha < \Omega} \mathscr{E}_\alpha$, we have proved that $\mathscr{A} \subset \mathscr{S}$. It is trivial that $\mathscr{E} \subset \mathscr{A}$, and so to complete the proof we need only to show that $\mathscr{A}$ is a σ-algebra. Since $\varnothing \in \mathscr{E}_0$, we have

$$X = (\varnothing' \cup \varnothing \cup \varnothing \cup \cdots) \in \mathscr{E}_1 \subset \mathscr{A}.$$

Now let $A \in \mathscr{A}$. There is an $\alpha < \Omega$ such that $A \in \mathscr{E}_\alpha$; hence $A' = (A' \cup A' \cup \cdots) \in \mathscr{E}_\alpha^* \subset \mathscr{E}_\beta$ for every $\beta > \alpha$, and therefore $A' \in \mathscr{A}$. Next let $(A_n)_{n=1}^\infty$ be a sequence of elements of $\mathscr{A}$. For each $n \in N$, there is an $\alpha_n < \Omega$ such that $A_n \in \mathscr{E}_{\alpha_n}$. Apply (4.49.iv) to find a $\beta < \Omega$ such that $\alpha_n < \beta$ for each $n \in N$. Then

$$\bigcup_{n=1}^\infty A_n \in \left(\bigcup_{n=1}^\infty \mathscr{E}_{\alpha_n}\right)^* \subset \mathscr{E}_\beta \subset \mathscr{A}.$$

Therefore $\mathscr{A}$ is a σ-algebra and $\mathscr{A} = \mathscr{S}$.

By hypothesis, we have $\overline{\overline{\mathscr{E}_0}} = \mathfrak{c} \geqq 2$. Considering the ways in which the sets $\bigcup_{n=1}^\infty A_n \in \mathscr{E}_1$ can be formed [at most $2\mathfrak{c}$ choices for each A_n], we see that $\overline{\overline{\mathscr{E}_1}} \leqq (2\mathfrak{c})^{\aleph_0} = \mathfrak{c}^{\aleph_0}$ [(4.32), (4.34), (4.24.vii), and (4.24.xi)]. Now suppose that $\overline{\overline{\mathscr{E}_\beta}} \leqq \mathfrak{c}^{\aleph_0}$ for all β such that $1 \leqq \beta < \alpha$, where $1 < \alpha < \Omega$. Then $\overline{\overline{\bigcup_{\beta<\alpha} \mathscr{E}_\beta}} \leqq \mathfrak{c}^{\aleph_0} \aleph_0 = \mathfrak{c}^{\aleph_0}$ (4.32), and so, arguing as above, $\overline{\overline{\mathscr{E}_\alpha}} \leqq (\mathfrak{c}^{\aleph_0})^{\aleph_0} = \mathfrak{c}^{\aleph_0}$ [(4.24.viii) and (4.31)]. It follows by transfinite induction that $\overline{\overline{\mathscr{E}_\alpha}} \leqq \mathfrak{c}^{\aleph_0}$ for every α such that $0 \leqq \alpha < \Omega$. Therefore

$$\overline{\overline{\mathscr{S}}} = \overline{\overline{\mathscr{A}}} = \overline{\overline{\bigcup_{\alpha<\Omega} \mathscr{E}_\alpha}} \leqq \mathfrak{c}^{\aleph_0} \aleph_1 = \mathfrak{c}^{\aleph_0}$$

[(4.50) and (4.32)]. □

(10.24) Exercise. Notation is as in (10.16).

(a) Find $\mathscr{S}(\mathscr{E})$ if $\mathscr{E} = \{\varnothing\}$.

(b) Find $\mathscr{S}(\mathscr{E})$ if $\mathscr{E} = \{U_1, U_2, \ldots, U_m\}$, where the U_j are nonvoid and pairwise disjoint and $U_1 \cup \cdots \cup U_m = X$. What is $\overline{\overline{\mathscr{S}(\mathscr{E})}}$?

(c) Find $\mathscr{S}(\mathscr{E})$ if $\mathscr{E}$ is an arbitrary finite family of subsets of X. What is $\overline{\overline{\mathscr{S}(\mathscr{E})}}$?

(d) Find $\mathscr{S}(\mathscr{E})$ if $\mathscr{E}$ is the family of all finite subsets of X [X can be finite or infinite]. What is $\overline{\overline{\mathscr{S}(\mathscr{E})}}$?

(10.25) Corollary. *If X is a topological space with a countable base $\mathscr{C}$ for its topology, then $\overline{\overline{\mathscr{B}(X)}} \leqq \mathfrak{c}$.*

Proof. The definition of base (6.10) shows that every open subset of X is in the family $\mathscr{C}_1$ [notation is borrowed with evident changes from (10.23)]. This proves that $\mathscr{S}(\mathscr{C}) = \mathscr{B}(X)$; applying (10.23) and (4.34), we obtain

$$\overline{\overline{\mathscr{B}(X)}} \leqq (\aleph_0)^{\aleph_0} = \mathfrak{c}. \quad \square$$

The problem of finding non μ-measurable sets for a given outer measure μ can be terribly complicated. Even for outer measures constructed as in § 9, no general facts or methods are known. For Lebesgue outer measure λ on $\mathscr{P}(R)$, however, it is reasonably simple to find non λ-measurable sets. We begin with a simple fact about λ.

(10.26) Definition. For subsets A and B of R, let $A + B = \{x + y : x \in A, y \in B\}$, $A - B = \{x - y : x \in A, y \in B\}$, and $-A = \{-x : x \in A\}$. For $x \in R$, the set $\{x\} + A$ will be written $x + A$, and is called the *translate of A by x*. The sets AB, A^{-1} [if $0 \notin A$], AB^{-1}, and xA are defined analogously.

(10.27) Lemma. *For all $x \in R$ and $A \subset R$, the equalities $\lambda(x + A) = \lambda(A) = \lambda(-A)$ obtain.*

Proof. By (9.24) and (9.23), we have

$$\lambda(A) = \inf\{\lambda(U) : U \text{ is open in } R \text{ and } A \subset U\}$$
$$= \inf\left\{\sum_{n=1}^{\infty} (b_n - a_n) : A \subset \bigcup_{n=1}^{\infty}]a_n, b_n[\right\}.$$

Since the inclusions

$$A \subset \bigcup_{n=1}^{\infty}]a_n, b_n[\,,$$
$$x + A \subset \bigcup_{n=1}^{\infty}]a_n + x, b_n + x[\,, \tag{1}$$
$$-A \subset \bigcup_{n=1}^{\infty}]-b_n, -a_n[$$

are mutually equivalent, and since the three unions of intervals in (1) have the same Lebesgue measure, the lemma is proved. □

(10.28) Theorem. *Let T be a λ-measurable subset of R such that $\lambda(T) > 0$. Then T contains a subset E that is not λ-measurable. In fact, E can be chosen to have the following property. If $\mathscr{A}$ is a σ-algebra of subsets of R such that $\mathscr{M}_\lambda \subset \mathscr{A}$ and $x + A \in \mathscr{A}$ whenever $A \in \mathscr{A}$ and $x \in R$ [for example $\mathscr{A} = \mathscr{M}_\lambda$], and if μ is a countably additive measure on $\mathscr{A}$ such that $\mu(A) = \lambda(A)$ for all $A \in \mathscr{M}_\lambda$ and $\mu(x + A) = \mu(A)$ for all $A \in \mathscr{A}$ and $x \in R$, then $E \notin \mathscr{A}$.*

Proof. Since $T = \bigcup_{n=1}^{\infty} (T \cap [-n, n])$, (9.21.iv) shows that

$$0 < \lambda(T) \leq \sum_{n=1}^{\infty} \lambda(T \cap [-n, n]).$$

Hence there is a $p \in N$ such that

$$0 < \lambda(T \cap [-p, p]) \leq \lambda([-p, p]) = 2p < \infty,$$

and we lose nothing in supposing that $0 < \lambda(T) < \infty$ and that $T \subset [-p, p]$. Since countable sets have λ-measure 0 (9.34), we must have

$$\overline{\overline{T}} > \aleph_0 .\text{[1]}$$

[1] By (10.30) *infra*, T contains a compact set F of positive λ-measure. By (9.34), F is uncountable, and (6.65) and (6.66) imply that $\overline{\overline{F}} = \mathfrak{c}$. Hence $\overline{\overline{T}} = \mathfrak{c}$.

Now let D be any countably infinite subset of T (4.15) and let H be the smallest additive subgroup of R that contains D. That is, H consists of all finite sums $\sum_{k=1}^{m} n_k d_k$, where the n_k's are integers and the d_k's are in D; from this it is clear that H is countable.

Consider the cosets $\{t + H : t \in R\}$. Since H is a subgroup of R, these cosets are pairwise disjoint; *i.e.*, for any t_1 and t_2, the cosets $t_1 + H$ and $t_2 + H$ are disjoint or identical. Let $\{t_\gamma : \gamma \in \Gamma\}$ be chosen in R so that the sets $t_\gamma + H$; $\gamma \in \Gamma$, are all of the distinct cosets of H, *i.e.*, $\gamma_1 \neq \gamma_2$ implies $(t_{\gamma_1} + H) \cap (t_{\gamma_2} + H) = \varnothing$, and for each $t \in R$, there is a $\gamma \in \Gamma$ such that $t + H = t_\gamma + H$. Let $\Gamma_0 = \{\gamma \in \Gamma : (t_\gamma + H) \cap T \neq \varnothing\}$. For each $\gamma \in \Gamma_0$, choose just one element $x_\gamma \in (t_\gamma + H) \cap T$, and let $E = \{x_\gamma : \gamma \in \Gamma_0\}$. [Since T is contained in $\bigcup \{t_\gamma + H : \gamma \in \Gamma_0\}$ and $\overline{\overline{t_\gamma + H}} = \aleph_0$, we must have

$$\overline{\overline{\Gamma_0}} = \overline{\overline{T}} > \aleph_0 \,.]$$

In finding E we have twice made an uncountable number of arbitrary choices, and to do this we must invoke the axiom of choice[1].

To prove that E possesses the pathological properties ascribed to it, define the set J to be $H \cap (T - T)$. Since $D - D \subset J \subset H$, it is obvious that $\overline{\overline{J}} = \aleph_0$. We claim that $(y_1 + E) \cap (y_2 + E) = \varnothing$ for distinct $y_1, y_2 \in J$. If not, then there are distinct $x_1, x_2 \in E$ such that $y_1 + x_1 = y_2 + x_2$. Since $y_1, y_2 \in H$, this implies

$$x_2 = x_1 + (y_1 - y_2) \in x_1 + H,$$

a contradiction to the definition of E, as x_1 and x_2 lie in disjoint cosets of H. Hence the family $\{y + E : y \in J\}$ is pairwise disjoint. Now assume that E is in the σ-algebra $\mathscr{A}$. If $\mu(E) = 0$, our hypotheses on μ imply that

$$\mu(J + E) = \mu\Big(\bigcup_{y \in J} (y + E)\Big) = \sum_{y \in J} \mu(y + E) = \sum_{y \in J} \mu(E) = 0 \,. \tag{1}$$

If $\mu(E) > 0$, a similar reckoning gives

$$\mu(J + E) = \infty \,. \tag{2}$$

Both (1) and (2) are impossible. To see this, we first prove that

$$T \subset J + E \,. \tag{3}$$

In fact, if $v \in T$, then $v \in t_\gamma + H = x_\gamma + H$ for some $\gamma \in \Gamma_0$, and so

[1] Every example of a non λ-measurable set has been constructed by using the axiom of choice. A recent announcement by R. Solovay [Notices Amer. Math. Soc. 12, 217 (1965)] indicates that without the axiom of choice, non λ-measurable sets cannot be obtained at all.

$v = x_\gamma + h$ for some $h \in H$. Thus

$$h = v - x_\gamma \in (T - T) \cap H = J ,$$

which proves (3). If (1) holds, then (3) implies that

$$\mu(T) \leqq \mu(J + E) = 0 .$$

Since $\lambda(T)$ is positive and $\mu(T) = \lambda(T)$, this is a contradiction, and (1) cannot hold. It is also obvious that

$$J + E = (H \cap (T - T)) + E \subset (T - T) + T \subset [-3p, 3p] ,$$

so that

$$\mu(J + E) \leqq \mu([-3p, 3p]) = \lambda([-3p, 3p]) = 6p < \infty .$$

Thus (2) is impossible, and the assumption that $E \in \mathscr{A}$ must be rejected. □

(10.29) Remarks. (a) There exists a *finitely* additive measure μ on $\mathscr{P}(R)$ such that $\mu(A) = \lambda(A)$ for all $A \in \mathscr{M}_\lambda$ and $\mu(x + A) = \mu(A)$ for $x \in R$ and $A \subset R$. This was first proved by S. BANACH [Fund. Math. **4**, 7–33 (1923)]. The construction is sketched in (20.40) *infra*. A far-reaching generalization of BANACH'S result appears in HEWITT and ROSS, *Abstract Harmonic Analysis I* [Springer-Verlag, Heidelberg, 1963], pp. 242–245, to which interested readers are referred.

(b) Countably additive extensions μ of Lebesgue measure to very large σ-algebras $\mathscr{M}$ of subsets of R have been found, retaining the property that $\mu(x + A) = \mu(A)$. One can make $2^{\mathfrak{c}}$ new sets μ-measurable, and in fact there is a family $\mathscr{D} \subset \mathscr{M}$ such that $\overline{\overline{\mathscr{D}}} = 2^{\mathfrak{c}}$ and $\mu(D_1 \triangle D_2) = 1$ for distinct $D_1, D_2 \in \mathscr{D}$. Such extensions are implicit in a construction given by KAKUTANI and OXTOBY [Ann. of Math. (2) **52**, 580–590 (1950)]. They are given explicitly in a construction by HEWITT and ROSS [Math. Annalen, to appear].

(c) For an interpretation of (b) in terms of a certain metric space, see (10.45) and (10.47) below.

We return to our outer measures ι on locally compact Hausdorff spaces, proving some useful facts about ι-measurable sets.

(10.30) Theorem. *Let X be a locally compact Hausdorff space and let ι be as in* § 9. *Let A be an ι-measurable subset of X such that $A \subset \bigcup_{n=1}^{\infty} B_n$ for some sequence $(B_n)_{n=1}^{\infty}$ of sets such that $\iota(B_n) < \infty$ for all n. Then*

$$\iota(A) = \sup\{\iota(F) : F \textit{ is compact, } F \subset A\} .$$

Proof. (I) Suppose first that $\iota(A) < \infty$. Let ε be any positive number. By (9.24) there is an open set V such that $A \subset V$ and $\iota(V) < \iota(A) + \frac{1}{4}\varepsilon$. Since $\iota(V) = \iota(A) + \iota(V \cap A')$, we have $\iota(V \cap A') < \frac{1}{4}\varepsilon$. Using (9.26), select a compact subset E of V such that $\iota(V \cap E') < \frac{1}{2}\varepsilon$. Using (9.24) again, choose an open set W such that $V \cap A' \subset W \subset V$ and $\iota(W) < \frac{1}{2}\varepsilon$. The set $F = E \cap W'$ is compact. It is clear that $F \subset A$, for

$$E \cap W' \subset E \cap (V' \cup A) \subset V \cap (V' \cup A) = A .$$

We have

$$\begin{aligned} \iota(A \cap F') &= \iota(A \cap (E' \cup W)) \leqq \iota(A \cap E') + \iota(A \cap W) \\ &\leqq \iota(V \cap E') + \iota(W) < \frac{1}{2}\varepsilon + \frac{1}{2}\varepsilon = \varepsilon . \end{aligned}$$

Using the ι-measurability of A, we see that $\iota(F) = \iota(A) - \iota(A \cap F') > \iota(A) - \varepsilon$. Since ε is arbitrary, the theorem follows for $\iota(A) < \infty$.

(II) Suppose that $\iota(A) = \infty$. In view of (9.24), we may suppose that the sets B_n in our hypothesis are ι-measurable [in fact, open]. Write $A_n = A \cap (B_1 \cup \cdots \cup B_n)$ for $n \in N$ and $A_0 = \varnothing$. Then A_n is ι-measurable, $\iota(A_n) < \infty$, $A_n \subset A_{n+1}$, and $\bigcup_{n=1}^{\infty} A_n = A$.[1]

By (10.11), $(X, \mathscr{M}_\iota, \iota)$ is a measure space; hence (10.13) implies that

$$\infty = \iota(A) = \lim_{n\to\infty} \iota(A_n) . \tag{1}$$

Using part (I), choose for each $n \in N$ a compact set F_n such that $F_n \subset A_n$ and $\iota(F_n) \geqq \frac{1}{2}\iota(A_n)$. It is plain from (1) that

$$\lim_{n\to\infty} \iota(F_n) = \lim_{n\to\infty} \iota(A_n) = \infty = \iota(A). \quad \square$$

(10.31) Theorem. *Let X be a locally compact Hausdorff space and let ι be as in § 9. For $A \subset X$, the following statements are equivalent:*

(i) *A is ι-measurable;*

(ii) *$\iota(U) \geqq \iota(U \cap A) + \iota(U \cap A')$ for all open sets U such that $\iota(U) < \infty$;*

(iii) *$A \cap U$ is ι-measurable for each open set U such that $\iota(U) < \infty$;*

(iv) *$A \cap F$ is ι-measurable for every compact set F.*

Proof. Theorem (10.20) shows that each compact set F is ι-measurable, since it is closed; and so (i) implies (iv).

Suppose that (iv) holds and let U be an open set such that $\iota(U) < \infty$. Theorem (10.30) shows that for each $n \in N$ there is a compact set $F_n \subset U$ such that $\iota(F_n) > \iota(U) - \frac{1}{n}$. Let $F = \bigcup_{n=1}^{\infty} F_n$. Then we have: $F \subset U$; F is ι-measurable; and $\iota(F) \geqq \iota(F_n) > \iota(U) - \frac{1}{n}$ for each $n \in N$.

---

[1] A set that is the union of a countable family of sets of finite measure is called *σ-finite* [cf. (10.3)]. Recall also: a set that is the union of a countable number of compact sets is called *σ-compact*.

It follows that $\iota(F) = \iota(U)$ and $\iota(U \cap F') = 0$. Hence

$$A \cap U = A \cap [F \cup (U \cap F')] = (A \cap F) \cup (A \cap U \cap F')$$
$$= \left(\bigcup_{n=1}^{\infty} (A \cap F_n)\right) \cup (A \cap U \cap F'),$$

and so, since $\iota(A \cap U \cap F') = 0$, $A \cap U$ is a countable union of ι-measurable sets. Therefore (iv) implies (iii).

Next suppose that (iii) holds and let U be open of finite ι-measure. Then U and $A \cap U$ are both ι-measurable, and so $U \cap A' = U \cap (U \cap A)'$ is ι-measurable. Thus

$$\iota(U) = \iota((U \cap A) \cup (U \cap A')) = \iota(U \cap A) + \iota(U \cap A'),$$

i.e., (iii) implies (ii).

Suppose that (ii) holds, and let T be an arbitrary subset of X. In establishing (10.5), we may suppose that $\iota(T) < \infty$, since otherwise $\iota(T) = \infty \geqq \iota(T \cap A) + \iota(T \cap A')$. For a given $\varepsilon > 0$, choose an open set U such that $T \subset U$ and $\iota(U) < \iota(T) + \varepsilon$. Then (ii) implies that

$$\iota(T) + \varepsilon > \iota(U) \geqq \iota(U \cap A) + \iota(U \cap A') \geqq \iota(T \cap A) + \iota(T \cap A').$$

Since ε is arbitrary, we are through. □

(10.32) Corollary. *If A is locally ι-null, then A is ι-measurable and $\iota(A) = 0$ or ∞.*

Proof. For each compact set F we have $\iota(F \cap A) = 0$ (9.29), and so $F \cap A$ is ι-measurable (10.7). It follows from (10.31) that A is ι-measurable. Suppose that $\iota(A) < \infty$. Applying (10.30), we have $\iota(A) = \sup\{\iota(F): F$ is compact and $F \subset A\} = 0$. □

(10.33) Remark. Since for some choices of X and ι there exist locally ι-null sets which are not ι-null [see (9.41.e)], (10.32) shows that (10.30) cannot in general be strengthened to admit all ι-measurable sets. However if X is a countable union of compact sets [*e.g.* $X = R^n$], then every ι-measurable subset of X satisfies the hypothesis of (10.30).

(10.34) Theorem. *Let X be a locally compact Hausdorff space and let ι be as in* § 9. *For every σ-finite, ι-measurable subset A of X, there are subsets B and C of X such that B is σ-compact, C is a Borel set, the inclusions $B \subset A \subset C$ obtain, and $\iota(C \cap B') = 0$.*

Proof. (I) Suppose that $\iota(A) < \infty$. For each $n \in N$, there is a compact set $F_n \subset A$ such that $\iota(F_n) > \iota(A) - \frac{1}{n}$. Let $B = \bigcup_{n=1}^{\infty} F_n$. For each n we have $\iota(F_n) \leqq \iota(B) \leqq \iota(A)$, and so $\iota(B) = \iota(A)$. Next, for each $n \in N$, select an open set $U_n \supset A$ such that $\iota(U_n) < \iota(A) + \frac{1}{n}$. Let $C = \bigcap_{n=1}^{\infty} U_n$; then C is a G_δ set, and hence clearly a Borel set. It is clear that $\iota(C) = \iota(A)$. Using the ι-measurability of A, we have $\iota(C \cap A')$

$= \iota(C) - \iota(A) = 0$, $\iota(A \cap B') = \iota(A) - \iota(B) = 0$, and so also

$$\iota(C \cap B') = \iota(C \cap A') + \iota(A \cap B') = 0 .$$

(II) If $\iota(A) = \infty$, then, as in the proof of (10.30), write $A = \bigcup_{n=1}^{\infty} A_n$ where each A_n is measurable and has finite measure. By case (I), there are σ-compact sets B_n and G_δ sets C_n, $n = 1, 2, 3, \ldots$, such that $B_n \subset A_n \subset C_n$ and $\iota(C_n \cap B_n') = 0$. Let $B = \bigcup_{n=1}^{\infty} B_n$; B is clearly σ-compact. We have

$$A \cap B' = \left(\bigcup_{n=1}^{\infty} A_n\right) \cap \left(\bigcup_{k=1}^{\infty} B_k\right)' = \left(\bigcup_{n=1}^{\infty} A_n\right) \cap \left(\bigcap_{k=1}^{\infty} B_k'\right) \subset \bigcup_{n=1}^{\infty} (A_n \cap B_n') ,$$

and so

$$\iota(A \cap B') \leqq \iota\left(\bigcup_{n=1}^{\infty} (A_n \cap B_n')\right) \leqq \sum_{n=1}^{\infty} \iota(A_n \cap B_n') = 0 .$$

Now let $C = \bigcup_{n=1}^{\infty} C_n$; C is clearly a Borel set. The argument given above to prove that $\iota(A \cap B') = 0$ can be used to prove that $\iota(C \cap A') = 0$, and as in part (I) it follows that $\iota(C \cap B') = 0$. □

The functional $\bar{I}$ of § 9 satisfies the inequality $\bar{I}(f+g) \leqq \bar{I}(f) + \bar{I}(g)$ for all $f, g \in \mathfrak{F}^+$ (9.18). It is possible to exhibit functionals I and functions f and g such that strict inequality holds [see (10.41)]; *i.e.*, $\bar{I}$ is not in general additive on $\mathfrak{F}^+$. However $\bar{I}$ is additive on special classes of functions, and we now exhibit one such class.

(10.35) Theorem. *Let A and B be disjoint ι-measurable sets and let α and β be nonnegative real numbers. Then we have*

(i) $\bar{I}(\alpha \xi_A + \beta \xi_B) = \alpha \bar{I}(\xi_A) + \beta \bar{I}(\xi_B)$.

Proof. By the subadditivity of $\bar{I}$, it obviously suffices to prove that

$$\bar{I}(\alpha \xi_A + \beta \xi_B) \geqq \alpha \bar{I}(\xi_A) + \beta \bar{I}(\xi_B) . \tag{1}$$

The inequality (1) is easy to verify if $\alpha = 0$ or $\beta = 0$, or if $\iota(A) = 0$ or $\iota(A) = \infty$, or if $\iota(B) = 0$ or $\iota(B) = \infty$. We leave these verifications to the reader, and prove (1) under the hypothesis that $\alpha\beta > 0$, $0 < \iota(A) < \infty$, and $0 < \iota(B) < \infty$.

(I) Suppose that A and B are compact. By (6.80), there is a continuous real-valued function φ on X such that $\varphi(A) = \{0\}$ and $\varphi(B) = \{1\}$. The sets $\left\{x \in X : \varphi(x) < \frac{1}{3}\right\}$ and $\left\{x \in X : \varphi(x) > \frac{2}{3}\right\}$ are open disjoint sets containing A and B, respectively. Since A and B have finite measure, they are contained in open sets having finite measure. Taking the intersections of these open sets with those defined by φ, we obtain open sets U_0 and V_0 such that $U_0 \supset A$ and $V_0 \supset B$, $U_0 \cap V_0 = \varnothing$, $0 < \iota(U_0) < \infty$, and $0 < \iota(V_0) < \infty$. We have

$$\bar{I}(\alpha \xi_A + \beta \xi_B) \leqq \alpha \bar{I}(\xi_A) + \beta \bar{I}(\xi_B) = \alpha \iota(A) + \beta \iota(B) < \infty .$$

Now choose $\varepsilon > 0$. There is a function $f \in \mathfrak{M}^+$ such that $f \geqq \alpha\xi_A + \beta\xi_B$ and

$$\bar{I}(f) - \frac{1}{3}\varepsilon < \bar{I}(\alpha\xi_A + \beta\xi_B)\,.$$

Choose $\delta > 0$ such that

$$0 < \delta < \min\left\{\frac{\varepsilon}{3\,\iota(U_0)}, \frac{\varepsilon}{3\,\iota(V_0)}, \alpha, \beta\right\}.$$

For all $x \in A$, we have $f(x) \geqq \alpha$. By the lower semicontinuity of f, there is an open set U such that $A \subset U \subset U_0$ and $f(x) > \alpha - \delta$ for all $x \in U$. Similarly, there is an open set V such that $B \subset V \subset V_0$ and $f(x) > \beta - \delta$ for all $x \in V$. Thus we have $f \geqq (\alpha - \delta)\,\xi_U + (\beta - \delta)\,\xi_V$; therefore

$$\begin{aligned}\bar{I}(f) &\geqq \bar{I}((\alpha-\delta)\,\xi_U + (\beta-\delta)\,\xi_V) = (\alpha-\delta)\,\bar{I}(\xi_U) + (\beta-\delta)\,\bar{I}(\xi_V)\\ &= (\alpha-\delta)\,\iota(U) + (\beta-\delta)\,\iota(V) \geqq \alpha\iota(A) + \beta\iota(B) - \delta(\iota(U) + \iota(V))\\ &\geqq \alpha\iota(A) + \beta\iota(B) - \delta(\iota(U_0) + \iota(V_0)) > \alpha\iota(A) + \beta\iota(B) - \frac{2}{3}\varepsilon\,.\end{aligned}$$

Summarizing, we have shown that

$$\alpha\bar{I}(\xi_A) + \beta\bar{I}(\xi_B) - \varepsilon < \bar{I}(f) - \frac{1}{3}\varepsilon < \bar{I}(\alpha\xi_A + \beta\xi_B)\,.$$

Since ε is arbitrary, it follows that (1) holds if A and B are compact.

(II) We suppose now that A and B are arbitrary ι-measurable sets of finite positive measure. Choose $\varepsilon > 0$. Using (10.30), choose compact sets E and F such that $E \subset A$ and $F \subset B$, $\alpha\iota(E) > \alpha\iota(A) - \frac{1}{2}\varepsilon$, and $\beta\iota(F) > \beta\iota(B) - \frac{1}{2}\varepsilon$. Using part (I) for the compact sets E and F, we have

$$\begin{aligned}\bar{I}(\alpha\xi_A + \beta\xi_B) \geqq \bar{I}(\alpha\xi_E + \beta\xi_F) &= \alpha\iota(E) + \beta\iota(F) > \alpha\iota(A) + \beta\iota(B) - \varepsilon\\ &= \alpha\bar{I}(\xi_A) + \beta\bar{I}(\xi_B) - \varepsilon\,. \quad \square\end{aligned}$$

We close this section with a large collection of exercises. A few of these [for example (10.37)] are actually needed for subsequent theorems in the main text. The rest of the exercises illustrate and extend the theory in various directions, and we trust that all serious readers will work through most of them.

(10.36) Exercise. Let X be an arbitrary set and $\mathscr{A}$ an algebra of subsets of X [$\mathscr{A}$ need not be a σ-algebra]. Let μ be a countably additive measure on $\mathscr{A}$ in the sense of (10.3). Define a set function $\bar{\mu}$ on $\mathscr{P}(X)$ as follows: for $T \subset X$, let

$$\bar{\mu}(T) = \inf\left\{\sum_{n=1}^{\infty} \mu(A_n) : T \subset \bigcup_{n=1}^{\infty} A_n \text{ and } A_1, A_2, \ldots, A_n, \ldots \in \mathscr{A}\right\}.$$

(a) Prove that $\bar{\mu}$ is an outer measure on $\mathscr{P}(X)$.
(b) Prove that $\bar{\mu}$ is equal to μ on the algebra $\mathscr{A}$.
(c) Prove that all elements of $\mathscr{A}$ are measurable with respect to $\bar{\mu}$.

(d) Prove that μ can be extended to a countably additive measure defined on a σ-algebra of subsets of X that contains $\mathscr{A}$. [The fact that this can be done is called *E. Hopf's extension theorem.*]

(10.37) Exercise. Let X and $\mathscr{A}$ be as in (10.36). Let γ be a set function on $\mathscr{A}$ satisfying the following conditions:

(i) $0 \leqq \gamma(A) \leqq \infty$ for all $A \in \mathscr{A}$;

(ii) $\gamma(A \cup B) = \gamma(A) + \gamma(B)$ for $A, B \in \mathscr{A}$ and $A \cap B = \varnothing$;

(iii) if $A_1, A_2, \ldots \in \mathscr{A}$, if $A_1 \supset A_2 \supset \cdots \supset A_n \supset \cdots$, and $\bigcap_{n=1}^{\infty} A_n = \varnothing$, then $\lim_{n\to\infty} \gamma(A_n) = 0$.

Define $\bar{\gamma}$ just as in (10.36). Prove that (a), (b), (c), and (d) of (10.36) hold for the set functions γ and $\bar{\gamma}$. [This is another version of E. Hopf's extension theorem.]

(10.38) Exercise. Let $(X, \mathscr{A}, \mu)$ be a measure space. Prove that μ can be extended to a measure $\bar{\mu}$ on a σ-algebra $\bar{\mathscr{A}}$ such that every subset of every set of μ-measure 0 is $\bar{\mu}$-measurable and has $\bar{\mu}$-measure 0.

The following exercise will be needed in the sequel to prove two important theorems of the main text [(20.56) and (20.57)], and so we spell out the proof in some detail.

(10.39) Exercise. Let X, $\mathscr{A}$, μ, and $\bar{\mu}$ be as in (10.36). Let $\mathscr{S} = \mathscr{S}(\mathscr{A})$ be the smallest σ-algebra of subsets of X that contains the algebra $\mathscr{A}$. Suppose that $(X, \mathscr{S}, \nu)$ is a measure space such that $\nu(A) = \mu(A)$ for all $A \in \mathscr{A}$. Prove the following.

(a) If $B \in \mathscr{S}$, then $\bar{\mu}(B) \geqq \nu(B)$. [Hint. Let $\mathscr{A}_\sigma$ denote the family of all countable unions of sets in $\mathscr{A}$. If $A = \bigcup_{n=1}^{\infty} A_n \in \mathscr{A}_\sigma$ where $\{A_n\}_{n=1}^{\infty} \subset \mathscr{A}$, then $A = A_1 \cup \bigcup_{n=2}^{\infty} (A_n \cap A_1' \cap A_2' \cap \cdots \cap A_{n-1}')$ is a disjoint union of sets in $\mathscr{A}$, and so $\nu(A) = \bar{\mu}(A)$. Thus

$$\bar{\mu}(B) = \inf\{\bar{\mu}(A) : B \subset A \in \mathscr{A}_\sigma\} = \inf\{\nu(A) : B \subset A \in \mathscr{A}_\sigma\} \geqq \nu(B).]$$

(b) If $F \in \mathscr{S}$ and $\bar{\mu}(F) < \infty$, then $\nu(F) = \bar{\mu}(F)$. [Hint. For $\varepsilon > 0$, choose $A \in \mathscr{A}_\sigma$ such that $F \subset A$ and $\bar{\mu}(A) < \bar{\mu}(F) + \varepsilon$. Then use (a) to show that

$$\bar{\mu}(F) \leqq \bar{\mu}(A) = \nu(A) = \nu(F) + \nu(A \cap F') \leqq \nu(F) + \bar{\mu}(A \cap F') < \nu(F) + \varepsilon.]$$

(c) If there is a sequence $(F_n)_{n=1}^{\infty} \subset \mathscr{A}$ such that $\mu(F_n) < \infty$ for all n and $X = \bigcup_{n=1}^{\infty} F_n$, then $\nu(E) = \bar{\mu}(E)$ for all $E \in \mathscr{S}$, *i.e.*, the extension of μ to $\mathscr{S}$ is unique. [Hint. We may suppose that $F_n \cap F_m = \varnothing$ for $n \neq m$. Then by (b) we have

$$\nu(E) = \sum_{n=1}^{\infty} \nu(E \cap F_n) = \sum_{n=1}^{\infty} \bar{\mu}(E \cap F_n) = \bar{\mu}(E).]$$

(d) If the σ-finiteness hypothesis in (c) fails, then μ may have more than one extension to $\mathcal{S}$. [Hint. Let $X = [0, 1[$ and let $\mathcal{A}$ be the algebra of all finite unions of intervals of the form $[a, b[\subset [0, 1[$. Define μ on $\mathcal{A}$ by $\mu(\varnothing) = 0$ and $\mu(A) = \infty$ if $A \neq \varnothing$. Show that there are exactly $2^{\mathfrak{c}}$ countably additive measures on the Borel sets of $[0, 1[$ that agree with μ on $\mathcal{A}$.]

(10.40) Exercise. An outer measure μ on $\mathcal{P}(X)$ is said to be *regular* if for each $E \subset X$, there exists a μ-measurable set $A \subset X$ such that $E \subset A$ and $\mu(A) = \mu(E)$.

(a) Prove that any outer measure obtained from a measure on an algebra of sets as in (10.36) is a regular outer measure.

(b) Prove that if X is a locally compact Hausdorff space and ι is as in § 9, then ι is a regular outer measure.

(c) Let $X = \{0, 1\}$. Construct an irregular outer measure on $\mathcal{P}(X)$.

(d) Let μ be a regular outer measure on $\mathcal{P}(X)$, let $E \subset X$, and let $\mathcal{A}$ be the smallest σ-algebra containing $\{E\} \cup \mathcal{M}_\mu$. Prove that if μ is finitely additive on $\mathcal{A}$, then $E \in \mathcal{M}_\mu$.

(e) Notice that, since λ is a regular outer measure on R, none of the extensions of λ [as a measure on $\mathcal{M}_\lambda$] mentioned in (10.29.b) can agree with the outer measure λ except at sets in $\mathcal{M}_\lambda$.

(10.41) Exercise. Let A be a subset of $[0, 1]$ that is not λ-measurable, and let $B = [0, 1] \cap A'$. Prove that $\overline{S}(\xi_A + \xi_B) < \overline{S}(\xi_A) + \overline{S}(\xi_B)$, where S is the Riemann integral.

(10.42) Exercise. Let X and Y be topological spaces. Prove the following.

(a) If f is a continuous function from X into Y and if $B \in \mathcal{B}(Y)$, then $f^{-1}(B) \in \mathcal{B}(X)$. [Consider the family of all sets B for which the assertion holds.]

(b) If $A \in \mathcal{B}(X)$ and $B \in \mathcal{B}(Y)$, then $A \times B \in \mathcal{B}(X \times Y)$. [Recall the definition of the product topology (6.41) and use (a).]

(c) Generalize (b) to products of countably many topological spaces.

(10.43) Exercise [H. Steinhaus]. Let T be a λ-measurable set in R such that $\lambda(T) > 0$. Prove that the set $T - T$ contains an interval $[-\alpha, \alpha]$ $(\alpha > 0)$. The following steps may be useful.

(a) If U and V are open in R and have finite λ-measure, the function $x \to \lambda((x + U) \cap V)$ is continuous on R. [Begin with intervals and use (6.59) for general U and V.]

(b) If A and B are λ-measurable of finite λ-measure, then $x \to \lambda((x + A) \cap B)$ is continuous. [For $U \supset A$ and $V \supset B$, prove first that

$$|\lambda((x + U) \cap V) - \lambda((x + A) \cap B)| \leqq \lambda(U \cap A') + \lambda(V \cap B').]$$

(c) The set $T - T$ contains an interval $[-\alpha, \alpha]$. [The function $x \to \lambda((x + T) \cap T)$ is positive at 0, and if $(x + T) \cap T \neq \varnothing$, then $x \in T - T$.]

(10.44) Exercise. Generalize (10.43) to $A + B$, where A, B are in $\mathscr{M}_\lambda$ and both have positive measure.

(10.45) Exercise. Let $(X, \mathscr{M}, \mu)$ be a measure space. Define φ on $[0, \infty]$ by letting $\varphi(t) = 1 - \exp(-t)$ if $0 \leqq t < \infty$ and putting $\varphi(\infty) = 1$. For $A, B \in \mathscr{M}$, define

$$\varrho(A, B) = \varphi(\mu(A \triangle B)) .$$

(a) Identifying sets A and B for which $\mu(A \triangle B) = 0$, prove that $(\mathscr{M}, \varrho)$ is a complete metric space.

(b) Show that the mappings from $\mathscr{M} \times \mathscr{M}$ to $\mathscr{M}$ with values $A \cup B$, $A \triangle B$, and $A \cap B$ at (A, B) are continuous. Show also that $A \to A'$ is continuous from $\mathscr{M}$ to $\mathscr{M}$.

(10.46) Exercise. Prove that the metric space $(\mathscr{M}, \varrho)$ defined in (10.45) is not compact in the case that $X = [0, 1]$, $\mathscr{M} = \mathscr{B}([0, 1])$, and $\mu = \lambda$.

(10.47) Exercise. (a) Let X be a locally compact Hausdorff space and let ι be a measure on X as in §9. Prove that if there exists a countable base for the topology of X, then the metric space $(\mathscr{M}_\iota, \varrho)$ defined as in (10.45) is separable.

(b) Note that the metric space $(\mathscr{M}_\lambda, \varrho)$ has a countable dense subset, where $\mathscr{M}_\lambda$ is the σ-algebra of Lebesgue measurable subsets of R and λ is the measure used to define ϱ, as in (10.45). Find the smallest cardinal number of a dense subset of $(\mathscr{M}, \varrho)$ for the invariant extension μ of Lebesgue measure described in (10.29.b).

(10.48) Exercise. Let X be a metric space with metric ϱ and let μ be any outer measure on $\mathscr{P}(X)$ such that if $A, B \subset X$, $A \neq \varnothing$, $B \neq \varnothing$, and $\varrho(A, B) > 0$, then $\mu(A \cup B) = \mu(A) + \mu(B)$. Such outer measures are called *metric outer measures*. Let U be an open proper subset of X and let A be a nonvoid subset of U. For each $n \in N$ define $A_n = \left\{x \in A : \varrho(x, U') \geqq \frac{1}{n}\right\}$. Prove that:

(a) $\lim\limits_{n \to \infty} \mu(A_n) = \mu(A)$

[consider the sets $D_{2n} = A_{2n} \cap A'_{2n-1}$ and $D_{2n+1} = A_{2n+1} \cap A'_{2n}$];

(b) U is μ-measurable;

(c) $\mathscr{B}(X) \subset \mathscr{M}_\mu$.

(10.49) Exercise: Construction of a class of outer measures. Let X be a separable metric space, let $\mathscr{O}$ be the family of all open sets in X, and let p be a positive real number. For each $\varepsilon > 0$ let $\mathscr{O}_\varepsilon = \{U \in \mathscr{O}:$

$\operatorname{diam} U \leqq \varepsilon\} \cup \{\varnothing\}$. For each $E \subset X$, define

$$\mu_{p,\varepsilon}(E) = \inf\left\{\sum_{n=1}^{\infty} (\operatorname{diam} U_n)^p : U_n \in \mathcal{O}_\varepsilon,\ E \subset \bigcup_{n=1}^{\infty} U_n\right\},$$

where we define $\operatorname{diam} \varnothing = 0$.

(a) Prove that $\mu_{p,\varepsilon}(E)$ is nondecreasing as ε decreases.

Define $\mu_p(E) = \lim\limits_{\varepsilon \downarrow 0} \mu_{p,\varepsilon}(E)$ for each $E \subset X$.

(b) Prove that μ_p is a metric outer measure on X.

(c) Prove that if $\mu_p(E) < \infty$ and $q > p$, then $\mu_q(E) = 0$.

The set function μ_p is called the *Hausdorff p-dimensional* [*outer*] *measure* on X. For $E \subset X$ we define the *Hausdorff dimension* of E to be the number $\sup\{p \in R : p > 0,\ \mu_p(E) = \infty\}$, where we let $\sup \varnothing = 0$.

(10.50) Exercise. Let R have its usual metric. We consider Hausdorff measures μ_p on R [see (10.49)]. For $E \subset R$, let $\dim E$ be the Hausdorff dimension of E. Prove that

(a) $\mu_1 = \lambda$;

(b) $\dim U = 1$ for all nonvoid open sets $U \subset R$;

(c) $\dim E = 0$ implies $\lambda(E) = 0$;

(d) if P is CANTOR's ternary set, then $\dim P = \dfrac{\log 2}{\log 3}$ [consider the sets P_n of (6.62)];

(e) there is an uncountable subset E of R such that $\dim E = 0$.

(10.51) Exercise: Another class of outer measures. Let (X, ϱ) be a metric space. For a nonvoid set $E \subset X$ and $t > 0$, define $n(E, t)$ as follows:

$$n(E, t) = 1 \quad \text{if } \varrho(x, x') \leqq t \quad \text{for all} \quad x, x' \in E;$$

$n(E, t) = \sup\{\overline{\overline{F}} : F \subset E,\ F \text{ is finite},\ \varrho(x, x') > t \text{ for distinct } x, x' \in F\}$ if this supremum is finite;

$n(E, t) = \infty$ in all other cases.

Define $n(\varnothing, t)$ as zero. Let φ be a real-valued, positive, strictly decreasing function defined on $]0, 1]$ such that $\lim\limits_{t \downarrow 0} \varphi(t) = \infty$. For all $E \subset X$, define

$$\operatorname{ext}_\varphi(E) = \varliminf_{t \downarrow 0} \frac{n(E, t)}{\varphi(t)}.$$

For all $E \subset X$, define

$$\nu_\varphi(E) = \inf\left\{\sum_{k=1}^{\infty} \operatorname{ext}_\varphi(A_k)\right\},$$

where the infimum is taken over all countable, pairwise disjoint families of sets $\{A_k\}_{k=1}^{\infty}$ such that $\bigcup\limits_{k=1}^{\infty} A_k = E$.

(a) Prove that ν_φ is a metric outer measure as defined in (10.48). [Hint. The only nontriviality is showing that $\nu_\varphi(A \cup B) = \nu_\varphi(A) + \nu_\varphi(B)$

if $\varrho(A, B) > 0$. This follows from the equality $n(A \cup B, t) = n(A, t) + n(B, t)$, which is valid for $t < \varrho(A, B)$.]

(b) Compute ν_φ for $X = R$ with the usual metric for R and $\varphi(t) = \frac{1}{t}$.

(c) Compare the outer measures ν_φ [with suitable φ!] with Hausdorff p-dimensional measures.

(d) Prove that $\nu_\varphi(E) = 0$ if E is countable.

(e) Prove that $\nu_\varphi(E) = \nu_\varphi(F)$ if there is an isometry of E onto F. [An *isometry* is a mapping ψ of one metric space onto another such that $\varrho(x, y) = \varrho'(\psi(x), \psi(y))$, ϱ and ϱ' being the metrics on the two spaces.]

(10.52) Exercise. Let α be any real-valued nondecreasing function on R and let λ_α be the Lebesgue-Stieltjes measure on R induced by the Riemann-Stieltjes integral as in § 9. Prove that $\lambda_\alpha(\{x\}) = 0$ for $x \in R$ if and only if α is continuous at x.

(10.53) Exercise. Let α and λ_α be as in (10.52) and suppose that α is continuous. Prove the following assertions.

(a) For each $\varepsilon > 0$ there exists a nowhere dense perfect set $A \subset [0, 1]$ such that $\lambda_\alpha(A) > \lambda_\alpha([0, 1]) - \varepsilon$.

(b) There exists an F_σ set $B \subset [0, 1]$ such that B is of first category and $\lambda_\alpha(B) = \lambda_\alpha([0, 1])$.

(c) There exists a G_δ set of second category contained in $[0, 1]$ having λ_α-measure zero.

(10.54) Exercise. In this exercise, we first sketch the construction of a subset B of R measurable for no measure λ_α with continuous α.

(a) Prove that every uncountable closed subset F of R has cardinal number $\mathfrak{c}$. [Use (6.65) and (6.66).]

(b) [F. BERNSTEIN]. Prove that there is a subset B of R such that $B \cap F \neq \varnothing$ and $B' \cap F \neq \varnothing$ for every uncountable closed subset F of R. [Hints. There are just $\mathfrak{c}$ open subsets of R and hence just $\mathfrak{c}$ uncountable closed subsets. Let $\omega_\mathfrak{c}$ be the smallest ordinal number with corresponding cardinal number $\mathfrak{c}$ [use (4.47) to show that $\omega_\mathfrak{c}$ exists]. Let $\{F_\eta : \eta < \omega_\mathfrak{c}\}$ be a well ordering of all uncountable closed subsets of R. Define B by transfinite recursion and the axiom of choice, as follows. Let x_0 and y_0 be any two distinct points in F_0. Suppose that x_γ and y_γ have been defined for all $\gamma < \eta$, where $\eta < \omega_\mathfrak{c}$. The set $A_\eta = \{x_\gamma : \gamma < \eta\} \cup \{y_\gamma : \gamma < \eta\}$ has cardinal number $< \mathfrak{c}$, because $\omega_\mathfrak{c}$ is the *smallest* ordinal number of cardinal $\mathfrak{c}$. Hence the set $F_\eta \cap A'_\eta$ has cardinal number $\mathfrak{c}$. Let x_η and y_η be any two distinct points in the set $F_\eta \cap A'_\eta$. Finally let $B = \{x_\eta : \eta < \omega_\mathfrak{c}\}$. It is clear that $B \cap F_\eta \neq \varnothing$ and that $B' \cap F_\eta \neq \varnothing$ for all $\eta < \omega_\mathfrak{c}$.]

(c) Prove that B is non λ_α-measurable if α is continuous and $\lambda_\alpha \neq 0$. [Hints. Assume that B is λ_α-measurable. Then by (10.30), we have $\lambda_\alpha(B) = \sup\{\lambda_\alpha(F) : F \text{ is compact}, F \subset B\}$. The only compact subsets of B are countable, and since $\lambda_\alpha(\{x\}) = 0$ for all $x \in R$, it follows that

$\lambda_\alpha(B) = 0$. Similarly $\lambda_\alpha(B') = 0$, and so if B is λ_α-measurable, λ_α is the zero measure.]

Throughout the remainder of this exercise, assume the continuum hypothesis, *i.e.*, $\aleph_1 = \mathfrak{c}$ [see (4.49) and (4.50)].

(d) Prove that there exists an indexing $\{C_\eta : 0 \leqq \eta < \Omega\}$ of the family of all nowhere dense closed subsets of $[0, 1]$ by the set P_Ω of all countable ordinal numbers.

Define [by transfinite recursion and the axiom of choice] a set $S = \{x_\eta : 0 \leqq \eta < \Omega\}$ as follows: let $x_0 \in [0, 1] \cap C_0'$ and

$$x_\eta \in [0, 1] \cap (\bigcup_{\theta < \eta} (C_\theta \cup \{x_\theta\}))'.$$

(e) Prove that $\overline{\overline{S}} = \mathfrak{c}$ and that $S \cap C_\eta$ is countable for all $\eta \in P_\Omega$.

(f) Prove that $\lambda_\alpha(S) = 0$ for all Lebesgue-Stieltjes measures λ_α such that α is continuous.

(10.55) Remark. The set S defined above is not a Borel set in R. In fact, it is known that each uncountable Borel set in a complete separable metric space contains a nonvoid perfect set [see W. Sierpiński, *loc. cit.* (10.21.b), p. 228].

(10.56) Exercise. (a) Let $(X, \mathscr{A}, \mu)$ be a measure space such that $0 < \mu(X) < \infty$ and μ assumes only a finite number of distinct positive values. Prove that $X = E_1 \cup \cdots \cup E_n \cup F$, where the summands are $\mathscr{A}$-measurable and pairwise disjoint and have the following properties. There is a nondecreasing sequence $(\alpha_k)_{k=1}^n$ of positive numbers such that if $A \in \mathscr{A}$ and $A \subset E_k$, then $\mu(A) = 0$ or $\mu(A) = \alpha_k$; $\mu(E_k) = \alpha_k$; and $\mu(F) = 0$. [Hint. Let α_1 be the least positive value assumed by μ and let E_1 be any set in $\mathscr{A}$ for which $\mu(E_1) = \alpha_1$. Consider E_1' and proceed by induction.]

(b) Let X be an uncountable set, $\mathscr{A}$ the family $\{A : A \subset X,\ A$ is countable or A' is countable$\}$, and μ on $\mathscr{A}$ defined by $\mu(A) = 1$ if A' is countable and $\mu(A) = 0$ if A is countable. Prove that $(X, \mathscr{A}, \mu)$ is a measure space. Use this example to show that the decomposition described in (*a*) need not be unique.

(c) Let $(X, \mathscr{A}, \mu)$ be a measure space such that $0 < \mu(X) < \infty$ and μ assumes infinitely many distinct values. Show that there is an infinite pairwise disjoint family $\{A_n\}_{n=1}^\infty \subset \mathscr{A}$ such that $0 < \mu(A_n) < \infty$ for all n.

(d) Let $(X, \mathscr{A}, \mu)$ be a measure space such that every $A \in \mathscr{A}$ such that $\mu(A) = \infty$ contains a set $B \in \mathscr{A}$ such that $0 < \mu(B) < \infty$. Then every such A contains a set $C \in \mathscr{A}$ such that $\mu(C) = \infty$ and C is the union of a countable number of sets of finite measure. [Hints. Let $\alpha = \sup\{\mu(B) : B \in \mathscr{A}, B \subset A, \mu(B) < \infty\}$. Let $(B_n)_{n=1}^\infty$ be a nondecreasing sequence of sets in $\mathscr{A}$ such that $\mu(B_n) < \infty$, $B_n \subset A$, and $\lim_{n\to\infty} \mu(B_n) = \alpha$. Let $C = \bigcup_{n=1}^{\infty} B_n$. The assumption $\mu(C) < \infty$ leads at once to a contradiction.]

(10.57) Exercise. Let X be a set and let $\mathscr{L}$ be a family of subsets of X such that $A \cup B \in \mathscr{L}$ and $A \cap B \in \mathscr{L}$ if $A, B \in \mathscr{L}$. Suppose also that $X \in \mathscr{L}$ and $\varnothing \in \mathscr{L}$. Such a family $\mathscr{L}$ is called a *lattice of sets* [with *unit* and *zero*]. Let $\mathscr{D}$ be the family of all proper differences of sets in $\mathscr{L}$, *i.e.*, $\mathscr{D} = \{B \cap A' : A, B \in \mathscr{L}, A \subset B\}$. Finally, let $\mathscr{U}$ be the family of all finite disjoint unions of sets in $\mathscr{D}$. Prove the following.

(a) If $D_1, D_2 \in \mathscr{D}$, then $D_1 \cap D_2 \in \mathscr{D}$. [If $D_j = B_j \cap A_j'$ $(j = 1, 2)$, then $D_1 \cap D_2 = (B_1 \cap B_2) \cap ((A_1 \cap B_2) \cup (A_2 \cap B_1))'$.]

(b) If $U_1, U_2 \in \mathscr{U}$, then $U_1 \cap U_2 \in \mathscr{U}$.

(c) If $U \in \mathscr{U}$, then $U' \in \mathscr{U}$. [Use induction on the number n, where $U = D_1 \cup D_2 \cup \cdots \cup D_n$.]

(d) The family $\mathscr{U}$ is the smallest algebra of subsets of X containing $\mathscr{L}$.

Let $\mathscr{A}$ be a σ-algebra of subsets of X such that $\mathscr{L} \subset \mathscr{A}$ and let μ and ν be two measures defined on $\mathscr{A}$.

(e) If $\mu(A) = \nu(A)$ for all $A \in \mathscr{L}$ and if there exists a sequence $(A_n) \subset \mathscr{L}$ such that $\bigcup_{n=1}^{\infty} A_n = X$ and $\mu(A_n) < \infty$ for all $n \in N$, then $\mu(E) = \nu(E)$ for all $E \in \mathscr{S}(\mathscr{L})$. [First show that μ and ν agree on $\mathscr{U}$ and then use (10.39).]

(10.58) Exercise. Use (10.57) to prove the following.

(a) If X is a topological space and μ and ν are two finite measures defined on $\mathscr{B}(X)$ that agree on (1) the family of all open sets, or (2) the family of all closed sets, or (3) the family of all compact sets [in the case that X is σ-compact], then μ and ν agree on $\mathscr{B}(X)$.

(b) If X is a metric space and μ is a finite measure defined on $\mathscr{B}(X)$, then $\mu(E) = \inf\{\mu(U) : U \text{ is open}, E \subset U\}$ for all $E \in \mathscr{B}(X)$. [Define $\nu(E) = \inf\{\mu(U) : U \text{ is open}, E \subset U\}$ for all $E \subset X$. Show that ν is a metric outer measure (10.48) and use (a).]

§ 11. Measurable functions

(11.1) Introduction. As was pointed out in (10.40.d), the outer measures ι constructed in § 9 from nonnegative linear functionals I need not be even finitely additive on all sets. However, we have learned that they are in fact countably additive on their σ-algebras of measurable sets. In like manner, we cannot expect that the extensions $\overline{\overline{I}}$ should be finitely additive on all nonnegative functions [see (10.41)]. In this section we construct a large class of functions on which the functionals $\overline{\overline{I}}$ are countably additive. [The countable additivity will be proved in § 12.] These so-called *measurable functions* bear a relationship to the family of all functions which is analogous in many ways to the relation between measurable sets and the family of all sets.

Throughout this section, X will denote an arbitrary set and $\mathscr{A}$ will denote an arbitrary σ-algebra of subsets of X. The ordered pair $(X, \mathscr{A})$ is called a *measurable space*[1].

(11.2) Definition. Let f be an extended real-valued function defined on X. Suppose that $f^{-1}(]a, \infty]) \in \mathscr{A}$ for every $a \in R$, *i.e.*, $\{x \in X : a < f(x) \leqq \infty\} \in \mathscr{A}$ for all real numbers a. Then f is said to be an $\mathscr{A}$-*measurable function.* [The reader should notice that this definition closely resembles the definition of lower semicontinuity (7.21.d).] If X is a topological space and $\mathscr{A}$ is the σ-algebra $\mathscr{B}(X)$ of Borel sets, then any $\mathscr{B}(X)$-measurable function is said to be *Borel measurable.* If $X = R$ and $\mathscr{A} = \mathscr{M}_\lambda$, then an $\mathscr{M}_\lambda$-measurable function is called a *Lebesgue measurable function.* [Notice that the definition of measurable function depends in no way upon any measure, but only upon a particular σ-algebra.]

(11.3) Theorem. *Let D be any dense subset of R [that is, $D^- = R$]. The following conditions on an extended real-valued function f with domain X are equivalent:*

(i) *f is $\mathscr{A}$-measurable;*
(ii) $f^{-1}(]a, \infty]) \in \mathscr{A}$ *for all* $a \in D$;
(iii) $f^{-1}([a, \infty]) \in \mathscr{A}$ *for all* $a \in D$;
(iv) $f^{-1}([-\infty, a[) \in \mathscr{A}$ *for all* $a \in D$;
(v) $f^{-1}([-\infty, a]) \in \mathscr{A}$ *for all* $a \in D$.

Proof. It is trivial that (i) implies (ii). To see that (ii) implies (iii), let $a \in D$ and let (a_n) be a strictly increasing sequence in D such that $a_n \to a$. Then we have $f^{-1}([a, \infty]) = \bigcap_{n=1}^{\infty} f^{-1}(]a_n, \infty])$. To see that (iii) implies (iv), observe that $f^{-1}([-\infty, a[) = (f^{-1}([a, \infty]))'$. The proof that (iv) implies (v) is similar to the proof that (ii) implies (iii). It remains only to show that (v) implies (i). For $a \in R$, choose a strictly decreasing sequence (b_n) in D such that $b_n \to a$. Then

$$f^{-1}(]a, \infty]) = \left(\bigcap_{n=1}^{\infty} f^{-1}([-\infty, b_n])\right)' . \quad \square$$

(11.4) Theorem. *Let f be an extended real-valued function having domain X. Then f is $\mathscr{A}$-measurable if and only if*

(i) $f^{-1}(\{-\infty\})$ *and* $f^{-1}(\{\infty\})$ *are both in* $\mathscr{A}$,

and

(ii) $f^{-1}(B) \in \mathscr{A}$ *for every* $B \in \mathscr{B}(R)$.

Proof. Since $]a, \infty[\in \mathscr{B}(R)$ and $]a, \infty] =]a, \infty[\cup \{\infty\}$ for every $a \in R$, it is clear that (i) and (ii) imply that f is $\mathscr{A}$-measurable.

---

[1] A purist might cavil at the term "measurable space", as there is absolutely no guarantee that a nontrivial measure exists on $\mathscr{A}$. We use the term *faute de mieux.*

Conversely, suppose that f is $\mathscr{A}$-measurable. Then

$$f^{-1}(\{-\infty\}) = \bigcap_{n=1}^{\infty} f^{-1}([-\infty, -n]) \in \mathscr{A} \text{ and } f^{-1}(\{\infty\}) = \bigcap_{n=1}^{\infty} f^{-1}(]n, \infty]) \in \mathscr{A}.$$

Thus (i) obtains. To prove (ii), let $\mathscr{S} = \{S \subset R : f^{-1}(S) \in \mathscr{A}\}$. We will show that $\mathscr{S}$ is a σ-algebra of subsets of R. Clearly $\varnothing \in \mathscr{S}$. If (S_n) is any sequence in $\mathscr{S}$, then $f^{-1}\left(\bigcup_{n=1}^{\infty} S_n\right) = \bigcup_{n=1}^{\infty} f^{-1}(S_n) \in \mathscr{A}$; thus countable unions of sets in $\mathscr{S}$ are again in $\mathscr{S}$. If $S \in \mathscr{S}$, then

$$f^{-1}(R \cap S') = [f^{-1}(S) \cup f^{-1}(\{-\infty\}) \cup f^{-1}(\{\infty\})]' \in \mathscr{A};$$

thus $\mathscr{S}$ is closed under complementation. It follows that $\mathscr{S}$ is a σ-algebra of subsets of R. We next show that $\mathscr{S}$ contains every open subset of R. Indeed, since f is $\mathscr{A}$-measurable, $R \in \mathscr{S}$, and (i) is true, we have $f^{-1}(]a, \infty])$ $\in \mathscr{A}$ and $f^{-1}([-\infty, b[) \in \mathscr{A}$ whenever $-\infty \leqq a < \infty$ and $-\infty < b \leqq \infty$. Thus if $]a, b[$ is any open interval of R, we have

$$f^{-1}(]a, b[) = f^{-1}([-\infty, b[) \cap f^{-1}(]a, \infty]) \in \mathscr{A}$$

and so $]a, b[\in \mathscr{S}$. It follows that all open subsets of R are in $\mathscr{S}$. Thus $\mathscr{S}$ is a σ-algebra containing $\mathscr{B}(R)$ and (ii) obtains. □

(11.5) Corollary. *Suppose that X is a topological space and that $\mathscr{A} \supset \mathscr{B}(X)$. Then all real-valued continuous functions and all extended real-valued lower [upper] semicontinuous functions defined on X are $\mathscr{A}$-measurable.*

(11.6) Remark. It is clear that if f is a real-valued Lebesgue measurable function on R, then $f^{-1}(B)$ is a Lebesgue measurable set whenever B is a Borel set. It is worth noting that even for certain real-valued *continuous* functions f on R there exist Lebesgue measurable sets A such that $f^{-1}(A)$ is not Lebesgue measurable. We sketch the construction of such a set. Let P be a nowhere dense perfect subset of $[0, 1]$ such that $\inf P = 0$, $\sup P = 1$, and $\lambda(P) > 0$ (10.53.a). Let C denote CANTOR's ternary set. Let $\mathscr{I}$ and $\mathscr{J}$ denote the families of component open subintervals of $[0, 1]$ that are complementary to P and C, respectively. Linearly order both $\mathscr{I}$ and $\mathscr{J}$ in the obvious way [$I_1 < I_2$ if I_1 lies to the left of I_2]. Then $\mathscr{I}$ and $\mathscr{J}$ are both of order type η (4.53), and so there exists an order-isomorphism φ from $\mathscr{I}$ onto $\mathscr{J}$. [If P is a Cantor-like set constructed as in (6.62), then φ can be defined explicitly by associating complementary intervals having like subscripts.] Define a function f from $[0, 1]$ onto $[0, 1]$ as follows: $f(0) = 0$; for $I \in \mathscr{I}$, define f linearly from I onto $\varphi(I)$ by mapping the lower [upper] endpoint of I to the lower [upper] endpoint of $\varphi(I)$ and joining with a line segment; and for $x \in P \cap \{0\}'$, define $f(x) = \sup\{f(t) : t < x, t \in \bigcup \mathscr{I} = [0, 1] \cap P'\}$. Then f is a continuous one-to-one [strictly increasing!] function from $[0, 1]$ onto $[0, 1]$ and $f(P) = C$. Let S be a non Lebesgue measurable subset of P (10.28) and let $A = f(S)$. Then we have $A \subset C$, so that $\lambda(A) = 0$ and $A \in \mathscr{M}_\lambda$. However $f^{-1}(A) = S \notin \mathscr{M}_\lambda$. [Note that the above construction can be used to show that any two nowhere dense compact perfect subsets of R are homeomorphic.]

We conclude from (11.4) that the set A in the above example is not a Borel set. This is yet another proof that there exist Lebesgue measurable sets that are not Borel sets [cf. (10.21.a)].

Let f, A, and S be as above and let $g = \xi_A$. It is clear that $g \circ f = \xi_S$ (on $[0, 1]$). Notice that f and g are both Lebesgue measurable and that $g \circ f$ is not. Thus the composition of two measurable functions need not be measurable. We do however have the following theorem.

(11.7) Theorem. *Let φ be any extended real-valued function defined on $R^\#$ such that $\varphi^{-1}([a, \infty]) \cap R$ is a Borel set for all real a,* i.e., *φ is $\mathscr{B}(R^\#)$-measurable. Let f be $\mathscr{A}$-measurable. Then $\varphi \circ f$ is $\mathscr{A}$-measurable.*

Proof. We have

$$\begin{aligned}(\varphi \circ f)^{-1}([a, \infty]) &= f^{-1}(\varphi^{-1}([a, \infty])) \\ &= f^{-1}((\varphi^{-1}([a, \infty]) \cap R) \cup A_+ \cup A_-) \\ &= f^{-1}(\varphi^{-1}([a, \infty]) \cap R) \cup f^{-1}(A_+) \cup f^{-1}(A_-)\,,\end{aligned}$$

where $A_+ = \{\infty\} \cap \varphi^{-1}([a, \infty])$ and $A_- = \{-\infty\} \cap \varphi^{-1}([a, \infty])$. Since $\varphi^{-1}([a, \infty]) \cap R$ is a Borel set by hypothesis, $f^{-1}(\varphi^{-1}([a, \infty]) \cap R)$ is in $\mathscr{A}$, by (11.4). It is easy to see that $f^{-1}(A_+)$ and $f^{-1}(A_-)$ are also in $\mathscr{A}$; therefore $(\varphi \circ f)^{-1}([a, \infty]) \in \mathscr{A}$, and so $\varphi \circ f$ is $\mathscr{A}$-measurable. □

(11.8) Theorem. *If f is $\mathscr{A}$-measurable, then the following assertions hold.*

(i) *The function $f + \alpha$ is $\mathscr{A}$-measurable for all real α.*

(ii) *The function αf is $\mathscr{A}$-measurable for all real α.*

(iii) *Let*

$$h(x) = \begin{cases} |f(x)|^p & \text{if } f(x) \text{ is finite}\,, \\ \beta_- & \text{if } f(x) = -\infty\,, \\ \beta_+ & \text{if } f(x) = \infty\,, \end{cases}$$

where β_- and β_+ are arbitrary extended real numbers and p is any positive real number. Then h is $\mathscr{A}$-measurable.

(iv) *Let*

$$h(x) = \begin{cases} [f(x)]^m & \text{if } f(x) \text{ is finite}\,, \\ \beta_- & \text{if } f(x) = -\infty\,, \\ \beta_+ & \text{if } f(x) = \infty\,, \end{cases}$$

where m is a positive integer and β_- and β_+ are arbitrary extended real numbers. Then h is $\mathscr{A}$-measurable.

(v) *Let $h = \frac{1}{f}$ where f is finite and not zero, and let h assume constant but arbitrary values β_+, β_-, and β_0 on the sets $\{x \in X : f(x) = \infty\}$,*

$\{x \in X : f(x) = -\infty\}$, *and* $\{x \in X : f(x) = 0\}$, *respectively. Then* h *is* $\mathscr{A}$*-measurable.*

Proof. In each case, we define a suitable function φ such that the function in question is equal to $\varphi \circ f$ and apply (11.7). For (i), let

$$\varphi(t) = \begin{cases} t + \alpha & \text{if } t \in R, \\ \pm\infty & \text{if } t = \pm\infty. \end{cases}$$

To prove (ii), let

$$\varphi(t) = \begin{cases} -\infty & \text{if } t = -\infty, \\ \alpha t & \text{if } t \in R, \\ \infty & \text{if } t = \infty, \end{cases}$$

if $\alpha > 0$, and

$$\varphi(t) = \begin{cases} \infty & \text{if } t = -\infty, \\ \alpha t & \text{if } t \in R, \\ -\infty & \text{if } t = \infty, \end{cases}$$

if $\alpha < 0$; if $\alpha = 0$, the assertion is trivial.

For (iii), let $\varphi(\pm\infty) = \beta_{\pm}$ and $\varphi(t) = |t|^p$ for real t. Since φ is continuous on R, it is clear that $\varphi^{-1}([a, \infty]) \cap R$ is a Borel set for all real a.

The proof of (iv) is similar, with $\varphi(t) = t^m$ for real t and $\varphi(\pm\infty) = \beta_{\pm}$.

To prove (v), let $\varphi(t) = \frac{1}{t}$ for $t \neq 0, \infty, -\infty$, and let $\varphi(0) = \beta_0$, $\varphi(\pm\infty) = \beta_{\pm}$. □

(11.9) Lemma. *Let* f *and* g *be* $\mathscr{A}$*-measurable. Then the sets*

(i) $\{x \in X : f(x) > g(x)\}$,

(ii) $\{x \in X : f(x) \geqq g(x)\}$,

and

(iii) $\{x \in X : f(x) = g(x)\}$

are in $\mathscr{A}$.

Proof. We have

$$\{x \in X : f(x) > g(x)\} = \bigcup_{u \in Q} (\{x \in X : f(x) > u\} \cap \{x \in X : g(x) < u\}),$$

and from this identity the measurability of (i) follows. The set (ii) is the complement of the set (i) with the rôles of f and g interchanged, and so it too is measurable. The set (iii) is the intersection of two measurable sets of type (ii), and so is measurable. □

(11.10) Theorem. *Let* f, g *be* $\mathscr{A}$*-measurable. Let* $h(x) = f(x) + g(x)$ *for all* $x \in X$ *such that* $f(x) + g(x)$ *is defined and let* h *have any fixed value* β *[an extended real number] elsewhere. Then* h *is* $\mathscr{A}$*-measurable.*

Proof. For any real number a, we have

$$\begin{aligned} h^{-1}(]a, \infty]) &= \{x \in X : f(x) + g(x) > a\} \cup A_\beta \\ &= \{x \in X : f(x) > a - g(x)\} \cup A_\beta, \end{aligned}$$

where

$$A_\beta = \begin{cases} (\{x : f(x) = \infty\} \cap \{x : g(x) = -\infty\}) \\ \cup\, (\{x : f(x) = -\infty\} \cap \{x : g(x) = \infty\}) \quad \text{if} \quad a < \beta, \\ \varnothing \quad \text{if} \quad a \geqq \beta\,. \end{cases}$$

The set $\{x \in X : f(x) > a - g(x)\}$ is in $\mathscr{A}$ by (11.9); therefore, since A_β is also in $\mathscr{A}$, the set $h^{-1}(]a, \infty])$ is in $\mathscr{A}$. □

(11.11) Theorem. *Let f and g be $\mathscr{A}$-measurable functions. Let h be defined on X by*

$$h(x) = \begin{cases} f(x)\, g(x) & \text{if} \quad x \notin A \\ \beta & \text{if} \quad x \in A\,, \end{cases}$$

where β is an arbitrary extended real number, and $A = \{x \in X : f(x) = \infty$ and $g(x) = -\infty\} \cup \{x \in X : f(x) = -\infty$ and $g(x) = \infty\}$. Then h is an $\mathscr{A}$-measurable function.

Proof. Consider $a \in R$ such that $a > 0$, and let

$$A_\beta = \begin{cases} A & \text{if} \quad a < \beta\,, \\ \varnothing & \text{if} \quad a \geqq \beta\,. \end{cases}$$

We have $h^{-1}(]a, \infty]) = \{x \in X : h(x) > a\} = A_\beta \cup \{x \in X : f(x) = \infty$ and $g(x) > 0\} \cup \{x \in X : f(x) > 0$ and $g(x) = \infty\} \cup \{x \in X : f(x) < 0$ and $g(x) = -\infty\} \cup \{x \in X : f(x) = -\infty$ and $g(x) < 0\} \cup \Big\{x \in X : f(x)$ and $g(x)$ are finite and $\frac{1}{4}\,[(f(x) + g(x))^2 - (f(x) - g(x))^2] > a\Big\}$. Applying (11.4), (11.8) and (11.10), we see that $h^{-1}(]a, \infty]) \in \mathscr{A}$. Similar expressions hold for $a < 0$ and $a = 0$, and it follows that h is $\mathscr{A}$-measurable. □

We next study limits of sequences of measurable functions.

(11.12) Theorem. *Let (f_n) be a sequence of $\mathscr{A}$-measurable functions defined on X. Then the four functions $\inf\limits_n f_n$, $\sup\limits_n f_n$, $\varliminf\limits_{n \to \infty} f_n$, and $\varlimsup\limits_{n \to \infty} f_n$ [defined pointwise as in (7.1)] are all $\mathscr{A}$-measurable.*

Proof. It follows immediately from the identity $\{x \in X : \sup\limits_n f_n(x) > a\} = \bigcup\limits_{n=1}^{\infty} \{x \in X : f_n(x) > a\}$ that $\sup\limits_n f_n$ is $\mathscr{A}$-measurable. The $\mathscr{A}$-measurability of $\inf\limits_n f_n$ follows at once from the identity $\inf\limits_n f_n(x) = -\sup\limits_n(-f_n(x))$ [recall that $-(\infty) = -\infty$ and $-(-\infty) = \infty$]. The rest follows from the first two results and the identities

$$\varliminf_{n \to \infty} f_n(x) = \sup_k \big(\inf_{n \geqq k} f_n(x)\big)$$

and

$$\varlimsup_{n \to \infty} f_n(x) = \inf_k \big(\sup_{n \geqq k} f_n(x)\big)\,. \; □$$

(11.13) Corollary. *Let $f_1, \ldots, f_m$ be $\mathscr{A}$-measurable. Then the functions*

$$\max\{f_1, \ldots, f_m\} \quad \text{and} \quad \min\{f_1, \ldots, f_m\}$$

[defined pointwise] are $\mathscr{A}$-measurable.

Proof. Define $f_n = f_m$ for all $n > m$ and apply (11.12). □

(11.14) Corollary. *If (f_n) is a sequence of $\mathscr{A}$-measurable functions defined on X and $\lim_{n\to\infty} f_n(x)$ exists in $R^{\#}$ for all $x \in X$, then $\lim_{n\to\infty} f_n$ is $\mathscr{A}$-measurable.*

Proof. Since $\lim_{n\to\infty} f_n = \varliminf_{n\to\infty} f_n = \varlimsup_{n\to\infty} f_n$, (11.12) applies. □

We now consider the concept of measurability for complex-valued [finite] functions.

(11.15) Definition. A complex-valued function f defined on X is said to be *$\mathscr{A}$-measurable* if both $\mathrm{Re} f$ and $\mathrm{Im} f$ are $\mathscr{A}$-measurable.

(11.16) Theorem. *Let f be a complex-valued function defined on X. Then the following statements are equivalent:*

(i) *f is $\mathscr{A}$-measurable;*

(ii) *$f^{-1}(U) \in \mathscr{A}$ for each open $U \subset K$;*

(iii) *$f^{-1}(B) \in \mathscr{A}$ for every $B \in \mathscr{B}(K)$.*

Proof. Let $f_1 = \mathrm{Re} f$ and $f_2 = \mathrm{Im} f$. Then $f = f_1 + i f_2$. First suppose that (i) is true and let $V = \{s + it \in K : a < s < b,\ c < t < d\}$ where $\{a, b, c, d\} \subset Q$. Then $f^{-1}(V) = f_1^{-1}(]a, b[) \cap f_2^{-1}(]c, d[) \in \mathscr{A}$. Next, let U be any open subset of K. There exists a sequence (V_n) of rational rectangles of the form V above such that $U = \bigcup_{n=1}^{\infty} V_n$. It follows that $f^{-1}(U) = \bigcup_{n=1}^{\infty} f^{-1}(V_n) \in \mathscr{A}$. Thus (i) implies (ii).

Now suppose that (ii) is true and set $\mathscr{S} = \{S \subset K : f^{-1}(S) \in \mathscr{A}\}$. As in the proof of (11.4), we see that $\mathscr{S}$ is a σ-algebra of subsets of K. Also $\mathscr{S}$ contains all open subsets of K, and so $\mathscr{B}(K) \subset \mathscr{S}$. Thus (iii) follows and therefore (ii) implies (iii).

Finally, suppose that (iii) is true. For $a \in R$, let $A_1 = \{s + it \in K : s > a\}$ and $A_2 = \{s + it \in K : t > a\}$. Then $f_j^{-1}(]a, \infty]) = f_j^{-1}(]a, \infty[) = f^{-1}(A_j) \in \mathscr{A}$ because $A_j \in \mathscr{B}(K)$ $(j = 1, 2)$. Thus f_1 and f_2 are $\mathscr{A}$-measurable, and so (i) is true. □

(11.17) Theorem. *Let f and g be complex-valued, $\mathscr{A}$-measurable functions on X, let $\alpha \in K$, let $m \in N$, and let p be a positive real number. Then all of the following functions are $\mathscr{A}$-measurable on X: $f + \alpha$; αf; $|f|^p$; f^m; $\frac{1}{f}$, if $f(x) \neq 0$ for all $x \in X$; $f + g$; fg.*

Proof. These results all follow immediately from Definition (11.15) by applying (11.8), (11.10), and (11.11). □

(11.18) Theorem. *Let (f_n) be a sequence of $\mathscr{A}$-measurable complex-valued functions on X and suppose that $\lim_{n\to\infty} f_n(x) = f(x) \in K$ for each $x \in X$. Then f is $\mathscr{A}$-measurable.*

Proof. Apply (11.15) and (11.14). □

(11.19) Remark. Theorems (11.18) and (11.14) both require that the sequence in question converge for every $x \in X$. However, a large portion of our work will deal with the case in which there is some specific measure μ defined on $\mathscr{A}$, the functions in question are defined only μ-almost everywhere, and the convergence of sequences is only μ-a.e[1]. Thus we would like a corresponding theorem for this case. Such a theorem will require some additional hypothesis about μ, for consider the case that $X = R$, $\mathscr{A} = \mathscr{B}(R)$, $\mu = \lambda$, P = CANTOR's ternary set, $A \subset P$, $A \notin \mathscr{B}(R)$, $f = \xi_A$, and $f_n = 0$ for all $n \in N$. Then each f_n is $\mathscr{B}(R)$-measurable and f is *not* $\mathscr{B}(R)$-measurable, but $f_n(x) \to f(x)$ for all $x \in R \cap P'$, *i.e.*, $f_n \to f$ λ-a.e. To avoid such irritating situations, it is enough to consider complete measures, defined as follows.

(11.20) Definition. Suppose that μ is a measure defined on $\mathscr{A}$ and that $B \in \mathscr{A}$ whenever $A \in \mathscr{A}$, $\mu(A) = 0$, and $B \subset A$, *i.e.*, all subsets of sets of measure zero are measurable. Then μ is said to be a *complete measure* and $(X, \mathscr{A}, \mu)$ is called a *complete measure space.*

Theorem (10.7) implies that if μ is an outer measure on X, then $(X, \mathscr{M}_\mu, \mu)$ is a complete measure space. We gain much and lose little [as the next theorem shows] by restricting our attention to complete measure spaces.

(11.21) Theorem. *Let $(X, \mathscr{A}, \mu)$ be any measure space. Define $\bar{\mathscr{A}} = \{E \cup A : E \in \mathscr{A},\ A \subset B$ for some $B \in \mathscr{A}$ such that $\mu(B) = 0\}$ and define $\bar{\mu}$ on $\bar{\mathscr{A}}$ by the rule $\bar{\mu}(E \cup A) = \mu(E)$. Then $\bar{\mathscr{A}}$ is a σ-algebra, $\bar{\mu}$ is a complete measure on $\bar{\mathscr{A}}$, and $(X, \bar{\mathscr{A}}, \bar{\mu})$ is a complete measure space. This measure space is called the* completion *of $(X, \mathscr{A}, \mu)$ and $\bar{\mu}$ is called the* completion of μ.

Proof. Exercise.

(11.22) Definition. If $E \in \mathscr{A}$ and $\mathscr{A}_E = \{F \in \mathscr{A} : F \subset E\}$, then $\mathscr{A}_E$ is plainly a σ-algebra of subsets of E and $(E, \mathscr{A}_E)$ is a measurable space. A function defined on E will be called *$\mathscr{A}$-measurable* if it is $\mathscr{A}_E$-measurable.

(11.23) Theorem. *Let $(X, \mathscr{A}, \mu)$ be a complete measure space and let f be an $\mathscr{A}$-measurable function defined μ-a.e. on X. Suppose that g is a function defined μ-a.e. on X such that $f = g$ μ-a.e. Then g is $\mathscr{A}$-measurable.*

---

[1] The term "μ-almost everywhere" and its abbreviation "μ-a.e." were defined in (9.29) for the case in which μ is a measure ι on a locally compact Hausdorff space. The extension to arbitrary measure spaces $(X, \mathscr{A}, \mu)$ is immediate.

Proof. Let $A = \{x \in (\operatorname{dom} f) \cap (\operatorname{dom} g) : f(x) = g(x)\}$. Then $\mu(A') = 0$ and all subsets of A' are in $\mathscr{A}$. We suppose that f and g are extended real-valued, the complex case being similar. For $a \in R$, we have

$$\begin{aligned} g^{-1}(]a, \infty]) &= (g^{-1}(]a, \infty]) \cap A) \cup (g^{-1}(]a, \infty]) \cap A') \\ &= (f^{-1}(]a, \infty]) \cap A) \cup (g^{-1}(]a, \infty]) \cap A') \in \mathscr{A} . \quad \square \end{aligned}$$

(11.24) Theorem. *Let $(X, \mathscr{A}, \mu)$ be a complete measure space and let (f_n) be a sequence of $\mathscr{A}$-measurable functions each of which is defined μ-a.e. on X. Suppose that f is defined μ-a.e. on X and that $\lim_{n\to\infty} f_n(x) = f(x)$ μ-a.e. on X. Then f is $\mathscr{A}$-measurable.*

Proof. Define A as the set

$$(\operatorname{dom} f) \cap \left(\bigcap_{n=1}^{\infty} \operatorname{dom} f_n\right) \cap \{x \in X : f_n(x) \to f(x)\} .$$

It is obvious that $A \in \mathscr{A}$ and that $\mu(A') = 0$. For each $n \in N$, define

$$g_n(x) = \begin{cases} f_n(x) & \text{if } x \in A, \\ 0 & \text{if } x \in A', \end{cases}$$

and define

$$g(x) = \begin{cases} f(x) & \text{if } x \in A, \\ 0 & \text{if } x \in A'. \end{cases}$$

Theorem (11.23) implies that g_n is $\mathscr{A}$-measurable for all $n \in N$. Clearly $g_n(x) \to g(x)$ for all $x \in X$. Applying (11.14) or (11.18), we see that g is $\mathscr{A}$-measurable. Again by (11.23), f is $\mathscr{A}$-measurable. $\square$

Mathematical analysis is heavily concerned with convergence of sequences and series of functions. Indeed, a main goal of analysis is the approximation of complicated functions by means of simple functions. [The terms "approximation", "complicated", and "simple" have different meanings in different situations.] Up to this point we have met in this book two kinds of convergence: pointwise [almost everywhere] and uniform. We now introduce a third kind of convergence and prove some theorems that show a number of relationships among these three kinds of convergence.

(11.25) Definition. Let $(X, \mathscr{A}, \mu)$ be a measure space and let f and $(f_n)_{n=1}^{\infty}$ be $\mathscr{A}$-measurable functions on X. They may be either extended real- or complex-valued. Suppose that for every $\delta > 0$, we have

$$\lim_{n\to\infty} \mu(\{x \in X : |f(x) - f_n(x)| \geqq \delta\}) = 0 .$$

Then (f_n) is said to *converge in measure* [or *in probability*] to f. We write: $f_n \to f$ in measure.

(11.26) Theorem [F. Riesz]. *Let $(X, \mathscr{A}, \mu)$ be a measure space and let f and (f_n) be $\mathscr{A}$-measurable functions such that $f_n \to f$ in measure. Then there exists a subsequence (f_{n_k}) such that $f_{n_k} \to f$ μ-a.e.*

Proof. Choose $n_1 \in N$ such that $\mu(\{x \in X : |f(x) - f_{n_1}(x)| \geqq 1\}) < \frac{1}{2}$. Suppose that $n_1, n_2, \ldots, n_k$ have been chosen. Then choose n_{k+1} such that $n_{k+1} > n_k$ and

$$\mu\left(\left\{x \in X : |f(x) - f_{n_{k+1}}(x)| \geqq \frac{1}{k+1}\right\}\right) < \frac{1}{2^{k+1}} .$$

Let $A_j = \bigcup_{k=j}^{\infty} \left\{x : |f(x) - f_{n_k}(x)| \geqq \frac{1}{k}\right\}$ for each $j \in N$. Clearly we have $A_1 \supset A_2 \supset \cdots$. Next let $B = \bigcap_{j=1}^{\infty} A_j$. Since $\mu(A_1) < \sum_{k=1}^{\infty} \frac{1}{2^k} < \infty$, it follows from (10.15) that

$$\mu(B) = \lim_{j\to\infty} \mu(A_j) \leqq \lim_{j\to\infty} \sum_{k=j}^{\infty} \frac{1}{2^k} = \lim_{j\to\infty} \frac{1}{2^{j-1}} = 0 ,$$

that is, $\mu(B) = 0$. Next, let $x \in B' = \bigcup_{j=1}^{\infty} A_j'$. Then there is a j_x such that

$$x \in A_{j_x}' = \bigcap_{k=j_x}^{\infty} \left\{y \in X : |f(y) - f_{n_k}(y)| < \frac{1}{k}\right\} .$$

Given $\varepsilon > 0$, choose k_0 such that $k_0 \geqq j_x$ and $\frac{1}{k_0} \leqq \varepsilon$. Then $k \geqq k_0$ implies that $|f(x) - f_{n_k}(x)| < \frac{1}{k} \leqq \varepsilon$. This proves that $f_{n_k}(x) \to f(x)$ for all $x \in B'$. □

(11.27) Note. There exist sequences of functions that converge in measure and do not converge a.e. For example, let $X = [0, 1]$, $\mathscr{A} = \mathscr{M}_\lambda$, $\mu = \lambda$, and, for each $n \in N$, define

$$f_n = \xi_{\left[\frac{j}{2^k}, \frac{j+1}{2^k}\right]} \quad \text{where} \quad n = 2^k + j, 0 \leqq j < 2^k .$$

Thus $f_1 = \xi_{[0,1]}$, $f_2 = \xi_{\left[0, \frac{1}{2}\right]}, \ldots, f_{10} = \xi_{\left[\frac{1}{4}, \frac{3}{8}\right]}, \ldots$. It is clear that $\lambda(\{x : |f_n(x)| \geqq \delta\}) \leqq \frac{1}{2^k} \to 0$ as $n \to \infty$ for every $\delta > 0$. Thus $f_n \to 0$ in measure. On the other hand, if $x \in [0, 1]$, the sequence $(f_n(x))$ contains an infinite number of 0's and an infinite number of 1's. Thus the sequence of functions (f_n) converges *nowhere* on $[0, 1]$.

(11.28) Exercise. Find a subsequence of the sequence (f_n) of (11.27) that converges to zero λ-a.e. on $[0, 1]$. Can you find a subsequence (f_{n_k}) of (f_n) that converges to zero *everywhere* on $[0, 1]$?

(11.29) Note. We have already seen several instances in which finiteness or σ-finiteness of a measure space is an essential hypothesis: cf. for example (10.15), (10.30), (10.34), and (10.58). Our next two theorems are stated for finite measure spaces.

(11.30) Lemma. *Let $(X, \mathscr{A}, \mu)$ be a finite measure space and let f and (f_n) be $\mathscr{A}$-measurable functions that are defined and finite μ-a.e. on X. Suppose that $f_n \to f$ μ-a.e. on X. Then for each pair of positive real numbers*

δ and ε, there exist a set $J \in \mathscr{A}$ and an integer $n_0 \in N$ such that $\mu(J') < \varepsilon$ and $|f(x) - f_n(x)| < \delta$ for all $x \in J$ and $n \geqq n_0$.

Proof. Let $E = \{x \in X : f(x)$ is finite, $f_n(x)$ is finite for all $n \in N$, $f_n(x) \to f(x)\}$. By hypothesis, $\mu(E') = 0$. For each $m \in N$, let $E_m = \{x \in E : |f(x) - f_n(x)| < \delta$ for all $n \geqq m\}$. We have $E_1 \subset E_2 \subset \cdots$ and $\bigcup_{m=1}^{\infty} E_m = E$. Therefore $E_1' \supset E_2' \supset \cdots$ and $\bigcap_{m=1}^{\infty} E_m' = E'$. Since $\mu(E_1') \leqq \mu(X) < \infty$, it follows that $\lim_{m\to\infty} \mu(E_m') = \mu(E') = 0$. Thus choose $n_0 \in N$ such that $\mu(E_{n_0}') < \varepsilon$ and set $J = E_{n_0}$. □

(11.31) Theorem [Lebesgue]. *Let $(X, \mathscr{A}, \mu)$, f, and (f_n) be as in (11.30). Then $f_n \to f$ in measure.*

Proof. Choose arbitrary positive numbers δ and ε. For each $n \in N$, let $S_n(\delta) = \{x \in X : |f(x) - f_n(x)| \geqq \delta\}$. By (11.30), there exist $J \in \mathscr{A}$ and $n_0 \in N$ such that $\mu(J') < \varepsilon$ and $|f(x) - f_n(x)| < \delta$ for all $x \in J$ and for all $n \geqq n_0$. Thus $n \geqq n_0$ implies $S_n(\delta) \subset J'$. Therefore $n \geqq n_0$ implies $\mu(S_n(\delta)) \leqq \mu(J') < \varepsilon$. Since ε is arbitrary, it follows that $\lim_{n\to\infty} \mu(S_n(\delta)) = 0$, *i.e.*, $f_n \to f$ in measure. □

(11.32) Theorem [Egorov]. *Let $(X, \mathscr{A}, \mu)$, f, and (f_n) be as in (11.30). Then for each $\varepsilon > 0$ there exists a set $A \in \mathscr{A}$ such that $\mu(A') < \varepsilon$ and $f_n \to f$ uniformly on A.*

Proof. Choose a positive number ε. By (11.30), for each $m \in N$ there exist $J_m \in \mathscr{A}$ and $n_m \in N$ such that $\mu(J_m') < \frac{\varepsilon}{2^m}$ and $|f(x) - f_n(x)| < \frac{1}{m}$ for all $x \in J_m$ and all $n \geqq n_m$. Define A by $A = \bigcap_{m=1}^{\infty} J_m$. Then $A' = \bigcup_{m=1}^{\infty} J_m'$, and so

$$\mu(A') \leqq \sum_{m=1}^{\infty} \mu(J_m') < \sum_{m=1}^{\infty} \frac{\varepsilon}{2^m} = \varepsilon .$$

Also $n \geqq n_m$ implies that

$$\begin{aligned} \sup_{x\in A} |f(x) - f_n(x)| &\leqq \sup_{x \in J_m} |f(x) - f_n(x)| \\ &\leqq \frac{1}{m} \end{aligned}$$

for every $m \in N$. Thus $f_n \to f$ uniformly on A. □

(11.33) Caution. Theorems (11.31) and (11.32) depend heavily on the hypothesis that $\mu(X) < \infty$. For example, suppose that $X = R$, $\mathscr{A} = \mathscr{M}_\lambda$, $\mu = \lambda$, $f_n = \xi_{[n,n+1]}$, and $f = 0$. Then $f_n(x) \to f(x) = 0$ for all $x \in R$. But $\lambda(\{x \in R : |f(x) - f_n(x)| \geqq 1\}) = \lambda([n, n+1]) = 1 \not\to 0$, and so $f_n \not\to f$ in measure. Also, if $A \in \mathscr{M}_\lambda$ and $\lambda(A') < 1$, then for each $n \in N$ there exists $x_n \in A \cap [n, n+1]$, and so $|f(x_n) - f_n(x_n)| = 1$, *i.e.*, $f_n \not\to f$ uniformly on A.

The remainder of this section is devoted to an investigation of the structure of measurable functions. As usual, $(X, \mathscr{A})$ is an arbitrary measurable space.

(11.34) Definition. A *simple function* s on X is a function that assumes only a finite number of values. If $\operatorname{rng} s = \{\alpha_1, \ldots, \alpha_n\}$ and $A_k = \{x \in X : s(x) = \alpha_k\}$ $(1 \leqq k \leqq n)$, then it is obvious that $s = \sum_{k=1}^{n} \alpha_k \xi_{A_k}$.

(11.35) Theorem. *Let f be any complex- or extended real-valued $\mathscr{A}$-measurable function defined on X. Then there exists a sequence (s_n) of finite-valued, $\mathscr{A}$-measurable, simple functions defined on X such that $|s_1| \leqq |s_2| \leqq \cdots \leqq |s_n| \leqq \cdots$ and $s_n(x) \to f(x)$ for each $x \in X$. If f is bounded, then the functions s_n can be chosen so that the convergence is uniform. If $f \geqq 0$, the sequence (s_n) can be chosen so that $0 \leqq s_1 \leqq s_2 \leqq \cdots \leqq f$.*

Proof. First consider the case $f \geqq 0$. For each $n \in N$ and for $1 \leqq k \leqq n \cdot 2^n$, let

$$A_{n,k} = \left\{x \in X : \frac{k-1}{2^n} \leqq f(x) < \frac{k}{2^n}\right\},$$

and

$$B_n = \{x \in X : f(x) \geqq n\}.$$

Define

$$s_n(x) = \sum_{k=1}^{n \cdot 2^n} \frac{k-1}{2^n} \xi_{A_{n,k}}(x) + n \xi_{B_n}(x).$$

It is clear that all of the sets $A_{n,k}$ and B_n are in $\mathscr{A}$, and so each s_n is an $\mathscr{A}$-measurable simple function. It is also easy to see that

$$0 \leqq s_1 \leqq s_2 \leqq \cdots \leqq f, \quad |s_n| \leqq n, \quad \text{and} \quad |f(x) - s_n(x)| < \frac{1}{2^n}$$

for all $x \in B_n'$. It follows that $\lim_{n \to \infty} s_n(x) = f(x)$ for every $x \in X$. Moreover, if there exists $\beta \in R$ such that $|f| \leqq \beta$, then $\sup_{x \in X} |f(x) - s_n(x)| \leqq \frac{1}{2^n}$ for all $n \geqq \beta$; therefore $s_n \to f$ uniformly on X if f is bounded.

Now consider the general case. If f is extended real-valued, define $f^+ = \max\{f, 0\}$ and $f^- = -\min\{f, 0\}$. Then $f^+ \geqq 0$, $f^- \geqq 0$, and $f = f^+ - f^-$. If f is complex-valued, we may write

$$f = f_1 - f_2 + i(f_3 - f_4)$$

where each $f_j \geqq 0$. In either of these cases we apply the results of the preceding paragraph to the nonnegative extended real-valued functions making up f. We leave the details to the reader. □

(11.36) Theorem [N. N. Luzin]. *Let X be a locally compact Hausdorff space and let ι and $\mathscr{M}_\iota$ be as in §§ 9, 10. Suppose that $E \in \mathscr{M}_\iota$, $\iota(E) < \infty$, and that f is a complex-valued $\mathscr{M}_\iota$-measurable function on X such that $f(x) = 0$ for all $x \in E'$. Then for each $\varepsilon > 0$ there exists a function $g \in \mathfrak{C}_{00}(X)$*

such that $\iota(\{x \in X : f(x) \neq g(x)\}) < \varepsilon$. *Moreover if* $\|f\|_u < \infty$, *then* g *can be selected so that* $\|g\|_u \leq \|f\|_u$.

Proof. Let $\varepsilon > 0$ be given. Use (9.24) to select an open set U such that $E \subset U$ and $\iota(U) < \iota(E) + \frac{\varepsilon}{4}$. The set U is fixed throughout the proof.

(I) Suppose that $f = \xi_A$. Then $A \subset E$ and A is ι-measurable. Since $\iota(A) < \infty$, we may apply (10.30) to find a compact set F such that $F \subset A$ and $\iota(A \cap F') < \frac{\varepsilon}{2}$. Next use (9.24) and (6.79) to produce an open set V with compact closure such that $F \subset V \subset U$ and $\iota(V \cap F') < \frac{\varepsilon}{2}$. Use (6.80) to obtain a continuous function g from X into $[0, 1]$ such that $g(x) = 1$ for all $x \in F$ and $g(x) = 0$ for all $x \in V'$. Then we have: $g \in \mathfrak{C}_{00}(X)$, $g = 0$ on U', and $\{x \in X : f(x) \neq g(x)\} \subset (V \cap F') \cup (A \cap F')$. It follows that

$$\iota(\{x \in X : f(x) \neq g(x)\}) < \frac{\varepsilon}{2} + \frac{\varepsilon}{2} = \varepsilon,$$

and so the proof is complete if $f = \xi_A$.

(II) Consider next the case that f is a simple function, say $f = \sum_{k=1}^{n} \alpha_k \xi_{A_k}$ where each A_k is ι-measurable. We may [and do] suppose that $A_k \subset E$ for each k. Next apply (I) to find functions $g_k \in \mathfrak{C}_{00}(X)$ such that $g_k = 0$ on U' and $\iota(\{x \in X : \xi_{A_k}(x) \neq g_k(x)\}) < \frac{\varepsilon}{n}$ for $1 \leq k \leq n$. Define $g = \sum_{k=1}^{n} \alpha_k g_k$. Then we have: $g \in \mathfrak{C}_{00}(X)$; $g = 0$ on U'; and $\{x \in X : f(x) \neq g(x)\} \subset \bigcup_{k=1}^{n} \{x \in X : \xi_{A_k}(x) \neq g_k(x)\}$. Thus $\iota(\{x \in X : f(x) \neq g(x)\}) < \sum_{k=1}^{n} \frac{\varepsilon}{n} = \varepsilon$.

(III) Now consider the general case. Apply (11.35) to obtain a sequence $(s_n)_{n=1}^{\infty}$ of $\mathscr{M}_\iota$-measurable, complex-valued, simple functions such that $s_n = 0$ on E' for each $n \in N$ and $s_n(x) \to f(x)$ for all $x \in X$. For each $n \in N$, apply (II) to obtain a function $g_n \in \mathfrak{C}_{00}(X)$ such that $g_n = 0$ on U' and if $A_n = \{x \in X : s_n(x) \neq g_n(x)\}$, then $\iota(A_n) < \frac{\varepsilon}{5^n}$. Let $A = \bigcup_{n=1}^{\infty} A_n$. It is clear that $A \subset U$ and that

$$\iota(A) \leq \sum_{n=1}^{\infty} \iota(A_n) < \sum_{n=1}^{\infty} \frac{\varepsilon}{5^n} = \frac{\varepsilon}{4}.$$

Clearly $s_n(x) = g_n(x)$ for all $x \in A'$ and all $n \in N$. Thus $g_n(x) \to f(x)$ for all $x \in A'$. Since $\iota(U \cap A') < \infty$, Egorov's theorem (11.32) shows that there is an ι-measurable set $B \subset U \cap A'$ such that $\iota(U \cap A' \cap B') < \frac{\varepsilon}{4}$ and $g_n \to f$ uniformly on B. Next use (10.30) to select a compact set

$F \subset B$ such that $\iota(B \cap F') < \frac{\varepsilon}{4}$. Plainly $g_n \to f$ uniformly on $F \cup U'$ [recall that $g_n = f = 0$ on U']. Since each g_n is continuous, it follows from (7.9) that f is continuous on $F \cup U'$ [in the relative topology]. Applying the TIETZE extension theorem (7.40), for the compact set F and its open superset U, we find a function $g \in \mathfrak{C}_{00}(X)$ such that $f(x) = g(x)$ for all $x \in F \cup U'$. We have

$$\{x \in X: f(x) \neq g(x)\} \subset U \cap F' = A \cup (U \cap A' \cap B') \cup (B \cap F'),$$

and so

$$\iota(\{x \in X: f(x) \neq g(x)\}) \leqq \iota(A) + \iota(U \cap A' \cap B') + \iota(B \cap F')$$
$$< \frac{\varepsilon}{4} + \frac{\varepsilon}{4} + \frac{\varepsilon}{4} < \varepsilon.$$

Finally, suppose that $\|f\|_u < \infty$ and $\|g\|_u > \|f\|_u$. This is certainly possible. In this case we tamper a little with g to obtain the desired conclusion. Let $S = \{z \in K: |z| \leqq \|f\|_u\}$ and $T = \{z \in K: |z| \leqq \|g\|_u\}$. Define a mapping φ from T onto S as follows:

$$\varphi(z) = \begin{cases} z & \text{if } z \in S \\ \frac{z}{|z|} \cdot \|f\|_u & \text{if } z \in T \cap S'. \end{cases}$$

It is easy to see that φ is continuous and that $|\varphi(z)| = \|f\|_u$ for $z \in T \cap S'$. Now let $h = \varphi \circ g$. Then h is continuous and $h(x) = 0$ whenever $g(x) = 0$; thus $h \in \mathfrak{C}_{00}(X)$. Also it is evident that $\|h\|_u = \|f\|_u$ and $\{x \in X: f(x) \neq h(x)\} \subset \{x \in X: f(x) \neq g(x)\}$. □

(11.37) Exercise: Measures on measurable subsets. Let $(X, \mathscr{A}, \mu)$ be a measure space and E a set in $\mathscr{A}$. Let $\mathscr{A}_E$ be as in (11.22), and let μ_E be the restriction of μ to $\mathscr{A}_E$. Prove that $(E, \mathscr{A}_E, \mu_E)$ is a measure space.

(11.38) Exercise: Measures on image sets. (a) Let $(X, \mathscr{A}, \mu)$ be a measure space and τ a mapping of X *onto* a set Y. Let $\mathscr{B}$ be the family of all subsets B of Y such that $\tau^{-1}(B) \in \mathscr{A}$. For $B \in \mathscr{B}$, let $\nu(B) = \mu(\tau^{-1}(B))$. Prove that $(Y, \mathscr{B}, \nu)$ is a measure space.

(b) State and prove an analogue of (a) for outer measures.

(11.39) Exercise: Measures on nonmeasurable subsets. (a) Let $(X, \mathscr{A}, \mu)$ be a measure space and let Y be a subset of X such that if $B \subset Y'$ and $B \in \mathscr{A}$, then $\mu(B) = 0$. Let $\mathscr{A}^\dagger$ be the family of all sets $Y \cap M$ for $M \in \mathscr{A}$. For $M^\dagger \in \mathscr{A}^\dagger$, define $\mu^\dagger(M^\dagger)$ as $\mu(M)$ for an arbitrary $M \in \mathscr{A}$ such that $Y \cap M = M^\dagger$. Prove that $\mu^\dagger$ is well defined, that $\mathscr{A}^\dagger$ is a σ-algebra of subsets of Y, and that $(Y, \mathscr{A}^\dagger, \mu^\dagger)$ is a measure space.

(b) Consider the measure space $(R, \mathscr{M}_\lambda, \lambda)$. Find a subset Y of R that is not λ-measurable and which satisfies the condition of part (a). [Hint. Use a set B as constructed in (10.54.b).]

(c) Prove that $[0, 1]$ contains a subset D admitting a measure μ on $\mathscr{B}(D)$ such that $\mu(D) = 1$, $\mu(F) = 0$ for all compact sets $F \subset D$, and $\mu(Y) = \inf\{\mu(U) : U$ is open in the topology of D, $U \supset Y\}$ for all $Y \in \mathscr{B}(D)$. That is, $(D, \mathscr{B}(D), \mu)$ is irregular, although "outer" regular. [Hints. Construct a subset D of $[0, 1]$ such that $D \cap F \neq \varnothing \neq D' \cap F$ for every uncountable closed set $F \subset [0, 1]$: see (10.54.b). Then D is non λ-measurable, and so $0 < \lambda(D) \leqq 1$. Let X be a G_δ set such that $D \subset X \subset [0, 1]$ and $\lambda(X) = \lambda(D)$. Then every λ-measurable subset of $X \cap D'$ has λ-measure 0, and we may take μ to be $\left(\frac{1}{\lambda(D)} \lambda\right)^{\dagger}$ as in part (a). All of the claims made for μ are easy to verify. Compare this result with (10.58).]

(11.40) Exercise: Extending a measure. Let $(X, \mathscr{A}, \mu)$ be a measure space. Let $\mathscr{S}$ be a subfamily of $\mathscr{P}(X)$ such that:

(i) $P \in \mathscr{S} \cap \mathscr{A}$ implies $\mu(P) = 0$;

(ii) $P_1, P_2, P_3, \ldots \in \mathscr{S}$ imply $\bigcup_{n=1}^{\infty} P_n \in \mathscr{S}$.

Let $\mathscr{A}^*$ be the family of all subsets A^* of X such that the symmetric difference $A^* \triangle A$ is in $\mathscr{S}$ for some $A \in \mathscr{A}$. For such a set A^*, let $\mu^*(A^*) = \mu(A)$.

(a) Prove that $\mathscr{A}^*$ is a σ-algebra of subsets of X.

(b) Prove that μ^* is well defined on $\mathscr{A}^*$ and that $(X, \mathscr{A}^*, \mu^*)$ is a measure space.

(c) In what sense is $(X, \mathscr{A}^*, \mu^*)$ an extension of $(X, \mathscr{A}, \mu)$?

(d) Find a simple hypothesis on $\mathscr{S}$ necessary and sufficient for $(X, \mathscr{A}^*, \mu^*)$ to be a complete measure space. Prove your assertion!

(11.41) Exercise. Let $(X, \mathscr{A}, \mu)$ be a finite measure space. Let $(f_n)_{n=1}^{\infty}$ be a sequence of extended real-valued, $\mathscr{A}$-measurable functions on X. Suppose that $\lim_{n\to\infty} f_n(x) = f(x)$ for μ-almost all $x \in X$, where f is extended real-valued and $\mathscr{A}$-measurable. Define $\arctan(\infty) = \frac{\pi}{2}$ and $\arctan(-\infty) = -\frac{\pi}{2}$, and consider the functions $\arctan \circ f_n$ and $\arctan \circ f$. Now formulate and prove an analogue of EGOROV's theorem (11.32) on uniform convergence of f_n to f except on sets of arbitrarily small measure.

(11.42) Exercise. Let $(X, \mathscr{A}, \mu)$ be a σ-finite measure space and let f and (f_n) be $\mathscr{A}$-measurable complex-valued functions that are defined μ-a.e. on X. Suppose that $f_n \to f$ μ-a.e. on X. Prove that there exists a set $H \in \mathscr{A}$ and a family $\{E_k\}_{k=1}^{\infty} \subset \mathscr{A}$ such that $X = H \cup \bigcup_{k=1}^{\infty} E_k$, $\mu(H) = 0$, and $f_n \to f$ uniformly on each E_k. [Use EGOROV's theorem (11.32).]

(11.43) Exercise. (a) Find a sequence $(f_n) \subset \mathfrak{C}^r([0, 1])$ and a real-valued function f such that $f_n(x) \to f(x)$ for all $x \in [0, 1]$ but $f_n \to f$

uniformly on no subinterval of $[0, 1]$. [Make f discontinuous on a dense set.]

(b) Use the sequence constructed in (a) to show that the conclusion of (11.42) fails for the measure space $([0, 1], \mathscr{P}([0, 1]), \nu)$ where ν is counting measure on $[0, 1]$ (10.4.a). [Show that the E_k's can be taken closed and apply BAIRE's category theorem (6.54).]

(c) Show that there is a sequence $(f_n) \subset \mathfrak{C}^r([0, 1])$ such that $f_n(x) \to 0$ for every $x \in [0, 1]$ but $f_n \to 0$ uniformly on no subinterval of $[0, 1]$. [For each $n \in N$, let F_n be the set of all numbers in $]0, 1]$ having the form $\frac{k}{2^m}$ for an integer k and an integer $m \in \{0, 1, \ldots, n\}$. Let f_n be zero on F_n. For $\frac{k}{2^m} \in F_n$, where k is odd, let $f_n\left(\frac{k}{2^m} - \frac{1}{2^{n+1}}\right) = \frac{1}{2^m}$. Let f_n be linear in all subintervals of $[0, 1]$ where it is not yet defined.]

(11.44) Exercise. Let X be a locally compact Hausdorff space and let ι be a measure on X as in § 9. Suppose that f is a complex-valued $\mathscr{M}_\iota$-measurable function on X such that $\{x \in X : f(x) \neq 0\}$ is σ-finite with respect to ι. Prove that there exists a Borel measurable function g on X such that $|g| \leqq |f|$ and $\iota(\{x \in X : f(x) \neq g(x)\}) = 0$. [Use (11.35) and (10.34).]

(11.45) Exercise. Let $(X, \mathscr{A}, \mu)$ be a finite measure space. Suppose that f and $(f_n)_{n=1}^{\infty}$ are $\mathscr{A}$-measurable complex-valued functions on X. Prove that $f_n \to f$ in measure if and only if each subsequence of (f_n) admits a subsubsequence that converges to f μ-a.e.

(11.46) Exercise. Let X be a topological space. A family $\mathfrak{C}$ of complex-valued functions on X is said to be *closed under pointwise limits* if $f \in \mathfrak{C}$ whenever f is a complex-valued function on X and, for some sequence $(f_n) \subset \mathfrak{C}$, $f(x) = \lim\limits_{n\to\infty} f_n(x)$ for all $x \in X$. The family $\mathfrak{R}(X)$ of all *Baire functions on* X is defined to be the intersection of all families $\mathfrak{C}$ of complex-valued functions on X such that $\mathfrak{C}$ contains all complex-valued continuous functions on X and $\mathfrak{C}$ is closed under pointwise limits. Notice that K^X is such a class $\mathfrak{C}$.

(a) Prove that every Baire function is Borel measurable.

Let $\mathfrak{R}_0$ be the set of all complex-valued continuous functions on X. If α is an ordinal number such that $0 < \alpha < \Omega$ [see (4.49)], define $\mathfrak{R}_\alpha$ to be the family of all functions f such that f is the pointwise limit of some sequence $(f_n) \subset \bigcup \{\mathfrak{R}_\beta : \beta \text{ is an ordinal number}, \beta < \alpha\}$. The functions in $\mathfrak{R}_\alpha$ are known as the *Baire functions of type* α.

(b) Prove that $\mathfrak{R}(X) = \bigcup\limits_{\alpha < \Omega} \mathfrak{R}_\alpha$ [compare the proof of (10.23)].

(c) Prove that if f and g are Baire functions, then so are $f + g$, fg, and $|f|$; if f and g are real-valued Baire functions, then so are $\max\{f, g\}$ and $\min\{f, g\}$ [use (b) and transfinite induction].

Let $\mathscr{B}_0(X)$ be the smallest σ-algebra of subsets of X that contains all sets of the form $\{x \in X : f(x) = 0\}$, where f is a continuous complex-valued function on X. The sets in $\mathscr{B}_0(X)$ are called the *Baire sets of* X.

(d) Prove that $f \in \mathfrak{R}(X)$ if and only if f is a complex-valued function on X and f is $\mathscr{B}_0(X)$-measurable. [For the "if" statement, first show that $\{E \subset X : \xi_E \in \mathfrak{R}(X)\} = \mathscr{B}_0(X)$, and then use (11.35). For the "only if" statement use (b) and transfinite induction.]

(c) Prove that $\mathscr{B}_0(X) = \mathscr{B}(X)$ if X is a metric space [use (6.86)].

§ 12. The abstract Lebesgue integral

This is perhaps the most important single section in the entire book. In it we construct, and study the remarkable properties of, the Lebesgue integral on an arbitrary measure space. It turns out that this integral is equal to the functional $\bar{I}$ for all nonnegative measurable functions when the measure space is $(X, \mathscr{M}_\iota, \iota)$. Throughout the present section, $(X, \mathscr{A}, \mu)$ denotes an arbitrary measure space, except where further restrictions are explicitly stated. The symbol $\mathfrak{S}$ denotes all simple, $\mathscr{A}$-measurable functions on X that are complex- or extended real-valued; $\mathfrak{S}^+$ is as usual the set of all nonnegative functions in $\mathfrak{S}$.

(12.1) Definition. A *measurable dissection of* X is any finite, pairwise disjoint family $\{A_1, A_2, \ldots, A_n\} \subset \mathscr{A}$ such that $\bigcup_{k=1}^{n} A_k = X$.

(12.2) Definition. Let f be any function from X into $[0, \infty]$. Define

$$L(f) = \sup\left\{\sum_{k=1}^{n} \inf\{f(x) : x \in A_k\}\, \mu(A_k) : \{A_1, \ldots, A_n\} \text{ is a measurable dissection of } X\right\}.$$

Since one or more A_k's may be $\varnothing$, we must define $\inf \varnothing$: as a matter of convenience we set $\inf \varnothing = 0$.

For an extended real-valued function f we define [just as in (11.35)]

$$f^+ = \max\{f, 0\} \quad \text{and} \quad f^- = -\min\{f, 0\}.$$

Notice that $f^+ \geqq 0$, $f^- \geqq 0$, and $f = f^+ - f^-$. We define $L(f) = L(f^+) - L(f^-)$ provided that at least one of the numbers $L(f^+)$ and $L(f^-)$ is finite. If $L(f^+) = L(f^-) = \infty$, then we do not define $L(f)$. The number $L(f)$ [when defined] is called the *Lebesgue integral* [or simply the *integral*] *of* f.

(12.3) Examples.

(a) If $\alpha \in R^\#$ and $f(x) = \alpha$ for all $x \in X$, then $L(f) = \alpha\mu(X)$.

(b) If $f(x) = \infty$ for $x \in E$ and $\mu(E) > 0$, then $L(f) = \infty$ if it is defined.

(c) If $X = [0, 1]$, $\mu = \lambda$, and f is nonnegative and Riemann integrable on $[0, 1]$, then $L(f) \geqq S(f; [0, 1])$. This inequality is all but trivial: each lower Darboux sum for f is less than or equal to one of the numbers of which $L(f)$ is the supremum. [Actually the equality $L(f) = S(f; [0, 1])$ holds: see (12.51.f) *infra.*]

(12.4) Theorem. *Let $f \in \mathfrak{S}^+$, say $f = \sum_{k=1}^{n} \alpha_k \xi_{E_k}$, where the E_k's are pairwise disjoint and in $\mathscr{A}$. Then $L(f)$ exists and $L(f) = \sum_{k=1}^{n} \alpha_k \mu(E_k)$.*

Proof. We may suppose that the pairwise disjoint family $\{E_k\}_{k=1}^n$ covers X. Since $\inf\{f(x) : x \in E_k\} = \alpha_k$, we have $L(f) \geqq \sum_{k=1}^{n} \alpha_k \mu(E_k)$. Now let $\{B_j\}_{j=1}^m$ be an arbitrary measurable dissection of X. Then

$$\begin{aligned}\sum_{j=1}^{m} \inf\{f(x) : x \in B_j\}\, \mu(B_j) &= \sum_{j=1}^{m} \sum_{k=1}^{n} \inf\{f(x) : x \in B_j\}\, \mu(E_k \cap B_j)\\ &\leqq \sum_{j=1}^{m} \sum_{k=1}^{n} \inf\{f(x) : x \in E_k \cap B_j\}\, \mu(E_k \cap B_j)\\ &= \sum_{j=1}^{m} \sum_{k=1}^{n} \alpha_k\, \mu(E_k \cap B_j)\\ &= \sum_{k=1}^{n} \alpha_k\, \mu(E_k)\,.\end{aligned}$$

Thus we obtain $L(f) \leqq \sum_{k=1}^{n} \alpha_k \mu(E_k)$ and hence $L(f) = \sum_{k=1}^{n} \alpha_k \mu(E_k)$. □

(12.5) Theorem. *Let f and g be any nonnegative functions on X. If $f(x) \leqq g(x)$ for all $x \in X$, then $L(f) \leqq L(g)$.*

Proof. Trivial.

(12.6) Theorem. *Let f be a nonnegative measurable function. If $\mu(\{x \in X : f(x) > 0\})$ is positive, then $L(f)$ is positive.*

Proof. We will find a set $A \in \mathscr{A}$ and a positive number α such that $\mu(A) > 0$ and $f(x) \geqq \alpha$ for all $x \in A$. It will then follow that

$$\begin{aligned}L(f) &\geqq \inf\{f(x) : x \in A\}\, \mu(A) + \inf\{f(x) : x \in A'\}\, \mu(A')\\ &\geqq \alpha\, \mu(A) > 0\,.\end{aligned}$$

For each positive integer n, let $A_n = \left\{x \in X : f(x) \geqq \frac{1}{n}\right\}$. We have $A_1 \subset A_2 \subset \cdots \subset A_n \subset \cdots$ and $\bigcup_{n=1}^{\infty} A_n = \{x \in X : f(x) > 0\}$. By (10.13), we have $\lim_{n\to\infty} \mu(A_n) = \mu\left(\bigcup_{n=1}^{\infty} A_n\right) > 0$. Hence there is a positive integer n_0 for which $\mu(A_{n_0}) > 0$ and $f(x) \geqq \frac{1}{n_0}$ on A_{n_0}. □

(12.7) Theorem. *Let f and g be in $\mathfrak{S}^+$. Then we have $L(f+g) = L(f) + L(g)$.*

Proof. Write $f = \sum_{j=1}^{m} \alpha_j \xi_{A_j}$ and $g = \sum_{k=1}^{n} \beta_k \xi_{B_k}$, where $\bigcup_{j=1}^{m} A_j = \bigcup_{k=1}^{n} B_k = X$. Then $f + g = \sum_{j=1}^{m} \sum_{k=1}^{n} (\alpha_j + \beta_k)\, \xi_{(A_j \cap B_k)}$. Thus by (12.4) we have

$$
\begin{aligned}
L(f+g) &= \sum_{j=1}^{m} \sum_{k=1}^{n} (\alpha_j + \beta_k)\, \mu(A_j \cap B_k) \\
&= \sum_{j=1}^{m} \sum_{k=1}^{n} \alpha_j\, \mu(A_j \cap B_k) + \sum_{k=1}^{n} \sum_{j=1}^{m} \beta_k\, \mu(A_j \cap B_k) \\
&= \sum_{j=1}^{m} \alpha_j\, \mu\left(A_j \cap \left(\bigcup_{k=1}^{n} B_k\right)\right) + \sum_{k=1}^{n} \beta_k\, \mu\left(B_k \cap \left(\bigcup_{j=1}^{m} A_j\right)\right) \\
&= \sum_{j=1}^{m} \alpha_j\, \mu(A_j) + \sum_{k=1}^{n} \beta_k\, \mu(B_k) \\
&= L(f) + L(g). \quad \square
\end{aligned}
$$

(12.8) Theorem. *If $f \geqq 0$ and $t \in R$, then we have $L(tf) = tL(f)$.*

The proof is easy and is omitted.

Our immediate aim is to establish the extremely important identity

$$L\left(\sum_{n=1}^{\infty} f_n\right) = \sum_{n=1}^{\infty} L(f_n)$$

for all sequences (f_n) of nonnegative $\mathscr{A}$-measurable functions. We begin with a lemma.

(12.9) Lemma. *Let f be any extended real-valued function on X and suppose that $E = \{x \in X : f(x) \neq 0\}$ is an $\mathscr{A}$-measurable set. Let $\mathscr{A}_E$ and μ_E be as in* (11.22) *and* (11.37). *Then, if $L(f)$ exists, we have $L(f) = L_E(f)$, where L_E is the integral for the measure space $(E, \mathscr{A}_E, \mu_E)$.*

Proof. First suppose that $f \geqq 0$ and let γ be any real number such that $\gamma < L(f)$. There exists a measurable dissection $\{A_1, \ldots, A_m\}$ of X satisfying the inequality

$$\gamma < \sum_{k=1}^{m} \inf\{f(x) : x \in A_k\}\, \mu(A_k).$$

Using the fact that E is $\mathscr{A}$-measurable, we have

$$
\begin{aligned}
\gamma &< \sum_{k=1}^{m} \inf\{f(x) : x \in A_k\}\, \mu(A_k \cap E) + \sum_{k=1}^{m} \inf\{f(x) : x \in A_k\}\, \mu(A_k \cap E') \\
&\leqq \sum_{k=1}^{m} \inf\{f(x) : x \in A_k \cap E\}\, \mu(A_k \cap E) \\
&\leqq L_E(f).
\end{aligned}
$$

[Note that $A_k \cap E' \neq \varnothing$ implies $\inf\{f(x) : x \in A_k\} = 0$.] It follows that $L(f) \leqq L_E(f)$.

Next suppose that $\gamma < L_E(f)$, and let $\{B_k\}_{k=1}^{m}$ be any measurable dissection of E such that

$$\gamma < \sum_{k=1}^{m} \inf\{f(x) : x \in B_k\} \mu(B_k) .$$

We have

$$\gamma < \sum_{k=1}^{m} \inf\{f(x) : x \in B_k\} \mu(B_k) + 0 \cdot \mu(E') \leqq L(f) ,$$

and so $L_E(f) \leqq L(f)$. The assertion for arbitrary functions follows immediately. □

The following result, which looks harmless enough, is the key to the proof that L is countably additive.

(12.10) Theorem. *Let $(g_n)_{n=1}^{\infty}$ be any nondecreasing sequence in $\mathfrak{S}^+$. Suppose that $h \in \mathfrak{S}^+$ and that $\lim_{n\to\infty} g_n \geqq h$. Then we have $\lim_{n\to\infty} L(g_n) \geqq L(h)$.*

Proof. The theorem is trivial if $\mu(X) = 0$, and so we suppose throughout that $0 < \mu(X) \leqq \infty$. Let $h = \gamma_1 \xi_{E_1} + \gamma_2 \xi_{E_2} + \cdots + \gamma_m \xi_{E_m}$, where the E_k's are pairwise disjoint, $X = \bigcup_{k=1}^{m} E_k$, and $0 \leqq \gamma_1 < \gamma_2 < \cdots < \gamma_m \leqq \infty$. Suppose that $\gamma_1 = 0$; then by (12.9) we have $L(h) = L_{E_1'}(h)$. Supposing that the theorem is established for the case $\gamma_1 > 0$ and letting E_1' take the rôle of X, we have

$$L(h) = L_{E_1'}(h) \leqq \lim_{n\to\infty} L_{E_1'}(g_n) \leqq \lim_{n\to\infty} L(g_n) .$$

It thus suffices to prove the theorem under the assumption that $\gamma_1 > 0$.

Case (I): $\mu(X)$ and γ_m are finite. For any $\delta > 0$, choose $\varepsilon > 0$ satisfying the inequality

$$\varepsilon \leqq \min\left\{\frac{\delta}{2\mu(X)}, \gamma_1\right\} .$$

For every positive integer n, let $S_n = \{x \in X : g_n(x) > h(x) - \varepsilon\}$. Since $\lim_{n\to\infty} g_n \geqq h$, we have $X = \bigcup_{n=1}^{\infty} S_n$. The sequence (g_n) is nondecreasing, and so the sequence $S_1, S_2, \ldots$ is nondecreasing. From these facts and the countable additivity of μ, we find (10.13) that

$$\lim_{n\to\infty} \mu(S_n) = \mu(X)$$

and that

$$\lim_{n\to\infty} \mu(S_n') = 0 .$$[1]

We also have $L(g_n) \geqq L(g_n \xi_{S_n}) \geqq L((h - \varepsilon)\xi_{S_n}) = L(h\xi_{S_n}) - \varepsilon L(\xi_{S_n})$. The relations $h = h\xi_{S_n} + h\xi_{S_n'} \leqq h\xi_{S_n} + \gamma_m \xi_{S_n'}$ imply that $L(h) \leqq$

[1] Countable additivity of μ is used in Case (I) only to establish this relation. However, countable additivity is essential: the theorem fails for μ's that are finitely but not countably additive.

$L(h\xi_{S_n}) + \gamma_m \mu(S'_n)$. Hence

$$\begin{aligned} L(g_n) &\geqq L(h) - \gamma_m \mu(S'_n) - \varepsilon \mu(S_n) \\ &\geqq L(h) - \gamma_m \mu(S'_n) - \varepsilon \mu(X) \\ &\geqq L(h) - \gamma_m \mu(S'_n) - \frac{1}{2}\delta . \end{aligned}$$

If n is so large that $\gamma_m \mu(S'_n) < \frac{\delta}{2}$, then we have $L(g_n) > L(h) - \delta$. The inequality $\lim_{n\to\infty} L(g_n) \geqq L(h)$ follows, as δ is arbitrary.

Case (II): γ_m is finite and $\mu(X) = \infty$. We plainly have $L(h) \geqq \gamma_1 \mu(X) = \infty$. Let ε be any number such that $0 < \varepsilon < \gamma_1$, and define S_n for $n \in N$ just as in Case (I). For $x \in S_n$, we have $g_n(x) > h(x) - \varepsilon \geqq \gamma_1 - \varepsilon$. Therefore the relations $L(g_n) \geqq L(g_n \xi_{S_n}) \geqq L((\gamma_1 - \varepsilon)\xi_{S_n}) = (\gamma_1 - \varepsilon)\mu(S_n)$ obtain, and (10.13) implies that

$$\lim_{n\to\infty} L(g_n) \geqq (\gamma_1 - \varepsilon) \lim_{n\to\infty} \mu(S_n) = (\gamma_1 - \varepsilon)(\infty) = \infty = L(h) .$$

Case (III): $\mu(E_m)$ is positive and $\gamma_m = \infty$. Here we have $L(h) \geqq \gamma_m \mu(E_m) = \infty$. Choose any real number $\gamma > \gamma_{m-1}$, and let $h_\gamma = \gamma_1 \xi_{E_1} + \cdots + \gamma_{m-1}\xi_{E_{m-1}} + \gamma \xi_{E_m}$. By Cases (I) and (II), we have $\lim_{n\to\infty} L(g_n) \geqq L(h_\gamma) \geqq \gamma \mu(E_m)$. Since γ can be arbitrarily large, it follows that $\lim_{n\to\infty} L(g_n) = \infty = L(h)$.

Case (IV): $\gamma_m = \infty$ and $\mu(E_m) = 0$. Here we have $L(h) = \sum_{j=1}^{m-1} \gamma_j \mu(E_j)$. Let $B = E_1 \cup E_2 \cup \cdots \cup E_{m-1}$. Then

$$g_n \geqq g_n \xi_B \quad \text{and} \quad \lim_{n\to\infty} g_n \xi_B \geqq h\xi_B = \sum_{j=1}^{m-1} \gamma_j \xi_{E_j} .$$

Since $\gamma_{m-1} < \infty$, Case (I) or Case (II) applies to $(g_n \xi_B)$ and $h\xi_B$, so that $\lim_{n\to\infty} L(g_n) \geqq \lim_{n\to\infty} L(g_n \xi_B) \geqq L(h\xi_B) = L(h)$. □

(12.11) Theorem. *Let (g_n) be a nondecreasing sequence of functions in $\mathfrak{S}^+$. Then we have $\lim_{n\to\infty} L(g_n) = L(\lim_{n\to\infty} g_n)$.*

Proof. Let $\lim_{n\to\infty} g_n = \varphi$, and let γ be any real number such that $\gamma < L(\varphi)$. There exists a measurable dissection $\{A_1, A_2, \ldots, A_m\}$ of X such that

$$\gamma < \sum_{k=1}^{m} \inf\{\varphi(x) : x \in A_k\}\, \mu(A_k) = L\left(\sum_{k=1}^{m} \alpha_k \xi_{A_k}\right) \leqq \lim_{n\to\infty} L(g_n) ,$$

where $\alpha_k = \inf\{\varphi(x) : x \in A_k\}$. Here we have used (12.10). Since γ is arbitrary, we infer that $L(\varphi) \leqq \lim_{n\to\infty} L(g_n)$. The reverse inequality is immediate. □

(12.12) Theorem. *Let f and g be nonnegative $\mathscr{A}$-measurable functions. Then*

$$L(f+g) = L(f) + L(g)\,.$$

Proof. Let (f_n) and (g_n) be nondecreasing sequences of nonnegative simple functions with limits f and g respectively (11.35). The sequence $(f_n + g_n)$ increases to $f + g$, and so by (12.11) and the additivity of L on nonnegative simple functions (12.7), we have

$$L(f) + L(g) = \lim_{n\to\infty} L(f_n) + \lim_{n\to\infty} L(g_n) = \lim_{n\to\infty} L(f_n + g_n) = L(f+g)\,. \ \square$$

(12.13) Theorem. *Let f be a nonnegative $\mathscr{A}$-measurable function on X. Then $L(f) = 0$ if and only if $f = 0$ μ-a.e.*

Proof. Let $E = \{x \in X : f(x) > 0\}$. If $L(f) = 0$, then it follows from (12.6) that $\mu(E) = 0$, *i.e.*, $f = 0$ a.e. Conversely, suppose that $\mu(E) = 0$. Then

$$0 \leqq L(f) \leqq L(\infty \cdot \xi_E) = \infty \cdot \mu(E) = 0\,. \ \square$$

(12.14) Theorem. *Let f and g be $\mathscr{A}$-measurable, extended real-valued functions on X such that $f = g$ μ-a.e. and $L(f)$ is defined. Then $L(g)$ is defined and $L(g) = L(f)$.*

Proof. Let $E = \{x \in X : f(x) \neq g(x)\}$. By hypothesis $\mu(E) = 0$.

Case (I): $f \geqq 0$, $g \geqq 0$. Apply (12.12) and (12.13) to obtain

$$L(f) = L(f\xi_E) + L(f\xi_{E'}) = L(f\xi_{E'}) = L(g\xi_{E'}) = L(g\xi_E) + L(g\xi_{E'}) = L(g)\,.$$

Case (II): general case. For $x \in X$ such that $f(x) = g(x)$, we have $f^+(x) = \max\{f(x), 0\} = \max\{g(x), 0\} = g^+(x)$ and $g^-(x) = -\min\{g(x), 0\} = -\min\{f(x), 0\} = f^-(x)$. Therefore

$$\{x \in X : f^+(x) \neq g^+(x)\} \cup \{x \in X : f^-(x) \neq g^-(x)\} \subset E\,,$$

and so $f^+ = g^+$ a.e. and $f^- = g^-$ a.e. Applying Case (I) twice, we conclude that $L(f) = L(f^+) - L(f^-) = L(g^+) - L(g^-) = L(g)$. $\square$

(12.15) Theorem. *Let f be an extended real-valued, $\mathscr{A}$-measurable function defined on X and suppose that $L(f)$ is defined and finite. Then $\mu(\{x \in X : f(x) = \pm\infty\}) = 0$, i.e., f is finite μ-a.e.*

Proof. Let $A = \{x \in X : f(x) = \infty\}$ and $B = \{x \in X : f(x) = -\infty\}$. By the definition of L we have

$$\infty \cdot \mu(A) + \inf\{f^+(x) : x \in A'\}\,\mu(A') \leqq L(f^+) < \infty$$

and

$$\infty \cdot \mu(B) + \inf\{f^-(x) : x \in B'\}\,\mu(B') \leqq L(f^-) < \infty\,.$$

It follows that $\mu(A) = \mu(B) = 0$. $\square$

(12.16) Remarks. Let f be an extended real-valued, $\mathscr{A}$-measurable function defined on X, let E be any set in $\mathscr{A}$, and let α be any extended real number. Let f_1 be the function on X such that

$$\text{(i)} \qquad f_1(x) = \begin{cases} \alpha & \text{if } x \in E, \\ f(x) & \text{if } x \in E'. \end{cases}$$

It is obvious from (11.2) that f_1 is $\mathscr{A}$-measurable. If $\mu(E) = 0$ and $L(f)$ is defined, then (12.14) shows that $L(f_1)$ is defined and that $L(f_1) = L(f)$. If $L(f)$ is finite, we use (12.15) and the value $\alpha = 0$ [say] in (i) to replace f by a finite-valued function f_1 equal to f a.e. and having the same integral as f. Thus we lose nothing in dealing with $\mathscr{A}$-measurable functions having finite integrals if we suppose that these functions are finite-valued. It is also convenient at times to consider functions defined only almost everywhere. The definition follows.

(12.17) Definition. Let E be a set in $\mathscr{A}$ such that $\mu(E') = 0$. Let $\mathscr{A}_E$ be as in (11.37), and let f be an $\mathscr{A}_E$-measurable, extended real-valued function defined on E. Let f_0 be any extended real-valued, $\mathscr{A}$-measurable function on X such that $f_0(x) = f(x)$ for $x \in E$ [*e.g.*, $f_0(x) = 0$ for $x \in E'$]. Let $L(f) = L(f_0)$ if $L(f_0)$ is defined, and leave $L(f)$ undefined if $L(f_0)$ is undefined. [It is immediate from (12.14) that $L(f)$ is uniquely determined by the definition just given.] We shall frequently in the sequel encounter functions that are defined only on sets E as above and are $\mathscr{A}_E$-measurable. To avoid tedious repetition, we shall call such functions $\mathscr{A}$-measurable, although this is not really correct, and we will whenever convenient think of these functions as being extended over all of X so as to be $\mathscr{A}$-measurable.

We now introduce a very important space of functions.

(12.18) Definition. We define $\mathfrak{L}_1^r(X, \mathscr{A}, \mu)$ as the set of all $\mathscr{A}$-measurable real-valued functions f defined μ-a.e. on X such that $L(f)$ exists and is finite. Where confusion seems impossible, we will write $\mathfrak{L}_1^r$ for $\mathfrak{L}_1^r(X, \mathscr{A}, \mu)$.

The functional L is ordinarily written in integral notation:

$$L(f) = \int_X f(x)\, d\mu(x) = \int_X f(t)\, d\mu(t) = \int_X f\, d\mu = \int f\, d\mu .$$

We will adopt this notation in dealing with $\mathscr{A}$-measurable functions. In case $X = [a, b]$ and $\mu = \lambda$, we write $\int_a^b f(x)\, dx$, $\int_a^b f(t)\, dt$, etc., for $\int_{[a,b]} f\, d\lambda$. The notations $\int_{-\infty}^{\infty} f(x)\, dx$, $\int_a^{\infty} f(x)\, dx$, and $\int_{-\infty}^{b} f(x)\, dx$ are self-explanatory.

(12.19) Theorem. *Let $f \in \mathfrak{L}_1^r$, and let $f = f_1 - f_2$, where $f_1 \geq 0$, $f_2 \geq 0$ and $f_1, f_2 \in \mathfrak{L}_1^r$. Then*

$$\int_X f\, d\mu = \int_X f_1\, d\mu - \int_X f_2\, d\mu .$$

Proof. By definition, we have $\int\limits_X f\,d\mu = \int\limits_X f^+\,d\mu - \int\limits_X f^-\,d\mu$. Since $f_1 - f_2 = f^+ - f^-$, we have $f_1 + f^- = f^+ + f_2$. From this equality and (12.12) we infer that

$$\int\limits_X f_1\,d\mu + \int\limits_X f^-\,d\mu = \int\limits_X f^+\,d\mu + \int\limits_X f_2\,d\mu. \quad \square$$

(12.20) Theorem. *For $f, g \in \mathfrak{L}_1^r$ and $\alpha, \beta \in R$, we have*

$$\int\limits_X (\alpha f + \beta g)\,d\mu = \alpha \int\limits_X f\,d\mu + \beta \int\limits_X g\,d\mu\,.$$

That is, the mapping $f \to \int\limits_X f\,d\mu$ is a linear functional on $\mathfrak{L}_1^r$.

The proof is easy and is omitted.

(12.21) Theorem [Lebesgue]. *Let (f_n) be a sequence of nonnegative, extended real-valued, $\mathscr{A}$-measurable functions on X. Then*

$$\int\limits_X \left(\sum_{j=1}^{\infty} f_j\right) d\mu = \sum_{j=1}^{\infty} \int\limits_X f_j\,d\mu\,.$$

Proof. For every positive integer m, we have

$$\sum_{j=1}^{\infty} f_j \geqq \sum_{j=1}^{m} f_j;$$

therefore

$$\int\limits_X \left(\sum_{j=1}^{\infty} f_j\right) d\mu \geqq \int\limits_X \left(\sum_{j=1}^{m} f_j\right) d\mu = \sum_{j=1}^{m} \int\limits_X f_j\,d\mu\,,$$

and consequently

$$\int\limits_X \left(\sum_{j=1}^{\infty} f_j\right) d\mu \geqq \sum_{j=1}^{\infty} \int\limits_X f_j\,d\mu\,.$$

For every positive integer n, let $(s_n^{(k)})_{k=1}^{\infty}$ be a nondecreasing sequence of functions in $\mathfrak{S}^+$ with limit f_n. For $k \in N$, let $g_k = s_1^{(k)} + s_2^{(k)} + \cdots + s_k^{(k)}$. The sequence $(g_k)_{k=1}^{\infty}$ is obviously nondecreasing. If $m \leqq k$, then we have

$$s_1^{(k)} + s_2^{(k)} + \cdots + s_m^{(k)} \leqq g_k \leqq f_1 + \cdots + f_k \leqq \sum_{j=1}^{\infty} f_j\,.$$

Taking the limit with respect to k, we find that

$$f_1 + f_2 + \cdots + f_m \leqq \lim_{k\to\infty} g_k \leqq \sum_{j=1}^{\infty} f_j$$

for each m. Taking the limit with respect to m, we obtain

$$\lim_{k\to\infty} g_k = \sum_{j=1}^{\infty} f_j.$$

Now (12.11) implies that

$$\int_X \left(\sum_{j=1}^{\infty} f_j\right) d\mu = \lim_{k\to\infty} \int_X g_k\, d\mu \leqq \lim_{k\to\infty} \int_X (f_1 + f_2 + \cdots + f_k)\, d\mu$$

$$= \lim_{k\to\infty} \sum_{j=1}^{k} \int_X f_j\, d\mu = \sum_{j=1}^{\infty} \int_X f_j\, d\mu. \quad \square$$

(12.22) B. Levi's Theorem. *Let $(f_k)_{k=1}^{\infty}$ be a nondecreasing sequence of extended real-valued, $\mathcal{A}$-measurable functions on X such that $\int_X f_k^- \, d\mu < \infty$ for some k. Then*

$$\lim_{k\to\infty} \int_X f_k\, d\mu = \int_X (\lim_{k\to\infty} f_k)\, d\mu\,.$$

Proof. We may suppose with no loss of generality that $\int_X f_1^- \, d\mu < \infty$, and in view of (12.16) that no f_k assumes the value $-\infty$. If any $\int_X f_k\, d\mu$ is equal to ∞, the result is trivial. Otherwise, for $k \in N$, we define

$$g_k(x) = \begin{cases} f_{k+1}(x) - f_k(x) & \text{if} \quad f_k(x) < \infty\,, \\ \infty \quad \text{otherwise}\,. \end{cases}$$

Then we have $\lim_{n\to\infty} f_n = \lim_{n\to\infty} \left(f_1 + \sum_{k=1}^{n-1} g_k\right) = f_1 + \sum_{k=1}^{\infty} g_k$, and so by (12.21) and (12.19),

$$\int_X (\lim_{n\to\infty} f_n)\, d\mu = \int_X f_1\, d\mu + \sum_{k=1}^{\infty} \left(\int_X f_{k+1}\, d\mu - \int_X f_k\, d\mu\right) = \lim_{n\to\infty} \int_X f_n\, d\mu. \quad \square$$

(12.23) Fatou's Lemma. *Let $(f_n)_{n=1}^{\infty}$ be a sequence of nonnegative, extended real-valued, $\mathcal{A}$-measurable functions on X.* Then

$$\int_X \underline{\lim}_{n\to\infty} f_n\, d\mu \leqq \underline{\lim}_{n\to\infty} \int_X f_n\, d\mu\,.$$

Proof. For every positive integer k, let $g_k = \inf\{f_k, f_{k+1}, \ldots\}$. Plainly g_k is $\mathcal{A}$-measurable, (g_k) is nondecreasing, and $g_k \leqq f_k$. The hypotheses of (12.22) obtain for $(g_k)_{k=1}^{\infty}$, and so we have

$$\int_X \underline{\lim}_{n\to\infty} f_n\, d\mu = \int_X \lim_{k\to\infty} g_k\, d\mu = \lim_{k\to\infty} \int_X g_k\, d\mu \leqq \underline{\lim}_{n\to\infty} \int_X f_n\, d\mu\,. \quad \square$$

(12.24) Lebesgue's Dominated Convergence Theorem. *Let $(f_n)_{n=1}^{\infty}$ be a sequence of extended real-valued, $\mathcal{A}$-measurable functions each defined a.e. on X, and suppose that there is a function $s \in \mathfrak{L}_1^+$ such that for each n, the inequality $|f_n(x)| \leqq s(x)$ holds a.e. on X. Then*

(i) $$\int_X \underline{\lim}_{n\to\infty} f_n\, d\mu \leqq \underline{\lim}_{n\to\infty} \int_X f_n\, d\mu$$

and

(ii) $$\int_X \overline{\lim}_{n\to\infty} f_n\, d\mu \geqq \overline{\lim}_{n\to\infty} \int_X f_n\, d\mu\,.$$

If $\lim_{n\to\infty} f_n(x)$ *exists for* μ*-almost all* $x \in X$, *then* $\lim_{n\to\infty} \int_X f_n \, d\mu$ *exists and*

(iii) $$\int_X \lim_{n\to\infty} f_n \, d\mu = \lim_{n\to\infty} \int_X f_n \, d\mu .$$

Proof. It is obvious that all f_n are in $\mathfrak{L}_1^r$. Hence all f_n, and s, are finite a.e. on X. Let $A = \{x \in X : f_n(x) \text{ is } \pm\infty \text{ or } |f_n(x)| > s(x) \text{ for some } n \in N\}$, let $B = \{x \in X : f_n(x) \text{ is undefined for some } n \in N\}$, and let $C = \{x \in X : s(x) \text{ is infinite or is undefined}\}$. Let $f_n(x) = s(x) = 0$ on $A \cup B \cup C$. Since $\mu(A \cup B \cup C) = 0$, (12.14) shows that none of the integrals appearing in the statement of the theorem has been changed by this definition. Furthermore we have $|f_n(x)| \leq s(x) < \infty$ for all $n \in N$ and all $x \in X$.

The sequence $(s + f_n)_{n=1}^{\infty}$ consists of nonnegative functions. Applying FATOU's lemma (12.23), we find

$$\int_X s \, d\mu + \int_X \underline{\lim}_{n\to\infty} f_n \, d\mu = \int_X \left[\underline{\lim}_{n\to\infty} (s + f_n)\right] d\mu$$

$$\leq \underline{\lim}_{n\to\infty} \int_X (s + f_n) \, d\mu = \int_X s \, d\mu + \underline{\lim}_{n\to\infty} \int_X f \, d\mu .$$

Thus (i) holds. [The reader will note that the function $\underline{\lim}_{n\to\infty} f_n$ occurring in (i) is defined only a.e., but is equal a.e. to the function $\underline{\lim}_{n\to\infty} f_n$ with the f_n defined everywhere as in the proof.] The inequality (ii) is proved in like manner, starting with the sequence $(s - f_n)_{n=1}^{\infty}$ and using the equality $\overline{\lim}_{n\to\infty} \alpha_n = - \underline{\lim}_{n\to\infty} (-\alpha_n)$.

Finally, if $\lim_{n\to\infty} f_n$ exists a.e. on X, (i) and (ii) imply that

$$\underline{\lim}_{n\to\infty} \int_X f_n \, d\mu \geq \int_X \lim_{n\to\infty} f_n \, d\mu \geq \overline{\lim}_{n\to\infty} \int_X f_n \, d\mu .$$

Hence $\lim_{n\to\infty} \int_X f_n \, d\mu$ exists and (iii) holds. □

(12.25) Note. The presence of the dominating function s in the above theorem is of the utmost importance. If no such function exists, the conclusion may fail. For example, let $X = R$, $\mu = \lambda$, and $f_n = n\xi_{\left]0, \frac{1}{n}\right]}$. Then $\int_R f_n \, d\lambda = n \cdot \frac{1}{n} = 1$ for all $n \in N$ while $\lim_{n\to\infty} f_n(x) = 0$ for all $x \in R$. That is, $\lim_{n\to\infty} \int_R f_n \, d\lambda = 1 \neq 0 = \int_R \lim_{n\to\infty} f_n \, d\lambda$.

We next extend our integral to complex-valued functions.

(12.26) Definition. Let $\mathfrak{L}_1(X, \mathscr{A}, \mu)$ [written for brevity as $\mathfrak{L}_1$] denote the set of all complex-valued functions f such that f is defined μ-a.e. on X, $\operatorname{Re} f \in \mathfrak{L}_1^r$, and $\operatorname{Im} f \in \mathfrak{L}_1^r$. For $f \in \mathfrak{L}_1$, we define $\int_X f \, d\mu = \int_X \operatorname{Re} f \, d\mu + i \int_X \operatorname{Im} f \, d\mu$. Functions in $\mathfrak{L}_1$ are sometimes called *integrable* or *summable*.

(12.27) Theorem. *Let $f, g \in \mathfrak{L}_1$ and $\alpha, \beta \in K$. Then $\alpha f + \beta g \in \mathfrak{L}_1$ and $\int_X (\alpha f + \beta g)\, d\mu = \alpha \int_X f\, d\mu + \beta \int_X g\, d\mu$,* i.e., *$\mathfrak{L}_1$ is a complex linear space and $\int_X \cdots d\mu$ is a linear functional on $\mathfrak{L}_1$.*

This theorem follows at once by considering real and imaginary parts and applying previous results.

(12.28) Theorem. *Let f be a complex-valued $\mathscr{A}$-measurable function on X. Then*

(i) *$f \in \mathfrak{L}_1$ if and only if $|f| \in \mathfrak{L}_1$,*

and

(ii) *if $f \in \mathfrak{L}_1$, then $\left|\int_X f\, d\mu\right| \leqq \int_X |f|\, d\mu$.*

Proof. Conclusion (i) follows directly from the inequalities

$$|f| \leqq |\mathrm{Re} f| + |\mathrm{Im} f| \leqq 2|f|$$

and the fact that $|g| = g^+ + g^-$ for real-valued g's. To prove (ii), repeat the argument of (9.4). □

(12.29) Note. To find necessary and sufficient conditions for equality in the inequality $\left|\int_X f\, d\mu\right| \leqq \int_X |f|\, d\mu$, consider $h \in \mathfrak{L}_1$. We then ask when the equality

$$\left|\int h\, d\mu\right| = \int |h|\, d\mu$$

holds. It clearly suffices to have $h = \exp(i\alpha)\, |h|$ where α is any real number. We now show that this condition is also necessary. Suppose then that $\int h\, d\mu = \exp(i\beta) \left|\int h\, d\mu\right|$ for a real number β, and define

$$\varphi = \exp(-i\beta)\, h = \varphi_1 + i\varphi_2,$$

where φ_1 and φ_2 are real-valued functions. We have

$$\int \varphi\, d\mu = \exp(-i\beta) \int h\, d\mu = \exp(-i\beta)\exp(i\beta) \left|\int h\, d\mu\right| = \int |h|\, d\mu\,.$$

Hence

$$\int \varphi\, d\mu = \int \varphi_1\, d\mu + i \int \varphi_2\, d\mu = \int [\varphi_1^2 + \varphi_2^2]^{\frac{1}{2}}\, d\mu\,,$$

and therefore

$$\int \varphi\, d\mu = \int \varphi_1\, d\mu \leqq \int |\varphi_1|\, d\mu \leqq \int [\varphi_1^2 + \varphi_2^2]^{\frac{1}{2}}\, d\mu = \int \varphi\, d\mu\,.$$

Hence we have $\varphi_2 = 0$ a.e. and so $\varphi = \varphi_1$ a.e. Since $\int \varphi\, d\mu = \int |\varphi|\, d\mu$, we have $\varphi \geqq 0$ a.e. Thus the equality $\varphi = \exp(-i\beta) h \geqq 0$ holds a.e. and from this we conclude that $h = \exp(i\beta)\varphi_1 = \exp(i\beta)\, |h|$ a.e.

(12.30) Lebesgue's Dominated Convergence Theorem [complex form]. *Let (f_n) be a sequence in $\mathfrak{L}_1$ such that $\lim_{n\to\infty} f_n(x)$ exists μ-a.e. on X. Suppose that there exists a function $s \in \mathfrak{L}_1^r$ such that $|f_n| \leqq s$ μ-a.e. for each $n \in N$.*

Then $\lim\limits_{n\to\infty} f_n \in \mathfrak{L}_1$ *and*

$$\lim_{n\to\infty} \int_X f_n \, d\mu = \int_X \lim_{n\to\infty} f_n \, d\mu .$$

Proof. Let $f(x) = \lim\limits_{n\to\infty} f_n(x)$ whenever this limit exists. Clearly f is defined a.e. on X and is $\mathscr{A}$-measurable. Also $|f(x) - f_n(x)| \leqq |f(x)| + |f_n(x)| \leqq 2s(x)$ for all $n \in N$ and $\lim\limits_{n\to\infty} |f(x) - f_n(x)| = 0$ a.e. Thus, by (12.24) and (12.28.ii), we have

$$\left|\int_X f \, d\mu - \int_X f_n \, d\mu\right| \leqq \int_X |f - f_n| \, d\mu \to \int_X 0 \, d\mu = 0 . \quad \square$$

(12.31) Definition. Let f be any function for which $\int_X f \, d\mu$ is defined. For each $E \in \mathscr{A}$ we define

$$\int_E f \, d\mu = \int_X \xi_E f \, d\mu .$$

It is easy to see that

$$\int_E f \, d\mu = \int_E f \, d\mu_E ,$$

where μ_E is the measure μ restricted to the σ-algebra $\mathscr{A}_E$ (11.37).

(12.32) Corollary. *Let f be in $\mathfrak{L}_1$, let $(A_n)_{n=1}^\infty$ be a pairwise disjoint sequence in $\mathscr{A}$, and write $A = \bigcup\limits_{n=1}^\infty A_n$. Then*

$$\int_A f \, d\mu = \sum_{n=1}^\infty \int_{A_n} f \, d\mu .$$

Proof. Define $g_n = f\xi_{A_1} + \cdots + f\xi_{A_n}$. Then

$$|g_n| \leqq |f| \in \mathfrak{L}_1$$

and

$$\lim_{n\to\infty} g_n(x) = f(x)\,\xi_A(x) \quad \text{a.e.}$$

From (12.30) we have

$$\begin{aligned} \int_A f \, d\mu &= \int_X f\xi_A \, d\mu = \int_X \lim_{n\to\infty} g_n \, d\mu = \lim_{n\to\infty} \int_X g_n \, d\mu \\ &= \lim_{n\to\infty} \sum_{k=1}^n \int_{A_k} f \, d\mu \\ &= \sum_{k=1}^\infty \int_{A_k} f \, d\mu. \quad \square \end{aligned}$$

(12.33) Corollary. *Let (f_n) be a sequence of complex-valued $\mathscr{A}$-measurable functions on X such that $\sum\limits_{n=1}^\infty |f_n| \in \mathfrak{L}_1$ [or, equivalently, $\sum\limits_{n=1}^\infty \int_X |f_n| \, d\mu < \infty$]. Then $\sum\limits_{n=1}^\infty f_n$ is in $\mathfrak{L}_1$ and $\displaystyle\int_X \sum_{n=1}^\infty f_n \, d\mu = \sum_{n=1}^\infty \int_X f_n \, d\mu$.*

Proof. Exercise.

LEBESGUE'S theorem on dominated convergence (12.30), and its cousins (12.21)–(12.24), are used very frequently in analysis. It is not too much to say that Fourier analysis, for example, depends upon (12.30). We shall take up some of these applications in the sequel; for the moment we content ourselves with a simple though nonobvious corollary of (12.22).

(12.34) Theorem. *Let $f \in \mathfrak{L}_1(X, \mathscr{A}, \mu)$. For every $\varepsilon > 0$ there exists a $\delta > 0$ depending only on ε and f such that for all $E \in \mathscr{A}$ satisfying $\mu(E) < \delta$, we have*

$$\int_E |f|\, d\mu < \varepsilon .$$

Proof. For $n = 1, 2, \ldots$, let

$$\psi_n(x) = \begin{cases} |f(x)| & \text{if } \ |f(x)| \leqq n, \\ n & \text{otherwise}. \end{cases}$$

Then (ψ_n) is a nondecreasing sequence of $\mathscr{A}$-measurable functions and $\lim\limits_{n\to\infty} \psi_n = |f|$. By (12.22), we have

$$\lim_{n\to\infty} \int_X \psi_n\, d\mu = \int_X \lim_{n\to\infty} \psi_n\, d\mu = \int_X |f|\, d\mu .$$

Select n so that $\int\limits_X (|f| - \psi_n)\, d\mu < \frac{1}{2}\varepsilon$. Setting $\delta = \frac{\varepsilon}{2n}$ and choosing any $E \in \mathscr{A}$ such that $\mu(E) < \delta$, we have

$$\int_E \psi_n\, d\mu \leqq \int_E n\, d\mu = n\,\mu(E) < \frac{1}{2}\varepsilon .$$

It follows that

$$\begin{aligned} \Big|\int_E f\, d\mu\Big| &\leqq \int_E |f|\, d\mu = \int_E (|f| - \psi_n)\, d\mu + \int_E \psi_n\, d\mu \\ &< \int_X (|f| - \psi_n)\, d\mu + \frac{1}{2}\varepsilon < \frac{1}{2}\varepsilon + \frac{1}{2}\varepsilon = \varepsilon \end{aligned}$$

for all $E \in \mathscr{A}$ such that $\mu(E) < \delta$. □

We now return to the functionals I, $\bar{I}$, and $\bar{\bar{I}}$ of § 9. We wish to show that $\bar{\bar{I}}$ is actually an integral.

(12.35) Theorem. *Let X be a locally compact Hausdorff space, let I be a nonnegative linear functional on $\mathfrak{C}_{00}(X)$, $\bar{\bar{I}}$ as in* (9.15), *and ι as in* (9.19). *Then $(X, \mathscr{M}_\iota, \iota)$ is a measure space* (10.20); *and for every nonnegative $\mathscr{M}_\iota$-measurable function f on X, we have*

(i) $$\bar{\bar{I}}(f) = \int_X f\, d\iota .$$

Proof. Let (s_n) be the sequence of simple functions defined in terms of f as in (11.35):

$$s_n = \sum_{k=1}^{n\cdot 2^n} \frac{k-1}{2^n}\, \xi_{A_{n,k}} + n\,\xi_{B_n} ,$$

where $A_{n,k} = \left\{x \in X : \frac{k-1}{2^n} \leq f(x) < \frac{k}{2^n}\right\}$ and $B_n = \{x \in X : f(x) \geq n\}$.
By (10.35) and (12.4), we have

$$\bar{\bar{I}}(s_n) = \int_X s_n d\iota \quad \text{for all} \quad n \in N .$$

By (9.17), we have

$$\lim_{n\to\infty} \bar{\bar{I}}(s_n) = \bar{\bar{I}}(f) .$$

By (12.11), we have

$$\lim_{n\to\infty} \int_X s_n d\iota = \int_X f d\iota .$$

Combining these equalities, we have (i). □

Theorem (12.35) is a generalized form of one of the most famous and most important theorems of modern analysis; we now state it.

(12.36) F. Riesz's Representation Theorem. *Let X be a locally compact Hausdorff space and let I be a nonnegative linear functional on $\mathfrak{C}_{00}(X)$. Then there is a measure space $(X, \mathscr{M}_\iota, \iota)$, where $\mathscr{M}_\iota$ contains all Borel sets, such that*

(i) $$I(f) = \int_X f(x)\, d\iota(x)$$

for all $f \in \mathfrak{C}_{00}(X)$.

Proof. This is a special case of (12.35), since $\bar{\bar{I}}(f) = I(f)$ for all nonnegative f in $\mathfrak{C}_{00}(X)$, and I and $\int_X \cdots d\iota$ are linear functionals. □

(12.37) Remark. The importance of Riesz's representation theorem lies in the countable additivity of the integral, as described in (12.21) to (12.24) and (12.30). Frequently we encounter functionals I on $\mathfrak{C}_{00}(X)$ that are nonnegative and linear. Riesz's theorem shows that we can write I as a countably additive integral; and from this useful consequences often ensue.

(12.38) Remark. In (12.36) there is no statement that the measure ι corresponding to a given functional I is unique. In fact, in some cases there are distinct measures ι and η defined on $\mathscr{B}(X)$ such that $\int_X f d\iota = \int_X f d\eta$ for all $f \in \mathfrak{C}_{00}(X)$ [see (12.58) *infra*]. However, this phenomenon does not occur if we restrict our attention to *regular* measures.

(12.39) Definition. Let X be a locally compact Hausdorff space and let μ be a measure defined on a σ-algebra $\mathscr{A}$ of subsets of X such that $\mathscr{A}$ contains $\mathscr{B}(X)$, the Borel sets of X. Then μ is called a *regular measure* if:

(i) $\mu(F) < \infty$ for all compact sets $F \subset X$;

(ii) $\mu(A) = \inf\{\mu(U) : U \text{ is open in } X, A \subset U\}$ for all $A \in \mathscr{A}$;

(iii) $\mu(U) = \sup\{\mu(F) : F \text{ is compact}, F \subset U\}$ for all open sets $U \subset X$.

It follows from (10.20), (9.27), (9.24), and (9.26) that every measure ι defined as in § 9 is a regular measure on $\mathscr{M}_\iota$. [Cf. the different but related definition of regular *outer* measure in (10.40) *supra.*]

(12.40) Theorem. *Let μ be a regular measure defined on a σ-algebra $\mathscr{A}$ of subsets of a locally compact Hausdorff space X. Then*

$$\mu(A) = \sup\{\mu(F) : F \text{ is compact, } F \subset A\}$$

for every $A \in \mathscr{A}$ that is σ-finite with respect to μ.

Proof. Repeat verbatim the proof of (10.30) with ι replaced by μ. □

(12.41) Theorem. *Let X be a locally compact Hausdorff space and let μ and ν be regular measures defined on σ-algebras $\mathscr{M}_\mu$ and $\mathscr{M}_\nu$ respectively. Suppose that $\int\limits_X f\,d\mu = \int\limits_X f\,d\nu$ for all $f \in \mathfrak{C}_{00}^+(X)$. Then $\mu(E) = \nu(E)$ for all $E \in \mathscr{M}_\mu \cap \mathscr{M}_\nu$.*

Proof. Let F be any nonvoid compact subset of X. Use (12.39.ii) to find sequences (U_n) and (V_n) of open sets containing F such that $\mu(U_1) < \infty$, $\nu(V_1) < \infty$, $\mu(U_n) \to \mu(F)$, and $\nu(V_n) \to \nu(F)$. For each $n \in N$, set $W_n = \bigcap\limits_{k=1}^{n} (U_k \cap V_k)$. Then each W_n is open, $W_1 \supset W_2 \supset \cdots \supset F$, $\nu(W_n) \to \nu(F)$, and $\mu(W_n) \to \mu(F)$. For each $n \in N$, use (6.80) to obtain a function $f_n \in \mathfrak{C}_{00}^+(X)$ such that $f_n(X) \subset [0, 1]$, $f_n(F) = \{1\}$, and $f_n(W_n')$ $\subset \{0\}$. Next let $g_n = \min\{f_1, \ldots, f_n\}$. It is clear that $g_n \in \mathfrak{C}_{00}^+(X)$, that $g_1 \geqq g_2 \geqq \cdots$, and that $\lim\limits_{n\to\infty} g_n(x) = \xi_F(x)$ for all $x \in \left(\bigcap\limits_{n=1}^{\infty} W_n \cap F'\right)'$, *i.e.* μ-a.e. and ν-a.e. Since $\int\limits_X g_1\,d\mu \leqq \mu(W_1) < \infty$ and $\int\limits_X g_1\,d\nu \leqq \nu(W_1) < \infty$ Theorem (12.24) applies to yield

$$\mu(F) = \int\limits_X \xi_F\,d\mu = \lim_{n\to\infty} \int\limits_X g_n\,d\mu = \lim_{n\to\infty} \int\limits_X g_n\,d\nu = \int\limits_X \xi_F\,d\nu = \nu(F)\,.$$

Thus $\mu(F) = \nu(F)$ for all compact sets $F \subset X$.

For open sets $U \subset X$, we have $\mu(U) = \sup\{\mu(F) : F \text{ is compact}, F \subset U\} = \sup\{\nu(F) : F \text{ is compact}, F \subset U\} = \nu(U)$. For arbitrary set $E \in \mathscr{M}_\mu \cap \mathscr{M}_\nu$, we have $\mu(E) = \inf\{\mu(U) : U \text{ is open}, E \subset U\} = \inf\{\nu(U) : U \text{ is open}, E \subset U\} = \nu(E)$. □

(12.42) Theorem. *Let X be a locally compact Hausdorff space an[d] let μ be a regular measure defined on a σ-algebra $\mathscr{A}$ of subsets of X such th[at] $(X, \mathscr{A}, \mu)$ is a complete measure space. Suppose that $E \in \mathscr{A}$ if and onl[y] if $E \cap F \in \mathscr{A}$ for every compact set $F \subset X$. Define I on $\mathfrak{C}_{00}(X)$ by $I(f)$ $= \int\limits_X f\,d\mu$ and let ι be the measure constructed from I as in § 9. Then $\mathscr{A} = \mathscr{M}_\iota$ and $\mu(E) = \iota(E)$ for all $E \in \mathscr{M}_\iota$.*

Proof. Let E be in $\mathscr{A}$ and suppose that $\mu(E) < \infty$. Since μ is regular, there exist sequences of sets (F_n) and (U_n) such that $F_n \subset E \subset U_n$, each F_n is compact, each U_n is open, $\mu(F_n) \to \mu(E)$, and $\mu(U_n) \to \mu(E)$. Let $A = \bigcup_{n=1}^{\infty} F_n$ and $B = \bigcap_{n=1}^{\infty} U_n$. Then A and B are Borel sets, $A \subset E \subset B$, $\mu(A) = \mu(E) = \mu(B)$, and $\mu(B \cap A') = 0$. It follows from (12.35) and (12.41) that $\iota(A) = \iota(B)$ and $\iota(B \cap A') = 0$. Therefore, since ι is complete, $E = A \cup [E \cap (B \cap A')] \in \mathscr{M}_\iota$. This argument proves:

$$E \in \mathscr{A} \quad \text{and} \quad \mu(E) < \infty \quad \text{imply} \quad E \in \mathscr{M}_\iota\,. \tag{1}$$

Repeating the above argument with the rôles of μ and ι reversed, we have

$$E \in \mathscr{M}_\iota \quad \text{and} \quad \iota(E) < \infty \quad \text{imply} \quad E \in \mathscr{A}\,. \tag{2}$$

Next consider any E in $\mathscr{A}$ and any compact set $F \subset X$. By hypothesis we have $E \cap F \in \mathscr{A}$, and of course $\mu(E \cap F) < \infty$. Therefore $E \cap F$ is in $\mathscr{M}_\iota$ for all compact sets F. Applying (10.31.iv), we infer that $E \in \mathscr{M}_\iota$, and so we have proved that $\mathscr{A} \subset \mathscr{M}_\iota$. A very similar argument proves the reversed inclusion, and so we have $\mathscr{A} = \mathscr{M}_\iota$. Finally, (12.41) shows that $\mu(E) = \iota(E)$ for all $E \in \mathscr{M}_\iota$. □

(12.43) Remark. The hypothesis that $E \in \mathscr{A}$ if and only if $E \cap F \in \mathscr{A}$ for all compact sets $F \subset X$ is essential to prove (12.42). For example, let $X = R_d \times R$, define I on $\mathfrak{C}_{00}(X)$ as in (9.41), and let ι be the measure induced by I. Let ν be the restriction of ι to $\mathscr{B}(X)$ and let $(X, \mathscr{A}, \mu)$ be the completion of the measure space $(X, \mathscr{B}(X), \nu)$ [see (11.21)]. Then $I(f) = \int_X f\,d\mu$ for all $f \in \mathfrak{C}_{00}(X)$, but we also have $\mathscr{A} \subsetneqq \mathscr{M}_\iota$. We construct a set $A \in \mathscr{M}_\iota \cap \mathscr{A}'$ as follows. Let φ be a one-to-one mapping of R onto $\mathscr{B}(R)$ (10.25). Define $A = \bigcup_{x \in R} \{(x, y) : y \in \varphi(x)\}$. The set A is in $\mathscr{M}_\iota$ because $A \cap F \in \mathscr{B}(X) \subset \mathscr{M}_\iota$ for all compact $F \subset X$. Assuming that $A \in \mathscr{A}$, we have $A = B \cup C$ where C meets only countably many lines $V_x = \{(x, y) : y \in R\}$ and $B \in \mathscr{B}(X)$. Then there exists an ordinal number $\alpha < \Omega$ such that $B \in \mathscr{E}_\alpha$ [where $\mathscr{E}_0$ is the family of all open subsets of X and succeeding $\mathscr{E}_\alpha$'s are defined as in the proof of (10.23)]. Thus $B \cap V_x \in \mathscr{E}_{\alpha+2}$ for all vertical lines V_x. It follows that $\varphi(x) \in \mathscr{F}_{\alpha+2}$ [where $\mathscr{F}_0$ is the family of all open subsets of R] for all $x \in R$ except possibly those countably many x's for which $C \cap V_x \neq \varnothing$. But this contradicts the known fact that $\mathscr{B}(R) \cap \mathscr{F}'_\beta \neq \varnothing$ for all $\beta < \Omega$ [see K. Kuratowski, C. R. Paris, vol. 176 (1923), 229; also see W. Sierpiński, Fund. Math., vol. 6 (1924), 39]. Thus $A \notin \mathscr{A}$.

Next we note an important property of integrals with respect to Lebesgue measure.

(12.44) Theorem. *Let f be a Lebesgue measurable function R, let t be a real number, and let f_t be the translate of f by t (8.14). Then*

(i) $$\int_R f_t(x)\,dx = \int_R f(-x)\,dx = \int_R f(x)\,dx$$

whenever any of these integrals is defined.

Proof. For $f \geqq 0$, the translation and inversion invariance of the Riemann integral makes it clear that

$$\overline{S}(f_t) = \overline{S}(f^\star) = \overline{S}(f)\,. \tag{1}$$

[Use (8.15.iv), (8.15.v), (9.8), and (9.15).] Now call on (12.35); and refer to (12.2) and (12.26) for conditions under which the integrals in (i) are defined. □

(12.45) Continuous images of measures. This construction requires some preliminary explanation, although the basic idea is simple enough. Let X and Y be locally compact Hausdorff spaces, and let φ be a continuous mapping of X *on*to Y. Suppose that we are given a measure μ on X in the sense of § 9, and make the hypothesis that

(i) $\varphi^{-1}(F)$ is compact in X for every compact subset F of Y

or that

(ii) $\mu(X)$ is finite.

Consider an arbitrary function $f \in \mathfrak{C}_{00}(Y)$. It is easy to see that the composite function $f \circ \varphi$ is in $\mathfrak{L}_1(X, \mathscr{M}_\mu, \mu)$, since it is in $\mathfrak{C}_{00}(X)$ if (i) holds and is a bounded continuous function in any case and so is in $\mathfrak{L}_1(X, \mathscr{M}_\mu, \mu)$ if $\mu(X)$ is finite. Therefore the mapping

(iii) $$f \to \int_X f \circ \varphi(x)\,d\mu(x)$$

is a nonnegative linear functional on $\mathfrak{C}_{00}(Y)$. Accordingly there is a measure ν in the sense of § 9 on Y such that

(iv) $$\int_X f \circ \varphi(x)\,d\mu(x) = \int_Y f(y)\,d\nu(y)$$

for all $f \in \mathfrak{C}_{00}(Y)$. The measure ν is called *the image of the measure μ under the continuous mapping φ.*

(12.46) Theorem. *Notation is as in* (12.45). *For all σ-finite ν-measurable subsets B of Y, we have*

(i) $$\nu(B) = \mu(\varphi^{-1}(B)) = \int_X \xi_B \circ \varphi(x)\,d\mu(x).$$

For every function $f \in \mathfrak{L}_1(Y, \mathscr{M}_\nu, \nu)$, $f \circ \varphi$ is in $\mathfrak{L}_1(X, \mathscr{M}_\mu, \mu)$ and we have

(ii) $$\int_Y f(y)\,d\nu(y) = \int_X f \circ \varphi(x)\,d\mu(x).$$

Proof. We proceed in steps. Suppose first that U is a nonvoid open subset of Y. Let $\mathfrak{K}$ be the set of all functions f in $\mathfrak{C}_{00}^+(Y)$ such that $f \leqq \xi_U$. URYSOHN's theorem (6.80) implies that $\sup\{f : f \in \mathfrak{K}\} = \xi_U$

and it is obvious that $\mathfrak{K}$ is directed upward in the sense of (9.11). Taking note of (12.35) and applying (9.11) twice, we find that

$$\begin{aligned}\nu(U) &= \int_Y \xi_U(y)\, d\nu(y) = \sup\left\{\int_Y f(y)\, d\nu(y) : f \in \mathfrak{K}\right\}\\ &= \sup\left\{\int_X f\circ\varphi(x)\, d\mu(x) : f \in \mathfrak{K}\right\}\\ &= \int_X \sup\{f\circ\varphi(x) : f\in\mathfrak{K}\}\, d\mu(x) = \int_X \xi_U\circ\varphi(x)\, d\mu(x)\\ &= \mu(\varphi^{-1}(U)). \qquad (1)\end{aligned}$$

Next suppose that B is an arbitrary subset of Y. Theorem (9.24) and (1) show that

$$\begin{aligned}\nu(B) &= \inf\{\nu(U) : U \text{ is open in } Y \text{ and } U \supset B\}\\ &= \inf\{\mu(\varphi^{-1}(U)) : U \text{ is open in } Y \text{ and } U \supset B\}\\ &\geqq \mu(\varphi^{-1}(B)). \qquad (2)\end{aligned}$$

In particular, if $\nu(B) = 0$, then (i) holds for B. If F is a compact subset of Y, then F is contained in an open set U such that U^- is compact (6.79). Thus $\nu(U)$ is finite (9.27), and so

$$\nu(U) - \nu(U \cap F') = \nu(F). \qquad (3)$$

Apply (1) to (3) and note that $U \cap F'$ is open:

$$\begin{aligned}\nu(F) &= \mu(\varphi^{-1}(U)) - \mu(\varphi^{-1}(U\cap F'))\\ &= \mu\big(\varphi^{-1}(U) \cap (\varphi^{-1}(U\cap F'))'\big)\\ &= \mu(\varphi^{-1}(F)). \qquad (4)\end{aligned}$$

Every σ-finite ν-measurable set B can be written as

$$B = \left(\bigcup_{n=1}^{\infty} F_n\right) \cup P, \qquad (5)$$

where $F_1 \subset F_2 \subset \cdots \subset F_n \subset \cdots$, each F_n is compact, P is disjoint from $\bigcup_{n=1}^{\infty} F_n$, and $\nu(P) = 0$. This follows readily from (10.34), and we omit the details. Applying (4) to (5) and using (10.13), we obtain

$$\begin{aligned}\nu(B) &= \nu\left(\bigcup_{n=1}^{\infty} F_n\right) = \lim_{n\to\infty} \nu(F_n) = \lim_{n\to\infty} \mu(\varphi^{-1}(F_n))\\ &= \mu\left(\bigcup_{n=1}^{\infty} \varphi^{-1}(F_n)\right) = \mu\left(\varphi^{-1}\left(\bigcup_{n=1}^{\infty} F_n\right)\right) \leqq \mu(\varphi^{-1}(B)). \qquad (6)\end{aligned}$$

Now (i) follows from (6) and (2).

Let us show that $\varphi^{-1}(B) \in \mathscr{M}_\mu$ for all $B \in \mathscr{M}_\nu$. If B is σ-finite, use (5) to write

$$\varphi^{-1}(B) = \bigcup_{n=1}^{\infty} \varphi^{-1}(F_n) \cup \varphi^{-1}(P);$$

each $\varphi^{-1}(F_n)$ is closed, because φ is continuous, and $\mu(\varphi^{-1}(P)) = 0$

because of (2). If B is not σ-finite, let E be any compact subset of X. We have

$$E \cap \varphi^{-1}(B) = E \cap \varphi^{-1}(\varphi(E) \cap B)$$

since φ is single-valued. The set $\varphi(E) \cap B$ is plainly ν-measurable and finite for ν, so that $\varphi^{-1}(\varphi(E) \cap B)$ is μ-measurable, as was just proved. Thus $E \cap \varphi^{-1}(B)$ is in $\mathscr{M}_\mu$, and (10.31) implies that $\varphi^{-1}(B)$ is in $\mathscr{M}_\mu$.

It is now obvious that for every ν-measurable complex function f on Y, the function $f \circ \varphi$ on X is μ-measurable. Using (i), it is easy to establish (ii). Consider $f \in \mathfrak{L}_1(Y, \mathscr{M}_\nu, \nu)$; we may suppose that $f \geqq 0$. By (11.35) there is a nondecreasing sequence $(s_n)_{n=1}^\infty$ of simple, ν-measurable functions such that $\lim\limits_{n\to\infty} s_n(y) = f(y)$ everywhere on Y. Plainly

$$\int_Y s_n(y)\, d\nu(y) \leqq \int_Y f(y)\, d\nu(y) < \infty\,. \tag{7}$$

Write $s_n = \sum\limits_{k=1}^m \alpha_k \xi_{B_k}$, where $0 < \alpha_1 < \cdots < \alpha_m$. Then $\nu(B_k)$ is finite for all k, as (7) proves, and (12.4) and (i) imply that

$$\begin{aligned}\int_Y s_n(y)\, d\nu(y) &= \sum_{k=1}^m \alpha_k \nu(B_k) = \sum_{k=1}^m \alpha_k\, \mu(\varphi^{-1}(B_k)) \\ &= \int_X \left(\sum_{k=1}^m \alpha_k \xi_{\varphi^{-1}(B_k)}(x)\right) d\mu(x) \\ &= \int_X s_n \circ \varphi(x)\, d\mu(x)\,. \end{aligned} \tag{8}$$

Take the limit as $n \to \infty$ of both ends of (8), and cite B. LEVI's theorem (12.22). This proves (ii). □

We close this section with a large collection of exercise. Of these (12.48), (12.51), (12.54), and (12.63) are important either for later applications or for understanding the theory expounded up to this point. We hope that all readers will work through at least these exercises, and that most readers will work through all of them.

(12.47) Exercise. Let $(X, \mathscr{A}, \mu)$ be any finite measure space and let $\mathfrak{S}$ be the set of all complex-valued $\mathscr{A}$-measurable functions on X. For $f, g \in \mathfrak{S}$ define

$$\varrho(f, g) = \int_X \frac{|f-g|}{1+|f-g|}\, d\mu\,.$$

Prove the following assertions.

(a) $\varrho(f, g) = 0$ if and only if $f = g$ a.e.

(b) $\varrho(f, g) = \varrho(g, f)$.

(c) $\varrho(f, h) \leqq \varrho(f, g) + \varrho(g, h)$.

(d) If $(f_n)_{n=1}^\infty$ satisfies $\lim\limits_{m, n\to\infty} \varrho(f_n, f_m) = 0$, then there exists a complex-valued measurable function g such that $\lim\limits_{n\to\infty} \varrho(f_n, g) = 0$.

That is, identifying functions equal a.e., we have defined a complete metric space $\mathfrak{F}$.

(e) For $f \in \mathfrak{F}$ and $(f_n) \subset \mathfrak{F}$ we have $\varrho(f_n, f) \to 0$ if and only if $f_n \to f$ in measure. For this reason we call ϱ the *metric of convergence in measure.*

(12.48) Exercise. Let $(X, \mathscr{A}, \mu)$ be any measure space and let f be a nonnegative, real-valued, bounded, $\mathscr{A}$-measurable function on X. Let $\alpha = \inf\{f(x) : x \in X\}$ and $\beta = \sup\{f(x) : x \in X\}$. For $n \in N$ and $j = 1, 2, \ldots, n-1$, let

$$A_j = \left\{x \in X : \alpha + \frac{(j-1)\,(\beta-\alpha)}{n} \leqq f(x) < \alpha + \frac{j(\beta-\alpha)}{n}\right\}$$

and let

$$A_n = \left\{x \in X : \alpha + \frac{(n-1)\,(\beta-\alpha)}{n} \leqq f(x) \leqq \beta\right\}.$$

The *Lebesgue sums for f* are defined as the numbers

$$s_n = \sum_{j=1}^{n} \left(\alpha + \frac{(j-1)\,(\beta-\alpha)}{n}\right) \mu(A_j)\,.$$

Prove that $\lim\limits_{n\to\infty} s_n = \int\limits_X f\,d\mu$.

Next suppose that $\mu(X) < \infty$. Let f be any bounded, real-valued, $\mathscr{A}$-measurable function on X. Define s_n as above. Prove that $\lim\limits_{n\to\infty} s_n = \int\limits_X f\,d\mu$.

(12.49) Exercise. Let $(X, \mathscr{A}, \mu)$ be any finite measure space and let $(f_n)_{n=1}^{\infty}$ and f be complex-valued, $\mathscr{A}$-measurable functions on X such that $f_n \to f$ a.e. Suppose that there exists $\beta \in R$ such that $|f_n| \leqq \beta$ a.e. for all $n \in N$. Use Egorov's theorem, not (12.21)–(12.24) or (12.30), to prove that

$$\lim_{n\to\infty} \int\limits_X f_n\,d\mu = \int\limits_X f\,d\mu\,.$$

(12.50) Exercise. Let $(X, \mathscr{A}, \mu)$ be a σ-finite measure space. Let g be an $\mathscr{A}$-measurable function on X such that $fg \in \mathfrak{L}_1$ for every $f \in \mathfrak{L}_1$. Prove that there is a number $\alpha \in R$ such that $\mu(\{x \in X : |g(x)| > \alpha\}) = 0$. Give an example to show that this conclusion may fail if the hypothesis of σ-finiteness is dropped.

(12.51) Exercise. Let $-\infty < a < b < \infty$ and let f be any bounded real-valued function on $[a, b]$. For each $\delta > 0$ and $x \in [a, b]$, define

$$m_\delta(x) = \inf\{f(t) : t \in [a, b] \cap\,]x-\delta, x+\delta[\}$$

and

$$M_\delta(x) = \sup\{f(t) : t \in [a, b] \cap\,]x-\delta, x+\delta[\}\,,$$

and define

$$m(x) = \lim_{\delta\downarrow 0} m_\delta(x),\quad M(x) = \lim_{\delta\downarrow 0} M_\delta(x)\,.$$

(a) Prove that f is continuous at x if and only if $m(x) = M(x)$.

Let $(\Delta_j)_{j=1}^{\infty}$ be a sequence of subdivisions of $[a, b]$, say

$$\Delta_j = \{a = x_0^{(j)} < x_1^{(j)} < \cdots < x_{n_j}^{(j)} = b\},$$

such that

$$\lim_{j\to\infty} \max\{x_k^{(j)} - x_{k-1}^{(j)} : 1 \leqq k \leqq n_j\} = 0 .$$

Let

$$m_k^{(j)} = \inf\{f(t) : x_{k-1}^{(j)} \leqq t \leqq x_k^{(j)}\},\ M_k^{(j)} = \sup\{f(t) : x_{k-1}^{(j)} \leqq t \leqq x_k^{(j)}\},$$

$$\varphi_j = \sum_{k=1}^{n_j} m_k^{(j)} \xi_{]x_{k-1}^{(j)}, x_k^{(j)}[}, \quad \text{and} \quad \psi_j = \sum_{k=1}^{n_j} M_k^{(j)} \xi_{]x_{k-1}^{(j)}, x_k^{(j)}[} .$$

Prove that:

(b) if $x \in [a, b]$ and x is distinct from all $x_k^{(j)}$, then $\lim_{j\to\infty} \varphi_j(x) = m(x)$ and $\lim_{j\to\infty} \psi_j(x) = M(x)$;

(c) m and M are Lebesgue measurable on $[a, b]$;

(d) if $L(f, \Delta_j)$ and $U(f, \Delta_j)$ are the lower and upper Darboux sums respectively [defined with $\alpha(x) = x$] for the function f corresponding to the subdivision Δ_j, then

$$\lim_{j\to\infty} L(f, \Delta_j) = \int_a^b m(x)\,dx$$

and

$$\lim_{j\to\infty} U(f, \Delta_j) = \int_a^b M(x)\,dx;$$

(e) f is Riemann integrable on $[a, b]$ if and only if f is continuous a.e., *i.e.*, $\lambda(\{x \in [a, b] : f \text{ is discontinuous at } x\}) = 0$;

(f) if f is Riemann integrable on $[a, b]$, then $f \in \mathfrak{L}_1([a, b], \mathscr{M}_\lambda, \lambda)$ and

$$S(f; [a, b]) = \int_a^b f(x)\,dx ,$$

where S denotes the Riemann integral (8.6). [The reader should note that the foregoing applies only to *bounded* functions on *finite* intervals.]

(12.52) Exercise. Find a real-valued function f on $[0, 1]$ such that f is continuous on $]0, 1]$, $\lim_{\delta\downarrow 0} \int_{[\delta,1]} f\,d\lambda$ is finite, and the Lebesgue integral of f over $[0, 1]$ is not defined.

(12.53) Exercise. Find a bounded, real-valued, Lebesgue measurable function f on $[0, 1]$ such that $\int_{[0,1]} |f - g|\,d\lambda > 0$ for every Riemann integrable function g on $[0, 1]$.

(12.54) Exercise. Let $-\infty < a < b < \infty$ and let f be a Lebesgue measurable function on $[a, b]$ such that $\int_a^x f(t)\,dt = 0$ for every $x \in [a, b]$. Prove that $f = 0$ λ-a.e. [This exercise is needed for the proof of Theorem

(16.34) *infra*. Hints are as follows. Since $\int_a^b f(t)\,dt$ exists and is finite, $|f|$ is in $\mathfrak{L}_1([a, b], \mathscr{M}_\lambda, \lambda)$. It is evident from (6.59), (12.32), and our hypothesis that $\int_U f\,d\lambda = 0$ for all open subsets U of $[a, b]$. Use (9.24) to infer that $\int_A f\,d\lambda = 0$ for all λ-measurable sets $A \subset [a, b]$. From this the identity $f = 0$ λ-a.e. is immediate.]

(12.55) Exercise. Let X be a locally compact Hausdorff space such that every open subset of X is σ-compact. Suppose that μ is a measure defined on $\mathscr{B}(X)$ such that $\mu(F) < \infty$ for all compact sets $F \subset X$. Prove that μ is a regular measure. [First consider the case that X is compact. Consider the family $\mathscr{R} = \{E \in \mathscr{B}(X) : \mu(E) = \inf\{\mu(U) : U$ is open, $U \supset E\}$ and $\mu(E) = \sup\{\mu(F) : F$ is compact, $F \subset E\}\}$.]

(12.56) Exercise. Let μ be a measure defined on $\mathscr{B}(R)$ such that $\mu([0, 1]) = 1$ and $\mu(E + x) = \mu(E)$ for every $E \in \mathscr{B}(R)$ and $x \in R$. Prove that $\mu(E) = \lambda(E)$ for all $E \in \mathscr{B}(R)$. [First prove that $\mu(\{x\}) = 0$ for all $x \in R$. Next show that $\mu(]a, b[) = b - a$ for all $a < b$ in R. Use (12.55)].

(12.57) Exercise. Let $(X, \mathscr{A}, \mu)$ be an arbitrary measure space and let f and $(f_n)_{n=1}^\infty$ be complex-valued measurable functions on X. Suppose that $f_n \to f$ in measure and that there exists a function $g \in \mathfrak{L}_1^r$ such that $|f_n| \leq g$ a.e. for all $n \in N$. Prove that $\lim_{n\to\infty} \int_X |f - f_n|\,d\mu = 0$. [Assume that $\overline{\lim_{n\to\infty}} \int_X |f - f_n|\,d\mu = \alpha > 0$ and choose a subsequence (f_{n_k}) such that $\lim_{k\to\infty} \int_X |f - f_{n_k}|\,d\mu = \alpha$. Then use (11.26)].

(12.58) Exercise. Let I be the nonnegative linear functional on $\mathfrak{C}_{00}(R_d \times R)$ defined in (9.41). Define η on $\mathscr{B}(R_d \times R)$ by the rule $\eta(E) = \sum_{x\in R} \lambda(\{y : (x, y) \in E\})$. Prove that:

(a) η is a measure on $\mathscr{B}(R_d \times R)$;

(b) $I(f) = \int_{R_d \times R} f\,d\eta$ for all $f \in \mathfrak{C}_{00}(R_d \times R)$;

(c) η is not a regular measure.

(12.59) Exercise. Let X be a locally compact Hausdorff space and let μ be a regular measure defined on a σ-algebra $\mathscr{A}$ of subsets of X [of course $\mathscr{A} \supset \mathscr{B}(X)$] such that $\mu(\{x\}) = 0$ for all $x \in X$. Suppose that B is in $\mathscr{A}$ and that

$$\mu(B) = \sup\{\mu(F) : F \text{ is compact, } F \subset B\} < \infty .$$

Prove that for each $\alpha \in [0, \mu(B)]$ there exists a σ-compact set $A \subset B$ such that $\mu(A) = \alpha$.

(12.60) Exercise. Let $(X, \mathscr{A}, \mu)$ be a measure space as described in (10.56.a). Let f be an $\mathscr{A}$-measurable function on X. Prove that f is constant on each E_k except for a set of μ-measure zero, and accordingly write the integral $\int_X f\,d\mu$ as a certain finite sum.

(12.61) Exercise. Let $(X, \mathscr{A}, \mu)$ be a measure space that is degenerate in the sense of (10.3): $\mu(A) = 0$ or $\mu(A) = \infty$ for all $A \in \mathscr{A}$. Show that every function $f \in \mathfrak{L}_1(X, \mathscr{A}, \mu)$ vanishes except on a set of μ-measure zero and that $\int_X |f|\,d\mu = 0$. [This unpleasant property justifies the term "degenerate" for the measure spaces under consideration.]

(12.62) Exercise. Let X be a locally compact, σ-compact Hausdorff space and let μ be a regular measure defined on a σ-algebra $\mathscr{A}$ of subsets of X $[\mathscr{A} \supset \mathscr{B}(X)]$. Let f be an $\mathscr{A}$-measurable function on X such that $f(X) \subset [0, \infty]$ and such that $f\xi_F \in \mathfrak{L}_1(X, \mathscr{A}, \mu)$ for all compact subsets F of X. [Such an f is called *locally μ-integrable.*] Define the set-function ν on $\mathscr{A}$ by

$$\nu(A) = \int_A f(x)\,d\mu(x)\,.$$

Prove that ν is a regular measure on $\mathscr{A}$.

(12.63) Exercise: Integrals on the completion of a measure space. Let $(X, \mathscr{A}, \mu)$ be a measure space and $(X, \bar{\mathscr{A}}, \bar{\mu})$ its completion (11.21).

(a) Let $\bar{f}$ be a complex- or extended real-valued $\bar{\mathscr{A}}$-measurable function defined on X. Prove that there is an $\mathscr{A}$-measurable function f such that $f = \bar{f}$ $\bar{\mu}$-almost everywhere on X. [Hints. Suppose that $\bar{f}$ is extended real-valued. Use (11.35) to find a sequence $(\bar{s}_n)$ of real-valued, $\bar{\mathscr{A}}$-measurable, simple functions such that $\lim\limits_{n\to\infty} \bar{s}_n = \bar{f}$ everywhere. Each $\bar{s}_n$ has the form $\sum\limits_k \alpha_{n,k}\xi_{A_{n,k}}$ with $A_{n,k} \in \bar{\mathscr{A}}$ and the $A_{n,k}$'s pairwise disjoint. Each $A_{n,k}$ is contained in a set $B_{n,k} \in \mathscr{A}$ such that $\bar{\mu}(B_{n,k} \cap A'_{n,k}) = 0$. Let $s_n = \sum\limits_k \alpha_{n,k}\xi_{B_{n,k}}$ and define f as $\varliminf\limits_{n\to\infty} s_n$.]

(b) Let $\bar{f}$ be a function in $\mathfrak{L}_1(X, \bar{\mathscr{A}}, \bar{\mu})$. Prove that there is a function f in $\mathfrak{L}_1(X, \mathscr{A}, \mu)$ such that

$$\int_X |f - \bar{f}|\,d\bar{\mu} = 0$$

and

$$\int_X f\,d\mu = \int_X \bar{f}\,d\bar{\mu}\,.$$

[It suffices to consider nonnegative functions $\bar{f}$. By adding sets of μ-measure 0, it is easy to make the simple functions s_n of part (a) into a nondecreasing sequence. By (12.22) we then find

$$\int_X \bar{f}\,d\bar{\mu} = \lim_{n\to\infty} \int_X \bar{s}_n\,d\bar{\mu} = \lim_{n\to\infty} \int_X s_n\,d\mu = \int_X f\,d\mu\,.$$

The other equality is obvious.]

(c) Let X be a locally compact Hausdorff space and $(X, \mathscr{M}_\iota, \iota)$ as in §§ 9 and 10. Suppose that X is σ-finite with respect to ι. Let $\mathscr{B}(X)$

denote the Borel sets of X. Prove that $(X, \mathscr{M}_\iota, \iota)$ is the completion of $(X, \mathscr{B}(X), \iota)$. What does this tell you about Borel measurable functions and arbitrary $\mathscr{M}_\iota$-measurable functions? [Use part (a).]

(d) Drop the hypothesis in part (c) that X be σ-finite. Let f be an arbitrary function in $\mathfrak{L}_1(X, \mathscr{M}_\iota, \iota)$. Prove that there is a Borel measurable function f_1 on X such that $f_1 = f$ ι-almost everywhere and such that $|f_1| \leqq |f|$. [Consider a subset Y of X σ-finite with respect to ι such that f vanishes on Y' [Y can be chosen to be open if you like], and then argue as in part (c).]

(12.64) Exercise. Let $(X, \mathscr{A}, \mu)$ be a measure space and let f be a complex-valued $\mathscr{A}$-measurable function on X. Prove that $f \in \mathfrak{L}_1(X, \mathscr{A}, \mu)$ if and only if there exists a sequence (s_n) of simple functions such that $(s_n) \subset \mathfrak{L}_1$, $s_n \to f$ in measure, and

$$\lim_{m,n\to\infty} \int_X |s_m - s_n| \, d\mu = 0 .$$

In this case we have

$$\int_X f \, d\mu = \lim_{n\to\infty} \int_X s_n \, d\mu .$$

[If one first defines $\int_X s \, d\mu$ for complex-valued, $\mathscr{A}$-measurable, simple functions, then the above facts can be used to define $\mathfrak{L}_1$ and the integral on $\mathfrak{L}_1$. This approach is useful when dealing with functions with values in a Banach space. It does not depend directly on the ordering of the real numbers as our definition of the integral does.]

CHAPTER FOUR

Function Spaces and Banach Spaces

The theory of integration developed in Chapter Three enables us to define certain spaces of functions that have remarkable properties and are of enormous importance in analysis as well as in its applications. We have already, in § 7, considered spaces whose points are functions. In § 7, we considered only the uniform norm $\| \|_u$ [see (7.3)] to define the distance between two functions. The present chapter is concerned with norms that are defined in one way or another from *integrals*. The most important such norms are defined and studied in § 13. These special norms lead us very naturally to study abstract Banach spaces, to which § 14 is devoted. While we are not concerned with Banach spaces *per se*, it is an inescapable fact that many results can be proved as easily for all Banach spaces [perhaps with some additional property] as for the special Banach spaces defined in §§ 7 and 13. Our desires both for economy of effort and for clarity of exposition dictate that we treat these results in general Banach spaces. In § 15, we give a strictly computational construction of the conjugate spaces of the function spaces $\mathfrak{L}_p (1 < p < \infty)$. We have chosen this construction because of its elementary nature and also because we think that manipulation of inequalities is something that every student of analysis should learn. In § 16, we consider Hilbert spaces, which are $\mathfrak{L}_2$ spaces looked at abstractly, and also give some concrete examples and illustrations.

All of the sections of this chapter are important, and the reader is advised to study them all.

§ 13. The spaces $\mathfrak{L}_p$ $(1 \leqq p < \infty)$

As usual, we begin with a definition.

(13.1) Definition. Let p be a positive real number, and let $(X, \mathscr{A}, \mu)$ be an arbitrary measure space. Let f be a complex-valued $\mathscr{A}$-measurable function defined μ-a.e. on X such that $|f|^p \in \mathfrak{L}_1^r$. We then say that $f \in \mathfrak{L}_p(X, \mathscr{A}, \mu)$, and we define the symbol $\|f\|_p$ by

$$\|f\|_p = \left[\int_X |f|^p \, d\mu\right]^{\frac{1}{p}}.$$

Where no confusion seems possible, we will write $\mathfrak{L}_p$ for $\mathfrak{L}_p(X, \mathscr{A}, \mu)$.[1] For $p \geqq 1$ and $f \in \mathfrak{L}_p$, we call $\|f\|_p$ the *norm of* f or the $\mathfrak{L}_p$*-norm of* f.[2] The symbols L^p, L_p, and $\mathfrak{L}^p$ are employed by some writers to denote $\mathfrak{L}_p$.

For $1 \leqq p < \infty$, the function $f \to \|f\|_p$ on $\mathfrak{L}_p$ satisfies all the axioms for a norm set down in (7.5), except for the positivity requirement: $\|f\|_p > 0$ if $f \neq 0$. [If $\mathscr{A}$ contains a nonvoid set E such that $\mu(E) = 0$, then $\xi_E \neq 0$ but $\|\xi_E\|_p = 0$.] The only nontrivial fact is the triangle inequality (7.5.iii), which for $\mathfrak{L}_p$ is

$$\|f + g\|_p \leqq \|f\|_p + \|g\|_p .$$

We shall first prove this inequality, paying attention to possible equalities and also obtaining some other inequalities useful in the sequel.

(13.2) Theorem [Young's **inequality**]. *Let* φ *be a continuous, real-valued, strictly increasing function defined on* $[0, \infty[$ *such that* $\lim_{u \to \infty} \varphi(u) = \infty$ *and* $\varphi(0) = 0$. *Let* $\psi = \varphi^{-1}$. *For all* $x \in [0, \infty[$ *define*

$$\Phi(x) = \int_0^x \varphi(u)\, du$$

and

$$\Psi(x) = \int_0^x \psi(v)\, dv .$$

Then $a, b \in [0, \infty[$ *imply*

$$ab \leqq \Phi(a) + \Psi(b)$$

and equality obtains if and only if $b = \varphi(a)$.

Proof. A formal proof can be given using the fact that $\int_0^c \varphi(u)\, du + \int_0^{\varphi(c)} \psi(v)\, dv = c\,\varphi(c)$ for all $c \geqq 0$. However, interpreting the integrals as areas, we render the result obvious by the accompanying Fig. 6. □

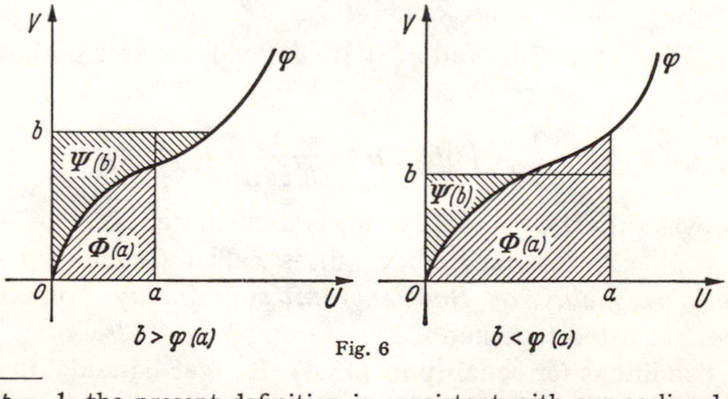

Fig. 6

---

[1] For $p = 1$, the present definition is consistent with our earlier definition of $\mathfrak{L}_1$ given in (12.26), in view of the assertion (12.28.i).

[2] For $0 < p < 1$ and all but a few measure spaces, the function $f \to \|f\|_p$ on $\mathfrak{L}_p$ is *not* a norm in the sense of (7.5). See (13.25.c) for a discussion.

For any positive real number p such that $p \neq 1$, define $p' = \frac{p}{p-1}$ $\left[\text{thus } \frac{1}{p} + \frac{1}{p'} = 1\right]$.

(13.3) Corollary. *For $p > 1$ and a and b any nonnegative real numbers, we have*

(i) $$ab \leq \frac{a^p}{p} + \frac{b^{p'}}{p'}.$$

Equality holds in (i) *if and only if $a^p = b^{p'}$.*

Proof. For $u \in [0, \infty[$, define $\varphi(u) = u^{p-1}$; φ is continuous and strictly increasing, $\lim\limits_{u\to\infty} \varphi(u) = \infty$, and $\varphi(0) = 0$. The inverse ψ of φ is given by $\psi(v) = v^{\frac{1}{p-1}}$. We have $\Phi(a) = \int_0^a \varphi(u)\,du = \int_0^a u^{p-1}\,du = \frac{a^p}{p}$ and $\Psi(b) = \int_0^b \psi(v)\,dv = \int_0^b v^{\frac{1}{p-1}}\,dv = \frac{b^{p'}}{p'}$ [the Lebesgue integral and the Riemann integral agree on Riemann integrable functions (12.51.f)]. The corollary follows at once from (13.2). □

(13.4) Theorem $\left[\textbf{Hölder's inequality for } p > 1\right]$. *Let $f \in \mathfrak{L}_p$ and $g \in \mathfrak{L}_{p'}$, where $p > 1$. Then $fg \in \mathfrak{L}_1$, and we have*

(i) $$\left|\int fg\,d\mu\right| \leq \int |fg|\,d\mu$$

and

(ii) $$\int |fg|\,d\mu \leq \|f\|_p \|g\|_{p'};$$

and so also

(iii) $$\left|\int fg\,d\mu\right| \leq \|f\|_p \|g\|_{p'}.$$

Proof. We first prove (ii). [Note that (ii) and (12.28.ii) imply (i).] If f or g is zero μ-a.e., then (ii) is trivial. Otherwise, using (13.3), we have

$$\frac{|f(u)|}{\|f\|_p} \frac{|g(u)|}{\|g\|_{p'}} \leq \frac{1}{p} \frac{|f(u)|^p}{\|f\|_p^p} + \frac{1}{p'} \frac{|g(u)|^{p'}}{\|g\|_{p'}^{p'}}$$

for all u in X such that $f(u)$ and $g(u)$ are defined, *i.e.*, for μ-almost all u. Thus we have

$$\frac{1}{\|f\|_p \|g\|_{p'}} \int |fg|\,d\mu \leq \frac{1}{p\|f\|_p^p} \int |f|^p\,d\mu + \frac{1}{p'\|g\|_{p'}^{p'}} \int |g|^{p'}\,d\mu = \frac{1}{p} + \frac{1}{p'} = 1,$$

and this proves (ii). The inequality (iii) is immediate. □

For $p = p' = 2$, the inequality (ii) is called *Cauchy's inequality*, or *Schwarz's inequality*, or *Bunyakovskiĭ's inequality*; sometimes the three names are listed together.

(13.5) Conditions for equality in (13.4). To get equality in (13.4.ii), it is clearly necessary and sufficient that we have

$$\frac{|f(u)|}{\|f\|_p} \frac{|g(u)|}{\|g\|_{p'}} = \frac{1}{p} \frac{|f(u)|^p}{\|f\|_p^p} + \frac{1}{p'} \frac{|g(u)|^{p'}}{\|g\|_{p'}^{p'}}$$

for almost all $u \in X$. By (13.3), this happens if and only if $\frac{|f|^p}{\|f\|_p^p} = \frac{|g|^{p'}}{\|g\|_{p'}^{p'}}$ almost everywhere. Thus equality obtains in (13.4.ii) if and only if there are nonnegative real numbers A and B, not both zero, such that

$$A\,|f|^p = B\,|g|^{p'}$$

almost everywhere.

The reader can easily formulate from this and (12.29) a necessary and sufficient condition that equality hold in (13.4.iii).

(13.6) Theorem [Hölder's inequality for $0 < p < 1$]. *Let $0 < p < 1$ and let f and g be functions in $\mathfrak{L}_p^+$ and $\mathfrak{L}_{p'}^+$, respectively. Then we have*

(i) $$\int fg\,d\mu \geqq \left(\int f^p\,d\mu\right)^{\frac{1}{p}} \left(\int g^{p'}\,d\mu\right)^{\frac{1}{p'}}$$

unless $\int g^{p'}\,d\mu = 0$ [note that $p' < 0$].

Proof. In the case we are concerned with, we have $0 < \int g^{p'}\,d\mu < \infty$, and since $p' < 0$, this implies that $g(x) > 0$ for almost all $x \in X$. Let $q = \frac{1}{p}$, and define $\varphi = g^{-\frac{1}{q}}$ and $\psi = g^{\frac{1}{q}} f^{\frac{1}{q}}$. It is easy to see that $\varphi^{q'} = g^{p'}$, and so $\varphi \in \mathfrak{L}_{q'}$. If $\int fg\,d\mu = \infty$, then (i) holds trivially. Otherwise fg is in $\mathfrak{L}_1$, and so ψ is in $\mathfrak{L}_q$. Applying (13.4) with p replaced by q, we have

$$\int f^p\,d\mu = \int \varphi\psi\,d\mu \leqq \left(\int \psi^q\,d\mu\right)^{\frac{1}{q}} \left(\int \varphi^{q'}\,d\mu\right)^{\frac{1}{q'}} = \left(\int fg\,d\mu\right)^p \left(\int g^{p'}\,d\mu\right)^{\frac{1}{q'}}.$$

It follows immediately that

$$\int fg\,d\mu \geqq \left(\int f^p\,d\mu\right)^{\frac{1}{p}} \left(\int g^{p'}\,d\mu\right)^{-\frac{1}{q'p}},$$

and since $-\frac{1}{q'p} = \frac{1}{p'}$, the theorem follows. □

(13.7) Theorem [Minkowski's inequality]. *For $1 \leqq p < \infty$ and $f, g \in \mathfrak{L}_p$, we have*

(i) $$\|f + g\|_p \leqq \|f\|_p + \|g\|_p.$$

Proof. Suppose first that $p > 1$. We have

$$\begin{aligned} |f + g|^p &\leqq (|f| + |g|)^p \leqq [2\max\{|f|, |g|\}]^p \\ &= 2^p \max\{|f|^p, |g|^p\} \leqq 2^p (|f|^p + |g|^p). \end{aligned}$$

This crude estimate shows that $|f + g|^p \in \mathfrak{L}_1$, *i.e.*, $f + g \in \mathfrak{L}_p$. Thus (13.4) implies that

$$\begin{aligned} \|f+g\|_p^p &= \int |f+g|^p\,d\mu \leqq \int |f+g|^{p-1}|f|\,d\mu + \int |f+g|^{p-1}|g|\,d\mu \\ &\leqq \left(\int |f|^p d\mu\right)^{\frac{1}{p}} \left(\int |f+g|^{(p-1)p'} d\mu\right)^{\frac{1}{p'}} + \left(\int |g|^p d\mu\right)^{\frac{1}{p}} \left(\int |f+g|^{(p-1)p'} d\mu\right)^{\frac{1}{p'}} \\ &= (\|f\|_p + \|g\|_p)\,\|f+g\|_p^{\frac{p}{p'}}. \end{aligned}$$

The inequality

$$\|f+g\|_p^{p-\frac{p}{p'}} \leqq \|f\|_p + \|g\|_p$$

thus holds. Observing that $p - \frac{p}{p'} = 1$, we obtain MINKOWSKI'S inequality for $p > 1$. Since $\int |f+g|\,d\mu \leqq \int |f|\,d\mu + \int |g|\,d\mu$, the inequality is trivial for $p = 1$. □

We now give conditions for equality in MINKOWSKI'S inequality.

(13.8) Theorem. *For $p = 1$, we obtain equality in* (13.7.i) *if and only if there is a positive measurable function ϱ such that*

$$f(x)\,\varrho(x) = g(x)$$

almost everywhere on the set $\{x : f(x)\,g(x) \neq 0\}$. Equality obtains for $1 < p < \infty$ if and only if $Af = Bg$, where A and B are nonnegative real numbers such that $A^2 + B^2 > 0$.

Proof. Exercise.

(13.9) Theorem. *For $0 < p < 1$ and $f, g \in \mathfrak{L}_p^+$, we have*

(i) $$\|f+g\|_p \geqq \|f\|_p + \|g\|_p\,.$$

Proof. The estimate given in (13.7) for $|f+g|^p$ shows again that $f + g \in \mathfrak{L}_p$. To prove (i), use (13.6.i) and the argument of (13.7). □

We next describe the exact sense in which $\mathfrak{L}_p$ is a normed linear space ($p \geqq 1$).

(13.10) Theorem. *For $1 \leqq p < \infty$, $\mathfrak{L}_p$ is a normed linear space over K, where we agree that $f = g$ means $f(x) = g(x)$ for μ-almost all $x \in X$. [Alternatively, let $\mathfrak{N} = \{f \in \mathfrak{L}_p : f(x) = 0$ a.e. on $X\}$; then $\mathfrak{N}$ is a closed linear subspace of $\mathfrak{L}_p$. What we call $\mathfrak{L}_p$ upon identifying functions that are equal a.e. is really $\mathfrak{L}_p/\mathfrak{N}$.]*

Proof. It is trivial that $\|\alpha f\|_p = |\alpha|\,\|f\|_p$. All other necessary verifications have been made. □

The following theorem is of vital importance in many applications of integration theory. A very special case, the RIESZ-FISCHER theorem, was regarded as sensational when it was first enunciated in 1906. Now, as we will see, the general theorem is not hard to prove.

(13.11) Theorem. *For $1 \leqq p < \infty$, $\mathfrak{L}_p$ is a complex Banach space,* i.e., *in the metric $\varrho(f, g) = \|f - g\|_p$, $\mathfrak{L}_p$ is a complete metric space.*

Proof. Let $(f_n)_{n=1}^{\infty}$ be a Cauchy sequence in $\mathfrak{L}_p$, *i.e.*, (f_n) has the property that $\lim\limits_{m,n\to\infty} \|f_n - f_m\|_p = 0$. The sequence of numbers $(f_n(x))_{n=1}^{\infty}$ may converge at no point $x \in X$ [the sequence (f_n) constructed in (11.27) serves as an example of this phenomenon.] However, we can find a subsequence of (f_n) that does converge μ-almost everywhere. In fact, choose $(f_{n_k})_{k=1}^{\infty}$ as any subsequence of (f_n) such that $n_1 < n_2 < \cdots$

$< n_k < \cdots$ and $\sum_{k=1}^{\infty} \|f_{n_{k+1}} - f_{n_k}\|_p = \alpha < \infty$. This is possible: *e.g.*, we can select increasing n_k's such that $\|f_m - f_{n_k}\|_p < 2^{-k}$ for all $m \geqq n_k$. Now define

$$g_k = |f_{n_1}| + |f_{n_2} - f_{n_1}| + \cdots + |f_{n_{k+1}} - f_{n_k}|, \quad \text{for} \quad k = 1, 2, 3, \ldots.$$

It is clear that

$$\|g_k^p\|_1 = \|g_k\|_p^p = (\| \, |f_{n_1}| + |f_{n_2} - f_{n_1}| + \cdots + |f_{n_{k+1}} - f_{n_k}| \, \|_p)^p$$
$$\leqq \left(\|f_{n_1}\|_p + \sum_{j=1}^{k} \|f_{n_{j+1}} - f_{n_j}\|_p \right)^p \leqq (\|f_{n_1}\|_p + \alpha)^p < \infty .$$

Let $g = \lim_{k\to\infty} g_k$. By B. LEVI's theorem (12.22), and the above estimate, we have

$$\int g^p \, d\mu = \int \lim_{k\to\infty} g_k^p \, d\mu = \lim_{k\to\infty} \int g_k^p \, d\mu < \infty .$$

Hence g is in $\mathfrak{L}_p$; *i.e.*,

$$\int \left[|f_{n_1}| + \sum_{j=1}^{\infty} |f_{n_{j+1}} - f_{n_j}| \right]^p d\mu < \infty .$$

The nonnegative integrand above must be finite μ-a.e., and so the series $\sum_{j=1}^{\infty} |f_{n_{j+1}}(x) - f_{n_j}(x)|$ converges μ-a.e. Obviously the series

$$f_{n_1}(x) + \sum_{j=1}^{\infty} (f_{n_{j+1}}(x) - f_{n_j}(x))$$

also converges μ-a.e. The kth partial sum of this series is $f_{n_{k+1}}(x)$, and so the sequence $(f_{n_k}(x))_{k=1}^{\infty}$ converges to a complex number $f(x)$ for all $x \in A$, where $A \in \mathscr{A}$ and $\mu(A') = 0$. Define $f(x)$ as 0 for all $x \in A'$. It is easy to see that f is $\mathscr{A}$-measurable, and obviously f is complex-valued on X.

We will show that f is the limit in $\mathfrak{L}_p$ of the sequence (f_n), and this will of course prove that $\mathfrak{L}_p$ is complete in the metric induced by the $\mathfrak{L}_p$-norm. Given $\varepsilon > 0$, let l be so large that

$$\|f_s - f_t\|_p < \varepsilon \quad \text{for} \quad s, t \geqq n_l .$$

Then for $k \geqq l$ and $m > n_l$, we have

$$\|f_m - f_{n_k}\|_p < \varepsilon .$$

By FATOU'S lemma (12.23), we have

$$\int |f - f_m|^p \, d\mu = \int \underline{\lim}_{k\to\infty} |f_{n_k} - f_m|^p \, d\mu$$
$$\leqq \underline{\lim}_{k\to\infty} \int |f_{n_k} - f_m|^p \, d\mu \leqq \varepsilon^p .$$

Thus for each $m > n_l$, the function $f - f_m$ is in $\mathfrak{L}_p$, and so $f = f - f_m + f_m$

is in $\mathfrak{L}_p$; and

$$\lim_{n\to\infty} \|f - f_n\|_p = 0 . \quad \square$$

(13.12) Remark. The function spaces $\mathfrak{L}_p^r$ [real-valued $\mathscr{A}$-measurable functions defined μ-a.e. on X such that $\|f\|_p = (\int |f|^p \, d\mu)^{\frac{1}{p}} < \infty$] are real normed linear spaces for $1 \leqq p < \infty$, and they too are complete. The proofs are very like the proofs for the complex spaces $\mathfrak{L}_p$.

(13.13) Example. Let D be any nonvoid set and consider all complex-valued functions f on D such that $\sum_{x \in D} |f(x)|^p < \infty$, where $0 < p < \infty$. [Recall that $\sum_{x \in D} |f(x)|^p = \sup\{\sum_{x\in F} |f(x)|^p : F$ is a finite subset of $D\}$.] If $\mathscr{A}$ is all subsets of D and μ the counting measure defined in (10.4.a), then these functions are the elements of $\mathfrak{L}_p(D, \mathscr{A}, \mu)$. Custom dictates that this space be designated by $l_p(D)$, and if $D = N$, simply by l_p. If $1 \leqq p < \infty$, then $l_p(D)$ is a complete metric space in which the metric is obtained from the norm

$$\|f\|_p = \Big(\sum_{x\in D} |f(x)|^p\Big)^{\frac{1}{p}} .$$

The HÖLDER and MINKOWSKI inequalities take the forms

$$\sum_{x\in D} |f(x)\, g(x)| \leqq \Big(\sum_{x\in D} |(f(x)|^p\Big)^{\frac{1}{p}} \Big(\sum_{x\in D} |g(x)|^{p'}\Big)^{\frac{1}{p'}}$$

and

$$\Big(\sum_{x\in D} |f(x) + g(x)|^p\Big)^{\frac{1}{p}} \leqq \Big(\sum_{x\in D} |f(x)|^p\Big)^{\frac{1}{p}} + \Big(\sum_{x\in D} |g(x)|^p\Big)^{\frac{1}{p}} ,$$

respectively. If D is finite, say $D = \{1, 2, \ldots, n\}$, then the foregoing produces the l_p norm and its corresponding metric on K^n and R^n. The distance between two points $(x_1, x_2, \ldots, x_n)$ and $(y_1, y_2, \ldots, y_n)$ is $\left(\sum_{j=1}^{n} |x_j - y_j|^p\right)^{\frac{1}{p}}$. For $p = 2$, we obtain the classical Euclidean metric. The topologies induced by the l_p metrics on K^n and R^n are all the same [cf. (6.17)].

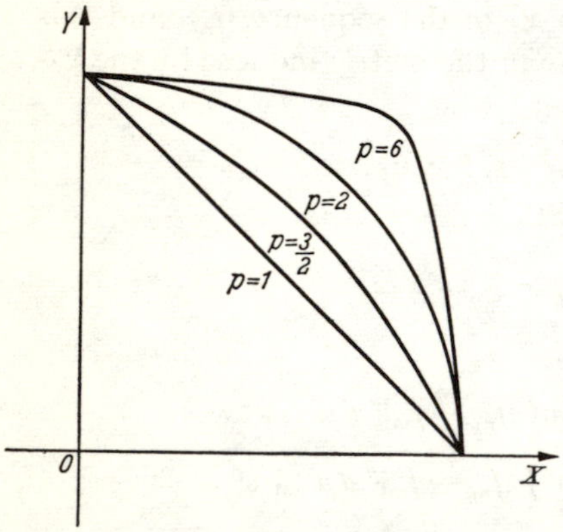

Fig. 7.

The first quadrant of the unit balls in R^2 for various values of p are sketched in Fig. 7.

(13.14) Examples. The spaces $\mathfrak{L}_p([0, 1])$ and $\mathfrak{L}_p(R)$, where $0 < p < \infty$ [it is understood that $\mu = \lambda$

and $\mathscr{A} = \mathscr{M}_\lambda$], are very important function spaces in both pure and applied analysis.

(13.15) Example. For $p = 2$, we obtain the famous function space $\mathfrak{L}_2(X, \mathscr{A}, \mu)$. In this case $p = p' = 2$, and so HÖLDER'S inequality takes the form

$$\int |fg| \, d\mu \leqq \|f\|_2 \, \|g\|_2$$

for $f, g \in \mathfrak{L}_2$. Consider the mapping that takes $\mathfrak{L}_2 \times \mathfrak{L}_2$ into K by the rule

$$(f, g) \to \int f \bar{g} \, d\mu = \langle f, g \rangle ,$$

where the equality defines $\langle f, g \rangle$. This mapping has the following properties:

$$\begin{aligned} \langle f_1 + f_2, g \rangle &= \langle f_1, g \rangle + \langle f_2, g \rangle; \\ \langle \alpha f, g \rangle &= \alpha \langle f, g \rangle \quad \text{for} \quad \alpha \in K; \\ \overline{\langle f, g \rangle} &= \langle g, f \rangle; \\ \langle f, f \rangle &> 0 \quad \text{for} \quad f \neq 0 . \end{aligned}$$

We infer from these identities [or directly] that

$$\begin{aligned} \langle f, g_1 + g_2 \rangle &= \langle f, g_1 \rangle + \langle f, g_2 \rangle , \\ \langle f, \alpha g \rangle &= \bar{\alpha} \langle f, g \rangle , \end{aligned}$$

and

$$\langle 0, f \rangle = \langle f, 0 \rangle = 0 .$$

The spaces $\mathfrak{L}_2$ can be described in abstract terms, as follows.

(13.16) Definition. Let H be a linear space over K having an inner product

$$(x, y) \to \langle x, y \rangle \in K$$

mapping $H \times H$ into K such that

$$\begin{aligned} \langle x + y, z \rangle &= \langle x, z \rangle + \langle y, z \rangle , \\ \langle \alpha x, y \rangle &= \alpha \langle x, y \rangle \quad \text{for} \quad \alpha \in K , \\ \overline{\langle x, y \rangle} &= \langle y, x \rangle , \\ \langle x, x \rangle &> 0 \quad \text{if} \quad x \neq 0 . \end{aligned}$$

[The other properties of $\langle \ , \ \rangle$ listed for $\mathfrak{L}_2$ in (13.15) can be proved for H from the above relations.] Then H is called an *inner product space* or a *pre-Hilbert space*. For $x \in H$, define

$$\|x\| = \langle x, x \rangle^{\frac{1}{2}} .$$

The inequalities

$$|\langle x, y \rangle| \leqq \|x\| \cdot \|y\|$$

and

$$\|x + y\| \leqq \|x\| + \|y\|$$

can be proved. Thus H is a normed linear space. If H is complete relative to this norm, then H is called a *Hilbert space*. There is a very extensive theory of Hilbert spaces. We will take up the rudiments of this theory in § 16 *infra*. One of the most striking facts of this theory is that every Hilbert space is identifiable *qua* Hilbert space with some $l_2(D)$. Thus in particular every $\mathfrak{L}_2$ space can be identified with some space $l_2(D)$. We will deal with this identification problem in (16.29).

We now return to the spaces $\mathfrak{L}_p$, establishing a few more simple facts.

(13.17) Theorem. *If $\mu(X) < \infty$ and $0 < p < q < \infty$, then $\mathfrak{L}_q \subset \mathfrak{L}_p$, and the inequality*

$$\|f\|_p \leqq \|f\|_q (\mu(X))^{\frac{1}{p} - \frac{1}{q}}$$

holds for $f \in \mathfrak{L}_q$.

Proof. Let $r = \frac{q}{p} > 1$. For any $f \in \mathfrak{L}_q$, we have

$$\int |f|^p \, d\mu \leqq \left(\int |f|^{pr} \, d\mu\right)^{\frac{1}{r}} \left(\int 1^{r'} \, d\mu\right)^{\frac{1}{r'}} = \left(\int |f|^q \, d\mu\right)^{\frac{p}{q}} \left(\mu(X)\right)^{\frac{q-p}{q}}.$$

It follows that $f \in \mathfrak{L}_p$, and that

$$\|f\|_p \leqq \|f\|_q \left(\mu(X)\right)^{\frac{q-p}{pq}} = \|f\|_q \left(\mu(X)\right)^{\frac{1}{p} - \frac{1}{q}}. \quad \square$$

(13.18) Theorem. *If $0 < p < q < \infty$, then $l_p(D) \subset l_q(D)$; and the inclusion is proper if D is infinite.*

Proof. Suppose that $f \in l_p(D)$; then we have

$$\sum_{x \in D} |f(x)|^q = \sum_{x \in D} |f(x)|^p \, |f(x)|^{q-p} \leqq A^{q-p} \sum_{x \in D} |f(x)|^p ,$$

where A is a constant such that $|f(x)| < A$ for all $x \in D$. The reader should find it easy to construct an example illustrating that the inclusion is proper if D is infinite. $\square$

(13.19) Theorem. *If $f \in \mathfrak{L}_p \cap \mathfrak{L}_q$, where $0 < p < q < \infty$, and if $p < r < q$, then $f \in \mathfrak{L}_r$. Also, the function φ defined by*

$$\varphi(r) = \log\left(\|f\|_r^r\right)$$

on $[p, q]$ is convex, i.e., $0 < \alpha < 1$ *implies*

$$\varphi(\alpha p + (1 - \alpha) q) \leqq \alpha \varphi(p) + (1 - \alpha) \varphi(q) .$$

Proof. Let $r = \alpha p + (1 - \alpha) q$, $0 < \alpha < 1$. Using Hölder's inequality with $\frac{1}{\alpha}$ $\left[\text{note that } \left(\frac{1}{\alpha}\right)' = \frac{1}{1-\alpha}\right]$, we have

$$\begin{aligned} \int |f|^r \, d\mu = \int |f|^{\alpha p + (1-\alpha) q} \, d\mu &\leqq \left(\int |f|^{\alpha p \cdot \frac{1}{\alpha}} \, d\mu\right)^{\alpha} \left(\int |f|^{(1-\alpha) q \, \frac{1}{1-\alpha}} \, d\mu\right)^{1-\alpha} \\ &= \left(\int |f|^p \, d\mu\right)^{\alpha} \left(\int |f|^q \, d\mu\right)^{1-\alpha} . \end{aligned}$$

[Note that the functions $|f|^{\alpha p}$ and $|f|^{(1-\alpha)q}$ are in $\mathfrak{L}_{\frac{1}{\alpha}}$ and $\mathfrak{L}_{\left(\frac{1}{\alpha}\right)'}$, respectively.]

Hence we have $f \in \mathfrak{L}_{\alpha p+(1-\alpha)q}$, and

$$\|f\|_{\alpha p+(1-\alpha)q}^{\alpha p+(1-\alpha)q} \leqq (\|f\|_p^p)^\alpha (\|f\|_q^q)^{1-\alpha} .$$

Taking the logarithm of both sides of this inequality, we have

$$\varphi(\alpha p + (1-\alpha) q) \leqq \alpha\varphi(p) + (1-\alpha)\varphi(q) ,$$

i.e., φ is convex. □

(13.20) Theorem. *Consider any $\mathfrak{L}_p$, $1 \leqq p < \infty$. For every $f \in \mathfrak{L}_p$ and every $\varepsilon > 0$, there exists a simple function $\sigma \in \mathfrak{L}_p$ such that $|\sigma| \leqq |f|$ and $\|\sigma - f\|_p < \varepsilon$. In particular, $\mathfrak{S} \cap \mathfrak{L}_p$ is dense in $\mathfrak{L}_p$.*

Proof. Note first that $\mathfrak{S} \cap \mathfrak{L}_p \subset \mathfrak{L}_p$. Suppose next that $f \geqq 0$, $f \in \mathfrak{L}_p$. According to (11.35) there exists a nondecreasing sequence (s_n) of nonnegative simple functions such that $s_n(x) \to f(x)$ μ-a.e. For each $n \in N$ we have

$$|f - s_n|^p \leqq |f + s_n|^p \leqq |2f|^p = 2^p f^p \in \mathfrak{L}_1 .$$

It follows from LEBESGUE'S theorem on dominated convergence (12.24) that

$$\lim_{n\to\infty} \int |f - s_n|^p \, d\mu = \int \lim_{n\to\infty} |f - s_n|^p \, d\mu = 0 ,$$

and so we can choose $s_n \in \mathfrak{S} \cap \mathfrak{L}_p$ so that $s_n \leqq f$ and $\|s_n - f\|_p$ is arbitrarily small. For an arbitrary $f \in \mathfrak{L}_p$, write $f = f_1 - f_2 + i(f_3 - f_4)$, where $f_j \in \mathfrak{L}_p^+$ and $f_1 f_2 = f_3 f_4 = 0$. For $\varepsilon > 0$, choose $\sigma_j \in \mathfrak{L}_p^+ \cap \mathfrak{S}$ such that $\sigma_j \leqq f_j$ and $\|\sigma_j - f_j\|_p < \frac{\varepsilon}{4}$ $(j \in \{1, 2, 3, 4\})$. Define σ as $\sigma_1 - \sigma_2 + i(\sigma_3 - \sigma_4)$; obviously σ is in $\mathfrak{S} \cap \mathfrak{L}_p$. Also we have

$$\|f - \sigma\|_p \leqq \sum_{j=1}^{4} \|f_j - \sigma_j\|_p < \varepsilon$$

and

$$|\sigma|^2 = (\sigma_1 - \sigma_2)^2 + (\sigma_3 - \sigma_4)^2 = \sum_{j=1}^{4} \sigma_j^2 \leqq \sum_{j=1}^{4} f_j^2 = |f|^2 . \quad □$$

(13.21) Theorem. *Let X be a locally compact Hausdorff space, let ι be any measure on X as in § 9, and let $\mathcal{M}_\iota$ be the σ-algebra of ι-measurable sets. Then $\mathfrak{C}_{00}(X) \subset \mathfrak{L}_p(X, \mathcal{M}_\iota, \iota)$ and $\mathfrak{C}_{00}(X)$ is dense in $\mathfrak{L}_p(X, \mathcal{M}_\iota, \iota)$ for $1 \leqq p < \infty$. That is, if $f \in \mathfrak{L}_p(X, \mathcal{M}_\iota, \iota)$ and $\varepsilon > 0$, then there exists a function $\varphi \in \mathfrak{C}_{00}(X)$ such that $\|f - \varphi\|_p < \varepsilon$. Moreover, if f is bounded, then φ can be chosen so that $\|\varphi\|_u \leqq \|f\|_u$.*

Proof. If $\varphi \in \mathfrak{C}_{00}(X)$, then there exists a compact set F [say the support of φ] such that $|\varphi|^p \leqq \|\varphi\|_u^p \xi_F$. Since $\iota(F) < \infty$, it follows that $\varphi \in \mathfrak{L}_p$.

Let $f \in \mathfrak{L}_p(X, \mathcal{M}_\iota, \iota)$ and let $\varepsilon > 0$ be given. Apply (13.20) to obtain a function $\sigma \in \mathfrak{S} \cap \mathfrak{L}_p$ such that $|\sigma| \leqq |f|$ and $\|f - \sigma\|_p < \frac{\varepsilon}{2}$. Since σ takes on only finitely many complex values, it is clear that $\|\sigma\|_u < \infty$.

If $\sigma = 0$, then $\sigma \in \mathfrak{C}_{00}(X)$ and the proof is complete. Thus suppose that $\|\sigma\|_u = M > 0$.

Let $E = \{x \in X : \sigma(x) \neq 0\}$. Since $\sigma \in \mathfrak{S} \cap \mathfrak{L}_p$, we have $\iota(E) < \infty$. We apply LUZIN's theorem (11.36) to obtain a function $\varphi \in \mathfrak{C}_{00}$ such that $\|\varphi\|_u \leqq M$ and if $A = \{x \in X : \varphi(x) \neq \sigma(x)\}$, then $\iota(A) < \left(\frac{\varepsilon}{4M}\right)^p$. We therefore have

$$\|f-\varphi\|_p \leqq \|f-\sigma\|_p + \|\sigma-\varphi\|_p < \frac{\varepsilon}{2} + \left(\int_X |\sigma-\varphi|^p \, d\iota\right)^{\frac{1}{p}}$$

$$= \frac{\varepsilon}{2} + \left(\int_A |\sigma-\varphi|^p \, d\iota\right)^{\frac{1}{p}} \leqq \frac{\varepsilon}{2} + \left(\int_A (2M)^p \, d\iota\right)^{\frac{1}{p}}$$

$$= \frac{\varepsilon}{2} + 2M \cdot \iota(A)^{\frac{1}{p}} < \varepsilon. \quad \square$$

(13.22) Definition. A *step function on* R is any function of the form $\sum_{k=1}^{n} \alpha_k \xi_{I_k}$ where $\alpha_1, \ldots, \alpha_n$ are complex numbers and each I_k is a bounded interval [open, closed, or half-open].

(13.23) Theorem. *Let ι be any measure on R as in § 9 and let $1 \leqq p < \infty$. Then the step functions on R form a dense subset of $\mathfrak{L}_p(R, \mathscr{M}_\iota, \iota)$.*

Proof. Clearly each step function is bounded and vanishes off of a compact set, and so each step function is in $\mathfrak{L}_p$.

Let $f \in \mathfrak{L}_p$ and $\varepsilon > 0$ be given. Use (13.20) to find a simple function $\sigma \in \mathfrak{L}_p$ such that $\|f-\sigma\|_p < \frac{\varepsilon}{2}$, say $\sigma = \sum_{j=1}^{m} \beta_j \xi_{B_j}$, where $B_j \in \mathscr{M}_\iota$ for all j. Since $\sigma \in \mathfrak{L}_p$, we may suppose that $\iota(B_j) < \infty$ for all j. Let $\delta = \frac{1}{2}\left(\frac{\varepsilon}{2\left(1 + \sum_{j=1}^{m} |\beta_j|\right)}\right)^p$, and fix j in $\{1, 2, \ldots, m\}$. Use (9.24) to obtain an open set U_j such that $B_j \subset U_j$ and $\iota(U_j) < \iota(B_j) + \delta$. It follows from (6.59) that $U_j = \bigcup_{k=1}^{\infty} I_{j,k}$, where the $I_{j,k}$'s are pairwise disjoint open intervals. We have $\sum_{k=1}^{\infty} \iota(I_{j,k}) = \iota(U_j) < \infty$, and so we may choose k_0 such that $\sum_{k=k_0+1}^{\infty} \iota(I_{j,k}) < \delta$. Let $V_j = \bigcup_{k=1}^{k_0} I_{j,k}$ and $W_j = U_j \cap V_j'$. We see at once that $\iota(W_j) < \delta$, $V_j \triangle B_j \subset (U_j \cap B_j') \cup W_j$, and $\iota(V_j \triangle B_j) \leqq \iota(U_j \cap B_j') + \iota(W_j) < 2\delta$. Therefore

$$\|\xi_{B_j} - \xi_{V_j}\|_p = \|\xi_{V_j \triangle B_j}\|_p < (2\delta)^{\frac{1}{p}}.$$

Next set $s = \sum_{j=1}^{m} \beta_j \xi_{V_j}$. Since each V_j is the union of a finite number of intervals, s is a step function. Also

$$\|\sigma - s\|_p = \left\| \sum_{j=1}^{m} \beta_j (\xi_{B_j} - \xi_{V_j}) \right\|_p$$
$$\leqq \sum_{j=1}^{m} |\beta_j| \cdot \|\xi_{B_j} - \xi_{V_j}\|_p < (2\delta)^{\frac{1}{p}} \left(\sum_{j=1}^{m} |\beta_j| \right) < \frac{\varepsilon}{2},$$

and so

$$\|f - s\|_p \leqq \|f - \sigma\|_p + \|\sigma - s\|_p < \varepsilon . \quad \square$$

The norm in the spaces $\mathfrak{L}_p(R, \mathscr{M}_\lambda, \lambda)$ has an interesting and useful continuity property, which we now establish. Recall the definition (8.14) of the translate f_h of a function f on R by a real number h.

(13.24) Theorem. *Let $X = R$ and let λ be as usual Lebesgue measure. Let p be a real number such that $1 \leqq p < \infty$, and let f be any function in $\mathfrak{L}_p(R, \mathscr{M}_\lambda, \lambda)$. Then*

(i) $$\lim_{h \downarrow 0} \|f_h - f\|_p = \lim_{h \uparrow 0} \|f_h - f\|_p = 0 .$$

Proof. Let ε be an arbitrary positive number. Applying (13.21), choose $\varphi \in \mathfrak{C}_{00}(R)$ such that $\|\varphi - f\|_p < \frac{\varepsilon}{3}$. Then φ is uniformly continuous (7.18), and also there is a positive real number α such that $\varphi(t) = 0$ if $|t| \geqq \alpha$. Choose $\delta > 0$ so that $\delta < 1$ and

$$|\varphi(t + h) - \varphi(t)| < \frac{\varepsilon}{3} \left(\frac{1}{2(\alpha + 1)} \right)^{\frac{1}{p}} \quad \text{for} \quad |h| < \delta .$$

Then $|h| < \delta$ implies that

$$\int_R |\varphi(t+h) - \varphi(t)|^p dt = \int_{-(\alpha+1)}^{\alpha+1} |\varphi(t+h) - \varphi(t)|^p dt < \left(\frac{\varepsilon}{3}\right)^p,$$

which is to say that

$$\|\varphi_h - \varphi\|_p < \frac{\varepsilon}{3} .$$

It is clear from (12.44) that

$$\|\varphi_h - f_h\|_p = \|\varphi - f\|_p \quad \text{for all} \quad h \in R ,$$

and so we have

$$\|f_h - f\|_p \leqq \|f_h - \varphi_h\|_p + \|\varphi_h - \varphi\|_p + \|\varphi - f\|_p < \varepsilon$$

if $|h| < \delta$. $\square$

(13.25) Exercise. Let p be a real number such that $0 < p < 1$. Let $(X, \mathscr{A}, \mu)$ be any measure space, and let f, g be functions in $\mathfrak{L}_p(X, \mathscr{A}, \mu)$.

(a) Prove that

(i) $$\|f + g\|_p \leqq 2^{\frac{1}{p} - 1} (\|f\|_p + \|g\|_p) .$$

[Hints. For $0 \leqq t < \infty$, prove that

$$1 + t^p \geqq (1+t)^p .$$

This implies that

$$\|f+g\|_p^p \leqq \|f\|_p^p + \|g\|_p^p .$$

Next look at the function $\psi(t) = \left(1 + t^{\frac{1}{p}}\right)\left(1+t\right)^{-\frac{1}{p}}$. The function ψ has exactly one minimum in $[0, \infty[$, at $t = 1$. Computing this minimum, one finds that (i) can be established.]

(b) Prove that $\varrho(f, g) = \|f - g\|_p^p$ is a metric on $\mathfrak{L}_p$ under which $\mathfrak{L}_p$ is a complete metric space. [Hint. Imitate the proof of (13.11).]

(c) Suppose that X contains two disjoint $\mathscr{A}$-measurable sets A and B each of finite positive μ-measure. Prove that $\| \ \|_p$ is not a norm, *i.e.*, that there are functions $f, g \in \mathfrak{L}_p$ for which $\|f+g\|_p > \|f\|_p + \|g\|_p$. [Write q for the number $\frac{1}{p}$. Note that $(1+t)^q > 1 + t^q$ for all positive real numbers t. Let $f = \alpha\xi_A$ and $g = \beta\xi_B$, where α and β are positive real numbers. Now computing the $\mathfrak{L}_p$-norms of f, g, and $f+g$, and using the inequality just noted, one can choose α and β so as to solve the present problem.]

(d) In this part, let p be any positive number. Suppose that no two sets in $\mathscr{A}$ of finite positive μ-measure are disjoint. Determine completely the structure of $\mathfrak{L}_p(X, \mathscr{A}, \mu)$. From this determination, show that $\mathfrak{L}_p(X, \mathscr{A}, \mu)$ is a trivial normed linear space for all $p > 0$. [Hints. Suppose that there are no sets at all of finite positive μ-measure. Then every $\mathfrak{L}_p$ reduces to $\{0\}$. Suppose next that there is some set of finite positive μ-measure. Then, under our hypothesis, every $\mathfrak{L}_p$ reduces to K alone.]

(13.26) Exercise: Generalized Hölder's inequality. Let $\alpha_1, \alpha_2, \ldots, \alpha_n$ be positive real numbers such that $\sum_{j=1}^{n} \alpha_j = 1$. For $f_1, f_2, \ldots, f_n$ in $\mathfrak{L}_1^+$, we have

$$f_1^{\alpha_1} f_2^{\alpha_2} \cdots f_n^{\alpha_n} \in \mathfrak{L}_1^+ ,$$

and

$$\int_X \left(f_1^{\alpha_1} f_2^{\alpha_2} \cdots f_n^{\alpha_n}\right) d\mu \leqq \|f_1\|_1^{\alpha_1} \|f_2\|_1^{\alpha_2} \cdots \|f_n\|_1^{\alpha_n} .$$

(13.27) Exercise. Write out carefully and prove the conditions for equality in Minkowski's inequality for $p = 1$ and also for $1 < p < \infty$.

(13.28) Exercise. Let X be a locally compact Hausdorff space and ι and $\mathscr{M}_\iota$ as usual. Suppose that $\iota(X) > 0$ and $\iota(\{x\}) = 0$ for all $x \in X$. Let p be any positive number.

(a) Find a function $f \in \mathfrak{L}_p$ such that $f \notin \mathfrak{L}_{p+\delta}$ for all $\delta > 0$.

[Hint. Use (12.59) and the fact that $\sum_{n=1}^{\infty} \frac{2^{n\delta}}{n^2} = \infty$ for all $\delta > 0$.]

(b) Find a function f on X such that $f \in \mathfrak{L}_{p-\delta}$ for all $\delta > 0$ and $f \notin \mathfrak{L}_p$. [Recall that $\sum_{n=1}^{\infty} n^{-\alpha}$ converges if $\alpha > 1$ and diverges if $0 < \alpha \leqq 1$.]

(c) Find a nonnegative real-valued function that is in no $\mathfrak{L}_p$.

(13.29) Exercise. Consider the set $[0, \infty[$ and Lebesgue measure λ on it. For every $p > 0$, find a function f on $[0, \infty[$ such that $f \in \mathfrak{L}_p$ and $f \notin \mathfrak{L}_q$ if $p \neq q$. [Hint. Consider the function g such that $g(x) = \frac{1}{x(1 + |\log(x)|)^2}$.]

(13.30) Exercise. Let $(X, \mathscr{A}, \mu)$ be a finite measure space and let f be any bounded measurable function on X. Prove that

$$\lim_{p \to \infty} \|f\|_p = \inf\{\alpha \in R : \alpha > 0, \mu(\{x \in X : |f(x)| > \alpha\}) = 0\}.$$

(13.31) Exercise. Let p be a real number such that $1 \leqq p < \infty$, and let f be a function in $\mathfrak{L}_p(R)$ such that f is uniformly continuous. Prove that $f \in \mathfrak{C}_0(R)$. Show by examples that each $\mathfrak{L}_p(R)$ contains an unbounded continuous function.

(13.32) Exercise. Let $(X, \mathscr{A}, \mu)$ be a measure space such that $\mu(X) = 1$ and let f be a function in $\mathfrak{L}_1^+(X, \mathscr{A}, \mu)$. Define $\log(0)$ as $-\infty$.

(a) Prove that

(i) $$\int_X \log f(x)\, d\mu(x) \leqq \log\Big(\int_X f(x)\, d\mu(x)\Big).$$

[Hint. Check the inequality $\log(t) \leqq t - 1$ for $0 \leqq t < \infty$. Replace t by $\frac{1}{\|f\|_1} f(x)$ and integrate.]

(b) Prove that equality holds in (i) if and only if f is a constant function a. e. [Hint. Check that $\log(t) < t - 1$ if $t \neq 1$.]

(c) Prove that

(ii) $$\lim_{r \downarrow 0} \|f\|_r = \exp\Big[\int_X \log f(x)\, d\mu(x)\Big].$$

[Hints. Show that $(f^r - 1)/r$ decreases to $\log f$ as $r \downarrow 0$, and apply the dominated convergence theorem to prove that

$$\lim_{r \downarrow 0} r^{-1}\Big(\int_X f^r d\mu - 1\Big) = \int_X \log f\, d\mu.$$

Using (i), show that

$$\frac{1}{r}\Big[\int_X f^r d\mu - 1\Big] \geqq \frac{1}{r} \log \int_X f^r d\mu \geqq \frac{1}{r} \int_X \log(f^r)\, d\mu = \int_X \log f\, d\mu.\Big]$$

(13.33) Exercise. Let $(X, \mathscr{A}, \mu)$ be a measure space and let p be any positive real number. Prove that if f and $(f_n)_{n=1}^{\infty}$ are in $\mathfrak{L}_p(X, \mathscr{A}, \mu)$ and $\|f - f_n\|_p \to 0$, then $f_n \to f$ in measure. Find a sequence $(f_n)_{n=1}^{\infty} \subset \mathfrak{L}_p([0, 1], \mathscr{M}_\lambda, \lambda)$ such that $f_n \to 0$ in measure but $\|f_n\|_p \not\to 0$.

(13.34) Exercise: Convex functions. Let I be an interval of R. A real-valued function Φ defined on I is said to be *convex* if whenever $a < b$ in I and $0 \leqq t \leqq 1$ we have

$$\Phi(ta + (1-t)\,b) \leqq t\Phi(a) + (1-t)\,\Phi(b)\,,$$

i.e., on the interval $[a, b]$ the graph of Φ is never above the chord [line segment] joining the points $(a, \Phi(a))$ and $(b, \Phi(b))$. Let Φ be a convex function.

(a) Prove that if $t_1, \ldots, t_n$ are positive real numbers and $\{x_1, \ldots, x_n\} \subset I$, then

$$\text{(i)} \qquad \Phi\left(\frac{t_1 x_1 + \cdots + t_n x_n}{t_1 + \cdots + t_n}\right) \leqq \frac{t_1 \Phi(x_1) + \cdots + t_n \Phi(x_n)}{t_1 + \cdots + t_n}\,.$$

[Use induction.]

(b) Prove that Φ is continuous on the interior I° of I and show by an example that Φ may be discontinuous at the endpoints of I.

(c) Prove that if c is in I°, then there exists a real number α such that $\Phi(u) \geqq \alpha(u - c) + \Phi(c)$ for all $u \in I$, *i.e.*, the line through $(c, \Phi(c))$ having slope α is always below or on the graph of Φ.

(d) Prove the following generalization of inequality (i). Let $(X, \mathscr{A}, \mu)$ be a finite measure space. If $f \in \mathfrak{L}_1^r(X, \mathscr{A}, \mu)$, if $f(X) \subset I$, and if $\Phi \circ f \in \mathfrak{L}_1^r(X, \mathscr{A}, \mu)$, then

$$\text{(ii)} \qquad \Phi\left[\frac{1}{\mu(X)} \int_X f\, d\mu\right] \leqq \frac{1}{\mu(X)} \int_X (\Phi \circ f)\, d\mu\,.$$

Inequalities (i) and (ii) are known as *JENSEN's inequalities*. [Hints. Let $c = \frac{1}{\mu(X)} \int_X f\, d\mu$. Show that $c \in I$. For the case that $c \in I^\circ$, let α be as in (c). Then

$$\Phi \circ f(x) \geqq \alpha(f(x) - c) + \Phi(c) \quad \text{for all} \quad x \in X\,.$$

Integrate both sides of this inequality. The other case is straightforward.]

(13.35) Exercise. Let φ be a real-valued nondecreasing function defined on an interval $[a, b[\subset R$. For $a \leqq x < b$, let $\Phi(x) = \int_a^x \varphi(u)\, du$. Prove that Φ is convex on $[a, b[$.

(13.36) Exercise. Let Φ and Ψ be as in YOUNG's inequality (13.2). Let $(X, \mathscr{A}, \mu)$ be a σ-finite measure space and let $\mathfrak{L}_\Phi^\dagger$ be the set of all complex-valued $\mathscr{A}$-measurable functions on X such that $\Phi \circ |f| \in \mathfrak{L}_1^+(X, \mathscr{A}, \mu)$.

(a) If Φ increases too rapidly we may have $f \in \mathfrak{L}_\Phi^\dagger$ and $2f \notin \mathfrak{L}_\Phi^\dagger$. Give such an example. [Try $\varphi(u) = \exp(u) - 1$ and $f(t) = \log(t^{-\frac{1}{2}})$.]

Let $\mathfrak{L}_\Phi$ be the set of all complex-valued $\mathscr{A}$-measurable functions f on X such that

$$\|f\|_\Phi = \sup\{\, \|fg\|_1 : g \in \mathfrak{L}_\Psi^\dagger,\ \int_X \Psi \circ |g|\, d\mu \leqq 1\} < \infty\,.$$

Prove that:

(b) $\mathfrak{L}_\Phi^\dagger \subset \mathfrak{L}_\Phi$ [use (13.2)];

(c) $\mathfrak{L}_\Phi$ is a complex linear space;

(d) $\|\ \|_\Phi$ is a norm on $\mathfrak{L}_\Phi$, where functions equal a.e. are identified;

(e) with the norm $\|\ \|_\Phi$, $\mathfrak{L}_\Phi$ is a Banach space. [First suppose that $\mu(X) < \infty$. Prove that if $\|f_n - f_m\|_\Phi \to 0$, then $\|f_n - f_m\|_1 \to 0$.]

The spaces $\mathfrak{L}_\Phi$ are called *Birnbaum-Orlicz spaces*. For further information about these spaces, the reader should consult A. C. Zaanen, *Linear Analysis*, Vol. I [New York: Interscience Publishers, 1953].

(13.37) Exercise. Define Φ on $[0, \infty[$ by $\Phi(t) = 0$ if $0 \leqq t \leqq 1$ and $\Phi(t) = t \cdot \log t$ if $t \geqq 1$. The Birnbaum-Orlicz space $\mathfrak{L}_\Phi(X, \mathscr{A}, \mu)$ [see (13.36)] is often denoted $\mathfrak{L} \log^+ \mathfrak{L}$. Prove that for the measure space $([0, 1], \mathscr{M}_\lambda, \lambda)$ we have

$$\mathfrak{L}_p \subset \mathfrak{L} \log^+ \mathfrak{L} \subset \mathfrak{L}_1$$

for all $p > 1$. The space $\mathfrak{L} \log^+ \mathfrak{L}$ arises quite naturally in Fourier analysis. See for example A. Zygmund, *Trigonometric Series*, 2nd Edition. [2 Vols. Cambridge: Cambridge University Press, 1959], and also Theorem (21.80) *infra*.

(13.38) Exercise: Vitali's convergence theorem. Let $(X, \mathscr{A}, \mu)$ be a measure space and let $1 \leqq p < \infty$. Let $(f_n)_{n=1}^\infty$ be a sequence in $\mathfrak{L}_p(X, \mathscr{A}, \mu)$ and let f be an $\mathscr{A}$-measurable function such that f is finite μ-a.e. and $f_n \to f$ μ-a.e. Then $f \in \mathfrak{L}_p(X, \mathscr{A}, \mu)$ and $\|f - f_n\|_p \to 0$ if and only if:

(i) for each $\varepsilon > 0$, there exists a set $A_\varepsilon \in \mathscr{A}$ such that $\mu(A_\varepsilon) < \infty$ and $\int_{A'_\varepsilon} |f_n|^p\, d\mu < \varepsilon$ for all $n \in N$;

and

(ii) $$\lim_{\mu(E) \to 0} \int_E |f_n|^p\, d\mu = 0$$

uniformly in n, *i.e.*, for each $\varepsilon > 0$ there is a $\delta > 0$ such that $E \in \mathscr{A}$ and $\mu(E) < \delta$ imply $\int_E |f_n|^p\, d\mu < \varepsilon$ for all $n \in N$.

Prove this theorem. [To prove the necessity of (i), let $\varepsilon > 0$ be given, choose $n_0 \in N$ such that $\|f - f_n\|_p < \varepsilon$ for all $n \geqq n_0$, choose $B_\varepsilon, C_\varepsilon \in \mathscr{A}$ of finite measure such that $\int_{B'_\varepsilon} |f|^p\, d\mu < \varepsilon$ and $\int_{C'_\varepsilon} |f_n|^p\, d\mu < \varepsilon$ for $n = 1, \ldots, n_0$. Then put $A_\varepsilon = B_\varepsilon \cup C_\varepsilon$. The necessity of (ii) is proved similarly by using (12.34). Next suppose that (i) and (ii) hold. Use (i), Fatou's lemma and Minkowski's inequality to reduce the problem to the case that $\mu(X) < \infty$.

For $\varepsilon > 0$, let δ be as in (ii). Use EGOROV's theorem to find $B \in \mathscr{A}$ such that $\mu(B) < \delta$ and $f_n \to f$ uniformly on B'. Use FATOU's lemma to prove that $\int_B |f|^p \, d\mu < \varepsilon$. Then use MINKOWSKI's inequality to show that $\int_X |f - f_n|^p \, d\mu < 3^p \varepsilon$ for all large n. Thus conclude that $f = (f - f_n) + f_n \in \mathfrak{L}_p$ and $\|f - f_n\|_p \to 0$.

VITALI's convergence theorem has considerable theoretical importance and can also be frequently applied to prove other useful theorems. The next exercise is also useful for applications [see for example (20.58) *infra*] and so we provide plentiful hints for its proof.

(13.39) Exercise. Let $(X, \mathscr{A}, \mu)$, p, (f_n), and f be as in (13.38). Suppose that $f_n \to f$ μ-a.e. For each $(n, k) \in N \times N$ let $B_{n,k} = \{x \in X: |f_n(x)|^p \geqq k\}$.

(a) Suppose that condition (13.38.i) holds [as it does, for example, if $\mu(X) < \infty$]. Prove that the following four assertions are equivalent:

(i) $f \in \mathfrak{L}_p$ and $\|f - f_n\|_p \to 0$;

(ii) if $(E_k)_{k=1}^{\infty} \subset \mathscr{A}$, $E_1 \supset E_2 \supset \cdots$, and $\bigcap_{k=1}^{\infty} E_k = \varnothing$, then $\lim_{k\to\infty} \int_{E_k} |f_n|^p \, d\mu = 0$ uniformly in n;

(iii) $\lim_{k\to\infty} \int_{B_{n,k}} |f_n|^p \, d\mu = 0$ uniformly in n[1];

(iv) condition (13.38.ii) holds.

[Hints. Assertions (i) and (iv) are equivalent by (13.38). To show that (i) implies (ii), consider $\varepsilon > 0$ and $n_0 \in N$ such that $\|f_n - f\|_p < \varepsilon$ for all $n \geqq n_0$. Then for $n \geqq n_0$, we have

$$\left(\int_{E_k} |f_n|^p \, d\mu\right)^{\frac{1}{p}} \leqq \left(\int_{E_k} |f|^p \, d\mu\right)^{\frac{1}{p}} + \left(\int_{E_k} |f_n - f|^p \, d\mu\right)^{\frac{1}{p}}$$
$$< \left(\int_{E_k} |f|^p \, d\mu\right)^{\frac{1}{p}} + \varepsilon; \tag{1}$$

now apply dominated convergence to $(|f|^p \xi_{E_k})_{k=1}^{\infty}$ to show that (1) is less than 2ε for $k \geqq k_0$ and all $n \geqq n_0$. If $n \in \{1, \ldots, n_0\}$, then

$$\int_{E_k} |f_n|^p \, d\mu \leqq \int_{E_k} \max\{|f_1|^p, \ldots, |f_{n_0}|^p\} \, d\mu,$$

and dominated convergence implies (ii).

Next suppose that (ii) holds, and write $E_k = \bigcup_{n=k}^{\infty} B_{n,k}$. Plainly $E_1 \supset E_2 \supset \cdots$, and $\overline{\lim_{n\to\infty}} |f_n(x)| = \infty$ on $\bigcap_{k=1}^{\infty} E_k$. Hence $\mu\left(\bigcap_{k=1}^{\infty} E_k\right)$ is 0; write $F_k = E_k \cap \left(\bigcap_{k=1}^{\infty} E_k\right)'$. Use (ii) to choose a k_0 such that for $k \geqq k_0$

[1] A sequence $(|f_n|^p)_{n=1}^{\infty}$ satisfying (iii) is said to be *uniformly integrable*.

and for all n,

$$\int_{F_k} |f_n|^p \, d\mu < \varepsilon .$$

For $n \geqq k_0$, we have $B_{n,k_0} \subset E_{k_0}$ and so for $n \geqq k_0$ and $k \geqq k_0$, it follows that

$$\int_{B_{n,k}} |f_n|^p \, d\mu \leqq \int_{B_{n,k_0}} |f_n|^p \, d\mu \leqq \int_{E_{k_0}} |f_n|^p \, d\mu = \int_{F_{k_0}} |f_n|^p \, d\mu < \varepsilon .$$

For $n \in \{1, \ldots, k_0 - 1\}$, we have $|f_n|^p \leq \max\{|f_1|^p, \ldots, |f_{k_0-1}|^p\} = g$, and so $\int_{B_{n,k}} |f_n|^p \, d\mu \leqq \int_{B_k} g \, d\mu$, where $B_k = \bigcup_{n=1}^{k_0-1} B_{n,k} = \{x \in X : g(x) \geqq k\}$. Thus dominated convergence applies, and so (iii) holds if (ii) holds.

Finally, suppose that (iii) holds. Choose k_0 so large that if $k \geqq k_0$ and $n \in N$, we have

$$\int_{B_{n,k}} |f_n|^p \, d\mu < \varepsilon^p .$$

If $E \in \mathscr{A}$ and $\mu(E) < k_0^{-1} \varepsilon^p$, then

$$\left(\int_E |f_n|^p \, d\mu\right)^{\frac{1}{p}} \leqq \left(\int_{E \cap B_{n,k_0}} |f_n|^p \, d\mu\right)^{\frac{1}{p}} + \left(\int_{E \cap B'_{n,k_0}} |f_n|^p \, d\mu\right)^{\frac{1}{p}} < \varepsilon + \varepsilon .$$

Hence (iv) holds if (iii) holds.]

(b) Prove that condition (ii) of part (a) implies conditions (13.38.i) and (13.38.ii).

(13.40) Discussion. We conclude this section with a study of a concept of convergence in $\mathfrak{L}_p$ spaces different from norm convergence. Thus far we have considered four important concepts of convergence for sequence of functions: uniform [unif.]; pointwise almost everywhere [a.e.]; in measure [meas.]; and in the $\mathfrak{L}_p$ norm [mean-p]. We have also expended considerable effort in examining the relationships among these four types of convergence. Let us summarize our main results. It is trivial that [unif.] implies [a.e.], in fact "everywhere". Obvious examples show that the converse fails. Nevertheless, it is easy to infer from (9.6) that if X is a compact Hausdorff space and (f_n) is a *monotone* sequence or directed family in $\mathfrak{C}^r(X)$ that converges pointwise to a function $f \in \mathfrak{C}^r(X)$, then $f_n \to f$ uniformly. [This fact is called DINI's *theorem.*] Our most useful result in this direction is EGOROV's theorem (11.32). The relationship between [a.e.] and [meas.] was thoroughly examined in (11.26), (11.27), (11.31), and (11.33). RIESZ's theorem (11.26) is often valuable in weakening an hypothesis of [a.e.] to [meas.] [see (13.45) *infra* and (12.57)]. We have a number of theorems on interchanging the order of limit and integral, *viz.* (12.21)–(12.24), (12.30), (13.38), and (13.39). These theorems can all be regarded as relating [a.e.] to [mean-p]. The relation between [meas.] and [mean-p] is set down in (13.33). Plainly [unif.] is much stronger than either [meas.] or [mean-p] on *finite* measure spaces, but for infinite measure spaces there is no implication running either way.

We now introduce a fifth kind of convergence for functions in $\mathfrak{L}_p$ spaces, and will study its relations with the notions studied previously.

(13.41) Definition. Let $(X, \mathscr{A}, \mu)$ be a measure space, let $1 \leqq p < \infty$, and let f and $(f_n)_{n=1}^{\infty}$ be functions in $\mathfrak{L}_p(X, \mathscr{A}, \mu)$. If $p > 1$, then (f_n) is said to *converge* to f *weakly* [*in* $\mathfrak{L}_p$] if

$$\lim_{n\to\infty} \int_X f_n g \, d\mu = \int_X f g \, d\mu$$

for every $g \in \mathfrak{L}_{p'}$. If $p = 1$, then (f_n) is said to *converge* to f *weakly* [*in* $\mathfrak{L}_1$] if

$$\lim_{n\to\infty} \int_X f_n g \, d\mu = \int_X f g \, d\mu$$

for every bounded $\mathscr{A}$-measurable function g on X.

(13.42) Theorem. *Notation is as in* (13.41). *If* $\|f - f_n\|_p \to 0$, *then* $f_n \to f$ *weakly.*

Proof. This follows at once from Hölder's inequality. □

(13.43) Examples. We now give a few examples to show that there is no connection at all between weak convergence and the four kinds of convergence discussed in (13.40) [except for (13.42) of course] unless further hypotheses are imposed on either the sequence or the measure space. In all of these examples we use Lebesgue measure λ.

(a) For each $n \in N$, define f_n on $[0, 2\pi]$ by $f_n(x) = \cos(nx)$. Then $(f_n) \subset \mathfrak{L}_p([0, 2\pi])$ for each $p \geqq 1$. The Riemann-Lebesgue lemma [which we prove in (16.35) *infra*] shows that $f_n \to 0$ weakly for all $p \geqq 1$. Since $\int_0^{2\pi} f_n^2 \, d\lambda = \pi$ for all $n \in N$, (12.24) and (12.57) show that $f_n \to 0$ for none of the other four kinds of convergence.

(b) Take $f_n = n\xi_{\left[0, \frac{1}{n}\right]}$. Then $(f_n) \subset \mathfrak{L}_p([0, 1])$ for all $p \geqq 1$, $f_n \to 0$ a.e. and in measure, but $f_n \not\to 0$ weakly [take $g = \xi_{[0,1]}$].

(c) Let $f_n = \frac{1}{n} \xi_{[1, \exp(n)]}$. Then $(f_n) \subset \mathfrak{L}_p(R)$ for all $p \geqq 1$. Write $g(x) = \frac{1}{x}$ for $x \geqq 1$ and $g(x) = 0$ for $x < 1$. Then $g \in \mathfrak{L}_{p'}(R)$ for all $p > 1$, g is bounded, and $\int_R f g \, d\lambda = \frac{1}{n} \int_1^{\exp(n)} \frac{dx}{x} = 1$ for all $n \in N$. Thus $f_n \to 0$ uniformly on R but $f_n \not\to 0$ weakly in $\mathfrak{L}_p(R)$.

For finite measure spaces, we know that $\mathfrak{L}_p \subset \mathfrak{L}_1$ if $p \geqq 1$ (13.27), and so uniform convergence implies $\mathfrak{L}_p$-weak convergence for finite measure spaces.

In spite of the negative results just exhibited, we do have some positive results if our sequences (f_n) satisfy certain side conditions.

(13.44) Theorem. *Notation is as in (13.41). Suppose that* $1 < p < \infty$ *and that* $(\|f_n\|_p)_{n=1}^{\infty}$ *is a bounded sequence of numbers. If* $f_n \to f$ μ*-a.e., then* $f_n \to f$ *weakly in* $\mathfrak{L}_p$.

Proof. Choose $\alpha \in R$ such that $\|f_n\|_p \leqq \alpha$ for all $n \in N$. By FATOU's lemma (12.23), we have

$$\begin{aligned}\|f\|_p^p &= \int_X |f|^p\, d\mu = \int_X \lim_{n\to\infty} |f_n|^p\, d\mu \\ &\leqq \underline{\lim}_{n\to\infty} \int_X |f_n|^p\, d\mu \leqq \alpha^p .\end{aligned} \tag{1}$$

Let $\varepsilon > 0$ and $g \in \mathfrak{L}_{p'}$ be given. Use (12.34) to obtain $\delta > 0$ such that for all $E \in \mathscr{A}$ for which $\mu(E) < \delta$, we have

$$2\alpha\Big(\int_E |g|^{p'}\, d\mu\Big)^{\frac{1}{p'}} < \frac{\varepsilon}{3} . \tag{2}$$

Next select $A \in \mathscr{A}$ such that $\mu(A) < \infty$ and

$$2\alpha\Big(\int_{A'} |g|^{p'}\, d\mu\Big)^{\frac{1}{p'}} < \frac{\varepsilon}{3} . \tag{3}$$

Apply EGOROV's theorem (11.32) to obtain $B \in \mathscr{A}$ such that $B \subset A$, $\mu(A \cap B') < \delta$, and $f_n \to f$ uniformly on B. Finally, choose $n_0 \in N$ such that $n \geqq n_0$ implies

$$|f(x) - f_n(x)|\,(\mu(B))^{\frac{1}{p}}\,\|g\|_{p'} < \frac{\varepsilon}{3}$$

for all $x \in B$. Then $n \geqq n_0$ implies

$$\Big(\int_B |f - f_n|^p\, d\mu\Big)^{\frac{1}{p}}\,\|g\|_{p'} < \frac{\varepsilon}{3} . \tag{4}$$

Thus, combining (1), (2) [with $E = A \cap B'$], (3), and (4) and using HÖLDER's and MINKOWSKI's inequalities, we have

$$\begin{aligned}\Big|\int_X f g\, d\mu - \int_X f_n g\, d\mu\Big| &\leqq \int_X |f - f_n|\,|g|\, d\mu \\ &= \int_{A\cap B'} |f - f_n|\,|g|\, d\mu + \int_{A'} |f - f_n|\,|g|\, d\mu \\ &+ \int_B |f - f_n|\,|g|\, d\mu \leqq \|f - f_n\|_p \Big(\int_{A\cap B'} |g|^{p'}\, d\mu\Big)^{\frac{1}{p'}} \\ &+ \|f - f_n\|_p \Big(\int_{A'} |g|^{p'}\, d\mu\Big)^{\frac{1}{p'}} + \Big(\int_B |f - f_n|^p\, d\mu\Big)^{\frac{1}{p}}\,\|g\|_{p'} \\ &< \frac{\varepsilon}{3} + \frac{\varepsilon}{3} + \frac{\varepsilon}{3} = \varepsilon\end{aligned}$$

for all $n \geqq n_0$. □

(13.45) Corollary. *The hypothesis* $f_n \to f$ μ*-a.e. in (13.44) can be replaced by the hypothesis that* $f_n \to f$ *in measure.*

Proof. Assume that f_n does *not* converge weakly to f. Choose $g \in \mathfrak{L}_{p'}$ such that

$$\overline{\lim_{n\to\infty}} \left| \int_X fg\,d\mu - \int_X f_n g\,d\mu \right| = \alpha \neq 0\,.$$

Use (6.84) to find integers $n_1 < n_2 < \cdots$ such that

$$\lim_{k\to\infty} \left| \int_X (f - f_{n_k})\, g\,d\mu \right| = \alpha\,. \tag{1}$$

Next use (11.26) to find a subsequence $(f_{n_{k_j}})_{j=1}^{\infty}$ of $(f_{n_k})_{k=1}^{\infty}$ such that $\lim\limits_{j\to\infty} f_{n_{k_j}} = f$ μ-a.e. It follows from (13.44) that

$$\lim_{j\to\infty} \left| \int_X (f - f_{n_{k_j}})\, g\,d\mu \right| = 0\,.$$

But this equality is incompatible with (1). □

(13.46) Remark. Example (13.43.b) shows that neither (13.44) nor (13.45) is true for the case $p = 1$. However, if we replace the hypothesis that $(\|f_n\|_1)$ be a bounded sequence by the hypothesis that $\|f_n\|_1 \to \|f\|_1$, we get a much stronger conclusion.

(13.47) Theorem. *Notation is as in* (13.41). *Suppose that* $p = 1$, *that* $f_n \to f$ *μ-a.e., and that* $\|f_n\|_1 \to \|f\|_1$. *Then*

(i) $\int_E |f_n|\,d\mu \to \int_E |f|\,d\mu$ *for all* $E \in \mathscr{A}$,

(ii) $\|f - f_n\|_1 \to 0$,

and

(iii) $f_n \to f$ *weakly in* $\mathfrak{L}_1$.

Proof. Let $E \in \mathscr{A}$. Then FATOU's lemma (12.23) shows that

$$\begin{aligned}
\underline{\lim_{n\to\infty}} \int_E |f_n|\,d\mu &\geqq \int_E |f|\,d\mu = \int_X |f|\,d\mu - \int_{E'} |f|\,d\mu \\
&\geqq \int_X |f|\,d\mu - \underline{\lim_{n\to\infty}} \int_{E'} |f_n|\,d\mu \\
&= \overline{\lim_{n\to\infty}} \left(\int_X |f_n|\,d\mu - \int_{E'} |f_n|\,d\mu \right) = \overline{\lim_{n\to\infty}} \int_E |f_n|\,d\mu\,.
\end{aligned}$$

Hence $\lim\limits_{n\to\infty} \int_E |f_n|\,d\mu$ exists and (i) holds.

To prove (ii), let $\varepsilon > 0$ be given. Select $A \in \mathscr{A}$ such that $\mu(A) < \infty$ and $\int_{A'} |f|\,d\mu < \frac{\varepsilon}{5}$. Use (12.34) to obtain a $\delta > 0$ such that if $E \in \mathscr{A}$ and $\mu(E) < \delta$, then $\int_E |f|\,d\mu < \frac{\varepsilon}{5}$. Next apply EGOROV's theorem (11.32) to find $B \in \mathscr{A}$ such that $B \subset A$, $\mu(A \cap B') < \delta$, and $f_n \to f$ uniformly on B. Choose $n_0 \in N$ such that

$$\left[\sup_{x\in B} |f(x) - f_n(x)|\right] \cdot \mu(B) < \frac{\varepsilon}{5}$$

for all $n \geqq n_0$. Now apply (i) to get

$$\overline{\lim_{n\to\infty}} \int_X |f-f_n|\,d\mu = \overline{\lim_{n\to\infty}} \left[\int_{A'} |f-f_n|\,d\mu + \int_{A\cap B'} |f-f_n|\,d\mu + \int_B |f-f_n|\,d\mu\right]$$

$$\leqq \overline{\lim_{n\to\infty}} \left[\int_{A'} |f|\,d\mu + \int_{A'} |f_n|\,d\mu + \int_{A\cap B'} |f|\,d\mu + \int_{A\cap B'} |f_n|\,d\mu\right] + \frac{\varepsilon}{5}$$

$$= 2\int_{A'} |f|\,d\mu + 2\int_{A\cap B'} |f|\,d\mu + \frac{\varepsilon}{5} < \frac{2\varepsilon}{5} + \frac{2\varepsilon}{5} + \frac{\varepsilon}{5} = \varepsilon\,.$$

Since ε is arbitrary, we have proved (ii). Conclusion (iii) follows at once from (ii). □

(13.48) Note. Theorem (13.42) admits a partial converse involving no hypothesis about pointwise convergence or convergence in measure. This converse is easy to prove once certain inequalities are established, and we postpone it to (15.17) *infra*.

(13.49) Exercise. Notation is as in (13.41). Suppose that $\|f_n\|_p \to \|f\|_p$. Prove the following.

(a) If $f_n \to f$ μ-a.e., then $\|f - f_n\|_p \to 0$.

(b) If $f_n \to f$ in measure, then $\|f - f_n\|_p \to 0$. [Recall that $1 \leqq p < \infty$.]

(13.50) Exercise. Notation is as in (13.41). Suppose that $f_n \to f$ μ-a.e. and suppose that there is a function $g \in \mathfrak{L}_1^+$ such that $|f_n|^p \leqq g$ for all $n \in N$. Prove that for every $\varepsilon > 0$ there is a set $B \in \mathscr{A}$ such that $\mu(B') < \varepsilon$ and $f_n \to f$ uniformly on B. [Take a hard look at the proof of (11.30) and then proceed as in (11.32).]

(13.51) Exercise. Let $(X, \mathscr{A}, \mu)$ be a measure space such that $\{x\} \in \mathscr{A}$ and $\mu(\{x\}) > 0$ for all $x \in X$. Let $f, f_n \in \mathfrak{L}_1(X, \mathscr{A}, \mu)$ for $n = 1, 2, \ldots$. Prove the following.

(a) If $f_n \to f$ weakly in $\mathfrak{L}_1$, then $f_n(x) \to f(x)$ for all $x \in X$.

(b) The converse of (i) is false except when X is a finite set.

(c) If $f_n \to f$ weakly, then $\|f - f_n\|_1 \to 0$. [Write $f_0 = f$ and note that $\bigcup_{n=0}^{\infty} \{x \in X : f_n(x) \neq 0\}$ is a countable set. Use (a) and (13.47).]

(d) For $p > 1$, find a sequence $(f_n) \subset l_p$ such that $f_n \to 0$ weakly but $\|f_n\|_p \not\to 0$. [See (13.13).]

§ 14. Abstract Banach spaces

We have already defined Banach spaces (7.7) and have met several specific examples: $\mathfrak{C}(X)$ and $\mathfrak{C}_0(X)$ in § 7 and $\mathfrak{L}_p(X, \mathscr{A}, \mu)$ in § 13. In the present section we give a short introduction to the abstract theory of Banach spaces and prove some important theorems about these spaces. For a thorough treatment of the subject, the reader is invited to consult the treatise *Linear Operators Part I* by Nelson

DUNFORD and JACOB T. SCHWARTZ [New York: Interscience Publishers, 1958]. Throughout this section F will denote either the field R or the field K.

(14.1) Definition. Let A and B be linear spaces over F. A function T from A into B is called a *linear transformation* [or *linear operator*] if

(i) $T(x+y) = T(x) + T(y)$

and

(ii) $T(\alpha x) = \alpha T(x)$

for all $x, y \in A$ and $\alpha \in F$. If A and B are normed linear spaces, a linear transformation T from A into B is said to be *bounded* if there exists a nonnegative real number M such that

(iii) $\|T(x)\| \leqq M\|x\|$ for all $x \in A$

[*i.e.*, T is bounded on the unit sphere of A]. In this case we define the *norm of* T to be the infimum of the set of all M's that satisfy (iii), and we write $\|T\|$ for the norm of T. This norm is called the *operator norm.*

(14.2) Theorem. *Let A and B be normed linear spaces and let T be a bounded linear transformation from A into B. Then*

$$\|T\| = \sup\left\{\frac{\|T(x)\|}{\|x\|} : x \in A, x \neq 0\right\} = \sup\{\|T(x)\| : x \in A, \|x\| = 1\} = \sup\{\|T(x)\| : x \in A, \|x\| \leqq 1\},$$

and

$$\|T(x)\| \leqq \|T\| \cdot \|x\| \quad \text{for all} \quad x \in A .$$

Proof. Exercise.

(14.3) Theorem. *Let A and B be normed linear spaces and let T be a linear transformation from A into B. Then the following three statements are mutually equivalent:*

(i) *T is bounded;*

(ii) *T is uniformly continuous on A;*

(iii) *T is continuous at some point of A.*

The continuity statements are, as always in this section, understood to be relative to the metric topologies induced on A and B by their respective norms.

Proof. If (i) holds, we have

$$\|T(x) - T(y)\| = \|T(x-y)\| \leqq \|T\| \cdot \|x-y\|$$

for all $x, y \in A$, and so (ii) follows. Trivially (ii) implies (iii). Next suppose that (iii) holds, say T is continuous at $x_0 \in A$. Then there exists a $\delta > 0$ such that $\|T(x) - T(x_0)\| \leqq 1$ whenever $\|x - x_0\| \leqq \delta$. Therefore $\|x\| \leqq 1$ implies $\|(\delta x + x_0) - x_0\| \leqq \delta$, and so

$$\|Tx\| = \frac{1}{\delta}\|T(\delta x + x_0) - T(x_0)\| \leqq \frac{1}{\delta} .$$

Thus T is bounded and $\|T\| \leqq \frac{1}{\delta}$. □

(14.4) Theorem. *Let A and B be normed linear spaces over F and let $\mathfrak{B}(A, B)$ denote the set of all bounded linear transformations from A into B. Then, with pointwise linear operations and the operator norm, $\mathfrak{B}(A, B)$ is a normed linear space. Moreover $\mathfrak{B}(A, B)$ is a Banach space if B is a Banach space.*

Proof. We prove only that $\mathfrak{B}(A, B)$ is complete if B is complete; the other verifications are routine and we omit them. Suppose that B is complete and let (T_n) be a Cauchy sequence in $\mathfrak{B}(A, B)$. For $x \in A$ we have $\|T_n(x) - T_m(x)\| \leqq \|T_n - T_m\| \cdot \|x\|$, and so $(T_n(x))$ is a Cauchy sequence in B. Thus for each $x \in A$, there is a vector $T(x) \in B$ such that $\|T_n(x) - T(x)\| \to 0$. This defines a mapping T from A into B. For $x, y \in A$, we have $\|T(x+y) - [T(x) + T(y)]\| \leqq \|T(x+y) - T_n(x+y)\| + \|T_n(x) - T(x)\| + \|T_n(y) - T(y)\| \to 0$, and thus $T(x+y) = T(x) + T(y)$. We prove similarly that $T(\alpha x) = \alpha T(x)$ for all $x \in A$ and $\alpha \in F$. Thus T is linear.

Since (T_n) is a Cauchy sequence, there is a positive constant β such that $\|T_n\| \leqq \beta$ for all $n \in N$. For $\|x\| \leqq 1$, we have $\|T(x)\| \leqq \|T(x) - T_n(x)\| + \|T_n(x)\| \leqq 1 + \beta$ [for all large n], and hence T is bounded, *i.e.*, $T \in \mathfrak{B}(A, B)$. It remains only to show that $\|T - T_n\| \to 0$. Let $\varepsilon > 0$ be given. Choose an integer p so large that $m, n \geqq p$ implies $\|T_m - T_n\| < \frac{\varepsilon}{2}$. Next let $x \in A$ be such that $\|x\| \leqq 1$ and choose $m_x \in N$ such that $m_x \geqq p$ and $\|T(x) - T_{m_x}(x)\| < \frac{\varepsilon}{2}$. Then $n \geqq p$ implies that

$$\|T(x) - T_n(x)\| \leqq \|T(x) - T_{m_x}(x)\| + \|T_{m_x} - T_n\| < \frac{\varepsilon}{2} + \frac{\varepsilon}{2} = \varepsilon .$$

It follows that $n \geqq p$ implies

$$\|T - T_n\| = \sup\{\|T(x) - T_n(x)\| : \|x\| \leqq 1\} \leqq \varepsilon . \quad \square$$

(14.5) Remark. The reader should notice the similarity between the above proof and the proof of (7.9).

(14.6) Definition. Let E be a linear space over F. A *linear functional* on E is a linear transformation from E into F [where F is regarded as a one-dimensional linear space over F]. If E is a normed linear space [and the absolute value is used as a norm on F], let E^* denote the space of all bounded linear functionals on E, *i.e.*, $E^* = \mathfrak{B}(E, F)$. Since F is complete, it follows from (14.4) that E^* is a Banach space. The space E^* is called the *conjugate* [*adjoint, dual*] *space of* E. The conjugate space E^{**} of the space E^* is called the *second conjugate space of* E, etc.

(14.7) Discussion. Let E be a normed linear space. There is a so-called *natural mapping of* E *into* E^{**} defined as follows. For $x \in E$, define $\hat{x}$ on E^* by the rule $\hat{x}(f) = f(x)$. Simple computations show that each $\hat{x}$ is a linear functional on E^* and that the mapping $x \to \hat{x}$ is a linear transformation. Also

$$\sup\{|\hat{x}(f)| : f \in E^*, \|f\| \leqq 1\} = \sup\{|f(x)| : f \in E^*, \|f\| \leqq 1\}$$
$$\leqq \sup\{\|f\| \, \|x\| : f \in E^*, \|f\| \leqq 1\} \leqq \|x\|$$

for each $x \in E$. Thus the mapping $x \to \hat{x}$ is a bounded linear transformation from E into E^{**} of norm $\leqq 1$. Several questions arise. (1) Is this mapping one-to-one? (2) Does it preserve norms? (3) Is it onto E^{**}? (4) Indeed, are there any nonzero elements in E^*? In general none of these questions have obvious answers; however we are able to answer (1), (2), and (4) with the aid of the Hahn-Banach theorem, which is next on our program. Question (3) will be answered in the exercises.

(14.8) Definition. Let E be a real linear space. A real-valued function p defined on E is said to be a *sublinear functional* if

(i) $p(x+y) \leqq p(x) + p(y)$

and

(ii) $p(\alpha x) = \alpha p(x)$

for all $x, y \in E$ and all *positive* real numbers α. Notice that a norm is a sublinear functional.

(14.9) Hahn-Banach Theorem. *Let E be a real linear space and let M be a linear subspace of E. Suppose that p is a sublinear functional defined on E and that f is a linear functional defined on M such that $f(x) \leqq p(x)$ for every $x \in M$. Then there exists a linear functional g defined on E such that g is an extension of f* [i.e., $f \subset g$] *and $g(x) \leqq p(x)$ for every $x \in E$.*

Proof. Let $\mathfrak{Z}$ be the set of all real functions h such that $\operatorname{dom} h$ is a linear subspace of E, h is linear, $f \subset h$, and $h(x) \leqq p(x)$ for all $x \in \operatorname{dom} h$. Notice that $f \in \mathfrak{Z}$. Partially order $\mathfrak{Z}$ by $\subset$ [recall that a function is a set of ordered pairs]. Let $\mathfrak{C}$ be any chain in $\mathfrak{Z}$ and let $h = \bigcup \mathfrak{C}$. Then, with the help of (2.19), we see that $h \in \mathfrak{Z}$. Applying ZORN'S lemma (3.10), we see also that $\mathfrak{Z}$ has a maximal member, say g. To complete the proof we need only show that $\operatorname{dom} g = E$. Assume that this is false, and let y be any element in $E \cap (\operatorname{dom} g)'$. Let $G = \operatorname{dom} g$ and define $H = \{x + \alpha y : x \in G, \alpha \in R\}$. Clearly H is a linear subspace of E and $G \subsetneqq H$. Let c be a fixed, but arbitrary, real number and define h on H by

$$h(x + \alpha y) = g(x) + \alpha c \,.$$

Then h is well-defined since if $x_1 + \alpha_1 y = x_2 + \alpha_2 y$, where $x_1, x_2 \in G$ and $\alpha_1, \alpha_2 \in R$, then $(\alpha_1 - \alpha_2)\, y = x_2 - x_1 \in G$ so that $\alpha_1 = \alpha_2$ and $x_1 = x_2$. Clearly h is a linear functional and $g \subsetneqq h$. If we can select c in such a way that $h(x) \leqq p(x)$ for all $x \in H$, then we will have $h \in \mathfrak{Z}$, which contradicts the maximality of g and will complete our proof. The remainder of our proof is therefore devoted to showing that c can be so selected.

Our requirement is that $g(x) + \alpha c = h(x + \alpha y) \leqq p(x + \alpha y)$ for all $x \in G$, $\alpha \in R$. By the linearity of g and the sublinearity of p, this is equivalent to the two requirements

$$g\left(\frac{x}{\alpha}\right) + c \leqq p\left(\frac{x}{\alpha} + y\right) \quad \text{for} \quad x \in G \quad \text{and} \quad \alpha > 0\,,$$

and

$$g\left(\frac{x}{\alpha}\right) + c \geqq -p\left(-\frac{x}{\alpha} - y\right) \quad \text{for} \quad x \in G \quad \text{and} \quad \alpha < 0\,.$$

Therefore it is sufficient to have

$$g(u) - p(u - y) \leqq c \leqq -g(v) + p(v + y)$$

for all $u, v \in G$. But we do have

$$g(u) + g(v) = g(u + v) \leqq p(u + v) \leqq p(u - y) + p(v + y)$$

for all $u, v \in G$. Write

$$a = \sup\{g(u) - p(u - y) : u \in G\}$$

and

$$b = \inf\{-g(v) + p(v + y) : v \in G\}\,.$$

It is clear that $a \leqq b$. Taking c to be any real number such that $a \leqq c \leqq b$, we complete our construction. □

(14.10) Remark. The crux of the Hahn-Banach theorem is that the extended functional is still majorized by p. If this requirement were not made we could obtain an extension of f simply by taking any Hamel basis for M, enlarging it to a Hamel basis for E, defining g arbitrarily on the new basis vectors, and defining g to be linear on E.

(14.11) Corollary. *Let E be a real normed linear space and let M be a linear subspace of E. If $f \in M^*$, then there exists $g \in E^*$ such that $f \subset g$ and $\|g\| = \|f\|$.*

Proof. Define p on E by $p(x) = \|f\| \cdot \|x\|$. Then p is a sublinear functional on E and we have $f(x) \leqq |f(x)| \leqq p(x)$ for all $x \in M$. Apply (14.9) to obtain a linear functional g on E such that $f \subset g$ and $g(x) \leqq p(x)$ for all $x \in E$. Clearly $g \in E^*$ and $\|g\| \leqq \|f\|$. But we also have

$$\begin{aligned} \|g\| &= \sup\{|g(x)| : x \in E,\ \|x\| \leqq 1\} \\ &\geqq \sup\{|g(x)| : x \in M,\ \|x\| \leqq 1\} \\ &= \sup\{|f(x)| : x \in M,\ \|x\| \leqq 1\} = \|f\|\,. \end{aligned}$$

Thus $\|g\| = \|f\|$. □

(14.12) Theorem [BOHNENBLUST-SOBCZYK-SUHOMLINOV]. *Let E be a complex normed linear space and let M be a linear subspace of E. If $f \in M^*$, then there exists $g \in E^*$ such that $f \subset g$ and $\|g\| = \|f\|$.*

Proof. For each $x \in M$, write $f(x) = f_1(x) + i f_2(x)$ where f_1 and f_2 are real-valued. An easy computation shows that f_1 and f_2 are *real* linear functionals on M, *i.e.*, $f_j(x + y) = f_j(x) + f_j(y)$ and $f_j(\alpha x) = \alpha f_j(x)$

for $\alpha \in R$. It is also obvious that $|f_j(x)| \leqq |f(x)| \leqq \|f\| \cdot \|x\|$, and so f_1 and f_2 are bounded and $\|f_j\| \leqq \|f\|$. Now, regarding E and M as real linear spaces [simply ignore multiplication by all but real scalars], we apply (14.11) to obtain a bounded real linear functional g_1 on E such that $f_1 \subset g_1$ and $\|g_1\| = \|f_1\|$. Next define g on E by the rule

$$g(x) = g_1(x) - ig_1(ix) .$$

It is easy to see that g is a complex linear functional, *e.g.*, $ig(x) = ig_1(x) + g_1(ix) = g_1(ix) - ig_1(i(ix)) = g(ix)$. To see that $f \subset g$, notice that for $x \in M$ we have

$$\begin{aligned} g_1(ix) + if_2(ix) &= f_1(ix) + if_2(ix) = f(ix) = if(x) \\ &= -f_2(x) + if_1(x) = -f_2(x) + ig_1(x), \end{aligned}$$

so that $g_1(ix) = -f_2(x)$ and therefore $g(x) = g_1(x) - ig_1(ix) = f_1(x) + if_2(x) = f(x)$. We need only show that g is bounded and that $\|g\| = \|f\|$. Let $x \in E$ be arbitrary and write $g(x) = r\exp(i\theta)$ where $r \geqq 0$ and $\theta \in R$. Then we have

$$\begin{aligned} |g(x)| = r = \exp(-i\theta)\, g(x) &= g(\exp(-i\theta)x) = g_1(\exp(-i\theta)x) \\ &\leqq \|g_1\| \cdot \|x\| = \|f_1\| \cdot \|x\| \leqq \|f\| \cdot \|x\| . \end{aligned}$$

This proves that g is bounded and that $\|g\| \leqq \|f\|$. As in (14.11), it is obvious that $\|f\| \leqq \|g\|$. Therefore $\|g\| = \|f\|$. □

(14.13) Corollary. *Let E be a normed linear space and let S be a linear subspace of E. Suppose that $z \in E$ and* $\operatorname{dist}(z, S) = d > 0$. *Then there exists $g \in E^*$ such that $g(S) = \{0\}$, $g(z) = d$, and $\|g\| = 1$. In particular, if $S = \{0\}$, then we have $g(z) = \|z\|$.*

Proof. Let $M = \{x + \alpha z : x \in S,\ \alpha \in F\}$. Then M is a linear subspace of E. Define f on M by $f(x + \alpha z) = \alpha d$. Clearly f is a well-defined linear functional on M such that $f(S) = \{0\}$ and $f(z) = d$. Also $\|f\| = \sup\left\{\frac{|\alpha d|}{\|x + \alpha z\|} : x + \alpha z \in M,\ \|x + \alpha z\| \neq 0\right\} = \sup\left\{\frac{d}{\|-y + z\|} : y \in S\right\} = \frac{d}{d} = 1$. Apply (14.11) if $F = R$ or (14.12) if $F = K$ to obtain the required functional $g \in E^*$. □

We now return to the mapping $x \to \hat{x}$ discussed in (14.7).

(14.14) Theorem. *Let E be a normed linear space and let π be the natural mapping of E into E^{**}: $\pi(x)(f) = f(x)$. Then π is a norm-preserving linear transformation from E into E^{**}. Consequently π is one-to-one.*

Proof. We have already observed in (14.7) that π is a bounded linear transformation from E into E^{**} and that $\|\pi\| \leqq 1$. Let x be any nonzero element of E. According to (14.13), there is an element $g \in E^*$ such that $\|g\| = 1$ and $g(x) = \|x\|$. Thus

$$\|x\| = g(x) \leqq \sup\{|f(x)| : f \in E^*,\ \|f\| = 1\} = \|\pi(x)\| \leqq \|x\| ,$$

that is,

$$\|\pi(x)\| = \|x\| .$$

Clearly $\|\pi(0)\| = 0 = \|0\|$. We have thus proved that π preserves norms. Consequently $x \neq y$ in E implies that $\|\pi(x) - \pi(y)\| = \|\pi(x - y)\| = \|x - y\| \neq 0$, and so $\pi(x) \neq \pi(y)$. □

(14.15) Remark. In view of (14.14), a normed linear space E is indistinguishable *qua* normed linear space from the subspace $\pi(E)$ of E^{**}. The mapping π need not be *onto* E^{**} [see (14.26)]. In case $\pi(E) = E^{**}$, the space E is called *reflexive*. Since E^{**} is complete and π is an isometry, every reflexive normed linear space is a Banach space. In § 15 we will show that every $\mathfrak{L}_p$ space $(1 < p < \infty)$ is reflexive.

We next present three theorems which, together with the Hahn-Banach theorem, are often regarded as the cornerstones of functional analysis. These are the open-mapping theorem, the closed-graph theorem, and the uniform boundedness principle. Several applications of these theorems will be given in the corollaries and the exercises. Unlike the Hahn-Banach theorem, these three theorems require completeness.

(14.16) Open mapping theorem [BANACH]. *Let A and B be Banach spaces and let T be a bounded linear transformation from A* onto *B. Then $T(U)$ is open in B for each open subset U of A.*

Proof. For each $\varepsilon > 0$, define $A_\varepsilon = \{x \in A : \|x\| < \varepsilon\}$ and $B_\varepsilon = \{y \in B : \|y\| < \varepsilon\}$. Let $\varepsilon > 0$ be given. We will show that there exists a $\delta > 0$ such that $T(A_\varepsilon) \supset B_\delta$. For each $n \in N$, let $\varepsilon_n = \frac{\varepsilon}{2^n}$. It is clear that if n is fixed, then

$$A = \bigcup_{j=1}^{\infty} jA_{\varepsilon_n}$$

[we define jA_{ε_n} as in (5.6.f)], and so we also have

$$B = T(A) = \bigcup_{j=1}^{\infty} T(jA_{\varepsilon_n}) .$$

Since B is complete, the Baire category theorem (6.54) implies that not every $T(jA_{\varepsilon_n})$, $j = 1, 2, \ldots$, is nowhere dense. Thus there is a $j_n \in N$ such that $[T(j_n A_{\varepsilon_n})]^-$ has nonvoid interior. But

$$[T(A_{\varepsilon_n/2})]^- = \frac{1}{2j_n} [T(j_n A_{\varepsilon_n})]^- ,$$

and αW is open in B if $\alpha \neq 0$ and W is open in B. Thus there exists a nonvoid open set $V_n \subset [T(A_{\varepsilon_n/2})]^-$. It follows that

$$[T(A_{\varepsilon_n})]^- \supset [T(A_{\varepsilon_n/2}) - T(A_{\varepsilon_n/2})]^- \supset [T(A_{\varepsilon_n/2})]^- - [T(A_{\varepsilon_n/2})]^- \supset V_n - V_n .$$ [1]

[1] For subsets C and D of B, we write $C - D = \{x - y : x \in C,\ y \in D\}$: see (5.6.f).

Since $0 \in V_n - V_n$ and $V_n - V_n = \cup \{V_n - x : x \in V_n\}$ is open in B, there exists a $\delta_n > 0$ such that

$$B_{\delta_n} \subset V_n - V_n \subset [T(A_{\varepsilon_n})]^- . \tag{1}$$

We may suppose that $\delta_n < \frac{1}{n}$ for every $n \in N$. We will now show that $B_{\delta_1} \subset T(A_\varepsilon)$.

To this end, let y be any element of B_{δ_1}. We must find an $x \in A_\varepsilon$ such that $T(x) = y$. By (1), there exists $x_1 \in A_{\varepsilon_1}$ such that $\|y - T(x_1)\| < \delta_2$, so that $y - T(x_1) \in B_{\delta_2}$. In view of (1) there exists $x_2 \in A_{\varepsilon_2}$ such that $\|y - T(x_1) - T(x_2)\| < \delta_3$. Continuing by finite induction, we find a sequence $(x_n)_{n=1}^\infty$ such that for each $n \in N$, x_n is in A_{ε_n} and

$$\left\| y - \sum_{k=1}^{n} T(x_k) \right\| < \delta_{n+1} . \tag{2}$$

Let $z_n = x_1 + x_2 + \cdots + x_n$. For $m < n$, we have

$$\|z_n - z_m\| \leq \sum_{k=m+1}^{n} \|x_k\| < \sum_{k=m+1}^{\infty} \frac{\varepsilon}{2^k} = \frac{\varepsilon}{2^m} ,$$

which has limit 0 as $m \to \infty$. Thus (z_n) is a Cauchy sequence in A; since A is complete, there is an $x \in A$ such that $\|x - z_n\| \to 0$. It is clear that

$$\|x\| = \lim_{n\to\infty} \|z_n\| \leq \sum_{k=1}^{\infty} \|x_k\| < \sum_{k=1}^{\infty} \frac{\varepsilon}{2^k} = \varepsilon ,$$

and so x is in A_ε. It follows from (2) that $\|y - T(z_n)\| \to 0$. Since T is continuous, we have $\|T(z_n) - T(x)\| \to 0$, and so $y = T(x)$.

Finally, let U be any nonvoid open subset of A, and let y be any element of $T(U)$. Then there exists $x \in U$ such that $T(x) = y$. Since U is open, there is an $\varepsilon > 0$ such that $x + A_\varepsilon \subset U$. Applying our previous result, we find that there is a $\delta > 0$ for which $B_\delta \subset T(A_\varepsilon)$. Therefore $y + B_\delta \subset T(x) + T(A_\varepsilon) = T(x + A_\varepsilon) \subset T(U)$. Thus y is an interior point of $T(U)$, and $T(U)$ is open. □

(14.17) Corollary. *If A and B are Banach spaces and T is a one-to-one continuous linear transformation from A onto B, then T^{-1} is continuous.*

Proof. If U is open in $A = \text{rng}\, T^{-1}$, then $(T^{-1})^{-1}(U) = T(U)$ is open in $B = \text{dom}\, T^{-1}$. □

(14.18) Corollary. *Let E be a linear space over F and suppose that $\| \ \|$ and $\| \ \|'$ are two Banach space norms for E. Then the metric topologies induced on E by $\| \ \|$ and $\| \ \|'$ are identical if and only if there exists a positive constant α such that*

$$\alpha \|x\| \geqq \|x\|'$$

for all $x \in E$.

Proof. Consider the identity mapping on E as a linear transformation from the Banach space $(E, \| \|)$ onto the Banach space $(E, \| \|')$. We leave the details as an exercise. □

(14.19) Lemma. *Let A and B be normed linear spaces. Then $A \times B$, with coordinatewise linear operations and the norm*

$$\|(x, y)\| = \|x\| + \|y\|$$

is a normed linear space. Moreover $A \times B$ is complete if and only if both A and B are complete.

Proof. Exercise.

(14.20) Definition. Let A and B be normed linear spaces. A linear transformation $T: A \to B$ is said to have a *closed graph* if whenever $x_n \to x$ in A and $T(x_n) \to y$ in B we have $T(x) = y$, *i.e.*, T, as a set of ordered pairs, is a closed set in $A \times B$.

It is trivial that if T is continuous, then T has a closed graph. The converse is not always true. However the converse is true if A and B are *Banach* spaces.

(14.21) Closed graph theorem. *Let A and B be Banach spaces and let T be a linear transformation from A into B such that T has a closed graph. Then T is continuous.*

Proof. Let $G = \{(x, T(x)) : x \in A\}$ be the graph of T [actually G is T]. Then G is a closed linear subspace of the Banach space $A \times B$, and so G is a Banach space. Let P_1 and P_2 be the projections of G into A and B respectively, *i.e.*, $P_1(x, T(x)) = x$ and $P_2(x, T(x)) = T(x)$ for all $x \in A$. We have

$$\|P_1(x, T(x))\| = \|x\| \leqq \|(x, T(x))\|$$

and

$$\|P_2(x, T(x))\| = \|T(x)\| \leqq \|(x, T(x))\|$$

and therefore P_1 and P_2 are continuous linear transformations. Since T is single-valued and $\operatorname{dom} T = A$, P_1 is one-to-one and onto A. It follows from (14.17) that P_1^{-1} is continuous. Clearly $T = P_2 \circ P_1^{-1}$, and so T is continuous. □

(14.22) Lemma. *Let B be a Banach space and let I be a nonvoid set. Let Γ denote the set of all functions γ from I into B such that* $\sup\{\|\gamma(\iota)\| : \iota \in I\} < \infty$ *and let $\|\gamma\|$ denote this supremum. Then Γ, with pointwise linear operations and the above norm, is a Banach space.*

The proof is almost the same as that given in (7.9) and we therefore omit it.

(14.23) Theorem: Uniform boundedness principle. *Let A and B be Banach spaces and let $\{T_\iota : \iota \in I\}$ be a nonvoid family of bounded linear transformations from A into B such that*

$$\sup\{\|T_\iota(x)\| : \iota \in I\} < \infty$$

for every $x \in A$. *Then*

$$\sup\{\|T_\iota\| : \iota \in I\} < \infty .$$

Proof. Let Γ be as in (14.22). Define a mapping $S : A \to \Gamma$ by

$$S(x)(\iota) = T_\iota(x) \quad \text{for} \quad x \in A, \iota \in I .$$

Our hypothesis that the family $\{T_\iota : \iota \in I\}$ is pointwise bounded shows that $S(x) \in \Gamma$ for each $x \in A$. Clearly S is linear. We show that S is continuous by using the closed graph theorem. Thus suppose that $x_n \to x$ in A and that $S(x_n) \to \gamma$ in Γ. For each $\iota \in I$, we have

$$\begin{aligned} \|\gamma(\iota) - S(x)(\iota)\| &\leqq \|\gamma(\iota) - S(x_n)(\iota)\| + \|S(x_n)(\iota) - S(x)(\iota)\| \\ &\leqq \|\gamma - S(x_n)\| + \|T_\iota(x_n) - T_\iota(x)\| . \end{aligned}$$

The last expression has limit 0 as $n \to \infty$ because T_ι is continuous. Therefore $\gamma(\iota) = S(x)(\iota)$ for all $\iota \in I$, and so $\gamma = S(x)$. This proves that the graph of S is closed, and hence S is continuous, *i.e.*, $\|S\| = \sup\{\|S(x)\| : \|x\| \leqq 1\} < \infty$. Since

$$\begin{aligned} \|T_\iota\| &= \sup\{\|T_\iota(x)\| : \|x\| \leqq 1\} \\ &= \sup\{\|S(x)(\iota)\| : \|x\| \leqq 1\} \\ &\leqq \sup\{\|S(x)\| : \|x\| \leqq 1\} = \|S\| \end{aligned}$$

for every $\iota \in I$, we conclude that

$$\sup\{\|T_\iota\| : \iota \in I\} \leqq \|S\| < \infty . \quad \square$$

(14.24) Corollary [Banach-Steinhaus theorem]. *Let* A *and* B *be Banach spaces and let* $(T_n)_{n=1}^{\infty}$ *be a pointwise convergent sequence of bounded linear transformations from* A *into* B. *Then the mapping* $T : A \to B$ *defined by*

$$T(x) = \lim_{n\to\infty} T_n(x)$$

is a bounded linear transformation.

Proof. It is obvious that T is linear. It is also clear that for each $x \in A$ we have

$$\sup\{\|T_n(x)\| : n \in N\} < \infty .$$

It follows from (14.23) that there exists a positive constant M such that $\|T_n\| \leqq M$ for all $n \in N$. Thus $x \in A$ implies

$$\|T(x)\| = \lim_{n\to\infty} \|T_n(x)\| \leqq M\|x\| ,$$

and so T is bounded and $\|T\| \leqq M$. $\square$

(14.25) Exercise. Let D be a nonvoid set and let $c_0(D)$ denote the set of all complex-valued functions f defined on D such that for each $\varepsilon > 0$ the set $\{x \in D : |f(x)| \geqq \varepsilon\}$ is finite. Thus $c_0(D) = \mathfrak{C}_0(D)$ where D is equipped with the discrete topology [see (7.12)]. Define linear operations pointwise on $c_0(D)$, and for $f \in c_0(D)$ define $\|f\|_u = \sup\{|f(x)| : x \in D\}$. Prove the following.

(a) $c_0(D)$ is a Banach space.

(b) $c_0(D)$ is separable if and only if D is countable.

(c) If φ is a bounded linear functional on $c_0(D)$, then there exists a function $g \in l_1(D)$ [see (13.13)] such that

$$\varphi(f) = \sum_{x \in D} f(x)\, g(x)$$

for all $f \in c_0(D)$.

(d) The mapping $\varphi \rightarrow g$ described in (c) is a norm-preserving isomorphism from $c_0(D)^*$ onto $l_1(D)$.

(14.26) Exercise. Let D be a nonvoid set and let $l_\infty(D)$ denote the set of all bounded complex-valued functions defined on D. [The space $l_\infty(D)$ is denoted $m(D)$ by some writers.] Define linear operations pointwise in $l_\infty(D)$, and for $f \in l_\infty(D)$ define

$$\|f\|_u = \sup\{|f(x)| : x \in D\}.$$

Notice that $l_\infty(D) = \mathfrak{C}(D)$ where D has the discrete topology. Notation is as in (14.25). Prove the following.

(a) If $\varphi \in l_1(D)^*$, then there exists a g in $l_\infty(D)$ such that

$$\varphi(f) = \sum_{x \in D} f(x)\, g(x)$$

for all $f \in l_1(D)$.

(b) The mapping $\varphi \rightarrow g$ described in (a) is a norm-preserving isomorphism from $l_1(D)^*$ onto $l_\infty(D)$.

(c) If D is infinite, then $c_0(D)$ is not reflexive. [Compute the natural mapping of $c_0(D)$ into $l_\infty(D)$ explicitly.]

The *density character* of a topological space X is the smallest cardinal number such that there exists a dense subset of X having that cardinal number.

(d) If D is infinite and $\overline{\overline{D}} = \mathfrak{d}$, then both $c_0(D)$ and $l_1(D)$ have density character $\mathfrak{d}$, but $l_\infty(D)$ has density character $2^{\mathfrak{d}}$.

(e) If D is infinite, then there exists an element $\varphi \in l_\infty(D)^*$ such that $\varphi(f) = 0$ for all $f \in c_0(D)$ and $\|\varphi\| = 1$. [Use an extension theorem.]

(f) If $l_1(D)$ is reflexive, then D is finite. [Use (b) and (e). For an explicit computation of $l_\infty(D)^*$, see (20.27)—(20.35) *infra*.]

(14.27) Exercise. Let E be a real linear space and let $P \subset E$ be such that

(i) $x, y \in P$ and $\alpha, \beta \geqq 0$ imply $\alpha x + \beta y \in P$;

(ii) $x \in P$ and $-x \in P$ imply $x = 0$.

Then P is called a *convex cone*. For $x, y \in E$, define $x \leqq y$ to mean that $y - x \in P$. Prove that $\leqq$ is a partial ordering on E. Let S be a linear subspace of E such that for all $x \in E$, $(x + S) \cap P \neq \varnothing$ if and only if

$(-x + S) \cap P \neq \varnothing$. Suppose that f is a linear functional on S such that $x \in S$ and $x \geqq 0$ imply $f(x) \geqq 0$. Prove that f can be extended to a linear functional g on E such that $x \in E$ and $x \geqq 0$ imply $g(x) \geqq 0$. [Use the Hahn-Banach theorem, defining

$$p(x) = \inf\{f(y) : y \in S, y \geqq x\}$$

for all x in the linear span of $S \cup P$. Alternatively, give a direct proof using ZORN's lemma.] This result is known as *KREĬN's extension theorem for nonnegative linear functionals.*

(14.28) Exercise. Prove that there exists no sequence $(c_n)_{n=1}^{\infty}$ of complex numbers such that an infinite series $\sum_{n=1}^{\infty} a_n$ of complex numbers converges absolutely if and only if $(c_n a_n)_{n=1}^{\infty}$ is a bounded sequence. [Assume such a sequence exists with $c_n \neq 0$ for all n. Consider the mapping $T : l_\infty(N) \to l_1(N)$ given by $T(f)(n) = \frac{f(n)}{c_n}$ and use the open mapping theorem.]

(14.29) Exercise. Let E be a normed linear space and let D be a nonvoid subset of E such that

$$\sup\{|f(x)| : x \in D\} < \infty$$

for each $f \in E^*$. Prove that $\sup\{\|x\| : x \in D\} < \infty$. [Consider $\pi(D) \subset E^{**}$.]

(14.30) Exercise. Let E be a Banach space and let A and B be closed linear subspaces of E such that $A \cap B = \{0\}$. Prove the following.

(a) If $A + B$ is closed in E, then the mapping $x + y \to x$ for $x \in A$, $y \in B$ is a continuous linear mapping of $A + B$ onto A.

(b) For $x \in A, y \in B$ let $\|x + y\|' = \|x\| + \|y\|$. Then $\| \ \|'$ is a complete norm for $A + B$.

(c) The set $A + B$ is closed in E if and only if $\| \ \|$ and $\| \ \|'$ induce the same topology on $A + B$.

(14.31) Exercise. (a) Let E be a normed linear space and let M be a closed linear subspace of E. Suppose that $z \in E \cap M'$. Let $S = \{x + \alpha z: x \in M, \alpha \in F\}$. Thus S is the smallest linear subspace of E containing M and z. Prove that S is closed in E. [Define f on S by $f(x + \alpha z) = \alpha$. Show that $f \in S^*$ and $\|f\| \leqq \frac{1}{\operatorname{dist}(z, M)}$. Then use the fact that F is complete.]

(b) Prove that every finite-dimensional linear subspace of E is closed in E. [Use (a) and induction].

(14.32) Exercise. Prove that there exists no Banach space of algebraic dimension $\aleph_0$. [Use (14.31) and the Baire category theorem.] For an arbitrary nonzero cardinal number $\mathfrak{m}$ [finite or infinite], construct a normed linear space of algebraic dimension $\mathfrak{m}$.

(14.33) Exercise. Let A and B be Banach spaces and let T be a linear transformation from A into B such that $g \circ T \in A^*$ for every $g \in B^*$. Prove that T is continuous.

(14.34) Exercise. Let A and B be normed linear spaces and let T be a bounded linear transformation from A into B. For $g \in B^*$, define $T^*(g) = g \circ T$. [The transformation T^* is frequently called the *adjoint of* T.] Prove that:

(a) T^* is a bounded linear transformation from B^* into A^*;

(b) $\|T^*\| = \|T\|$;

(c) T^* is one-to-one if and only if $T(A)$ is dense in B;

(d) T is one-to-one if and only if $T^*(B^*)$ separates points of A.

(14.35) Exercise. Let A be a normed linear space.

(a) Prove that there exists a Banach space B and a norm-preserving linear transformation $T: A \to B$ such that $T(A)$ is dense in B. [Consider the natural mapping of A into A^{**}.]

(b) Prove that if B_1 and B_2 are any two Banach spaces having the property ascribed to B in part (a), then there exists a norm-preserving linear transformation from B_1 onto B_2.

(14.36) Exercise. Let I be a nonvoid set, and for each $\iota \in I$, let E_ι be a normed linear space over F. Let p be a real number $\geqq 1$. Let E be the set of all $x = (x_\iota) \in \underset{\iota \in I}{\times} E_\iota$ such that $\sum_{\iota \in I} \|x_\iota\|^p < \infty$, and for all $x \in E$, let $\|x\| = \left[\sum_{\iota \in I} \|x_\iota\|^p\right]^{\frac{1}{p}}$. With linear operations $(x + y)(\iota) = x(\iota) + y(\iota)$ and $(\alpha x)(\iota) = \alpha(x(\iota))$ and the norm just defined, E is a normed linear space. The space E is a Banach space if and only if each E_ι is. Prove the preceding two assertions.

(14.37) Exercise. Let E be a finite-dimensional linear space and let $\| \ \|_1$ and $\| \ \|_2$ be norms on E. Prove that there are positive numbers α and β such that

$$\alpha \|x\|_1 \leqq \|x\|_2 \leqq \beta \|x\|_1$$

for all $x \in E$. Thus all pairs of norms on E are "equivalent", and all norms make E a Banach space.

(14.38) Exercise. Let E be a normed linear space and let M be a closed linear subspace of E. Consider the quotient space $E/M = \{x + M : x \in E\}$, where linear operations are defined by

$$(x + M) + (y + M) = (x + y) + M$$

and

$$\alpha(x + M) = (\alpha x) + M .$$

Define the *quotient norm* on E/M by the rule

(i) $\|x + M\| = \inf\{\|x + m\| : m \in M\}$.

Prove the following.

(a) Formula (i) defines a norm on E/M.

(b) If E is a Banach space, then so is E/M. [For a given Cauchy sequence in E/M, choose a subsequence $(x_n + M)_{n=1}^{\infty}$ such that $\|x_k - x_n + M\| < 2^{-n}$ for $k \geq n$. Then choose $z_n \in M$ such that $\|x_{n+1} - x_n + z_n\| < 2^{-n}$ for each n. Let $y_n = x_n + z_{n-1} + z_{n-2} + \cdots + z_1$. Prove that (y_n) is a Cauchy sequence in E, let $y_n \to y$, and prove that $x_n + M \to y + M$ in E/M.]

(c) The natural mapping φ of E onto E/M: $\varphi(x) = x + M$, is a bounded linear transformation with $\|\varphi\| \leq 1$, and φ sends open sets onto open sets.

(d) If E is a Banach space, then φ maps the open unit ball of E onto the open unit ball of E/M.

(e) If M and E/M are both complete, then so is E.

(14.39) Exercise. Let E be a separable Banach space, let $(x_n)_{n=1}^{\infty}$ be a countable dense subset of E, and let $B = \{f \in E^* : \|f\| \leq 1\}$. For $f, g \in B$, define

$$\varrho(f, g) = \sum_{n=1}^{\infty} \frac{1}{2^n} \frac{|f(x_n) - g(x_n)|}{1 + |f(x_n) - g(x_n)|} .$$

Prove that

(a) ϱ is a metric for B,

and

(b) with the metric ϱ, B is a compact metric space.

Let P be CANTOR's ternary set.

(c) Prove that there exists a norm-preserving linear transformation T from E onto a closed subspace of $\mathfrak{C}(P)$, where $\mathfrak{C}(P)$ has the uniform norm. [Use (b) and (6.100) to obtain a continuous mapping φ from P onto B. Define $T(x)(t) = \varphi(t)(x)$, for $x \in E$ and $t \in P$.]

(d) Prove (c) with P replaced by $[0, 1]$.

(14.40) Exercise. Let E be a normed linear space and let S be a linear subspace of E that is dense in E. Suppose that f is a bounded linear functional on S. Without recourse to the Hahn-Banach theorem, prove that there exists a unique $g \in E^*$ such that $g(x) = f(x)$ for all $x \in S$. Prove also that $\|g\| = \|f\|$.

§ 15. The conjugate space of $\mathfrak{L}_p$ $(1 < p < \infty)$

In this section we construct the conjugate spaces of an important, if special, class of Banach spaces. Throughout this section $(X, \mathscr{A}, \mu)$ denotes a fixed but arbitrary measure space and p a fixed but arbitrary real number such that $1 < p < \infty$. We abbreviate $\mathfrak{L}_p(X, \mathscr{A}, \mu)$ as $\mathfrak{L}_p$. Recall that $p' = p/(p-1)$ (§ 13).

(15.1) Theorem. *Let $g \in \mathfrak{L}_{p'}$ and define L_g on $\mathfrak{L}_p$ by the rule*

$$L_g(f) = \int_X f\bar{g}\, d\mu \quad \text{for all} \quad f \in \mathfrak{L}_p .$$

Then $L_g \in \mathfrak{L}_p^*$ and $\|L_g\| = \|g\|_{p'}$.

Proof. By HÖLDER'S inequality (13.4) we have

$$|L_g(f)| \leqq \|f\|_p \, \|g\|_{p'} .$$

Thus L_g is a bounded linear functional on $\mathfrak{L}_p$ [the linearity of L_g is obvious], and

$$\|L_g\| \leqq \|g\|_{p'} .$$

In fact, equality holds here, *i.e.*, $\|L_g\| = \|g\|_{p'}$. To see this, let f be $|g|^{p'-1}\operatorname{sgn}(g)$; then we have $|f|^p = |g|^{p'}$. Thus f is in $\mathfrak{L}_p$, and

$$\|f\|_p = \|g\|_{p'}^{\frac{p'}{p}} = \|g\|_{p'}^{p'-1} .$$

Hence the equalities

$$\begin{aligned} L_g(f) &= \int f\bar{g}\, d\mu = \int |g|^{p'-1}\operatorname{sgn}(g)\, \bar{g}\, d\mu = \int |g|^{p'}\, d\mu \\ &= \|g\|_{p'}^{p'} = \|g\|_{p'}^{p'-1} \|g\|_{p'} = \|f\|_p \, \|g\|_{p'} \end{aligned}$$

hold, and so also $\|L_g\| \geqq \|g\|_{p'}$. Hence we have $\|L_g\| = \|g\|_{p'}$.[1] □

(15.2) Remark. Our goal in this section is to prove that *every* bounded linear functional on $\mathfrak{L}_p$ has the form L_g for some $g \in \mathfrak{L}_{p'}$. It will then follow that $\mathfrak{L}_p^*$ and $\mathfrak{L}_{p'}$ are indistinguishable as Banach spaces, even though, naturally, they consist of quite different entities. Our proof is elementary, making use of no sophisticated facts. Only techniques of the calculus are needed.

(15.3) Lemma. *Suppose that $p \geqq 2$. Then the inequality*

(i) $\left(\frac{1+x}{2}\right)^p + \left(\frac{1-x}{2}\right)^p \leqq \frac{1}{2}(1+x^p)$

obtains for all $x \in [0, 1]$.

Proof. Define $F(x) = \left(\frac{1+x}{2}\right)^p + \left(\frac{1-x}{2}\right)^p - \frac{1}{2}(1+x^p)$. We must show that $F(x) \leqq 0$ for $0 \leqq x \leqq 1$. Since $F(0) = 2^{-1}(2^{-p+2} - 1)$ and $p \geqq 2$, we have $F(0) \leqq 0$. For $0 < x \leqq 1$, it is convenient to consider the function Φ defined by

$$\Phi(x) = \frac{2^p}{x^p} F(x) . \tag{1}$$

Thus

$$\Phi(x) = \left[\left(\frac{1}{x} + 1\right)^p + \left(\frac{1}{x} - 1\right)^p - 2^{p-1}\left(\frac{1}{x^p} + 1\right)\right];$$

clearly $\Phi(1) = 0$. Let us prove that $\Phi'(x) \geqq 0$ for $0 < x < 1$. This deriv-

[1] We could just as well have used g as $\bar{g}$ in the definition of the functional L_g. For the case $p = 2$, $\bar{g}$ is more natural, as we will see in § 16, and so we keep it here.

ative is

$$\Phi'(x) = -\frac{p}{x^{p+1}}\left[(1+x)^{p-1} + (1-x)^{p-1} - 2^{p-1}\right]. \tag{2}$$

Write $\alpha = p-1$ [note that $\alpha \geqq 1$], and consider the function Ψ defined by

$$\Psi(x) = (1+x)^{\alpha} + (1-x)^{\alpha} - 2^{\alpha}.$$

We have

$$\Psi'(x) = \alpha(1+x)^{\alpha-1} - \alpha(1-x)^{\alpha-1} \geqq 0 \quad \text{for} \quad 0 < x < 1.$$

Thus Ψ is a nondecreasing function on [0, 1], and since $\Psi(1) = 0$, the mean value theorem implies that $\Psi(x) \leqq 0$ for $0 \leqq x \leqq 1$. Going back to (2), we infer that $\Phi'(x) \geqq 0$ for $0 < x < 1$, and since $\Phi(1) = 0$, $\Phi(x)$ is nonpositive for $0 < x < 1$. The definition (1) shows that $F(x)$ is also nonpositive for $0 < x < 1$. □

(15.4) Lemma. *Let z and w be complex numbers, and suppose that $p \geqq 2$. Then we have*

(i) $$\left|\frac{z+w}{2}\right|^p + \left|\frac{z-w}{2}\right|^p \leqq \frac{|z|^p}{2} + \frac{|w|^p}{2}.$$

Proof. If $w = 0$, the inequality becomes $\frac{|z|^p}{2^{p-1}} \leqq \frac{|z|^p}{2}$, which holds since $p - 1 \geqq 1$. Thus we may suppose that $|z| \geqq |w| > 0$. The inequality (i) is equivalent to

$$\left|\frac{1}{2}\left(1+\frac{w}{z}\right)\right|^p + \left|\frac{1}{2}\left(1-\frac{w}{z}\right)\right|^p \leqq \frac{1}{2}\left(1+\left|\frac{w}{z}\right|^p\right), \tag{1}$$

which we will now prove. The inequality (1) can be written in the form

$$\left|\frac{1+r\exp(i\theta)}{2}\right|^p + \left|\frac{1-r\exp(i\theta)}{2}\right|^p \leqq \frac{1}{2}(1+r^p) \tag{2}$$

where $0 < r \leqq 1$ and $0 \leqq \theta < 2\pi$. If $\theta = 0$, the inequality (2) is just (15.3.i). The proof will be complete if we show that the left side of (2) is a maximum when $\theta = 0$, for fixed r. Clearly we may consider only θ such that $0 \leqq \theta \leqq \frac{\pi}{2}$. We must show that the function g defined by

$$g(\theta) = |1 + r\exp(i\theta)|^p + |1 - r\exp(i\theta)|^p$$

has a maximum on $\left[0, \frac{\pi}{2}\right]$ at $\theta = 0$. We have

$$g(\theta) = [1 + r^2 + 2r\cos(\theta)]^{\frac{p}{2}} + [1 + r^2 - 2r\cos(\theta)]^{\frac{p}{2}}$$

and so

$$\begin{aligned} g'(\theta) &= \frac{p}{2}(1 + r^2 + 2r\cos(\theta))^{\frac{p}{2}-1}(-2r\sin(\theta)) \\ &\quad + \frac{p}{2}(1 + r^2 - 2r\cos(\theta))^{\frac{p}{2}-1}(2r\sin(\theta)) \\ &= -pr\sin(\theta)\left[(1+r^2+2r\cos(\theta))^{\frac{p}{2}-1} - (1+r^2-2r\cos(\theta))^{\frac{p}{2}-1}\right]. \end{aligned} \tag{3}$$

Since $p \geqq 2$, it is clear from (3) that $g'(\theta) \leqq 0$ for $\theta \in \left[0, \frac{\pi}{2}\right]$. Therefore the function g is nonincreasing in $\left[0, \frac{\pi}{2}\right]$, *i.e.*, g assumes its maximum value at 0. □

(15.5) Clarkson's inequality for $p \geqq 2$. *For $p \geqq 2$ and $f, g \in \mathfrak{L}_p$, we have*

(i) $\left\|\frac{f+g}{2}\right\|_p^p + \left\|\frac{f-g}{2}\right\|_p^p \leqq \frac{1}{2}\|f\|_p^p + \frac{1}{2}\|g\|_p^p$.

Proof. We may suppose that f and g assume complex values and are defined μ-a.e. [(12.18) and (12.26)]. Then for all $x \in X$ such that $f(x)$ and $g(x)$ are defined, (15.4.i) implies that

$$\left|\frac{f(x)+g(x)}{2}\right|^p + \left|\frac{f(x)-g(x)}{2}\right|^p \leqq \frac{|f(x)|^p}{2} + \frac{|g(x)|^p}{2}. \tag{1}$$

Integrating both sides of (1) over X, we obtain (i). □

There is an analogue of (15.5.i) for $1 < p < 2$, which we establish next. The inequality and its proof, for some reason, are more complicated than for $p \geqq 2$.

(15.6) Lemma. *Suppose that $1 < p \leqq 2$. Then the inequality*

(i) $(1+x)^{p'} + (1-x)^{p'} \leqq 2(1+x^p)^{\frac{1}{p-1}}$

obtains for all $x \in [0, 1]$.

Proof. The result is trivial if $p = 2$. Thus we suppose that $1 < p < 2$. For $x = 0$ and for $x = 1$, (i) becomes an equality. As u runs from 0 to 1, the function $\frac{1-u}{1+u}$ decreases [strictly] from 1 to 0. Hence our desired inequality (i) is equivalent to

$$\left(1+\frac{1-u}{1+u}\right)^{p'} + \left(1-\frac{1-u}{1+u}\right)^{p'} \leqq 2\left(1+\left(\frac{1-u}{1+u}\right)^p\right)^{\frac{1}{p-1}} \tag{1}$$

for $0 < u < 1$. Multiplying both sides of (1) by $(1+u)^{p'}$, we obtain

$$2^{p'}(1+u^{p'}) \leqq 2\left[(1+u)^p + (1-u)^p\right]^{\frac{1}{p-1}}. \tag{2}$$

Raising both sides of (2) to the $(p-1)^{st}$ power, we get

$$(1+u^{p'})^{p-1} \leqq \frac{1}{2}\left[(1+u)^p + (1-u)^p\right], \tag{3}$$

for $0 < u < 1$. It is clear that the steps going from (i) to (3) are reversible, so that we need only to prove (3). Expanding in power series, we have

$$\begin{aligned}
&\frac{1}{2}\left[(1+u)^p + (1-u)^p\right] - (1+u^{p'})^{p-1} \\
&= \frac{1}{2}\left[\sum_{k=0}^{\infty}\binom{p}{k}u^k + \sum_{k=0}^{\infty}\binom{p}{k}(-1)^k u^k\right] - \sum_{k=0}^{\infty}\binom{p-1}{k}u^{p'k} \\
&= \sum_{k=0}^{\infty}\left[\binom{p}{2k}u^{2k} - \binom{p-1}{k}u^{p'k}\right] \\
&= \sum_{k=1}^{\infty}\left[\binom{p}{2k}u^{2k} - \binom{p-1}{2k-1}u^{p'(2k-1)} - \binom{p-1}{2k}u^{p'2k}\right]. \qquad (4)
\end{aligned}$$

As shown in (7.25), the last line of (4) converges absolutely and uniformly for $u \in [0, 1]$. We will show that each term $[\cdots]$ in this series is nonnegative. Plainly this will prove (3). The kth term is

$$\begin{aligned}
&\frac{p(p-1)(p-2)\cdots(p-(2k-1))}{(2k)!}u^{2k} - \frac{(p-1)(p-2)\cdots(p-(2k-1))}{(2k-1)!}u^{p'(2k-1)} \\
&\qquad - \frac{(p-1)(p-2)\cdots(p-2k)}{(2k)!}u^{p'2k} \\
&= \frac{p(p-1)(2-p)\cdots(2k-1-p)}{(2k)!}u^{2k} - \frac{(p-1)(2-p)(3-p)\cdots(2k-1-p)}{(2k-1)!}u^{p'(2k-1)} \\
&\qquad + \frac{(p-1)(2-p)\cdots(2k-p)}{(2k)!}u^{p'2k} \\
&= u^{2k}\frac{(2-p)(3-p)\cdots(2k-p)}{(2k-1)!} \\
&\qquad \times\left[\frac{p(p-1)}{(2k)(2k-p)} - \frac{(p-1)}{(2k-p)}u^{p'(2k-1)-2k} + \frac{(p-1)}{(2k)}u^{p'2k-2k}\right].
\end{aligned}$$

The first factor here is obviously positive. Rewrite the expression in brackets as

$$\begin{aligned}
&\left[\frac{1}{\frac{2k-p}{p-1}} - \frac{1}{\frac{2k}{p-1}} - \frac{1}{\frac{2k-p}{p-1}}u^{\frac{2k-p}{p-1}} + \frac{1}{\frac{2k}{p-1}}u^{\frac{2k}{p-1}}\right] \\
&\qquad = \left[\frac{1-u^{\frac{2k-p}{p-1}}}{\frac{2k-p}{p-1}} - \frac{1-u^{\frac{2k}{p-1}}}{\frac{2k}{p-1}}\right]. \qquad (5)
\end{aligned}$$

An elementary argument [which the reader should carry out] shows that for any $u > 0$ the function with values $\frac{1-u^t}{t}$, $0 < t < \infty$, is decreasing as a function of t. Since $\frac{2k-p}{p-1} < \frac{2k}{p-1}$, it follows that (5) is positive. □

(15.7) Theorem. *Let z and w be complex numbers, and suppose that $1 < p \leqq 2$. Then we have*

(i) $$|z+w|^{p'} + |z-w|^{p'} \leqq 2(|z|^p + |w|^p)^{\frac{1}{p-1}}.$$

Proof. If $z = 0$ or $w = 0$, (i) is obvious. Otherwise, we may suppose that $0 < |z| \leqq |w|$. The desired inequality is thus equivalent to the inequality

$$\left|1 + \frac{z}{w}\right|^{p'} + \left|-1 + \frac{z}{w}\right|^{p'} \leqq 2\left(\left|\frac{z}{w}\right|^p + 1\right)^{\frac{1}{p-1}}. \qquad (1)$$

Write (1) in the form

$$|1 + r\exp(i\theta)|^{p'} + |-1 + r\exp(i\theta)|^{p'} \leqq 2(r^p + 1)^{\frac{1}{p-1}}, \qquad (2)$$

where $\frac{z}{w} = r\exp(i\theta)$, $0 < r \leqq 1$, and $0 \leqq \theta < 2\pi$. For $\theta = 0$, the inequality (2) is just (15.6.i). Just as in the proof of (15.4.2), one shows that the expression on the left in (2) attains its maximum on $\left[0, \frac{\pi}{2}\right]$ at $\theta = 0$. Thus (2) holds for all θ. □

(15.8) Clarkson's inequality for $1 < p < 2$. *For functions f and g in $\mathfrak{L}_p$, the inequality*

(i) $$\left\|\frac{f+g}{2}\right\|_p^{p'} + \left\|\frac{f-g}{2}\right\|_p^{p'} \leqq \left[\frac{1}{2}\|f\|_p^p + \frac{1}{2}\|g\|_p^p\right]^{\frac{1}{p-1}}$$

holds.

Proof. By Minkowski's inequality for $0 < p < 1$ (13.9), we have

$$\left\|\left|\frac{f+g}{2}\right|^{p'}\right\|_{p-1} + \left\|\left|\frac{f-g}{2}\right|^{p'}\right\|_{p-1} \leqq \left\|\left|\frac{f+g}{2}\right|^{p'} + \left|\frac{f-g}{2}\right|^{p'}\right\|_{p-1}. \qquad (1)$$

The left side of (1) is the left side of (i), since $\| |h|^{p'} \|_{p-1} = \|h\|_p^{p'}$ for any $h \in \mathfrak{L}_p$. The right side is

$$\left[\int \left(\left|\frac{f+g}{2}\right|^{p'} + \left|\frac{f-g}{2}\right|^{p'}\right)^{p-1} d\mu\right]^{\frac{1}{p-1}},$$

which by (15.7) is less than or equal to

$$\left[\int 2^{p-1}\left(\left|\frac{f}{2}\right|^p + \left|\frac{g}{2}\right|^p\right) d\mu\right]^{\frac{1}{p-1}} = \left[\frac{1}{2}\|f\|_p^p + \frac{1}{2}\|g\|_p^p\right]^{\frac{1}{p-1}}. \quad □$$

Throughout (15.9)–(15.11), p is fixed and greater than 1, $\mathfrak{L}_p$ denotes an arbitrary $\mathfrak{L}_p(X, \mathscr{A}, \mu)$, and L is an arbitrary bounded linear functional on $\mathfrak{L}_p$ different from 0.

(15.9) Theorem. *There is a function $\varphi_0 \in \mathfrak{L}_p$ such that $\|\varphi_0\|_p = 1$ and $L(\varphi_0) = \|L\|$, that is, L assumes a maximum absolute value on the unit ball of $\mathfrak{L}_p$.*

Proof. The definition (14.1) of $\|L\|$ shows that there is a sequence $(\varphi'_n)_{n=1}^{\infty}$ in $\mathfrak{L}_p$ such that $\|\varphi'_n\|_p = 1$, $|L(\varphi'_n)| > \frac{1}{2}\|L\|$, and $\lim_{n\to\infty} |L(\varphi'_n)| = \|L\|$. Let $\varphi_n = \operatorname{sgn}[\overline{L(\varphi'_n)}]\,\varphi'_n$. Then we obviously have:

$$L(\varphi_n) = |L(\varphi'_n)| > \frac{1}{2}\|L\| > 0\,; \tag{1}$$

$$\|\varphi_n\|_p = 1; \tag{2}$$

$$\lim_{n\to\infty} L(\varphi_n) = \|L\|\,. \tag{3}$$

We will show that (φ_n) is a Cauchy sequence in $\mathfrak{L}_p$. In the contrary case, there are a positive number α and subsequences $(\varphi_{n_k})_{k=1}^{\infty}$ and $(\varphi_{m_k})_{k=1}^{\infty}$ such that $\|\varphi_{n_k} - \varphi_{m_k}\|_p > \alpha$ for $k = 1, 2, \ldots$. For $p \geqq 2$, we use CLARKSON's inequality (15.5) to write

$$\left\|\frac{\varphi_{m_k}+\varphi_{n_k}}{2}\right\|_p^p + \left\|\frac{\varphi_{m_k}-\varphi_{n_k}}{2}\right\|_p^p \leqq \frac{1}{2}\|\varphi_{n_k}\|_p^p + \frac{1}{2}\|\varphi_{m_k}\|_p^p = 1\,. \tag{4}$$

For $1 < p < 2$, we use CLARKSON's inequality (15.8) to write

$$\left\|\frac{\varphi_{m_k}+\varphi_{n_k}}{2}\right\|_p^{p'} + \left\|\frac{\varphi_{m_k}-\varphi_{n_k}}{2}\right\|_p^{p'} \leqq \left[\frac{1}{2}\|\varphi_{m_k}\|_p^p + \frac{1}{2}\|\varphi_{n_k}\|_p^p\right]^{\frac{1}{p-1}} = 1\,. \tag{5}$$

For $p \geqq 2$, the inequality (4) implies that

$$\left\|\frac{\varphi_{m_k}+\varphi_{n_k}}{2}\right\|_p^p < 1 - \left(\frac{\alpha}{2}\right)^p, \tag{6}$$

and for $1 < p < 2$, (5) implies that

$$\left\|\frac{\varphi_{m_k}+\varphi_{n_k}}{2}\right\|_p^{p'} < 1 - \left(\frac{\alpha}{2}\right)^{p'}. \tag{7}$$

From (6) and (7) we can find, for each $p > 1$, a number $\beta \in]0, 1[$ that is independent of k and such that

$$\left\|\frac{\varphi_{m_k}+\varphi_{n_k}}{2}\right\|_p < 1 - \beta \tag{8}$$

for $k = 1, 2, \ldots$. Consider the sequence of functions $(g_k)_{k=1}^{\infty}$ defined by

$$g_k = \frac{\varphi_{m_k}+\varphi_{n_k}}{\|\varphi_{m_k}+\varphi_{n_k}\|_p}\,. \tag{9}$$

No denominator in (9) is zero, for otherwise we would have $\varphi_{n_k} = -\varphi_{m_k}$, and hence the equality $L(\varphi_{n_k}) = -L(\varphi_{m_k})$ would hold, contradicting (1). For $k = 1, 2, \ldots$, (8) and (9) show that

$$\begin{aligned} L(g_k) &= \frac{1}{\left\|\frac{\varphi_{m_k}+\varphi_{n_k}}{2}\right\|_p}\left[\frac{1}{2}L(\varphi_{m_k}) + \frac{1}{2}L(\varphi_{n_k})\right] \\ &> \frac{1}{1-\beta}\left[\frac{1}{2}L(\varphi_{m_k}) + \frac{1}{2}L(\varphi_{n_k})\right]. \end{aligned} \tag{10}$$

By (3) we have $\lim_{k\to\infty} L(\varphi_{m_k}) = \lim_{k\to\infty} L(\varphi_{n_k}) = \|L\|$. Thus (10) implies that

$$\varliminf_{k\to\infty} L(g_k) \geqq \frac{1}{1-\beta}\|L\|.$$

Since $\|g_k\|_p = 1$, this is an evident contradiction. Therefore (φ_n) is a Cauchy sequence in $\mathfrak{L}_p$ and so has a limit φ_0 in $\mathfrak{L}_p$ (13.11). It is clear from (3) that $\lim_{n\to\infty} L(\varphi_n) = L(\varphi_0) = \|L\|$. □

(15.10) Lemma[1]. *Let E be a complex normed linear space, and let L be a nonzero bounded linear functional on E for which there exists $g \in E$ satisfying the conditions $\|g\| = 1$ and $L(g) = \|L\|$. Consider the function*

(i) $t \to \|g + tf\| = \psi_f(t)$

defined on R, where f is any element of E. If ψ_f and ψ_{-if} are differentiable at $t = 0$, then we have

(ii) $\frac{1}{\|L\|} L(f) = \psi_f'(0) + i\psi_{-if}'(0)$.

Proof. Suppose without loss of generality that $\|L\| = 1$. For any complex number z, we have

$$L\big(g + z(f - L(f)\,g)\big) = L(g) + z\big(L(f) - L(f)\,L(g)\big) = L(g) = 1\,.$$

Since $|L(h)| \leqq \|h\|$ for all $h \in E$, it follows that

$$\|g + z(f - L(f)\,g)\| \geqq 1 \quad \text{for all} \quad z \in K\,. \tag{1}$$

For each $t \in R$ different from $\frac{-1}{L(f)}$, write $g + tf$ in the form $g + tf = (1 + tL(f))\left[g + t\frac{1}{1+tL(f)}(f - L(f)\,g)\right]$. The norm of the expression in brackets is greater than or equal to 1 for all t, and for $t = 0$ it is equal to 1. Hence

$$\begin{aligned}\|g + tf\| - \|g\| &\geqq |1 + tL(f)| - 1\\ &= \left[(1 + t\,\mathrm{Re}(L(f)))^2 + (t\,\mathrm{Im}(L(f)))^2\right]^{\frac{1}{2}} - 1\\ &\geqq 1 + t\,\mathrm{Re}(L(f)) - 1 = t\,\mathrm{Re}(L(f))\,,\end{aligned}$$

which implies that

$$\frac{\|g + tf\| - \|g\|}{t} \geqq \mathrm{Re}(L(f)) \quad \text{if} \quad t > 0\,, \tag{2}$$

and

$$\frac{\|g + tf\| - \|g\|}{t} \leqq \mathrm{Re}(L(f)) \quad \text{if} \quad t < 0\,. \tag{3}$$

Both (2) and (3) follow trivially from (1) if $L(f) = 0$. It follows that

$$\psi_f'(0) = \mathrm{Re}(L(f)) \quad \text{for all} \quad f \in E\,. \tag{4}$$

---

[1] This lemma is due to E. J. McShane [Proc. Amer. Math. Soc. 1 (1950), p. 402].

Applying (4) to the function $-if$, we obtain

$$\psi'_{-if}(0) = \mathrm{Re}(L(-if)) = \mathrm{Im}(L(f)); \tag{5}$$

and (4) and (5) imply (ii). □

(15.11) Theorem [F. RIESZ]. *Let L be a bounded linear functional on $\mathfrak{L}_p$ $(1 < p < \infty)$. There is a function $h \in \mathfrak{L}_{p'}$ such that $L(f) = \int_X f\bar{h}\,d\mu$ for all $f \in \mathfrak{L}_p$.*

Proof. The result is trivial for $L = 0$, so we suppose that $L \neq 0$. Using (15.9), select a function $g \in \mathfrak{L}_p$ such that $L(g) = \|L\|$ and $\|g\|_p = 1$. We want to apply (15.10), and to do this we must show that the function

$$t \to \|tf + g\|_p = \psi_f(t)$$

is differentiable at $t = 0$ for every $f \in \mathfrak{L}_p$. Let $\omega(t) = \psi_f^p(t) = \int_X |tf + g|^p\,d\mu$. Writing $f = f_1 + if_2$ and $g = g_1 + ig_2$, we have

$$|tf + g|^p = \left[(tf_1 + g_1)^2 + (tf_2 + g_2)^2\right]^{\frac{p}{2}},$$

and so almost everywhere on X we have

$$\frac{d}{dt}|tf + g|^p = p\,|tf + g|^{p-2}\left[(tf_1 + g_1)\,f_1 + (tf_2 + g_2)\,f_2\right] \tag{1}$$

for all t. [If $1 < p < 2$ and the points $x \in X$ and $t \in R$ are such that $tf(x) + g(x) = 0$, then the first factor in the above expression for $\frac{d}{dt}|tf + g|^p$ is undefined, and the second factor is zero. In this case, as the reader can check, the derivative is actually zero.] For every $t \neq 0$, we have

$$\frac{\omega(t) - \omega(0)}{t} = \int_X \frac{|tf + g|^p - |g|^p}{t}\,d\mu\,. \tag{2}$$

Using the mean value theorem and (1) to rewrite the integrand in (2), we have

$$\frac{\omega(t) - \omega(0)}{t} = \int_X p\,|t'f + g|^{p-2}\,[(t'f_1 + g_1)\,f_1 + (t'f_2 + g_2)\,f_2]\,d\mu \tag{3}$$

where $0 < |t'| < |t|$ *and t' is a function of $x \in X$.* [If $1 < p < 2$ and $t'f(x) + g(x) = 0$, then the integrand is zero.] Since $(t'f_j + g_j) \leqq |t'f + g|$ and $f_j \leqq |f|$, the absolute value of the integrand in (3) is less than or equal to $2p\,|t'f + g|^{p-1}\,|f|$. If $|t| \leqq 1$, then we have $2p\,|t'f + g|^{p-1}\,|f| \leqq 2p\,(|f| + |g|)^{p-1}|f|$. The functions $|f|$ and $|g|$ are both in $\mathfrak{L}_p$, and so $(|f| + |g|)^{p-1}$ is in $\mathfrak{L}_{p'}$, and $(|f| + |g|)^{p-1}|f|$ is in $\mathfrak{L}_1$, by HÖLDER'S inequality (13.4). Thus for all $|t| \leqq 1$, the integrand in (3) is less than or equal to the fixed function $2p\,(|f| + |g|)^{p-1}|f|$, which is in $\mathfrak{L}_1$. LEBESGUE'S theorem on dominated convergence (12.24) implies that

$$\lim_{t \to 0} \int_X \frac{|g + tf|^p - |g|^p}{t} \, d\mu = \int_X p \, |g|^{p-2} [g_1 f_1 + g_2 f_2] \, d\mu \, . \tag{4}$$

[If $1 < p < 2$ and $g(x) = 0$, then the integrand in (4) and in the following integrals is zero.] Combining (2) and (4), we see that $\omega'(0)$ exists and that

$$\omega'(0) = \int_X p \, |g|^{p-2} [g_1 f_1 + g_2 f_2] \, d\mu \, . \tag{5}$$

Consequently $\psi_f'(0)$ also exists. Using (5), we write

$$\psi_f'(0) = \frac{1}{p} \left(\int_X |g|^p d\mu \right)^{\frac{1}{p} - 1} \cdot \omega'(0) = \frac{1}{p} \|g\|_p^{1-p} \, \omega'(0)$$

$$= \int_X |g|^{p-2} \, [g_1 f_1 + g_2 f_2] \, d\mu \, . \tag{6}$$

Lemma (15.10) and (6) imply that

$$L(f) = \|L\| \, (\psi_f'(0) + i \psi_{-if}'(0))$$

$$= \|L\| \int_X |g|^{p-2} ((g_1 f_1 + g_2 f_2) + i (g_1 f_2 - g_2 f_1)) \, d\mu = \|L\| \cdot \int_X |g|^{p-2} \bar{g} f \, d\mu \, .$$

The theorem follows when we set $h = \|L\| \cdot |g|^{p-1} \operatorname{sgn}(g)$; *i.e.*,

$$L(f) = \int_X f \bar{h} \, d\mu \, . \quad \square$$

(15.12) Theorem. *Let $(X, \mathscr{A}, \mu)$ be an arbitrary measure space and let p be a real number such that $1 < p < \infty$. Then the mapping T defined by*

$$T(g) = L_{\bar{g}}$$

[see (15.1)] is a norm-preserving linear transformation from $\mathfrak{L}_{p'}$ onto $\mathfrak{L}_p^*$. Thus, as Banach spaces, $\mathfrak{L}_{p'}$ and $\mathfrak{L}_p^*$ are isomorphic.

Proof. The fact that T is a norm-preserving mapping from $\mathfrak{L}_{p'}$ *into* $\mathfrak{L}_p^*$ is (15.1). It follows from (15.11) that T is *onto* $\mathfrak{L}_p^*$. It is trivial that T is linear. Since T is linear and norm-preserving, T is one-to-one. $\square$

(15.13) Exercise [J. A. Clarkson]. Let $(Y, \mathscr{A}, \mu)$ be a measure space such that $\mathscr{A}$ contains two disjoint sets of finite positive measure. There is a [unique] least positive number c such that

$$\frac{1}{c} \leqq \frac{\|f + g\|_p^2 + \|f - g\|_p^2}{2 (\|f\|_p^2 + \|g\|_p^2)} \leqq c$$

for all $f, g \in \mathfrak{L}_p$ $(1 < p < \infty)$ such that $\|f\|_p$ and $\|g\|_p$ are not both zero. Prove that c exists and that $c = 2^{\frac{|2-p|}{p}}$. Also, the constants c and $\frac{1}{c}$ are attained.

(15.14) Exercise. (a) Let $(X, \mathscr{A}, \mu)$ be a measure space, let p be a real number such that $p > 1$, and let f be an $\mathscr{A}$-measurable function on X such that:

(i) $\{x \in X : f(x) \neq 0\}$ is the union of a countable number of sets in $\mathscr{A}$ having finite measure;

(ii) $fg \in \mathfrak{L}_1(X, \mathscr{A}, \mu)$ for all $g \in \mathfrak{L}_p(X, \mathscr{A}, \mu)$.

Then f is in $\mathfrak{L}_{p'}(X, \mathscr{A}, \mu)$. [Hints. Construct a sequence of functions $(f_n)_{n=1}^{\infty}$ such that $(|f_n|)_{n=1}^{\infty}$ is nondecreasing, $|f_n| \to |f|$ everywhere, and each f_n vanishes except on a set of finite measure. Then use (12.22), (15.1), and (14.23) to infer that $f \in \mathfrak{L}_{p'}$.]

(b) [E. B. Leach]. Let $(X, \mathscr{A}, \mu)$ be a measure space and suppose that every set A in $\mathscr{A}$ such that $\mu(A) = \infty$ contains a set $B \in \mathscr{A}$ for which $0 < \mu(B) < \infty$. Let f be any $\mathscr{A}$-measurable function on X for which (ii) above holds. Then f is in $\mathfrak{L}_{p'}(X, \mathscr{A}, \mu)$. [Hints. Let $A_n = \left\{x \in X : |f(x)| \geq \frac{1}{n}\right\}$. If $\mu(A_m) = \infty$, use (10.56.d) to find a subset C of A_m such that $C \in \mathscr{A}$, $\mu(C) = \infty$, and C is σ-finite. Then $f\xi_C$ satisfies (i) above and also (ii), since f satisfies (ii). It follows that $f\xi_C \in \mathfrak{L}_{p'}$, a contradiction. Hence f satisfies (i), and (a) applies.]

(15.15) Exercise. (a) Let $(X, \mathscr{A}, \mu)$ be the measure space described in (10.56.b). Show that the conclusion of (15.13) fails for this measure space for each p such that $1 < p < \infty$.

(b) Let $(X, \mathscr{A}, \mu)$ be a measure space for which there is a set $D \in \mathscr{A}$ such that $\mu(D) = \infty$ and no $\mathscr{A}$-measurable subset of D has finite positive measure. Prove that there is an $\mathscr{A}$-measurable, nonnegative, real-valued function f on X such that (15.14.ii) holds and f is in no $\mathfrak{L}_r$ $(0 < r < \infty)$.

(15.16) Exercise. Let E be a [real or complex] normed linear space such that for all $\varepsilon > 0$ and $x, y \in E$ such that $\|x\| = \|y\| = 1$ and $\|x-y\| > \varepsilon$, the inequality

(i) $$\left\|\frac{1}{2}(x+y)\right\| \leq (1-\delta)$$

obtains, where $\delta = \delta(\varepsilon)$ is independent of x and y and $0 < \delta < 1$. Such spaces are called *uniformly convex* [by some writers *uniformly rotund*].

(a) Let E be a uniformly convex Banach space and L a bounded linear functional on E. Prove that there is an $x \in E$ such that $\|x\| = 1$ and $L(x) = \|L\|$. [Imitate the proof of (15.9), noting that (15.9.8) is simply the assertion that $\mathfrak{L}_p$ is uniformly convex.]

(b) Prove that a uniformly convex Banach space is reflexive. [Use McShane's lemma (15.10).]

(c) Let E be a uniformly convex normed linear space, let S be a proper linear subspace of E that is complete in the norm on E[1], and let x

---

[1] For example, S can be any closed subspace of E if E is a Banach space, or any finite-dimensional subspace of an arbitrary E [see (14.31.b) and (14.37)].

be any element of $E \cap S'$. Show that there is one and only one element $y_0 \in S$ such that

$$\|y_0 - x\| = \inf\{\|y - x\| : y \in S\}.$$

(15.17) Exercise: Weak convergence and the RADON-RIESZ theorem. In (13.41), we defined weak convergence for sequences of functions in $\mathfrak{L}_p$ and thereafter studied relations between weak convergence and other sorts of convergence. Weak convergence can be defined for sequences (x_n) in any normed linear space E, as follows. The sequence (x_n) *converges weakly to* $x \in E$ if $f(x_n) \to f(x)$ for all f in the conjugate space E^* of E. Theorem (15.11) shows that the present definition is consistent with the definition offered in (13.41) for $E = \mathfrak{L}_p$ $(1 < p < \infty)$: for $E = \mathfrak{L}_1$, see (20.19) *infra*.

(a) Let E be a normed linear space [over R or K] with the property that if (x_n) is a sequence in E, $x \in E$, $\|x_n\| = 1$, and $\|x\| = 1$, then the relation

$$\left\|\frac{x_n + x}{2}\right\| \to 1$$

implies

$$\|x_n - x\| \to 0.$$

[Such spaces are called *locally uniformly convex*.]. Let (y_n) be a sequence in E and y an element of E such that:

(i) $\|y_n\| \to \|y\|$;

(ii) $y_n \to y$ weakly.

Prove that $\|y_n - y\| \to 0$. [If $\|y\| = 0$, the assertion is trivial. Otherwise, write $x_n = \|y_n\|^{-1} y_n$ and $x = \|y\|^{-1} y$. Then $x_n \to x$ weakly and $\|x_n\| = \|x\| = 1$. Also if $\|x_n - x\| \to 0$, it follows that $\|y_n - y\| \to 0$. Hence we need only to show that $\|x_n - x\| \to 0$. Assume the contrary. Then by local uniform convexity, there is a subsequence $(x_{n_k})_{k=1}^{\infty}$ such that

$$\left\|\frac{1}{2}(x_{n_k} + x)\right\| < \alpha < 1 \quad (k = 1, 2, 3, \ldots). \tag{1}$$

By (14.13), there is an $f \in E^*$ such that $\|f\| = 1$ and $f(x) = 1$. For this f, (1) implies that

$$\frac{1}{2}|f(x_{n_k}) + 1| < \alpha,$$

so x_n does *not* converge weakly to x.]

(b) [RADON-RIESZ theorem]. Let p be a real number such that $1 < p < \infty$ and let $(X, \mathscr{A}, \mu)$ be a measure space. Write $\mathfrak{L}_p$ for $\mathfrak{L}_p(X, \mathscr{A}, \mu)$. Let (f_n) be a sequence of functions in $\mathfrak{L}_p$ and f a function in $\mathfrak{L}_p$ such that $f_n \to f$ weakly and $\|f_n\|_p \to \|f\|_p$. Prove that $\|f_n - f\|_p \to 0$. [Use CLARKSON'S inequalities (15.5) and (15.8) to show that $\mathfrak{L}_p$ is locally uniformly convex. Then apply part (a)[1].]

---

[1] This short proof of the RADON-RIESZ theorem, and part (a), were kindly suggested to us by Professor IRVING GLICKSBERG.

(15.18) Exercise. Define f and f_n, $n = 1, 2, \ldots$, on $[0, 2\pi]$ by the rules: $f(x) = 1$, $f_n(x) = 1 + \sin(nx)$. Notice that $f, f_n \in \mathfrak{L}_1([0, 2\pi], \mathscr{M}_\lambda, \lambda)$ for all $n \in N$. Prove that:

(a) $f_n \to f$ weakly in $\mathfrak{L}_1$ [use (16.35)];
(b) $\|f\|_1 = \|f_n\|_1 = 2\pi$ for all $n \in N$;
(c) $f_n \not\to f$ in measure;
(d) $\|f_n - f\|_1 \not\to 0$;
(e) $f_n \not\to f$ a.e.;
(f) $\mathfrak{L}_1([0, 2\pi], \mathscr{M}_\lambda, \lambda)$ is not locally uniformly convex.

§ 16. Abstract Hilbert spaces

(16.1) Inner product spaces. Recall (13.16) that an *inner product space* is a linear space H over K together with a mapping $(x, y) \to \langle x, y\rangle$ of $H \times H$ into K such that:

$$\langle x + y, z\rangle = \langle x, z\rangle + \langle y, z\rangle;$$
$$\langle \alpha x, y\rangle = \alpha\langle x, y\rangle;$$
$$\langle y, x\rangle = \overline{\langle x, y\rangle};$$
$$\langle x, x\rangle > 0 \quad \text{if} \quad x \neq 0;$$

[for all $x, y, z \in H$ and $\alpha \in K$]. The above relations imply trivially that:

$$\langle 0, y\rangle = \langle y, 0\rangle = 0;$$
$$\langle x, \alpha y\rangle = \bar{\alpha}\langle x, y\rangle;$$
$$\langle x, y + z\rangle = \langle x, y\rangle + \langle x, z\rangle;$$

for all $x, y, z \in H$ and $\alpha \in K$.

An inner product space over R is defined similarly; in this case we have $\langle y, x\rangle = \langle x, y\rangle$. For many reasons complex inner product spaces are more useful in analysis than are real inner product spaces.

(16.2) Theorem [Inequality of Cauchy-Bunyakovskiĭ-Schwarz]. *For $x, y \in H$ we have*

(i) $|\langle x, y\rangle|^2 \leqq \langle x, x\rangle \langle y, y\rangle$.

Equality obtains in (i) *if and only if x and y are linearly dependent.*

Proof. If $y = 0$, equality obtains and we have $0x = 1y$. Thus suppose that $y \neq 0$ and let $\gamma \in K$. Then we have

$$0 \leqq \langle x - \gamma y, x - \gamma y\rangle = \langle x, x\rangle - \gamma\langle y, x\rangle - \bar{\gamma}\langle x, y\rangle + |\gamma|^2 \langle y, y\rangle. \quad (1)$$

Setting $\gamma = \dfrac{\langle x, y\rangle}{\langle y, y\rangle}$ in (1), we obtain

$$0 \leqq \langle x, x\rangle - \frac{\langle x, y\rangle\langle y, x\rangle}{\langle y, y\rangle} - \frac{\overline{\langle x, y\rangle}\langle x, y\rangle}{\langle y, y\rangle} + \frac{|\langle x, y\rangle|^2\langle y, y\rangle}{\langle y, y\rangle^2}$$
$$= \langle x, x\rangle - \frac{|\langle x, y\rangle|^2}{\langle y, y\rangle}.$$

It is clear that strict inequality holds throughout this computation unless $x - \gamma y = 0$.

If $\alpha x = \beta y$, where $|\alpha| + |\beta| \neq 0$, it is easy to see that the two sides of (i) are equal. □

(16.3) Theorem. *Let* $\|x\| = \langle x, x\rangle^{\frac{1}{2}}$. *With this definition of norm, an inner product space H is a normed linear space.*

Proof. All of the verifications that $\| \ \|$ is a norm on H are trivial except for the triangle inequality. Evidently

$$\|x+y\|^2 = \langle x+y, x+y\rangle = \langle x, x\rangle + \langle x, y\rangle + \langle y, x\rangle + \langle y, y\rangle$$
$$= \|x\|^2 + 2\,\mathrm{Re}\langle x, y\rangle + \|y\|^2 .$$

Applying (16.2), we have

$$2\,\mathrm{Re}\langle x, y\rangle \leqq 2\,|\langle x, y\rangle| \leqq 2\|x\| \cdot \|y\| .$$

Therefore $\|x+y\|^2 \leqq (\|x\| + \|y\|)^2$, and so $\|x+y\| \leqq \|x\| + \|y\|$. □

(16.4) Exercise. Find a necessary and sufficient condition that $\|x+y\| = \|x\| + \|y\|$.

(16.5) Exercise: The polar identity. Prove that if H is a *complex* inner product space, then

$$4\langle x, y\rangle = \|x+y\|^2 - \|x-y\|^2 + i\|x+iy\|^2 - i\|x-iy\|^2$$

for all $x, y \in H$.

(16.6) Exercise. Let E be a complex normed linear space. Prove that there exists an inner product on E that induces the given norm as in (16.3) if and only if the given norm satisfies the *parallelogram law*

(i) $$\|x+y\|^2 + \|x-y\|^2 = 2(\|x\|^2 + \|y\|^2) \quad \text{for all} \quad x, y \in E .$$

[If (i) holds, define $\langle x, y\rangle$ by $\mathrm{Re}\langle x, y\rangle = \frac{1}{2}(\|x+y\|^2 - \|x\|^2 - \|y\|^2)$ and $\mathrm{Im}\langle x, y\rangle = \frac{1}{2}(\|x+iy\|^2 - \|x\|^2 - \|y\|^2)$. Then (16.1) is easy to check. The reader should draw a diagram to show that (i) is really a well-known elementary fact about sides and diagonals of a parallelogram. This exercise shows that inner product spaces are the normed linear spaces in which all two-dimensional subspaces "look like" Euclidean spaces.]

(16.7) Definition. An inner product space which, with the norm defined in (16.3), is a Banach space is called a *Hilbert space*. An incomplete inner product space is sometimes called a *pre-Hilbert space*.

(16.8) Examples. (a) If $(X, \mathscr{A}, \mu)$ is any measure space, then $\mathfrak{L}_2(X, \mathscr{A}, \mu)$ is a Hilbert space [(13.11) and (13.15)].

(b) The space $\mathfrak{C}_{00}(R)$ with

$$\langle f, g\rangle = \int_{-\infty}^{\infty} f(t)\,\overline{g(t)}\,dt$$

is an incomplete inner product space. To see this, approximate $\xi_{[0,1]}$ by the sequence $(g_n)_{n=1}^{\infty}$ of continuous functions defined as follows: $g_n(x) = 0$ for $-\infty < x \leqq -\frac{1}{n}$ and for $1 + \frac{1}{n} \leqq x < \infty$; $g_n(x) = 1$ for $0 \leqq x \leqq 1$; and g_n is linear on $\left[-\frac{1}{n}, 0\right]$ and $\left[1, 1+\frac{1}{n}\right]$. The sequence (g_n) converges to $\xi_{[0,1]}$ in $\mathfrak{L}_2$, and so is a Cauchy sequence. However, it is clear that (g_n) converges to no function in $\mathfrak{C}_{00}(R)$.

(c) The sequence space $l_2(N)$, consisting of all complex sequences $x = (x_n)$ such that $\sum_{n=1}^{\infty} |x_n|^2 < \infty$, is a Hilbert space in which $\langle (x_n), (y_n) \rangle = \sum_{n=1}^{\infty} x_n \bar{y}_n$. [As we have observed before (13.13), $l_2(N)$ is actually the space $\mathfrak{L}_2(X, \mathscr{A}, \mu)$ in which X is the set N, $\mathscr{A}$ is all subsets of X, and μ is counting measure.]

(d) The space consisting of all sequences $x = (x_n)$ such that x_n is ultimately zero is an incomplete inner product space [the inner product is that induced by l_2]. For example, the sequence $(x^{(m)})_{m=1}^{\infty}$ in which $x^{(m)} = \left(1, \frac{1}{2}, \ldots, \frac{1}{m}, 0, 0, \ldots\right)$ converges in l_2, but its limit has no zero terms. The reader should note the analogy between this space and $\mathfrak{C}_{00}(R)$; indeed if the positive integers N are given the discrete topology, then the space being considered is $\mathfrak{C}_{00}(N)$.

We now take up a notion that lies ready to hand in any inner product space, and which we will use later to classify all Hilbert spaces.

(16.9) Definition. Let H be an inner product space. If $\langle x, y \rangle = 0$ for elements x and y of H, then x and y are said to be *orthogonal to each other*, and we write $x \perp y$. If E is a subset of H such that $\langle x, y \rangle = 0$ for all $x, y \in E$ such that $x \neq y$, then E is said to be *orthogonal*. If in addition $\|x\| = 1$ for all $x \in E$, then E is called *orthonormal*. If E and F are subsets of H such that $x \perp y$ for all $x \in E$ and $y \in F$, then we say that E and F are *orthogonal*, and we write $E \perp F$.

The sets $\varnothing$ and $\{x\}$ are orthogonal for all $x \in H$. The vector 0 is orthogonal to every vector in H.

(16.10) Theorem. *If* $\{z_1, \ldots, z_n\}$ *is an orthogonal set in* H, *then*

$$\|z_1 + \cdots + z_n\|^2 = \|z_1\|^2 + \cdots + \|z_n\|^2 .$$

Proof. We have

$$\left\|\sum_{k=1}^{n} z_k\right\|^2 = \left\langle \sum_{j=1}^{n} z_j, \sum_{k=1}^{n} z_k \right\rangle = \sum_{j=1}^{n} \sum_{k=1}^{n} \langle z_j, z_k \rangle = \sum_{j=1}^{n} \langle z_j, z_j \rangle$$
$$= \sum_{j=1}^{n} \|z_j\|^2 . \quad \square$$

(16.11) Definition. Let E be a normed linear space and let $(x_n)_{n=1}^{\infty}$ be a sequence of elements of E. We say that the series $\sum_{n=1}^{\infty} x_n$ *converges*, and we write $\sum_{n=1}^{\infty} x_n = x$, if there exists an $x \in E$ such that $\lim_{p\to\infty} \left\| x - \sum_{n=1}^{p} x_n \right\| = 0$.

(16.12) Theorem. *Let H be a Hilbert space and let $\{z_n\}_{n=1}^{\infty}$ be an orthogonal set in H,* i.e., *$z_n \perp z_m$ for $n \neq m$. Then $\sum_{n=1}^{\infty} z_n$ converges if and only if $\sum_{n=1}^{\infty} \|z_n\|^2 < \infty$. If $\sum_{n=1}^{\infty} z_n = z$, then $\|z\|^2 = \sum_{n=1}^{\infty} \|z_n\|^2$.*

Proof. For $n > m$, it is plain that

$$\left\| \sum_{k=1}^{n} z_k - \sum_{k=1}^{m} z_k \right\|^2 = \left\| \sum_{k=m+1}^{n} z_k \right\|^2 = \sum_{k=m+1}^{n} \|z_k\|^2 = \sum_{k=1}^{n} \|z_k\|^2 - \sum_{k=1}^{m} \|z_k\|^2 .$$

Thus $\left(\sum_{k=1}^{n} z_k \right)_{n=1}^{\infty}$ is a Cauchy sequence in H if and only if $\left(\sum_{k=1}^{n} \|z_k\|^2 \right)_{n=1}^{\infty}$ is a Cauchy sequence in R. This proves our first assertion.

Now suppose that $\sum_{k=1}^{\infty} z_k = z$. Writing $s_n = \sum_{k=1}^{n} z_k$, we have $\|z - s_n\| \to 0$ and $\|s_n\|^2 = \sum_{k=1}^{n} \|z_k\|^2$. Thus $(\|z\| + \|s_n\|)_{n=1}^{\infty}$ is a bounded sequence of numbers. Also we have $|\,\|z\| - \|s_n\|\,| \leq \|z - s_n\| \to 0$, and so

$$\begin{aligned} \left| \|z\|^2 - \sum_{k=1}^{n} \|z_k\|^2 \right| &= |\,\|z\|^2 - \|s_n\|^2| \\ &= (\|z\| + \|s_n\|)\,|\,\|z\| - \|s_n\|\,| \to 0, \end{aligned}$$

i. e.,

$$\lim_{n\to\infty} \sum_{k=1}^{n} \|z_k\|^2 = \|z\|^2 . \quad \square$$

(16.13) Theorem. *Let H be an inner product space and let E be an orthogonal subset of H not containing 0. Then E is linearly independent.*

Proof. Suppose that $\{x_1, \ldots, x_n\} \subset E$, that $\alpha_1, \ldots, \alpha_n$ are scalars, and that $\sum_{k=1}^{n} \alpha_k x_k = 0$. Then for each $j \in \{1, \ldots, n\}$ we have $0 = \langle 0, x_j \rangle = \left\langle \sum_{k=1}^{n} \alpha_k x_k, x_j \right\rangle = \sum_{k=1}^{n} \alpha_k \langle x_k, x_j \rangle = \alpha_j \|x_j\|^2$. But $x_j \neq 0$, and so $\|x_j\|^2 \neq 0$. Therefore all α_j's are zero. $\square$

(16.14) Definition. Let E be an arbitrary orthogonal set in an inner product space H. For $x \in H$ and $z \in E$, we define the *Fourier coefficient of x with respect to z* to be the number $\langle x, z \rangle$.

In the real Hilbert space R^3 [3-dimensional Euclidean space], the vectors (1, 0, 0), (0, 1, 0), and (0, 0, 1) form an orthonormal set, and the Fourier coefficient of a vector x with respect to each of these unit vectors is simply the length of the perpendicular projection of x in the direction determined by that unit vector.

The following example motivates the use of the term "Fourier coefficient" in (16.14).

(16.15) Example. Consider the space $\mathfrak{L}_2\left([-\pi, \pi], \mathscr{M}_\lambda, \frac{1}{2\pi}\lambda\right)$. In this space, the inner product is

$$\langle f, g\rangle = \frac{1}{2\pi}\int_{-\pi}^{\pi} f(t)\,\overline{g(t)}\,dt$$

for $f, g \in \mathfrak{L}_2$. For $n \in Z$, the functions $\chi_n : t \to \exp(int)$ are defined and continuous on R, and so are in $\mathfrak{L}_2([-\pi, \pi])$. We will show that these functions form an orthonormal set. For each $n \in Z$, we have $\|\chi_n\|^2 = \frac{1}{2\pi}\int_{-\pi}^{\pi} |\chi_n(t)|^2\,dt = \frac{1}{2\pi}\int_{-\pi}^{\pi} 1\,dt = 1$. If $m \neq n$ in Z, we have $\langle \chi_m, \chi_n\rangle = \frac{1}{2\pi}\int_{-\pi}^{\pi} \chi_m(t)\,\overline{\chi_n(t)}\,dt = \frac{1}{2\pi}\int_{-\pi}^{\pi} \exp(i(m-n)t)\,dt = \frac{1}{2\pi}\int_{-\pi}^{\pi} \cos((m-n)t)\,dt + \frac{i}{2\pi}\int_{-\pi}^{\pi} \sin((m-n)t)\,dt = 0$. [Use elementary calculus to evaluate these integrals.] Hence $\{\chi_n\}_{n=-\infty}^{\infty}$ is an orthonormal set. For $f \in \mathfrak{L}_1([-\pi, \pi])$, the classical n^{th} Fourier coefficient of f is the number

$$\hat{f}(n) = \frac{1}{2\pi}\int_{-\pi}^{\pi} f(t)\exp(-int)\,dt = \langle f, \chi_n\rangle .$$

For an inner product space H and an orthonormal set $\{z_1, \ldots, z_n\} \subset H$, one often wants to know how closely an element $x \in H$ can be approximated in the metric of H by a linear combination of the z_k's. This question is completely answered by the following theorem.

(16.16) Theorem. *Let H be an inner product space, let $\{z_1, \ldots, z_n\}$ be an orthonormal set in H, and let x be any element of H. Then the function f defined on K^n by*

(i) $f(\alpha_1, \ldots, \alpha_n) = \left\| x - \sum_{k=1}^{n} \alpha_k z_k \right\|$

attains an absolute minimum value at one and only one point of K^n, viz., $\alpha_k = \langle x, z_k\rangle$ $(k = 1, \ldots, n)$. *Furthermore, the inequality*

(ii) $\sum_{k=1}^{n} |\langle x, z_k\rangle|^2 \leqq \|x\|^2$

holds.

Proof. We have

$$\left\|x - \sum_{k=1}^{n} \alpha_k z_k\right\|^2 = \|x\|^2 - \sum_{k=1}^{n} \overline{\alpha_k} \langle x, z_k\rangle - \sum_{k=1}^{n} \alpha_k \langle z_k, x\rangle + \sum_{k=1}^{n} |\alpha_k|^2$$

$$= \|x\|^2 + \sum_{k=1}^{n} \left[|\langle x, z_k\rangle|^2 - \bar{\alpha}_k \langle x, z_k\rangle - \alpha_k \overline{\langle x, z_k\rangle} + |\alpha_k|^2\right] - \sum_{k=1}^{n} |\langle x, z_k\rangle|^2$$

$$= \|x\|^2 + \sum_{k=1}^{n} |\langle x, z_k\rangle - \alpha_k|^2 - \sum_{k=1}^{n} |\langle x, z_k\rangle|^2 .$$

Hence $f(\alpha_1, \ldots, \alpha_n)$ is a minimum if and only if $\alpha_k = \langle x, z_k\rangle$, $k = 1, \ldots, n$; and in this case we see that

$$0 \leqq \|x\|^2 - \sum_{k=1}^{n} |\langle x, z_k\rangle|^2 . \quad \square$$

(16.17) Theorem [Bessel's inequality]. *Let E be a nonvoid orthonormal set in an inner product space H, and let $x \in H$. Then we have*

(i) $\sum_{z \in E} |\langle x, z\rangle|^2 \leqq \|x\|^2$,

and so $\{z \in E : \langle x, z\rangle \neq 0\}$ is countable.

Proof. The inequality (16.16.ii) implies that for each nonvoid finite set $F \subset E$ we have

$$\sum_{z \in F} |\langle x, z\rangle|^2 \leqq \|x\|^2 .$$

Therefore

$$\sum_{z \in E} |\langle x, z\rangle|^2 = \sup\Big\{\sum_{z \in F} |\langle x, z\rangle|^2 : F \neq \varnothing,\ F \text{ finite},\ F \subset E\Big\} \leqq \|x\|^2 . \quad \square$$

(16.18) Theorem. *Let $\{z_k\}_{k=1}^{\infty}$ be an orthonormal set in a Hilbert space H. For every $x \in H$, the vector $y = \sum_{k=1}^{\infty} \langle x, z_k\rangle z_k$ exists in H and $x - y$ is orthogonal to every z_k.*

Proof. The existence of y follows from (16.12) and Bessel's inequality. Let $m \in N$. We must show that $\langle x - y, z_m\rangle = 0$. For each $n \in N$ let $y_n = \sum_{k=1}^{n} \langle x, z_k\rangle z_k$. Then for all $n \geqq m$, we have

$$\begin{aligned} |\langle x - y, z_m\rangle| &\leqq |\langle x - y_n, z_m\rangle| + |\langle y_n - y, z_m\rangle| \\ &\leqq \left|\langle x, z_m\rangle - \sum_{k=1}^{n} \langle x, z_k\rangle \langle z_k, z_m\rangle\right| + \|y_n - y\| \, \|z_m\| \\ &= 0 + \|y_n - y\| . \end{aligned}$$

We have used the fact that $\{z_k\}$ is orthonormal. Since $\|y_n - y\| \to 0$, it follows that $\langle x - y, z_m\rangle = 0$. $\square$

We shall now investigate the problem of writing an arbitrary element of an inner product space as a limit of linear combinations of elements of an orthonormal set. We first make a definition.

(16.19) Definition. An orthonormal set E in an inner product space H is said to be *complete* if the only vector orthogonal to all elements of E is 0.

(16.20) Theorem. *Every inner product space that is not* $\{0\}$ *contains a complete orthonormal set. In fact, every orthonormal subset of* H *is contained in a complete orthonormal set.*

Proof. We use TUKEY's lemma (3.8). Let A be any nonvoid orthonormal subset of H; for example, $A = \{\|x\|^{-1}x\}$ where $x \neq 0$, $x \in H$. Next let $\mathscr{F} = \{B : B \subset H, A \cup B \text{ is an orthonormal set}\}$. To test orthonormality one tests only two vectors at a time, and so it is clear that $\mathscr{F}$ is of finite character. Also $A \in \mathscr{F}$, so that $\mathscr{F}$ is nonvoid. By TUKEY's lemma, $\mathscr{F}$ has a maximal member E. It is obvious that $E \supset A$. We assert that E is complete. Assume that $y \neq 0$ and that $y \perp E$. Set $z = \|y\|^{-1} y$. Then $E \cup \{z\} \in \mathscr{F}$ and $E \subsetneqq E \cup \{z\}$. This contradicts the maximality of E. Thus $y \perp E$ implies $y = 0$. □

The above proof is not constructive; it gives no clue as to how to construct a complete orthonormal set in any given inner product space. There are methods of actually constructing complete orthonormal sets provided that they are not too large. We now take up this construction.

(16.21) Lemma. *Let* S *be a dense subset of an inner product space* H. *If* $x \perp S$, *then* $x = 0$.

Proof. Choose a sequence (y_n) in S such that $\|x - y_n\| \to 0$. Then

$$\|x\|^2 = \langle x, x\rangle - \langle x, y_n\rangle = \langle x, x - y_n\rangle \leqq \|x\| \cdot \|x - y_n\| \to 0 .$$

Therefore $\|x\| = 0$. □

(16.22) The Gram-Schmidt orthonormalization process. Let H be an inner product space and let $\{y_1, \ldots, y_n, \ldots\}$ be a finite or countably infinite linearly independent subset of H. Let $z_1 = y_1$ and set

$$u_1 = \|z_1\|^{-1} z_1 ,$$

and

$$z_2 = y_2 - \langle y_2, u_1\rangle u_1 .$$

The vector z_2 is not zero, for y_2 is not a multiple of y_1. Define

$$u_2 = \|z_2\|^{-1} z_2 .$$

We have

$$\langle u_2, u_1\rangle = \|z_2\|^{-1} \langle z_2, u_1\rangle = 0 .$$

Thus the set $\{u_1, u_2\}$ is orthonormal and it spans the same 2-dimensional subspace as $\{y_1, y_2\}$.

We will define inductively an orthonormal set $\{u_1, \ldots, u_n, \ldots\}$ such that for each positive integer k, the set $\{u_1, \ldots, u_k\}$ spans the same subspace as $\{y_1, \ldots, y_k\}$. Thus suppose that $\{u_1, \ldots, u_n\}$ has been constructed and that $\operatorname{span}\{u_1, \ldots, u_k\} = \operatorname{span}\{y_1, \ldots, y_k\}$ for $k = 1, \ldots, n$.

If $\{y_1, y_2, \ldots, y_n\}$ is *all* of the y's, stop the construction. If y_{n+1} exists, let

$$z_{n+1} = y_{n+1} - \sum_{k=1}^{n} \langle y_{n+1}, u_k \rangle u_k .$$

Now z_{n+1} is not the zero vector, because $y_{n+1} \notin \operatorname{span}\{y_1, \ldots, y_n\} = \operatorname{span}\{u_1, \ldots, u_n\}$. Next define

$$u_{n+1} = \|z_{n+1}\|^{-1} z_{n+1} .$$

For $1 \leq j \leq n$, we have

$$\begin{aligned} \langle u_{n+1}, u_j \rangle &= \|z_{n+1}\|^{-1} \left\langle y_{n+1} - \sum_{k=1}^{n} \langle y_{n+1}, u_k \rangle u_k, u_j \right\rangle \\ &= \|z_{n+1}\|^{-1} \big(\langle y_{n+1}, u_j \rangle - \langle y_{n+1}, u_j \rangle \cdot \langle u_j, u_j \rangle \big) = 0 . \end{aligned}$$

Thus the set $\{u_1, \ldots, u_{n+1}\}$ is orthonormal; it remains to show that it spans the same subspace as $\{y_1, \ldots, y_{n+1}\}$. It is obvious from the definitions of z_{n+1} and u_{n+1} that y_{n+1} is a linear combination of $u_1, \ldots, u_{n+1}$. Therefore

$$\{y_{n+1}\} \cup \big(\operatorname{span}\{y_1, \ldots, y_n\}\big) \subset \operatorname{span}\{u_1, \ldots, u_{n+1}\} ,$$

and so

$$\operatorname{span}\{y_1, \ldots, y_{n+1}\} \subset \operatorname{span}\{u_1, \ldots, u_{n+1}\} .$$

Similarly u_{n+1} is a linear combination of the vectors y_{n+1} and $u_1, \ldots, u_n$. By the inductive hypothesis, it follows that $u_{n+1} \in \operatorname{span}\{y_1, \ldots, y_{n+1}\}$, and therefore $\operatorname{span}\{u_1, \ldots, u_{n+1}\} \subset \operatorname{span}\{y_1, \ldots, y_{n+1}\}$. Thus these two subspaces are the same. [Alternatively, we could have reversed the first inclusion by a dimensionality argument.]

The Gram-Schmidt process yields an orthonormal set which is essentially unique. More precisely, if $\{u_1, \ldots, u_n\}$ is to span the same subspace as $\{y_1, \ldots, y_n\}$ for each n, then we must have $u_1 = \frac{\gamma}{\|y_1\|} y_1$, where $|\gamma| = 1$. Hence the choice of u_1 is unique up to a multiplicative constant of absolute value 1. Having defined $\{u_1, \ldots, u_n\}$, we must take $u_{n+1} \in \operatorname{span}\{u_1, \ldots, u_n, y_{n+1}\}$. Thus for some complex numbers $\beta_1, \ldots, \beta_n$, α, we have

$$u_{n+1} = \alpha y_{n+1} + \beta_1 u_1 + \cdots + \beta_n u_n .$$

The number α cannot be zero since, if it were, then $u_{n+1} \in \operatorname{span}\{u_1, \ldots, u_n\} = \operatorname{span}\{y_1, \ldots, y_n\}$, from which it would follow that $y_{n+1} \in \operatorname{span}\{u_1, \ldots, u_{n+1}\} = \operatorname{span}\{y_1, \ldots, y_n\}$. This would contradict the linear independence of the y_k's. Hence

$$\frac{1}{\alpha} u_{n+1} = y_{n+1} + \delta_1 u_1 + \cdots + \delta_n u_n ,$$

and so also

$$0 = \left\langle \frac{1}{\alpha} u_{n+1}, u_j \right\rangle = \langle y_{n+1}, u_j \rangle + \delta_j \,,$$

and

$$\delta_j = - \langle y_{n+1}, u_j \rangle \quad (1 \leqq j \leqq n) \,.$$

Thus we have

$$\frac{1}{\alpha} u_{n+1} = y_{n+1} - \sum_{k=1}^{n} \langle y_{n+1}, u_k \rangle u_k \,,$$

just as we defined z_{n+1}. The number α is now determined from this last equality by taking norms and noting that $\|u_{n+1}\| = 1$. Clearly α, and hence u_{n+1}, is unique up to a multiplicative factor of absolute value 1.

(16.23) Theorem. *Let H be an inner product space, not $\{0\}$, that contains a countable dense subset D. Then H contains a countable complete orthonormal set that is obtained from D by the Gram-Schmidt process.*

Proof. Suppose, as we may, that $0 \notin D$, and enumerate D as $(x_n)_{n=1}^{\infty}$. Define $y_1 = x_{n_1}$ where $n_1 = 1$. Suppose that $y_1 = x_{n_1}, \ldots, y_k = x_{n_k}$ have been defined and are linearly independent, and that $n_1 < n_2 < \cdots < n_k$. If there is no $j > n_k$ such that $\{y_1, \ldots, y_k, x_j\}$ is linearly independent, stop the process. Otherwise let n_{k+1} be the smallest such j and define $y_{k+1} = x_{n_{k+1}}$. We have thus defined a finite or countably infinite linearly independent set $\{y_1, y_2, \ldots\} \subset D$. Let S be the smallest linear subspace of H containing $\{y_1, y_2, \ldots\}$. It is clear that $D \subset S$, since if $x_j \in D$, then x_j is a linear combination of $y_1, \ldots, y_k$, where k is chosen so that $n_k \leqq j < n_{k+1}$ [or else y_k is the last y selected and $n_k \leqq j$]. Hence S is dense in H. Let $\{u_1, u_2, \ldots\}$ be the orthonormal set obtained from the set $\{y_1, y_2, \ldots\}$ by the Gram-Schmidt process. We will show that $\{u_1, u_2, \ldots\}$ is complete. Suppose that $x \in H$ and $\langle x, u_n \rangle = 0$ for all n. Then $\left\langle x, \sum_{n=1}^{p} \alpha_n u_n \right\rangle = 0$ for all finite linear combinations of the u_n's, and so $\langle x, y \rangle = 0$ for all $y \in S$. It follows from (16.21) that $x = 0$. □

(16.24) Corollary. *Let $n \in N$. Then an inner product space H is indistinguishable [as an inner product space] from K^n if and only if the algebraic dimension of H is n. [In K^n we define $\langle (z_1, \ldots, z_n), (w_1, \ldots, w_n) \rangle = \sum_{j=1}^{n} z_j \overline{w}_j$.]*

Proof. It is clear that if H has dimension n, then the process of choosing the y's in the previous proof stops with $\{y_1, \ldots, y_n\}$. Thus we obtain a complete orthonormal set $\{u_1, \ldots, u_n\} \subset H$. The mapping of H onto K^n given by

$$\sum_{j=1}^{n} \alpha_j u_j \to (\alpha_1, \ldots, \alpha_n)$$

preserves all inner product space structure, *i.e.*, it is one-to-one, onto, linear, and preserves inner products [hence norms as well]. The converse is trivial. □

(16.25) Example. By way of illustration, we work out a certain classical orthonormal set. For each integer $n \geqq 0$, define f_n on R by

$$f_n(x) = x^n \exp\left[-\frac{x^2}{2}\right].$$

Since $\sum_{k=1}^{\infty} \frac{k^{2n}}{\exp(k^2)} < \infty$ for all $n \geqq 0$, each f_n is evidently in the Hilbert space $\mathfrak{L}_2(R, \mathscr{M}_\lambda, \lambda)$. Since each polynomial has only finitely many roots, the set $\{f_n\}_{n=0}^{\infty}$ is linearly independent over K. For each integer $n \geqq 0$, define

$$H_n(x) = (-1)^n \exp[x^2] \exp^{(n)}[-x^2],$$

where the superscript $^{(n)}$ denotes the nth derivative of the function $x \to \exp[-x^2]$. The functions $(H_n)_{n=0}^{\infty}$ are clearly all polynomials. They are called the *Hermite polynomials.* The first three Hermite polynomials are

$$H_0(x) = 1,$$
$$H_1(x) = 2x,$$
$$H_2(x) = 4x^2 - 2.$$

One can go on computing them as long as patience will permit. Next let

$$\varphi_n(x) = \exp\left[-\frac{x^2}{2}\right] H_n(x);$$

these functions are called the *Hermite functions*. They are all in $\mathfrak{L}_2(R, \mathscr{M}_\lambda, \lambda)$, and, as we will now show, they are an orthogonal set. First we have

$$\varphi_n''(x) = (-1)^n \Big\{(x^2 + 1) \exp\left[\frac{x^2}{2}\right] \exp^{(n)}[-x^2] + 2x \exp\left[\frac{x^2}{2}\right] \exp^{(n+1)}[-x^2] + \exp\left[\frac{x^2}{2}\right] \exp^{(n+2)}[-x^2]\Big\}. \quad (1)$$

Using LEIBNIZ's rule for finding the derivatives of a product, we have

$$\begin{aligned}\exp^{(n+2)}[-x^2] &= \{-2x \exp[-x^2]\}^{(n+1)} \\ &= \sum_{k=0}^{n+1} \binom{n+1}{k} (-2x)^{(k)} \exp^{(n+1-k)}[-x^2] \\ &= (-2x) \exp^{(n+1)}[-x^2] + (n+1)(-2) \exp^{(n)}[-x^2].\end{aligned}$$

Substituting this expression in (1), we obtain

$$\varphi_n''(x) = (-1)^n \exp\left[\frac{x^2}{2}\right] \{(x^2+1)\exp^{(n)}[-x^2] + 2x\exp^{(n+1)}[-x^2]$$
$$+(-2x\exp^{(n+1)}[-x^2] - 2(n+1)\exp^{(n)}[-x^2])\}$$
$$= (-1)^n \exp\left[\frac{x^2}{2}\right] \exp^{(n)}[-x^2](x^2-2n-1) = (x^2-2n-1)\varphi_n(x).$$

Thus every φ_n satisfies the differential equation

$$\varphi_n''(x) = (x^2-2n-1)\varphi_n(x).$$

Hence for every pair of nonnegative integers m and n, we have

$$\varphi_n''\varphi_m - \varphi_n\varphi_m'' = (x^2-2n-1)\varphi_n\varphi_m - (x^2-2m-1)\varphi_n\varphi_m = 2(m-n)\varphi_n\varphi_m.$$

If $m \neq n$, we have

$$\int_{-\infty}^{\infty} \varphi_n\varphi_m\,dx = \frac{1}{2(m-n)} \int_{-\infty}^{\infty} [\varphi_n''\varphi_m - \varphi_n\varphi_m'']\,dx$$
$$= \frac{1}{2(m-n)} \lim_{A\to\infty} \left\{ \varphi_m\varphi_n' \Big|_{-A}^{A} - \int_{-A}^{A} \varphi_n'\varphi_m'\,dx - \varphi_m'\varphi_n \Big|_{-A}^{A} + \int_{-A}^{A} \varphi_n'\varphi_m'\,dx \right\} = 0.$$

[This computation requires LEBESGUE'S theorem on dominated convergence.] Thus the set $\{\varphi_n\}_{n=0}^{\infty}$ is orthogonal.

To normalize the φ_n's, we now compute $\int_{-\infty}^{\infty} \varphi_n^2\,dx$. We begin by establishing the equality

$$H_n' = 2nH_{n-1} \quad (n = 1, 2, 3, \ldots). \tag{2}$$

We have

$$H_n'(x) = (-1)^n \{2x\exp[x^2]\exp^{(n)}[-x^2] + \exp[x^2]\exp^{(n+1)}[-x^2]\}.$$

Computing as before, we find that

$$\exp^{(n+1)}[-x^2] = -2x\exp^{(n)}[-x^2] - 2n\exp^{(n-1)}[-x^2],$$

and therefore

$$H_n'(x) = (-1)^n \{2x\exp[x^2]\exp^{(n)}[-x^2]$$
$$+ \exp[x^2](-2x\exp^{(n)}[-x^2] - 2n\exp^{(n-1)}[-x^2])\}$$
$$= (-1)^{n-1} 2n\exp[x^2]\exp^{(n-1)}[-x^2] = 2nH_{n-1}.$$

This establishes (2). To evaluate our integrals, we first observe that

$$\int_{-\infty}^{\infty} \varphi_0^2(x)\,dx = \int_{-\infty}^{\infty} \exp[-x^2]\,dx = \pi^{\frac{1}{2}},$$

as is well known. [See (21.60) *infra.*] Next, we have

$$\int_{-\infty}^{\infty} \varphi_n^2(x)\,dx = \int_{-\infty}^{\infty} \exp[-x^2]\,H_n^2(x)\,dx$$

$$= \int_{-\infty}^{\infty} \exp[-x^2]\,H_n(x)\exp[x^2]\,(-1)^n \exp^{(n)}[-x^2]\,dx$$

$$= (-1)^n \int_{-\infty}^{\infty} H_n(x)\exp^{(n)}[-x^2]\,dx$$

$$= \lim_{A\to\infty}\Big\{(-1)^n H_n(x)\exp^{(n-1)}[-x^2]\big|_{-A}^{A} + (-1)^{n-1}\int_{-A}^{A} H_n'(x)\exp^{(n-1)}[-x^2]\,dx\Big\}$$

$$= 2n \int_{-\infty}^{\infty} (-1)^{n-1} H_{n-1}(x)\exp^{(n-1)}[-x^2]\,dx$$

$$= 2n \int_{-\infty}^{\infty} \varphi_{n-1}^2(x)\,dx\,.$$

This establishes the recursive formula

$$\int_{-\infty}^{\infty} \varphi_n^2(x)\,dx = 2n \int_{-\infty}^{\infty} \varphi_{n-1}^2(x)\,dx\,,$$

and it follows that

$$\int_{-\infty}^{\infty} \varphi_n^2(x)\,dx = \pi^{\frac{1}{2}}\,2^n n!$$

for $n = 0, 1, 2, \ldots$. Hence the functions $\{\psi_n\}_{n=0}^{\infty}$ given by $\psi_n(x) = \left(\pi^{\frac{1}{2}}\,2^n\,n!\right)^{-\frac{1}{2}} \varphi_n(x) = \left(\pi^{\frac{1}{2}}\,2^n n!\right)^{-\frac{1}{2}} (-1)^n \exp\left[\frac{x^2}{2}\right] \exp^{(n)}[-x^2]$ form an orthonormal set. The functions $\{\psi_n\}_{n=0}^{\infty}$ are obtained from $\left\{x^n \exp\left[-\frac{x^2}{2}\right]\right\}_{n=0}^{\infty}$ by the Gram-Schmidt process. This follows readily from the fact that H_n has degree n for all n, and from the essential uniqueness pointed out in (16.22).

(16.26) Theorem. *Let H be a Hilbert space. The following properties of an orthonormal subset E of H are equivalent.*

(i) *The set E is complete.*

(ii) *For each $x \in H$, we have $x = \sum_{z\in E} \langle x, z\rangle z$* [Fourier series].[1]

(iii) *For all $x \in H$, we have $\|x\|^2 = \sum_{z\in E} |\langle x, z\rangle|^2$* [PARSEVAL's identity].

---

[1] The equality (ii) means that the right side has only a countable number of nonzero terms and that for every enumeration of these terms the resulting series converges to x as in (16.11). The equalities (iii) and (iv) have analogous meanings.

(iv) *For all $x, y \in H$, we have* $\langle x, y\rangle = \sum_{z\in E} \langle x, z\rangle \overline{\langle y, z\rangle}$ [PARSEVAL'S identity].

(v) *The smallest subspace of H containing E is dense in H.*

Proof. Suppose that (i) holds and let $x \in H$. According to BESSEL'S inequality (16.17), there are only countably many $z \in E$ such that $\langle x, z\rangle \neq 0$; enumerate these as (z_n). By (16.18), the vector $y = \sum_n \langle x, z_n\rangle z_n = \sum_{z\in E} \langle x, z\rangle z$ exists and $x - y$ is orthogonal to E. Since E is complete, it follows that $x - y = 0$; and so (ii) holds.

To show that (ii) implies (iv), let $x, y \in H$ be given and let (z_k) be an enumeration of all $z \in E$ such that $\langle x, z\rangle \neq 0$ or $\langle y, z\rangle \neq 0$. Let

$$x_n = \sum_{k=1}^{n} \langle x, z_k\rangle z_k$$

and

$$y_n = \sum_{k=1}^{n} \langle y, z_k\rangle z_k .$$

Then we have

$$\begin{aligned}\left|\langle x, y\rangle - \sum_{k=1}^{n} \langle x, z_k\rangle \overline{\langle y, z_k\rangle}\right| &= |\langle x, y\rangle - \langle x_n, y_n\rangle| \\ &\leq |\langle x, y\rangle - \langle x_n, y\rangle| + |\langle x_n, y\rangle - \langle x_n, y_n\rangle| \\ &\leq \|x - x_n\| \cdot \|y\| + \|x_n\| \cdot \|y_n - y\| \to 0\end{aligned}$$

because $\|x - x_n\| \to 0$, $\|y_n - y\| \to 0$, and $\|x_n\| \to \|x\|$. Thus $\langle x, y\rangle = \sum_{k=1}^{\infty} \langle x, z_k\rangle \overline{\langle y, z_k\rangle} = \sum_{z\in E} \langle x, z\rangle \overline{\langle y, z\rangle}$; hence (iv) is established.

It is obvious that (iv) implies (iii). if (iii) holds and $\langle x, z\rangle = 0$ for all $z \in E$, then it is plain that $\|x\| = 0$. Thus (iii) implies (i). This completes the proof that (i), (ii), (iii), and (iv) are pairwise equivalent.

Plainly (ii) implies (v). Finally we show that (v) implies (i). Suppose that $x \in H$ and $\langle x, z\rangle = 0$ for all $z \in E$. Then it is clear that $\langle x, y\rangle = 0$ for all y in the linear span of E. It follows from (v) and (16.21) that $x = 0$. □

(16.27) Theorem. *Any two complete orthonormal sets in a Hilbert space H have the same cardinal number.*

Proof. Ignoring a trivial case, we suppose that $H \neq \{0\}$. Let A and B be any two complete orthonormal sets in H. If A is finite, it follows from (16.26.ii) and (16.13) that A is a Hamel basis for H over K. Since B is linearly independent, (3.26) shows that B is contained in a Hamel basis C, so that $\overline{\overline{B}} \leq \overline{\overline{C}}$, and $\overline{\overline{C}} = \overline{\overline{A}}$ by (4.58). Thus B is also finite, and is also a Hamel basis. Another reference to (4.58) shows that $\overline{\overline{B}} = \overline{\overline{A}}$.

The case in which A and B are infinite remains to be treated. For each $a \in A$, let $B_a = \{b \in B : \langle a, b\rangle \neq 0\}$. Then B_a is countable for all

$a \in A$. For any $b \in B$ we have $1 = \|b\|^2 = \sum_{a \in A} |\langle b, a \rangle|^2$ (16.26.iii), and so there is some $a \in A$ such that $\langle a, b \rangle \neq 0$, *i.e.*, $b \in B_a$. Thus $B = \bigcup_{a \in A} B_a$. It follows that $\overline{\overline{B}} \leq \aleph_0 \overline{\overline{A}} = \overline{\overline{A}}$. Interchanging the rôles of A and B in this argument, we see also that $\overline{\overline{A}} \leq \overline{\overline{B}}$. It follows from the Schröder-Bernstein theorem that $\overline{\overline{A}} = \overline{\overline{B}}$. □

(16.28) Definition. Let H be a Hilbert space. If $H \neq \{0\}$, we define the *orthogonal dimension of* H to be the [unique!] cardinal number of any complete orthonormal set in H. If $H = \{0\}$, we say that H has *orthogonal dimension zero.*

(16.29) Theorem. *Let H be a nonzero Hilbert space. Then there exists a set D and a linear transformation T from H onto $l_2(D)$ that preserves inner products [hence norms as well]. Also $\overline{\overline{D}}$ is the orthogonal dimension of H.*

Proof. Let D be an arbitrary complete orthonormal set in H; then $\overline{\overline{D}}$ is the orthogonal dimension of H. For $x \in H$, let $T(x)$ be that function on D such that

$$[T(x)](z) = \langle x, z \rangle$$

for all $z \in D$. Then T maps H into $l_2(D)$, for $\sum_{z \in D} |\langle x, z \rangle|^2 < \infty$ by BESSEL's inequality. Also, for $x, y \in H$, we have

$$[T(x+y)](z) = \langle x + y, z \rangle = \langle x, z \rangle + \langle y, z \rangle = [T(x)](z) + [T(y)](z)$$

for all $z \in D$; that is, $T(x+y) = T(x) + T(y)$. Similarly $T(\alpha x) = \alpha T(x)$ for all $\alpha \in K$. Thus T is linear. Using PARSEVAL's identity (16.26.iv), we have

$$\langle T(x), T(y) \rangle = \sum_{z \in D} [T(x)(z)] \overline{[T(y)(z)]} = \sum_{z \in D} \langle x, z \rangle \overline{\langle y, z \rangle} = \langle x, y \rangle ,$$

and so T preserves inner products. Finally, we show that T is *onto* $l_2(D)$. If $f \in l_2(D)$, then $\sum_{z \in D} |f(z)|^2 < \infty$, and it follows from (16.12) that $x = \sum_{z \in D} f(z) z$ is in H. Let (z_n) be an enumeration of all $z \in D$ such that $f(z) \neq 0$ or $\langle x, z \rangle \neq 0$. For a fixed p and any $m \geq p$, we have

$$|\langle x, z_p \rangle - f(z_p)| = \left| \langle x, z_p \rangle - \sum_{n=1}^{m} f(z_n) \cdot \langle z_n, z_p \rangle \right|$$

$$\leq \left\| x - \sum_{n=1}^{m} f(z_n) z_n \right\| \cdot \|z_p\| \to 0 \quad \text{as} \quad m \to \infty .$$

Therefore $f(z) = \langle x, z \rangle = T(x)(z)$ for all $z \in D$; hence $f = T(x)$. □

(16.30) Remark. It follows from (16.29) that every $\mathfrak{L}_2(X, \mathscr{A}, \mu)$ is completely determined *qua* Hilbert space by a single cardinal number, and in fact is indistinguishable from a certain $l_2(D)$. No characterization of this kind for $\mathfrak{L}_p(X, \mathscr{A}, \mu)$, $p \neq 2$, is known to the authors. We also see that there exists just one [complex] Hilbert space for each cardinal number, *i.e.*, a Hilbert space is completely determined by its orthogonal dimension.

(16.31) Theorem. *Let H be a Hilbert space and let f be a bounded linear functional on H. Then there exists a unique $y \in H$ such that*

$$f(x) = \langle x, y \rangle$$

for every $x \in H$. Moreover

$$\|f\| = \|y\| .$$

Proof. Since H may be identified with an $l_2(D)$, the theorem follows from (15.12). □

(16.32) Theorem. *The functions $\chi_n : t \to \exp(int)$ $(n \in Z)$ form a complete orthonormal set in $\mathfrak{L}_2\left([-\pi, \pi], \mathscr{M}_\lambda, \frac{1}{2\pi}\lambda\right)$.*

Proof. We have already shown (16.15) that $\{\chi_n\}_{n \in Z}$ is an orthonormal set in $\mathfrak{L}_2$. We now prove that it is complete. Let $T = \{z \in K : |z| = 1\} = \{\exp(it) : -\pi \leqq t \leqq \pi\}$. For every integer n, the function

$$\exp(it) \to \exp(int)$$

is continuous on T. Let $\mathfrak{T}$ denote the set of all functions of the form

$$\exp(it) \to \sum_{k=-n}^{n} \alpha_k \exp(ikt)$$

where $\alpha_k \in K$. These functions are called, for an obvious reason, *trigonometric polynomials*. We have proved in (7.35.b) that $\mathfrak{T}$ is uniformly dense in $\mathfrak{C}(T)$. It is plain that this implies that any function $f \in \mathfrak{C}([-\pi, \pi])$ such that $f(-\pi) = f(\pi)$ can be uniformly approximated [arbitrarily closely] by functions in $\mathfrak{T}$ [regarded as functions of $t \in [-\pi, \pi]$]. Now let $\varphi \in \mathfrak{L}_2([-\pi, \pi])$ and let ε be a positive real number. Use (13.21) to select a function $g \in \mathfrak{C}([-\pi, \pi])$ such that $\|\varphi - g\|_2 \leqq \frac{\varepsilon}{3}$. Next, by changing the values of g on an interval $[\pi - \delta, \pi]$ for an appropriate $\delta > 0$ [if necessary], it is easy to find a function $f \in \mathfrak{C}([-\pi, \pi])$ such that $f(-\pi) = f(\pi)$ and $\|g - f\|_2 < \frac{\varepsilon}{3}$. Finally, choose a function $p \in \mathfrak{T}$ such that $\|f - p\|_u < \frac{\varepsilon}{3}$. Then

$$\begin{aligned}
\|\varphi - p\|_2 &\leqq \|\varphi - g\|_2 + \|g - f\|_2 + \|f - p\|_2 \\
&< \frac{2}{3}\varepsilon + \left(\frac{1}{2\pi} \int_{-\pi}^{\pi} |f - p|^2 \, d\lambda\right)^{\frac{1}{2}} \\
&\leqq \left(\frac{1}{2\pi} \int_{-\pi}^{\pi} \left(\frac{\varepsilon}{3}\right)^2 d\lambda\right)^{\frac{1}{2}} + \frac{2}{3}\varepsilon \\
&= \varepsilon .
\end{aligned}$$

Thus $\mathfrak{T}$ is dense in $\mathfrak{L}_2([-\pi, \pi])$. By (16.26.v), $\{\chi_n\}_{n\in Z}$ is complete. □

(16.33) Definition. For $f \in \mathfrak{L}_1([-\pi, \pi], \mathscr{M}_\lambda, \lambda)$, let

$$\hat{f}(n) = \frac{1}{2\pi} \int_{-\pi}^{\pi} f(t) \exp(-int)\, dt \quad (n \in Z).$$

The number $\hat{f}(n)$ is called the nth *Fourier coefficient of* f. The function $\hat{f}$ on Z is called the *Fourier transform of* f.

(16.34) Theorem. *Let* $f \in \mathfrak{L}_1([-\pi, \pi])$. *If the identity* $\hat{f} = 0$ *holds, then* $f = 0$ *in* $\mathfrak{L}_1$.

Proof. Let $\mathfrak{T}$ be the set of all trigonometric polynomials on T, defined as in the proof of (16.32). If $\hat{f} = 0$, then it is obvious that

$$\int_{-\pi}^{\pi} f(t)\, p(t)\, dt = 0$$

for all $p \in \mathfrak{T}$. Let x be a fixed number in $]-\pi, \pi]$. It is easy to construct a nondecreasing sequence $(g_n)_{n=1}^{\infty}$ of nonnegative continuous functions such that:

$$g_n(-\pi) = g_n(\pi) = 0;$$

$$\lim_{n\to\infty} g_n(t) = \xi_{]-\pi, x[}(t) \quad \text{for all} \quad t \in [-\pi, \pi].$$

By (7.35.b), there exists for each $n \in N$ a polynomial $p_n \in \mathfrak{T}$ such that $\|g_n - p_n\|_u < \frac{1}{n}$. Then we have

$$f(t)\, \xi_{]-\pi, x[}(t) = \lim_{n\to\infty} f(t)\, p_n(t) \quad \text{for all} \quad t \in [-\pi, \pi],$$

and plainly

$$|f(t)\, p_n(t)| \leq |f(t)| \left(|g_n(t)| + \frac{1}{n}\right) \leq 2|f(t)|.$$

LEBESGUE's dominated convergence theorem (12.30) shows that

$$\int_{-\pi}^{x} f(t)\, dt = \int_{-\pi}^{\pi} f(t)\, \xi_{]-\pi, x[}(t)\, dt = \lim_{n\to\infty} \int_{-\pi}^{\pi} f(t)\, p_n(t)\, dt = 0.$$

From (12.54) we infer that $f = 0$ a.e. on $[-\pi, \pi]$. □

(16.35) Riemann-Lebesgue Lemma. *Let* $f \in \mathfrak{L}_1([-\pi, \pi])$. *Then* $\lim_{|n|\to\infty} \hat{f}(n) = 0$.

Proof. Let $\varepsilon > 0$ be given. Use (13.21) to choose $g \in \mathfrak{C}([-\pi, \pi])$ such that $\|f - g\|_1 < \frac{\varepsilon}{2}$. Clearly $g \in \mathfrak{L}_2([-\pi, \pi])$. By BESSEL's inequality we have $\sum_{n=-\infty}^{\infty} |\hat{g}(n)|^2 \leq \|g\|_2^2$. Thus there exists a positive integer p

such that $|\hat{g}(n)| < \frac{\varepsilon}{2}$ whenever $|n| \geqq p$. Hence $|n| \geqq p$ implies

$$\begin{aligned} |\hat{f}(n)| &\leqq |\hat{f}(n) - \hat{g}(n)| + |\hat{g}(n)| \\ &< \left| \frac{1}{2\pi} \int_{-\pi}^{\pi} (f(x) - g(x)) \exp(-inx)\, dx \right| + \frac{\varepsilon}{2} \\ &\leqq \frac{1}{2\pi} \int_{-\pi}^{\pi} |f(x) - g(x)|\, dx + \frac{\varepsilon}{2} \\ &< \frac{\varepsilon}{2} + \frac{\varepsilon}{2} = \varepsilon. \quad \square \end{aligned}$$

(16.36) Remark. There is an analogous theory of Fourier transforms for the space $\mathfrak{L}_1(R, \mathscr{M}_\lambda, \lambda) = \mathfrak{L}_1(R)$. In this case, one defines

$$\hat{f}(y) = (2\pi)^{-\frac{1}{2}} \int_R f(x) \exp(-iyx)\, dx$$

for each $y \in R$. Once again the equality $\hat{f} = 0$ implies $f = 0$ a.e. Also, $\lim_{|y| \to \infty} \hat{f}(y) = 0$. These transforms are important in several parts of analysis. Using them, it is possible to prove that the Hermite functions are a complete orthonormal set in $\mathfrak{L}_2(R)$; for a short proof, see (21.64.b) *infra*.

(16.37) Theorem [Parseval's identity]. *For* $f \in \mathfrak{L}_2([-\pi, \pi])$, *the equality*

$$\text{(i)} \qquad \frac{1}{2\pi} \int_{-\pi}^{\pi} |f(t)|^2 dt \quad \sum_{n=-\infty}^{\infty} |\hat{f}(n)|^2$$

holds.

Proof. This is an immediate consequence of (16.26), (16.32), and the definition (16.33) of $\hat{f}(n)$. $\square$

(16.38) Remark. There is a generalization of (16.37) for $\mathfrak{L}_p([-\pi, \pi])$, $1 < p < 2$. In this case, the inequality

$$\left(\sum_{n=-\infty}^{\infty} |\hat{f}(n)|^{p'} \right)^{\frac{1}{p'}} < \left(\frac{1}{2\pi} \int_{-\pi}^{\pi} |f(t)|^p dt \right)^{\frac{1}{p}}$$

holds, unless $f(x) = \alpha \exp(imx)$ for some $\alpha \in K$ and $m \in Z$. This is a nontrivial fact, and its proof is fairly sophisticated. See for example Edwin Hewitt and I. I. Hirschman jr., Amer. J. Math. **76**, 839–852 (1954).

(16.39) Riesz-Fischer theorem. *Let* $(\alpha_n)_{n=-\infty}^{\infty}$ *be in* $l_2(Z)$, i.e., $\sum_{n=-\infty}^{\infty} |\alpha_n|^2 < \infty$. *Then there is a function* $f \in \mathfrak{L}_2([-\pi, \pi])$ *such that* $\hat{f}(n) = \alpha_n$, *for all* $n \in Z$.

Proof. This is a trivial consequence of the completeness of $\mathfrak{L}_2$ (13.11). Let

$$f_k(t) = \sum_{j=-k}^{k} \alpha_j \exp(ijt) \quad (k = 1, 2, \ldots).$$

Then the equality $\|f_k - f_l\|_2^2 = \sum_{j=-l}^{-k-1} |\alpha_j|^2 + \sum_{j=k+1}^{l} |\alpha_j|^2$ shows that (f_k) is a Cauchy sequence. Let f be the limit in $\mathfrak{L}_2$ of (f_k). We clearly have $\hat{f}(n) = \lim_{k\to\infty} \hat{f}_k(n) = \alpha_n$. □

(16.40) Theorem. *Let H be a Hilbert space and let $\mathfrak{B}(H)$ denote the space of all bounded linear operators [transformations] from H into H. Then $\mathfrak{B}(H)$ is a Banach algebra with unit, where we take composition for multiplication. Moreover, for each $T \in \mathfrak{B}(H)$ there exists a unique $T^* \in \mathfrak{B}(H)$ such that*

$$\langle T(x), y\rangle = \langle x, T^*(y)\rangle$$

for all $x, y \in H$. Also $T^{**} = T$ and $\|T^*\| = \|T\|$. The operator T^* is called the adjoint of T.

Proof. We have proved in (14.4) that $\mathfrak{B}(H)$ is a Banach space. For $T_1, T_2 \in \mathfrak{B}(H)$ and $x \in H$, we have

$$\|T_1 T_2(x)\| \leqq \|T_1\| \cdot \|T_2(x)\| \leqq \|T_1\| \cdot \|T_2\| \cdot \|x\| .$$

Therefore $T_1 T_2 \in \mathfrak{B}(H)$ and

$$\|T_1 T_2\| \leqq \|T_1\| \cdot \|T_2\| .$$

It is easy to see that $\mathfrak{B}(H)$ is an algebra over K. Thus $\mathfrak{B}(H)$ is a Banach algebra.

Let $T \in \mathfrak{B}(H)$. For each $y \in H$, define f_y on H by

$$f_y(x) = \langle T(x), y\rangle .$$

Then we have

$$|f_y(x)| = |\langle T(x), y\rangle| \leqq \|T\| \cdot \|x\| \cdot \|y\|$$

for all $x \in H$. Obviously f_y is linear. Thus f_y is a bounded linear functional on H and $\|f_y\| \leqq \|T\| \cdot \|y\|$. It follows from (16.31) that there exists a *unique* element $T^*(y) \in H$ such that $\langle T(x), y\rangle = f_y(x) = \langle x, T^*(y)\rangle$ for all $x \in H$ and

$$\|T^*(y)\| = \|f_y\| \leqq \|T\| \cdot \|y\| . \tag{1}$$

This defines T^* on H. A simple computation shows that T^* is linear, and (1) implies that T^* is bounded and

$$\|T^*\| \leqq \|T\| . \tag{2}$$

Applying the above results to T^*, we have $\langle T^*(x), y\rangle = \overline{\langle y, T^*(x)\rangle} = \overline{\langle T(y), x\rangle} = \langle x, T(y)\rangle$ for all $x, y \in H$. Since the adjoint is unique, this implies that $T = (T^*)^* = T^{**}$. We next apply the inequality (2) with T^* taking the rôle of T to obtain $\|T\| = \|T^{**}\| \leqq \|T^*\|$. Thus $\|T\| = \|T^*\|$. □

(16.41) Theorem. *Let H and $\mathfrak{B}(H)$ be as in* (16.40). *Then the mapping $T \to T^*$ of $\mathfrak{B}(H)$ into $\mathfrak{B}(H)$ has the following properties:*

(i) $(T_1 + T_2)^* = T_1^* + T_2^*$;

(ii) $(\alpha T)^* = \bar{\alpha} T^*$ *for* $\alpha \in K$;

(iii) $(T_1 T_2)^* = T_2^* T_1^*$;

(iv) $T^{**} = T$;

(v) $\|T^* T\| = \|T\|^2$.

Proof. Equality (iv) was established in (16.40). Equalities (i)—(iii) follow from the uniqueness of the adjoint and the following computations:

$$\begin{aligned}\langle (T_1 + T_2)(x), y\rangle &= \langle T_1(x), y\rangle + \langle T_2(x), y\rangle \\ &= \langle x, T_1^*(y)\rangle + \langle x, T_2^*(y)\rangle \\ &= \langle x, (T_1^* + T_2^*)(y)\rangle;\end{aligned}$$

$$\begin{aligned}\langle (\alpha T)(x), y\rangle &= \alpha\langle T(x), y\rangle = \alpha\langle x, T^*(y)\rangle \\ &= \langle x, \bar{\alpha} T^*(y)\rangle;\end{aligned}$$

$$\langle (T_1 T_2)(x), y\rangle = \langle T_1(T_2(x)), y\rangle = \langle T_2(x), T_1^*(y)\rangle = \langle x, T_2^* T_1^*(y)\rangle .$$

We next prove (v). First we have $\|T^* T\| \leqq \|T^*\| \cdot \|T\| = \|T\|^2$. Next, for all $x \in H$ we have

$$\|T(x)\|^2 = \langle T(x), T(x)\rangle = \langle x, T^* T(x)\rangle \leqq \|x\| \cdot \|T^* T(x)\| \leqq \|x\|^2 \cdot \|T^* T\| ;$$

hence $\|T(x)\| \leqq \|T^* T\|^{\frac{1}{2}} \cdot \|x\|$. Therefore $\|T\| \leqq \|T^* T\|^{\frac{1}{2}}$, and so $\|T\|^2 \leqq \|T^* T\|$. □

(16.42) Exercise. Let H be a Hilbert space and let M be a closed linear subspace of H. Prove that there exists a closed linear subspace $M^\perp$ of H such that $M \perp M^\perp$ and $H = M \oplus M^\perp$, *i.e.*, $M + M^\perp = H$ and $M \cap M^\perp = \{0\}$. [Hint. Consider a complete orthonormal set in M; extend it to a complete orthonormal set in H; and let $M^\perp$ be the closed linear subspace spanned by the added orthonormal vectors.]

(16.43) Exercise. Let H be an inner product space and let A be a nonvoid subset of H that is complete in the norm metric of H, and also has the property that $\frac{1}{2}(x + y) \in A$ if $x, y \in A$. Prove the following statements.

(a) The set A is closed in H.

(b) For $x_1, \ldots, x_n \in A$ and positive real numbers $\alpha_1, \ldots, \alpha_n$ such that $\sum_{k=1}^{n} \alpha_k = 1$, the element $\sum_{k=1}^{n} \alpha_k x_k$ is in A. [This is the property of *convexity*.]

(c) Every finite-dimensional linear subspace of H is complete in its norm metric, *i.e.*, it is a Hilbert space.

(16.44) Exercise. Let H and A be as in (16.43) and let z be any element of H. Prove that there is a unique element $y_0 \in A$ such that

$$\text{(i)} \qquad \|y_0 - z\| = \inf\{\|y - z\| : y \in A\}.$$

[The right side of (i) is the *distance from* $\{z\}$ *to* A as defined in (6.87)]. [Hints. Cf. (15.16.c), noting that H is uniformly convex; or apply the parallelogram law directly. Choose a sequence (y_n) of elements in A such that $\lim_{n \to \infty} \|y_n - z\| = \inf\{\|y - z\| : y \in A\}$. Apply uniform convexity to $y_n - z$ and $y_m + z$ to infer that (y_n) is a Cauchy sequence and so has a limit y_0 in A. Uniqueness is proved similarly.]

(16.45) Exercise. Let H be an inner product space and let $\{x_n\}_{n=1}^{\infty}$ be a set of vectors in H. Let S_n be the linear space spanned by $\{x_1, \ldots, x_n\}$. Suppose that x_n is the element of S_n nearest to x_{n+1} for $n = 1, 2, 3, \ldots$. [This element exists by (16.43.c) and (16.44).] The set $\{x_n\}_{n=1}^{\infty}$ is called a *martingale in the wide sense*. Write $y_1 = x_1$ and $y_n = x_n - x_{n-1}$ for $n > 1$. Prove the following.

(a) The vectors y_n are pairwise orthogonal and $x_n = y_1 + \cdots + y_n$.

(b) The inequalities $\|x_1\| \leq \|x_2\| \leq \cdots \leq \|x_n\| \leq \cdots$ hold.

(c) If $\{z_n\}_{n=1}^{\infty}$ is an orthogonal set, then $\{z_1 + \cdots + z_n\}_{n=1}^{\infty}$ is a martingale in the wide sense.

(16.46) Exercise. Let H be a Hilbert space and let $T \in \mathfrak{B}(H)$. Prove:

(a) if $\langle T(x), y \rangle = 0$ for all $x, y \in H$, then $T = 0$;

(b) if $\langle T(x), x \rangle = 0$ for all $x \in H$, then $T = 0$;

(c) $T^*T = TT^*$ if and only if $\|Tx\| = \|T^*x\|$ for all $x \in H$.

(16.47) Exercise. Let H be a Hilbert space and let M be a closed linear subspace of H. As in (16.42), we have $H = M \oplus M^{\perp}$. Prove that:

(a) for each $x \in H$, there exists a unique $(y, z) \in M \times M^{\perp}$ such that $x = y + z$; define $P(x) = y$;

(b) $P \in \mathfrak{B}(H)$;

(c) $P^2 = P$;

(d) $P = P^*$;

(e) $P(H) = M$;

(f) $P^{-1}(0) = M^{\perp}$.

The operator P is known as the *projection of H onto M*.

Prove that if $T \in \mathfrak{B}(H)$ satisfies $T^2 = T$ and $T = T^*$, then T is the projection of H onto some closed subspace of H.

(16.48) Exercise: Construction of a particular projection operator. (a) Let $(X, \mathscr{A}, \mu)$ be a measure space such that $0 < \mu(X) < \infty$; write $\mathfrak{L}_2$ for $\mathfrak{L}_2(X, \mathscr{A}, \mu)$. Let $\mathfrak{F}$ be a dense linear subspace of $\mathfrak{L}_2$ closed under the formation of complex conjugates and containing only bounded functions. Let $\mathfrak{M}$ be a closed linear subspace of $\mathfrak{L}_2$ such that $f \in \mathfrak{M}$

and $g \in \mathfrak{F}$ imply $gf \in \mathfrak{M}$ [*i.e.*, $\mathfrak{M}$ is *invariant under multiplication by functions in* $\mathfrak{F}$]. Let P be the projection of $\mathfrak{L}_2$ onto $\mathfrak{M}$ as defined in (16.47). Prove that $P(\varphi) = P(1) \cdot \varphi$ for all $\varphi \in \mathfrak{L}_2$ and that $P(1) = \xi_E$ for some set $E \in \mathscr{A}$. Thus $\mathfrak{M}$ is exactly the set of all functions in $\mathfrak{L}_2$ that vanish on E'. Note that every such set is plainly a closed subspace of $\mathfrak{L}_2$ invariant under multiplication by functions in $\mathfrak{F}$. [Hints. Consider any $h \in \mathfrak{M}^\perp$, $f \in \mathfrak{M}$, and $g \in \mathfrak{F}$. We have

$$\int_X f\overline{g h}\, d\mu = \int_X (f\bar{g})\, \bar{h}\, d\mu = 0$$

and so $\mathfrak{M}^\perp$ too is invariant under multiplication by functions in $\mathfrak{F}$. Since $1 - P(1) \in \mathfrak{M}^\perp$, it follows that $P\big(g(1 - P(1))\big) = 0$ and so $P(g) = P(gP(1)) = gP(1)$ for all $g \in \mathfrak{F}$. As $\mathfrak{F}$ is dense in $\mathfrak{L}_2$, we infer that $P(\varphi) = \varphi P(1)$ for all $\varphi \in \mathfrak{L}_2$. Since $P(1) = P(1)^2$, we have $P(1) = \xi_E$.]

(b) Let $(X, \mathscr{A}, \mu)$ be a σ-finite measure space, and let $\mathfrak{M}$ and $\mathfrak{F}$ be subspaces of $\mathfrak{L}_2(X, \mathscr{A}, \mu)$ just as in part (a). Prove that the projection operator P of $\mathfrak{L}_2(X, \mathscr{A}, \mu)$ onto $\mathfrak{M}$ has the form $P(f) = \xi_E f$ for some $E \in \mathscr{A}$.[1] [Write $X = \bigcup_{n=1}^{\infty} F_n$, where the F_n's are pairwise disjoint sets in $\mathscr{A}$ of finite measure. Apply (a) to each subspace $\xi_{F_n} \mathfrak{L}_2(X, \mathscr{A}, \mu)$ and add.]

(16.49) Exercise. Let H be the real Hilbert space $l_2^r = l_2^r(N)$. Define $C = \left\{f \in H : |f(n)| \leqq \frac{1}{n} \text{ for all } n \in N\right\}$. Prove that C is a compact, nowhere dense subset of H. The metric space C is known as the *Hilbert parallelotope.*

(16.50) Exercise. Let x, y, z be elements of an inner product space H such that $\|x\| = \|y\| = \|z\|$. Prove that

(i) $|\langle x, x\rangle\langle z, y\rangle - \langle x, y\rangle\langle z, x\rangle|^2 \leqq [\langle x, x\rangle^2 - |\langle y, x\rangle|^2][\langle x, x\rangle^2 - |\langle z, x\rangle|^2]$.

[Hint. You may obviously suppose that $\|x\| = \|y\| = \|z\| = 1$. Use the Gram-Schmidt process to replace H by K^3, x by $(1, 0, 0)$, y by $(\alpha, \beta, 0)$, and z by $(\gamma, \delta, \varepsilon)$, where $|\alpha|^2 + |\beta|^2 = |\gamma|^2 + |\delta|^2 + |\varepsilon|^2 = 1$. Then (i) becomes nearly trivial.]

(16.51) Exercise. Let x, y, z and H be as in (16.50). Prove that

(i) $\langle x, x\rangle^2 (|\langle y, z\rangle|^2 + |\langle y, x\rangle|^2 + |\langle z, x\rangle|^2)$

$$\leqq \langle x,x\rangle^4 + \langle x,x\rangle\langle z,y\rangle\langle y,x\rangle\langle x,z\rangle + \langle x,x\rangle\langle y,z\rangle\langle x,y\rangle\langle z,x\rangle.$$

(16.52) Exercise. Prove that a Hilbert space H is separable if and only if the orthogonal dimension of H is $\leqq \aleph_0$ [see (16.29)].

(16.53) Exercise. Let $(X, \mathscr{A}, \mu)$ be a measure space such that $0 < \mu(X) < \infty$ and let $(\mathscr{A}, \varrho)$ be the metric space defined in (10.45). Let $\mathfrak{m}$ be the smallest cardinal number of a dense subset of $(\mathscr{A}, \varrho)$.

[1] For a yet more general result, see (19.76).

Let $\mathfrak{d}$ be the orthogonal dimension of the Hilbert space $\mathfrak{L}_2(X, \mathscr{A}, \mu)$. Prove the following assertions. If $\mathfrak{d}$ is finite, then $\mathfrak{m} = 2^{\mathfrak{d}}$. If $\mathfrak{d}$ is infinite, then $\mathfrak{m} = \mathfrak{d}$. [Hints. Consider first the case in which μ assumes only finitely many values, and use (10.56.a) and (12.60) to prove that $\mathfrak{m} = 2^{\mathfrak{d}}$. If μ assumes infinitely many values, then (10.56.c) shows that $\mathfrak{d}$ is infinite. In this case, the inequalities $\mathfrak{m} \leqq \mathfrak{d}$ and $\mathfrak{d} \leqq \mathfrak{m}$ are proved by simple arguments.]

(16.54) Exercise. Let H be the real Hilbert space $l_2^r(R)$. For each $t \in R$ let u_t be that element of H defined on R by $u_t(s) = \delta_{t,s}$ [KRONECKER'S delta]. Define $X = \{f \in H : f = \alpha u_t$ for some $t \in R$, $0 \leqq \alpha \leqq 1\}$. Then X, with the metric of H, is a metric space. Prove that there exists a one-to-one mapping φ of X onto the French railroad space (D, ϱ) [see (6.13.e)] such that both φ and φ^{-1} are uniformly continuous.

(16.55) Exercise. Let H be a Hilbert space and let $T \in \mathfrak{B}(H)$. Write $\alpha(T) = \sup\{|\langle Tx, x\rangle| : x \in H, \|x\| = 1\}$.

(a) Prove that $\alpha(T) = \|T\|$ if $T = T^*$. [Use the identity

$$4\|Tx\|^3 = \langle T(\|Tx\|x + Tx), \|Tx\|x + Tx\rangle - \langle T(\|Tx\|x - Tx), \|Tx\|x - Tx\rangle$$

and the parallelogram law.]

(b) Find a T for which $\alpha(T) \neq \|T\|$.

(16.56) Exercise. Use (16.42) to give a proof of (16.31) that does not invoke the results of § 15. $\Big[$For $f \in H^*$, let $M = f^{-1}(\{0\})$. If $M = H$, let $y = 0$. If $M \neq H$, let z be a nonzero element of $M^\perp$ and set $y = \overline{f(z)}\,\|z\|^{-2}z$. Note that $\left(x - \frac{f(x)}{f(z)}\,z\right) \in M$ for all $x \in H$.$\Big]$

(16.57) Exercise. Let H be a Hilbert space and let $T \in \mathfrak{B}(H)$. Prove that $[T^{-1}(\{0\})] \perp [\overline{T^*(H)}]$ and $H = [T^{-1}(\{0\})] + [\overline{T^*(H)}]$. [See (16.42). Show that $\overline{T^*(H)}^{\perp} = T^{-1}(\{0\})$.]

(16.58) Exercise. Let H be a Hilbert space and let $T \in \mathfrak{B}(H)$ be such that $T = T^*$. Suppose that there exists a positive constant c such that $\|T(x)\| \geqq c\|x\|$ for all $x \in H$. Prove that

(a) T is one-to-one,

(b) $T(H) = H$,

and

(c) $T^{-1} \in \mathfrak{B}(H)$.

[Use (16.57) and (14.17).]

CHAPTER FIVE

Differentiation

This chapter contains first a brief but reasonably complete treatment of the theory of differentiation for complex-valued functions defined on intervals of the line. Section 17 is severely classical, containing examples and LEBESGUE'S famous theorem on differentiation of functions of finite variation. In § 18, we explore the conditions under which the classical equality

$$f(b) - f(a) = \int_a^b f'(t)\,dt$$

is valid. This exploration leads to interesting and perhaps unexpected measure-theoretic ideas, which have little to do with differentiation and which have applications in extraordinarily diverse fields. The main result in this direction is the LEBESGUE-RADON-NIKODÝM theorem, which we examine thoroughly in § 19 and apply to the decomposition of measures on R. In § 20, we present several other applications of the LEBESGUE-RADON-NIKODÝM theorem to problems in abstract analysis. Sections 17 and 18 are important, and should be studied by all readers. The same is true of § 19, up to and including (19.24). The remainder of § 19 may be omitted by readers pressed for time. Of § 20, (20.1)–(20.5) and (20.41)–(20.52) are topics important for every student. The remainder of § 20 is in our opinion interesting but less vital, and it too may be omitted by readers pressed for time.

§ 17. Differentiable and nondifferentiable functions

This section deals solely with functions defined on intervals of R. While reasonably elementary, it is an indispensable introduction to the more sophisticated matters considered in §§ 18 and 19. Throughout this section, "almost everywhere" means "λ-almost everywhere" and "measurable" means "$\mathscr{M}_\lambda$-measurable". As usual, we begin with some definitions.

(17.1) Definition. Let $a \in R$ and $\delta > 0$. If φ is a real-valued function defined on $]a, a+\delta[$, define

$$\varliminf_{h \downarrow a} \varphi(h) = \sup\{\inf\{\varphi(h) : a < h < t\} : a < t \leqq a + \delta\}$$

and

$$\varlimsup_{h \downarrow a} \varphi(h) = \inf\{\sup\{\varphi(h) : a < h < t\} : a < t \leqq a + \delta\}.$$

These two extended real numbers are called the *lower right limit* and the *upper right limit of φ at a* respectively. If φ is a real-valued function defined on $]a-\delta, a[$, define the *lower left limit* and the *upper left limit of φ at a* to be the extended real numbers

$$\underline{\lim}_{h \uparrow a} \varphi(h) = \sup\{\inf\{\varphi(h) : t < h < a\} : a - \delta \leqq t < a\}$$

and

$$\overline{\lim}_{h \uparrow a} \varphi(h) = \inf\{\sup\{\varphi(h) : t < h < a\} : a - \delta \leqq t < a\}$$

respectively.

(17.2) Definition. Let $a \in R$ and $\delta > 0$. If f is a real-valued function defined on $[a, a+\delta[$, define

$$D_+ f(a) = \underline{\lim}_{h \downarrow 0} \frac{f(a+h) - f(a)}{h}$$

and

$$D^+ f(a) = \overline{\lim}_{h \downarrow 0} \frac{f(a+h) - f(a)}{h}.$$

If f is a real-valued function defined on $]a-\delta, a]$, define

$$D_- f(a) = \underline{\lim}_{h \uparrow 0} \frac{f(a+h) - f(a)}{h}$$

and

$$D^- f(a) = \overline{\lim}_{h \uparrow 0} \frac{f(a+h) - f(a)}{h}.$$

These four extended real numbers are known as the *Dini derivates of f at a*; $D_+ f(a)$ is the *lower right derivate*, $D^+ f(a)$ is the *upper right derivate*, $D_- f(a)$ is the *lower left derivate*, and $D^- f(a)$ is the *upper left derivate*.

(17.3) Remarks. The inequalities

$$(D_+ f)(a) \leqq (D^+ f)(a)$$

and

$$(D_- f)(a) \leqq (D^- f)(a)$$

obviously hold. Also it is easy to see that $(D^+ f)(a)$ $[(D_+ f)(a)]$ is the largest [smallest] limit of a sequence $\left(\frac{f(a+h_n) - f(a)}{h_n}\right)$, where $h_n > 0$ and $\lim_{n \to \infty} h_n = 0$. Similar statements hold for $(D^- f)(a)$ and $(D_- f)(a)$.

(17.4) Definition. If $(D^+ f)(a) = (D_+ f)(a)$, then f is said to have a *right derivative at a*, and we write $f'_+(a)$ for the common value $(D^+ f)(a) = (D_+ f)(a)$. The *left derivative of f at a* is defined analogously, and is written $f'_-(a)$. If $f'_+(a)$ and $f'_-(a)$ exist and are equal, then f is said to be *differentiable at a*, or to have a *derivative at a*, and we write $f'(a)$ for the common value $f'_+(a) = f'_-(a)$. The number $f'(a)$ is called the *derivative of f at a*. Notice that our definition does not exclude ∞ or $-\infty$ as a value for $f'(a)$. For example, if $f(x) = x^{\frac{1}{3}}$ $(x \in R)$, then $f'(0)$ exists and $f'(0) = \infty$.

(17.5) Definition. If f is a complex-valued function defined on $[a, a + \delta[$, we say that f is *right differentiable at a* or has a *right derivative at a* if

$$\lim_{h \downarrow 0} \frac{f(a + h) - f(a)}{h}$$

exists and is a complex number. Left derivatives and two-sided derivatives [which are simply called *derivatives*] are defined similarly. It is obvious that f has a derivative at a if and only if $\mathrm{Re} f$ and $\mathrm{Im} f$ have *finite* derivatives at a. Then

$$f'(a) = (\mathrm{Re} f)'(a) + i(\mathrm{Im} f)'(a) .$$

(17.6) Remarks. (a) There is a slight solecism in Definitions (17.4) and (17.5), since a real-valued function is certainly complex-valued, and $\pm\infty$ are admitted as values of f' if f is real-valued. The point should cause no trouble, however, and we will not bother with the special terminology that would be needed to remove the difficulty.

(b) If $f'_+(a)$ exists and is finite, then $\lim\limits_{h \downarrow 0} f(a + h) = f(a)$. Similarly, if $f'_-(a)$ exists and is finite, then $\lim\limits_{h \uparrow 0} f(a + h) = f(a)$. Hence if f has a *finite* derivative at a, then it is continuous at a. [Is this true if $f'(a) = \infty$ or $-\infty$?] The following two theorems show that the converse of this statement fails in a striking way.

(17.7) Theorem. *Let a be an odd positive integer and b a real number such that $0 < b < 1$. Suppose also that $ab > 1 + \frac{3\pi}{2}$. Let f be the function defined on R by*

$$f(x) = \sum_{k=0}^{\infty} b^k \cos[a^k \pi x] .$$

Then f is continuous and bounded on R, and f has a finite derivative at no point[1].

Proof. For each k we have $|b^k \cos[a^k \pi x]| \leqq b^k$ for all $x \in R$. Thus the series defining f converges absolutely and $|f(x)| \leqq \sum\limits_{k=0}^{\infty} b^k = \frac{1}{1 - b}$

---

[1] The investigation of the relationship between continuity and differentiability has a long history. The function f defined here was constructed by WEIERSTRASS (*ca.* 1875). It was minutely examined by HARDY in 1916 [Trans. Amer. Math. Soc. 17, 301—325 (1916)]. Among other things, HARDY was able to show that f has the stated properties if $ab \geqq 1$. In the same paper, he showed that the continuous function

$$g(x) = \sum_{n=1}^{\infty} \frac{\sin[n^2 \pi x]}{n^2}$$

is nowhere differentiable; this is considerably more difficult to prove than the corresponding statement for f. RIEMANN had conjectured many years earlier that g is nowhere differentiable.

for all $x \in R$. Also, if $f_n(x) = \sum_{k=0}^{n-1} b^k \cos[a^k \pi x]$, we have

$$\|f - f_n\|_u \leqq \sum_{k=n}^{\infty} b^k = \frac{b^n}{1-b} \to 0$$

as $n \to \infty$. It follows as in (7.9) that f is continuous on R.

We will show that f is differentiable at no point of R. Let $x \in R$ be fixed. For $n \in N$ and $h > 0$, write

$$\frac{f(x+h) - f(x)}{h} = \sum_{k=0}^{n-1} b^k \frac{\cos[a^k \pi (x+h)] - \cos[a^k \pi x]}{h}$$
$$+ \sum_{k=n}^{\infty} b^k \frac{\cos[a^k \pi (x+h)] - \cos[a^k \pi x]}{h} = S_n + R_n .$$

Using the mean value theorem, we write

$$\frac{\cos[a^k \pi (x+h)] - \cos[a^k \pi x]}{h} = - a^k \pi \sin[a^k \pi (x + h')] , \qquad (1)$$

where $0 < h' < h$. The absolute value of the right side of (1) is less than or equal to $a^k \pi$, and so

$$|S_n| \leqq \sum_{k=0}^{n-1} a^k b^k \pi = \pi \frac{a^n b^n - 1}{ab - 1} < \frac{\pi a^n b^n}{ab - 1} . \qquad (2)$$

Now write

$$a^n x = \alpha_n + \beta_n ,$$

where α_n is an integer, and $-\frac{1}{2} \leqq \beta_n < \frac{1}{2}$. Let

$$h_n = \frac{1 - \beta_n}{a^n} .$$

Since $\frac{3}{2} \geqq 1 - \beta_n > \frac{1}{2}$, we have $\frac{3}{2a^n} \geqq \frac{1-\beta_n}{a^n} > \frac{1}{2a^n}$ and so

$$\frac{2a^n}{3} \leqq \frac{1}{h_n} < 2a^n .$$

We now estimate $|R_n|$. For $k \geqq n$, consider

$$a^k \pi (x + h_n) = a^{k-n} a^n \pi (x + h_n) = a^{k-n} \pi (a^n x + 1 - \beta_n) = a^{k-n} \pi (1 + \alpha_n) .$$

Since a is odd, the equalities

$$\cos[a^k \pi (x + h_n)] = \cos[a^{k-n} \pi (1 + \alpha_n)] = (-1)^{1+\alpha_n}$$

hold. We also have

$$- \cos[\pi a^k x] = - \cos[\pi a^{k-n} a^n x] = - \cos[\pi a^{k-n} (\alpha_n + \beta_n)]$$
$$= - \cos[\pi a^{k-n} \alpha_n] \cos[\pi a^{k-n} \beta_n] = (-1)^{1+\alpha_n} \cos[\pi a^{k-n} \beta_n] .$$

Upon setting $h = h_n$, we find that $|R_n|$ becomes

$$|R_n| = \left| \sum_{k=n}^{\infty} b^k \frac{(-1)^{1+\alpha_n} + (-1)^{1+\alpha_n} \cos[\pi a^{k-n} \beta_n]}{h_n} \right|$$

$$= \left| \frac{(-1)^{1+\alpha_n}}{h_n} \right| \cdot \sum_{k=n}^{\infty} b^k (1 + \cos[\pi a^{k-n} \beta_n]) \geqq \frac{b^n}{h_n} \geqq \frac{2a^n b^n}{3},$$

since $\cos \pi \beta_n \geqq 0$. Combining this estimate for $|R_n|$ with (2), we obtain

$$\left| \frac{f(x + h_n) - f(x)}{h_n} \right| \geqq |R_n| - |S_n| > \frac{2a^n b^n}{3} - \frac{\pi a^n b^n}{ab-1} = (ab)^n \left[\frac{2}{3} - \frac{\pi}{ab-1} \right].$$

Since $ab > 1 + \frac{3}{2}\pi$, $\left[\frac{2}{3} - \frac{\pi}{ab-1}\right]$ is a positive constant, and it follows that

$$\lim_{n \to \infty} \left| \frac{f(x + h_n) - f(x)}{h_n} \right| = \infty.$$

Since $\lim_{n \to \infty} h_n = 0$, it is clear that at least one right derivate of f at x is infinite. □

Our next theorem gives an indirect proof that continuous nowhere differentiable functions exist. It shows actually that in a certain sense most continuous functions are nowhere differentiable.

(17.8) Theorem. *Consider the real Banach space* $\mathfrak{C}^r = \mathfrak{C}^r([0, 1])$. *Let* $\mathfrak{D} = \{f \in \mathfrak{C}^r : D^+ f(x)$ *and* $D_+ f(x)$ *are both finite for some* $x \in [0, 1[\}$. *Then* $\mathfrak{D}$ *is of first category in the complete metric space* $\mathfrak{C}^r$. *Thus the set of all continuous functions on* $[0, 1]$ *which have at least one infinite right derivate at every point of* $[0, 1[$ *is dense in* $\mathfrak{C}^r([0, 1])$.

Proof. For each integer $n > 1$, let $\mathfrak{E}_n = \Big\{ f \in \mathfrak{C}^r :$ there exists an $x \in \left[0, 1 - \frac{1}{n}\right]$ such that $\left| \frac{f(x+h) - f(x)}{h} \right| \leqq n$ for all $h \in \left]0, \frac{1}{n}\right] \Big\}$. We first show that $\mathfrak{D} = \bigcup_{n=2}^{\infty} \mathfrak{E}_n$. Obviously $\bigcup_{n=2}^{\infty} \mathfrak{E}_n \subset \mathfrak{D}$. Let $f \in \mathfrak{D}$. Then there exists an $x \in [0, 1[$ and a constant $\alpha > 0$ such that for some δ, $0 < \delta < 1 - x$, the inequality

$$\left| \frac{f(x + h) - f(x)}{h} \right| < \alpha$$

holds for every $h \in]0, \delta[$. Select an integer n for which $n > \max\left\{\frac{1}{\delta}, \alpha\right\}$. It is plain that $f \in \mathfrak{E}_n$, and so $\mathfrak{D} \subset \bigcup_{n=2}^{\infty} \mathfrak{E}_n$.

Next we show that each $\mathfrak{E}_n$ is closed in $\mathfrak{C}^r$. To this end fix n, let $f \in \overline{\mathfrak{E}_n}$, and choose a sequence $(f_k)_{k=1}^{\infty}$ of functions in $\mathfrak{E}_n$ such that $\|f - f_k\|_u \to 0$. For each k, choose $x_k \in \left[0, 1 - \frac{1}{n}\right]$ to correspond to f_k as in the definition of $\mathfrak{E}_n$. Since $\left[0, 1 - \frac{1}{n}\right]$ is compact, the sequence (x_k) has a convergent subsequence whose corresponding subsequence of

(f_k) converges in norm to f; we denote these subsequences again by (x_k) and (f_k) respectively. Let $x = \lim_{k \to \infty} x_k$; clearly $x \in \left[0, 1 - \frac{1}{n}\right]$. Now fix $h \in \left]0, \frac{1}{n}\right]$ and let $\varepsilon > 0$ be arbitrary. Choose k so large that

$$\|f - f_k\|_u < \frac{h\varepsilon}{4}, \ |f(x_k) - f(x)| < \frac{h\varepsilon}{4}, \ \text{and} \ |f(x_k + h) - f(x + h)| < \frac{h\varepsilon}{4}.$$

Then $|f(x+h) - f(x)| \leqq |f(x+h) - f(x_k + h)| + |f(x_k + h) - f_k(x_k + h)| + |f_k(x_k + h) - f_k(x_k)| + |f_k(x_k) - f(x_k)| + |f(x_k) - f(x)| < \frac{h\varepsilon}{4} + \frac{h\varepsilon}{4} + nh + \frac{h\varepsilon}{4} + \frac{h\varepsilon}{4} = h(n + \varepsilon)$. Since ε is arbitrary, it follows that

$$\left|\frac{f(x+h) - f(x)}{h}\right| \leqq n$$

for all $h \in \left]0, \frac{1}{n}\right]$; hence $f \in \mathfrak{E}_n$. It follows that $\mathfrak{E}_n$ is closed.

We now show that every $\mathfrak{E}_n$ has void interior. Assume the contrary, *i.e.*, that there exist an n, an $f \in \mathfrak{E}_n$, and an $\varepsilon > 0$ such that $\mathfrak{B}_\varepsilon(f) = \{\varphi \in \mathfrak{C}^r : \|f - \varphi\|_u < \varepsilon\} \subset \mathfrak{E}_n$. By WEIERSTRASS's approximation theorem (7.31), there is a polynomial p such that $\|p - f\|_u < \varepsilon$. Let $\delta = \varepsilon - \|p - f\|_u$. Then we have $\mathfrak{B}_\delta(p) \subset \mathfrak{B}_\varepsilon(f) \subset \mathfrak{E}_n$. Next we construct a function $g \in \mathfrak{C}^r$ such that: $\|g\|_u < \delta$, $g'_+(x)$ exists, and $|g'_+(x)| > n + \|p'\|_u$ for all $x \in [0, 1[$. [Such g's clearly exist; *e.g.*, let g be a nonnegative "sawtooth" function on $[0, 1]$ with maximum $\frac{\delta}{2}$ and slopes greater in absolute value than the constant $n + \|p'\|_u$.] Then $g + p \in \mathfrak{B}_\delta(p)$, and we also have

$$|(g + p)'_+| = |g'_+ + p'| \geqq |g'_+| - |p'| \geqq |g'_+| - \|p'\|_u > n$$

at all points of $[0, 1[$. Thus $g + p \notin \mathfrak{E}_n$. This contradiction proves that $\mathfrak{E}_n^\circ = \varnothing$ for all n.

We conclude that each set $\mathfrak{E}_n$ is nowhere dense in $\mathfrak{C}^r$. Therefore $\mathfrak{D} = \bigcup_{n=2}^{\infty} \mathfrak{E}_n$ is of first category in $\mathfrak{C}^r$. Since $\mathfrak{C}^r$ is a complete metric space, it follows from the Baire category theorem (6.54) that $\mathfrak{C}^r \cap \mathfrak{D}'$ is dense in $\mathfrak{C}^r$.[1] □

The technique used in the proof of (17.7) is important. Many existence proofs throughout analysis and set-theoretic topology are carried out in just this way.

We next examine the extent to which a function can have different right and left derivatives.

---

[1] Many writers have made constructions of this sort. Our construction is taken from S. BANACH, Studia Math. 3, 174—179 (1931). See also K. KURATOWSKI, *Topologie I*, Deuxième Édition, Monografie Matematyczne, Tom XX, Warszawa-Wrocław, 1948, pp. 326—328.

(17.9) Theorem. *Let $]a, b[$ be any open interval of R and let f be an arbitrary real-valued function defined on $]a, b[$. Then there exist only countably many points $x \in]a, b[$ such that $f'_-(x)$ and $f'_+(x)$ both exist [they may be infinite] and are not equal.*

Proof. Let $A = \{x \in]a, b[: f'_-(x) \text{ exists}, f'_+(x) \text{ exists}, f'_+(x) < f'_-(x)\}$ and let $B = \{x \in]a, b[: f'_-(x) \text{ exists}, f'_+(x) \text{ exists}, f'_+(x) > f'_-(x)\}$. For each $x \in A$, choose a rational number r_x such that $f'_+(x) < r_x < f'_-(x)$. Next choose rational numbers s_x and t_x such that $a < s_x < x < t_x < b$,

$$\frac{f(y) - f(x)}{y - x} > r_x \quad \text{if} \quad s_x < y < x\,, \tag{1}$$

and

$$\frac{f(y) - f(x)}{y - x} < r_x \quad \text{if} \quad x < y < t_x\,. \tag{2}$$

Combining (1) and (2), we have

$$f(y) - f(x) < r_x(y - x) \tag{3}$$

whenever $y \neq x$ and $s_x < y < t_x$. Thus we obtain a function φ from A into the countable set Q^3, defined by $\varphi(x) = (r_x, s_x, t_x)$. We will prove that A is countable by showing that φ is one-to-one. Assume that there are distinct x and y in A such that $\varphi(x) = \varphi(y)$. Then $]s_y, t_y[=]s_x, t_x[$, and x and y are both in this interval. It follows from (3) that

$$f(y) - f(x) < r_x(y - x)$$

and

$$f(x) - f(y) < r_y(x - y)\,.$$

Since $r_x = r_y$, adding these two inequalities yields $0 < 0$. This is a contradiction, so that φ is one-to-one and A is countable. Similar reasoning proves that B is countable. □

Our next goal is to prove H. LEBESGUE's famous theorem that a monotone function has a finite derivative almost everywhere. The main tool used in the proof is a remarkable theorem of VITALI, which we present next. VITALI's theorem has numerous applications in classical analysis, particularly in the theory of differentiation.

(17.10) Definition. Let $E \subset R$. A family $\mathscr{V}$ of closed intervals of R, each having positive length, is called a *Vitali cover of E* if for each $x \in E$ and each $\varepsilon > 0$ there exists an interval $I \in \mathscr{V}$ such that $x \in I$ and $\lambda(I) < \varepsilon$, *i.e.*, each point of E is in arbitrarily short intervals of $\mathscr{V}$.

(17.11) VITALI's covering theorem. *Let E be an arbitrary subset of R and let $\mathscr{V}$ be any [nonvoid] Vitali cover of E. Then there exists a pairwise disjoint countable family $\{I_n\} \subset \mathscr{V}$ such that*

$$\lambda(E \cap (\bigcup_n I_n)') = 0\,.$$

Moreover, if $\lambda(E) < \infty$, then for each $\varepsilon > 0$ there exists a pairwise disjoint *finite family* $\{I_1, \ldots, I_p\} \subset \mathscr{V}$ *such that*

$$\lambda\Big(E \cap \Big(\bigcup_{n=1}^{p} I_n\Big)'\Big) < \varepsilon .$$

Proof[1]. Case I: $\lambda(E) < \infty$. Choose an open set V such that $E \subset V$ and $\lambda(V) < \infty$. Let $\mathscr{V}_0 = \{I \in \mathscr{V} : I \subset V\}$. Plainly $\mathscr{V}_0$ is a Vitali cover of E. Let $I_1 \in \mathscr{V}_0$. If $E \subset I_1$, the construction is complete. Otherwise we continue by induction as follows. Suppose that $I_1, I_2, \ldots, I_n$ have been selected and are pairwise disjoint. If $E \subset \bigcup_{k=1}^{n} I_k$, the construction is complete. Otherwise, write

$$A_n = \bigcup_{k=1}^{n} I_k, \quad U_n = V \cap A_n' .$$

Clearly A_n is closed, U_n is open, and $U_n \cap E \neq \varnothing$. Let

$$\delta_n = \sup\{\lambda(I) : I \in \mathscr{V}_0, I \subset U_n\} . \tag{1}$$

Choose $I_{n+1} \in \mathscr{V}_0$ such that $I_{n+1} \subset U_n$ and $\lambda(I_{n+1}) > \frac{1}{2}\,\delta_n$. If our process does not stop after a finite number of steps [in which case there is nothing left to prove], then it yields an infinite sequence $(I_n)_{n=1}^{\infty}$ of pairwise disjoint members of $\mathscr{V}_0$. Let $A = \bigcup_{n=1}^{\infty} I_n$. We must show that $\lambda(E \cap A') = 0$. For each n, let J_n be the closed interval having the same midpoint as I_n and such that

$$\lambda(J_n) = 5\lambda(I_n) .$$

We have

$$\begin{aligned} \lambda\Big(\bigcup_{n=1}^{\infty} J_n\Big) &\leq \sum_{n=1}^{\infty} \lambda(J_n) = 5 \sum_{n=1}^{\infty} \lambda(I_n) \\ &= 5\lambda(A) \leq 5\lambda(V) < \infty . \end{aligned} \tag{2}$$

Theorem (10.15) shows that

$$\lim_{p\to\infty} \lambda\Big(\bigcup_{n=p}^{\infty} J_n\Big) = 0 .$$

Thus, to prove that $\lambda(E \cap A') = 0$, it suffices to prove that $E \cap A' \subset \bigcup_{n=p}^{\infty} J_n$ for every $p \in N$. Fix $p \in N$ and let $x \in E \cap A'$. Then we have $x \in E \cap A_p' \subset U_p$, and so there exists an $I \in \mathscr{V}_0$ such that $x \in I \subset U_p$. It is evident that $\delta_n < 2\lambda(I_{n+1})$, and (2) shows that $\lambda(I_n) \to 0$ as $n \to \infty$. Hence there is an integer n such that $\delta_n < \lambda(I)$. Thus, by (1), there exists an integer n such that $I \not\subset U_n$; let q be the smallest such integer. It is obvious that $p < q$. We infer that

$$I \cap A_q \neq \varnothing \quad \text{and} \quad I \cap A_{q-1} = \varnothing .$$

[1] We give the ingenious proof of this theorem due to S. Banach [Fund. Math. 6, 170—188 (1924)].

It follows that

$$I \cap I_q \neq \varnothing \tag{3}$$

and, since $I \subset U_{q-1}$, we have

$$\lambda(I) \leqq \delta_{q-1} < 2\lambda(I_q) . \tag{4}$$

Since $\lambda(J_q) = 5\lambda(I_q)$, (3) and (4) show that

$$I \subset J_q \subset \bigcup_{n=p}^{\infty} J_n ,$$

so that $x \in \bigcup_{n=p}^{\infty} J_n$. Hence we have $E \cap A' \subset \bigcup_{n=p}^{\infty} J_n$, which implies that $\lambda(E \cap A') = 0$.

Now let $\varepsilon > 0$ be given and choose an integer p so large that

$$\sum_{n=p+1}^{\infty} \lambda(I_n) < \varepsilon .$$

Then

$$E \cap A'_p \subset (E \cap A') \cup \left(\bigcup_{n=p+1}^{\infty} I_n \right),$$

and so

$$\lambda(E \cap A'_p) \leqq 0 + \lambda\left(\bigcup_{n=p+1}^{\infty} I_n \right) < \varepsilon .$$

Thus the proof is finished if $\lambda(E) < \infty$.

Case II: $\lambda(E) = \infty$. For each $n \in Z$, let $E_n = E \cap]n, n+1[$ and let $\mathscr{V}_n = \{I \in \mathscr{V} : I \subset]n, n+1[\}$. Clearly $\mathscr{V}_n$ is a Vitali cover of E_n. Apply Case I to find a countable pairwise disjoint family $\mathscr{I}_n \subset \mathscr{V}_n$ such that $\lambda(E_n \cap (\cup \mathscr{I}_n)') = 0$ for each $n \in Z$. Let $\mathscr{I} = \bigcup_{n=-\infty}^{\infty} \mathscr{I}_n$. Then $\mathscr{I}$ is a countable pairwise disjoint subfamily of $\mathscr{V}$ and

$$E \cap (\cup \mathscr{I})' \subset Z \cup \left[\bigcup_{n=-\infty}^{\infty} \left(E_n \cap (\cup \mathscr{I}_n)' \right) \right] .$$

We see that

$$\lambda(E \cap (\cup \mathscr{I})') \leqq \lambda(Z) + \sum_{n=-\infty}^{\infty} 0 = 0 . \quad \square$$

(17.12) Theorem [Lebesgue]. *Let $[a, b]$ be a closed interval in R and let f be a real-valued monotone function on $[a, b]$. Then f has a finite derivative almost everywhere on $[a, b]$.*

Proof. We suppose that f is nondecreasing [otherwise consider $-f$]. Let $E = \{x : a \leqq x < b,\ D_+f(x) < D^+f(x)\}$. We will first show that $\lambda(E) = 0$. For every pair of positive rational numbers u and v such that $u < v$, let

$$E_{u,v} = \{x \in E : D_+f(x) < u < v < D^+f(x)\} .$$

Clearly $E = \cup \{E_{u,v} : u, v \in Q, 0 < u < v\}$. Since this is a countable union, it suffices to show that $\lambda(E_{u,v}) = 0$ for all $0 < u < v$ in Q. Assume the contrary: that there exist positive rational numbers u and v, $u < v$, such that $\lambda(E_{u,v}) = \alpha > 0$. Let ε be such that

$$0 < \varepsilon < \frac{\alpha(v-u)}{u+2v}\,.$$

Choose an open set $U \supset E_{u,v}$ such that $\lambda(U) < \alpha + \varepsilon$. For each $x \in E_{u,v}$, there exist arbitrarily small positive numbers h such that $[x, x+h] \subset U \cap [a,b]$ and

$$f(x+h) - f(x) < uh\,. \tag{1}$$

The family $\mathscr{V}$ of all such closed intervals is a Vitali cover of $E_{u,v}$, and so, by (17.11), there exists a finite, pairwise disjoint subfamily $\{[x_i, x_i+h_i]\}_{i=1}^{m}$ of $\mathscr{V}$ such that

$$\lambda\left(E_{u,v} \cap \left(\bigcup_{i=1}^{m} [x_i, x_i + h_i]\right)'\right) < \varepsilon\,.$$

Let $V = \bigcup_{i=1}^{m}]x_i, x_i + h_i[$. Then we have

$$\lambda(E_{u,v} \cap V') < \varepsilon\,. \tag{2}$$

The inclusion $V \subset U$ implies that

$$\sum_{i=1}^{m} h_i = \lambda(V) \leqq \lambda(U) < \alpha + \varepsilon\,,$$

and so (1) yields the inequalities

$$\sum_{i=1}^{m} (f(x_i+h_i) - f(x_i)) < u \sum_{i=1}^{m} h_i < u(\alpha+\varepsilon)\,. \tag{3}$$

Again, for all $y \in E_{u,v} \cap V$, there exist arbitrarily small positive numbers k such that $[y, y+k] \subset V$ and

$$f(y+k) - f(y) > vk\,. \tag{4}$$

The family of all such closed intervals is a Vitali cover of $E_{u,v} \cap V$, and so there is a finite, pairwise disjoint family $\{[y_j, y_j + k_j]\}_{j=1}^{n}$ of such intervals with the property that

$$\lambda\left(E_{u,v} \cap V \cap \left(\bigcup_{j=1}^{n} [y_j, y_j + k_j]\right)'\right) < \varepsilon\,.$$

This inequality together with (2) implies that

$$\alpha = \lambda(E_{u,v}) \leqq \lambda(E_{u,v} \cap V') + \lambda(E_{u,v} \cap V) < \varepsilon + \left(\varepsilon + \sum_{j=1}^{n} k_j\right). \tag{5}$$

Next, using (4) and (5), we have

$$v(\alpha - 2\varepsilon) < v \sum_{j=1}^{n} k_j < \sum_{j=1}^{n} (f(y_j+k_j) - f(y_j))\,. \tag{6}$$

Since $\bigcup_{j=1}^{n} [y_j, y_j + k_j] \subset \bigcup_{i=1}^{m} [x_i, x_i + h_i]$ and f is nondecreasing, we also have

$$\sum_{j=1}^{n} (f(y_j + k_j) - f(y_j)) \leqq \sum_{i=1}^{m} (f(x_i + h_i) - f(x_i)) . \tag{7}$$

Combining (6), (7), and (3) gives

$$v(\alpha - 2\varepsilon) < u(\alpha + \varepsilon) ,$$

which contradicts our choice of ε. Thus $\lambda(E) = 0$, and so $f'_+(x)$ exists a.e. on $[a, b]$. Similarly $f'_-(x)$ exists a.e. on $[a, b]$. Now apply (17.9) to see that $f'(x)$ exists a.e. on $[a, b]$.

It remains only to show that the set F of points x in $]a, b[$ for which $f'(x) = \infty$ has measure zero. Let β be an arbitrary positive number. For each $x \in F$, there exist arbitrarily small positive numbers h such that $[x, x + h] \subset]a, b[$ and

$$f(x + h) - f(x) > \beta h . \tag{8}$$

By VITALI's theorem (17.11), there exists a countable pairwise disjoint family $\{[x_n, x_n + h_n]\}$ of these intervals such that

$$\lambda\big(F \cap (\bigcup_n [x_n, x_n + h_n])'\big) = 0.$$

From this fact and (8) we obtain

$$\beta\lambda(F) \leqq \beta \sum_n h_n < \sum_n (f(x_n + h_n) - f(x_n)) \leqq f(b) - f(a) .$$

Thus

$$\beta\lambda(F) < f(b) - f(a) \quad \text{for all} \quad \beta \in R ,$$

which implies that $\lambda(F) = 0$. □

(17.13) Question. Suppose that $\lambda(A) = 0$, $A \subset [a, b]$. Is it possible to find a monotone function f on $[a, b]$ such that f' exists exactly on $A' \cap]a, b[$? The complete answer seems to be unknown.

(17.14) Definition. Let f be a complex-valued function defined on $[a, b] \subset R$. Define

$$V_a^b f = \sup \left\{ \sum_{k=1}^{n} |f(x_k) - f(x_{k-1})| : a = x_0 < x_1 < \cdots < x_n = b \right\} .$$

The extended real number $V_a^b f$ is called the *total variation of f over* $[a, b]$. If $V_a^b f < \infty$, then f is said to be of *finite variation* [or *bounded variation*] *over* $[a, b]$.

(17.15) Remarks. (a) The function f has finite variation if and only if the functions Ref and Imf have finite variation.

(b) The equality $V_a^b f + V_b^c f = V_a^c f$ holds for $a < b < c$.

(c) The function $x \to V_a^x f$ is nondecreasing.

(17.16) Theorem [Jordan decomposition theorem]. *A real-valued function of finite variation is the difference of two nondecreasing functions.*

Proof. Write $f(x) = V_a^x f - (V_a^x f - f(x))$, where we define $V_a^a f = 0$. Evidently the function $x \to V_a^x f$ is nondecreasing. The function

$$x \to V_a^x f - f(x)$$

is also nondecreasing, for if $x' > x$, then

$$V_a^{x'} f - f(x') - (V_a^x f - f(x)) = V_x^{x'} f - (f(x') - f(x)) \geqq 0. \quad \square$$

(17.17) Theorem [LEBESGUE]. *A complex-valued function of finite variation has a finite derivative a.e.*

Proof. This is an immediate consequence of (17.16), (17.15.a), (17.12), and (17.5). □

(17.18) Theorem [FUBINI][1]. *Let $(f_n)_{n=1}^{\infty}$ be a sequence of nondecreasing [or nonincreasing] real-valued functions on an interval $[a, b]$ such that $\sum_{n=1}^{\infty} f_n(x) = s(x)$ exists and is finite in $[a, b]$. Then*

(i)
$$s'(x) = \sum_{n=1}^{\infty} f_n'(x)$$

a.e. in $]a, b[$.

Proof. There is no harm in supposing [and we do] that all f_n are nondecreasing. Also, by considering the functions $f_n - f_n(a)$, we may suppose that $f_n \geqq 0$. Thus $s = \sum_{n=1}^{\infty} f_n$ is nonnegative and nondecreasing. The derivative $s'(x)$ exists and is finite for almost all $x \in]a, b[$, as (17.12) shows.

Consider then the partial sums $s_n = f_1 + f_2 + \cdots + f_n$, and the remainders $r_n = s - s_n$. Each f_j has a finite derivative a.e.; hence there is a set $A \subset]a, b[$ such that $\lambda(A' \cap]a, b[) = 0$,

$$s_n'(x) = f_1'(x) + f_2'(x) + \cdots + f_n'(x) < \infty$$

for all $x \in A$ and all n, and $s'(x)$ exists and is finite for $x \in A$. For any $x \in]a, b[$ and every $h > 0$ such that $x + h \in]a, b[$, it follows from the equality

$$\frac{s(x+h) - s(x)}{h} = \frac{s_n(x+h) - s_n(x)}{h} + \frac{r_n(x+h) - r_n(x)}{h}$$

that

$$\frac{s_n(x+h) - s_n(x)}{h} \leqq \frac{s(x+h) - s(x)}{h};$$

and this inequality implies that $s_n'(x) \leqq s'(x)$ for all $x \in A$. The inequality $s_n'(x) \leqq s_{n+1}'(x)$ is clear, and so we have

$$s_n'(x) \leqq s_{n+1}'(x) \leqq s'(x)$$

[1] This is not the theorem ordinarily called "FUBINI's theorem", which deals with product measures and integrals and will be taken up in Chapter Six.

for $x \in A$ and $n = 1, 2, \ldots$. Hence

$$\lim_{n\to\infty} s'_n(x) = \sum_{j=1}^{\infty} f'_j(x)$$

exists a.e., and it remains to show that $\lim_{n\to\infty} s'_n(x) = s'(x)$ a.e. Since the sequence $(s'_n(x))_{n=1}^{\infty}$ is nondecreasing for each $x \in A$, it suffices to show that (s'_n) admits a subsequence converging a.e. to s'. To this end, let $n_1, n_2, \ldots, n_k, \ldots$ be an increasing sequence of integers such that

$$\sum_{k=1}^{\infty} [s(b) - s_{n_k}(b)] < \infty .$$

For each n_k and for every $x \in]a, b[$, we have

$$0 \leqq s(x) - s_{n_k}(x) \leqq s(b) - s_{n_k}(b) .$$

The terms on the left side of this inequality are bounded by the terms of a convergent series of nonnegative terms. Hence $\sum_{k=1}^{\infty} [s(x) - s_{n_k}(x)]$ converges. The terms of this series are monotone functions that have finite derivatives a.e. Therefore the argument used above to prove that $\sum_{j=1}^{\infty} f'_j(x)$ converges a.e. also proves that $\sum_{k=1}^{\infty} [s'(x) - s'_{n_k}(x)]$ converges a.e.; and of course it follows that $\lim_{k\to\infty} s'_{n_k}(x) = s'(x)$ a.e. □

We close this section with a long collection of exercises. A number are merely illustrative examples; several are minor theorems with sketched proofs [(17.24), (17.25), (17.26), (17.27), (17.31), (17.36), (17.37)]; and (17.33) and (17.34) are needed for later theorems of the main text. The reader should bear these facts in mind when doing the exercises.

(17.19) Exercise. Let ψ be LEBESGUE'S singular function, defined in (8.28). Compute all of the derivates of ψ at each point of [0, 1].

(17.20) Exercise. Define the function φ on R by

$$\varphi(x) = \begin{cases} x & \text{if } 0 \leqq x < \frac{1}{2}, \\ 1 - x & \text{if } \frac{1}{2} \leqq x < 1, \end{cases}$$

$$\varphi(x + k) = \varphi(x) \quad \text{for all} \quad k \in Z .$$

Let

$$f(x) = \sum_{n=1}^{\infty} 2^{-n} \varphi(2^n x) .$$

Prove that f is continuous on R. Compute all four derivates of f at each dyadic rational point. Prove that f fails to have a finite derivative at every point not a dyadic rational.

(17.21) Exercise. Let $\{a_n\}_{n=1}^{\infty}$ be a set of distinct points in the interval $[a, b]$. Let $(u_n)_{n=1}^{\infty}$ and $(v_n)_{n=1}^{\infty}$ be sequences of real numbers such that $\sum_{n=1}^{\infty} |u_n| < \infty$ and $\sum_{n=1}^{\infty} |v_n| < \infty$. Define

$$f_n(x) = \begin{cases} 0 & \text{if} \quad x < a_n, \\ u_n & \text{if} \quad x = a_n, \\ v_n & \text{if} \quad x > a_n. \end{cases}$$

Prove that $s(x) = \sum_{n=1}^{\infty} f_n(x)$ has a finite derivative a.e., and that $s'(x) = 0$ a.e. [Hint. The function s has finite variation; find each V_a^x by using the numbers $|u_n|$ and $|v_n|$. Then apply (17.18).]

(17.22) Exercise. Find a real-valued *strictly* increasing function f on R such that $f'(x) = 0$ a.e.

(17.23) Exercise. Let $f \in \mathfrak{C}^r([a, b])$. Suppose that there exist real constants $\alpha < \beta$ such that

$$\alpha \leqq D^+ f(x) \leqq \beta$$

for all $x \in [a, b[$. Prove that

$$h\alpha \leqq f(x+h) - f(x) \leqq h\beta$$

if $a \leqq x < x + h \leqq b$. [Hint. Assuming that

$$f(x_0 + h_0) - f(x_0) < \gamma h_0 < \alpha h_0$$

and writing $h_1 = \sup\{h : 0 < h < h_0, f(x_0 + h) - f(x_0) \geqq \gamma h\}$, show that $D^+ f(x_0 + h_1) < \alpha$.]

(17.24) Exercise. Let f be a function in $\mathfrak{C}^r([a, b])$ and let c be a number in $]a, b[$. Suppose that $D^+ f(c)$ is finite and that $D^+ f$ is continuous at c. Prove that $f'(c)$ exists. [Use (17.23).]

(17.25) Exercise. Let f be a real-valued nondecreasing function on $[a, b]$. Suppose that $u \geqq 0$ and $E \subset [a, b]$ are such that for each $x \in E$, there exists some derivate of f at x which does not exceed u. Prove that $\lambda(f(E)) \leqq u\lambda(E)$. [Consider an appropriate Vitali cover of $f(A)$, where $A = \{x \in E : f(x) \neq f(y) \text{ for all } y \in [a, b] \cap \{x\}'\}$. Notice that $f(E \cap A')$ is countable.]

(17.26) Exercise. Let f be as in (17.25). Suppose that $v \geqq 0$ and $F \subset [a, b]$ are such that for each $x \in F$, some derivate of f at x is greater than or equal to v. Prove that $\lambda(f(F)) \geqq v\lambda(F)$. [Consider an appropriate Vitali cover of B, where $B = \{x \in F : f \text{ is continuous at } x\}$. Notice that $F \cap B'$ is countable.]

(17.27) Exercise. Let f be a real-valued function defined on $[a, b]$. Suppose that $c \geqq 0$ and $E \subset [a, b]$ are such that $f'(x)$ exists and $|f'(x)| \leqq c$ for all $x \in E$. Prove that $\lambda(f(E)) \leqq c\lambda(E)$. [Consider a Vitali cover of $f(E)$ by intervals $[f(x), f(x+h)]$ such that $f([x, x+h]) \subset [f(x), f(x+h)]$.]

(17.28) Exercise. Let E be a subset of R that is the union of a family of quite arbitrary intervals, each being open, closed, or half open and half closed. Prove that E is Lebesgue measurable. [Use VITALI's theorem.]

(17.29) Exercise. Let α and β be positive real numbers. Define f on $[0, 1]$ by $f(x) = x^\alpha \sin(x^{-\beta})$ $(0 < x \leqq 1)$, $f(0) = 0$. Prove that f is of finite variation on $[0, 1]$ if and only if $\alpha > \beta$.

(17.30) Exercise. Prove or disprove the following statement. If f is a function in $\mathfrak{C}^r([0, 1])$, then there exist $a, b \in R$ such that $0 \leqq a < b \leqq 1$ and f is of finite variation on $[a, b]$.

(17.31) Exercise. A function f defined on an interval I of R is said to satisfy a *Lipschitz condition of order* $\alpha > 0$ if there exists a constant $M \geqq 0$ such that

$$|f(x) - f(y)| \leqq M |x - y|^\alpha$$

for all $x, y \in I$. We write $f \in \mathfrak{Lip}_\alpha(I)$. Prove the following.

(a) If $\alpha > 1$ and $f \in \mathfrak{Lip}_\alpha(I)$, then f is a constant.

(b) If $0 < \alpha < 1$, then there exists a function $f \in \mathfrak{Lip}_\alpha([0, 1])$ such that f has infinite variation over $[0, 1]$.

(c) There exists a continuous function of finite variation on $[0, 1]$ which satisfies *no* Lipschitz condition.

(d) If $f \in \mathfrak{C}^r([a, b])$, then $f \in \mathfrak{Lip}_1([a, b])$ if and only if $D^+ f$ is bounded on $[a, b[$. [Hint. Use (17.23).]

(17.32) Exercise. Let f be a complex-valued function of finite variation on $[a, b]$. Suppose that f is continuous at $c \in [a, b]$. Prove that the function $g : x \to V_a^x f$, where $g(a) = 0$, is continuous at c.

(17.33) Exercise. Let f be a function in $\mathfrak{C}^r([a, b])$. For each subdivision $\Delta = \{a = x_0 < x_1 < \cdots < x_n = b\}$ of $[a, b]$, define $|\Delta| = \max\{x_k - x_{k-1} : 1 \leqq k \leqq n\}$ and

$$\omega_k = \max\{f(x) : x_{k-1} \leqq x \leqq x_k\} - \min\{f(x) : x_{k-1} \leqq x \leqq x_k\}$$

for $k = 1, \ldots, n$. Prove that

$$V_a^b f = \lim_{|\Delta| \to 0} \sum_{k=1}^{n} \omega_k .$$

(17.34) Exercise. Let f be a function in $\mathfrak{C}^r([a, b])$. For each $y \in R$, let $A_y = \{x \in [a, b] : f(x) = y\}$. Define ν on R by

$$\nu(y) = \begin{cases} \overline{\overline{A}}_y & \text{if } A_y \text{ is finite}, \\ \infty & \text{if } A_y \text{ is infinite}. \end{cases}$$

Prove that the function ν is Lebesgue measurable and that

$$\int_R \nu(y)\, dy = V_a^b f .$$

[Hint. Use (17.33). Let $\Delta_1 \subset \Delta_2 \subset \cdots$ be a sequence of subdivisions of $[a, b]$ such that $|\Delta_n| \to 0$, say $\Delta_n = \{a = x_0^{(n)} < \cdots < x_{m_n}^{(n)} = b\}$. For each n, define

$$v_n = \sum_{k=1}^{m_n} \xi_{B_{n,k}},$$

where $B_{n,k} = f([x_{k-1}^{(n)}, x_k^{(n)}])$. Prove that $v_n(y) \to v(y)$ for almost all $y \in R$ and apply B. LEVI's monotone convergence theorem.] The function v is known as the *Banach indicatrix of* f.

(17.35) Exercise. For an interval $[a, b] \subset R$, let $\mathfrak{V}([a, b])$ denote the set of all complex-valued functions f on $[a, b]$ such that $V_a^b f < \infty$ and $f(a) = 0$. For $f \in \mathfrak{V}([a, b])$, define $\|f\| = V_a^b f$. Prove (a)–(c) and answer (d) and (e).

(a) With pointwise operations $\mathfrak{V}([a, b])$ is a complex linear algebra.
(b) For $f \in \mathfrak{V}([a, b])$, the inequality $\|f\|_u \leqq \|f\|$ holds.
(c) With the above norm, $\mathfrak{V}([a, b])$ is a Banach space.
(d) Is it true that $\|fg\| \leqq \|f\| \cdot \|g\|$ for all $f, g \in \mathfrak{V}([a, b])$?
(e) Find the least cardinal number of a dense subset of $\mathfrak{V}([a, b])$ in the topology defined by the variation norm.

(17.36) Exercise. Let x be a real number and let f be a real-valued function defined in a neighborhood of x. The *upper* [*lower*] *first* and *second symmetric derivatives* of f at x are defined to be the limits superior [inferior] of the expressions

$$\frac{f(x+h) - f(x-h)}{2h} \tag{1}$$

and

$$\frac{f(x+h) + f(x-h) - 2f(x)}{h^2} \tag{2}$$

as $h \downarrow 0$, respectively. These derivatives are denoted by $\overline{D}_1 f(x)$ and $\overline{D}_2 f(x)$ [$\underline{D}_1 f(x)$ and $\underline{D}_2 f(x)$] respectively. If $\overline{D}_1 f(x) = \underline{D}_1 f(x)$ [$\overline{D}_2 f(x) = \underline{D}_2 f(x)$], we call this common value the *first* [*second*] *symmetric derivative of f at x* and denote it by $D_1 f(x)$ [$D_2 f(x)$]. Prove the following.

(a) If $f'(x)$ exists, then so does $D_1 f(x)$ and they are equal.
(b) The converse of (a) is false.
(c) If f' exists and is finite in a neighborhood of x and $f''(x)$ is finite, then $D_2 f(x)$ exists and is equal to $f''(x)$. [Use the mean value theorem on (2) as a function of h.]
(d) $D_2 f(x)$ may exist even when f is continuous only at x.

(17.37) Exercise: More on convex functions. Let I be an open interval in R and let f be a convex function [see (13.34)] defined on I.

(a) Prove that $f'_+(x)$ and $f'_-(x)$ exist and are finite for all $x \in I$, also that f'_+ and f'_- are nondecreasing functions and $f'_- \leqq f'_+$ on I. Thus f' exists and is finite a.e. on I. [Hints. For $x < y < z$, we have

$y = \frac{z-y}{z-x} x + \frac{y-x}{z-x} z$ and so $f(y) \leqq \frac{z-y}{z-x} f(x) + \frac{y-x}{z-x} f(z)$. Hence

$$\frac{f(y)-f(x)}{y-x} \leqq \frac{f(z)-f(x)}{z-x} \leqq \frac{f(z)-f(y)}{z-y}. \tag{1}$$

From (1) our assertions follow easily.]

(b) Prove that f is in $\mathfrak{Lip}_1([a, b])$ for all closed bounded subintervals $[a, b]$ of I. [For $a < x < y < b$, (1) implies that

$$\frac{f(x)-f(a)}{x-a} \leqq \frac{f(y)-f(x)}{y-x} \leqq \frac{f(b)-f(y)}{b-y},$$

and from this and (a) it is easy to see that

$$\left|\frac{f(y)-f(x)}{y-x}\right| \leqq \max\{|f'_+(a)|, |f'_-(b)|\}.]$$

(c) Let g be a real-valued function on an open interval I in R. Prove that g is convex if and only if g is continuous and $\overline{D}_2 g \geqq 0$ on I [see (17.36)]. [First suppose that $\overline{D}_2 g > 0$ on I, assume that g is not convex on I, and find a point x such that $\overline{D}_2 g(x) \leqq 0$. Next consider the functions $g_n(x) = g(x) + \frac{1}{n} x^2$, $n = 1, 2, \ldots$.]

§ 18. Absolutely continuous functions

In this section, we identify the class of functions F of the form $F(x) = \int_a^x f(t)\,dt$ for $f \in \mathfrak{L}_1([a, b])$. We also identify the functions on intervals of R that are integrals of their derivatives. This study leads directly to some classical facts in the theory of Fourier series, which we also take up. As in § 17, "almost everywhere" means "λ-almost everywhere", and "measurable" means "$\mathcal{M}_\lambda$-measurable". We begin with some simple theorems.

(18.1) Theorem. *Let $f \in \mathfrak{L}_1([a, b])$ and define F on $[a, b]$ by*

$$F(x) = \int_a^x f(t)\,dt.$$

[*The function F is called the* indefinite integral *of f.*] *Then F is uniformly continuous and has finite variation, and $V_a^b F = \int_a^b |f(t)|\,dt$. A similar assertion holds for $f \in \mathfrak{L}_1(R)$ and $F(x) = \int_{-\infty}^x f(t)\,dt$.* [*If φ is a complex-valued function on R, then we define $V_{-\infty}^{\infty}(\varphi) = \lim_{A\to\infty} V_{-A}^{A}\varphi$; $V_{-\infty}^{a}(\varphi)$ and $V_a^{\infty}(\varphi)$ are defined similarly.*]

Proof. For $x' > x$, the equality $|F(x') - F(x)| = \left|\int_x^{x'} f(t)\,dt\right|$ holds; therefore it is clear from (12.34) that F is uniformly continuous. If $a = x_0 < x_1 < \cdots < x_n = b$, then we have

$$\sum_{k=1}^{n} |F(x_k) - F(x_{k-1})| = \sum_{k=1}^{n} \left|\int_{x_{k-1}}^{x_k} f(t)\,dt\right| \leqq \sum_{k=1}^{n} \int_{x_{k-1}}^{x_k} |f(t)|\,dt = \int_a^b |f(t)|\,dt\,.$$

Hence the inequality $V_a^b F \leqq \int_a^b |f(t)|\,dt$ holds, and so F has finite variation.

To prove the reversed inequality, first recall that step functions

$$\sigma = \sum_{k=1}^{n} \alpha_k\, \xi_{[x_{k-1}, x_k[} \quad (a = x_0 < x_1 < \cdots < x_n = b) \tag{1}$$

are dense in $\mathfrak{L}_1([a, b])$ (13.23). Consider the function $\operatorname{sgn}\bar{f}$; for every positive integer m, select a step function σ_m of the form (1) such that

$$\|\sigma_m - \operatorname{sgn}\bar{f}\|_1 < \frac{1}{m}\,. \tag{2}$$

Since $|\operatorname{sgn}\bar{f}(x)| = 1$ or 0 for every x, it is easy to see that the inequality (2) is only improved by replacing every α_k such that $|\alpha_k| > 1$ by the number $\alpha_k|\alpha_k|^{-1}$. There is thus no harm in supposing that $|\sigma_m(x)| \leqq 1$ for all $x \in [a, b]$ and $m \in N$. Clearly $\sigma_m \to \operatorname{sgn}\bar{f}$ in measure, and so by (11.26) there is a subsequence (σ_{m_j}) of (σ_m) such that

$$\lim_{j\to\infty} \sigma_{m_j}(t) = \operatorname{sgn}\bar{f}(t) \quad \text{a.e. in } [a, b]\,.$$

We then infer from LEBESGUE's theorem on dominated convergence (12.30) that

$$\int_a^b |f(t)|\,dt = \int_a^b f(t)\operatorname{sgn}\bar{f}(t)\,dt = \lim_{j\to\infty} \int_a^b f(t)\,\sigma_m(t)\,dt\,. \tag{3}$$

Since σ_{m_j} has the form (1), the absolute value of the last integral in (3) has the form

$$\begin{aligned} \left|\sum_{k=1}^{n} \alpha_k \int_{x_{k-1}}^{x_k} f(t)\,dt\right| &= \left|\sum_{k=1}^{n} \alpha_k (F(x_k) - F(x_{k-1}))\right| \\ &\leqq \sum_{k=1}^{n} |\alpha_k| \cdot |F(x_k) - F(x_{k-1})| \\ &\leqq \sum_{k=1}^{n} |F(x_k) - F(x_{k-1})| \leqq V_a^b F\,. \end{aligned} \tag{4}$$

Combining (3) and (4), we have

$$\int_a^b |f(t)|\,dt \leqq V_a^b F\,. \quad \square$$

The foregoing theorem shows that indefinite integrals are continuous and have finite variation. We wish now to show that the derivative of an indefinite integral is the integrated function [a.e.!]. To prove this, we need a preliminary, which is of some interest in its own right.

(18.2) Theorem. *Let A be an arbitrary subset of R. Then*

$$\text{(i)}\quad \lim_{k\downarrow 0} \frac{\lambda(A\cap\,]x, x+k[)}{k} = \lim_{h\downarrow 0} \frac{\lambda(A\cap\,]x-h, x[)}{h} = \lim_{h,k\downarrow 0} \frac{\lambda(A\cap\,]x-h, x+k[)}{h+k} = 1$$

for almost all $x \in A$. If A is λ-measurable, the limits in (i) *are equal to zero for almost all $x \in A'$.*[1]

Proof. With no harm done, we can [and do] suppose that A is bounded. There are bounded open sets U_n, $n = 1, 2, \ldots$, such that

$$U_1 \supset U_2 \supset \cdots \supset U_n \supset \cdots \supset A$$

and $\lambda(U_n) - 2^{-n} < \lambda(A)$. Let $a = \inf U_1$, and consider the functions

$$\varphi_n(x) = \lambda(U_n \cap\,]a, x[)$$

and

$$\varphi(x) = \lambda(A \cap\,]a, x[)\,.$$

For $x \in U_n$ and sufficiently small positive h, it is clear that

$$\frac{\varphi_n(x+h) - \varphi_n(x)}{h} = \frac{\varphi_n(x) - \varphi_n(x-h)}{h} = 1;$$

hence $\varphi_n'(x)$ exists for all $x \in U_n$ and

$$\varphi_n'(x) = 1\,.$$

We want to apply FUBINI's theorem (17.18) to the sum

$$(\varphi_1 - \varphi) + (\varphi_2 - \varphi) + \cdots + (\varphi_n - \varphi) + \cdots;$$

we first show that each $\varphi_n - \varphi$ is monotone. For $x' > x$, we have

$$\begin{aligned} &\varphi_n(x') - \varphi(x') - (\varphi_n(x) - \varphi(x)) \\ &\quad = \lambda(U_n \cap [x, x'[) - \lambda(A \cap\,]a, x'[) + \lambda(A \cap\,]a, x[) \\ &\quad \geqq \lambda(U_n \cap [x, x'[) - \lambda(A \cap [x, x'[) \geqq 0 \end{aligned}$$

because

$$\lambda(A \cap\,]a, x'[) \leqq \lambda(A \cap\,]a, x[) + \lambda(A \cap [x, x'[)$$

and

$$A \cap [x, x'[\, \subset U_n \cap [x, x'[;$$

thus $\varphi_n - \varphi$ is monotone. Now let $b = \sup U_1$; then

$$\varphi_n(b) - \varphi(b) = \lambda(U_n) - \lambda(A) < 2^{-n},$$

---

[1] Points x for which the relations (i) hold are called *points of density of A.*

and so for $a \leqq x \leqq b$ we have

$$\sum_{n=1}^{\infty} (\varphi_n(x) - \varphi(x)) \leqq \sum_{n=1}^{\infty} (\varphi_n(b) - \varphi(b)) \leqq \sum_{n=1}^{\infty} 2^{-n} < \infty .$$

Let

$$s(x) = \sum_{n=1}^{\infty} (\varphi_n(x) - \varphi(x)) .$$

By (17.18) and (17.12), the relations

$$s'(x) = \sum_{n=1}^{\infty} (\varphi_n'(x) - \varphi'(x)) < \infty$$

hold for almost all x in $]a, b[$, and so also we have

$$\lim_{n\to\infty} \varphi_n'(x) = \varphi'(x)$$

a.e. in $]a, b[$. Thus $\varphi'(x) = 1$ on $\bigcap_{n=1}^{\infty} U_n$ except on a set of λ-measure zero, and this implies the first assertion of the theorem.[1]

If A is λ-measurable, then

$$\begin{aligned} 1 &= \frac{\lambda(A \cap]x-h, x+k[)}{h+k} + \frac{\lambda(A' \cap]x-h, x+k[)}{h+k} \\ &= \psi_A(x) + \psi_{A'}(x) \end{aligned}$$

for all h, k. As h and k go to 0, $\psi_{A'}(x)$ goes to 1 for almost all $x \in A'$ [apply the first part of the theorem to the set A']. Hence $\psi_A(x)$ goes to zero a.e. on A'. □

(18.3) Theorem. *Let $f \in \mathfrak{L}_1([a, b])$, and let F be as in* (18.1). *Then the equality*

(i) $F'(x) = f(x)$

holds for almost all $x \in]a, b[$.

Proof. If $f = \xi_A$, where A is a measurable subset of $]a, b[$, then $F(x) = \lambda(]a, x[\cap A)$; and (18.2) shows that $F'(x) = \xi_A(x)$ a.e. in $]a, b[$. Next, let $s = \sum_{k=1}^{n} \alpha_k \xi_{A_k}$ be a nonnegative simple measurable function, so that

$$S(x) = \int_a^x s(t)\, dt = \sum_{k=1}^{n} \alpha_k \int_a^x \xi_{A_k}(t)\, dt .$$

Theorem (18.2) implies that

$$S'(x) = s(x) \quad \text{a.e. in }]a, b[. \tag{1}$$

For a nonnegative function f in $\mathfrak{L}_1$, let $(s_n)_{n=1}^{\infty}$ be a nondecreasing sequence of simple measurable functions such that $\lim_{n\to\infty} s_n(x) = f(x)$ for all

---

[1] We have actually proved a little more than claimed in the theorem. We have $\varphi'(x) = 1$ a.e. on the set $\bigcap_{n=1}^{\infty} U_n$; if A is nonmeasurable, the nonmeasurable set $\left(\bigcap_{n=1}^{\infty} U_n\right) \cap A'$ does not have measure 0.

$x \in [a, b]$ (11.35). Write $S_n(x) = \int_a^x s_n(t)\,dt$ for $n \in N$; B. LEVI's theorem (12.22) shows that

$$F(x) = \int_a^x f(t)\,dt = \lim_{n\to\infty} \int_a^x s_n(t)\,dt = \lim_{n\to\infty} S_n(x)$$

$$= S_1(x) + \sum_{n=1}^{\infty} [S_{n+1}(x) - S_n(x)] \tag{2}$$

for all $x \in [a, b]$. Each function $S_{n+1} - S_n$ is the integral of a nonnegative function and so is nondecreasing. FUBINI's theorem (17.18) applied to (2) gives us the equalities

$$F'(x) = S_1'(x) + \sum_{n=1}^{\infty} [S_{n+1}'(x) - S_n'(x)]$$

$$= \lim_{n\to\infty} S_n'(x) \quad \text{a.e. in }]a, b[\,, \tag{3}$$

and (1) gives us

$$\lim_{n\to\infty} S_n'(x) = \lim_{n\to\infty} s_n(x) \quad \text{a.e. in }]a, b[\,. \tag{4}$$

Combining (3) and (4), we obtain (i).

Finally, if f is an arbitrary function in $\mathfrak{L}_1([a, b])$, write

$$f = (f_1 - f_2) + i(f_3 - f_4)$$

where $f_j \in \mathfrak{L}_1^+$ and apply (i) for nonnegative functions. □

Theorem (18.3) can be sharpened considerably, as the next two assertions show.

(18.4) Lemma [LEBESGUE]. *Let f be a function in $\mathfrak{L}_1([a, b])$. Then there is a set $E \subset\,]a, b[$ such that $\lambda(E' \cap [a, b]) = 0$ and*

(i) $\lim\limits_{h\downarrow 0} \int_x^{x+h} |f(t) - \alpha|\,dt = \lim\limits_{h\downarrow 0} \int_{x-h}^{x} |f(t) - \alpha|\,dt = |f(x) - \alpha|$

for all $\alpha \in K$ and all $x \in E$.

Proof. Let $\{\beta_n\}_{n=1}^{\infty}$ be any countable dense subset of K. The functions g_n defined by

$$g_n(t) = |f(t) - \beta_n| \quad (n \in N)$$

are in $\mathfrak{L}_1([a, b])$. By (18.3), there are sets $E_n \subset\,]a, b[$ such that $\lambda(E_n' \cap [a, b]) = 0$ and

$$\lim_{h\downarrow 0} \frac{1}{h} \int_x^{x+h} g_n(t)\,dt = \lim_{h\downarrow 0} \frac{1}{h} \int_{x-h}^{x} g_n(t)\,dt = g_n(x)$$

for all $x \in E_n$. Let E be the intersection $\bigcap_{n=1}^{\infty} E_n$; clearly $\lambda(E' \cap [a, b]) = 0$. For $\varepsilon > 0$ and $\alpha \in K$, select an n such that $|\beta_n - \alpha| < \frac{\varepsilon}{3}$. Then we have

$$\big||f(t) - \alpha| - |f(t) - \beta_n|\big| \leqq |\beta_n - \alpha| < \frac{\varepsilon}{3} \quad \text{for all} \quad t \in [a, b]\,.$$

It follows that

$$\left|\frac{1}{h}\int_x^{x+h} |f(t)-\alpha|\,dt - \frac{1}{h}\int_x^{x+h} |f(t)-\beta_n|\,dt\right| \leqq \frac{1}{h}\int_x^{x+h} \frac{\varepsilon}{3}\,dt = \frac{\varepsilon}{3},$$

and in turn that

$$\left|\frac{1}{h}\int_x^{x+h} |f(t)-\alpha|\,dt - |f(x)-\alpha|\right|$$

$$\leqq \left|\frac{1}{h}\int_x^{x+h} |f(t)-\alpha|\,dt - \frac{1}{h}\int_x^{x+h} |f(t)-\beta_n|\,dt\right|$$

$$+\left|\frac{1}{h}\int_x^{x+h} g_n(t)\,dt - g_n(x)\right| + |\beta_n-\alpha| < \frac{\varepsilon}{3}+\frac{\varepsilon}{3}+\frac{\varepsilon}{3} = \varepsilon$$

if $x \in E$ and $0 < h < h_0$, where h_0 depends on ε and n. But n depends only on ε and α. Thus we conclude that

$$\lim_{h\downarrow 0}\frac{1}{h}\int_x^{x+h} |f(t)-\alpha|\,dt = |f(x)-\alpha|$$

for all $x \in E$ and $\alpha \in K$. A similar argument shows that

$$\lim_{h\downarrow 0}\frac{1}{h}\int_{x-h}^{x} |f(t)-\alpha|\,dt = |f(x)-\alpha|$$

for all $x \in E$ and $\alpha \in K$. □

(18.5) Theorem [Lebesgue]. *Let $f \in \mathfrak{L}_1([a, b])$. Then we have*

(i) $\lim\limits_{h\downarrow 0}\frac{1}{h}\int_0^h |f(x+t)+f(x-t)-2f(x)|\,dt = 0$

for almost all $x \in]a, b[$.

Proof. For fixed $x \in]a, b[$, write

$$\frac{1}{h}\int_0^h |f(x+t)+f(x-t)-2f(x)|\,dt$$

$$\leqq \frac{1}{h}\int_x^{x+h} |f(t)-f(x)|\,dt + \frac{1}{h}\int_{x-h}^{x} |f(t)-f(x)|\,dt\,.$$

Applying (18.4) with $\alpha = f(x)$, we see that (i) holds for almost all $x \in]a, b[$. □

(18.6) Definition. Suppose that $f \in \mathfrak{L}_1([a, b])$ and that $x \in]a,b[$. Then x is called a *Lebesgue point for* f if (18.5.i) holds. The set of all Lebesgue points for f is called the *Lebesgue set for* f.

It is obvious that every point of $]a, b[$ at which f is continuous is a Lebesgue point for f. However f need not be continuous anywhere, and yet almost every point is a Lebesgue point. The Lebesgue set plays an important rôle in the theory of Fourier series and integrals, as we shall see in (18.29) and (21.43).

We now inquire into the "reverse" [not really a converse] of Theorem (18.3). Given a continuous function φ that is differentiable a.e., is it true that

$$\varphi(x) - \varphi(a) = \int_a^x \varphi'(t)\,dt?$$

That is, does the fundamental theorem of the calculus hold with the Riemann integral replaced by the Lebesgue integral and differentiability replaced by differentiability a.e.? The following examples answer this question with an emphatic "no".

(18.7) Example. As in (8.28), let ψ be LEBESGUE'S singular function. Then ψ is continuous on $[0, 1]$ and it is clear that $\psi'(x) = 0$ for all x in $[0, 1]$ that are not in P [CANTOR's ternary set]. Thus $\psi'(x) = 0$ a.e. It follows that

$$\psi(1) - \psi(0) = 1 \neq 0 = \int_0^1 \psi'(x)\,dx\,.$$

The next example seems not as geometrically obvious as (18.7), but it is much more dramatic.

(18.8) Example [Adapted from RIESZ-NAGY]. We will exhibit a real-valued function F on $[0, 1]$ such that $F(0) = 0$, $F(1) = 1$, F is continuous and *strictly* increasing, and $F'(x) = 0$ a.e. We define continuous functions F_n, $n = 0, 1, 2, \ldots$, inductively as follows. First, let $(t_n)_{n=1}^{\infty}$ be any sequence of numbers in $]0, 1[$. Let $F_0(x) = x$, and define $F_1(0) = 0$, $F_1(1) = 1$, and $F_1\left(\frac{1}{2}\right) = \frac{1 - t_1}{2} \cdot 0 + \frac{1 + t_1}{2} \cdot 1$; and define F_1 to be linear on $\left[0, \frac{1}{2}\right]$ and $\left[\frac{1}{2}, 1\right]$. Suppose that $F_0, F_1, \ldots, F_n$ have been defined. Then we define:

$$F_{n+1}\left(\frac{k}{2^n}\right) = F_n\left(\frac{k}{2^n}\right) \quad \text{for} \quad k = 0, 1, \ldots, 2^n;$$

$$F_{n+1}\left(\frac{2k+1}{2^{n+1}}\right) = \frac{1 - t_{n+1}}{2} F_n\left(\frac{k}{2^n}\right) + \frac{1 + t_{n+1}}{2} F_n\left(\frac{k+1}{2^n}\right)$$

for $k = 0, 1, \ldots, 2^n - 1$;

and define F_{n+1} to be linear in the intervals $\left[\frac{k}{2^{n+1}}, \frac{k+1}{2^{n+1}}\right]$ for $k = 0$, $1, \ldots, 2^{n+1} - 1$. The functions F_n are plainly continuous. They are also strictly increasing. Indeed, for $0 < t < 1$ and any α, β such that $\alpha < \beta$, the inequalities

$$\beta - \left(\frac{1-t}{2}\alpha + \frac{1+t}{2}\beta\right) = \frac{1-t}{2}(\beta - \alpha) > 0$$

and

$$\frac{1-t}{2}\alpha + \frac{1+t}{2}\beta - \alpha = \frac{1+t}{2}(\beta - \alpha) > 0$$

hold. These inequalities show that if F_n is strictly increasing, then

$$F_{n+1}\left(\frac{k}{2^n}\right) < F_{n+1}\left(\frac{2k+1}{2^{n+1}}\right) < F_{n+1}\left(\frac{k+1}{2^n}\right)$$

for $k = 0, 1, \ldots, 2^n - 1$. The piecewise linearity of F_{n+1} proves that it too is strictly increasing. Also, if $\alpha < \beta$, we have

$$\frac{1-t}{2}\alpha + \frac{1+t}{2}\beta - \frac{\alpha+\beta}{2} = \frac{t}{2}(\beta - \alpha) > 0,$$

and from this inequality it follows that

$$F_n\left(\frac{2k+1}{2^{n+1}}\right) < F_{n+1}\left(\frac{2k+1}{2^{n+1}}\right)$$

for $k = 0, 1, \ldots, 2^n - 1$. Hence, again by linearity, the inequality

$$F_n(x) \leqq F_{n+1}(x)$$

holds for all $x \in [0, 1]$. Thus the sequence $(F_n(x))_{n=1}^{\infty}$ converges for all $x \in [0, 1]$; let

$$F(x) = \lim_{n \to \infty} F_n(x).$$

It is clear that F is nondecreasing. Actually it is strictly increasing. For, if $x < x'$ and k and n are such that $x < \frac{k}{2^n} < x'$, then we have

$$F(x) \leqq F\left(\frac{k}{2^n}\right) = F_n\left(\frac{k}{2^n}\right) < F_n(x') \leqq F(x').$$

We next consider any sequence of pairs of numbers $(\alpha_n, \beta_n)_{n=0}^{\infty}$ satisfying the following conditions:

$$\alpha_n \leqq \alpha_{n+1} \quad \text{and} \quad \beta_{n+1} \leqq \beta_n \ (n = 0, 1, 2, \ldots); \tag{1}$$

$$\alpha_n = \frac{k_n}{2^n} \quad \text{and} \quad \beta_n = \frac{k_n+1}{2^n}, \tag{2}$$

where $k_n \in \{0, 1, \ldots, 2^n - 1\}$ $(n = 0, 1, 2, \ldots)$. Thus we have $\alpha_n = \alpha_{n+1}$ and $\beta_{n+1} = \beta_n - \frac{1}{2^{n+1}}$, or $\beta_{n+1} = \beta_n$ and $\alpha_{n+1} = \alpha_n + \frac{1}{2^{n+1}}$. In the first case, we go to the *left* in proceeding from (α_n, β_n) to $(\alpha_{n+1}, \beta_{n+1})$; in the second, we go to the *right*.

Let k be a fixed nonnegative integer. Suppose that we go to the right in going from (α_k, β_k) to $(\alpha_{k+1}, \beta_{k+1})$. Then we have

$$\begin{aligned} F(\beta_{k+1}) - F(\alpha_{k+1}) &= F_{k+1}(\beta_{k+1}) - F_{k+1}(\alpha_{k+1}) \\ &= F_k(\beta_k) - \left\{\frac{1-t_{k+1}}{2} F_k(\alpha_k) + \frac{1+t_{k+1}}{2} F_k(\beta_k)\right\} \\ &= \frac{1-t_{k+1}}{2}(F(\beta_k) - F(\alpha_k)) . \end{aligned} \tag{3}$$

If we go to the left in going from (α_k, β_k) to $(\alpha_{k+1}, \beta_{k+1})$, then a like computation shows that

$$F(\beta_{k+1}) - F(\alpha_{k+1}) = \frac{1+t_{k+1}}{2}(F(\beta_k) - F(\alpha_k)) . \tag{4}$$

A simple induction shows that

$$F(\beta_n) - F(\alpha_n) = \prod_{k=1}^{n} \left(\frac{1+\varepsilon_k t_k}{2}\right),$$

where the ε_k's are 1 or -1. Thus

$$|F(\beta_n) - F(\alpha_n)| \leqq \prod_{k=1}^{n} \left(\frac{1+t_k}{2}\right),$$

and if all t_k are less than a number less than 1, F is obviously continuous. [We will not bother with exploring necessary conditions for the continuity of F.]

We now look at the derivates of F. Consider any dyadic rational $\frac{l}{2^p}$ such that $0 \leqq \frac{l}{2^p} < 1$. Define a sequence (α_n, β_n) satisfying (1) and (2) for which $\alpha_p = \alpha_{p+1} = \cdots = \frac{l}{2^p}$: we care not what $\alpha_1, \beta_1, \ldots, \alpha_{p-1}, \beta_{p-1}$ are. Then we must have

$$\beta_{p+s} = \frac{l}{2^p} + \frac{1}{2^{p+s}} \quad \text{for} \quad s = 0, 1, 2, \ldots .$$

Applying (4), we see that

$$\begin{aligned} \frac{F(\beta_{p+s}) - F\left(\frac{l}{2^p}\right)}{\beta_{p+s} - \frac{l}{2^p}} &= 2^{p+s} \prod_{j=1}^{s} \frac{1}{2}(1+t_{p+j})(F(\beta_p) - F(\alpha_p)) \\ &= 2^p (F(\beta_p) - F(\alpha_p)) \prod_{j=1}^{s} (1+t_{p+j}) . \end{aligned}$$

If the series $\sum_{n=1}^{\infty} t_n$ diverges, it follows that $D^+F\left(\frac{l}{2^p}\right) = \infty$ $\left[\log(1+t) \geqq \frac{t}{2}\right.$ for $\left.0 \leqq t \leqq 1\right]$. Similarly we have $D_-F\left(\frac{l}{2^p}\right) = 0$ if $0 < \frac{l}{2^p} \leqq 1$ and $\prod_{n=1}^{\infty}(1-t_n) = 0$. For these two results it is sufficient that $\overline{\lim_{n\to\infty}}\, t_n$ be positive. Consider next a point $x = \sum_{k=1}^{\infty} \frac{x_k}{2^k}$, where $x_k = 0$ or $x_k = 1$, and each value is assumed for infinitely many k's, *i.e.*, x is not a dyadic rational. For each n, there is a unique l such that $\frac{l}{2^n} < x < \frac{l+1}{2^n}$; let $\alpha_n = \frac{l}{2^n}$ and $\beta_n = \frac{l+1}{2^n}$. In fact these numbers are given by

$$\alpha_n = \frac{l}{2^n} = \frac{x_1}{2} + \cdots + \frac{x_n}{2^n} < x < \frac{x_1}{2} + \cdots + \frac{x_n}{2^n} + \frac{1}{2^n} = \frac{l+1}{2^n} = \beta_n .$$

We will compute $\dfrac{F(\beta_n) - F(\alpha_n)}{\frac{1}{2^n}}$. If $x_n = 0$, then $\alpha_n = \alpha_{n-1}$ and

$$\beta_n = \alpha_{n-1} + \frac{1}{2^n} = \frac{\alpha_{n-1} + \beta_{n-1}}{2} .$$

If $x_n = 1$, then $\beta_n = \beta_{n-1}$, and

$$\alpha_n = \alpha_{n-1} + \frac{1}{2^n} = \frac{\alpha_{n-1} + \beta_{n-1}}{2} .$$

In the first case we have

$$\frac{F(\beta_n) - F(\alpha_n)}{\frac{1}{2^n}} = 2^n \left\{ \frac{1 - t_n}{2} F_{n-1}(\alpha_{n-1}) + \frac{1 + t_n}{2} F_{n-1}(\beta_{n-1}) - F_{n-1}(\alpha_{n-1}) \right\}$$

$$= 2^n \left\{ \frac{1 + t_n}{2} \cdot (F_{n-1}(\beta_{n-1}) - F_{n-1}(\alpha_{n-1})) \right\} .$$

In the second case the factor $\frac{1 + t_n}{2}$ is replaced by $\frac{1 - t_n}{2}$ in the preceding line, and so we obtain

$$\frac{F(\beta_n) - F(\alpha_n)}{\frac{1}{2^n}} = 2^n \frac{(1 + (-1)^{x_n} t_n)}{2} \cdot (F_{n-1}(\beta_{n-1}) - F_{n-1}(\alpha_{n-1}))$$

$$= \cdots = \prod_{k=1}^{n} (1 + (-1)^{x_k} t_k) \tag{5}$$

for all n. We know that the function F has a finite derivative a.e., and hence the limit of the product in (5) exists, is finite, and is equal to $F'(x)$ for almost all x. The ratio $\dfrac{\prod_{k=1}^{n+1} (1 + (-1)^{x_k} t_k)}{\prod_{k=1}^{n} (1 + (-1)^{x_k} t_k)}$ is $1 \pm t_{n+1}$, and hence it converges to 1 if and only if $\lim_{n\to\infty} t_n = 0$. Thus if $\overline{\lim_{n\to\infty}}\, t_n > 0$, the product $\prod_{k=1}^{n} (1 + (-1)^{x_k} t_k)$ cannot converge to a positive finite number, and so $F'(x) = 0$ for almost all x.

We summarize. Given a sequence $(t_n)_{n=1}^{\infty}$ with values in $]0, 1[$, we have constructed a real-valued function F on $[0, 1]$ having the following properties:

(i) $F(0) = 0$, $F(1) = 1$, F is strictly increasing;

(ii) if $\overline{\lim_{n\to\infty}}\, t_n < 1$, then F is continuous;

(iii) if $\overline{\lim_{n\to\infty}}\, t_n > 0$ and x is a dyadic rational in $]0, 1[$, then $D^+F(x) = \infty$ and $D_-F(x) = 0$;

(iv) if $\overline{\lim_{n\to\infty}}\, t_n > 0$, then $F'(x) = 0$ for almost all $x \in]0, 1[$.

Thus if $0 < \overline{\lim_{n\to\infty}}\, t_n < 1$, then F is continuous, strictly increasing, and $F' = 0$ a.e. The reader should sketch the first few approximants F_n to F for a special choice of (t_n), say $t_n = \frac{1}{2}$ for all n, to see what is going on.

(18.9) Note. The construction in (18.8) proves also a curious measure-theoretic fact. If $(t_n)_{n=1}^{\infty}$ is a sequence of numbers in $]0, 1[$ not having limit 0, then

$$\prod_{k=1}^{\infty} (1 + (-1)^{x_k} t_k) = 0$$

for almost all numbers $x = \sum_{k=1}^{\infty} \frac{x_k}{2^k}$.

We now identify the class of functions that are indefinite integrals of functions in $\mathfrak{L}_1$.

(18.10) Definition. Let f be a complex-valued function defined on a subinterval J of R.[1] Suppose that for every $\varepsilon > 0$, there is a $\delta > 0$ such that

(i) $\sum_{k=1}^{n} |f(d_k) - f(c_k)| < \varepsilon$

for every finite, pairwise disjoint, family $\{]c_k, d_k[\}_{k=1}^{n}$ of open subintervals of J for which

(ii) $\sum_{k=1}^{n} (d_k - c_k) < \delta$.

Then f is said to be *absolutely continuous on* J.

(18.11) Examples. (a) Theorem (12.34) shows that the indefinite integral of a function in $\mathfrak{L}_1([a, b])$ is absolutely continuous. Our next project will be to prove that every absolutely continuous function is an indefinite integral.

(b) LEBESGUE's singular function ψ is *not* absolutely continuous. We can enclose CANTOR's ternary set P in a union $\bigcup_{k=1}^{\infty}]a_k, b_k[$ of pairwise disjoint open intervals such that $\sum_{k=1}^{\infty} (b_k - a_k)$ is arbitrarily small. Extend ψ so that $\psi(x) = 0$ for $x < 0$ and $\psi(x) = 1$ for $x > 1$. Then it is easy to

[1] Recall that by (6.1) J can be open, closed, or half-open, and that J can be bounded or unbounded.

see that $\sum_{k=1}^{\infty}(\psi(b_k) - \psi(a_k)) = 1$, and so $\sum_{k=1}^{n}(\psi(b_k) - \psi(a_k)) \geqq \frac{1}{2}$ for sufficiently large n, while $\sum_{k=1}^{n}(b_k - a_k)$ is arbitrarily small.

(c) None of the functions F of (18.8) is absolutely continuous. This is most easily seen from Theorem (18.15) *infra.*

We first set down some elementary properties of absolutely continuous functions.

(18.12) Theorem. *Any complex-valued absolutely continuous function f defined on $[a, b]$ has finite variation on $[a, b]$.*

Proof. Let $\delta > 0$ satisfy the conditions of Definition (18.10) for $\varepsilon = 1$. Let n be any integer such that $n > \frac{b-a}{\delta}$, and subdivide $[a, b]$ by points $a = x_0 < x_1 < \cdots < x_n = b$ such that $x_k - x_{k-1} = \frac{b-a}{n} < \delta$ for $k = 1, 2, \ldots, n$. From our choice of δ it follows that $V_{x_{k-1}}^{x_k} f \leqq 1$ for all k. Thus $V_a^b f = \sum_{k=1}^{n} V_{x_{k-1}}^{x_k} f \leqq n$. □

(18.13) Theorem. *Any absolutely continuous function f on $[a, b]$ is continuous, and can be written as*

(i) $f = f_1 - f_2 + i(f_3 - f_4)$,

where the f_j are real, nondecreasing, and absolutely continuous on $[a, b]$.

Proof. If f is absolutely continuous, then the continuity of f and the absolute continuity of $\operatorname{Im} f$ and $\operatorname{Re} f$ are obvious. For a real-valued, absolutely continuous function g, write $g_1(x) = V_a^x g$. Then $g = g_1 - (g_1 - g)$, and the proof will be complete upon showing that g_1 is absolutely continuous. [Note that g_1 and $g_1 - g$ are nondecreasing (17.16).] For an arbitrary $\varepsilon > 0$, let $\delta > 0$ be so small that $\sum_{k=1}^{n} |g(d_k) - g(c_k)| < \frac{\varepsilon}{2}$ whenever the pairwise disjoint intervals $]c_k, d_k[$ are such that

$$\sum_{k=1}^{n} (d_k - c_k) < \delta . \tag{1}$$

Let $\{]c_k, d_k[\}_{k=1}^{n}$ be a fixed system of pairwise disjoint intervals satisfying (1). Since g has finite variation, there is for each $k \in \{1, 2, \ldots, n\}$ a subdivision $c_k = a_0^{(k)} < a_1^{(k)} < \cdots < a_{l_k}^{(k)} = d_k$ such that

$$V_{c_k}^{d_k} g < \sum_{j=0}^{l_k - 1} |g(a_{j+1}^{(k)}) - g(a_j^{(k)})| + \frac{\varepsilon}{2n} .$$

Hence we have

$$\sum_{k=1}^{n} |g_1(d_k) - g_1(c_k)| = \sum_{k=1}^{n} V_{c_k}^{d_k} g < \sum_{k=1}^{n} \sum_{j=0}^{l_k - 1} |g(a_{j+1}^{(k)}) - g(a_j^{(k)})| + \frac{\varepsilon}{2}$$
$$< \frac{\varepsilon}{2} + \frac{\varepsilon}{2} = \varepsilon ;$$

and so g_1 is absolutely continuous. □

(18.14) Theorem. *If f is a real-valued, nondecreasing function on $[a, b]$, then f' is Lebesgue measurable and*

(i) $\int_a^b f'(x)\,dx \leqq f(b) - f(a)$.

If g is a complex-valued function of finite variation on $[a, b]$, then $g' \in \mathfrak{L}_1([a, b])$.

Proof. For $x > b$, let $f(x) = f(b)$. Let

$$f_n(x) = \frac{f\left(x + \frac{1}{n}\right) - f(x)}{\frac{1}{n}}$$

for $n = 1, 2, 3, \ldots$ and $a \leqq x \leqq b$. Then (f_n) is a sequence of nonnegative measurable functions and

$$\lim_{n\to\infty} f_n(x) = f'(x)$$

for almost all $x \in]a, b[$ [see (17.12)]; thus f' is measurable. By FATOU's lemma (12.23) and (12.44) we have

$$\begin{aligned}
\int_a^b f'(x)\,dx &= \int_a^b \lim_{n\to\infty} f_n(x)\,dx \leqq \varliminf_{n\to\infty} \int_a^b f_n(x)\,dx \\
&= \varliminf_{n\to\infty} n \int_a^b \left[f\left(x + \frac{1}{n}\right) - f(x)\right] dx \\
&= \varliminf_{n\to\infty} \left[n \int_b^{b+\frac{1}{n}} f(x)\,dx - n \int_a^{a+\frac{1}{n}} f(x)\,dx\right] \\
&\leqq \varliminf_{n\to\infty} \left[n \int_b^{b+\frac{1}{n}} f(b)\,dx - n \int_a^{a+\frac{1}{n}} f(a)\,dx\right] \\
&= f(b) - f(a)\,.
\end{aligned}$$

This proves our first assertion. The second assertion plainly follows from the first and the fact that g can be expressed as a linear combination of four nondecreasing functions. □

(18.15) Theorem. *Let f be an absolutely continuous complex-valued function on $[a, b]$ and suppose that $f'(x) = 0$ a.e. in $]a, b[$. Then f is a constant.*

Proof. We lose no generality by supposing that f is real-valued, for otherwise we examine $\text{Re} f$ and $\text{Im} f$ separately. We will show that $f(c) = f(a)$ for all $c \in]a, b]$. Thus let $c \in]a, b]$ and $\varepsilon > 0$ be arbitrary. Select a number $\delta > 0$ corresponding to the given ε for which the condition in the definition of absolute continuity (18.10) is satisfied. Let

$E = \{x \in]a, c[: f'(x) = 0\}$. Clearly $\lambda(E) = c - a$. For each $x \in E$ there exist arbitrarily small $h > 0$ such that $[x, x+h] \subset]a, c[$ and

$$|f(x+h) - f(x)| < \frac{\varepsilon h}{c-a}. \tag{1}$$

The family of all such intervals $[x, x+h]$ is a Vitali cover of E, and so by VITALI's theorem (17.11) there exists a finite pairwise disjoint family $\{[x_k, x_k + h_k]\}_{k=1}^n$ of these intervals such that

$$\lambda\left(E \cap \left(\bigcup_{k=1}^n [x_k, x_k + h_k]\right)'\right) < \delta.$$

Then

$$\lambda(]a, c[) = \lambda(E) < \delta + \sum_{k=1}^n h_k. \tag{2}$$

We may [and do] suppose that $x_1 < x_2 < \cdots < x_n$. It follows from (2) that the sum of the lengths of the open intervals

$$]a, x_1[,]x_1 + h_1, x_2[, \ldots,]x_n + h_n, c[$$

complementary to $\bigcup_{k=1}^n [x_k, x_k + h_k]$ is less than δ, and so, in view of our choice of δ, we have

$$|f(a) - f(x_1)| + \sum_{k=1}^{n-1} |f(x_k + h_k) - f(x_{k+1})| + |f(x_n + h_n) - f(c)| < \varepsilon. \tag{3}$$

The inequalities (1) and (3) combine to yield

$$\begin{aligned} |f(a) - f(c)| &\leqq |f(a) - f(x_1)| + \sum_{k=1}^{n-1} |f(x_k + h_k) - f(x_{k+1})| \\ &\quad + |f(x_n + h_n) - f(c)| + \sum_{k=1}^{n} |f(x_k + h_k) - f(x_k)| \\ &< \varepsilon + \sum_{k=1}^{n} \frac{\varepsilon h_k}{c-a} \leqq 2\varepsilon. \end{aligned}$$

Since ε is arbitrary, it follows that $f(c) = f(a)$. □

(18.16) Theorem [Fundamental theorem of the integral calculus for Lebesgue integrals]. *Let f be a complex-valued, absolutely continuous function on $[a, b]$. Then $f' \in \mathfrak{L}_1([a, b])$ and*

(i) $f(x) = f(a) + \int_a^x f'(t)\,dt$

for every $x \in [a, b]$.

Proof. From (18.12) and (18.14) it follows that $f' \in \mathfrak{L}_1$. Let $g(x) = \int_a^x f'(t)\,dt$. Then g is absolutely continuous and, by (18.3), $g'(x) = f'(x)$ a.e. Thus the function $h = f - g$ is absolutely continuous and $h'(x) = f'(x) - g'(x) = 0$ a.e. It follows from (18.15) that h is a constant.

Therefore

$$f(x) = h(x) + g(x) = h(a) + \int_a^x f'(t)\,dt = f(a) + \int_a^x f'(t)\,dt$$

for all $x \in [a, b]$. □

(18.17) Theorem. *A function f on $[a, b]$ has the form*

$$f(x) = f(a) + \int_a^x \varphi(t)\,dt$$

for some $\varphi \in \mathfrak{L}_1([a, b])$ if and only if f is absolutely continuous on $[a, b]$. In this case we have $\varphi(x) = f'(x)$ a.e. on $]a, b[$.

Proof. This is just a summary of (18.3), (18.11.a), and (18.16). □

Indefinite integrals on R can be characterized in much the same way.

(18.18) Theorem. *A function f on R has the form*

(i) $f(x) = \int_{-\infty}^x \varphi(t)\,dt$

for some $\varphi \in \mathfrak{L}_1(R)$ if and only if f is absolutely continuous on $[-A, A]$ for all $A > 0$, $V_{-\infty}^{\infty} f$ is finite, and $\lim_{x \to -\infty} f(x) = 0$.

Proof. Suppose that f has the form (i). Then f is absolutely continuous on $[-A, A]$ by (12.34), and by (18.1) we have

$$V_{-\infty}^{\infty} f = \int_R |\varphi(t)|\,dt < \infty .$$

The dominated convergence theorem (12.30) implies that $\lim_{x \to -\infty} f(x) = 0$. Conversely, if f is absolutely continuous on $[-A, A]$ for all $A > 0$, (18.17) shows that

$$f(x) = f(-A) + \int_{-A}^x f'(t)\,dt$$

for all positive real A and x such that $x > -A$. Taking the limit as $A \to \infty$, we have

$$f(x) = \lim_{A \to \infty} \int_{-A}^x f'(t)\,dt . \tag{1}$$

Applying (12.22), (18.17), and (18.1), we have

$$\int_{-\infty}^{\infty} |f'(t)|\,dt = \lim_{n \to \infty} \int_{-n}^{n} |f'(t)|\,dt$$
$$= \lim_{n \to \infty} V_{-n}^{n} f \leq V_{-\infty}^{\infty} f < \infty .$$

Thus f' is in $\mathfrak{L}_1(R)$, and the dominated convergence theorem applied to the right side of (1) shows that $f(x) = \int_{-\infty}^x f'(t)\,dt$. □

The formula for integration by parts holds for absolutely continuous functions and Lebesgue integrals.

(18.19) Theorem. *Let f, g be functions in $\mathfrak{L}_1([a, b])$, let*

$$F(x) = \alpha + \int_a^x f(t)\, dt\,,$$

and let

$$G(x) = \beta + \int_a^x g(t)\, dt\,.$$

Then

(i) $\int_a^b G(t) f(t)\, dt + \int_a^b g(t) F(t)\, dt = F(b)\, G(b) - F(a)\, G(a)\,.$

Proof. The inequality

$$|F(v) G(v) - F(u) G(u)| \leqq \|F\|_u |G(v) - G(u)| + \|G\|_u |F(v) - F(u)|$$

shows that FG is absolutely continuous. Hence FG is differentiable a.e., and

$$(FG)' = FG' + F'G\,,$$

as an elementary calculation shows. By (18.3) we have $G' = g$ a.e. and $F' = f$ a.e.; thus (i) follows from (18.16). □

(18.20) Corollary. *Let f and g be absolutely continuous functions on $[a, b]$. Then*

(i) $\int_a^b f(t)\, g'(t)\, dt + \int_a^b f'(t)\, g(t)\, dt = f(b)\, g(b) - f(a)\, g(a)\,.$

Proof. This is just (18.19) rewritten with the aid of (18.17). □

(18.21) Corollary. *Let f and g be functions on R satisfying the conditions of* (18.18). *Then*

(i) $$\int_R f(t)\, g'(t)\, dt + \int_R f'(t)\, g(t)\, dt = \int_R f'(t)\, dt \cdot \int_R g'(t)\, dt = \lim_{x\to\infty} f(x) \cdot \lim_{x\to\infty} g(x)\,.$$

Proof. Take limits as $a \to -\infty$ and $b \to \infty$ in (18.20.i). □

(18.22) Note. Another famous integral formula is the elementary formula for integration by substitution:

(i) $\int_\alpha^\beta f(y)\, dy = \int_a^b f(\varphi(x))\, \varphi'(x)\, dx\,,$

where f is Riemann integrable and φ is a function with positive continuous derivative on $[\varphi^{-1}(\alpha), \varphi^{-1}(\beta)] = [a, b]$. A much more general formula is in fact true, which subsumes (i) as a very special case. The proof seems most easily carried out by using the LEBESGUE-RADON-NIKODÝM theorem (19.24), and we postpone it to (20.4). We continue here with some technical facts about absolutely continuous functions needed in (20.4) and (20.5).

(18.23) Theorem. *Let φ be a complex-valued, absolutely continuous function on $[a, b]$. For $[x, y] \subset [a, b]$, let*

$$\omega_\varphi(x, y) = \sup\{|\varphi(u) - \varphi(v)| : u, v \in [x, y]\}.$$

Then for every $\varepsilon > 0$, there is a $\delta > 0$ such that

(i) $\sum_{k=1}^{n} \omega_\varphi(c_k, d_k) < \varepsilon$

for every finite, pairwise disjoint, family $\{]c_k, d_k[\}_{k=1}^n$ of open subintervals of $[a, b]$ for which

(ii) $\sum_{k=1}^{n} (d_k - c_k) < \delta$.[1]

Proof. The function φ being continuous and the interval $[c_k, d_k]$ being compact, it is easy to see that $[c_k, d_k]$ contains points u_k and v_k such that $u_k < v_k$ and $|\varphi(u_k) - \varphi(v_k)| = \omega_\varphi(c_k, d_k)$. Since

$$\sum_{k=1}^{n} (v_k - u_k) \leqq \sum_{k=1}^{n} (d_k - c_k) < \delta,$$

we obtain (i) at once from (18.10.i). □

For Theorem (20.4), we will need another definition.

(18.24) Definition. Let g be a function with domain $[a, b] \subset R$ and range $[\alpha, \beta] \subset R$. If $\lambda(E) = 0$ implies $\lambda(g(E)) = 0$ for all $E \subset [a, b]$, then g is said to be an *N-function* or to *satisfy the condition N*.[2]

(18.25) Theorem. *Let φ be a continuous function of finite variation with domain $[a, b] \subset R$ and range $[\alpha, \beta] \subset R$. Then φ is an N-function if and only if φ is absolutely continuous.*[3]

Proof. Suppose that φ is absolutely continuous and that $\lambda(E) = 0$. Let ε be an arbitrary positive number and let δ be as in (18.23). Since $\lambda(\varphi(\{a, b\}))$ is trivially zero, we may suppose that $E \subset]a, b[$. We choose a family $\{]c_k, d_k[\}_{k=1}^\infty$ of pairwise disjoint open subintervals of $]a, b[$ such that $E \subset \bigcup_{k=1}^{\infty}]c_k, d_k[$ and $\sum_{k=1}^{\infty} (d_k - c_k) < \delta$. By (18.23), we have

$$\sum_{k=1}^{n} \omega_\varphi(c_k, d_k) < \varepsilon \quad \text{for all } n$$

and so

$$\sum_{k=1}^{\infty} \omega_\varphi(c_k, d_k) \leqq \varepsilon. \tag{1}$$

---

[1] Thus we can replace the condition (18.10.i) by the apparently stronger condition (18.23.i) in the definition of absolute continuity.

[2] This terminology, and the concept itself, are due to N. N. Luzin (1915); he thought of the property as a "null condition".

[3] This theorem is due to Banach [Fund. Math. 7, 225—236 (1925)]; we give his original proof.

Plainly we have

$$\varphi(E) \subset \varphi\left(\bigcup_{k=1}^{\infty}]c_k, d_k[\right) = \bigcup_{k=1}^{\infty} \varphi(]c_k, d_k[) . \tag{2}$$

It is also evident that

$$\lambda(\varphi(]c_k, d_k[)) = \lambda(\varphi([c_k, d_k])) = \omega_\varphi(c_k, d_k) .$$

Hence (2) and (1) imply that $\lambda(\varphi(E)) \leqq \varepsilon$. Since ε is arbitrary, $\lambda(\varphi(E))=0$ and φ is an N-function.

The converse is less obvious. Suppose that φ is an N-function and assume that φ is not absolutely continuous. By (18.23), there is a positive number ε_0 such that we can find a sequence

$$\{]c_1^{(1)}, d_1^{(1)}[, \ldots,]c_{l_1}^{(1)}, d_{l_1}^{(1)}[\} = \mathscr{D}_1,$$

$$\{]c_1^{(2)}, d_1^{(2)}[, \ldots,]c_{l_2}^{(2)}, d_{l_2}^{(2)}[\} = \mathscr{D}_2,$$

$$\ldots$$

$$\{]c_1^{(n)}, d_1^{(n)}[, \ldots,]c_{l_n}^{(n)}, d_{l_n}^{(n)}[\} = \mathscr{D}_n,$$

$$\ldots,$$

with the following properties. First, the intervals comprising each $\mathscr{D}_n$ are pairwise disjoint. Second, the inequalities

$$\sum_{k=1}^{l_n} \omega_\varphi(c_k^{(n)}, d_k^{(n)}) \geqq \varepsilon_0 \tag{3}$$

hold for all n. Third,

$$\sum_{n=1}^{\infty} \sum_{k=1}^{l_n} (d_k^{(n)} - c_k^{(n)}) < \infty . \tag{4}$$

For each n, and for $y \in [\alpha, \beta]$, let $N_n(y)$ be the *number* of intervals $]c_k^{(n)}, d_k^{(n)}[$ in $\mathscr{D}_n$ that have nonvoid intersection with $\varphi^{-1}(\{y\})$. As the intervals $]c_k^{(n)}, d_k^{(n)}[$ are pairwise disjoint, it is evident that

$$N_n(y) \leqq \nu(y) ; \tag{5}$$

here ν is the Banach indicatrix of φ, defined in (17.34). It is obvious that

$$N_n = \sum_{k=1}^{l_n} \xi_{\varphi(]c_k^{(n)}, d_k^{(n)}[)} .$$

Since $\varphi([c_k^{(n)}, d_k^{(n)}])$ is a closed interval [whose measure is $\omega_\varphi(c_k^{(n)}, d_k^{(n)})$] and since $(\varphi(]c_k^{(n)}, d_k^{(n)}[))' \cap \varphi([c_k^{(n)}, d_k^{(n)}])$ contains at most two points, N_n is Borel measurable [actually the pointwise limit of a sequence of continuous functions] and

$$\int_R N_n(y)\, dy = \sum_{k=1}^{l_n} \omega_\varphi(c_k^{(n)}, d_k^{(n)}) \geqq \varepsilon_0 . \tag{6}$$

Let A be the set $\{y : y \in [\alpha, \beta], \overline{\lim_{n\to\infty}}\, N_n(y) \neq 0\}$. Let $A_1 = \{y \in A : \nu(y) = \infty\}$. Since φ has finite variation by hypothesis, ν is in $\mathfrak{L}_1([\alpha, \beta])$ and so

$\lambda(A_1) = 0$. Consider any point $y_0 \in A \cap A_1'$. There exists a sequence $(x_{n_j})_{j=1}^{\infty}$ $(n_1 < n_2 < \cdots)$ of points in $[a, b]$ such that

$$x_{n_j} \in \bigcup_{k=1}^{l_{n_j}}]c_k^{(n_j)}, d_k^{(n_j)}[\quad \text{and} \quad \varphi(x_{n_j}) = y_0$$

$(j = 1, 2, \ldots)$. Since $y_0 \notin A_1$, only a finite number of the points x_{n_j} are distinct, and so there is an $x_0 \in [a, b]$ in infinitely many of the sets $\bigcup_{k=1}^{l_n}]c_k^{(n)}, d_k^{(n)}[$ such that $\varphi(x_0) = y_0$. Write

$$E = \bigcap_{j=1}^{\infty} \bigcup_{n=j}^{\infty} \left(\bigcup_{k=1}^{l_n}]c_k^{(n)}, d_k^{(n)}[\right).$$

It is obvious that $x_0 \in E$. From (4) and (10.15) we infer that $\lambda(E) = 0$. Since φ is an N-function, we have $\lambda(\varphi(E)) = 0$. We have proved above that $\varphi(E) \supset A \cap A_1'$, so that

$$\lambda(A) = \lambda(A \cap A_1') = 0. \tag{7}$$

The definition of A and (7) show that $\lim_{n\to\infty} N_n(y) = 0$ for almost all y in $[\alpha, \beta]$. As $N_n \leq \nu$ and $\nu \in \mathfrak{L}_1([\alpha, \beta])$, we infer from LEBESGUE's dominated convergence theorem (12.24) that

$$\lim_{n\to\infty} \int_R N_n(y)\, dy = 0. \tag{8}$$

Since (6) and (8) contradict each other, the proof is complete. □

The hypothesis in (18.25) that φ have finite variation is essential, as the following example shows.

(18.26) Example. Consider any interval $[a, b]$ and a perfect nowhere dense subset F of $[a, b]$ that contains both a and b. The measure $\lambda(F)$ may be zero or positive. Write the open set $[a, b] \cap F'$ as $\bigcup_{n=1}^{\infty}]a_n, b_n[$, where the intervals $]a_n, b_n[$ are pairwise disjoint and are enumerated in an arbitrary order. Let $c_n = \frac{1}{2}(a_n + b_n)$ and let $(t_n)_{n=1}^{\infty}$ be a sequence of positive numbers with limit zero. Define a function g on $[a, b]$ as follows:

$g(x) = 0$ for all $x \in F$;
$g(c_n) = t_n$ $(n = 1, 2, \ldots)$;
g is linear in $[a_n, c_n]$ and in $[c_n, b_n]$
$(n = 1, 2, \ldots)$.

It is easy to see that g is continuous. Also it is easy to see that $V_a^b g = 2 \sum_{k=1}^{\infty} t_k$. We leave both proofs to the reader. To see that g is an N-function, consider any set $E \subset [a, b]$ such that $\lambda(E) = 0$. Using (2.15.i), we have

$$g(E) = g(E \cap F) \cup \bigcup_{k=1}^{\infty} g(E \cap]a_k, b_k[).$$

Since g is linear on $[a_k, c_k]$ and on $[c_k, b_k]$, it is plain that

$$\lambda(g(E \cap]a_k, b_k[)) = 0\,,$$

and so

$$\lambda(g(E)) \leqq \lambda(\{0\}) + \sum_{k=1}^{\infty} \lambda(g(E \cap]a_k, b_k[)) = 0\,.$$

If $\sum_{k=1}^{\infty} t_k = \infty$, then g certainly fails to be absolutely continuous, since it has infinite variation.

(18.27) Discussion. We close this section by giving a famous application of (18.5) [which in fact led LEBESGUE to the definition (18.6) of the Lebesgue set]. Consider a function $f \in \mathfrak{L}_1([-\pi, \pi])$ and its Fourier coefficients $\hat{f}(n)$ (16.33). The uniqueness theorem (16.34) tells us that f is determined [as an element of $\mathfrak{L}_1([-\pi, \pi])$, of course] by the function $\hat{f}$ defined on Z. This theorem leaves untouched the problem of reconstructing f from $\hat{f}$. This problem is important not only for its own sake but also for applications to physics, chemistry, and engineering, since many data obtained by spectroscopy, X-ray analysis, and the like, are nothing other than Fourier coefficients of functions which one wishes to determine. The simplest way to try to recapture f from $\hat{f}$ is by means of the *Fourier series of f*, the partial sums of which are defined as

$$\text{(i)} \quad s_n f(x) = \sum_{k=-n}^{n} \hat{f}(k) \exp(ikx) \qquad (n = 0, 1, 2, \ldots)\,.$$

In order to rewrite (i) and other expressions to be defined shortly, we define f on the entire line R by periodicity: $f(x + 2k\pi) = f(x)$, for $x \in [-\pi, \pi[$ and $k \in Z$ [the number $f(\pi)$ has no importance for $f \in \mathfrak{L}_1([-\pi, \pi])$].

We then have:

$$\text{(ii)} \quad s_n f(x) = \sum_{k=-n}^{n} \frac{1}{2\pi} \int_{-\pi}^{\pi} f(t) \exp(-ikt)\, dt \cdot \exp(ikx)$$

$$= \frac{1}{2\pi} \int_{-\pi}^{\pi} f(t) \left[\sum_{k=-n}^{n} \exp(ik(x-t))\right] dt$$

$$= \frac{1}{2\pi} \int_{-\pi}^{\pi} f(x-t) \left[\sum_{k=-n}^{n} \exp(ikt)\right] dt\,.$$

[The reader should check the last equality in (ii).] It is elementary to show that

$$\text{(iii)} \quad \sum_{k=-n}^{n} \exp(ikt) = \begin{cases} \dfrac{\sin\left(\left(n+\frac{1}{2}\right)t\right)}{\sin\left(\frac{1}{2}t\right)} & \text{if} \quad \exp(it) \neq 1\,, \\ 2n+1 & \text{if} \quad \exp(it) = 1\,. \end{cases}$$

The function defined by (iii) is called the *Dirichlet kernel* and is denoted by $D_n(t)$. Thus we may write

$$\text{(iv)}\quad s_n f(x) = \frac{1}{2\pi} \int_{-\pi}^{\pi} f(x-t)\, D_n(t)\, dt\,.$$

For many functions, the sequence $s_n f$ does in fact converge to f.[1] For others it does not. To reconstruct f from $\hat{f}$, we follow FEJÉR[2] in taking the *arithmetic means* of the partial sums (i). Accordingly we define

$$\begin{aligned}\text{(v)}\quad \sigma_n f(x) &= \frac{1}{n+1}\left[s_0 f(x) + s_1 f(x) + \cdots + s_n f(x)\right]\\ &= \sum_{k=-n}^{n} \left(1 - \frac{|k|}{n+1}\right) \hat{f}(k) \exp(ikx)\,.\end{aligned}$$

Using (iv), we write

$$\text{(vi)}\quad \sigma_n f(x) = \frac{1}{2\pi} \int_{-\pi}^{\pi} f(x-t) \left[\frac{1}{n+1}(D_0(t) + D_1(t) + \cdots + D_n(t))\right] dt\,.$$

The expression $[\cdots]$ in (vi) is called the *Fejér kernel*; it is denoted by $K_n(t)$; and one easily proves that

$$\text{(vii)}\quad K_n(t) = \begin{cases} \dfrac{1}{n+1}\left[\dfrac{\sin\left(\frac{1}{2}(n+1)t\right)}{\sin\left(\frac{1}{2}t\right)}\right]^2 & \text{if } \sin\left(\frac{1}{2}t\right) \neq 0\,,\\[2ex] n+1 & \text{if } \sin\left(\frac{1}{2}t\right) = 0\,.\end{cases}$$

The reader can easily verify the following:

$$\text{(viii)}\quad K_n(-t) = K_n(t)\,;$$

$$\text{(ix)}\quad 0 \leqq K_n(t) \leqq n+1\,;$$

$$\text{(x)}\quad \frac{1}{2\pi}\int_{-\pi}^{\pi} K_n(t)\, dt = 1\,;$$

since $\sin(\theta) > \frac{2}{\pi}\theta$ for $0 < \theta < \frac{\pi}{2}$,

$$\text{(xi)}\quad K_n(t) \leqq \frac{\pi^2}{(n+1)\, t^2} \quad \text{for} \quad 0 < |t| \leqq \pi\,.$$

It follows trivially from (xi) that

$$\text{(xii)}\quad \lim_{n\to\infty} \int_{\delta}^{\pi} K_n(t)\, dt = 0 \ \text{ for } \delta \in\,]0, \pi[\,.$$

Our first inversion theorem is elementary.

---

[1] For all details of this fact, and indeed the whole theory of trigonometric series, the best guide is undoubtedly the classical work of ZYGMUND, *Trigonometric Series* [2 vols., Cambridge University Press, 1959].

[2] LEOPOLD FEJÉR (1880—1959) was a distinguished Hungarian mathematician.

(18.28) Theorem. *Let p be a real number such that $1 \leqq p < \infty$, and let f be a function in $\mathfrak{L}_p([-\pi, \pi])$. Then*

(i) $\lim_{n\to\infty} \|f - \sigma_n f\|_p = 0$.

Proof. We choose an auxiliary function g, which for $p > 1$ is an arbitrary function in $\mathfrak{L}_{p'}([-\pi, \pi])$ such that $\|g\|_{p'} \leqq 1$, and which for $p = 1$ is the function identically 1. We then write

$$\left|\frac{1}{2\pi}\int_{-\pi}^{\pi} (f(x) - \sigma_n f(x))\, g(x)\, dx\right|$$

$$= \left|\frac{1}{2\pi}\int_{-\pi}^{\pi} \left(f(x)\,\frac{1}{2\pi}\int_{-\pi}^{\pi} K_n(t)\, dt - \frac{1}{2\pi}\int_{-\pi}^{\pi} f(x-t)\, K_n(t)\, dt\right) g(x)\, dx\right|$$

$$\leqq \frac{1}{4\pi^2}\int_{-\pi}^{\pi}\int_{-\pi}^{\pi} |f(x) - f(x-t)|\, |g(x)|\, K_n(t)\, dt\, dx\,.^1 \tag{1}$$

We anticipate FUBINI's theorem (21.13) [which of course is proved without recourse to the present theorem] to reverse the order of integration in the last expression of (1). This produces

$$\frac{1}{4\pi^2}\int_{-\pi}^{\pi}\int_{-\pi}^{\pi} |f(x) - f(x-t)|\, |g(x)|\, dx\, K_n(t)\, dt\,. \tag{2}$$

Now use HÖLDER's inequality (13.4.ii) on the inner integral in (2):

$$\frac{1}{2\pi}\int_{-\pi}^{\pi} |f(x) - f(x-t)|\, |g(x)|\, dx \leqq \|f - f_{-t}\|_p \cdot \|g\|_{p'}$$

$$\leqq \|f - f_{-t}\|_p\,. \tag{3}$$

Going back to (1), we therefore have

$$\left|\frac{1}{2\pi}\int_{-\pi}^{\pi} (f(x) - \sigma_n f(x))\, g(x)\, dx\right| \leqq \frac{1}{2\pi}\int_{-\pi}^{\pi} K_n(t)\, \|f - f_{-t}\|_p dt$$

$$= \frac{1}{2\pi}\int_{|t|\leqq\delta} + \frac{1}{2\pi}\int_{|t|>\delta}$$

$$\leqq \sup\{\|f - f_{-t}\|_p : |t| \leqq \delta\}\, \frac{1}{2\pi}\int_{|t|\leqq\delta} K_n(t)\, dt$$

$$+ 2\|f\|_p \frac{1}{2\pi}\int_{|t|>\delta} K_n(t)\, dt\,. \tag{4}$$

By (13.24), the supremum in (4) is arbitrarily small if δ is sufficiently small. By (18.27.xii), the limit of the last expression as $n \to \infty$ is zero, no

[1] We will prove in § 21 that this iterated integral is well defined.

matter how small δ may be. That is, the first expression in (4) is arbitrarily small for n sufficiently large. This implies by (15.1) that

$$\lim_{n\to\infty} \|f - \sigma_n f\|_p = 0 \quad (p > 1)\,.$$

For $p = 1$, use (3) and repeat the argument with obvious changes. □ [1]

Our point in going through (18.28) was to lead up to the much subtler fact that $\sigma_n f$ converges to f not merely "in the mean" [*i.e.*, in the $\mathfrak{L}_p$ norm] but also *pointwise almost everywhere.*

(18.29) Theorem [LEBESGUE]. *Let f be a function in $\mathfrak{L}_1([-\pi, \pi])$. Then if x is in the Lebesgue set of f, we have*

(i) $\lim\limits_{n\to\infty} \sigma_n f(x) = f(x)\,.$

Proof. For brevity we write $f(x + t) + f(x - t) - 2f(x)$ as $\varphi(x, t)$. We also define

$$\Phi(x, t) = \int_0^t |\varphi(x, u)|\, du$$

and we write the number $\Phi(x, \pi)$ as a. Theorem (18.5) shows that $\frac{1}{t}\Phi(x, t) \to 0$ as $t \to 0$ [x is in the Lebesgue set of f!]. Consider any $\varepsilon > 0$ and choose $\alpha > 0$ so that $\left|\frac{1}{t}\Phi(x, t)\right| < \varepsilon$ for $|t| \leqq \alpha$. Next use (18.27.xi) to choose an integer $n_0 > \frac{1}{\alpha}$ so that

$$n \geqq n_0 \text{ and } \alpha \leqq t \leqq \pi \text{ imply } |K_n(t)| < \frac{\varepsilon}{a + 1}\,. \tag{1}$$

Note that

$$n \geqq n_0 \text{ implies } n\Phi\left(x, \frac{1}{n}\right) < \varepsilon\,. \tag{2}$$

It is easy to see that

$$\sigma_n f(x) - f(x) = \frac{1}{2\pi}\int_0^\pi \varphi(x, t)\, K_n(t)\, dt$$

and hence

$$\begin{aligned} 2\pi\, |\sigma_n f(x) - f(x)| &\leqq \int_0^\pi |\varphi(x, t)|\, K_n(t)\, dt \\ &= \int_0^{1/n} |\varphi(x, t)|\, K_n(t)\, dt + \int_{1/n}^{\alpha} |\varphi(x, t)|\, K_n(t)\, dt \\ &\quad + \int_{\alpha}^{\pi} |\varphi(x, t)|\, K_n(t)\, dt = S_1 + S_2 + S_3\,. \end{aligned} \tag{3}$$

---

[1] For $p > 1$, we even have $\|f - s_n f\|_p \to 0$. This is much harder to prove, and is typical of the intriguing and delicate results obtained in the theory of Fourier series. See ZYGMUND, *loc. cit.*, Chapter VII, Theorem (6.4).

Now suppose that $n \geqq n_0$. For S_3, (1) implies that

$$S_3 \leqq \int_{\alpha}^{\pi} |\varphi(x,t)| \frac{\varepsilon}{a+1} dt \leqq \frac{\varepsilon}{a+1} a < \varepsilon . \tag{4}$$

By (18.27.ix) and (2), we have

$$S_1 \leqq \int_0^{1/n} |\varphi(x,t)| (n+1) dt = (n+1) \Phi\left(\frac{1}{n}\right) \leqq 2n\Phi\left(\frac{1}{n}\right) < 2\varepsilon . \tag{5}$$

Using (18.27.xi), we have

$$S_2 = \int_{1/n}^{\alpha} |\varphi(x,t)| K_n(t) \, dt \leqq \int_{1/n}^{\alpha} |\varphi(x,t)| \frac{\pi^2}{n+1} \frac{1}{t^2} dt .$$

We now apply (18.19), with

$$g(t) = -2t^{-3} \text{ and } G(t) = n^2 + \int_{1/n} g(u) \, du = t^{-2}$$

This yields

$$\int_{1/n}^{\alpha} |\varphi(x,t)| t^{-2} dt = \Phi(x,\alpha) \alpha^{-2} - \Phi\left(x, \frac{1}{n}\right) n^2 + 2 \int_{1/n}^{\alpha} \Phi(x,t) t^{-3} dt .$$

Hence

$$\begin{aligned} S_2 &\leqq \frac{\pi^2}{n+1} \Phi(x,\alpha) \frac{1}{\alpha^2} + \frac{2\pi^2}{n+1} \int_{1/n}^{\alpha} \Phi(x,t) \, t^{-3} dt \\ &\leqq \frac{\pi^2}{n+1} \frac{\varepsilon}{\alpha} + \frac{2\pi^2}{n+1} \int_{1/n}^{\alpha} \Phi(x,t) \, t^{-3} dt \\ &< \pi^2 \varepsilon + S_4 . \end{aligned}$$

Finally we have

$$S_4 \leqq \frac{2\pi^2}{n+1} \int_{1/n}^{\alpha} \varepsilon t^{-2} \, dt = \frac{2\pi^2 \varepsilon}{n+1} \left[n - \frac{1}{\alpha} \right] < \frac{2\pi^2 \varepsilon}{n+1} \cdot n < 2\pi^2 \varepsilon$$

and hence

$$S_2 < \pi^2 \varepsilon + 2\pi^2 \varepsilon = 3\pi^2 \varepsilon . \tag{6}$$

The relations (3), (4), (5), and (6) show that $n \geqq n_0$ implies

$$2\pi |\sigma_n f(x) - f(x)| < 3\varepsilon + 3\pi^2 \varepsilon < 33\varepsilon. \quad \square$$

(18.30) Exercise. Let E be a subset of R [not *a priori* measurable] such that for some positive real number δ, the inequality $\lambda(E \cap I) \geqq \delta \lambda(I)$ holds for all intervals $I \subset R$. Prove that E' is a set of measure zero. [Use (18.2).]

(18.31) Exercise. Find a real-valued, absolutely continuous function on $[0, 1]$ that is monotone on no interval. [Construct a measurable set $A \subset [0, 1]$ such that $\lambda(A \cap I) > 0$ and $\lambda(A' \cap I) > 0$ for every interval $I \subset [0, 1]$. This can be done by using Cantor-like sets (6.62). Now integrate $\xi_A - \xi_{A'}$.]

(18.32) Exercise. Consider a positive real number α, and define a function f_α on $[0, 1]$ by

$$f_\alpha(0) = 0\,,$$

$$f_\alpha(x) = x^\alpha \cos(x^{-1}) \text{ for } 0 < x \leqq 1\,.$$

[Define x^α as $\exp(\alpha \log(x))$ with $\log(x)$ real.]

(a) For what α's does f_α have finite variation?

(b) For what α's is f_α absolutely continuous?

(c) For what α's does f_α have a finite derivative in $]0, 1[$ and a finite right derivative at 0?

(18.33) Exercise. (a) Let f be a continuous complex-valued function of finite variation on $[a, b]$ and suppose that f is absolutely continuous on $[a, c]$ for all c such that $a < c < b$. Prove that f is absolutely continuous on $[a, b]$.

(b) Show that part (a) fails for some continuous functions on $[a, b]$ having infinite variation on $[a, b]$.

(18.34) Exercise. Prove the following. A complex-valued function on $[a, b]$ is in $\mathfrak{Lip}_1([a, b])$ [see the definition in (17.31)] if and only if for every $\varepsilon > 0$ there is a $\delta > 0$ such that for all sequences $([a_k, b_k])_{k=1}^n$ of subintervals of $[a, b]$ for which

$$\sum_{k=1}^{n} (b_k - a_k) < \delta$$

holds, the inequality

$$\sum_{k=1}^{n} |f(b_k) - f(a_k)| < \varepsilon$$

holds. That is, f is "absolutely continuous with overlap permitted".

(18.35) Exercise. Let f be a nondecreasing function in $\mathfrak{C}^r([a, b])$ and let $E = \{x : a \leqq x < b,\ D^+ f(x) = \infty\}$. Prove that f is absolutely continuous if and only if $\lambda(f(E)) = 0$. [Use (18.25) and (17.25).]

(18.36) Exercise. Let f be a complex-valued function on $[a, b]$. Prove that f is absolutely continuous on $[a, b]$ if and only if there exists a sequence $(f_n)_{n=1}^\infty$ of functions in $\mathfrak{Lip}_1([a, b])$ such that $V_a^b(f - f_n) \to 0$. [If f is absolutely continuous, let $f_n(x) = \int\limits_a^x g_n(t)\,dt$, where $g_n(t) = f'(t)$ if $|f'(t)| \leqq n$ and $g_n(t) = 0$ otherwise.]

(18.37) Exercise. Let g be absolutely continuous on $[a, b]$, let $[\alpha, \beta] = \operatorname{rng} g$, and let f be absolutely continuous on $[\alpha, \beta]$.

(a) Prove that $f \circ g$ is absolutely continuous on $[a, b]$ if and only if $f \circ g$ is of finite variation on $[a, b]$. [Use (18.25).]

(b) Give an example to show that $f \circ g$ need not be absolutely continuous on $[a, b]$.

(18.38) Exercise. Let f be a complex-valued function defined on a closed interval $[\alpha, \beta]$. Prove that $f \in \mathfrak{Lip}_1([\alpha, \beta])$ if and only if $f \circ g$ is absolutely continuous on $[a, b]$ for all closed intervals $[a, b]$ and absolutely continuous functions g with $\operatorname{dom} g = [a, b]$ and $\operatorname{rng} g \subset [\alpha, \beta]$.

(18.39) Exercise. (a) Find a strictly increasing function $f \in \mathfrak{C}^r([0, 1])$ for which there is a subset A of $[0, 1]$ such that $\lambda(A) = 0$ and $\lambda(f(A)) = 1$. [Use (18.8).]

(b) Show that a function in $\mathfrak{C}^r([a, b])$ maps every measurable set onto a measurable set if and only if f is an N-function. [Hints. If f is an N-function and A is a measurable set, write A as a σ-compact set plus a set of measure zero. If f is not an N-function, apply (10.28).]

(18.40) Exercise. By a particular choice of the numbers t_n in (18.26), functions with quite extraordinary properties can be found.

(a) [Ruziewicz-Saks] Let F be as in (18.26). The structure of F makes it clear that the family of intervals $\{[a_1, b_1], \ldots, [a_n, b_n]\}$ are pairwise disjoint for all n: let ϱ_n be the maximum of the lengths of the open subintervals of $[a, b]$ that form the set $[a, b] \cap \left(\bigcup_{k=1}^{n} [a_k, b_k]\right)'$. It is obvious that $\lim_{n\to\infty} (b_n - a_n) = 0$. Since F is nowhere dense, a simple argument proves that $\lim_{n\to\infty} \varrho_n = 0$. In the definition of g in (18.26), let $t_k = \frac{1}{2}(b_k - a_k) + \varrho_k$. Then g has a finite derivative at no point of F. Thus if $\lambda(F) > 0$, we have an N-function that fails on a set of positive measure to have a derivative. [Hints. Show first that if g is differentiable at $x \in F$, then $g'(x) = 0$. Show next that a derivative of g is nonzero if $x = a_k$ or $x = b_k$. For other points $x \in F$ and for each index n, there is an interval $]a_{j(n)}, b_{j(n)}[$ among the intervals $]a_1, b_1[, \ldots,]a_n, b_n[$ at minimum distance from x [there are obviously at most two such intervals]. A moment's thought shows that $\lim_{n\to\infty} j(n) = \infty$, that

$$0 < |x - c_{j(n)}| < \varrho_{j(n)} + \frac{1}{2}(b_{j(n)} - a_{j(n)}),$$

and that $\lim_{n\to\infty} c_{j(n)} = x$. Hence we have

$$\frac{g(c_{j(n)}) - g(x)}{|c_{j(n)} - x|} > \frac{\frac{1}{2}(b_{j(n)} - a_{j(n)}) + \varrho_{j(n)}}{\frac{1}{2}(b_{j(n)} - a_{j(n)}) + \varrho_{j(n)}} = 1,$$

which shows that either $D^+g(x) \geqq 1$ or $D_-g(x) \leqq -1$. Thus g is differentiable nowhere on F.]

(b) Suppose that $\lambda(F) > 0$. Prove that $\sum_{k=1}^{\infty} \varrho_k = \infty$. $\Big[$If $\sum_{k=1}^{\infty} \varrho_k$ were finite, then g would have finite variation and would be differentiable almost everywhere.$\Big]$

(c) Consider CANTOR's ternary set $P \subset [0, 1]$, with the notation of (6.62). Write its complementary intervals in the order

$$I_{1,1}, I_{2,1}, I_{2,2}, I_{3,1}, I_{3,2}, I_{3,3}, I_{3,4}, \ldots, I_{n,1}, \ldots, I_{n,2^{n-1}}, \ldots .$$

Compute the sum $\sum_{k=1}^{\infty} \varrho_k$, where ϱ_k is defined as in (a). Why does this not conflict with (b)?

(18.41) Exercise. This exercise is somewhat demanding, but its part (d) is so elegant a result that we hope all readers will work through it. The closed interval $[a, b]$ is fixed throughout.

(a) Let A be a subset of $[a, b]$ of measure 0. There is an absolutely continuous nondecreasing function ψ on $[a, b]$ such that $\psi'(x) = \infty$ for all $x \in A$. [Hints. Let $(U_n)_{n=1}^{\infty}$ be a decreasing sequence of open supersets of A such that $\lambda(U_n) < 2^{-n}$. Let $\varphi_n = \sum_{k=1}^{n} \xi_{U_k}$, and $\varphi = \lim_{n\to\infty} \varphi_n$. Verify that $\varphi \in \mathfrak{L}_1^+([a, b])$ and that $\int_a^b \varphi(t)\,dt < 1$. Let $\psi(x) = \int_a^x \varphi(t)\,dt$, and $\psi_n(x) = \int_a^x \varphi_n(t)\,dt$. If $x \in A$ and $[x, x+h] \subset U_n$, then

$$\frac{\psi(x+h) - \psi(x)}{h} \geqq \frac{1}{h} \int_x^{x+h} \varphi_n(t)\,dt = n\,.$$

Thus $\psi'_+(x) = \infty$; similarly for $\psi'_-(x)$.]

(b) [A. ZYGMUND]. For $f \in \mathfrak{C}^r([a, b])$, define

$$E = \{x : a \leqq x < b, D^+f(x) \leqq 0\}\,.$$

Suppose that $f(E)$ contains no interval. Then f is nondecreasing. [Hints. If $f(c) > f(d)$, where $a \leqq c < d \leqq b$, choose any $y_0 \in]f(d), f(c)[$ and define x_0 as $\sup\{x : c \leqq x < d, f(x) \geqq y_0\}$. Show that $f(x_0) = y_0$, and hence that $D^+f(x_0) \leqq 0$. This implies that $f(E) \supset]f(d), f(c)[$, a contradiction to our hypothesis.]

(c) Consider any $f \in \mathfrak{C}^r([a, b])$. Suppose that D_+f is nonnegative almost everywhere on $[a, b[$. Suppose also that the set B, defined by $B = \{x : x \in [a, b[, D_+f(x) = -\infty\}$, is countable. Then f is nondecreasing. [Hints. Let $A = \{x : x \in]a, b[, D^+f(x)$ or $D_+f(x)$ is negative and finite$\}$. Let ψ be as in part (a) for the set A, let $\sigma(x) = \psi(x) + x$, and let $g = f + \varepsilon\sigma$, where ε is a positive number. Show that $D^+g(x) = \infty$ for $x \in A$ and that $D^+g(x)$ is positive for $x \in A' \cap B'$. Hence the set $E = \{x : D^+g(x) \leqq 0\}$ is contained in the countable set B and $g(E)$ can accordingly contain no interval. By part (b), g is nondecreasing. As ε is arbitrary, f too must be nondecreasing.]

(d) [Main result]. Let f be a function in $\mathfrak{C}([a, b])$. Suppose that $f'(x)$ exists and is finite for all but a countable set of x in $]a, b[$ and that $f' \in \mathfrak{L}_1([a, b])$. Then

$$f(x) - f(a) = \int_a^x f'(t)\,dt \quad \text{for} \quad a \leqq x \leqq b,$$

and in particular f is absolutely continuous. [A sketch of the proof follows. Consider $f \in \mathfrak{C}^r([a, b])$, the complex case being a trivial extension of this. For each positive integer n, define

$$g_n = \max\{f', -n\} \quad \text{and} \quad f_n(x) = \int_a^x g_n(t)\,dt.$$

Apply (12.24) to prove that

$$\lim_{n\to\infty} f_n(x) = \int_a^x f'(t)\,dt.$$

Next show that

$$D_+(f_n - f)(x) = f_n'(x) - f'(x) = g_n(x) - f'(x) \geqq 0$$

for almost all $x \in]a, b[$. Since

$$\frac{f_n(x+h) - f_n(x)}{h} \geqq \frac{1}{h}\int_x^{x+h} (-n)\,dt = -n,$$

one sees that $D_+(f_n - f)(x)$ is greater than $-\infty$ except on the countable set where f' is nonfinite or does not exist. By part (c), $f_n - f$ is nondecreasing. Thus

$$f_n(x) - f(x) \geqq f_n(a) - f(a) = -f(a),$$

and so

$$\int_a^x f'(t)\,dt = \lim_{n\to\infty} f_n(x) \geqq f(x) - f(a).$$

Replace f by $-f$ to reverse the last inequality.]

(18.42) Exercise. Let f be defined on $[0, 1]$ by $f(x) = x^2 \sin(x^{-2})$ for $0 < x \leqq 1$ and $f(0) = 0$. Prove that f has a finite derivative at all points of $]0, 1[$ and $f_+'(0) = 0$, but $f' \notin \mathfrak{L}_1([0, 1])$. Is f absolutely continuous on $[0, 1]$? Is f of finite variation on $[0, 1]$? [Compare this with (18.41.d). This example raises the following question. If f is a continuous function on $[a, b]$ such that f' is finite everywhere on $]a, b[$ but is not in $\mathfrak{L}_1([a, b])$, then how can f be reconstructed from f'? This problem is solved by making use of the *Denjoy integral*, which was invented for just this purpose. The reader who wishes to learn the details about this integral and other integrals more general than that of LEBESGUE is referred to S. SAKS, *Theory of the Integral*, 2nd Ed., Monografie Matematyczne, Warszawa-Łwow, 1937].

(18.43) Exercise: Integral representation of convex functions. Prove the following. Let I be an open interval in R and f a real-valued function on I. The function f is convex if and only if there are a nondecreasing function φ on I and a point $c \in I$ such that

$$f(x) = \begin{cases} f(c) + \int\limits_c^x \varphi(t)\,dt & \text{for} \quad x \geqq c\,, \\ f(c) - \int\limits_x^c \varphi(t)\,dt & \text{for} \quad x < c\,. \end{cases}$$

[The "if" part is a tiny extension of (13.35). For the "only if" part, combine (17.37.b), (18.16), and (17.37.a).]

(18.44) Exercise. Let f be a complex-valued function of period 2π defined on R such that $f \in \mathfrak{L}_1([-\pi, \pi])$. Suppose that f is continuous at every point of $[a, b]$. Prove that $(\sigma_n f)_{n=1}^{\infty}$ converges to f uniformly on $[a, b]$. [Use uniform continuity and the Fejér kernel as in (18.29).]

(18.45) Exercise. Use the uniform boundedness principle to prove that there exists a real-valued continuous function of period 2π defined on R whose Fourier series diverges at 0. [Let

$$\mathfrak{P} = \{f \in \mathfrak{C}^r([-\pi, \pi]) : f(-\pi) = f(\pi)\}\,.$$

Then $\mathfrak{P}$, with the uniform norm, is a real Banach space. For each $n \in N$, define $T_n : \mathfrak{P} \to R$ by $T_n f = s_n f(0) = \frac{1}{2\pi} \int\limits_{-\pi}^{\pi} f(t)\, D_n(t)\, dt$, where the notation is as in (18.27). Prove that $T_n \in \mathfrak{P}^*$ and that $\|T_n\| = \frac{1}{2\pi} \int\limits_{-\pi}^{\pi} |D_n(t)|\, dt$ for each n. Next show that $\lim\limits_{n\to\infty} \|T_n\| = \infty$. Finally conclude from (14.23) that there exists an $f \in \mathfrak{P}$ for which $\overline{\lim\limits_{n\to\infty}}\, |s_n f(0)| = \infty$.]

(18.46) Exercise. (a) Let χ be a complex-valued Lebesgue measurable function defined on R such that:

(i) χ is bounded;

(ii) χ is not identically zero;

and

(iii) $\chi(x + y) = \chi(x)\,\chi(y)$ for all $x, y \in R$.

Prove that there exists a real number α such that $\chi(x) = \exp(i\alpha x)$ for all $x \in R$. [Hints. Use (ii) and (iii) to prove that χ vanishes nowhere. Noting (i), define f on R by the rule $f(x) = \int\limits_0^x \chi(t)\,dt$. Choose $a \in R$ such that $f(a) \neq 0$, prove that $\chi(x)\, f(a) = f(x + a) - f(x)$ for all $x \in R$, and conclude that χ is absolutely continuous. These facts imply that f, and therefore χ, has a continuous derivative on R. Differentiate (iii) with respect to y and then set $y = 0$ to obtain $\chi'(x) = \chi(x)\chi'(0)$. Next set $\chi'(0) = i\alpha$, where $\alpha \in K$. Verify that $\frac{d}{dx}[\chi(x)\exp(-i\alpha x)] = 0$ for all $x \in R$. Conclude that there exists a $\beta \in K$ such that $\chi(x) = \beta\exp(i\alpha x)$ and show that $\chi(0) = 1$ so that $\beta = 1$. Finally, use (i) to prove that $\alpha \in R$. Functions χ satisfying (i)—(iii) are called *characters of* R.]

(b) Let ψ be a real-valued Lebesgue measurable function on R that satisfies (ii) and (iii) above [naturally with χ replaced by ψ]. Prove that $\psi(x) = \exp(\beta x)$ for a real number β. [It is elementary to show that $\psi(x) > 0$ for all x. Use (10.43) to show that ψ is continuous at 0 and use (iii) to show that ψ is continuous everywhere. Now integrate as in part (a).]

(c) Let ω be a complex-valued Lebesgue measurable function on R that satisfies (ii) and (iii). Prove that $\omega(x) = \exp(\gamma x)$ for some $\gamma \in K$. [Use part (a) on $\omega|\omega|^{-1}$ and part (b) on $|\omega|$.]

(d) Construct examples to show parts (a) and (b) fail if the hypothesis of measurability is dropped. [Use a Hamel basis for R over Q as in (5.46), and note that a discontinuous χ as in part (a) cannot be Lebesgue measurable; similarly for ψ's as in part (b).]

(18.47) Abel summability of Fourier series. Theorem (18.29) has an analogue for another classical summability method, and in fact the exceptional set for this method may be much smaller than the complement of the Lebesgue set. We sketch the construction and proof, leaving many details to the reader as exercises. All notation not explained here is as in (18.27). For $f \in \mathfrak{L}_1([-\pi, \pi])$ and for $0 < r < 1$, let

(i) $$\alpha_r f(x) = \sum_{k=-\infty}^{\infty} r^{|k|}\hat{f}(k)\exp(ikx)\,.$$

[Compare this with the definition of $\sigma_n f$ in (18.27.v).] The function $\alpha_r f$ is called the rth *Abel sum of the series* $\sum_{k=-\infty}^{\infty}\hat{f}(k)\exp(ikx)$. Using the uniform convergence of the series in (i), show that

(ii) $$\alpha_r f(x) = \frac{1}{2\pi}\int_{-\pi}^{\pi} f(x-t)\left[\sum_{k=-\infty}^{\infty} r^{|k|}\exp(ikt)\right]dt\,.$$

The expression $[\cdots]$ in (ii) is called the *Poisson kernel*, and is denoted $P(r, t)$. A simple computation shows that

(iii) $$P(r,t) = \frac{1-r^2}{1+r^2-2r\cos(t)}\,.$$

Also easy to verify are the relations

(iv) $P(r,t) = P(r,-t)$,

(v) $\frac{1-r}{1+r} \leq P(r,t) \leq \frac{1+r}{1-r}$,

(vi) $\frac{1}{2\pi}\int_{-\pi}^{\pi} P(r,t)\,dt = 1$,

(vii) $P'(r,t) = -\frac{(1-r^2)\,2r\sin(t)}{(1+r^2-2r\cos(t))^2}$

[differentiation with respect to t].

Consider the function $F(x) = \frac{1}{2\pi}\int_{-\pi}^{x} f(t)\,dt$, and consider any $x \in]-\pi,\pi[$ at which the symmetric derivative

$$\lim_{h\downarrow 0}\frac{F(x+h)-F(x-h)}{2h} = D_1F(x)$$

exists and is finite [see (17.36)]. We wish to prove that

(viii) $\lim_{r\uparrow 1} \alpha_r f(x) = D_1F(x)$.

In view of (17.36.a) and (18.3), this will prove that $\lim_{r\uparrow 1}\alpha_r f(x) = f(x)$ a.e. on $[-\pi,\pi]$ [note that the set where this occurs contains, perhaps properly, the Lebesgue set for f (18.5)]. By adding a constant to f [which disturbs nothing] we may suppose that $F(\pi) = 0$. Applying (18.19) to (ii), we see that

$$\begin{aligned}
\text{(ix)}\quad \alpha_r f(x) &= \frac{1}{2\pi}\int_{-\pi}^{\pi} F(t)\,P'(r,x-t)\,dt \\
&= \frac{1}{2\pi}\int_{-\pi}^{\pi} F(x-t)\,P'(r,t)\,dt \\
&= -\frac{1}{2\pi}\int_{-\pi}^{\pi} F(x+t)\,P'(r,t)\,dt \\
&= \frac{1}{2\pi}\int_{-\pi}^{\pi} \frac{F(x+t)-F(x-t)}{2\sin t}\,M(r,t)\,dt\,,
\end{aligned}$$

where the kernel $M(r,t)$ is defined by

(x) $M(r,t) = \frac{(1-r^2)\,2r\sin^2(t)}{(1+r^2-2r\cos(t))^2} = -\sin(t)\,P'(r,t)$.

The equality

$$P'(r,t) = \sum_{k=-\infty}^{\infty} r^{|k|}(ik)\exp(ikt)$$

holds because the infinite series converges uniformly in t. Therefore

$$-\sin(t)\,P'(r,t) = \frac{1}{2}\sum_{k=-\infty}^{\infty} r^{|k|}\left[-k\exp(i(k+1)t) + k\exp(i(k-1)t)\right],$$

so that

$$\text{(xi)}\quad \frac{1}{2\pi}\int_{-\pi}^{\pi} M(r,t)\,dt = r\,.$$

Since $1 + r^2 - 2r\cos(t) = |1 - r\exp(it)|^2$, it is easy to see from (x) that for every $\delta \in]0, \pi[$, the equality

$$\text{(xii)}\quad \lim_{r\uparrow 1}\left[\max\{M(r,t) : \delta \leqq |t| \leqq \pi\}\right] = 0$$

holds. Finally, for $|t|$ sufficiently small, $\left|\frac{F(x+t) - F(x-t)}{2\sin t} - D_1F(x)\right|$ is arbitrarily small. Combining this with (xii), (xi), and (ix), we obtain (viii) forthwith.

(18.48) Exercise: More on N-Functions. Let $[a, b]$ be a compact interval in R and let f be a real-valued function defined on $]a, b[$.

(a) Suppose that $E \subset]a, b[$ and $\beta \geqq 0$ are such that $D^+f(x) \leqq \beta$ and $D_-f(x) \geqq -\beta$ for every $x \in E$. Prove that $\lambda(f(E)) \leqq \beta\lambda(E)$. [Hints. For $\varepsilon > 0$ and $n \in N$, define $E_n = \left\{x \in E : f(t) - f(x) < (\beta + \varepsilon)|t - x| \text{ for all } t \in]a, b[\text{ for which } |t - x| < \frac{1}{n}\right\}$. Then $E_1 \subset E_2 \subset \cdots$ and $\bigcup_{n=1}^{\infty} E_n = E$. For each n, let $\{I_{n,k}\}_{k=1}^{\infty}$ be a cover of E_n by intervals of length $< \frac{1}{n}$ such that $\sum_{k=1}^{\infty} \lambda(I_{n,k}) < \lambda(E_n) + \varepsilon$. Then

$$\lambda(f(E_n)) \leqq \sum_{k=1}^{\infty} \lambda(f(E_n \cap I_{n,k})) < (\beta + \varepsilon) \sum_{k=1}^{\infty} \lambda(I_{n,k}) < (\beta + \varepsilon)(\lambda(E) + \varepsilon)$$

for all $n \in N$. Let n go to ∞ and use (9.17).]

(b) Suppose that f has a finite derivative at all but countably many points of $[a, b]$. Prove that f is an N-function. [Hint. Consider the sets $A_n = \{x \in]a, b[: f'(x) \text{ exists and } |f'(x)| \leqq n\}$ and use part (a).]

(c) Suppose that B is a Lebesgue measurable subset of $]a, b[$ and that f has a finite derivative at each point of B. Prove that f and f' are both Lebesgue measurable on B and that $\lambda(f(B)) \leqq \int_B |f'(x)|\,dx$. [Hints. For $\varepsilon > 0$ and $n \in N$, let $B_n = \{x \in B : (n-1)\varepsilon \leqq |f'(x)| < n\varepsilon\}$. Applying part (a), we have

$$\begin{aligned}\lambda(f(B)) &\leqq \sum_{n=1}^{\infty} \lambda(f(B_n)) \leqq \sum_{n=1}^{\infty} n\varepsilon\lambda(B_n) \\ &\leqq \sum_{n=1}^{\infty}\left[\int_{B_n} |f'(x)|\,dx + \varepsilon\lambda(B_n)\right] \\ &= \int_B |f'(x)|\,dx + \varepsilon\lambda(B)\,.]\end{aligned}$$

(d) Use part (c) [*not* (18.25)] to prove that if f is a continuous N-function of finite variation on $[a, b]$, then f is absolutely continuous on

$[a, b]$. [Hints. For $[c, d] \subset [a, b]$, let $B = \{x \in]c, d[: f'(x)$ exists and is finite$\}$, and let $A = [c, d] \cap B'$. Then $|f(d) - f(c)| \leqq \lambda(f([c,d])) = \lambda(f(B)) + \lambda(f(A)) = \lambda(f(B)) \leqq \int_c^d |f'(x)|\,dx$. Recall that $f' \in \mathfrak{L}_1([a, b])$.][1]

§ 19. Complex measures and the Lebesgue-Radon-Nikodým theorem

In this section we make a further study of the measure-theoretic significance of absolutely continuous functions and functions of finite variation. We begin by examining abstract analogues of some of the classical notions of §§ 17 and 18. We will then use our abstract results to obtain further information about the classical case.

The most useful generalization of the notion of indefinite integral seems to be the following. Let $(X, \mathscr{A}, \mu)$ be an arbitrary measure space and let f be any function in $\mathfrak{L}_1(X, \mathscr{A}, \mu)$. Define ν on $\mathscr{A}$ by

$$\nu(E) = \int_E f\,d\mu$$

for $E \in \mathscr{A}$. Clearly ν is complex-valued, $\nu(\varnothing) = 0$, and ν is countably additive (12.32). Thus ν enjoys two essential properties of a measure. Since ν can assume arbitrary complex values, it is not always a measure in the sense of (10.3). This leads us to define and study signed measures and complex measures.

(19.1) Definition. Let $(X, \mathscr{A})$ be an arbitrary measurable space. An extended real-valued function ν defined on $\mathscr{A}$ is called a *signed measure* if

(i) $\nu(\varnothing) = 0$

and

(ii) $\nu\left(\bigcup_{n=1}^{\infty} E_n\right) = \sum_{n=1}^{\infty} \nu(E_n)$

for all pairwise disjoint sequences $(E_n)_{n=1}^{\infty}$ of elements of $\mathscr{A}$. A complex-valued function ν defined on $\mathscr{A}$ that satisfies (i) and (ii) is called a *complex measure*.

(19.2) Note. It is implicit in the above definition that the infinite series appearing in (19.1.ii) must always be meaningful and must converge [or definitely diverge] to the value on the left side of the equality. In particular, a signed measure ν can assume at most one of the values ∞ and $-\infty$. For, if $\nu(E) = \infty$ and $\nu(F) = -\infty$, then the right side of the equality

$$\nu(E \cup F) = \nu(E \cap F') + \nu(E \cap F) + \nu(E' \cap F)$$

is undefined since it contains both ∞ and $-\infty$ among its three terms [see (6.1.b)].

[1] Note that we have here a short proof of (18.25).

Our first goal is to show that just as a function of finite variation can be expressed as a linear combination of four monotone functions, so a complex measure can be expressed as a linear combination of four measures. It is obvious that any complex measure ν can be expressed uniquely in the form $\nu = \nu_1 + i\nu_2$ where ν_1 and ν_2 are real-valued signed measures; simply let $\nu_1(E) = \operatorname{Re}\nu(E)$ and $\nu_2(E) = \operatorname{Im}\nu(E)$ for all $E \in \mathscr{A}$. We therefore take up signed measures first.

(19.3) Theorem. *Let ν be a signed measure on a measurable space $(X, \mathscr{A})$. We have:*

(i) *if $E, F \in \mathscr{A}$, $|\nu(E)| < \infty$, and $F \subset E$, then $|\nu(F)| < \infty$;*

(ii) *if $A_n \in \mathscr{A}$ $(n = 1, 2, 3, \ldots)$ and $A_1 \subset A_2 \subset \cdots \subset A_n \subset \cdots$, then*

$$\lim_{n\to\infty} \nu(A_n) = \nu\left(\bigcup_{n=1}^{\infty} A_n\right);$$

(iii) *if $A_n \in \mathscr{A}$ $(n = 1, 2, 3, \ldots)$, if $A_1 \supset A_2 \supset \cdots \supset A_n \supset \cdots$, and $|\nu(A_1)| < \infty$, then*

$$\lim_{n\to\infty} \nu(A_n) = \nu\left(\bigcap_{n=1}^{\infty} A_n\right).$$

Proof. To prove (i), observe that $\nu(E) = \nu(F) + \nu(E \cap F')$. In order that $\nu(E)$ be finite, it is necessary and sufficient that both summands on the right side be finite. Conclusions (ii) and (iii) are proved by repeating *verbatim* the proofs of (10.13) and (10.15), respectively. In the proof of (iii) we use (i) to write $\nu(A_1 \cap A_n') = \nu(A_1) - \nu(A_n)$. □

(19.4) Definition. Let ν be a signed measure on a measurable space $(X, \mathscr{A})$. A set $P \in \mathscr{A}$ is called a *nonnegative set for* ν if $\nu(P \cap E) \geqq 0$ for all $E \in \mathscr{A}$. A set $M \in \mathscr{A}$ is called a *nonpositive set for* ν if $\nu(M \cap E) \leqq 0$ for all $E \in \mathscr{A}$. Note that $\varnothing$ is both a nonnegative set and a nonpositive set for ν. If P is a nonnegative set for ν and P' [the complement of P in X] is a nonpositive set for ν, then the ordered pair (P, P') is called a *Hahn decomposition of X for ν*.

(19.5) Lemma. *Let ν be a signed measure on $(X, \mathscr{A})$ and suppose that E is a set in $\mathscr{A}$ such that $0 < \nu(E) < \infty$. Then there exists a set $S \in \mathscr{A}$ such that $S \subset E$, S is a nonnegative set for ν, and $\nu(S) > 0$.*

Proof. It follows from (19.3.i) that $|\nu(F)| < \infty$ for all $F \in \mathscr{A}$ such that $F \subset E$. Assume that no set S of the required sort exists. In particular, E is not a nonnegative set for ν. Let n_1 be the smallest positive integer for which there exists a set $F_1 \in \mathscr{A}$ such that $F_1 \subset E$ and $\nu(F_1) < -\frac{1}{n_1}$. Then we have

$$\nu(E \cap F_1') = \nu(E) - \nu(F_1) > \nu(E) > 0,$$

and so, by our assumption, $E \cap F_1'$ is not a nonnegative set for ν. As before, let n_2 be the smallest positive integer for which there exists a set

$F_2 \in \mathscr{A}$ such that $F_2 \subset E \cap F_1'$ and $\nu(F_2) < -\frac{1}{n_2}$. Then

$$\nu(E \cap (F_1 \cup F_2)') = \nu(E) - \nu(F_1) - \nu(F_2) > 0\,,$$

and so $E \cap (F_1 \cup F_2)'$ is not a nonnegative set for ν. Continuing this process, we obtain a sequence $(n_k)_{k=1}^{\infty}$ of minimal positive integers and a corresponding pairwise disjoint sequence $(F_k)_{k=1}^{\infty}$ of sets in $\mathscr{A}$ such that $\nu(F_k) < -\frac{1}{n_k}$ for each $k \in N$. Let $F = \bigcup_{k=1}^{\infty} F_k$. Then we have

$$\infty > \nu(E \cap F') = \nu(E) - \nu(F) = \nu(E) - \sum_{k=1}^{\infty} \nu(F_k) > \nu(E) + \sum_{k=1}^{\infty} \frac{1}{n_k} > 0\,.$$

Thus $\sum_{k=1}^{\infty} \frac{1}{n_k} < \infty$ and $E \cap F'$ is not a nonnegative set for ν. Choose $A \in \mathscr{A}$ such that $A \subset E \cap F'$ and $\nu(A) < 0$, and then choose k so large that $\nu(A) < -\frac{1}{n_k}$ and $n_k > 2$. Now

$$A \cup F_k \subset E \cap \left(\bigcup_{j=1}^{k-1} F_j\right)'$$

and

$$\nu(A \cup F_k) = \nu(A) + \nu(F_k) < -\frac{1}{n_k} - \frac{1}{n_k} < -\frac{1}{n_k - 1}\,.$$

This contradicts the minimality of n_k. We conclude that a set S of the required sort exists. □

(19.6) Hahn Decomposition Theorem. *Let ν be a signed measure on a measurable space $(X, \mathscr{A})$. There exists a Hahn decomposition of X for ν. Moreover, this decomposition is unique in the sense that if (P_1, P_1') and (P_2, P_2') are any two such decompositions, then $\nu(P_1 \cap E) = \nu(P_2 \cap E)$ and $\nu(P_1' \cap E) = \nu(P_2' \cap E)$ for every $E \in \mathscr{A}$.*

Proof. Since ν takes on at most one of the values ∞ and $-\infty$, we suppose that $\nu(E) < \infty$ for all $E \in \mathscr{A}$. In the other case, simply interchange the rôles of positive and negative by considering the signed measure $-\nu$.

Let $\alpha = \sup\{\nu(A) : A$ is a nonnegative set for $\nu\}$. Choose a sequence $(A_k)_{k=1}^{\infty}$ of nonnegative sets for ν such that $\lim_{k\to\infty} \nu(A_k) = \alpha$. Define $P = \bigcup_{k=1}^{\infty} A_k$ and $P_n = \bigcup_{k=1}^{n} A_k$ $(n = 1, 2, \ldots)$. It follows by induction on n that each P_n is a nonnegative set for ν and that $\nu(P_n) \geqq \nu(A_n)$. Applying (19.3.ii), we have $\nu(P \cap E) = \lim_{n\to\infty} \nu(P_n \cap E) \geqq 0$ for all $E \in \mathscr{A}$. Thus P is a nonnegative set for ν and $\nu(P) = \alpha$. [Note that $\alpha < \infty$.]

We next show that P' is a nonpositive set for ν. Assuming the contrary, choose a set $E \in \mathscr{A}$ such that $E \subset P'$ and $\nu(E) > 0$. Since ν does not assume the value ∞, we have $\nu(E) < \infty$. Apply (19.5) to obtain a set $S \subset E$ such that $S \in \mathscr{A}$, S is a nonnegative set for ν, and $\nu(S) > 0$. Then $S \cup P$ is a nonnegative set for ν and $\nu(S \cup P) = \nu(S) + \alpha > \alpha$; this violates the definition of α. Therefore P' is a nonpositive set for ν.

To prove our uniqueness assertion, suppose that (P_1, P_1') and (P_2, P_2') are two Hahn decompositions of X for ν, and select $E \in \mathscr{A}$. Since $E \cap P_1 \cap P_2'$ is a subset of both P_1 and P_2', we have $\nu(E \cap P_1 \cap P_2') = 0$; similarly $\nu(E \cap P_1' \cap P_2) = 0$. Thus we have $\nu(E \cap P_1) = \nu(E \cap (P_1 \cup P_2)) = \nu(E \cap P_2)$ and $\nu(E \cap P_1') = \nu(E \cap (P_1' \cup P_2')) = \nu(E \cap P_2')$. □

(19.7) Definition. Let ν be a signed measure on $(X, \mathscr{A})$ and let (P, P') be a Hahn decomposition of X for ν. Define ν^+, ν^-, and $|\nu|$ on $\mathscr{A}$ by:

$$\nu^+(E) = \nu(E \cap P);$$

$$\nu^-(E) = -\nu(E \cap P');$$

and

$$|\nu|(E) = \nu^+(E) + \nu^-(E)$$

for all $E \in \mathscr{A}$. The set functions ν^+, ν^-, and $|\nu|$ are called the *positive variation of* ν, the *negative variation of* ν, and the *total variation of* ν, respectively.

(19.8) Theorem. *Notation is as in* (19.7). *The set functions* ν^+, ν^-, *and* $|\nu|$ *are well-defined measures on* $(X, \mathscr{A})$. *Also we have*

(i) $\nu(E) = \nu^+(E) - \nu^-(E)$ *for all* $E \in \mathscr{A}$.[1]

The proof is very simple, and we omit it.

(19.9) Example. Let $(X, \mathscr{A}, \mu)$ be a measure space and let f be an $\mathscr{A}$-measurable, extended real-valued function on X for which $\int_X f\,d\mu$ is defined. Define ν on $\mathscr{A}$ by

$$\nu(E) = \int_E f\,d\mu .$$

Then we have

$$\nu^+(E) = \int_E f^+\,d\mu ,$$

$$\nu^-(E) = \int_E f^-\,d\mu ,$$

and

$$|\nu|(E) = \int_E |f|\,d\mu$$

for all $E \in \mathscr{A}$. If $P = \{x \in X : f(x) > 0\}$, then (P, P') is a Hahn decomposition of X for ν. Notice that the nonequality $|\nu|(E) \neq |\nu(E)|$ must occur for some sets $E \in \mathscr{A}$ if ν^+ and ν^- are nondegenerate.

(19.10) Theorem. *Let* ν *be a signed measure on* $(X, \mathscr{A})$. *Then*

(i) $|\nu|(E) = \sup\left\{\sum_{k=1}^{n} |\nu(E_k)| : \{E_1, \ldots, E_n\} \text{ is a measurable dissection of } E\right\}$

for every $E \in \mathscr{A}$.

[1] The expression $\nu = \nu^+ - \nu^-$ is known as the *Jordan decomposition of* ν, in analogy with (17.16).

Proof. Let E be any fixed set in $\mathscr{A}$, and let β denote the right side of (i). Then we have

$$\begin{aligned}\sum_{k=1}^{n} |\nu(E_k)| &= \sum_{k=1}^{n} |\nu^+(E_k) - \nu^-(E_k)| \\ &\leqq \sum_{k=1}^{n} (\nu^+(E_k) + \nu^-(E_k)) \\ &= \sum_{k=1}^{n} |\nu|(E_k) = |\nu|(E)\end{aligned}$$

for every measurable dissection $\{E_1, \ldots, E_n\}$ of E; hence $\beta \leqq |\nu|(E)$. Consider the dissection $\{E \cap P, E \cap P'\}$ where (P, P') is a Hahn decomposition of X for ν. We get

$$\beta \geqq |\nu(E \cap P)| + |\nu(E \cap P')| = \nu^+(E) + \nu^-(E) = |\nu|(E) . \quad \square$$

In view of (19.10), we make the following definition with no risk of inconsistency.

(19.11) Definition. Let ν be a complex measure on $(X, \mathscr{A})$. The *total variation of* ν is the function $|\nu|$ defined on $\mathscr{A}$ by the formula (19.10.i).[1]

(19.12) Theorem. *Notation is as in* (19.11). *The set-function* $|\nu|$ *is a measure on* $(X, \mathscr{A})$.

Proof. It is obvious that $|\nu|(\varnothing) = 0$. Thus we need only show that $|\nu|$ is countably additive. Let $(A_j)_{j=1}^{\infty}$ be a pairwise disjoint sequence of sets in $\mathscr{A}$ and let $A = \bigcup_{j=1}^{\infty} A_j$. Let β be an arbitrary real number such that $\beta < |\nu|(A)$. Choose a measurable dissection $\{E_1, \ldots, E_n\}$ of A such that $\beta < \sum_{k=1}^{n} |\nu(E_k)|$. Then we have

$$\begin{aligned}\beta &< \sum_{k=1}^{n} \left| \sum_{j=1}^{\infty} \nu(E_k \cap A_j) \right| \\ &\leqq \sum_{k=1}^{n} \sum_{j=1}^{\infty} |\nu(E_k \cap A_j)| \\ &= \sum_{j=1}^{\infty} \sum_{k=1}^{n} |\nu(E_k \cap A_j)| \\ &\leqq \sum_{j=1}^{\infty} |\nu|(A_j) .\end{aligned}$$

Since β is arbitrary, it follows that

$$|\nu|(A) \leqq \sum_{j=1}^{\infty} |\nu|(A_j) . \tag{1}$$

If $|\nu|(A) = \infty$ [which, as we shall see in (19.13.v), is impossible], then the

[1] Notice the similarity of this definition with our definition of $V_a^b f$ in (17.14).

reverse of this inequality is obvious. Thus suppose that $|\nu|(A) < \infty$. For any $j_0 \in N$ and any measurable dissection $\{B_1, \ldots, B_m\}$ of A_{j_0}, we have

$$\sum_{k=1}^{m} |\nu(B_k)| \leqq \Big|\nu\Big(\bigcup_{j \neq j_0} A_j\Big)\Big| + \sum_{k=1}^{m} |\nu(B_k)| \leqq |\nu|(A),$$

and so $|\nu|(A_j) < \infty$ for all $j \in N$. Let $\varepsilon > 0$ be arbitrary. For each j, choose a measurable dissection $\{E_{j,1}, \ldots, E_{j,n_j}\}$ of A_j such that

$$\sum_{k=1}^{n_j} |\nu(E_{j,k})| > |\nu|(A_j) - \frac{\varepsilon}{2^j}.$$

Then for all $m \in N$, we have

$$\begin{aligned}\sum_{j=1}^{m} |\nu|(A_j) &< \sum_{j=1}^{m} \left(\frac{\varepsilon}{2^j} + \sum_{k=1}^{n_j} |\nu(E_{j,k})|\right) \\ &< \varepsilon + \left|\nu\left(\bigcup_{j=m+1}^{\infty} A_j\right)\right| + \sum_{j=1}^{m} \sum_{k=1}^{n_j} |\nu(E_{j,k})| \\ &\leqq \varepsilon + |\nu|(A).\end{aligned}$$

Since ε is arbitrary, we have

$$\sum_{j=1}^{m} |\nu|(A_j) \leqq |\nu|(A)$$

for every $m \in N$, and so

$$\sum_{j=1}^{\infty} |\nu|(A_j) \leqq |\nu|(A). \tag{2}$$

Combining (1) and (2), we see that $|\nu|$ is countably additive. □

(19.13) Theorem. *Let ν be a complex measure on $(X, \mathscr{A})$ and let ν_1 and ν_2 be the real and imaginary parts of ν, respectively. Then:*

(i) *ν_1 and ν_2 are signed measures on $(X, \mathscr{A})$;*

and for all $E \in \mathscr{A}$ we have

(ii) $\nu(E) = \nu_1^+(E) - \nu_1^-(E) + i\nu_2^+(E) - i\nu_2^-(E)$;[1]

(iii) $|\nu|(E) \leqq \nu_1^+(E) + \nu_1^-(E) + \nu_2^+(E) + \nu_2^-(E)$;

(iv) $\sup\{|\nu(A)| : A \in \mathscr{A}, A \subset E\} \leqq |\nu|(E) \leqq 4 \cdot \sup\{|\nu(A)| : A \in \mathscr{A}, A \subset E\}$;

(v) $|\nu(E)| \leqq |\nu|(X) < \infty$;

and

(vi) $\nu_j^+(E) \leqq |\nu|(E)$ *and* $\nu_j^-(E) \leqq |\nu|(E)$ for $j = 1, 2$.

Thus ν_1^+, ν_1^-, ν_2^+, ν_2^-, and $|\nu|$ are finite measures on $(X, \mathscr{A})$.

Proof. Parts (i) and (ii) are obvious. Let (P_1, P_1') and (P_2, P_2') be Hahn decompositions of X for ν_1 and ν_2 respectively. For any measurable

[1] We call this the *Jordan decomposition of* ν, again in analogy with (17.16).

dissection $\{E_1, \ldots, E_n\}$ of E, we have

$$\sum_{k=1}^{n} |\nu(E_k)| \leqq \sum_{k=1}^{n} (\nu_1^+(E_k) + \nu_1^-(E_k) + \nu_2^+(E_k) + \nu_2^-(E_k))$$
$$= \nu_1^+(E) + \nu_1^-(E) + \nu_2^+(E) + \nu_2^-(E) .$$

Take the supremum over all $\{E_1, \ldots, E_n\}$ to obtain (iii). Next let $\alpha = \sup\{|\nu(A)| : A \in \mathscr{A}, A \subset E\}$. Applying (iii) and (19.7), we have

$$\begin{aligned} |\nu|(E) &\leqq \nu_1(E \cap P_1) + |\nu_1(E \cap P_1')| + \nu_2(E \cap P_2) + |\nu_2(E \cap P_2')| \\ &\leqq |\nu(E \cap P_1)| + |\nu(E \cap P_1')| + |\nu(E \cap P_2)| + |\nu(E \cap P_2')| \\ &\leqq 4\alpha . \end{aligned} \tag{1}$$

If $A \in \mathscr{A}$ and $A \subset E$, then $\{A, E \cap A'\}$ is a measurable dissection of E, and so $|\nu(A)| \leqq |\nu(A)| + |\nu(E \cap A')| \leqq |\nu|(E)$. Take the supremum over all such A's to obtain

$$\alpha \leqq |\nu|(E) . \tag{2}$$

Combine (1) and (2) to get (iv).

Conclusion (v) follows from (iv) and (iii) [recall that ν is complex-valued and that ∞ is not a complex number].

Finally, (vi) follows from (iv) because

$$\nu_j^+(E) = \nu_j(E \cap P_j) \leqq |\nu|(E)$$

and

$$\nu_j^-(E) = |\nu_j(E \cap P_j')| \leqq |\nu|(E) . \quad \square$$

The equality (19.13.ii) suggests our next definition, which also is useful in later arguments and constructions.

(19.14) Definition. Let $(X, \mathscr{A})$ be a measurable space, let μ_1 and μ_2 be complex measures on $(X, \mathscr{A})$, and let α_1 and α_2 be complex numbers. The set-function $\alpha_1\mu_1 + \alpha_2\mu_2$ on $\mathscr{A}$ is defined by

$$(\alpha_1\mu_1 + \alpha_2\mu_2)(E) = \alpha_1\mu_1(E) + \alpha_2\mu_2(E)$$

for all $E \in \mathscr{A}$. If μ is a signed measure on $(X, \mathscr{A})$ and $\alpha \in R$, then $\alpha\mu$ is the set-function on $\mathscr{A}$ such that

$$(\alpha\mu)(E) = \alpha(\mu(E)) .$$

If μ and ν are signed measures on $(X, \mathscr{A})$ and if the simultaneous equalities $\mu(E) = \infty$, $\nu(E) = -\infty$ or $\mu(E) = -\infty$, $\nu(E) = \infty$ hold for no $E \in \mathscr{A}$, then we define $\mu + \nu$ by

$$(\mu + \nu)(E) = \mu(E) + \nu(E) \quad (E \in \mathscr{A}) .$$

In this case, $\mu + \nu$ is said to be *defined*. If either of the interdicted pair of equalities holds for some $E \in \mathscr{A}$, then $\mu + \nu$ is undefined.

(19.15) Notes. (a) Notation is as in (19.14). It is all but obvious that $\alpha_1\mu_1 + \alpha_2\mu_2$ is a complex measure on $(X, \mathscr{A})$, that $\alpha\mu$ is a signed measure, and that $\mu + \nu$, if defined, is a signed measure.

(b) If μ_1 and μ_2 are measures on $(X, \mathscr{A})$ and α_1, α_2 are in $[0, \infty[$, then $\alpha_1\mu_1 + \alpha_2\mu_2$ is a measure on $(X, \mathscr{A})$.

(c) For notational convenience, we will usually write the Jordan decomposition (19.13.ii) of a complex measure ν as

$$\alpha_1\nu_1 + \alpha_2\nu_2 + \alpha_3\nu_3 + \alpha_4\nu_4 .$$

We now consider integration with respect to complex measures, beginning with a useful if rather obvious fact.

(19.16) Theorem. *Let ν be a complex measure on $(X, \mathscr{A})$ and let $\nu = \sum_{k=1}^{4} \alpha_k \nu_k$ be its Jordan decomposition. A complex-valued function f defined $|\nu|$-a.e. on X is in $\mathfrak{L}_1(X, \mathscr{A}, |\nu|)$ if and only if it is in $\mathfrak{L}_1(X, \mathscr{A}, \nu_k)$ for each $k \in \{1, 2, 3, 4\}$.*

Proof. The theorem is true if f is an $\mathscr{A}$-measurable, nonnegative, simple function $\left[\text{say } f = \sum_{j=1}^{m} \beta_j \xi_{E_j}\right]$, because

$$\begin{aligned}\int_X f d|\nu| &= \sum_{j=1}^{m} \beta_j |\nu|(E_j) \\ &\leq \sum_{j=1}^{m} \beta_j \left(\sum_{k=1}^{4} \nu_k(E_j)\right) \\ &= \sum_{k=1}^{4} \int_X f d\nu_k ,\end{aligned} \tag{1}$$

as (19.13.iii) shows. Furthermore,

$$\begin{aligned}\int_X f\, d\nu_k &= \sum_{j=1}^{m} \beta_j \nu_k(E_j) \\ &\leq \sum_{j=1}^{m} \beta_j |\nu|(E_j) = \int_X f d|\nu| ,\end{aligned} \tag{2}$$

by (19.13.vi). In the general case, let $(s_n)_{n=1}^{\infty}$ be a nondecreasing sequence of $\mathscr{A}$-measurable, nonnegative, simple functions which converges a.e. to $|f|$. [Note that $|\nu|(A) = 0$ if and only if $\nu_k(A) = 0$ for all k.] As usual, we apply B. LEVI's theorem (12.22) to (1) and to (2) with s_n in place of f to obtain

$$\int_X |f|\, d|\nu| \leq \sum_{k=1}^{4} \int_X |f|\, d\nu_k$$

and

$$\int_X |f|\, d\nu_k \leq \int_X |f|\, d|\nu|$$

for all k. □

We may now make the following definition.

(19.17) Definition. Notation is as in (19.16). For $f \in \mathfrak{L}_1(X, \mathscr{A}, |\nu|)$, define

$$\int_X f\, d\nu = \sum_{k=1}^{4} \alpha_k \int_X f\, d\nu_k .$$

(19.18) Theorem. *Let ν be a complex measure on $(X, \mathscr{A})$. If $f, g \in \mathfrak{L}_1(X, \mathscr{A}, |\nu|)$ and $\alpha \in K$, then*

(i) $\int_X (f + g)\, d\nu = \int_X f\, d\nu + \int_X g\, d\nu$

and

(ii) $\int_X \alpha f\, d\nu = \alpha \int_X f\, d\nu$.

Thus $\int_X \cdots d\nu$ is a linear functional on $\mathfrak{L}_1(X, \mathscr{A}, |\nu|)$.

This is proved by an obvious computation, which we leave to the reader.

We next define absolute continuity *for measures*. As we shall prove in (19.53), Borel measures on R absolutely continuous with respect to λ are just the Lebesgue-Stieltjes measures induced by absolutely continuous nondecreasing functions.

(19.19) Definition. Let $(X, \mathscr{A})$ be a measurable space and let μ and ν be signed or complex measures on $(X, \mathscr{A})$. We say that ν is *absolutely continuous with respect to* μ, and we write $\nu \ll \mu$, if $|\mu|(E) = 0$ implies $\nu(E) = 0$ for all $E \in \mathscr{A}$.

(19.20) Theorem. *Let μ and ν be complex or signed measures on $(X, \mathscr{A})$ and let $\sum_{k=1}^{4} \alpha_k \nu_k$ be the Jordan decomposition of ν. Then the following are equivalent:*

(i) $\nu \ll \mu$;

(ii) $\nu_k \ll \mu$ *for* $k \in \{1, 2, 3, 4\}$;

(iii) $|\nu| \ll \mu$.

Proof. We consider only the complex case. Suppose that $E \in \mathscr{A}$ and that $|\mu|(E) = 0$. Suppose that (i) holds. Since $|\mu|(F) = 0$ for all subsets F of E such that $F \in \mathscr{A}$, we also have $\nu(F) = 0$ for all such F. From (19.10.i), we infer that $|\nu|(E) = 0$. Thus (i) implies (iii). If (iii) holds, then $|\nu|(E) = 0$ and so (19.13.vi) shows that $\nu_k(E) = 0$ for all k. That is, (iii) implies (ii). Finally, it is obvious that (ii) implies (i), since $\nu = \sum_{k=1}^{4} \alpha_k \nu_k$. □

As noted in the introductory remarks to this section, if $(X, \mathscr{A}, \mu)$ is a measure space and $f \in \mathfrak{L}_1(X, \mathscr{A}, \mu)$, then the function ν defined on $\mathscr{A}$ by

$$\nu(A) = \int_A f \, d\mu \tag{1}$$

is a complex measure on $(X, \mathscr{A})$. It is clear that $\nu \ll \mu$. The LEBESGUE-RADON-NIKODÝM theorem asserts that if $\nu \ll \mu$ and certain other conditions are met, then ν has the form (1). We will present several avatars of this theorem. First, a technicality.

(19.21) Lemma. *Let μ and ν be measures on $(X, \mathscr{A})$ such that $\nu(E) \leqq \mu(E)$ for all $E \in \mathscr{A}$. If $p > 0$ and $f \in \mathfrak{L}_p(X, \mathscr{A}, \mu)$, it follows that $f \in \mathfrak{L}_p(X, \mathscr{A}, \nu)$ and $\int_X |f|^p d\nu \leqq \int_X |f|^p d\mu$.*

Proof. If $f \in \mathfrak{L}_p(X, \mathscr{A}, \mu)$, then there is a sequence $(\sigma_n)_{n=1}^\infty$ of $\mathscr{A}$-measurable, nonnegative, simple functions increasing to $|f|^p$, and so

$$\lim_{n\to\infty} \int_X \sigma_n d\mu = \int_X |f|^p d\mu < \infty .$$

It is clear that $\int_X \sigma_n d\nu \leqq \int_X \sigma_n d\mu$, and it follows that

$$\int_X |f|^p d\nu = \lim_{n\to\infty} \int_X \sigma_n d\nu \leqq \int_X |f|^p d\mu . \quad \square$$

The next lemma, which may at first glance appear rather strange, is actually the crucial step in our proof of the LEBESGUE-RADON-NIKODÝM theorem.

(19.22) Lemma. *Let μ and ν be finite measures on $(X, \mathscr{A})$ such that $\nu \ll \mu$. Then there is an $\mathscr{A}$-measurable function g on X such that $g(X) \subset [0, 1[$ and*

(i) $\int_X f(1-g)\, d\nu = \int_X fg \, d\mu$

for all $f \in \mathfrak{L}_2(X, \mathscr{A}, \mu + \nu)$.

Proof. For $f \in \mathfrak{L}_2(X, \mathscr{A}, \mu + \nu)$, define

$$L(f) = \int_X f \, d\nu . \tag{1}$$

Since μ and ν, and with them $\mu + \nu$, are finite, f is also in $\mathfrak{L}_1(X, \mathscr{A}, \mu + \nu)$; and so by (19.21), f is in $\mathfrak{L}_1(X, \mathscr{A}, \nu)$. Thus L is defined and finite on $\mathfrak{L}_2(X, \mathscr{A}, \mu + \nu)$. It is clear that $L(\alpha f + \beta g) = \alpha L(f) + \beta L(g)$ for all $\alpha, \beta \in K$ and $f, g \in \mathfrak{L}_2(X, \mathscr{A}, \mu + \nu)$. Inequality (13.4.iii) shows that

$$|L(f)| = \left|\int_X f \, d\nu\right| \leqq \Big(\int_X |f|^2 d\nu\Big)^{\frac{1}{2}} (\nu(X))^{\frac{1}{2}}$$
$$\leqq \Big(\int_X |f|^2 d(\mu + \nu)\Big)^{\frac{1}{2}} (\nu(X))^{\frac{1}{2}} = \|f\|_2 (\nu(X))^{\frac{1}{2}} .$$

[Here $\|f\|_2$ denotes the norm in $\mathfrak{L}_2(X, \mathscr{A}, \mu + \nu)$.] Thus L is a bounded linear functional on $\mathfrak{L}_2(X, \mathscr{A}, \mu + \nu)$, and so by (15.11), there is a function $h \in \mathfrak{L}_2(X, \mathscr{A}, \mu + \nu)$ such that

$$L(f) = \int_X f\bar{h}\, d(\mu + \nu)\ .^{1} \tag{2}$$

Actually h is real-valued and nonnegative $(\mu + \nu)$-a.e., as we now show. For any $f \in \mathfrak{L}_2(X, \mathscr{A}, \mu + \nu)$, we can write

$$L(f) = \int_X f \operatorname{Re} h\, d(\mu + \nu) - i \int_X f \operatorname{Im} h\, d(\mu + \nu)\,.$$

Assume that $\operatorname{Im} h$ fails to vanish $(\mu + \nu)$-a.e.; say the set

$$A = \{x : \operatorname{Im} h(x) > 0\}$$

satisfies $(\mu + \nu)(A) > 0$. Then $L(\xi_A) = \int_A \operatorname{Re} h\, d(\mu + \nu) - i \int_A \operatorname{Im} h\, d(\mu + \nu)$ is not real. By (1), L is obviously real-valued on real-valued functions. This is a contradiction. Similarly, if h were negative on a set B such that $(\mu + \nu)(B) > 0$, we would have $L(\xi_B) < 0$, which again contradicts the definition of L. Thus h is real and nonnegative $(\mu + \nu)$-a.e.; and we may suppose it to be so everywhere. The definition (1) of L and the representation (2) show that

$$\int_X f(1 - h)\, d\nu = \int_X fh\, d\mu \tag{3}$$

for all $f \in \mathfrak{L}_2(X, \mathscr{A}, \mu + \nu)$.

Next, let

$$E = \{x \in X : h(x) \geqq 1\}\,.$$

Since ξ_E is in $\mathfrak{L}_2(X, \mathscr{A}, \mu + \nu)$, we may apply (3) with $f = \xi_E$ to obtain

$$0 \leqq \mu(E) = \int_X \xi_E d\mu \leqq \int_X \xi_E h\, d\mu = \int_X \xi_E (1 - h) d\nu \leqq 0\,.$$

Thus we have $\mu(E) = 0$, and so $\nu(E) = 0$ also. [This is our only use of the hypothesis $\nu \ll \mu$.] Let $g = h\xi_{E'}$. Then $g(X) \subset [0, 1[$ and $g = h$ almost

[1] Theorem (15.11) is of course much more than we need to produce the representation (2): only the case $p = 2$ is needed. As is the case with many problems involving $\mathfrak{L}_p$ spaces, the case $p = 2$ is much the simplest, and there is in fact a proof of (15.11) for $p = 2$ couched in terms of abstract Hilbert spaces. This proof is sketched in (16.56). Using (16.56), we could then prove (19.22) and so also (19.24) and (19.27) without recourse to (15.11). It would be then possible to prove (15.11) for all $\mathfrak{L}_p(X, \mathscr{A}, \mu)$ such that $(X, \mathscr{A}, \mu)$ satisfies the hypotheses of (19.27). We prefer the proof given in (15.11), partly because it is completely general and partly because it is constructive and classical in spirit. In § 20 we construct the conjugate space of $\mathfrak{L}_1(X, \mathscr{A}, \mu)$, a process that apparently requires (19.27). Then the general case of (15.11) could be proved from (19.24).

everywhere with respect to both μ and ν. Thus (3) shows that

$$\int_X f(1-g)\,d\nu = \int_X fg\,d\mu$$

for every $f \in \mathfrak{L}_2(X, \mathscr{A}, \mu+\nu)$. □

(19.23) Theorem [LEBESGUE-RADON-NIKODÝM]. *Let $(X, \mathscr{A})$ be a measurable space and let μ and ν be finite measures on $(X, \mathscr{A})$ such that $\nu \ll \mu$. Then there exists a function $f_0 \in \mathfrak{L}_1^+(X, \mathscr{A}, \mu)$ such that*

(i) $\int_X f\,d\nu = \int_X ff_0\,d\mu$

for all nonnegative, extended real-valued, $\mathscr{A}$-measurable functions f on X. For $f \in \mathfrak{L}_1(X, \mathscr{A}, \nu)$, the function ff_0 is in $\mathfrak{L}_1(X, \mathscr{A}, \mu)$ and (i) *holds. In particular, we have*

(ii) $\nu(A) = \int_A f_0\,d\mu$

for all $A \in \mathscr{A}$.

Proof. First consider any bounded, nonnegative, $\mathscr{A}$-measurable function f. Let g be the function of (19.22.i). Since both g and f are bounded and $\mu+\nu$ is a finite measure, the function $(1+g+\cdots+g^{n-1})f$ is in $\mathfrak{L}_2(X, \mathscr{A}, \mu+\nu)$ for every positive integer n; and by (19.22) the equality

$$\int_X (1+g+g^2+\cdots+g^{n-1})\,f(1-g)\,d\nu$$
$$= \int_X (1+g+g^2+\cdots+g^{n-1})\,fg\,d\mu$$

holds. Since $0 \leq g(x) < 1$ for all $x \in X$, this equality can be written

$$\int_X (1-g^n)\,f\,d\nu = \int_X \frac{g}{1-g}(1-g^n)\,f\,d\mu\,. \tag{1}$$

The sequence of functions $(1-g^n)f$ increases to f as n goes to infinity. Using (12.22) to pass to the limit in both sides of (1), we have

$$\int_X f\,d\nu = \int_X \frac{g}{1-g}\,f\,d\mu\,. \tag{2}$$

Putting $f = 1$ in (2), we see that the function $\frac{g}{1-g}$ is in $\mathfrak{L}_1^+(X, \mathscr{A}, \mu)$; define f_0 as the function $\frac{g}{1-g}$.

If f is an unbounded, nonnegative, $\mathscr{A}$-measurable function, then we may write $f = \lim_{m\to\infty} f_m$, where $f_m = \min\{f, m\}$, and apply (12.22) to (2) to obtain (i). The other assertions of the theorem are now clear. □

(19.24) LEBESGUE-RADON-NIKODÝM Theorem. *Let μ and ν be σ-finite measures on $(X, \mathscr{A})$ such that $\nu \ll \mu$. Then there exists a nonnegative, finite-valued, $\mathscr{A}$-measurable function f_0 on X such that*

(i) $\int_X f\,d\nu = \int_X ff_0\,d\mu$

for all nonnegative, extended real-valued, $\mathscr{A}$-measurable functions f on X. For $f \in \mathfrak{L}_1(X, \mathscr{A}, \nu)$, the function ff_0 is in $\mathfrak{L}_1(X, \mathscr{A}, \mu)$, and (i) *holds. In particular, we have*

(ii) $\nu(A) = \int\limits_A f_0 \, d\mu$

for all $A \in \mathscr{A}$. Moreover, f_0 is unique in the sense that if g_0 is any nonnegative, extended real-valued, $\mathscr{A}$-measurable function for which (ii) *holds, then* $g_0 = f_0$ μ-a.e.

Proof. Let $\{A_n\}_{n=1}^{\infty}$ and $\{B_n\}_{n=1}^{\infty}$ be pairwise disjoint families of $\mathscr{A}$-measurable sets, each with union X, such that $\mu(A_n) < \infty$ and $\nu(B_n) < \infty$ for all n. The family $\mathscr{C} = \{A_m \cap B_n\}_{m,n=1}^{\infty}$ is pairwise disjoint, and its union is X. Also, each member of this family has finite ν and μ measure. Let $(E_n)_{n=1}^{\infty}$ be any arrangement of $\mathscr{C}$ into a sequence of sets. For each n, define μ_n and ν_n on $\mathscr{A}$ by $\mu_n(A) = \mu(A \cap E_n)$ and $\nu_n(A) = \nu(A \cap E_n)$. Then μ_n and ν_n are finite measures on $(X, \mathscr{A})$ and $\nu_n \ll \mu_n$ for each n; and so (19.23) applies. Thus, for each n, we obtain a nonnegative, finite-valued, $\mathscr{A}$-measurable function f_n on X such that

$$\int\limits_X f \, d\nu_n = \int\limits_X f f_n \, d\mu_n \tag{1}$$

for all nonnegative $\mathscr{A}$-measurable functions f on X. Let f_0 be the function on X which is equal to f_n on E_n for all $n \in N$. It is easy to see that f_0 is nonnegative, $\mathscr{A}$-measurable, and finite-valued. Also, by (12.21), if f is a nonnegative, $\mathscr{A}$-measurable function on X, then

$$\begin{aligned}\int\limits_X f \, d\nu = \sum_{n=1}^{\infty} \int\limits_X \xi_{E_n} f \, d\nu = \sum_{n=1}^{\infty} \int\limits_X f \, d\nu_n &= \sum_{n=1}^{\infty} \int\limits_X f f_n \, d\mu_n \\ &= \sum_{n=1}^{\infty} \int\limits_X \xi_{E_n} f f_0 \, d\mu = \int\limits_X f f_0 \, d\mu \,.\end{aligned}$$

This proves (i) for nonnegative, $\mathscr{A}$-measurable f; (i) for $f \in \mathfrak{L}_1(X, \mathscr{A}, \nu)$ and (ii) follow at once.

To prove the uniqueness of f_0, let g_0 be $\mathscr{A}$-measurable and satisfy (ii). Assume that there exists a set $E \in \mathscr{A}$ such that $\mu(E) > 0$ and $f_0(x) > g_0(x)$ for all $x \in E$. For some n, we have $\mu(E \cap E_n) > 0$; if $A = E \cap E_n$, then we have $\nu(A) < \infty$, $0 < \mu(A)$ and $f_0 - g_0 > 0$ on A. Applying (12.6) and (ii), we obtain

$$0 < \int\limits_A (f_0 - g_0) \, d\mu = \nu(A) - \nu(A) = 0 \,.$$

This contradiction shows that $f_0 \leqq g_0$ μ-a.e.; similarly $f_0 \geqq g_0$ μ-a.e. □

The most general form of the LEBESGUE-RADON-NIKODÝM theorem of any conceivable use deals with an arbitrary $\nu \ll \mu$ and a μ that can be decomposed in such a way that (19.24) can be applied to each piece. The definition is as follows.

(19.25) Definition. Let $(X, \mathscr{A}, \mu)$ be a measure space. Suppose that there is a subfamily $\mathscr{F}$ of $\mathscr{A}$ with the following properties:

(i) $0 \leqq \mu(F) < \infty$ for all $F \in \mathscr{F}$;

(ii) the sets in $\mathscr{F}$ are pairwise disjoint and $\bigcup \mathscr{F} = X$;

(iii) if $E \in \mathscr{A}$ and $\mu(E) < \infty$, then $\mu(E) = \sum_{F \in \mathscr{F}} \mu(E \cap F)$;[1]

(iv) if $S \subset X$ and $S \cap F \in \mathscr{A}$ for all $F \in \mathscr{F}$, then $S \in \mathscr{A}$.

Then $(X, \mathscr{A}, \mu)$ and μ itself are said to be *decomposable* and $\mathscr{F}$ is called a *decomposition of* $(X, \mathscr{A}, \mu)$.

Our general LEBESGUE-RADON-NIKODÝM theorem holds for a decomposable μ and an arbitrary ν such that $\nu \ll \mu$. We need the following technical lemma.

(19.26) Lemma. *Let $(X, \mathscr{A})$ be a measurable space, and let μ and ν be measures on $(X, \mathscr{A})$ such that $\mu(X) < \infty$ and $\nu \ll \mu$. Then there exists a set $E \in \mathscr{A}$ such that:*

(i) *for all $A \in \mathscr{A}$ such that $A \subset E$, $\nu(A) = 0$ or $\nu(A) = \infty$;*

(ii) *for all $A \in \mathscr{A}$ such that $A \subset E$, $\mu(A) = 0$ if $\nu(A) = 0$;*

(iii) *ν is σ-finite on E'.*

Proof. With an eye to proving (i), consider the family $\mathscr{D} = \{B \in \mathscr{A} : C \subset B \text{ and } C \in \mathscr{A} \text{ imply that } \nu(C) = 0 \text{ or } \nu(C) = \infty\}$. Note that $\varnothing \in \mathscr{D}$. Define α by

$$\alpha = \sup\{\mu(B) : B \in \mathscr{D}\};$$

it is obvious that $\alpha \leqq \mu(X) < \infty$. There is a nondecreasing sequence of sets $(B_n)_{n=1}^{\infty}$ in $\mathscr{D}$ such that $\lim_{n\to\infty} \mu(B_n) = \alpha$; let $D = \bigcup_{n=1}^{\infty} B_n$. Since μ is countably additive, it is clear that $\mu(D) = \alpha$ (10.13). The set D is in $\mathscr{D}$, for if $C \in \mathscr{A}$ and $C \subset D$, then

$$\nu(C) = \nu(C \cap B_1) + \sum_{n=2}^{\infty} \nu(C \cap (B_n \cap B'_{n-1})).$$

Since each set $C \cap (B_n \cap B'_{n-1})$ is in $\mathscr{A}$ and has ν measure 0 or ∞, the same is true of C.

Now consider the set D'. We will show that for every set $F \subset D'$ such that $F \in \mathscr{A}$ and $\nu(F) > 0$, there exists a set F_1 in $\mathscr{A}$ such that $F_1 \subset F$ and

$$0 < \nu(F_1) < \infty. \tag{1}$$

If $\nu(F) < \infty$, then (1) is trivial. Thus suppose that $\nu(F) = \infty$ and assume that $\nu(G) = 0$ or $\nu(G) = \infty$ for every subset G of F such that $G \in \mathscr{A}$. Under this assumption, it is clear that $F \cup D \in \mathscr{D}$. Since $\nu \ll \mu$ and

---

[1] This possibly uncountable sum is defined as the supremum of the sums $\sum_{F \in \mathscr{D}} \mu(E \cap F)$, where $\mathscr{D}$ runs through all finite subfamilies of $\mathscr{F}$.

$\nu(F) > 0$, we have $\mu(F) > 0$. But we also have $\mu(F \cup D) = \mu(F) + \mu(D) > \alpha$, and this is a contradiction since $F \cup D \in \mathscr{D}$. The existence of a set F_1 satisfying (1) follows.

We will next show that ν is σ-finite on D'. To this end, let

$$\mathscr{F} = \{F \in \mathscr{A} : F \subset D' \text{ and } \nu \text{ is } \sigma\text{-finite on } F\}.$$

There is a nondecreasing sequence $(F_n)_{n=1}^{\infty}$ in $\mathscr{F}$ such that $\lim_{n\to\infty} \mu(F_n) = \sup\{\mu(F) : F \in \mathscr{F}\} = \beta$; let $F = \bigcup_{n=1}^{\infty} F_n$. Since F is a countable union of sets on which ν is σ-finite, ν is also σ-finite on F; thus $F \in \mathscr{F}$. Also, the equality $\mu(F) = \beta$ follows from (10.13). We claim that $\nu(F' \cap D') = 0$. If not, then by the preceding paragraph, there exists a set $H \in \mathscr{A}$ such that $H \subset D' \cap F'$ and $0 < \nu(H) < \infty$; hence $F \cup H \in \mathscr{F}$ and $\mu(H) > 0$. However we have

$$\mu(F \cup H) = \mu(H) + \mu(F) > \mu(F) = \beta \geqq \mu(F \cup H).$$

This contradiction shows that $\nu(F' \cap D') = 0$, and so ν is σ-finite on D'.

Finally, we define the promised set E. Let

$$\mathscr{G} = \{B \in \mathscr{A} : B \subset D \text{ and } \nu(B) = 0\}.$$

There is a nondecreasing sequence of sets $(B_n)_{n=1}^{\infty}$ in $\mathscr{G}$ such that $\lim_{n\to\infty} \mu(B_n) = \sup\{\mu(B) : B \in \mathscr{G}\} = \gamma$. Let $G = \bigcup_{n=1}^{\infty} B_n$, and let $E = D \cap G'$. Since $\nu(G) = 0$, ν is σ-finite on $E' = D' \cup G$; *i.e.*, (iii) is satisfied. The assertion (i) is clear since $E \subset D$. To prove (ii), assume that there is a set $B \subset E$ such that $B \in \mathscr{A}$, $\nu(B) = 0$, and $\mu(B) > 0$. Then we must have $G \cup B \in \mathscr{G}$. But this is impossible since

$$\mu(G \cup B) = \mu(G) + \mu(B) > \mu(G) = \gamma \geqq \mu(G \cup B);$$

and so (ii) is proved. □

We can now prove our final version of the LEBESGUE-RADON-NIKODÝM theorem.

(19.27) LEBESGUE-RADON-NIKODÝM Theorem. *Let $(X, \mathscr{A}, \mu)$ be decomposable with decomposition $\mathscr{F}$, and let ν be any measure on $(X, \mathscr{A})$ such that $\nu \ll \mu$. There exists a nonnegative, extended real-valued, $\mathscr{A}$-measurable function f_0 on X [which can be chosen finite-valued on each $F \in \mathscr{F}$ where ν is σ-finite] with the following properties:*

(i) $\nu(A) = \int_A f_0 \, d\mu$

for all $A \in \mathscr{A}$ that are σ-finite with respect to μ;

(ii) $\int_X f \, d\nu = \int_X f f_0 \, d\mu$

for all nonnegative, extended real-valued, $\mathscr{A}$-measurable functions f on X such that $\{x \in X : f(x) > 0\}$ is σ-finite with respect to μ;

(iii) *if* $f \in \mathfrak{L}_1(X, \mathscr{A}, \nu)$ *and* $\{x \in X : f(x) \neq 0\}$ *is* σ*-finite with respect to* μ*, then* $f f_0 \in \mathfrak{L}_1(X, \mathscr{A}, \mu)$ *and* $\int_X f\,d\nu = \int_X f f_0\,d\mu$.

Also f_0 *is unique, in the sense that if* g_0 *is any nonnegative, extended real-valued,* $\mathscr{A}$*-measurable function on* X *for which*

(iv) $\nu(A) = \int_A g_0\,d\mu$

for all $A \in \mathscr{A}$ *such that* $\mu(A) < \infty$*, then* $f_0 \xi_E$ *and* $g_0 \xi_E$ *are equal* μ*-a.e. for all* $E \in \mathscr{A}$ *that are* σ*-finite with respect to* μ.

Proof. For each $F \in \mathscr{F}$, the restriction of ν to F is absolutely continuous with respect to the restriction of μ to F, and so by (19.26) there are sets D_F and E_F in $\mathscr{A}$ such that: $D_F \cap E_F = \varnothing$; $D_F \cup E_F = F$; (19.26.i) and (19.26.ii) hold for E_F; and ν is σ-finite on D_F. If ν is σ-finite on F, then we take $D_F = F$. Since ν is σ-finite on D_F and μ is finite on D_F, we can apply (19.24) to assert that there is a nonnegative, finite-valued, $\mathscr{A}$-measurable function $f_0^{(F)}$ defined on D_F such that the conclusions of (19.24) hold for μ and ν restricted to D_F. Now let f_0 be the function on X such that for all $F \in \mathscr{F}$,

$$f_0(x) = \begin{cases} f_0^{(F)}(x) & \text{if} \quad x \in D_F, \\ \infty & \text{if} \quad x \in E_F. \end{cases} \tag{1}$$

Plainly f_0 is finite-valued if each D_F is F. It is easy to see that f_0 is $\mathscr{A}$-measurable; we leave this to the reader. Let us also write D for $\bigcup_{F \in \mathscr{F}} D_F$ and E for $\bigcup_{F \in \mathscr{F}} E_F$.

To show that f_0 has all of the properties ascribed to it, we first consider a set $A \in \mathscr{A}$ for which $\mu(A) < \infty$. Condition (19.25.iii) ensures that

$$\mu(A) = \sum_{F \in \mathscr{F}_0} \mu(A \cap F), \tag{2}$$

the family $\mathscr{F}_0$ being a countable subfamily of $\mathscr{F}$. Condition (19.25.iii) also implies that

$$\mu(A \cap (\bigcup\{F : F \in \mathscr{F} \cap \mathscr{F}_0'\})) = 0,$$

and since $\nu \ll \mu$, we have

$$\nu(A \cap (\bigcup\{F : F \in \mathscr{F} \cap \mathscr{F}_0'\})) = 0,$$

so that

$$\nu(A) = \sum_{F \in \mathscr{F}_0} \nu(A \cap F). \tag{3}$$

It is clear from (1) that

$$\begin{aligned} \nu(A \cap F) &= \nu(A \cap D_F) + \nu(A \cap E_F) \\ &= \int_{A \cap D_F} f_0\,d\mu + \nu(A \cap E_F). \end{aligned} \tag{4}$$

By (19.26.i), the value $\nu(A \cap E_F)$ is either 0 or ∞; by (19.26.ii) and absolute continuity, $\nu(A \cap E_F)$ is zero if and only if $\mu(A \cap E_F)$ is zero.

Therefore

$$\nu(A \cap E_F) = \int\limits_{A \cap E_F} \infty \, d\mu = \int\limits_{A \cap E_F} f_0 \, d\mu \,. \tag{5}$$

Combining (3), (4), and (5), and harking back to (12.21), we find that

$$\nu(A) = \sum_{F \in \mathscr{F}_0} \nu(A \cap F) = \sum_{F \in \mathscr{F}_0} \int\limits_{A \cap F} f_0 \, d\mu = \int\limits_A f_0 \, d\mu \,. \tag{6}$$

Assertion (i) is now obvious, since both ends of (6) are countably additive.

The equality (ii) is proved by considering characteristic functions, then simple functions, and finally passing to the limit using (11.35) and (12.22). Equality (iii) follows upon writing f as a linear combination of functions in $\mathfrak{L}_1^+(X, \mathscr{A}, \nu)$.

It remains to prove the uniqueness of f_0. Let g_0 be as in (iv). Then for every $F \in \mathscr{F}$ and every $\mathscr{A}$-measurable subset A of D_F, we have

$$\nu(A) = \int\limits_A f_0 \, d\mu = \int\limits_A g_0 \, d\mu \,,$$

and so by (19.24), $f_0(x) = g_0(x)$ for μ-almost all $x \in D_F$. Now assume that there exists a set $A \in \mathscr{A}$ such that $A \subset E_F$, $\mu(A) > 0$, and

$$g_0(x) < \infty = f_0(x) \quad \text{for all} \quad x \in A.$$

From (19.26.ii), we see that $\nu(A) > 0$. For each $n \in N$, write

$$A_n = \{x \in A : g_0(x) < n\}.$$

Then $(A_n)_{n=1}^{\infty}$ is a nondecreasing sequence and $\bigcup\limits_{n=1}^{\infty} A_n = A$. Applying (10.13), we infer that

$$0 < \nu(A) = \lim_{n \to \infty} \nu(A_n) \,;$$

hence there exists an n such that $\nu(A_n) > 0$; since $A_n \subset E_F$, we have $\nu(A_n) = \infty$. A glance at (iv) reveals that

$$\infty = \nu(A_n) = \int\limits_{A_n} g_0 \, d\mu \leqq n \mu(A_n) < \infty \,.$$

This contradiction proves that $g_0 = f_0$ μ-a.e. on F. □

(19.28) Corollary. *Let $(X, \mathscr{A}, \mu)$ be a σ-finite measure space and let ν be any measure on $(X, \mathscr{A})$ such that $\nu \ll \mu$. Then* (19.27.i), (19.27.ii), (19.27.iii) *hold for all $A \in \mathscr{A}$, all nonnegative, extended real-valued, $\mathscr{A}$-measurable f, and all $f \in \mathfrak{L}_1(X, \mathscr{A}, \nu)$, respectively.*

Proof. Since X is the union of a countable family of sets of finite μ-measure, $(X, \mathscr{A}, \mu)$ is plainly decomposable; and the restrictions imposed in (19.27.i), (19.27.ii), and (19.27.iii) are no restrictions at all in the present case. □

(19.29) We next consider the LEBESGUE-RADON-NIKODÝM theorem for measure spaces $(X, \mathscr{M}_\iota, \iota)$, where X is a locally compact Hausdorff space and ι is a measure as in § 9. It turns out that every such measure space $(X, \mathscr{M}_\iota, \iota)$ is decomposable in the sense of (19.25), and so we will be able to apply (19.27). The discussion is somewhat technical, unavoidably so in our opinion.

(19.30) Theorem. *Let X be a locally compact Hausdorff space, and let $(X, \mathscr{M}_\iota, \iota)$ be a measure space constructed as in §§ 9 and 10. There exists a family $\mathscr{F}_0$ of subsets of X with the following properties:*

(i) *the sets in $\mathscr{F}_0$ are compact and have [finite!] positive measure;*

(ii) *the sets in $\mathscr{F}_0$ are pairwise disjoint;*

(iii) *if $F \in \mathscr{F}_0$, U is open, and $U \cap F \neq \varnothing$, then $\iota(U \cap F) > 0$;*

(iv) *if $E \in \mathscr{M}_\iota$ and $\iota(E) < \infty$, then $E \cap F$ is nonvoid for only a countable number of sets $F \in \mathscr{F}_0$;*

(v) *the set $D = X \cap (\cup \mathscr{F}_0)'$ is ι-measurable and is locally ι-null;*

(vi) *if Y is a subset of X such that $Y \cap F \in \mathscr{M}_\iota$ for all $F \in \mathscr{F}_0$, then $Y \in \mathscr{M}_\iota$.*

Proof. Let Ж be the collection of all families $\mathscr{F}$ of subsets of X enjoying the following properties:

(1) the sets in $\mathscr{F}$ are compact and have positive ι-measure;

(2) the sets in $\mathscr{F}$ are pairwise disjoint;

(3) if $F \in \mathscr{F}$ and U is an open subset of X for which $U \cap F \neq \varnothing$, then $\iota(U \cap F) \neq 0$.

Clearly Ж is nonvoid, since the void family satisfies (1)—(3) vacuously. It is also clear that Ж is a partially ordered set under inclusion: for $\mathscr{F}_1, \mathscr{F}_2 \in$ Ж we have $\mathscr{F}_1 \subset \mathscr{F}_2$ or we do not. If $Ж_0 \subset Ж$ and $Ж_0$ is linearly ordered by inclusion, then it is clear that $\cup\{\mathscr{F} : \mathscr{F} \in Ж_0\} \in Ж$. Thus ZORN's lemma implies that Ж contains a maximal family, which we call $\mathscr{F}_0$.

Let us show that $\mathscr{F}_0$ satisfies all of our conditions. Condition (i) holds because of (1) and the fact that compact sets have finite ι-measure (9.27). Condition (ii) is just (2), and (iii) is the same as (3). To verify (iv), consider any $E \in \mathscr{M}_\iota$ such that $\iota(E) < \infty$, and select an open set U such that $E \subset U$ and $\iota(U) < \infty$ [see (9.24)]. Assume that $E \cap F$ is nonvoid for an uncountable number of sets $F \in \mathscr{F}_0$. The same is then true of $U \cap F$, and property (iii) shows that $\iota(U \cap F) > 0$ for an uncountable number of $F \in \mathscr{F}_0$. Hence $\iota(U) = \infty$, and this contradiction proves (iv).

We next prove (v). Let U be any open set such that $\iota(U) < \infty$ and let $\mathscr{F}_1$ be the [countable] subfamily of $\mathscr{F}_0$ consisting of all F such that $\iota(U \cap F) > 0$. Then (iii) shows that $\mathscr{F}_1 = \{F \in \mathscr{F}_0 \colon U \cap F \neq \varnothing\}$. It is plain that

$$\iota(U) = \iota(U \cap (\cup \mathscr{F}_1)) + \iota(U \cap (\cup \mathscr{F}_1)')$$

since $\cup \mathscr{F}_1$ is σ-compact and hence ι-measurable; hence also

$$\iota(U) = \iota(U \cap (\cup \mathscr{F}_0)) + \iota(U \cap (\cup \mathscr{F}_0)'),$$

and (10.31) implies that $\bigcup \mathscr{F}_0$ and $(\bigcup \mathscr{F}_0)' = D$ are ι-measurable. Up to this point in the proof, we have not needed the maximality of $\mathscr{F}_0$. To prove that D is locally ι-null, we need this property. If D is not locally ι-null, then by definition there is a compact set C such that $\iota(C \cap D) > 0$, and from (10.30) and the fact that D is ι-measurable, we infer the existence of a compact set H such that $H \subset C \cap D$ and $\iota(H) > 0$. Consider the family $\mathscr{U}$ of all open sets U such that $\iota(U \cap H) = 0$. Then $\iota(H \cap (\bigcup \mathscr{U})) = 0$, for otherwise $H \cap (\bigcup \mathscr{U})$ would contain a compact set E of positive ι-measure (10.30), and E would be covered by a finite number of sets $U \cap H$, each of zero ι-measure. The set $H \cap (\bigcup \mathscr{U})'$ is compact and contained in D, and

$$\iota(H \cap (\bigcup \mathscr{U})') = \iota(H) - \iota(H \cap (\bigcup \mathscr{U})) = \iota(H) > 0 .$$

Also if V is open and $V \cap H \cap (\bigcup \mathscr{U})' \neq \varnothing$, then $V \notin \mathscr{U}$, so that

$$\iota(V \cap H \cap (\bigcup \mathscr{U})') = \iota(V \cap H) - \iota(V \cap H \cap (\bigcup \mathscr{U})) = \iota(V \cap H) > 0 .$$

Therefore we can adjoin $H \cap (\bigcup \mathscr{U})'$ to $\mathscr{F}_0$ and still preserve properties (1)—(3). This contradicts the maximality of $\mathscr{F}_0$.

It remains only to prove (vi), to do which we appeal to (10.31). Let U be an open set such that $\iota(U) < \infty$, let $\mathscr{F}_1 = \{F \in \mathscr{F}_0 : U \cap F \neq \varnothing\}$, and let Y be as in (vi). Then we write

$$\begin{aligned} U \cap Y &= (U \cap Y \cap D) \cup (U \cap Y \cap \bigcup \mathscr{F}_0) \\ &= (U \cap Y \cap D) \cup \bigcup_{F \in \mathscr{F}_1} (U \cap Y \cap F) . \end{aligned}$$

The set $U \cap Y \cap D$ is ι-measurable because it is locally ι-null (10.32) [note that every subset of a locally ι-null set is locally ι-null]. The set $\bigcup_{F \in \mathscr{F}_1} (U \cap Y \cap F)$ is a countable union of ι-measurable sets and so is ι-measurable. Hence $U \cap Y$ is ι-measurable, and (10.31) shows that Y is ι-measurable. □

(19.31) Corollary. *Let* $(X, \mathscr{M}_\iota, \iota)$ *be as in* (19.30). *The measure space* $(X, \mathscr{M}_\iota, \iota)$ *is decomposable in the sense of* (19.25).

Proof. Let $\mathscr{F}_0$ and D be as in (19.30) and let $\mathscr{F} = \mathscr{F}_0 \cup \{\{x\} : x \in D\}$. It is clear from (19.30.ii) and (19.30.v) that the sets in $\mathscr{F}$ are pairwise disjoint and that $\bigcup \mathscr{F} = X$; *i.e.*, (19.25.ii) holds for $\mathscr{F}$. Since each set in $\mathscr{F}$ is compact, (19.25.i) holds. Suppose that $E \in \mathscr{M}_\iota$ and that $\iota(E) < \infty$. Since D is locally ι-null, we have $\iota(E \cap D) = 0$, and so $\iota(E \cap \{x\}) = 0$ for all $x \in D$. Thus it follows from (19.30.iv) and the countable additivity of ι that

$$\begin{aligned} \sum_{F \in \mathscr{F}} \iota(E \cap F) &= \sum_{x \in D} \iota(E \cap \{x\}) + \sum_{F \in \mathscr{F}_0} \iota(E \cap F) \\ &= \iota(E \cap D) + \iota(E \cap (\bigcup \mathscr{F}_0)) = \iota(E) ; \end{aligned}$$

hence (19.25.iii) is satisfied by $\mathscr{F}$. Condition (19.25.iv) follows at once from (19.30.vi). □

We now present our version of the LEBESGUE-RADON-NIKODÝM theorem for locally compact Hausdorff spaces.

(19.32) Theorem. *Let X be a locally compact Hausdorff space and ι a measure on X as constructed in §§ 9 and 10. Let ν be any measure whatever on $(X, \mathscr{M}_\iota)$ such that $\nu \ll \iota$. Then all of the conclusions of (19.27) hold with μ replaced by ι and $\mathscr{A}$ by $\mathscr{M}_\iota$.*

Proof. By (19.31), $(X, \mathscr{M}_\iota, \iota)$ is decomposable. Now we have only to apply (19.27). □

(19.33) Remark. If X is a locally compact Hausdorff space and if ι and η are any two outer measures on X constructed from nonnegative linear functionals as in § 9 such that $\iota(E) = 0$ implies $\eta(E) = 0$, then $\mathscr{M}_\iota \subset \mathscr{M}_\eta$. To see this, let $A \in \mathscr{M}_\iota$ and let F be compact. According to (10.34), we have $A \cap F = B \cup E$ where $B \in \mathscr{B}(X)$ and $\iota(E) = 0$. Since $\mathscr{B}(X) \subset \mathscr{M}_\eta$ and $\eta(E) = 0$, we have $A \cap F \in \mathscr{M}_\eta$. It follows from (10.31) that $A \in \mathscr{M}_\eta$. Thus (19.27) holds for ι and η on $(X, \mathscr{M}_\iota)$.

(19.34) Note. Our labors throughout (19.25) to (19.33) would be unnecessary if all measures were σ-finite and all locally compact Hausdorff spaces σ-compact. One may well ask if the generality obtained in (19.27) and (19.33) is worth the effort. Many mathematicians believe it is not. But we feel it our duty to show the reader the most general theorems that we can reasonably produce and that he might reasonably need. Examples showing the failure of other plausible versions of the LEBESGUE-RADON-NIKODÝM theorem appear in Exercise (19.71).

(19.35). It is easy to extend the LEBESGUE-RADON-NIKODÝM theorem to the cases that μ and ν are signed measures or complex measures, by making use of Hahn and Jordan decompositions. We restrict our attention to just one important extension of this sort.

(19.36) Theorem. *Let $(X, \mathscr{A}, \mu)$ be a σ-finite measure space and let ν be a complex measure on $(X, \mathscr{A})$ such that $\nu \ll \mu$. Then there exists a unique $f_0 \in \mathfrak{L}_1(X, \mathscr{A}, \mu)$ such that*

(i) $\int_X f\, d\nu = \int_X f f_0\, d\mu$

for all $f \in \mathfrak{L}_1(X, \mathscr{A}, |\nu|)$ and

(ii) $\nu(A) = \int_A f_0\, d\mu$

for all $A \in \mathscr{A}$. Moreover,

(iii) $|\nu|(A) = \int_A |f_0|\, d\mu$

for all $A \in \mathscr{A}$, and in particular

(iv) $|\nu|(X) = \int_X |f_0|\, d\mu = \|f_0\|_1$.

Proof. Let $\nu = \sum_{k=1}^{4} \alpha_k \nu_k$ be the Jordan decomposition of ν [(19.13.ii) and (19.15.c)]. According to (19.13) and (19.20), ν_1, ν_2, ν_3, ν_4, and $|\nu|$ are finite measures on $(X, \mathscr{A})$, and each of them is absolutely continuous with respect to μ. By (19.24), there exist nonnegative, finite-valued, $\mathscr{A}$-measurable functions f_k on X such that $f \in \mathfrak{L}_1(X, \mathscr{A}, \nu_k)$ implies

$$\int_X f\, d\nu_k = \int_X f f_k \, d\mu \tag{1}$$

$(k \in \{1, 2, 3, 4\})$. Let $f_0 = \sum_{k=1}^{4} \alpha_k f_k$, where $\alpha_1 = 1$, $\alpha_2 = -1$, $\alpha_3 = i$, and $\alpha_4 = -i$. Of course f_0 is in $\mathfrak{L}_1(X, \mathscr{A}, \mu)$ since each f_k is in $\mathfrak{L}_1(X, \mathscr{A}, \mu)$ [set $f = 1$ in (1)]. Then if $f \in \mathfrak{L}_1(X, \mathscr{A}, |\nu|)$, (19.16) shows that $f \in \mathfrak{L}_1(X, \mathscr{A}, \nu_k)$ for all k, and so (1) and (19.17) yield

$$\int_X f\, d\nu = \sum_{k=1}^{4} \alpha_k \int_X f\, d\nu_k = \sum_{k=1}^{4} \alpha_k \int_X f f_k\, d\mu = \int_X f f_0\, d\mu$$

[here we have used the fact that $f f_k \in \mathfrak{L}_1(X, \mathscr{A}, \mu)$ for all k, which is evident from (1)]. This proves (i). The identity (ii) follows from (i) upon taking $f = \xi_A$, since $|\nu|$ is a finite measure.

To prove the uniqueness of f_0 in $\mathfrak{L}_1(X, \mathscr{A}, \mu)$, suppose that $h_0 \in \mathfrak{L}_1(X, \mathscr{A}, \mu)$ and that $\nu(A) = \int_A h_0\, d\mu$ for all $A \in \mathscr{A}$. Let $\nu = \sigma + i\tau$, where σ and τ are real-valued. Then

$$\int_A \operatorname{Re} h_0\, d\mu = \sigma(A) = \int_A \operatorname{Re} f_0\, d\mu$$

for all $A \in \mathscr{A}$. It follows, as in the uniqueness proof of (19.24), that $\operatorname{Re} h_0 = \operatorname{Re} f_0$ μ-a.e.; similarly $\operatorname{Im} h_0 = \operatorname{Im} f_0$ μ-a.e. Thus h_0 and f_0 are the same element of $\mathfrak{L}_1(X, \mathscr{A}, \mu)$.

Finally we prove (iii). Let $A \in \mathscr{A}$ be fixed. For an arbitrary measurable dissection $\{A_1, \ldots, A_n\}$ of A we have

$$\sum_{j=1}^{n} |\nu(A_j)| = \sum_{j=1}^{n} \Big|\int_{A_j} f_0\, d\mu\Big| \leqq \sum_{j=1}^{n} \int_{A_j} |f_0|\, d\mu = \int_A |f_0|\, d\mu\,.$$

Taking the supremum over all such dissections, we obtain

$$|\nu|(A) \leqq \int_A |f_0|\, d\mu\,. \tag{2}$$

To prove the reversed inequality, use (11.35) to choose a sequence $(\sigma_m)_{m=1}^{\infty}$ of $\mathscr{A}$-measurable simple functions such that $\sigma_m \to \xi_A \operatorname{sgn} \bar{f}_0$ μ-a.e. and $|\sigma_m| \leqq |\xi_A \operatorname{sgn} \bar{f}_0| \leqq 1$. Then

$$|f_0 \sigma_m| \leqq |f_0| \in \mathfrak{L}_1(X, \mathscr{A}, \mu)$$

for all m, and so LEBESGUE's dominated convergence theorem (12.30) implies that

$$\int_A |f_0|\, d\mu = \int_X f_0 \xi_A \operatorname{sgn} \bar{f}_0\, d\mu$$
$$= \lim_{m\to\infty} \int_X f_0 \sigma_m\, d\mu\,. \tag{3}$$

Each σ_m has the form $\sum_{j=1}^{n} \beta_j \xi_{A_j}$, where $\{A_1, \ldots, A_n\}$ is a measurable dissection of A and $|\beta_j| \leq 1$ for all j. Therefore

$$\left|\int_X f_0 \sigma_m \, d\mu\right| = \left|\sum_{j=1}^{n} \beta_j \int_{A_j} f_0 \, d\mu\right| \leq \sum_{j=1}^{n} \left|\int_{A_j} f_0 \, d\mu\right|$$

$$= \sum_{j=1}^{n} |\nu(A_j)| \leq |\nu|(A). \quad (4)$$

Combining (3) and (4), we have

$$\int_A |f_0| \, d\mu \leq |\nu|(A). \quad (5)$$

Now (2) and (5) together imply (iii). Setting $A = X$, we get (iv). □

(19.37) Note. The reader should find it illuminating to compare the statement and the proof of (19.36.iii) with the corresponding result about absolutely continuous *functions* (18.1).

(19.38) Corollary. *Let ν be a complex measure on $(X, \mathscr{A})$. Then there exists an $\mathscr{A}$-measurable function f_0 on X such that:*

(i) $|f_0| = 1$;

(ii) $\nu(A) = \int_A f_0 \, d|\nu|$ *for all* $A \in \mathscr{A}$;

and

(iii) $\int_X f \, d\nu = \int_X f f_0 \, d|\nu|$ *for all* $f \in \mathfrak{L}_1(X, \mathscr{A}, |\nu|)$.

Also,

(iv) $\left|\int_X f \, d\nu\right| \leq \int_X |f| \, d|\nu|$ *for all* $f \in \mathfrak{L}_1(X, \mathscr{A}, |\nu|)$.

Proof. Obviously we have $\nu \ll |\nu|$. Define f_0 as in (19.36); then (ii) and (iii) are immediate. Let $A = \{x \in X : |f_0(x)| < 1\}$ and $B = \{x \in X : |f_0(x)| > 1\}$. Apply (19.36.iii) to get

$$\int_A (1 - |f_0|) \, d|\nu| = 0 = \int_B (|f_0| - 1) \, d|\nu|.$$

From (12.6), we see that

$$|\nu|(A) = 0 = |\nu|(B).$$

Thus, with no harm done, we redefine f_0 on $A \cup B$ so that (i) holds. For $f \in \mathfrak{L}_1(X, \mathscr{A}, |\nu|)$ we have

$$\left|\int_X f \, d\nu\right| = \left|\int_X f f_0 \, d|\nu|\right| \leq \int_X |f f_0| \, d|\nu|$$

$$= \int_X |f| \, d|\nu|,$$

and so (iv) holds. □

We next consider a relationship between pairs of measures that is the antithesis of absolute continuity.

(19.39) Definition. Let μ and ν be measures, signed measures, or complex measures on $(X, \mathscr{A})$. We say that μ and ν are [*mutually*] *singular*, and we write $\mu \perp \nu$, if there exists a set $B \in \mathscr{A}$ such that $|\mu|(B) = 0$ and $|\nu|(B') = 0$. We also say that $\mu\,[\nu]$ is *singular with respect to* $\nu\,[\mu]$.

(19.40) Theorem. *Let μ, ν, and σ be complex or signed measures on $(X, \mathscr{A})$ such that $\nu + \sigma$ is defined, and let α be in K. Then:*

(i) *$\nu \ll \mu$ and $\sigma \ll \mu$ imply $\alpha\nu \ll \mu$ and $(\nu + \sigma) \ll \mu$;*

and

(ii) *$\nu \perp \mu$ and $\sigma \perp \mu$ imply $\alpha\nu \perp \mu$ and $(\nu + \sigma) \perp \mu$.*

Proof. Suppose that $\nu \ll \mu$ and $\sigma \ll \mu$, and let $E \in \mathscr{A}$ be such that $|\mu|(E) = 0$. Then $(\alpha\nu)(E) = \alpha\nu(E) = \alpha \cdot 0 = 0$ and $(\nu + \sigma)(E) = \nu(E) + \sigma(E) = 0$. Thus (i) holds.

Next suppose that $\nu \perp \mu$ and $\sigma \perp \mu$. Choose A and B in $\mathscr{A}$ such that $|\mu|(A) = |\mu|(B) = 0$, $|\nu|(A') = 0$, and $|\sigma|(B') = 0$. Let $C = A \cup B$. It is clear that

$$|\nu + \sigma|(C') \leqq (|\nu| + |\sigma|)(C') \leqq |\nu|(A') + |\sigma|(B') = 0$$

and that $|\mu|(C) \leqq |\mu|(A) + |\mu|(B) = 0$. Thus $(\nu + \sigma)$ and μ are singular. It is obvious that $\alpha\nu \perp \mu$. □

(19.41) Theorem. *Let μ and ν be complex or signed measures on $(X, \mathscr{A})$. Then the following are equivalent:*

(i) *$\mu \perp \nu$;*

(ii) *$|\mu| \perp |\nu|$;*

and

(iii) *$\mu_k \perp \nu_j$ for all $j, k \in \{1, 2, 3, 4\}$, where $\sum_{k=1}^{4} \alpha_k \mu_k$ and $\sum_{j=1}^{4} \alpha_j \nu_j$ are the Jordan decomposition of μ and ν, respectively.*

Proof. This follows easily from (19.40) and (19.13). We omit the details. □

Our next theorem shows that if a σ-finite measure μ is given on a measurable space, then all σ-finite measures on that space can be analyzed by considering only those that are absolutely continuous or singular with respect to μ.

(19.42) Lebesgue Decomposition Theorem. *Let $(X, \mathscr{A}, \mu)$ be a σ-finite measure space and let ν be a complex measure or a σ-finite signed measure on $(X, \mathscr{A})$. Then we have*

(i) $\nu = \nu_1 + \nu_2$,

where $\nu_1 \ll \mu$ and $\nu_2 \perp \mu$. Moreover the decomposition (i) *is unique; indeed, if the decomposition $\nu = \tilde{\nu}_1 + \tilde{\nu}_2$ has the same properties, then $\tilde{\nu}_1 = \nu_1$ and $\tilde{\nu}_2 = \nu_2$.*

Proof. In view of the Jordan decomposition of ν and because of (19.20), (19.40), and (19.41), it suffices to consider the case in which ν

is a measure. Thus suppose that ν is a σ-finite measure on $(X, \mathscr{A})$. The measure ν is absolutely continuous with respect to the measure $\mu + \nu$, and both are σ-finite. Hence, by (19.24), there exists a nonnegative, real-valued, $\mathscr{A}$-measurable function f_0 on X such that

$$\int_X f\,d\nu = \int_X f f_0\,d(\mu + \nu) \tag{1}$$

for all nonnegative $\mathscr{A}$-measurable functions f. We claim that $f_0 \leqq 1$ a.e. with respect to $\mu + \nu$. To see this, let $E = \{x : f_0(x) > 1\}$ and assume that $(\mu + \nu)(E) > 0$. We can write

$$E = \bigcup_{n=1}^{\infty} \left\{x : f_0(x) \geqq 1 + \frac{1}{n}\right\},$$

and from this equality it is clear that there exists a number $\alpha > 1$ such that $(\mu + \nu)(F) > 0$, where

$$F = \{x : f_0(x) \geqq \alpha > 1\}.$$

Since $\mu + \nu$ is σ-finite, there is a set $A \in \mathscr{A}$ such that $A \subset F$ and $0 < (\mu + \nu)(A) < \infty$. Putting $f = \xi_A$ in (1), we obtain

$$\nu(A) = \int_A f_0\,d\mu + \int_A f_0\,d\nu \geqq \alpha\mu(A) + \alpha\nu(A),$$

and so

$$(1 - \alpha)\nu(A) \geqq \alpha\mu(A).$$

Since $\alpha > 1$, the inequality $\nu(A) > 0$ implies that $\alpha\mu(A) < 0$; it follows that $\nu(A) = 0$. Hence we have $0 \geqq \alpha\mu(A)$, and so $\mu(A) = 0$ also. Thus the equality $(\mu + \nu)(A) = \mu(A) + \nu(A) = 0$ holds, and this is a contradiction. If we write $f_1 = \min\{f_0, 1\}$, then $0 \leqq f_1 \leqq 1$ and

$$\int_X f\,d\nu = \int_X f f_1\,d(\mu + \nu). \tag{2}$$

Now consider the set

$$B = \{x \in X : f_1(x) = 1\}.$$

For a set $C \in \mathscr{A}$ such that $C \subset B$, $\mu(C) < \infty$, and $\nu(C) < \infty$, we put $f = \xi_C$ in (2) to find that

$$\nu(C) = \int_C f_1\,d\mu + \int_C f_1\,d\nu = \mu(C) + \nu(C),$$

so that $\mu(C) = 0$. But μ is σ-finite on B, and it follows that $\mu(B) = 0$. Defining ν_2 on $\mathscr{A}$ by

$$\nu_2(A) = \nu(A \cap B),$$

we obtain $\nu_2(B') = 0$. Thus ν_2 and μ are mutually singular. [Obviously ν_2 is a measure on $(X, \mathscr{A})$.] Writing

$$\nu_1(A) = \nu(A \cap B')$$

for $A \in \mathscr{A}$, we obviously have $\nu = \nu_1 + \nu_2$.

We must show that ν_1 is absolutely continuous with respect to μ. To do this, first consider any $C \in \mathscr{A}$ such that $\mu(C) = 0$ and $\nu(C) < \infty$. We have

$$\nu_1(C) = \nu(C \cap B') = \int_{C\cap B'} f_1 d\mu + \int_{C\cap B'} f_1 d\nu = \int_{C\cap B'} f_1 d\nu,$$

and so

$$\int_{C\cap B'} (1 - f_1)\, d\nu = 0\,. \tag{3}$$

The function $1 - f_1$ is positive on B', and so the equality (3) implies that $\nu_1(C) = \nu(C \cap B') = 0$, as desired. For an arbitrary C in $\mathscr{A}$ such that $\mu(C) = 0$, write $C = \bigcup_{n=1}^{\infty} C_n$ where the C_n's are pairwise disjoint sets in $\mathscr{A}$ and $\nu(C_n) < \infty$ for all n. The case just considered applies to each C_n, and so $\nu_1(C_n) = 0$ for $n = 1, 2, \ldots$; and of course it follows by countable additivity that $\nu_1(C) = 0$.

Finally, we prove the uniqueness of the decomposition. Suppose that $\nu = \nu_1 + \nu_2 = \tilde{\nu}_1 + \tilde{\nu}_2$, where ν_1 and $\tilde{\nu}_1$ are absolutely continuous with respect to μ and ν_2 and $\tilde{\nu}_2$ are mutually singular with respect to μ. Let B and $\tilde{B}$ be sets in $\mathscr{A}$ such that $\mu(B) = \mu(\tilde{B}) = 0$ and $\nu_2(B') = \tilde{\nu}_2(\tilde{B}') = 0$. For a set C in $\mathscr{A}$ such that $C \subset B \cup \tilde{B}$ we have $\mu(C) = 0$, and so by the absolute continuity of ν_1 and $\tilde{\nu}_1$ the equality $\nu(C) = \nu_2(C) = \tilde{\nu}_2(C)$ holds. If $C \subset B' \cap \tilde{B}'$ and $C \in \mathscr{A}$, then the equality $\nu_2(C) = \tilde{\nu}_2(C) = 0$ holds. Hence for an arbitrary set $A \in \mathscr{A}$ we have

$$\begin{aligned}\nu_2(A) &= \nu_2(A \cap (B \cup \tilde{B})) + \nu_2(A \cap (B' \cap \tilde{B}')) \\ &= \tilde{\nu}_2(A \cap (B \cup \tilde{B})) + \tilde{\nu}_2(A \cap (B' \cap \tilde{B}')) = \tilde{\nu}_2(A)\,.\end{aligned}$$

Since $\nu_2 = \tilde{\nu}_2$ and every measure in sight is σ-finite, the equality $\nu_1 = \tilde{\nu}_1$ also holds. □ [1]

(19.43) Definition. The [essentially unique] function f_0 appearing in the LEBESGUE-RADON-NIKODÝM theorem [(19.23), (19.24), (19.27), (19.36)] is often called the *LEBESGUE-RADON-NIKODÝM derivative of ν with respect to μ*, and the notation $\frac{d\nu}{d\mu}$ is used to denote f_0. Also this relationship among μ, ν, and f_0 is sometimes denoted by the formulae

$$d\nu = f_0\, d\mu \quad \text{and} \quad \nu = f_0 \mu\,.$$

(19.44) Theorem [Chain Rule]. *Let μ_0, μ_1, and μ_2 be σ-finite measures on $(X, \mathscr{A})$ such that*

$$\mu_2 \ll \mu_1 \quad \textit{and} \quad \mu_1 \ll \mu_0.$$

Then

(i) $\mu_2 \ll \mu_0$

[1] There is a proof of (19.42) in S. SAKS, *loc. cit.* (18.42), which does not use the LEBESGUE-RADON-NIKODÝM theorem.

and

(ii) $\frac{d\mu_2}{d\mu_0} = \frac{d\mu_2}{d\mu_1} \cdot \frac{d\mu_1}{d\mu_0}$ μ_0-*a.e.*

Proof. Assertion (i) is trivial. Let $f_0 = \frac{d\mu_1}{d\mu_0}$ and $f_1 = \frac{d\mu_2}{d\mu_1}$. Assertion (ii) follows from the equalities

$$\int_X f \, d\mu_2 = \int_X f f_1 \, d\mu_1 = \int_X f f_1 f_0 \, d\mu_0 . \quad \square$$

We turn now to a detailed study of the relationship between absolute continuity for functions and for measures. We first show that the mappings

$$\alpha \to S_\alpha \to \lambda_\alpha$$

described in §§ 8 and 9 establish one-to-one correspondences between the set of all normalized nondecreasing functions α on R (8.20), the set of all nonnegative linear functionals on $\mathfrak{C}_{00}(R)$, and the set of all [regular] Borel measures on R.[1]

(19.45) Theorem. *Let ι be any regular Borel measure on R. Define α on R by the rule*

$$\text{(i)} \quad \alpha(t) = \begin{cases} \iota([0, t[) & \text{if } t > 0 , \\ 0 & \text{if } t = 0 , \\ -\iota([t, 0[) & \text{if } t < 0 . \end{cases}$$

Then α is a nondecreasing, real-valued, left-continuous function on R. Also, we have $\lim_{t\to-\infty} \alpha(t) > -\infty$ *if and only if* $\iota(]-\infty, 0[) < \infty$ *and* $\lim_{t\to\infty} \alpha(t) < \infty$ *if and only if* $\iota([0, \infty[) < \infty$.

Proof. If $0 < t_1 < t_2$, then

$$\alpha(t_2) - \alpha(t_1) = \iota([0, t_2[) - \iota([0, t_1[) = \iota([t_1, t_2[) \geqq 0 ,$$

and so $\alpha(t_2) \geqq \alpha(t_1)$. If $t_1 < 0 \leqq t_2$, then the inequalities $\alpha(t_1) \leqq 0 \leqq \alpha(t_2)$ hold, and so again $\alpha(t_1) \leqq \alpha(t_2)$. If $t_1 < t_2 < 0$, then

$$\alpha(t_2) - \alpha(t_1) = -\iota([t_2, 0[) + \iota([t_1, 0[) = \iota([t_1, t_2[) \geqq 0 .$$

Thus in all cases the relation $t_2 > t_1$ implies that $\alpha(t_2) \geqq \alpha(t_1)$.

To show that α is left continuous, consider first $t > 0$. Let $(\varepsilon_n)_{n=1}^{\infty}$ be any decreasing sequence of positive real number converging to 0 and such that $\varepsilon_1 < t$. We have $[0, t[= \bigcup_{n=1}^{\infty} [0, t - \varepsilon_n[$, and so

$$\alpha(t) = \iota([0, t]) = \lim_{n\to\infty} \iota([0, t - \varepsilon_n[) = \lim_{n\to\infty} \alpha(t - \varepsilon_n) .$$

[1] A *Borel measure* is of course a measure defined on the σ-algebra of Borel sets. In (19.45), we use the symbol "ι" to denote a regular (12.39) Borel measure, although ι in §§ 9 and 10 was defined as an outer measure on all subsets with a σ-algebra $\mathscr{M}_\iota$ of measurable sets. The distinction is wiped out by Theorem (19.48). It is worth while as well to note that if μ is a Borel measure on R and $\mu(F) < \infty$ for all compact sets $F \subset R$, then μ is automatically regular (12.55).

Since $\bigcap_{n=1}^{\infty} \left[-\frac{1}{n}, 0\right[= \varnothing$, we have $\lim_{n\to\infty} \alpha\left(-\frac{1}{n}\right) = \lim_{n\to\infty} \iota\left(\left[-\frac{1}{n}, 0\right[\right) = \iota(\varnothing) = 0$; and so α is left continuous at 0. Finally, if $t < 0$, then the equalities

$$\alpha\left(t-\frac{1}{n}\right) = -\iota\left(\left[t-\frac{1}{n}, 0\right[\right) = -\left[\iota\left(\left[t-\frac{1}{n}, t\right[\right) + \iota([t, 0[)\right]$$

show that $\alpha(t) - \alpha\left(t-\frac{1}{n}\right) = \iota\left(\left[t-\frac{1}{n}, t\right[\right)$. Since $\bigcap_{n=1}^{\infty}\left[t-\frac{1}{n}, t\right[= \varnothing$, it follows that $\lim_{n\to\infty} \iota\left(\left[t-\frac{1}{n}, t\right[\right) = 0$, and so α is left continuous at t.

By (10.13), the equalities $\lim_{t\to\infty} \alpha(t) = \iota([0, \infty[)$ and $-\lim_{t\to-\infty} \alpha(t) = \iota(]-\infty, 0])$ hold, and these equalities prove the last assertion of the theorem. □

(19.46) Remarks. The function α of (19.45) may fail to be right continuous. For example, if ι is the point mass defined by

$$\iota(A) = \begin{cases} 1 & \text{if} \quad 0 \in A, \\ 0 & \text{if} \quad 0 \notin A, \end{cases}$$

it is easy to see that the corresponding α is not right continuous at 0. The choice of the definition of α was in this respect arbitrary; it could as well have been defined so as to be right continuous and nondecreasing. Also, the choice that $\alpha(0) = 0$ is an arbitrary normalization. For finite measures ι [*i.e.*, $\iota(R) < \infty$], it is often more convenient to normalize α so that $\lim_{t\to-\infty} \alpha(t) = 0$. As in (8.20), we shall use the term *normalized nondecreasing function* to mean a nondecreasing function α on R that is left continuous and satisfies $\alpha(0) = 0$. Theorem (19.45) therefore defines a mapping $\iota \to \alpha$ of the set of all regular Borel measures on R *in*to the set of all normalized nondecreasing functions. We next show that this mapping is *on*to.

(19.47) Theorem. *Let α be a normalized nondecreasing function on R, let S_α be the Riemann-Stieltjes integral corresponding to α as in §8, and let λ_α be the Lebesgue-Stieltjes measure on R constructed from S_α in §9 see [in particular* (9.19)*]. If β is the normalized nondecreasing function constructed from λ_α in* (19.45)*, then $\beta = \alpha$. Thus $a < b$ in R implies*

(i) $\lambda_\alpha([a, b[) = \alpha(b) - \alpha(a)$.

Proof. The equalities $\beta(0) = \alpha(0) = 0$ are trivial. Take $t > 0$ in R. Then $\beta(t) = \lambda_\alpha([0, t[)$, and we want to show that this number is $\alpha(t)$. Consider any decreasing sequence $(\varepsilon_n)_{n=1}^{\infty}$ of positive numbers such that $\varepsilon_n \to 0$ and such that $0 < \varepsilon_n < \varepsilon_1 < t$. Let f_n be the function whose graph is pictured in Figure 8.

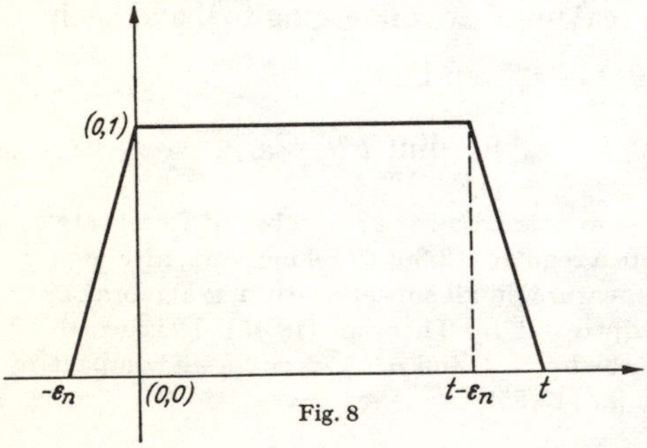

Fig. 8

The functions $(f_n)_{n=1}^{\infty}$ converge to $\xi_{[0,t[}$ everywhere, and

$$f_n \leqq \xi_{[-\varepsilon_1,t[} \in \mathfrak{L}_1(R, \mathscr{M}_{\lambda_\alpha}, \lambda_\alpha) \quad \text{for} \quad n = 1, 2, \ldots .$$

By (12.24) we have $\lim\limits_{n\to\infty} \int\limits_R f_n d\lambda_\alpha = \int\limits_R \lim\limits_{n\to\infty} f_n d\lambda_\alpha = \lambda_\alpha([0, t[) = \beta(t)$. By (12.36) the equality

$$\int\limits_R f_n d\lambda_\alpha = S_\alpha(f_n)$$

holds for $n = 1, 2, \ldots$. We complete the proof by showing that $\lim\limits_{n\to\infty} S_\alpha(f_n) = \alpha(t)$. If Δ_n is a subdivision of $[-\varepsilon_1, t]$ such that

$$\{-\varepsilon_n, 0, t - \varepsilon_n\} \subset \Delta,$$

then

$$\begin{aligned} S_\alpha(f_n) &\leqq U(f_n, \alpha, \Delta_n) \\ &\leqq (\alpha(t) - \alpha(t-\varepsilon_n)) + (\alpha(t-\varepsilon_n) - \alpha(0)) + (\alpha(0) - \alpha(-\varepsilon_n)) \\ &= \alpha(t) - \alpha(-\varepsilon_n) \to \alpha(t) - \alpha(0) = \alpha(t) \end{aligned}$$

and

$$S_\alpha(f_n) \geqq L(f_n, \alpha, \Delta_n) \geqq \alpha(t-\varepsilon_n) - \alpha(0) \to \alpha(t-) = \alpha(t).$$

[Recall that α is left continuous.] Thus

$$\alpha(t) = \lim_{n\to\infty} S_\alpha(f_n) = \lim_{n\to\infty} \int\limits_R f_n d\lambda_\alpha = \beta(t).$$

A similar argument shows that $\beta(t) = \alpha(t)$ for $t < 0$. Relation (i) now follows from the definition of β. □

(19.48) Theorem. *The mapping*

(i) $$\iota \to \alpha$$

defined in (19.45) *is a one-to-one mapping of the set of all regular Borel measures on R onto the set of all nondecreasing, real-valued, left continuous functions α on R such that $\alpha(0) = 0$. The inverse of this mapping is the mapping*

(ii) $$\alpha \to \lambda_\alpha.$$

Thus every regular Borel measure on R is a Lebesgue-Stieltjes measure.

Proof. Let α be given. Theorem (19.47) shows that α is the image of λ_α in (i) and so this mapping is onto. Suppose that α is also the image of the regular Borel measure ι. Now (19.47.i) and (19.45.i) show that

$$\lambda_\alpha([a, b[) = \alpha(b) - \alpha(a) = \iota([a, b[) \tag{1}$$

for all $a < b$ in R. Each open subset U of R can be expressed as a countable disjoint union of sets of the form $[a, b[$, *e.g.*, $]0, 1[= \bigcup\limits_{n=1}^{\infty} \left[\frac{1}{n+1}, \frac{1}{n}\right[$, and so (1) implies that $\lambda_\alpha(U) = \iota(U)$ for all open $U \subset R$. Since λ_α and ι are both regular, this implies that $\lambda_\alpha(E) = \iota(E)$ for all $E \in \mathscr{B}(R)$. Thus $\lambda_\alpha = \iota$ and the mapping (i) is one-to-one. The rest is clear. □

(19.49) Remark. There is a different proof of (19.48) which does not use the regularity of ι except for the requirement that $\iota([a, b[) < \infty$ for all $a < b$ in R, this being needed to define α. [Of course (12.35) implies that such an ι is regular.] This shows again that all Borel measures on R satisfying $\iota([a, b[) < \infty$ are Lebesgue-Stieltjes measures and hence are regular. The alternate proof runs as follows. Use (19.48.1) to show that ι and λ_α agree on the *algebra* of all finite disjoint unions of intervals of any of the forms $[a, b[$, $]-\infty, b[$, or $[a, \infty[$. Note that λ_α and ι are both σ-finite on this algebra, and finally use the uniqueness part of HOPF'S extension theorem (10.39) to infer that λ_α and ι agree on the σ-algebra generated by this algebra, namely $\mathscr{B}(R)$.

(19.50) Theorem. *Riemann-Stieltjes integrals S_α are the only nonnegative linear functionals on $\mathfrak{C}_{00}(R)$.*

Proof. Let I be any nonnegative linear functional on $\mathfrak{C}_{00}(R)$ and let ι be the measure constructed from I as in §§ 9 and 10. Use (19.48) to find an α such that $\iota = \lambda_\alpha$ on $\mathscr{B}(R)$. Finally infer from RIESZ'S representation theorem (12.36) that

$$I(f) = \int_R f\, d\iota = \int_R f\, d\lambda_\alpha = S_\alpha(f)$$

for all $f \in \mathfrak{C}_{00}(R)$. □ [1]

(19.51) Notation. To simplify our statements, we will write

$$\iota \to \alpha$$

to mean that ι is a regular Borel measure on R and that α is the normalized nondecreasing function on R obtained from ι in (19.45). Of course (19.48) shows that $\iota = \lambda_\alpha$.

(19.52) Theorem. *Let $\iota \to \alpha$. Then α is continuous if and only if $\iota(\{x\}) = 0$ for all $x \in R$. In fact, $\iota(\{x\}) = \lim\limits_{h \downarrow 0} \alpha(x+h) - \alpha(x)$ for all $x \in R$.*

Proof. For all $x \in R$, we have

$$\begin{aligned} \iota(\{x\}) &= \iota\left(\bigcap_{n=1}^{\infty} \left[x, x+\frac{1}{n}\right[\right) = \lim_{n\to\infty} \iota\left(\left[x, x+\frac{1}{n}\right[\right) \\ &= \lim_{n\to\infty} \alpha\left(x+\frac{1}{n}\right) - \alpha(x) = \lim_{h\downarrow 0} \alpha(x+h) - \alpha(x), \end{aligned}$$

as (19.47.i) shows. □

(19.53) Theorem. *The function α is absolutely continuous* (18.10) *in every interval $[-p, p]$, $p \in N$, if and only if the corresponding measure ι is absolutely continuous with respect to Lebesgue measure λ.*[2]

---

[1] Theorems (19.50) and (19.48), together with (12.56), prove that the Riemann integral is the only *invariant* nonnegative linear functional on $\mathfrak{C}_{00}(R)$, up to a positive constant. See (8.16) *supra*.

[2] The σ-algebra $\mathscr{M}_\lambda$ of course contains $\mathscr{B}(R)$. In applying the definition of absolute continuity of measures we here consider the measure spaces $(R, \mathscr{B}(R), \iota)$ and $(R, \mathscr{B}(R), \lambda)$.

Proof. Suppose first that α is absolutely continuous on $[-p, p]$ for all $p \in N$. Let A be any subset of R such that $\lambda(A) = 0$, and let ε be a positive number. For every $p \in N$, there exists $\delta(p) > 0$ such that $\sum_{k=1}^{n} (\alpha(b_k) - \alpha(a_k)) < \varepsilon$ whenever $\{]a_k, b_k[\}_{k=1}^{n}$ is a pairwise disjoint family of intervals such that $\bigcup_{k=1}^{n}]a_k, b_k[\subset [-p, p]$ and $\sum_{k=1}^{n} (b_k - a_k) < \delta(p)$. Since $\lambda(A \cap [-p, p]) = 0$, there are pairwise disjoint intervals $\{]a_k, b_k[\}_{k=1}^{\infty}$ such that $A \cap]-p, p[\subset \bigcup_{k=1}^{\infty}]a_k, b_k[\subset [-p, p]$ and $\sum_{k=1}^{\infty} (b_k - a_k) < \delta(p)$. By the choice of $\delta(p)$, we have

$$\sum_{k=1}^{\infty} (\alpha(b_k) - \alpha(a_k)) \leqq \varepsilon .$$

In view of (19.47) and the continuity of α, we also have $\alpha(b_k) - \alpha(a_k) = \iota(]a_k, b_k[)$ for all $k \in N$, and so

$$\varepsilon \geqq \sum_{k=1}^{\infty} (\alpha(b_k) - \alpha(a_k)) = \sum_{k=1}^{\infty} \iota(]a_k, b_k[= \iota\left(\bigcup_{k=1}^{\infty}]a_k, b_k[\right),$$

which implies that $\iota(A \cap]-p, p[) \leqq \varepsilon$. As ε is arbitrary, the equalities

$$\iota(A) = \lim_{p \to \infty} \iota(A \cap]-p, p[) = 0$$

follow.

Next suppose that ι is absolutely continuous with respect to λ [on Borel sets], and consider any interval $[-p, p]$ where $p \in N$. By (19.24), there is a nonnegative Borel measurable function f_0 such that

$$\iota(A) = \int_A f_0 \, d\lambda$$

for all Borel sets $A \subset R$. In particular $\int_{[-p,p]} f_0 \, d\lambda = \iota([-p, p]) < \infty$, and so $f_0 \in \mathfrak{L}_1([-p, p], \mathscr{B}(R), \lambda)$. Let $\varepsilon > 0$ be given. By (12.34) there is a $\delta > 0$ such that $\int_A f_0 \, d\lambda < \varepsilon$ for all Borel sets $A \subset [-p, p]$ such that $\lambda(A) < \delta$. Thus let $\{]a_k, b_k[\}_{k=1}^{n}$ be any pairwise disjoint family of open intervals with union $A \subset [-p, p]$ for which the inequality $\lambda(A) < \delta$ holds. Then we have

$$\int_A f_0 \, d\lambda < \varepsilon ,$$

and the integral is obviously equal to

$$\sum_{k=1}^{n} \iota([a_k, b_k[) = \sum_{k=1}^{n} (\alpha(b_k) - \alpha(a_k)) .$$

That is, α is absolutely continuous on $[-p, p]$. $\square$

We close this section by obtaining a classical decomposition of measures ι, and with it a corresponding decomposition for functions α. We begin by "skimming off" the discontinuous part of ι.

(19.54) Theorem. *For $x \in R$, let ε_x be the set function such that $\varepsilon_x(A) = \xi_A(x)$ for all $A \subset R$. Let $\{x_k\}_{k=1}^{\infty}$ be any countable subset of R, and let φ be a mapping of $\{x_k\}$ into $]0, \infty[$ such that $\Sigma\{\varphi(x_k) : |x_k| \leq n\} < \infty$ for $n = 1, 2, 3, \ldots$. Let $\iota(A) = \sum_{k=1}^{\infty} \varphi(x_k)\, \varepsilon_{x_k}(A)$ for all $A \subset R$. Then ι [on $\mathscr{B}(R)$] is the unique regular Borel measure obtained from the nonnegative linear functional*

$$I(f) = \sum_{k=1}^{\infty} \varphi(x_k)\, f(x_k)$$

on $\mathfrak{C}_{00}(R)$.

The proof is easy and is omitted.

(19.55) Definition. Measures ι defined as in (19.54) are called *purely discontinuous*. A measure ι such that $\iota(\{x\}) = 0$ for all $x \in R$ is called *continuous*.[1]

(19.56) Theorem. *If ι is purely discontinuous, then the corresponding α has derivative 0 a.e.*[2]

Proof. Let $\sigma_u = \xi_{]u,\infty[}$. For $t > 0$, we have

$$\begin{aligned}\alpha(t) = \iota([0, t[) &= \Sigma\{\iota(\{u\}) : 0 \leq u < t\} \\ &= \sum_{x_k \geq 0} \iota(\{x_k\})\, \sigma_{x_k}(t)\,.\end{aligned}$$

By (17.18), the equality $\alpha'(t) = \sum_{x_k \geq 0} \iota(\{x_k\})\, \sigma'_{x_k}(t)$ holds a.e. Since $\sigma'_{x_k}(t) = 0$ for $t \neq x_k$, we have $\alpha'(t) = 0$ a.e. for $t > 0$. For $t < 0$, consider the function $\tau_u = \xi_{]-\infty,u]}$. Then $\alpha(t) = -\sum_{x_k < 0} \iota(\{x_k\})\, \tau_{x_k}(t)$, and so $\alpha'(t) = 0$ a.e. for $t \leq 0$. □

(19.57) Theorem. *Let ι be any regular Borel measure on R. Then ι can be expressed in exactly one way in the form*

(i) $$\iota = \iota_c + \iota_d\,,$$

where ι_d is a purely discontinuous measure as in (19.54) *and ι_c is a continuous regular Borel measure.*

Proof. Let $D = \{x \in R : \iota(\{x\}) > 0\}$. Since $\iota([-n, n])$ is finite for each $n \in N$, D must be countable. If D is void, let $\iota_d = 0$ and $\iota_c = \iota$. Otherwise let $D = \{x_1, x_2, \ldots\}$ be an enumeration of D, where $x_j \neq x_k$ if $j \neq k$, and

---

[1] Theorem (19.54) and Definition (19.55) have obvious generalizations to $(X, \mathscr{M}_\iota, \iota)$ for an arbitrary locally compact Hausdorff space X. See (9.20) and (10.22).

[2] Wherever "almost everywhere", "a.e.", etc. is written with no further qualification, we mean with respect to Lebesgue measure.

define

$$\iota_d = \sum_k \iota(\{x_k\})\, \varepsilon_{x_k}.$$

It is obvious that ι_d is a purely discontinuous measure as in (19.54) and that

$$\iota_d(E) = \sum_{x \in E} \iota(\{x\}) \leqq \iota(E) \tag{1}$$

for every Borel set E. It is immediate from (1) that

$$\sum_k f(x_k)\, \iota(\{x_k\}) = \int_R f\, d\iota_d \leqq \int_R f\, d\iota \tag{2}$$

for every nonnegative Borel measurable function f on R. Define I_c on $\mathfrak{C}_{00}(R)$ by

$$I_c(f) = \int_R f\, d\iota - \int_R f\, d\iota_d. \tag{3}$$

Since $\iota(F)$ is finite for every compact subset F of R, both integrals on the right side of (3) are finite. Thus (2) shows that I_c is a nonnegative linear functional on $\mathfrak{C}_{00}(R)$. Let ι_c be the regular Borel measure on R constructed from I_c as in § 9. Theorem (12.36) shows that

$$\int_R f\, d\iota = \int_R f\, d\iota_c + \int_R f\, d\iota_d = \int_R f\, d(\iota_c + \iota_d)$$

for all $f \in \mathfrak{C}_{00}(R)$. The equality (i) follows from (19.54) and (12.41). For all $x \in R$, it is clear that $\iota_c(\{x\}) = \iota(\{x\}) - \iota_d(\{x\}) = 0$, and a moment's thought shows that the decomposition (i) of ι is unique. □

(19.58) Remark. Let ι be a regular Borel measure on R and let ι_c and ι_d be as in (19.57). Let the correspondences $\iota \to \alpha$, $\iota_c \to \alpha_c$, and $\iota_d \to \alpha_d$ be as in (19.51). It is easy to see from (19.45) that

$$\alpha = \alpha_c + \alpha_d.$$

The function α_c is continuous (19.52). The function α_d is called a *saltus function*; it has derivative zero a.e. (19.56) and a jump [or *saltus*] equal to $\alpha(x+) - \alpha(x) = \iota(\{x\})$ at each of its [countably many] discontinuities.

Next we turn to the question of singularity as regards measures and their corresponding functions α. First we need a lemma.

(19.59) Lemma. *Let μ and ν be σ-finite measures on a measurable space $(X, \mathscr{A})$. Then we have $\nu \perp \mu$ if and only if there exists no σ-finite measure $\tilde{\nu}$ on $(X, \mathscr{A})$ such that $\tilde{\nu} \neq 0$, $\tilde{\nu} \leqq \nu$, and $\tilde{\nu} \ll \mu$.*

Proof. If $\nu \perp \mu$, then $\nu = 0 + \nu$ is the *unique* Lebesgue decomposition of ν given in (19.42). Suppose that $\tilde{\nu}$ is a σ-finite measure on $(X, \mathscr{A})$ such that $\tilde{\nu} \leqq \nu$ and $\tilde{\nu} \ll \mu$. Then there is a σ-finite measure π on $(X, \mathscr{A})$ such that $\nu = \tilde{\nu} + \pi$ $\Big[$define $\pi(E) = \lim\limits_{n\to\infty} (\nu(E \cap A_n) - \tilde{\nu}(E \cap A_n))$, where

$A_1 \subset A_2 \subset \cdots$, $\tilde{\nu}(A_n) < \infty$, and $X = \bigcup_{n=1}^{\infty} A_n\Big]$. Thus

$$\nu = \tilde{\nu} + \pi_1 + \pi_2, \tag{1}$$

where $\pi_1 \ll \mu$ and $\pi_2 \perp \mu$. Since $\tilde{\nu} + \pi_1 \ll \mu$, (1) is a Lebesgue decomposition of ν. Therefore $\tilde{\nu} = 0$.

Suppose that ν and μ are not mutually singular. Use (19.42) to write $\nu = \nu_1 + \nu_2$, where $\nu_1 \ll \mu$ and $\nu_2 \perp \mu$. Then $\nu_1 \neq 0$ and $\nu_1 \leqq \nu$; hence ν_1 is the desired $\tilde{\nu}$. □

(19.60) Theorem. *Let ι be any regular Borel measure on R and let α be the normalized nondecreasing function associated with ι as in* (19.45). *Then ι and λ are mutually singular if and only if α' vanishes almost everywhere.*

Proof. Suppose first that ι and λ are not mutually singular. By (19.42), we have $\iota = \iota_1 + \iota_2$, where $\iota_1 \neq 0$, $\iota_1 \ll \lambda$, and $\iota_2 \perp \lambda$. Then ι_1 is a regular measure, as (19.49) shows.[1]

Next suppose that $\iota_1 \to \alpha_1$, *i.e.*, $\iota_1 = \lambda_{\alpha_1}$. Then α_1 is absolutely continuous on every interval $[-p, p]$ (19.53), and (19.45) and (18.16) imply that

$$\iota_1([0, x[) = \alpha_1(x) = \int_0^x \alpha_1'(t)\, dt \quad \text{for} \quad x > 0$$

and

$$\iota_1([x, 0[) = -\alpha_1(x) = \int_x^0 \alpha_1'(t)\, dt \quad \text{for } x < 0\,.$$

Since $\iota_1 \neq 0$ and ι_1 is regular, α_1' is positive on a Borel set E such that $\lambda(E) > 0$. If $\iota_2 \to \alpha_2$, then $\alpha = \alpha_1 + \alpha_2$, and so $\alpha' = \alpha_1' + \alpha_2' \geqq \alpha_1'$ a.e.; hence α' cannot vanish a.e. We have shown that $\alpha' = 0$ a.e. implies $\iota \perp \lambda$.

To prove the converse, suppose that $\alpha' > 0$ on a Borel set A of positive Lebesgue measure. By (18.14), α' is Lebesgue measurable on R and of course $\alpha' \geqq 0$ a.e. Define σ on $\mathscr{B}(R)$ by

$$\sigma(E) = \int_E \alpha'(t)\, dt\,.$$

It is obvious that

$$\sigma \ll \lambda \tag{1}$$

and that

$$\sigma(A) > 0\,. \tag{2}$$

---

[1] Here is an alternate proof. If E is a bounded Borel set, choose a decreasing sequence $(U_n)_{n=1}^{\infty}$ of bounded open sets such that $E \subset \bigcap_{n=1}^{\infty} U_n = B$, and $\lambda(B \cap E') = 0$. Then $\iota_1(B \cap E') = 0$, $\iota_1(U_1) \leqq \iota(U_1) < \infty$, and $\iota_1(E) = \iota_1(B) = \lim_{n\to\infty} \iota_1(U_n)$. Unbounded Borel sets of finite ι_1-measure are now easily dealt with. If U is open in R, choose an increasing sequence $(F_n)_{n=1}^{\infty}$ of compact sets such that $A = \bigcup_{n=1}^{\infty} F_n \subset U$ and $\lambda(U \cap A') = 0$. Then $\iota_1(U \cap A') = 0$ and $\iota_1(U) = \iota_1(A) = \lim_{n\to\infty} \iota_1(F_n)$. Thus ι_1 is regular.

From (18.14) we have

$$\sigma([a, b[) = \int_a^b \alpha'(t)\, dt \leqq \alpha(b) - \alpha(a) = \iota([a, b[) \tag{3}$$

for all $a < b$ in R. Like ι_1 in the preceding paragraph, σ is a regular Borel measure on R. It follows from (3), as in the proof of (19.48), that

$$\sigma(E) \leqq \iota(E) \tag{4}$$

for all $E \in \mathscr{B}(R)$. Combining (1), (2), (4), and (19.59), we see that $\iota \perp \lambda$ cannot obtain. Thus $\iota \perp \lambda$ implies $\alpha' = 0$ a.e. □

We now present our main decomposition theorem for regular Borel measures on R and their corresponding nondecreasing functions.

(19.61) Theorem. *Let ι be any regular Borel measure on R. Then ι can be expressed in just one way in the form*

(i) $$\iota = \iota_a + \iota_s + \iota_d,$$

where ι_a, ι_s, and ι_d are regular Borel measures on R, $\iota_a \ll \lambda$, $\iota_s \perp \lambda$, ι_d is purely discontinuous, and ι_s is continuous. If α, α_a, α_s, and α_d are the corresponding nondecreasing functions [see (19.45)], then:

(ii) $$\alpha = \alpha_a + \alpha_s + \alpha_d;$$

α_a is absolutely continuous on every compact interval; α_s is continuous; $\alpha_s' = 0$ a.e.; and α_d is a saltus function. Furthermore, we have

(iii) $$\iota_a(E) = \int_E \alpha'(t)\, dt$$

for all Borel sets E, and

(iv) $$\alpha_a(x) = \begin{cases} \int_0^x \alpha'(t)\, dt & \text{for } x \geqq 0, \\ -\int_x^0 \alpha'(t)\, dt & \text{for } x < 0. \end{cases}$$

Proof. We proved in (19.57) that $\iota = \iota_c + \iota_d$, where ι_c is regular and continuous and ι_d is purely discontinuous. We also proved that this decomposition is unique. In (19.58) we produced the decomposition $\alpha = \alpha_c + \alpha_d$, where α_c is continuous and α_d is a saltus function. In (19.56) we showed that $\alpha_d' = 0$ a.e.; hence $\alpha' = \alpha_c'$ a.e. Now define ι_a by formula (iii). Then ι_a is the regular measure σ that appears in the proof of (19.60). It is clear that $\iota_a \ll \lambda$. Let $\iota_a \to \alpha_a$. Then α_a is absolutely continuous on every compact interval (19.53) and $\iota_a(E) = \int_E \alpha_c'(t)\, dt$ for all $E \in \mathscr{B}(R)$. Applying (19.45) and (18.14), we have

$$\alpha_a(b) - \alpha_a(a) = \iota_a([a, b[) = \int_a^b \alpha_c'(t)\, dt \leqq \alpha_c(b) - \alpha_c(a) \tag{1}$$

whenever $a < b$ in R. Let α_s be the function $\alpha_c - \alpha_a$. Then α_s is continuous and, by (1), α_s is nondecreasing. Next let $\iota_s \to \alpha_s$, *i.e.*, $\iota_s = \lambda_{\alpha_s}$. Then ι_s is continuous (19.52) and regular and

$$\iota_c = \iota_a + \iota_s \,. \tag{2}$$

Formula (iv) follows at once from (iii) and the definition of α_a. From this fact and (18.3) we get $\alpha'_a = \alpha' = \alpha'_c$ a.e.; hence $\alpha'_s = 0$ a.e., and so $\iota_s \perp \lambda$ (19.60). Thus (2) gives the *unique* Lebesgue decomposition of ι_c with respect to λ. □

As usual, we close the section with a set of exercises. None of them is essential for subsequent work, but they all illustrate the theory in one way or another, and we recommend their study to the reader.

(19.62) Exercise. Let ν be a signed measure on $(X, \mathscr{A})$. Prove that

$$\nu^+(E) = \sup\{\nu(F) : F \in \mathscr{A}, F \subset E\}$$

and

$$\nu^-(E) = -\inf\{\nu(F) : F \in \mathscr{A}, F \subset E\}$$

for all $E \in \mathscr{A}$.

(19.63) Exercise. Let X be a nonvoid set and let $\mathscr{A}$ be an algebra of subsets of X. Suppose that ν is a real-valued set function defined on $\mathscr{A}$ such that $\nu(\varnothing) = 0$ and $\nu(A \cup B) = \nu(A) + \nu(B)$ if A and B are disjoint in $\mathscr{A}$. Let ν^+ and ν^- be defined as in (19.62).

(a) Prove that ν^+ and ν^- are finitely additive measures on $(X, \mathscr{A})$.

(b) Suppose that $\sup\{|\nu(A)| : A \in \mathscr{A}\} < \infty$. Prove that $\nu = \nu^+ - \nu^-$. [This is the Jordan decomposition of a finitely additive signed measure.]

(19.64) Exercise. Let $X = [-1, 1[$ and let $\mathscr{A}$ be the algebra of all finite disjoint unions of intervals of the form $[a, b[\subset [-1, 1[$. Let $f(x) = \frac{1}{x}$ for $x \neq 0$ and $f(0) = 0$. Define ν on $\mathscr{A}$ by

$$\nu\left(\bigcup_{k=1}^{n} [a_k, b_k[\right) = \sum_{k=1}^{n} f(b_k) - f(a_k) \,.$$

Prove that ν is well defined and satisfies the hypothesis of (19.63), but $\nu \neq \nu^+ - \nu^-$. Is there a Hahn decomposition of X for ν?

(19.65) Exercise. Let $X = [0, 1]$ and $\mathscr{A} = \mathscr{B}([0, 1])$. Define ν on $\mathscr{A}$ by $\nu(E) = \lambda(E) + i\lambda\left(E \cap \left[0, \frac{1}{2}\right]\right)$.

(a) Compute $|\nu|$ in terms of λ.

(b) Show that, in this case, strict inequality holds in (19.13.iii).

(c) Find a Borel measurable function g on $[0, 1]$ such that $|g| = 1$ and $\nu(E) = \int_E g\, d|\nu|$ for all $E \in \mathscr{A}$.

(19.66) Exercise. Let $(X, \mathscr{A})$ be a measurable space. Prove that the set of all complex measures on $(X, \mathscr{A})$ with setwise linear operations and $\|\nu\| = |\nu|(X)$ is a complex Banach space.

(19.67) Exercise. Let $(X, \mathscr{A})$ be a measurable space and μ and ν complex or signed measures on $\mathscr{A}$. Then ν is called *μ-continuous* if $\lim_{|\mu|(E)\to 0} \nu(E) = 0$ [that is, for every $\varepsilon > 0$ there is a $\delta > 0$ such that $|\mu|(E) < \delta$ implies $|\nu(E)| < \varepsilon$]. If ν is finite, prove that ν is μ-continuous if and only if it is absolutely continuous with respect to μ.

(19.68) Exercise. Let $(X, \mathscr{M}, \mu)$ be a measure space and $(\nu_n)_{n=1}^{\infty}$ a sequence of finite measures on $\mathscr{M}$ that are absolutely continuous with respect to μ. Suppose that $\lim_{n\to\infty} \nu_n(E) = \nu(E)$ exists and is finite for all $E \in \mathscr{M}$.

(a) Prove that the ν_n's are uniformly absolutely continuous with respect to μ; *i.e.*, $\lim_{\mu(E)\to 0} \nu_n(E) = 0$ uniformly in n.

(b) Prove that ν is a measure.

(c) Do (a) under the modified hypothesis that μ is a complex or signed measure on $\mathscr{M}$ and that the ν_n's are complex measures. Prove that, in this case, ν is a complex measure.

[Hints. Consider the metric space $(\mathscr{M}, \varrho)$ defined in (10.45). Show that each ν_n is well defined and continuous on this space. For given $\varepsilon > 0$, the families

$$\mathscr{M}_{m,n} = \left\{E \in \mathscr{M} : |\nu_n(E) - \nu_m(E)| \leqq \frac{\varepsilon}{3}\right\}, \quad m, n = 1, 2, 3, \ldots,$$

and

$$\mathscr{M}_p = \bigcap_{m,n \geqq p} \mathscr{M}_{m,n}, \qquad p = 1, 2, \ldots,$$

are thus closed. Apply the Baire category theorem to obtain an $\mathscr{M}_q$ having an interior point A. For a set $B \in \mathscr{M}$, write

$$\nu_n(B) = \nu_q(B) + [\nu_n(B) - \nu_q(B)]$$

and use the identity $\nu_k(B) = \nu_k(A \cup B) - \nu_k(A \cap B')$, $k = 1, 2, 3, \ldots$, to estimate $\nu_n(B)$. Use (a) to prove (b). To do (c), use μ to define a metric space analogous to $(\mathscr{M}, \varrho)$ and proceed as in (a) and (b).]

(19.69) Exercise. Let X be a nonvoid set and $\mathscr{M}$ a σ-algebra of subsets of X. Suppose that $(\nu_n)_{n=1}^{\infty}$ is a sequence of nonzero complex measures on $\mathscr{M}$ such that $\lim_{n\to\infty} \nu_n(E) = \nu(E)$ exists and is finite for all $E \in \mathscr{M}$. Prove that ν is a complex measure on $\mathscr{M}$. $\left[\text{Hint. Let } \mu(E) = \sum_{n=1}^{\infty} \frac{\nu_n(E)}{|\nu_n|(X)} 2^{-n},\right.$ show that each ν_n is absolutely continuous with respect to μ, and apply $\left.\text{(19.68).}\right]$

(19.70) Exercise. Let $\{\mu_\alpha\}_{\alpha \in I}$ be a family of measures on a σ-algebra $\mathscr{A}$. Prove that the set function μ given by

$$\mu(A) = \sum_{\alpha \in I} \mu_\alpha(A)$$

is a measure on $\mathscr{A}$.

(19.71) Exercise: Examples relating to the LEBESGUE-RADON-NIKODÝM Theorem.

(a) Let $(X, \mathscr{A}, \mu)$ be an arbitrary measure space and let ν on $\mathscr{A}$ be defined by: $\nu(A) = 0$ if $\mu(A) = 0$ and $\nu(A) = \infty$ if $\mu(A) > 0$. Prove that $(X, \mathscr{A}, \nu)$ is a measure space and that $\nu \ll \mu$. Find a function f_0 for which the conclusions of (19.24) are valid.

(b) Let X be a locally compact Hausdorff space admitting a nonzero measure ι as in § 9 such that $\iota(\{x\}) = 0$ for all $x \in X$. Let μ be counting measure (10.4.a), with its domain of definition restricted to $\mathscr{M}_\iota$. Show that $\iota \ll \mu$ and that there is no function f_0 on X for which (19.24.ii) obtains. Find an f_0 for which (19.24.ii) holds for all $A \in \mathscr{M}_\iota$ that are σ-finite with respect to μ.

(c) Let X be a locally compact Hausdorff space admitting a measure ι as in § 9 such that X is *not* σ-finite with respect to ι and such that $\iota(\{x\}) = 0$ for all $x \in X$. Prove that there is a subset of X that is locally ι-null but not ι-null. [Use (19.30), and choose a set S containing just one point in each $F \in \mathscr{F}_0$.]

(d) Let $X = R^2$. Impose first on X the topology $R_d \times R$ [refer to (9.41) for a description of this topology]. Let I be any nonnegative linear functional on $\mathfrak{C}_{00}(R_d \times R)$ such that $\iota(\{a\} \times R) > 0$ for all $a \in R$ and ι vanishes for points. [It is evident that many such I's exist.] Next repeat this construction with the topology of $R \times R_d$ on X; construct a nonnegative linear functional J on $\mathfrak{C}_{00}(R \times R_d)$ with corresponding measure η for which $\eta(R \times \{b\}) > 0$ for all $b \in R$ and η vanishes for points. Consider the measure space $(X, \mathscr{M}_\iota \cap \mathscr{M}_\eta, \iota + \eta)$. Observe that $\iota \ll \iota + \eta$. Prove that (19.27.i) holds for no function f_0 on X, even for all sets finite with respect to $\iota + \eta$.

(e) Generalize part (d) to products $X \times Y$, where X and Y are suitable locally compact Hausdorff spaces.

(19.72) Exercise. State and prove an analogue of FUBINI's theorem (17.18) on term-by-term differentiation, recast in terms of an infinite series of measures on a measurable space.

(19.73) Exercise. Suppose that μ and ν are σ-finite measures on $(X, \mathscr{A})$ such that $\mu \ll \nu$ and $\nu \ll \mu$. Prove that $\frac{d\nu}{d\mu} \neq 0$ a. e. and $\frac{d\mu}{d\nu} = 1 \Big/ \frac{d\nu}{d\mu}$ a. e. [Note that μ und ν have exactly the same sets of zero measure.]

(19.74) Exercise. Let $[a, b]$ be a closed interval in R and let f be a function of finite variation on $[a, b]$. Prove that

(a) $f = g + h$, where g is absolutely continuous on $[a, b]$ and $h' = 0$ a.e. on $[a, b]$, and

(b) the decomposition in (a) is unique except for additive constants.

(19.75) Exercise. There exist σ-finite measures on $\mathscr{B}(R)$ that are not Lebesgue-Stieltjes measures. Consider the following pathological example.

Let $(r_n)_{n=1}^{\infty}$ be an enumeration of Q and let g be the function such that

$$g(x) = x^{-\frac{1}{2}} \quad \text{if} \quad 0 < x < 1 \,,$$
$$g(x) = 0 \qquad \text{otherwise.}$$

Define f on R by the rule

$$f(x) = \sum_{n=1}^{\infty} 2^{-n} g(x + r_n) \,.$$

(a) Prove that $f \in \mathfrak{L}_1(R)$.

Define μ on $\mathscr{M}_\lambda$ by

$$\mu(E) = \int_E f^2 d\lambda \,.$$

Note that $\mu \ll \lambda$.

(b) Prove that $\mu([a, b]) = \infty$ for all $a < b$ in R.

(c) Prove that μ is σ-finite.

(d) Find a sequence $(F_n)_{n=1}^{\infty}$ of Borel subsets of R such that $\bigcup_{n=1}^{\infty} F_n = R$ and $\mu(F_n) < \infty$ for all $n \in N$. Can the F_n's be chosen so as to be compact?

(e) Why is this example not a counterexample to (19.48)?

(19.76) Exercise. Extend the result of (16.48.b) to subspaces of $\mathfrak{L}_2(X, \mathscr{A}, \mu)$, where $(X, \mathscr{A}, \mu)$ is an arbitrary decomposable measure space. That is, let $\mathfrak{M}$ and $\mathfrak{F}$ be as in (16.48.a). Then there is a set $E \in \mathscr{A}$ such that projection onto $\mathfrak{M}$ is just multiplication by ξ_E, and so $\mathfrak{M}$ is the set of all f in $\mathfrak{L}_2(X, \mathscr{A}, \mu)$ that vanish on E'.

(19.77) Exercise: Lebesgue decomposition for decomposable measures. Let $(X, \mathscr{A})$ be a measurable space and let μ and ν be decomposable measures on $(X, \mathscr{A})$.

(a) Prove that $\mu + \nu$ is a decomposable measure.

Modify the definition (19.39) of singularity as follows. Say that μ and ν are mutually singular if there is a set $B \in \mathscr{A}$ such that $\mu(B \cap E) = 0$ for all $E \in \mathscr{A}$ that are σ-finite with respect to μ and $\nu(B' \cap F) = 0$ for all $F \in \mathscr{A}$ that are σ-finite with respect to ν.

(b) State and prove an analogue of (19.24) for arbitrary decomposable measures μ and ν. Show that your result is a true generalization of (19.42).

§ 20. Applications of the LEBESGUE-RADON-NIKODÝM theorem

(20.1) Introduction. The LEBESGUE-RADON-NIKODÝM theorem has a large number of applications. It is very useful in establishing some well-known properties of integrals, in computing the conjugate spaces of various classical Banach spaces, in elucidating certain concepts of probability theory, and in studying product measures. We cannot give *all* known applications of the LEBESGUE-RADON-NIKODÝM theorem, or even any large part of them, and keep the text of reasonable size.

We choose four famous applications, which are either themselves important theorems in analysis or needed to establish such facts.

(20.2) Integration by substitution. We will apply the LEBESGUE-RADON-NIKODÝM theorem to prove a very general theorem on integration by substitution or by "change of variable". We wish to use the construction of continuous images of measures given in (12.45) and (12.46), and also (19.24). Hence we consider locally compact Hausdorff spaces X and Y and a continuous mapping φ of X onto Y. It is convenient to suppose that Y is σ-compact. [We could avoid this hypothesis, but at the cost of tedious complications.] Let μ and ϱ be measures on X and Y, respectively, in the sense of § 9. Suppose as in (12.45) that $\varphi^{-1}(F)$ is compact in X for F compact in Y, *or* that $\mu(X) < \infty$. Then define the measure ν on Y just as in (12.45). Since Y is σ-compact by hypothesis, ν must be σ-finite (9.27). Theorem (12.46) implies that

(i) $$\nu(B) = \mu(\varphi^{-1}(B))$$

for all ν-measurable subsets B of Y. With these preliminaries, we can state and prove our general theorem on integration by substitution.

(20.3) Theorem. *All notation is as in* (20.2). *Suppose that*

(i) $$\varrho(\varphi(E)) = 0 \text{ for all } E \subset X \text{ such that } \mu(E) = 0\,.$$

Then there is a nonnegative, real-valued, Borel measurable function w on X with the following properties. For every $f \in \mathfrak{L}_1(Y, \mathscr{M}_\varrho, \varrho)$, the function $(f \circ \varphi) w$ is in $\mathfrak{L}_1(X, \mathscr{M}_\mu, \mu)$, and

(ii) $$\int_Y f(y)\, d\varrho(y) = \int_X f \circ \varphi(x)\, w(x)\, d\mu(x)\,.$$

Furthermore, w can be taken to have the form $f_1 \circ \varphi$ for a Borel measurable function f_1 on Y.

Proof. Let ν be as in (20.2). Suppose that $B \subset Y$ and that $\nu(B) = 0$. Then as noted in (20.2.i), we have $0 = \nu(B) = \mu(\varphi^{-1}(B))$, and so our hypothesis (i) implies that $\varrho(B) = \varrho\big(\varphi(\varphi^{-1}(B))\big) = 0$. That is, $\varrho \ll \nu$. We apply Theorem (19.24) to the measurable space $(Y, \mathscr{B}(Y))$ and the measures ν and ϱ. Thus there exists a nonnegative, real-valued, Borel measurable function f_1 on Y such that

$$\int_Y g(y)\, d\varrho(y) = \int_Y g(y)\, f_1(y)\, d\nu(y) \tag{1}$$

for all Borel measurable functions g on Y for which the left side of (1) is defined. Also, if g is in $\mathfrak{L}_1(Y, \mathscr{B}(Y), \varrho)$, then $g f_1$ is in $\mathfrak{L}_1(Y, \mathscr{B}(Y), \nu)$.

We now appeal to (12.46.ii) to write

$$\int_Y g(y)\, f_1(y)\, d\nu(y) = \int_X g \circ \varphi(x)\, f_1 \circ \varphi(x)\, d\mu(x) \tag{2}$$

for all $g \in \mathfrak{L}_1(Y, \mathscr{B}(Y), \varrho)$. Since φ is continuous and f_1 is Borel measurable, it follows that $f_1 \circ \varphi$ is Borel measurable [see (10.42.a)]. Combining (1) and (2) and writing $w = f_1 \circ \varphi$, we obtain

$$\int_Y g(y)\, d\varrho(y) = \int_X g \circ \varphi(x)\, w(x)\, d\mu(x) \tag{3}$$

for all $g \in \mathfrak{L}_1(Y, \mathscr{B}(Y), \varrho)$.

Suppose that $B \subset Y$ and that $\varrho(B) = 0$. Let (U_n) be a decreasing sequence of open subsets of Y such that $U_n \supset B$, $\varrho(U_1) < \infty$, and $\lim_{n\to\infty} \varrho(U_n) = 0$. Write $A = \bigcap_{n=1}^{\infty} U_n$. Put $g = \xi_A$ in (3) and use (10.15) to obtain

$$0 = \varrho(A) = \int_X \xi_A \circ \varphi(x)\, w(x)\, d\mu(x)\,.$$

Thus $(\xi_A \circ \varphi)\, w$ vanishes μ-a.e. on X, and hence $(\xi_B \circ \varphi)\, w$ also vanishes μ-a.e. on X. If h is any $\mathscr{M}_\varrho$-measurable function on Y that vanishes except on B, we clearly have $(h \circ \varphi)\, w = (h \circ \varphi)\, (\xi_B \circ \varphi)\, w$, and so $(h \circ \varphi)\, w$ is $\mathscr{M}_\mu$-measurable and vanishes μ-a.e. on X. Finally, consider an arbitrary $f \in \mathfrak{L}_1(Y, \mathscr{M}_\varrho, \varrho)$. By (12.63), we can write

$$f = g + h\,,$$

where g is Borel measurable and h vanishes ϱ-a.e. on Y. Applying the last observations, we see that $(f \circ \varphi)\, w = (g \circ \varphi)\, w + (h \circ \varphi)\, w$, and so $(f \circ \varphi)\, w$ is the sum of a Borel measurable function and a function that vanishes μ-a.e. on X. Hence $(f \circ \varphi)\, w$ is $\mathscr{M}_\mu$-measurable. Furthermore, (3) yields

$$\int_Y f\, d\varrho = \int_Y g\, d\varrho = \int_X (g \circ \varphi)\, w\, d\mu = \int_X (f \circ \varphi)\, w\, d\mu\,. \quad \square$$

A classical case is that in which X and Y are closed intervals, as follows.

(20.4) Theorem. *Let $[a, b]$ be an interval in R and let φ be a nonconstant, real-valued, continuous N-function*[1] *defined on $[a, b]$. Let $[\alpha, \beta]$ be the image interval $\varphi([a, b])$. There is a nonnegative, real-valued, Borel measurable function w on $[a, b]$ such that for all $f \in \mathfrak{L}_1([\alpha, \beta], \mathscr{M}_\lambda, \lambda)$, $(f \circ \varphi)\, w$ is in $\mathfrak{L}_1([a, b], \mathscr{M}_\lambda, \lambda)$ and*

(i) $$\int_\alpha^\beta f(y)\, dy = \int_a^b f \circ \varphi(x)\, w(x)\, dx\,.$$

Furthermore, w can be taken to have the form $f_1 \circ \varphi$ for a certain Borel measurable function f_1 on $[\alpha, \beta]$.

Proof. In (20.3), we take $X = [a, b]$ and $Y = [\alpha, \beta]$. For μ we take Lebesgue measure λ on $[a, b]$, and for ϱ, Lebesgue measure λ on $[\alpha, \beta]$. Definition (18.24) ensures that our mapping function φ satisfies (20.3.i). Thus (20.3) applies. $\square$

[1] For the definition, see (18.24).

The function w in (20.4) can be rather complicated to compute [see (20.6)—(20.8)]. For *monotone* φ, however, it is simply $|\varphi'|$, as we now show. [This is the classical form of the theorem on integration by substitution.]

(20.5) Corollary. *Let φ be a monotone continuous N-function with domain $[a, b]$ and range $[\alpha, \beta]$ $(\alpha < \beta)$. Then φ is absolutely continuous, and the function w of (20.4) is equal to $|\varphi'|$ λ-almost everywhere on $[a, b]$. Thus for $f \in \mathfrak{L}_1([\alpha, \beta])$, we have $(f \circ \varphi)\,|\varphi'| \in \mathfrak{L}_1([a, b])$, and*

$$\text{(i)} \qquad \int_\alpha^\beta f(y)\,dy = \int_a^b f \circ \varphi(x)\,|\varphi'(x)|\,dx\,.$$

Proof. Theorem (18.25) shows that φ is absolutely continuous, since as a monotone function φ has finite variation. With no loss of generality, we suppose that φ is nondecreasing. Applying (20.4.i) to the function 1 on $[\alpha, \beta]$, we obtain

$$\varphi(b) - \varphi(a) = \int_\alpha^\beta dy = \int_a^b w(t)\,dt\,. \tag{1}$$

For every $x \in [a, b]$, the inclusion $\varphi^{-1}(\varphi([a, x])) \supset [a, x]$ holds, and hence by (20.4.i) applied to $\xi_{\varphi([a,x])}$ we have

$$\begin{aligned} \varphi(x) - \varphi(a) &= \int_\alpha^\beta \xi_{\varphi([a,x])}(y)\,dy \\ &= \int_a^b \xi_{\varphi([a,x])} \circ \varphi(t)\,w(t)\,dt \\ &= \int_a^b \xi_{\varphi^{-1}(\varphi([a,x]))}(t)\,w(t)\,dt \\ &\geqq \int_a^x w(t)\,dt\,. \end{aligned} \tag{2}$$

Similarly

$$\varphi(b) - \varphi(x) \geqq \int_x^b w(t)\,dt\,. \tag{3}$$

Inspection of (1), (2), and (3) shows that equality must hold in both (2) and (3). Since $w \in \mathfrak{L}_1([a, b])$ by (20.4) and

$$\varphi(x) = \int_a^x w(t)\,dt + \varphi(a)\,,$$

(18.3) shows that $\varphi'(x) = w(x)$ a.e. □

The following exercises illustrate the complications attendant on choosing functions w as in (20.4) for nonmonotone mapping functions φ.

(20.6) Exercise. Let φ be the function on $[0, 1]$ such that $\varphi(t) = \min\{t, 1-t\}$. Note that φ is absolutely continuous and hence an

N-function (18.25). Consider φ as a mapping of $[0, 1]$ onto $\left[0, \frac{1}{2}\right]$, to which (20.4) may be applied.

(a) Let v be a function in $\mathfrak{L}_1([0, 1], \mathcal{M}_\lambda, \lambda)$ with the property that

(i) $$\int_c^d v\,d\lambda + \int_{1-d}^{1-c} v\,d\lambda = d - c$$

for all $[c, d] \subset \left[0, \frac{1}{2}\right]$. Prove that

(ii) $$\int_0^{\frac{1}{2}} f(y)\,dy = \int_0^1 (f \circ \varphi(x))\,v(x)\,dx$$

for all $f \in \mathfrak{L}_1\left(\left[0, \frac{1}{2}\right], \mathcal{M}_\lambda, \lambda\right)$.

(b) If v is any function in $\mathfrak{L}_1([0, 1], \mathcal{M}_\lambda, \lambda)$ for which (ii) holds, prove that (i) holds for v.

(c) Infer that the function w in Theorem (20.4) is never unique if φ fails to be monotone.

(d) Is the function $|\varphi'|$ of (20.5) the only function [a.e. of course] for which (20.5.i) holds?

Extra complications appear with more complicated mapping functions φ, as the following exercise shows.

(20.7) Exercise. Let ψ be the function of period 1 on R such that $\psi(t) = t$ for $0 \leqq t \leqq \frac{1}{2}$ and $\psi(t) = 1 - t$ for $\frac{1}{2} \leqq t \leqq 1$. Define the function φ on $[0, 1]$ by the following rules:

for $2^{-1} \leqq t \leqq 1$, $\varphi(t) = 2^{-1}\psi(2t)$;
for $2^{-2} \leqq t \leqq 2^{-1}$, $\varphi(t) = 2^{-3}\psi(2^3 t)$;
for $2^{-3} \leqq t \leqq 2^{-2}$, $\varphi(t) = 2^{-5}\psi(2^5 t)$;
. . .
for $2^{-n} \leqq t \leqq 2^{-n+1}$, $\varphi(t) = 2^{-2n+1}\psi(2^{2n-1} t)$;
. . .
$\varphi(0) = 0$.

[Draw a sketch of the graph of φ.]

(a) Prove the following assertions. The variation $V_0^1 \varphi$ is equal to 1. The function φ is absolutely continuous; in fact, φ is in $\mathfrak{Lip}_1([0, 1])$. The derivative φ' assumes only the values ± 1.

(b) Consider φ as a mapping of $[0, 1]$ onto $\left[0, \frac{1}{4}\right]$ to which Theorem (20.4) may be applied. Find all of the $\mathcal{M}_\lambda$-measurable functions w on $[0, 1]$ for which (20.4.i) holds.

(20.8) Exercise. Let g be the function constructed in (18.40.a) *supra*. It is an N-function mapping $[0, 1]$ onto $[0, \beta]$, where

$$\beta = \max\left\{\frac{1}{2}(b_k - a_k) + \varrho_k\right\}_{k=1}^{\infty}.$$

Thus Theorem (20.4) can be applied, with $g = \varphi$. Since g' does not exist on F, no function w for which (20.4.i) holds can possibly be the derivative of g. Find all $\mathscr{M}_\lambda$-measurable functions on $[0, 1]$ for which (20.4.i) holds for $\varphi = g$.

(20.9) Note. There are obvious questions concerning classical transformations of integrals other than those treated in (20.4) and (20.5). For example, for a continuous mapping φ of R^n into R^n, the function w of (20.3) is in certain cases the absolute value of the *Jacobian* of the transformation φ. Lack of space and time compels us to omit this interesting subject, although its main outlines are clear enough from (20.3) and the n-dimensional Lebesgue integral to be defined in § 21.

In § 15 we computed the conjugate space for each of the Banach spaces $\mathfrak{L}_p(X, \mathscr{A}, \mu)$ for $1 < p < \infty$ and arbitrary measure spaces $(X, \mathscr{A}, \mu)$, but omitted all mention of this problem for spaces $\mathfrak{L}_1(X, \mathscr{A}, \mu)$. We now address ourselves to this computation. The results and the techniques employed here are quite different from those of § 15: our main tool is the LEBESGUE-RADON-NIKODÝM theorem.

(20.10) Remarks. Let $(X, \mathscr{A}, \mu)$ be any measure space. Let g be a bounded $\mathscr{A}$-measurable function on X and let L_g be defined on $\mathfrak{L}_1(X, \mathscr{A}, \mu)$ by the rule

(i) $$L_g(f) = \int_X f\bar{g}\, d\mu\,.$$

Obviously L_g is linear on $\mathfrak{L}_1$ and

$$|L_g(f)| \leqq \|g\|_u \|f\|_1$$

for all $f \in \mathfrak{L}_1$. Hence L_g is a bounded linear functional on $\mathfrak{L}_1$. If the function g is tampered with in any way whatever on a set of measure zero to obtain a new $\mathscr{A}$-measurable function h, then it is clear that $L_h = L_g$. Thus g need not be bounded to give rise to an element of $\mathfrak{L}_1^*$. It turns out in many important cases that all elements of $\mathfrak{L}_1^*$ indeed have the form L_g, where g is "essentially" bounded. We begin by making this idea precise.

(20.11) Definition. Let $(X, \mathscr{A}, \mu)$ be a measure space. A set $A \in \mathscr{A}$ is said to be *locally μ-null* if $\mu(A \cap E) = 0$ for all $E \in \mathscr{A}$ such that $\mu(E) < \infty$ [cf. (9.29)]. An $\mathscr{A}$-measurable function g on X is said to be *essentially bounded* if for some real number $a \geqq 0$ the set $\{x \in X : |g(x)| > a\}$ is locally μ-null. Let $\|g\|_\infty$ be the infimum of the set of all such numbers a. The number $\|g\|_\infty$ is called the *essential supremum of* $|g|$. This number is sometimes denoted by the symbols "ess sup$|g|$" or "vrai max$|g|$".

(20.12) Theorem. *If $(X, \mathscr{A}, \mu)$ is a σ-finite measure space and $A \in \mathscr{A}$, then A is locally μ-null if and only if $\mu(A) = 0$.*

Proof. Trivial.

(20.13) Theorem. *Let $(X, \mathscr{A}, \mu)$ be any measure space. Suppose that $f \in \mathfrak{L}_1(X, \mathscr{A}, \mu)$ and that g is essentially bounded. Then:*

(i) *the set $\{x \in X : |f(x)| > 0\}$ is σ-finite;*

(ii) $fg \in \mathfrak{L}_1$;

and

(iii) $$\left|\int_X fg\,d\mu\right| \leqq \|g\|_\infty \|f\|_1.$$

Proof. Let $E = \{x \in X : |f(x)| > 0\}$ and $E_n = \left\{x \in X : |f(x)| \geqq \frac{1}{n}\right\}$ for $n = 1, 2, \ldots$. Obviously $E = \bigcup_{n=1}^{\infty} E_n$ and $\mu(E_n) < \infty$ for all n. Thus (i) holds. Let a be any nonnegative real number such that the set

$$A = \{x \in X : |g(x)| > a\}$$

is locally μ-null. By (12.22), we have

$$\begin{aligned}\int_X |fg|\,d\mu &= \int_{E'} |fg|\,d\mu + \lim_{n\to\infty} \int_{E_n \cap A} |fg|\,d\mu + \int_{E\cap A'} |fg|\,d\mu \\ &\leqq 0 + 0 + a \int_{E\cap A'} |f|\,d\mu \\ &\leqq a \int_X |f|\,d\mu = a\|f\|_1.\end{aligned}$$

Taking the infimum over all such a, we obtain (ii) and (iii). □

(20.14) Theorem. *Let $(X, \mathscr{A}, \mu)$ be any measure space. Let g be a complex-valued, $\mathscr{A}$-measurable function on X.*

(i) *If g is essentially bounded, then the set $\{x \in X : |g(x)| > \|g\|_\infty\}$ is locally μ-null,* i.e., *the infimum in Definition* (20.11) *is attained.*

(ii) *The function g is essentially bounded if and only if there is a bounded $\mathscr{A}$-measurable function φ on X such that the set $\{x \in X : g(x) \neq \varphi(x)\}$ is locally μ-null.*

(iii) *If g is essentially bounded, then $\|g\|_\infty = \inf\{\|\varphi\|_u : \varphi$ is as in* (ii)}, *and this infimum is attained.*

Let $\mathfrak{L}_\infty(X, \mathscr{A}, \mu)$ denote the set of all essentially bounded $\mathscr{A}$-measurable functions on X, two functions being identified if they differ only on a locally μ-null set. Under pointwise linear operations and the norm $\|\ \|_\infty$, $\mathfrak{L}_\infty$ is a complex Banach space.

Proof. Assertion (i) follows from the evident fact that any countable union of locally μ-null sets is locally μ-null. Assertion (ii) is obvious.

To prove (iii), assume that there is a function φ as in (ii) such that $\|\varphi\|_u < \|g\|_\infty$. Then $\{x \in X : |g(x)| > \|\varphi\|_u\}$ is not locally μ-null and $g(x) \neq \varphi(x)$ at all points of this set. This contradicts the choice of φ, and so $\|\varphi\|_u \geqq \|g\|_\infty$ for all φ as in (ii). Let $B = \{x \in X : |g(x)| \leqq \|g\|_\infty\}$. Then $g\xi_B$ is bounded, $\mathscr{A}$-measurable, and equals g except on a locally μ-null subset of B'. Also, $\|g\xi_B\|_u = \|g\|_\infty$ [if $B = \varnothing$, then $\|g\|_\infty = 0$]. This proves (iii).

To prove that $\mathfrak{L}_\infty$ is a Banach space, first notice that (ii) shows that each equivalence class in $\mathfrak{L}_\infty$ contains a bounded function. Thus we may regard $\mathfrak{L}_\infty$ [as a linear space] as being the quotient space $\mathfrak{B}/\mathfrak{N}$, where $\mathfrak{B}$

is the linear space of all bounded, complex-valued, $\mathscr{A}$-measurable functions on X and $\mathfrak{N}$ is the linear subspace of all of those functions in $\mathfrak{B}$ that vanish except on a locally μ-null set. It is clear that $\mathfrak{B}$, with the uniform norm, is a Banach space and that $\mathfrak{N}$ is closed in $\mathfrak{B}$. Assertion (iii) shows that $\| \ \|_\infty$ is the quotient norm on $\mathfrak{B}/\mathfrak{N}$ [see (14.38)]. Thus (14.38.b) shows that $\mathfrak{L}_\infty$ is a Banach space. □

The reader will recall from (15.14.b) that, for certain measure spaces, if $p > 1$ and g is a measurable function such that $fg \in \mathfrak{L}_1$ for all $f \in \mathfrak{L}_p$, then $g \in \mathfrak{L}_{p'}$. The proof of this nontrivial fact suggested in (15.14) requires the uniform boundedness principle. The corresponding result for $p = 1$ and $p' = \infty$ is considerably easier to prove, as we shall now see.

(20.15) Theorem. *Let $(X, \mathscr{A}, \mu)$ be a measure space and let g be a complex-valued, $\mathscr{A}$-measurable function defined on X such that $\int_X |fg| \, d\mu < \infty$ for all $f \in \mathfrak{L}_1(X, \mathscr{A}, \mu)$. Then g is in $\mathfrak{L}_\infty(X, \mathscr{A}, \mu)$.*

Proof. Assume that $g \notin \mathfrak{L}_\infty$. Then there exist sequences of positive real numbers (α_n) and of $\mathscr{A}$-measurable sets (A_n) such that:

$$\alpha_1 < \alpha_2 < \cdots < \alpha_n < \cdots \quad \text{and} \quad \sum_{n=1}^{\infty} \frac{1}{\alpha_n} < \infty\,; \tag{1}$$

$$0 < \mu(A_n) < \infty\,; \tag{2}$$

and

$$\alpha_n < |g(x)| \leqq \alpha_{n+1} \quad \text{for} \quad x \in A_n. \tag{3}$$

To see this, let (β_n) be any sequence satisfying the conditions in (1). The assumption that $g \notin \mathfrak{L}_\infty$ implies that for each n the set

$$\{x \in X : |g(x)| > \beta_n\} = \bigcup_{j=1}^{\infty} \{x \in X : \beta_n < |g(x)| \leqq \beta_{n+j}\}$$

is not locally μ-null, and so for each n there exists a j such that $\{x \in X : \beta_n < |g(x)| \leqq \beta_{n+j}\}$ has a subset of finite positive μ-measure. Define the subsequence (α_n) of (β_n) by letting $\alpha_1 = \beta_1$, and, having defined α_n, letting α_{n+1} be the smallest value of β such that

$$\{x \in X : \alpha_n < |g(x)| \leqq \alpha_{n+1}\}$$

has a subset A_n of finite positive μ-measure. This establishes (1)—(3).

Let

$$f = \sum_{n=1}^{\infty} \frac{1}{\alpha_n \mu(A_n)} \xi_{A_n}.$$

Then we have $\int_X f \, d\mu = \sum_{n=1}^{\infty} \frac{1}{\alpha_n} < \infty$, and so $f \in \mathfrak{L}_1$. But we also have

$$\int_X |fg| \, d\mu = \sum_{n=1}^{\infty} \frac{1}{\alpha_n \mu(A_n)} \int_{A_n} |g| \, d\mu$$

$$\geqq \sum_{n=1}^{\infty} \frac{1}{\alpha_n \mu(A_n)} \int_{A_n} \alpha_n \, d\mu = \infty. \quad \square$$

We now return to our functionals L_g on $\mathfrak{L}_1$.

(20.16) Theorem. *Let $(X, \mathscr{A}, \mu)$ be any measure space and let g be an element of $\mathfrak{L}_\infty(X, \mathscr{A}, \mu)$. Define L_g on $\mathfrak{L}_1(X, \mathscr{A}, \mu)$ by the rule*

$$L_g(f) = \int_X f\bar{g}\, d\mu\,.$$

Then L_g is a bounded linear functional on $\mathfrak{L}_1(X, \mathscr{A}, \mu)$ and $\|L_g\| = \|g\|_\infty$.

Proof. Clearly L_g is linear; (20.13) shows that

$$|L_g(f)| \leqq \|g\|_\infty \|f\|_1$$

for all $f \in \mathfrak{L}_1$. Thus L_g is a bounded linear functional on $\mathfrak{L}_1$ and

$$\|L_g\| \leqq \|g\|_\infty\,. \tag{1}$$

It remains only to prove the reverse of this inequality. If $\|g\|_\infty = 0$, this is obvious. Thus suppose that $\|g\|_\infty > 0$ and let ε be an arbitrary real number such that $0 < \varepsilon < \|g\|_\infty$. Then the set $\{x \in X : |g(x)| > \|g\|_\infty - \varepsilon\}$ is not locally μ-null, and so it has a subset $E \in \mathscr{A}$ such that $0 < \mu(E) < \infty$. Define

$$f = \frac{1}{\mu(E)}\, \xi_E \operatorname{sgn}(g)\,.$$

It is obvious that $f \in \mathfrak{L}_1$ and that $\|f\|_1 \leqq 1$. Also, we have

$$L_g(f) = \int_X f\bar{g}\, d\mu = \frac{1}{\mu(E)} \int_E |g|\, d\mu$$
$$\geqq \|g\|_\infty - \varepsilon\,. \tag{2}$$

It follows from (1), (2), and the definition of $\|L_g\|$ that $\|L_g\| = \|g\|_\infty$. □

It is tempting to conjecture that every bounded linear functional on an $\mathfrak{L}_1$ space has the form L_g for some $g \in \mathfrak{L}_\infty$. This, however, is not the case for all measure spaces, as the following example shows.

(20.17) Example. Let $I = [0, 1]$ and let X be the unit square $I \times I$. Let $\mathscr{A}$ be the σ-algebra of Borel sets in X [usual topology] and define ν and μ on $\mathscr{A}$ by

$$\nu(E) = \sum_{x \in I} \lambda(\{y \in I : (x, y) \in E\})$$

and

$$\mu(E) = \nu(E) + \sum_{y \in I} \lambda(\{x \in I : (x, y) \in E\})\,.$$

[This construction is very like the construction in (19.71.d).] Define L on $\mathfrak{L}_1(X, \mathscr{A}, \mu)$ by

$$L(f) = \int_X f\, d\nu\,.$$

Since $\nu \leqq \mu$, it is plain that $L \in \mathfrak{L}_1^*(X, \mathscr{A}, \mu)$ and that $\|L\| \leqq 1$. Assume that there exists a function $g \in \mathfrak{L}_\infty(X, \mathscr{A}, \mu)$ such that $L = L_g$. For each fixed $y \in I$, set $H_y = \{(x, y) : x \in I\}$ and $f = \xi_{H_y} \operatorname{sgn}(g)$. Then f is in

$\mathfrak{L}_1(X, \mathscr{A}, \mu)$ and we have

$$\int_0^1 |g(x, y)| \, dx = \int_X f\bar{g} \, d\mu = L_g(f)$$
$$= L(f) = \int_X f \, d\nu$$
$$= 0 .$$

Since $y \in I$ is arbitrary, it follows that for each $y \in I$ we have $g(x, y) = 0$ for λ-almost all x. Anticipating FUBINI's theorem (21.12) [which of course depends in no way upon this example], it follows that there is a point $x_0 \in I$ such that $\int_0^1 |g(x_0, y)| \, dy = 0$. [In fact, λ-almost all x have this property.] Now let $V = \{(x_0, y) : y \in I\}$ and let $f = \xi_V$. Then f is in $\mathfrak{L}_1(X, \mathscr{A}, \mu)$, and we have

$$1 = \nu(V) = \int_X f \, d\nu = L(f) = L_g(f) = \int_X f\bar{g} \, d\mu$$
$$= \int_V \bar{g} \, d\mu \leqq \int_0^1 \bar{g}(x_0, y) \, dy = 0 .$$

This contradiction shows that no such g exists.

This example notwithstanding, it is true that for *decomposable* measure spaces (19.25), every $L \in \mathfrak{L}_1^*$ is an L_g for some $g \in \mathfrak{L}_\infty$. To prove this fact, we need the following lemma.

(20.18) Lemma. *Let $(X, \mathscr{A}, \mu)$ be a decomposable measure space and let $\mathscr{F}$ be a decomposition of $(X, \mathscr{A}, \mu)$ [see (19.25)]. Suppose that $f \in \mathfrak{L}_1(X, \mathscr{A}, \mu)$. Then there exists a countable subfamily $\mathscr{C}$ of $\mathscr{F}$ such that $f = 0$ μ-a.e. on $X \cap (\bigcup \mathscr{C})'$. Moreover if $(F_n)_{n=1}^\infty$ is any enumeration of $\mathscr{C}$, then we have*

(i) $$\lim_{p \to \infty} \left\| f - \sum_{n=1}^{p} f \xi_{F_n} \right\|_1 = 0$$

and

(ii) $$\int_X f \, d\mu = \sum_{n=1}^{\infty} \int_{F_n} f \, d\mu .$$

Proof. For $n \in N$, write

$$E_n = \left\{x \in X : |f(x)| > \frac{1}{n}\right\} \quad \text{and} \quad E = \{x \in X : f(x) \neq 0\} .$$

Then $E = \bigcup_{n=1}^{\infty} E_n$ and $\mu(E_n) < \infty$ for all $n \in N$. Let $\mathscr{C}_n = \{F \in \mathscr{F} : \mu(F \cap E_n) > 0\}$ and let $A_n = E_n \cap (\bigcup \mathscr{C}_n)'$. Since $\mu(E_n) = \sum_{F \in \mathscr{F}} \mu(F \cap E_n)$ (19.25.iii), it follows that $\mathscr{C}_n$ is a countable family and $\mu(A_n) = 0$. Let $\mathscr{C} = \bigcup_{n=1}^{\infty} \mathscr{C}_n$ and $A = \bigcup_{n=1}^{\infty} A_n$. Then $\mathscr{C}$ is countable, $\mu(A) = 0$, and f vanishes except on $A \cup (\cup \mathscr{C})$. Now let $(F_n)_{n=1}^{\infty}$ be as above. It is clear that $\lim_{p\to\infty} \sum_{n=1}^{p} f\xi_{F_n} = f$ μ-a.e. and $\left| f - \sum_{n=1}^{p} f\xi_{F_n} \right| \leqq |f| \in \mathfrak{L}_1$ for all $p \in N$. Thus (i) and (ii) follow from LEBESGUE's theorem on dominated convergence. □

(20.19) Theorem. *Let* $(X, \mathscr{A}, \mu)$ *be a decomposable measure space* (19.25). *Suppose that* L *is a bounded linear functional on* $\mathfrak{L}_1(X, \mathscr{A}, \mu)$. *Then there exists a function* $g \in \mathfrak{L}_\infty(X, \mathscr{A}, \mu)$ *such that* $L = L_g$ *as in* (20.16).

Proof. (I) Suppose that $\mu(X) < \infty$. For $E \in \mathscr{A}$, define

$$\nu(E) = L(\xi_E). \tag{1}$$

We claim that ν is a complex measure on $(X, \mathscr{A})$ such that $\nu \ll \mu$. Let $\{E_n\}_{n=1}^{\infty}$ be a pairwise disjoint family of sets in $\mathscr{A}$, let $E = \bigcup_{n=1}^{\infty} E_n$, and let $F_p = \bigcup_{n=p+1}^{\infty} E_n$ for each $p \in N$. We have $F_1 \supset F_2 \supset \cdots$ and $\bigcap_{p=1}^{\infty} F_p = \varnothing$; it follows from (10.15) and the hypothesis $\mu(X) < \infty$ that

$$\begin{aligned} \left| \nu(E) - \sum_{n=1}^{p} \nu(E_n) \right| &= \left| L(\xi_E) - L\left(\sum_{n=1}^{p} \xi_{E_n} \right) \right| = |L(\xi_{F_p})| \\ &\leqq \|L\| \cdot \|\xi_{F_p}\|_1 = \|L\| \cdot \mu(F_p) \to 0 \end{aligned}$$

as $p \to \infty$. Therefore

$$\nu(E) = \sum_{n=1}^{\infty} \nu(E_n),$$

and so ν is countably additive. If $\mu(E) = 0$, then $\xi_E = 0$ in $\mathfrak{L}_1$, and therefore $\nu(E) = L(\xi_E) = L(0) = 0$; hence $\nu \ll \mu$.

We infer from (19.36) that there exists a unique function $g \in \mathfrak{L}_1(X, \mathscr{A}, \mu)$ such that

$$\nu(E) = \int_E \bar{g}\, d\mu \tag{2}$$

and

$$|\nu|(E) = \int_E |g|\, d\mu \tag{3}$$

for all $E \in \mathscr{A}$. We assert that $|g| \leqq \|L\|$ μ-a.e. To see this, let

$$A = \{x \in X : |g(x)| > \|L\|\}$$

and assume that $\mu(A) > 0$. From (3) and (12.6) we have

$$|\nu|(A) = \int_A |g|\, d\mu > \int_A \|L\|\, d\mu = \|L\|\, \mu(A).$$

Thus there exists a measurable dissection $\{A_1, \ldots, A_n\}$ of A such that

$$\sum_{j=1}^{n} |\nu(A_j)| > \|L\|\, \mu(A).$$

Invoking (1), we obtain

$$\|L\|\,\mu(A) < \sum_{j=1}^{n} |\nu(A_j)| = \sum_{j=1}^{n} |L(\xi_{A_j})| \leqq \sum_{j=1}^{n} \|L\| \cdot \|\xi_{A_j}\|_1$$
$$= \|L\| \cdot \sum_{j=1}^{n} \mu(A_j) = \|L\|\,\mu(A)\,.$$

This contradiction shows that $|g| \leqq \|L\|$ μ-a.e., and so we suppose with no harm done that $|g(x)| \leqq \|L\|$ for all $x \in X$.

It obviously follows from (1) and (2) that

$$L(s) = \int_X s\bar{g}\,d\mu \tag{4}$$

for all complex-valued, $\mathscr{A}$-measurable, simple functions s defined on X. Now let $f \in \mathfrak{L}_1(X, \mathscr{A}, \mu)$, and use (11.35) to select a sequence $(s_n)_{n=1}^{\infty}$ of $\mathscr{A}$-measurable simple functions such that $|s_1| \leqq |s_2| \leqq \cdots \leqq |f|$ and $s_n(x) \to f(x)$ for all $x \in X$. Plainly $|s_n\bar{g}| \leqq \|L\| \cdot |f| \in \mathfrak{L}_1$ and $|f - s_n| \leqq 2|f| \in \mathfrak{L}_1$ for all $n \in N$; hence (12.30), (4), and the continuity of L imply that

$$L(f) = \lim_{n\to\infty} L(s_n) = \lim_{n\to\infty} \int_X s_n\bar{g}\,d\mu = \int_X f\bar{g}\,d\mu\,.$$

This proves the theorem for the case that $\mu(X) < \infty$.

(II) We now consider the general case. Let $\mathscr{F}$ be a decomposition of $(X, \mathscr{A}, \mu)$. For each $F \in \mathscr{F}$, define μ_F on $(X, \mathscr{A})$ by $\mu_F(E) = \mu(E \cap F)$ and define L_F on $\mathfrak{L}_1(X, \mathscr{A}, \mu_F)$ by $L_F(f) = L(f\xi_F)$. Plainly $(X, \mathscr{A}, \mu_F)$ is a finite measure space and L_F is a bounded linear functional on $\mathfrak{L}_1(X, \mathscr{A}, \mu_F)$ such that $\|L_F\| \leqq \|L\|$. Apply (I): for each $F \in \mathscr{F}$, there is a function $g_F \in \mathfrak{L}_\infty(X, \mathscr{A}, \mu_F)$ such that $|g_F(x)| \leqq \|L_F\| \leqq \|L\|$ for all $x \in X$ and

$$L(f\xi_F) = L_F(f) = \int_X f\bar{g}_F\,d\mu_F = \int_F f\bar{g}_F\,d\mu \tag{5}$$

for all $f \in \mathfrak{L}_1(X, \mathscr{A}, \mu_F)$. Since $\mathfrak{L}_1(X, \mathscr{A}, \mu) \subset \mathfrak{L}_1(X, \mathscr{A}, \mu_F)$, (5) plainly holds for all $f \in \mathfrak{L}_1(X, \mathscr{A}, \mu)$.

Next consider the functions g_F. All of them are defined everywhere on X, but the values of g_F on F' are of no consequence in determining L_F. In any event, we can [and do] define a function g on all of X by

$$g(x) = g_F(x) \quad \text{if} \quad x \in F \in \mathscr{F}\,.$$

It is clear that $g \in \mathfrak{L}_\infty(X, \mathscr{A}, \mu)$ [the $\mathscr{A}$-measurability of g follows from (19.25.iv)] and that $\|g\|_\infty \leqq \|g\|_u \leqq \|L\|$.

Let $f \in \mathfrak{L}_1(X, \mathscr{A}, \mu)$ and let $\{F_n\}_{n=1}^{\infty}$ be the countable subfamily of $\mathscr{F}$ described in (20.18). The continuity of L, (20.18), the boundedness of g, and (12.30) show that

$$L(f) = \lim_{p\to\infty} L\left(\sum_{n=1}^{p} f\xi_{F_n}\right) = \lim_{p\to\infty} \sum_{n=1}^{p} L(f\xi_{F_n})$$
$$= \lim_{p\to\infty} \sum_{n=1}^{p} \int_{F_n} f\bar{g}\,d\mu = \lim_{p\to\infty} \int_X \sum_{n=1}^{p} f\xi_{F_n}\bar{g}\,d\mu$$
$$= \int_X f\bar{g}\,d\mu\,. \quad \square$$

(20.20) Theorem. *Let* $(X, \mathscr{A}, \mu)$ *be a decomposable measure space* (19.25). *Then the mapping* T *defined by*

$$T(g) = L_{\bar{g}}$$

[see (20.16)*] is a norm-preserving linear mapping of* $\mathfrak{L}_\infty$ *onto the conjugate space* $\mathfrak{L}_1^*$. *Thus, as Banach spaces,* $\mathfrak{L}_\infty$ *and* $\mathfrak{L}_1^*$ *are isomorphic.*

Proof. The fact that T is a norm-preserving mapping from $\mathfrak{L}_\infty$ *into* $\mathfrak{L}_1^*$ is (20.16). It follows from (20.19) that T is *onto* $\mathfrak{L}_1^*$. It is trivial that T is linear. Since T is both linear and norm-preserving, it is one-to-one. $\square$

(20.21) Note. As we have shown in (20.17), the conclusion in (20.20) fails for some nondecomposable measure spaces. However J. SCHWARTZ has found a representation of $\mathfrak{L}_1^*(X, \mathscr{A}, \mu)$ for arbitrary $(X, \mathscr{A}, \mu)$ [Proc. Amer. Math. Soc. 2 (1951), 270–275], to which the interested reader is referred.

(20.22) Exercise. Let X be a locally compact Hausdorff space and let ι be an outer measure on $\mathscr{P}(X)$ as in § 9. Prove that the definitions of local ι-nullity given in (9.29) and in (20.11) are equivalent.

(20.23) Exercise. Let $(X, \mathscr{A}, \mu)$ be a degenerate measure space such that $\mu(X) = \infty$ [see (10.3) for the definition]. Is this measure space decomposable? Find $\mathfrak{L}_1$, $\mathfrak{L}_1^*$, and $\mathfrak{L}_\infty$ explicitly for this measure space.

(20.24) Exercise. Let $(X, \mathscr{A}, \mu)$ be any measure space and let $f \in \mathfrak{L}_1(X, \mathscr{A}, \mu)$. Define L on $\mathfrak{L}_\infty(X, \mathscr{A}, \mu)$ by

$$L(g) = \int_X g\bar{f}\,d\mu\,.$$

Prove that $L \in \mathfrak{L}_\infty^*$ and that $\|L\| = \|f\|_1$.

(20.25) Exercise. Prove that $\mathfrak{L}_1([0, 1])$ [with Lebesgue measure] is not reflexive by showing that not every $L \in \mathfrak{L}_\infty^*([0, 1])$ has the form described in (20.24). [Hint. Use the Hahn-Banach theorem to produce an $L \neq 0$ such that $L(g) = 0$ for all $g \in \mathfrak{L}_\infty$ for which g is essentially continuous, *i.e.*, $\|g - h\|_\infty = 0$ for some $h \in \mathfrak{C}([0, 1])$.]

(20.26) Exercise. (a) Prove that $\mathfrak{L}_\infty([0, 1])$ is not separable.

(b) Find necessary and sufficient conditions on a measure space that its $\mathfrak{L}_\infty$ space be separable. [Do not forget (20.23).]

Having found the conjugate space of $\mathfrak{L}_p(X, \mathscr{A}, \mu)$ for $1 < p < \infty$ and any measure space $(X, \mathscr{A}, \mu)$, and of $\mathfrak{L}_1(X, \mathscr{A}, \mu)$ for a large class of measure spaces, we find it natural to ask about the conjugate space of $\mathfrak{L}_\infty(X, \mathscr{A}, \mu)$. It turns out that each functional in $\mathfrak{L}_\infty^*$ can be represented as an integral with respect to a certain bounded, complex-valued, *finitely* additive measure on $\mathscr{A}$. We now sketch this representation. First we need some definitions.

(20.27) Definition. Let $(X, \mathscr{A}, \mu)$ be a measure space and let $\boldsymbol{F}(X, \mathscr{A}, \mu)$ denote the set of all complex-valued functions τ defined on $\mathscr{A}$ such that

(i) $\sup\{|\tau(A)| : A \in \mathscr{A}\} < \infty$;

(ii) $\tau(A \cup B) = \tau(A) + \tau(B)$ if $A, B \in \mathscr{A}$ and $A \cap B = \varnothing$;

and

(iii) $\tau(A) = 0$ if $A \in \mathscr{A}$ and A is locally μ-null.

For such a τ we define $|\tau|$ on $\mathscr{A}$ [just as in (19.10) and (19.11)] by the rule

$$|\tau|(A) = \sup\left\{\sum_{j=1}^{n} |\tau(A_j)| : \{A_1, \ldots, A_n\} \text{ is a measurable dissection of } A\right\}.$$

It is easy to show that $|\tau| \in \boldsymbol{F}(X, \mathscr{A}, \mu)$ [cf. (19.12)]. We define the *norm of* τ to be the number $\|\tau\| = |\tau|(X)$. One easily verifies that $\boldsymbol{F}(X, \mathscr{A}, \mu)$ is a complex normed linear space with this norm and with setwise linear operations.

It is an interesting fact that integrals can be defined for $\mathscr{A}$-measurable functions on X with respect to finitely additive measures τ as defined in (20.27). [Of course there is no analogue of the limit theorems (12.22) and (12.24).] We now outline the construction of integrals $\int \cdots d\tau$.

(20.28) Lemma. *Notation is as in* (20.27). *If* $f = \sum_{j=1}^{m} \alpha_j \xi_{A_j}$ *and* $g = \sum_{k=1}^{n} \beta_k \xi_{B_k}$ *are complex-valued simple functions on* X, *where* $\{A_1, \ldots, A_m\}$ *and* $\{B_1, \ldots, B_n\}$ *are measurable dissections of* X, *then for any* $\tau \in \boldsymbol{F}(X, \mathscr{A}, \mu)$ *we have*

(i) $$\left|\sum_{j=1}^{m} \alpha_j \tau(A_j) - \sum_{k=1}^{n} \beta_k \tau(B_k)\right| \leqq \|\tau\| \cdot \|f - g\|_u.$$

Proof. We write

$$\begin{aligned}\left|\sum_{j=1}^{m} \alpha_j \tau(A_j) - \sum_{k=1}^{n} \beta_k \tau(B_k)\right| &= \left|\sum_{j=1}^{m} \sum_{k=1}^{n} (\alpha_j - \beta_k)\, \tau(A_j \cap B_k)\right| \\ &\leqq \sum_{j=1}^{m} \sum_{k=1}^{n} |\alpha_j - \beta_k| \cdot |\tau(A_j \cap B_k)| \\ &\leqq \sum_{j=1}^{m} \sum_{k=1}^{n} \|f - g\|_u |\tau(A_j \cap B_k)| \\ &\leqq \|f - g\|_u \cdot \|\tau\|;\end{aligned}$$

the next to last inequality holds because either $A_j \cap B_k = \varnothing$, in which case the summand is zero, or there is an $x \in A_j \cap B_k$, in which case $|\alpha_j - \beta_k| = |f(x) - g(x)| \leqq \|f - g\|_u$. □

(20.29) Definition. Let $(X, \mathscr{A}, \mu)$ be a measure space and let $\tau \in \boldsymbol{F}(X, \mathscr{A}, \mu)$. If $s = \sum_{j=1}^{m} \alpha_j \xi_{A_j}$ is an $\mathscr{A}$-measurable simple function on X, we define

(i) $$\int_X s\, d\tau = \sum_{j=1}^{m} \alpha_j \tau(A_j)\,.$$

If f is a bounded, complex-valued $\mathscr{A}$-measurable function on X, use (11.35) to obtain a sequence $(s_n)_{n=1}^{\infty}$ of $\mathscr{A}$-measurable simple functions on X such that $\|f - s_n\|_u \to 0$. In view of Lemma (20.28), the sequence $(\int_X s_n d\tau)_{n=1}^{\infty}$ is a Cauchy sequence of complex numbers, and we define

(ii) $$\int_X f\, d\tau = \lim_{n\to\infty} \int_X s_n d\tau\,.$$

It is easy to see that this definition does *not* depend upon the particular sequence (s_n) [provided, of course, that it converges uniformly to f], and so the integral is well defined, and definitions (i) and (ii) are consistent. It is also easy to see that this definition agrees with Definition (19.17) in the case that τ is a complex measure.

(20.30) Theorem. *Let* $(X, \mathscr{A}, \mu)$ *and* τ *be as in* (20.29). *Let* f *and* g *be bounded* $\mathscr{A}$*-measurable functions on* X *and let* $\alpha \in K$. *Then*

(i) $$\int_X \alpha f\, d\tau = \alpha \int_X f\, d\tau\,,$$

(ii) $$\int_X (f + g)\, d\tau = \int_X f\, d\tau + \int_X g\, d\tau\,,$$

and

(iii) $$\left|\int_X f\, d\tau\right| \leqq \int_X |f|\, d|\tau|\,.$$

Proof. Exercise.

(20.31) Theorem. *Let* $(X, \mathscr{A}, \mu)$ *and* τ *be as in* (20.29). *Let* h *be a bounded* $\mathscr{A}$*-measurable function on* X *such that the set* $A = \{x \in X : h(x) \neq 0\}$ *is locally* μ*-null. Then* $\int_X h\, d\tau = 0$.

Proof. Using (20.30) and (20.27.iii), we find

$$\left|\int_X h\, d\tau\right| \leqq \int_X |h|\, d|\tau| = \int_X \xi_A |h|\, d|\tau| + \int_X \xi_{A'} |h|\, d|\tau|$$
$$\leqq \|h\|_u |\tau|(A) + 0 = 0\,. \quad □$$

(20.32) Definition. Let $(X, \mathscr{A}, \mu)$ and τ be as in (20.29) and let $g \in \mathfrak{L}_\infty(X, \mathscr{A}, \mu)$. Let f be a bounded function in $\mathfrak{L}_\infty$ such that $\|f - g\|_\infty = 0$. Define

$$\int_X g\, d\tau = \int_X f\, d\tau\,.$$

In view of (20.31), this definition does not depend on the particular bounded function f that is drawn from the $\mathfrak{L}_\infty$-class determined by g, and so the definition is unambiguous.

(20.33) Theorem. *Let $(X, \mathscr{A}, \mu)$ be any measure space and let $\tau \in \boldsymbol{F}(X, \mathscr{A}, \mu)$. Define L_τ on $\mathfrak{L}_\infty(X, \mathscr{A}, \mu)$ by the rule*

(i) $$L_\tau(g) = \int_X g \, d\tau \, .$$

Then L_τ is a bounded linear functional on $\mathfrak{L}_\infty$ and

(ii) $$\|L_\tau\| = \|\tau\| \, .$$

Proof. For $g \in \mathfrak{L}_\infty$, choose a bounded $f \in \mathfrak{L}_\infty$ such that $\|f - g\|_\infty = 0$ and $\|f\|_u = \|g\|_\infty$ (20.14). By (20.32), we have

$$|L_\tau(g)| = |\int_X g \, d\tau| = |\int_X f \, d\tau| \leqq \int_X |f| \, d|\tau| \leqq \|f\|_u \|\tau\|$$
$$= \|g\|_\infty \|\tau\| \, .$$

Thus $L_\tau \in \mathfrak{L}_\infty^*$ and

$$\|L_\tau\| \leqq \|\tau\| \, . \tag{1}$$

Let $\varepsilon > 0$ be given and select a measurable dissection $\{A_1, \ldots, A_n\}$ of X such that

$$\sum_{j=1}^{n} |\tau(A_j)| > \|\tau\| - \varepsilon \, .$$

For each j, let $\alpha_j = \operatorname{sgn}\left(\overline{\tau(A_j)}\right)$ and set $g = \sum_{j=1}^{n} \alpha_j \xi_{A_j}$. It is plain that $g \in \mathfrak{L}_\infty$, that $\|g\|_\infty = \|g\|_u \leqq 1$, and that

$$|L_\tau(g)| = |\int_X g \, d\tau| = \left|\sum_{j=1}^{n} \alpha_j \tau(A_j)\right| = \sum_{j=1}^{n} |\tau(A_j)| > \|\tau\| - \varepsilon \, .$$

Since ε is arbitrary, this shows that

$$\|L_\tau\| \geqq \|\tau\| \, . \tag{2}$$

Now combine (1) and (2) to get (ii). □

The converse of (20.33) holds: every bounded linear functional on $\mathfrak{L}_\infty$ is the integral with respect to a finitely additive measure. Plainly we get this measure by looking at characteristic functions of sets. The details follow.

(20.34) Theorem. *Let $(X, \mathscr{A}, \mu)$ be any measure space and let L be a bounded linear functional on $\mathfrak{L}_\infty(X, \mathscr{A}, \mu)$. Then there exists a $\tau \in \boldsymbol{F}(X, \mathscr{A}, \mu)$ such that $L = L_\tau$ as in* (20.33).

Proof. For each $A \in \mathscr{A}$, let

$$\tau(A) = L(\xi_A) \, . \tag{1}$$

We have

$$\sup\{|\tau(A)| : A \in \mathscr{A}\} \leqq \sup\{|L(g)| : g \in \mathfrak{L}_\infty, \|g\|_\infty \leqq 1\} = \|L\| \, .$$

Thus (20.27.i) holds for τ. Also, if A and B are disjoint sets in $\mathscr{A}$, then $\xi_A + \xi_B = \xi_{A\cup B}$, and so

$$\tau(A \cup B) = L(\xi_{A\cup B}) = L(\xi_A + \xi_B) = L(\xi_A) + L(\xi_B) = \tau(A) + \tau(B);$$

hence (20.27.ii) holds. Next, if $A \in \mathscr{A}$ and A is locally μ-null, then $\xi_A = 0$ in $\mathfrak{L}_\infty$ and therefore

$$\tau(A) = L(\xi_A) = L(0) = 0.$$

Thus (20.27.iii) holds, and so $\tau \in \boldsymbol{F}(X, \mathscr{A}, \mu)$. Let g be any function in $\mathfrak{L}_\infty$. Choose a bounded $f \in \mathfrak{L}_\infty$ such that $\|g - f\|_\infty = 0$ and $\|f\|_u = \|g\|_\infty$. Choose a sequence (s_n) of $\mathscr{A}$-measurable simple functions such that $\|f - s_n\|_u \to 0$. It is clear from (1), the linearity of L, and (20.29.i) that

$$L(s_n) = \int_X s_n\, d\tau \tag{2}$$

for each $n \in N$. Since $\|f - s_n\|_\infty \to 0$ and L is continuous on $\mathfrak{L}_\infty$, (2) and (20.29.ii) imply that

$$L(g) = L(f) = \lim_{n\to\infty} L(s_n) = \lim_{n\to\infty} \int_X s_n\, d\tau = \int_X f\, d\tau$$
$$= \int_X g\, d\tau = L_\tau(g). \quad \square$$

(20.35) Theorem. *Let $(X, \mathscr{A}, \mu)$ be an arbitrary measure space. Then the mapping T defined by*

$$T(\tau) = L_\tau$$

[see (20.33)] is a norm-preserving linear mapping of $\boldsymbol{F}(X, \mathscr{A}, \mu)$ onto the conjugate space $\mathfrak{L}_\infty^*$. Thus $\boldsymbol{F}$ is a Banach space and, as Banach spaces, $\boldsymbol{F}$ and $\mathfrak{L}_\infty^*$ are isomorphic.

Proof. The fact that T is a norm-preserving mapping from $\boldsymbol{F}$ *into* $\mathfrak{L}_\infty^*$ is (20.33). It is trivial that T is linear. Theorem (20.34) shows that T is *onto* $\mathfrak{L}_\infty^*$. Being both linear and norm-preserving, T is one-to-one and preserves Cauchy sequences. Since $\mathfrak{L}_\infty^*$ is complete, so is $\boldsymbol{F}$. $\square$

(20.36) Remark. Let X be an arbitrary nonvoid set. As in (7.3), let $\mathfrak{B}(X)$ denote the space of all bounded, complex-valued functions on X. This space has several other names: if X is regarded as a topological space with the discrete topology, then $\mathfrak{B}(X) = \mathfrak{C}(X)$ (7.8); if μ is the counting measure defined on $\mathscr{P}(X)$ (10.4.a), then $\mathfrak{B}(X) = \mathfrak{L}_\infty(X, \mathscr{P}(X), \mu)$; and in (14.26) this space was denoted by $l_\infty(X)$ [the conjugate space of $l_1(X)$]. In all cases the norm used on $\mathfrak{B}(X)$ has been the uniform norm. Thus Theorem (20.35) shows that the conjugate space $\mathfrak{B}(X)^*$ of the Banach space $\mathfrak{B}(X)$ is isometrically isomorphic to the space of all bounded, complex-valued, finitely additive measures defined on $\mathscr{P}(X)$.

(20.37) Exercise. Let X be a nonvoid set. Suppose that τ is a finitely additive measure defined on $\mathscr{P}(X)$ such that $\tau(X) = 1$ and $\tau(A) = 0$ or 1 for all $A \subset X$. Let $\mathscr{U} = \{A : A \subset X, \tau(A) = 1\}$.

(a) Prove that $\mathscr{U}$ has the following properties:

(i) $\varnothing \notin \mathscr{U}$;

(ii) if $A \in \mathscr{U}$ and $A \subset B \subset X$, then $B \in \mathscr{U}$;

(iii) if $A, B \in \mathscr{U}$, then $A \cap B \in \mathscr{U}$;

(iv) if $A \subset X$, then $A \in \mathscr{U}$ or $A' \in \mathscr{U}$.

Any family $\mathscr{U}$ that satisfies (i)—(iii) is called a *filter in* X. A filter in X satisfying (iv) is called an *ultrafilter in* X.

(b) Prove that if $\mathscr{V}$ is any ultrafilter in X and if σ is defined on $\mathscr{P}(X)$ by

$$\sigma(A) = \begin{cases} 1 & \text{if } A \in \mathscr{V}, \\ 0 & \text{if } A \notin \mathscr{V}, \end{cases}$$

then σ is a finitely additive measure on $(X, \mathscr{P}(X))$.

Thus we have set up a one-to-one correspondence between ultrafilters and finitely additive *zero-one measures*.

(c) Let $(X, \mathscr{A}, \mu)$ be any measure space and let τ be a finitely additive measure in $\boldsymbol{F}(X, \mathscr{A}, \mu)$. Prove that

$$\int_X fg\,d\tau = \int_X f\,d\tau \cdot \int_X g\,d\tau$$

for all $f, g \in \mathfrak{L}_\infty$ if and only if $\tau(A) = 0$ or 1 for all $A \in \mathscr{A}$.

(20.38) Exercise. Let X be a nonvoid set. A filter $\mathscr{U}$ in X is said to be *free* if $\bigcap \mathscr{U} = \varnothing$. All other filters are said to be *fixed*. Prove the following.

(a) If $\mathscr{U}$ is a fixed ultrafilter in X, then there is a point $p \in X$ such that $\mathscr{U} = \{A \in \mathscr{P}(X) : p \in A\}$.

(b) If X is finite, then every ultrafilter in X is fixed.

(c) If X is infinite, then there exists a free ultrafilter in X. [Let $\mathscr{F} = \{A \in \mathscr{P}(X) : A' \text{ is finite}\}$. Prove that $\mathscr{F}$ is a filter. Use ZORN's lemma to show that there is a maximal filter $\mathscr{V}$ containing $\mathscr{F}$. Prove that $\mathscr{V}$ is a free ultrafilter.]

(d) If $\mathscr{V}$ is a free ultrafilter in N and σ is as in (20.37), then $\sigma(F) = 0$ for all finite sets $F \subset N$, and so σ is not countably additive.

(e) If σ is as in (d), then for all $f \in l^r_\infty$ we have

$$\varliminf_{n\to\infty} f(n) \leq \int_N f\,d\sigma \leq \varlimsup_{n\to\infty} f(n)\,.$$

[First consider the case that $\lim\limits_{n\to\infty} f(n) = 0$. In general, find functions [sequences!] g and h such that $g \leq f \leq h$, $\varliminf\limits_{n\to\infty} f(n) = \lim\limits_{n\to\infty} g(n)$, and $\varlimsup\limits_{n\to\infty} f(n) = \lim\limits_{n\to\infty} h(n)$.]

(20.39) Exercise. Prove that there exists a nonnegative, real-valued, finitely additive measure τ defined on $\mathscr{P}([0, 1])$ such that $\tau(A) = \lambda(A)$ for all $A \in \mathscr{M}_\lambda$ for which $A \subset [0, 1]$. [Hints. Define L on $\{f : f$ is a bounded, real-valued, Lebesgue measurable function on $[0, 1]\}$ by $L(f) = \int_0^1 f(x)\,dx$.

Use KREIN's extension theorem (14.27) to extend L to a nonnegative linear functional on the space $\mathfrak{B}^r([0, 1])$ of all bounded real-valued functions on $[0, 1]$.]

(20.40) Exercise. (a) Prove that there exists a linear functional M on $\mathfrak{B}^r(R)$ such that for all $f \in \mathfrak{B}^r(R)$ and all $t \in R$, we have

(i) $$\inf\{f(x) : x \in R\} \leq M(f) \leq \sup\{f(x) : x \in R\}$$

and

(ii) $$M(f_t) = M(f)\,.$$ [1]

[Hints. Let $\mathfrak{H}$ be the linear subspace of $\mathfrak{B}^r(R)$ consisting of all finite sums of functions of the form $f_t - f$ for $f \in \mathfrak{B}^r(R)$ and $t \in R$. For

$$h = \sum_{k=1}^{n} [(f_k)_{t_k} - f_k] \in \mathfrak{H}\,,$$

prove that $\inf\{h(x) : x \in R\} \leq 0$. Assuming that this is false, choose $\varepsilon > 0$ such that $h(x) \geq \varepsilon$ for all $x \in R$. Let p be an arbitrary positive integer, and let Φ denote the set of all functions φ from $\{1, 2, \ldots, n\}$ into $\{1, 2, \ldots, p\}$. Clearly $\overline{\overline{\Phi}} = p^n$. For each $\varphi \in \Phi$, let $x(\varphi) = \varphi(1)t_1 + \varphi(2)t_2 + \cdots + \varphi(n)t_n$. Show that

$$\sum_{\varphi \in \Phi} [f_k(x(\varphi) + t_k) - f_k(x(\varphi))] \leq 2p^{n-1}\|f_k\|_u$$

for each $k \in \{1, 2, \ldots, n\}$, and then conclude that

$$p^n \varepsilon \leq \sum_{\varphi \in \Phi} h(x(\varphi)) \leq 2p^{n-1} \sum_{k=1}^{n} \|f_k\|_u\,.$$

This contradiction shows that no $h \in \mathfrak{H}$ has a positive lower bound. Next use (14.13) to obtain $M \in \mathfrak{B}^r(R)^*$ such that $M(1) = 1$, $\|M\| = 1$, and $M(h) = 0$ for all $h \in \mathfrak{H}$. Now (i) and (ii) follow easily.]

(b) Prove that there exists a [nonnegative, extended real-valued, finitely additive measure μ defined on $\mathscr{P}(R)$ such that

(iii) $\mu(A) = \lambda(A)$ for all $A \in \mathscr{M}_\lambda$

and

(iv) $\mu(A + t) = \mu(A)$ for all $A \subset R$ and all $t \in R$.

[Hints. Let τ be as in (20.39). Define ν on $\mathscr{P}(R)$ by

$$\nu(A) = \sum_{n=-\infty}^{\infty} \tau((A \cap [n, n+1[) - n)\,.$$

Show that ν is finitely additive and that $\nu(A) = \lambda(A)$ for all $A \in \mathscr{M}_\lambda$. For $A \subset R$, define f_A on R by the rule $f_A(t) = \nu(A + t)$. Let M be as in part (a) and define μ on $\mathscr{P}(R)$ by the rule

$$\mu(A) = \lim_{n \to \infty} M(\min\{f_A, n\})\,.$$

[1] As usual, f_t denotes the translate of f by t: $f_t(x) = f(x + t)$.

It is easy to prove (iii). To prove that μ is additive, use the inequalities $\min\{f_A + f_B, n\} \leqq \min\{f_A, n\} + \min\{f_B, n\} \leqq \min\{f_A + f_B, 2n\}$. To prove (iv), use the equality $f_{(A+t)}(x) = f_A(x+t)$.]

Our third application of the LEBESGUE-RADON-NIKODÝM theorem is to the study of yet another conjugate space. In (12.36), we saw that if X is a locally compact Hausdorff space, then every *nonnegative* linear functional I on $\mathfrak{C}_{00}(X)$ has the form $I(f) = \int\limits_X f\,d\iota$ for some regular Borel measure ι on X. This fact is useful in identifying the conjugate space of the complex Banach space $\mathfrak{C}_0(X)$.

(20.41) Definition. Let X be a locally compact Hausdorff space. A complex measure μ defined on the σ-algebra $\mathscr{B}(X)$ of Borel sets of X is said to be a *complex regular Borel measure on* X if for each $E \in \mathscr{B}(X)$ and each $\varepsilon > 0$ there exist a compact set F and an open set U such that $F \subset E \subset U$ and $|\mu(A)| < \varepsilon$ for all $A \in \mathscr{B}(X)$ such that $A \subset U \cap F'$. Let $\boldsymbol{M}(X)$ denote the set of all complex regular Borel measures on X.

(20.42) Note. If $\mu \in \boldsymbol{M}(X)$ and $\mu \geqq 0$, then, since all complex measures are bounded (19.13.v), μ is a finite measure. Theorem (12.40) thus shows that the definitions of regularity of μ given in (20.41) and in (12.39) are equivalent.

(20.43) Theorem. *Let X be a locally compact Hausdorff space, let $\mu, \nu \in \boldsymbol{M}(X)$, and let $\alpha \in K$. Then, operations being defined setwise, we have*

(i) $\alpha\mu \in \boldsymbol{M}(X)$

and

(ii) $\mu + \nu \in \boldsymbol{M}(X)$.

Thus $\boldsymbol{M}(X)$, with the norm defined by $\|\mu\| = |\mu|(X)$, is a complex normed linear space.

Proof. Exercise.

(20.44) Theorem. *Let X be a locally compact Hausdorff space, let μ be a complex measure defined on $\mathscr{B}(X)$, and let $\sum\limits_{j=1}^{4} \alpha_j\mu_j$ be the Jordan decomposition of μ as in* (19.15.c). *Then the following three statements are pairwise equivalent:*

(i) $\mu \in \boldsymbol{M}(X)$;

(ii) $|\mu| \in \boldsymbol{M}(X)$;

(iii) $\mu_j \in \boldsymbol{M}(X)$ *for* $j = 1, 2, 3, 4$.

Proof. We make heavy use of Theorem (19.13). The fact that (i) implies (ii) follows from the inequality

$$|\mu|(U \cap F') \leqq 4 \cdot \sup\{|\mu(A)| : A \in \mathscr{B}(X), A \subset U \cap F'\}.$$

The inequality $\mu_j(U \cap F') \leqq |\mu|(U \cap F')$ $(j = 1, 2, 3, 4)$ shows that

(ii) implies (iii). From the inequality $|\mu(A)| \leqq \sum_{j=1}^{4} \mu_j(A)$, we infer that (iii) implies (i). □

(20.45) Theorem. *Let X be a locally compact Hausdorff space, let $\mu \in \boldsymbol{M}(X)$, and define L_μ on $\mathfrak{C}_0(X)$ by the rule*

(i) $$L_\mu(f) = \int_X f \, d\mu .$$

Then L_μ is a bounded linear functional on the Banach space $\mathfrak{C}_0(X)$ and

(ii) $$\|L_\mu\| = \|\mu\| .$$

Proof. Each $f \in \mathfrak{C}_0$ is a bounded continuous function and so, since $|\mu|$ is a finite measure (19.13.v), f is in $\mathfrak{L}_1(X, \mathscr{B}(X), |\mu|)$. Thus L_μ is a linear functional on $\mathfrak{C}_0$. Also (19.38.iv) shows that if $f \in \mathfrak{C}_0$, then

$$|L_\mu(f)| \leqq \int_X |f| \, d|\mu| \leqq \|f\|_u \|\mu\| .$$

Thus $L_\mu \in \mathfrak{C}_0^*$ and

$$\|L_\mu\| \leqq \|\mu\| . \tag{1}$$

To prove the reverse of inequality (1), use (19.38) to obtain a complex-valued Borel measurable function f_0 on X such that $|f_0(x)| = 1$ for all $x \in X$ and

$$\int_X f \, d\mu = \int_X f f_0 d|\mu| \tag{2}$$

for all $f \in \mathfrak{L}_1(X, \mathscr{B}(X), |\mu|)$. Let $\varepsilon > 0$ be arbitrary. By (13.21), there exists a function $g \in \mathfrak{C}_{00}$ such that $\|g\|_u \leqq \|\bar{f}_0\|_u = 1$ and

$$\int_X |\bar{f}_0 - g| \, d|\mu| < \varepsilon . \tag{3}$$

Now (2) and (3) imply that

$$\|\mu\| = |\mu|(X) = \int_X \bar{f}_0 f_0 d|\mu| = \int_X \bar{f}_0 d\mu < \left|\int_X g \, d\mu\right| + \varepsilon = |L_\mu(g)| + \varepsilon . \tag{4}$$

Since ε is arbitrary and $\|g\|_u \leqq 1$, (4) implies that

$$\|\mu\| \leqq \|L_\mu\| . \tag{5}$$

Combine (1) and (5) to complete the proof. □

We will next show that every element of $\mathfrak{C}_0(X)^*$ has the form L_μ for some $\mu \in \boldsymbol{M}(X)$. First we need a lemma.

(20.46) Lemma. *Let X be a locally compact Hausdorff space and let $L \in \mathfrak{C}_0(X)^*$. Define I on $\mathfrak{C}_0^+$ by the rule*

(i) $I(f) = \sup\{|L(g)| : g \in \mathfrak{C}_0, |g| \leqq f\}$ *for all* $f \in \mathfrak{C}_0^+$.

Then I can be extended to a nonnegative linear functional in $\mathfrak{C}_0^*$ such that $\|I\| = \|L\|$.

Proof. Obviously if $f \in \mathfrak{C}_0^+$, then

$$I(f) \leqq \sup\{\|L\| \cdot \|g\|_u : g \in \mathfrak{C}_0, |g| \leqq f\} = \|L\| \cdot \|f\|_u < \infty . \quad (1)$$

Thus I is *real*-valued on $\mathfrak{C}_0^+$. It is also clear that

$$I(\alpha f) = \alpha I(f) \quad (2)$$

for $f \in \mathfrak{C}_0^+$ and $\alpha \geqq 0$. Let us show that I is additive on $\mathfrak{C}_0^+$. Let f_1, f_2 be in $\mathfrak{C}_0^+$. If $\varepsilon > 0$ is given, choose $g_1, g_2 \in \mathfrak{C}_0$ such that $|g_j| \leqq f_j$ and $|L(g_j)| > I(f_j) - \frac{\varepsilon}{2}$ $(j = 1, 2)$. Now write $L(g_j) = \beta_j |L(g_j)|$, where $\beta_j \in K$ and $|\beta_j| = 1$ $(j = 1, 2)$, and let $\beta = \beta_2 \bar{\beta}_1$. Then $|\beta| = 1$ and we have

$$\begin{aligned} I(f_1) + I(f_2) &< |L(g_1)| + |L(g_2)| + \varepsilon = \bar{\beta}_1 L(g_1) + \bar{\beta}_2 L(g_2) + \varepsilon \\ &= |\beta L(g_1) + L(g_2)| + \varepsilon = |L(\beta g_1 + g_2)| + \varepsilon \leqq I(f_1 + f_2) + \varepsilon \end{aligned}$$

[note that $|\beta g_1 + g_2| \leqq |g_1| + |g_2| \leqq f_1 + f_2$]. Thus

$$I(f_1) + I(f_2) \leqq I(f_1 + f_2) . \quad (3)$$

To prove the reversed inequality, choose $g \in \mathfrak{C}_0$ such that $|g| \leqq f_1 + f_2$ and $|L(g)| \geqq I(f_1 + f_2) - \varepsilon$. Let $h_1 = \min\{f_1, |g|\}$ and $h_2 = |g| - h_1$. It is plain that $h_1, h_2 \in \mathfrak{C}_0^+$, $h_1 \leqq f_1$, $h_2 \leqq f_2$, and $h_1 + h_2 = |g|$. Let $g_j = h_j \operatorname{sgn} g$ $(j = 1, 2)$. Then we have: $g_j \in \mathfrak{C}_0$; $|g_j| = h_j \leqq f_j$; and $g_1 + g_2 = g$. Therefore

$$\begin{aligned} I(f_1 + f_2) - \varepsilon \leqq |L(g)| = |L(g_1 + g_2)| &\leqq |L(g_1)| + |L(g_2)| \\ &\leqq I(f_1) + I(f_2) ; \end{aligned}$$

together with (3), this proves that

$$I(f_1 + f_2) = I(f_1) + I(f_2) . \quad (4)$$

We now extend I to $\mathfrak{C}_0$ in two steps. First, if $f \in \mathfrak{C}_0^r$, write $f = f^+ - f^-$, where as usual $f^+ = \max\{f, 0\}$ and $f^- = -\min\{f, 0\}$, and define

$$I(f) = I(f^+) - I(f^-) .$$

Notice that if g_1 and g_2 are *any* two functions in $\mathfrak{C}_0^+$ such that $f = g_1 - g_2$, then $f^+ + g_2 = g_1 + f^-$, and so (4) implies that $I(f) = I(g_1) - I(g_2)$. A simple computation shows that I is *real* linear on $\mathfrak{C}_0^r$. Next, for $f \in \mathfrak{C}_0$, define

$$I(f) = I(\operatorname{Re} f) + i I(\operatorname{Im} f) .$$

Another obvious computation shows that I is complex linear on $\mathfrak{C}_0$.

It is now clear [using (i)] that I is a nonnegative linear functional on $\mathfrak{C}_0$. Thus, just as in (9.4), for all $f \in \mathfrak{C}_0$ we have

$$|I(f)| \leqq I(|f|) .$$

This fact together with (1) yields

$$\|I\| = \sup\{|I(f)| : f \in \mathfrak{C}_0, \|f\|_u \leqq 1\} \leqq \sup\{I(|f|) : f \in \mathfrak{C}_0, \|f\|_u \leqq 1\} \leqq \|L\|.$$

To prove the reversed inequality, let $\varepsilon > 0$ be given and then select $g \in \mathfrak{C}_0$ such that $\|g\|_u \leqq 1$ and $|L(g)| > \|L\| - \varepsilon$. From (i) we have

$$\|L\| - \varepsilon < |L(g)| \leqq I(|g|) \leqq \|I\| .$$

Hence $\|I\| = \|L\|$. □

(20.47) Theorem. *Let X be a locally compact Hausdorff space and let $L \in \mathfrak{C}_0(X)^*$. Then there exists a complex measure $\mu \in \boldsymbol{M}(X)$ such that*

(i) $L(f) = \int_X f\, d\mu = L_\mu(f)$

for all $f \in \mathfrak{C}_0(X)$.

Proof. Let I be constructed from L as in (20.46). By F. RIESZ's representation theorem (12.36), there exists a measure ι on X as in § 9 such that

$$I(f) = \int_X f\, d\iota$$

for all $f \in \mathfrak{C}_{00}(X)$. It follows that

$$\begin{aligned}\iota(X) = \bar{I}(1) &= \sup\{I(f) : f \in \mathfrak{C}_{00}^+, f \leqq 1\}\\ &\leqq \sup\{|I(f)| : f \in \mathfrak{C}_0, \|f\|_u \leqq 1\}\\ &= \|I\| = \|L\| .\end{aligned}$$

Thus ι is a finite measure. We know (13.21) that $\mathfrak{C}_{00}$ is a dense linear subspace of $\mathfrak{L}_1(X, \mathscr{M}_\iota, \iota)$. It follows from (20.46.i) that, relative to the $\mathfrak{L}_1$ norm, L is a bounded linear functional of norm $\leqq 1$ on this dense subspace: in fact,

$$|L(f)| \leqq I(|f|) = \int_X |f|\, d\iota = \|f\|_1$$

for all $f \in \mathfrak{C}_{00}$. Thus, by (14.40) [or the Hahn-Banach theorem], L can be extended to a bounded linear functional L' on $\mathfrak{L}_1(X, \mathscr{M}_\iota, \iota)$ such that $\|L'\| \leqq 1$. Now apply (20.19) to obtain a function $g \in \mathfrak{L}_\infty(X, \mathscr{M}_\iota, \iota)$ such that $\|g\|_\infty = \|L'\| \leqq 1$ and

$$L'(f) = \int_X f\bar{g}\, d\iota$$

for all $f \in \mathfrak{L}_1(X, \mathscr{M}_\iota, \iota)$. Next define μ on $\mathscr{M}_\iota$ by the rule

$$\mu(E) = \int_E \bar{g}\, d\iota .$$

Then μ is a complex measure on $(X, \mathscr{M}_\iota)$, μ is absolutely continuous with respect to ι, and $\bar{g}$ is the unique LEBESGUE-RADON-NIKODÝM derivative of μ with respect to ι as in (19.36). [Actually $|\mu|$ is ι, but we do not need this fact here.] Now (19.36) implies that if $f \in \mathfrak{C}_0(X) \subset \mathfrak{L}_1(X, \mathscr{M}_\iota, |\mu|)$, then

$$\int_X f\, d\mu = \int_X f\bar{g}\, d\iota = L'(f) = L(f) ;$$

the last equality follows from the fact that L and L' are both continuous in the $\mathfrak{L}_1(X, \mathscr{M}_\iota, \iota)$-norm on $\mathfrak{C}_0$ and agree on the $\mathfrak{L}_1$-dense subspace $\mathfrak{C}_{00}$. All that remains is to show that μ is regular. Let $E \in \mathscr{M}_\iota$ and $\varepsilon > 0$ be given. Since ι is regular and finite, there exist a compact set F and an open set U such that $F \subset E \subset U$ and $\iota(U \cap F') < \varepsilon$. Thus if $A \in \mathscr{M}_\iota$ and $A \subset U \cap F'$, then

$$|\mu(A)| = |\int_A \bar{g}\, d\iota| \leq \iota(A) \leq \iota(U \cap F') < \varepsilon .$$

Hence $\mu \in \boldsymbol{M}(X)$ and the proof is complete. □

(20.48) Riesz Representation Theorem. *Let X be a locally compact Hausdorff space. Then the mapping T defined by*

$$T(\mu) = L_\mu$$

[see (20.45)] is a norm-preserving linear mapping of $\boldsymbol{M}(X)$ onto $\mathfrak{C}_0(X)^*$. Thus $\boldsymbol{M}(X)$ is a Banach space, and $\boldsymbol{M}(X)$ and $\mathfrak{C}_0(X)^*$ are isomorphic as Banach spaces.

Proof. The fact that T is a norm-preserving mapping from $\boldsymbol{M}$ *in*to $\mathfrak{C}_0^*$ is (20.45). It follows from (20.47) that T is *on*to $\mathfrak{C}_0^*$. It is trivial that T is linear. Since T is both linear and norm-preserving, T is one-to-one and preserves Cauchy sequences. Thus, since $\mathfrak{C}_0^*$ is complete, so is $\boldsymbol{M}$. □

(20.49) Exercise. Let X be a locally compact Hausdorff space and let $\mu \in \boldsymbol{M}(X)$. Let I be the nonnegative linear functional on $\mathfrak{C}_0(X)$ constructed from L_μ in (20.46). Prove that if ι is the measure corresponding to I as in (12.36), then $\iota = |\mu|$. [Show that $\int_X f\, d|\mu| = \sup\{|\int_X g\, d\mu| : g \in \mathfrak{C}_0, |g| \leq f\}$ for all $f \in \mathfrak{C}_0^+$.]

(20.50) Exercise. Let X be a locally compact Hausdorff space for which there exists a $\mu \in \boldsymbol{M}(X)$ such that $\mu(\{x\}) = 0$ for all $x \in X$ and $|\mu|(X) > 0$. Prove that the Banach space $\mathfrak{C}_0(X)$ is not reflexive. [Define Φ on $\boldsymbol{M}(X)$ by $\Phi(\nu) = \sum_{x \in X} \nu(\{x\})$.]

(20.51) Exercise. (a) Let X be a locally compact Hausdorff space and suppose that τ is a [nonnegative] finitely additive measure defined on $\mathscr{B}(X)$ such that τ is regular; *i.e.*, τ satisfies (i), (ii), and (iii) of Definition (12.39). Prove that τ is countably additive. [Use (12.36) to obtain a regular Borel measure ι on X such that $\int_X f\, d\tau = \int_X f\, d\iota$ for all $f \in \mathfrak{C}_{00}$. Then show that $\tau(E) = \iota(E)$ for every Borel set E.]

(b) Prove that any bounded, complex-valued, finitely additive measure τ defined on $\mathscr{B}(X)$ which is regular in the sense of (20.41) is countably additive, *i.e.*, $\tau \in \boldsymbol{M}(X)$.

In parts (c) and (d), let ι be a σ-finite regular Borel measure on X. Prove the following.

(c) If ν is a complex measure defined on $\mathscr{B}(X)$ such that $\nu \ll \iota$, then $\nu \in \boldsymbol{M}(X)$.

(d) If $\nu \in \boldsymbol{M}(X)$ and $\nu = \nu_1 + \nu_2$ is the Lebesgue decomposition of ν with respect to ι, then $\nu_j \in \boldsymbol{M}(X)$ $(j = 1, 2)$.

(20.52) Exercise. More on the structure of $\mathfrak{C}_{00}(X)$ and $\mathfrak{C}_0(X)$. (a) Let X be a nonvoid locally compact Hausdorff space, and let M be a *multiplicative linear functional* on $\mathfrak{C}_0(X)$. By this we mean that M satisfies (9.1.i) and (9.1.ii), that $M \neq 0$, and that $M(fg) = M(f)\,M(g)$ for all $f, g \in \mathfrak{C}_0(X)$. Prove that there is a point $a \in X$ such that $M(f) = E_a(f) = f(a)$ for all $f \in \mathfrak{C}_0(X)$. [Hints. It is convenient to prove first that M is continuous. In fact, we have $|M(f)| \leqq \|f\|_u$ for all $f \in \mathfrak{C}_0$. To see this, assume that $M(f) = \alpha$ where $|\alpha| > \|f\|_u$. Let $g = -\sum\limits_{k=1}^{\infty} \alpha^{-k} f^k$. The series converges uniformly and so $g \in \mathfrak{C}_0$. Check that $\alpha^{-1} f + g - \alpha^{-1} fg = 0$. Then one has $0 = M(0) = \alpha^{-1} M(f) + M(g) - \alpha^{-1} M(f) M(g) = 1 + M(g) - M(g) = 1$.

Now knowing that M is bounded, we apply (20.47) to write

$$M(f) = \int_X f\,d\mu = \int_X f\bar{g}\,d\iota$$

where $\mu \in \boldsymbol{M}(X)$, $\iota \in \boldsymbol{M}^+(X)$, $g \in \mathfrak{L}_\infty(X, \mathscr{B}(X), \iota)$, and $\|g\|_\infty \leqq 1$. If $A, B \in \mathscr{B}(X)$ and $\iota(A)$ and $\iota(B)$ are positive, then $A \cap B \neq \varnothing$. This follows readily from the regularity of ι and the identity $M(f_1 f_2) = M(f_1) M(f_2)$. Now it is easy to see that ι assumes only one positive value, since $\iota(A) = \iota(A \cap B') + \iota(A \cap B)$ and $\iota(B) = \iota(A' \cap B) + \iota(A \cap B)$. The family $\{F \subset X : F \text{ is compact}, \iota(F) > 0\}$ is nonvoid and has the finite intersection property and hence has nonvoid total intersection, say F_0. It is not hard to see that $\iota(F_0) > 0$ and that F_0 has no proper nonvoid subsets, *i.e.*, $F_0 = \{a\}$ for some $a \in X$. It follows immediately that $M(f) = f(a)$.]

(b) Prove that every multiplicative linear functional M on $\mathfrak{C}_{00}(X)$ has the form E_a for some $a \in X$. [Hints. Choose $f_0 \in \mathfrak{C}_{00}$ such that $M(f_0) = 1$. Let $U = \{x \in X : f_0(x) \neq 0\}$. Let $\mathfrak{F} = \{f \in \mathfrak{C}_{00}(X) : f(U') \subset \{0\}\}$. Then $\mathfrak{F}$ can be identified in an obvious way with $\mathfrak{C}_0(U)$ [note that U is locally compact]. The functional M is multiplicative and nonzero on $\mathfrak{F}$ and so part (a) shows that $M(f) = f(a)$ for all $f \in \mathfrak{F}$, a being a point in U. Given $g \in \mathfrak{C}_{00}(X)$, the function $g f_0$ is in $\mathfrak{F}$ and $M(g f_0) = M(g)$. Since $f_0(a) = 1$, the proof is complete.]

(c) Generalize part (a) as follows. Let $(X, \mathscr{A}, \mu)$ be a finite measure space such that $\mu(X) > 0$. Prove that $\int fg\,d\mu = \int f\,d\mu \int g\,d\mu$ for all $f \in \mathfrak{L}_1(X, \mathscr{A}, \mu)$ if and only if μ assumes only the values 0 and 1. [Compare this with (20 37.c).]

(d) Part (a) can be interpreted in the following way. A *left ideal* in an algebra A is a linear subspace I that is closed under left multiplication by arbitrary elements of A: $x \in I$ and $y \in A$ imply $yx \in I$. *Right ideals* are defined similarly. A set I that is both a left and a right ideal is called a *two-sided ideal*. In commutative algebras, the distinction between right and left ideals disappears, of course, and we use the term "ideal". An ideal is called *maximal* if the only ideal containing it properly is the entire algebra. In the commutative Banach algebra $\mathfrak{C}_0(X)$, every set $\{f \in \mathfrak{C}_0(X) : f(a) = 0\}$ is a closed maximal ideal. Furthermore, every maximal ideal [not assumed *a priori* to be closed] in $\mathfrak{C}_0(X)$ has this form. [All assertions are simple to verify except the last, for which the following hints are offered. If $\mathfrak{J}$ is an ideal in $\mathfrak{C}_0(X)$ and if there is a point $a \in X$ such that $f(a) = 0$ for all $f \in \mathfrak{J}$, then $\mathfrak{J}$ can be maximal only if $\mathfrak{J} = \{f \in \mathfrak{C}_0(X) : f(b) = 0\}$ for some $b \in X$. If $\mathfrak{J}$ is an ideal such that for all $a \in X$, $f(a) \neq 0$ for some $f \in \mathfrak{J}$, then a simple compactness argument, and the fact that $\bar{f}f \in \mathfrak{J}$ if $f \in \mathfrak{J}$, show that $\mathfrak{J} \supset \mathfrak{C}_{00}(X)$. Consider the algebra $\mathfrak{S} = \mathfrak{C}_0(X)/\mathfrak{J}$. It is a linear space over K and also an algebra with no proper ideals at all. This implies that $\mathfrak{S}$ is a field or is K as a linear space with all products equal to 0. Let τ denote the canonical mapping of $\mathfrak{C}_0(X)$ onto $\mathfrak{S}$: $\tau(f) = f + \mathfrak{J}$ for all $f \in \mathfrak{C}_0(X)$. If $\mathfrak{S}$ is a field, let h be a function in $\mathfrak{C}_0(X)$ such that $\tau(h)$ is the multiplicative unit of $\mathfrak{S}$. Since $\mathfrak{J}$ contains $\mathfrak{C}_{00}(X)$, there is an element $\varphi \in \mathfrak{J}$ such that $\|h - \varphi\|_u < 1$. Let $\psi = -\sum_{k=1}^{\infty} (h - \varphi)^k$. Then as in part (a), one can show that

$$h = \varphi - \psi\varphi + \psi h - \psi .$$

Clearly $\varphi - \psi\varphi$ is in $\mathfrak{J}$, and $\psi h - \psi$ is in $\mathfrak{J}$ because

$$\tau(\psi h - \psi) = \tau(\psi)\,\tau(h) - \tau(h) = \tau(h) - \tau(h) = 0 .$$

Hence h is in $\mathfrak{J}$, and so $\tau(h) = 0$, a contradiction. If $\mathfrak{S}$ is not a field, then all products in $\mathfrak{S}$ are 0, and in particular, if $f \in \mathfrak{C}_0^+$, then $f = f^{\frac{1}{2}} \cdot f^{\frac{1}{2}}$ is in $\mathfrak{J}$. From this it follows at once that $\mathfrak{J} = \mathfrak{C}_0$.]

(f) Part (e) admits the following extension. Let $\mathfrak{J}$ be any *closed* proper ideal in $\mathfrak{C}_0(X)$. Then there is a nonvoid closed subset F of X such that $\mathfrak{J} = \{f \in \mathfrak{C}_0(X) : f(F) = \{0\}\}$. [Hint. Use (6.80).]

(20.53) Discussion. Our last application of the LEBESGUE-RADON-NIKODÝM theorem deals with sequences of σ-algebras on a fixed set and with measures on them. We first generalize the LEBESGUE-RADON-NIKODÝM derivative defined in (19.43). Let X be a set, $\mathscr{A}$ a σ-algebra of subsets of X, μ a σ-finite measure on $\mathscr{A}$, and η a σ-finite *signed* measure on $\mathscr{A}$. Let $\eta = \eta_a + \eta_s$ be the [unique] Lebesgue decomposition of η with respect to μ [see (19.42)]. Let f be an extended real-valued, $\mathscr{A}$-measurable function on X. Then f is said to be a *derivative of* η *with respect to* μ if the following conditions hold:

(i) $\eta_a(A) = \int_A f\,d\mu$ for all $A \in \mathscr{A}$;

there exist sets $B, P \in \mathscr{A}$ such that

(ii) $\mu(B) = 0$ and $|\eta_s|(B') = 0$;

(iii) (P, P') is a Hahn decomposition for η_s;
(iv) $f = \infty$ on $B \cap P$ and $f = -\infty$ on $B \cap P'$.

It is obvious that this definition agrees with (19.43) in the case that $\eta \ll \mu$ [$\eta_s = 0$]. In this case we merely take $B = P = \varnothing$.

To see that such derivatives exist, let B be any set in $\mathscr{A}$ such that (ii) holds [$\eta_s \perp \mu$], let (P, P') be any Hahn decomposition for η_s (19.6), and let f_0 be any LEBESGUE-RADON-NIKODÝM derivative of η_a with respect to μ [apply (19.24) to η_a^+ and η_a^- to obtain f_0^+ and f_0^-]. Now define f by

$$f(x) = \begin{cases} f_0(x) & \text{for } x \in B', \\ \infty & \text{for } x \in B \cap P, \\ -\infty & \text{for } x \in B \cap P'. \end{cases}$$

Since $f = f_0$ μ-a.e., it is clear that f is a derivative.

The following lemma characterizing derivatives will be very useful. It has the advantage that it makes no mention of Lebesgue or Hahn decompositions.

(20.54) Lemma. *Let g be an extended real-valued, $\mathscr{A}$-measurable function on X. For each real number α, let $G_\alpha = \{x \in X : g(x) \geqq \alpha\}$ and $L_\alpha = \{x \in X : g(x) \leqq \alpha\}$. The function g is a derivative of η with respect to μ if and only if the following conditions obtain:*

(i) *for every real number α and every $A \in \mathscr{A}$, we have*

$$\eta(G_\alpha \cap A) \geqq \alpha\mu(G_\alpha \cap A);$$

(ii) *for every real number α and every $A \in \mathscr{A}$, we have*

$$\eta(L_\alpha \cap A) \leqq \alpha\mu(L_\alpha \cap A).$$

Proof. Suppose that g is a derivative of η with respect to μ and let B and P be as in (20.53) [with f replaced by g of course]. Then for all $A \in \mathscr{A}$ and $\alpha \in R$ we have

$$\begin{aligned} \eta(G_\alpha \cap A) &= \eta_a(G_\alpha \cap A) + \eta_s(G_\alpha \cap A) \\ &= \int_{G_\alpha \cap A} g\, d\mu + \eta_s(G_\alpha \cap A \cap B) \\ &\geqq \alpha\mu(G_\alpha \cap A) + \eta_s(G_\alpha \cap A \cap B). \end{aligned} \qquad (1)$$

Since $g \geqq \alpha > -\infty$ on G_α, it follows from (20.53.iv) that $G_\alpha \cap A \cap B \subset P$, and so $\eta_s(G_\alpha \cap A \cap B) \geqq 0$ (19.4). This fact and (1) imply (i). A similar argument, which we leave to the reader, establishes (ii).

Conversely, suppose that (i) and (ii) hold for the function g. Let f be any derivative of η with respect to μ as defined in (20.53), and let B and P be as in (20.53). We first show that $g = f$ μ-a.e. If this is false, then a moment's reflection reveals that either:

(a) there exist real numbers α and β and a set $F \in \mathscr{A}$ such that $0 < \mu(F) < \infty$, $|\eta_s|(F) = 0$, and $f(x) \leqq \alpha < \beta \leqq g(x)$ for all $x \in F$;

or

(b) there exist real numbers α and β and a set $F \in \mathscr{A}$ such that $0 < \mu(F) < \infty$, $|\eta_s|(F) = 0$, and $g(x) \leqq \alpha < \beta \leqq f(x)$ for all $x \in F$. Assume that (a) holds. By (i) we have

$$\eta(F) = \eta(G_\beta \cap F) \geqq \beta\mu(G_\beta \cap F) = \beta\mu(F)\,. \tag{2}$$

We also have

$$\eta(F) = \eta_a(F) = \int_F f\,d\mu \leqq \alpha\mu(F)\,. \tag{3}$$

An obvious contradiction ensues from (2) and (3), and so (a) must fail. Likewise (b) fails. Thus $g = f$ μ-a.e., and so (20.53.i) holds for g.

Now make the following definitions: $L = \{x \in X : g(x) = -\infty\}$; $G = \{x \in X : g(x) = \infty\}$; $B_0 = B \cap (L \cup G)$; and $P_0 = B \cap G$. We wish to show that (20.53.ii)–(20.53.iv) hold for B_0, P_0, and g. Since $B_0 \subset B$, it is obvious that $\mu(B_0) = 0$. It is also obvious that (20.53.iv) holds. Thus it suffices to show that $|\eta_s|(B_0') = 0$ and that (P_0, P_0') is a Hahn decomposition for η_s.

Since $\mu(B) = 0$ and $\eta_a \ll \mu$, we have $|\eta_a|(B) = 0$. Thus condition (ii) implies that for all $A \in \mathscr{A}$ and $\alpha \in R$, we have

$$\eta_s(L_\alpha \cap A \cap B) = \eta(L_\alpha \cap A \cap B) \leqq \alpha\mu(L_\alpha \cap A \cap B) = 0\,. \tag{4}$$

It is obvious that

$$L_1 \subset L_2 \subset \cdots \quad \text{and} \quad \bigcup_{n=1}^{\infty} L_n = G'\,;$$

hence (19.3.ii) and (4) imply that

$$\eta_s(G' \cap A \cap B) = \lim_{n\to\infty} \eta_s(L_n \cap A \cap B) \leqq 0$$

for all $A \in \mathscr{A}$. We conclude that $G' \cap B$ is a nonpositive set for η_s. Likewise we use (i) and the facts $G_{-1} \subset G_{-2} \subset \cdots$ and $\bigcup_{n=1}^{\infty} G_{-n} = L'$ to prove that $L' \cap B$ is a nonnegative set for η_s. Thus the set $L' \cap G' \cap B$ is both nonpositive and nonnegative for η_s, and so $|\eta_s|(L' \cap G' \cap B) = 0$. Now $B_0' = B' \cup [(L' \cap G') \cap B]$ and $|\eta_s|(B') = 0$; hence $|\eta_s|(B_0') = 0$.

It remains only to prove that (P_0, P_0') is a Hahn decomposition for η_s. It is plain that $G = \bigcap_{n=1}^{\infty} G_n$ and that $L = \bigcap_{n=1}^{\infty} L_{-n}$. Thus we may argue as above to prove that

$$\eta_s(G \cap A \cap B) \geqq 0 \quad \text{and} \quad \eta_s(L \cap A \cap B) \leqq 0$$

for all $A \in \mathscr{A}$ [this time we use (19.3.iii) and the fact that η_s is σ-finite]; we leave the details to the reader. Therefore $G \cap B$ is a nonnegative set for η_s and $L \cap B$ is a nonpositive set for η_s. Since P_0 is equal to $B \cap G$, P_0 is a nonnegative set for η_s. Since

$$P_0' = (B \cap L' \cap G') \cup (L \cap B) \cup B'$$

and since $|\eta_s|(B') = 0$ and $|\eta_s|(B \cap L' \cap G') = 0$, we see that P_0' is a nonpositive set for η_s. This completes our verification that g is a derivative of η with respect to μ if (i) and (ii) hold. □

(20.55) More discussion. The following notation and hypotheses are fixed throughout (20.55)–20.60). Let X be a set and $\mathscr{M}$ a σ-algebra of subsets of X. Let μ be a measure on $\mathscr{M}$ and η a signed measure on $\mathscr{M}$. Let $(\mathscr{M}_n)_{n=1}^{\infty}$ be a sequence of σ-algebras of subsets of X such that $\bigcup_{k=1}^{\infty} \mathscr{M}_n \subset \mathscr{M}$. We suppose that μ and $|\eta|$ are σ-finite on each of the σ-algebra $\mathscr{M}_n$.

For each $n \in N$, consider the measure spaces $(X, \mathscr{M}_n, \mu)$ and $(X, \mathscr{M}_n, \eta)$; that is, we restrict the domains of μ and η to $\mathscr{M}_n$. We write μ_n for μ on $\mathscr{M}_n$ and η_n for η on $\mathscr{M}_n$. Let f_n be a derivative of η_n with respect to μ_n.

We are concerned with pointwise limits of the sequence of functions (f_n) under two hypotheses concerning $(\mathscr{M}_n)$. Our first theorem is the following.

(20.56) Theorem. *Suppose that*

(i) $\mathscr{M}_1 \subset \mathscr{M}_2 \subset \mathscr{M}_3 \subset \cdots$,

and write $\mathscr{M}_\omega$ *for the smallest* σ*-algebra containing* $\bigcup_{n=1}^{\infty} \mathscr{M}_n$. *Let* μ_ω *and* η_ω *be* μ *and* η, *respectively, restricted to* $\mathscr{M}_\omega$. *Then both of the functions*

(ii) $\underline{f} = \underline{\lim}_{n\to\infty} f_n$ *and* $\bar{f} = \overline{\lim}_{n\to\infty} f_n$

are derivatives of η_ω *with respect to* μ_ω. *Thus:* $\lim_{n\to\infty} f_n(x)$ *exists for* μ*-almost all* $x \in X$; $\int_A (\lim_{n\to\infty} f_n)\, d\mu = \eta_a(A)$ *for all* $A \in \mathscr{M}_\omega$; *and the* μ_ω*-singular part of* η_ω *is confined to the set* $\{x \in X : \lim_{n\to\infty} f_n(x) = \pm\infty\}$, *as set forth in* (20.53).

Proof. Let α be a real number. In this proof, write $L_\alpha = \{x \in X : \underline{f}(x) \leqq \alpha\}$ and $G_\alpha = \{x \in X : \bar{f}(x) \geqq \alpha\}$. In view of (20.54), it suffices to prove that

$$\eta(L_\alpha \cap A) \leqq \alpha \mu(L_\alpha \cap A) \tag{1}$$

and

$$\eta(G_\alpha \cap A) \geqq \alpha \mu(G_\alpha \cap A) \tag{2}$$

for all $A \in \mathscr{M}_\omega$. [To obtain (20.54.i) and (20.54.ii) from (1) and (2), use the obvious inclusions

$$\{x \in X : \bar{f}(x) \geqq \alpha\} \supset \{x \in X : \underline{f}(x) \geqq \alpha\}$$

and

$$\{x \in X : \underline{f}(x) \leqq \alpha\} \supset \{x \in X : \bar{f}(x) \leqq \alpha\},$$

and replace A in (1) and (2) by its intersection with the smaller set.] Let $(\alpha_n)_{n=1}^{\infty}$ be a strictly decreasing sequence of real numbers with limit α.

For $n \in N$, let

$$H_n = \{x \in X : \inf\{f_{n+1}(x), f_{n+2}(x), \ldots\} < \alpha_n\},$$

$$H_{n,1} = \{x \in X : f_{n+1}(x) < \alpha_n\},$$

and

$$H_{n,p} = \{x \in X : \min\{f_{n+1}(x), \ldots, f_{n+p-1}(x)\} \geqq \alpha_n, f_{n+p}(x) < \alpha_n\}, \quad (p \geqq 2).$$

It is clear that $H_{n,p} \in \mathscr{M}_{n+p}$, that $\{H_{n,p}\}_{p=1}^{\infty}$ is a pairwise disjoint family, that $H_n = \bigcup_{p=1}^{\infty} H_{n,p}$, and that $L_\alpha = \bigcap_{n=1}^{\infty} H_n$. Let A be any set in the algebra $\bigcup_{n=1}^{\infty} \mathscr{M}_n$, so that $A \in \bigcap_{n=n_0}^{\infty} \mathscr{M}_n$ for some n_0. The set $H_{n,p} \cap A$ is in $\mathscr{M}_{n+p}$ for $n \geqq n_0$ and $p \geqq 1$. We assemble all of these facts and (20.54) to write

$$\eta(H_n \cap A) = \sum_{p=1}^{\infty} \eta(H_{n,p} \cap A) = \sum_{p=1}^{\infty} \eta_{n+p}(H_{n,p} \cap A)$$
$$\leqq \sum_{p=1}^{\infty} \alpha_n \mu_{n+p}(H_{n,p} \cap A) = \alpha_n \mu(H_n \cap A), \tag{3}$$

the relations (3) holding for all $n \geqq n_0$. If $|\eta|(A) < \infty$ and $\mu(A) < \infty$, then we take the limit as $n \to \infty$ in (3) to write

$$\eta(L_\alpha \cap A) = \lim_{n\to\infty} \eta(H_n \cap A) \leqq \lim_{n\to\infty} \alpha_n \mu(H_n \cap A) = \alpha \mu(L_\alpha \cap A). \tag{4}$$

Since $|\eta|$ and μ are σ-finite on $\mathscr{M}_1$, (4) holds for all $A \in \bigcup_{n=1}^{\infty} \mathscr{M}_n$.

Let $\{F_k\}_{k=1}^{\infty}$ be a pairwise disjoint family of sets in $\mathscr{M}_1$ such that $X = \bigcup_{k=1}^{\infty} F_k$ and μ and $|\eta|$ are finite on each F_k. Let ν_k be the set function on $\bigcup_{n=1}^{\infty} \mathscr{M}_n$ defined by

$$\nu_k(A) = \alpha \mu(F_k \cap L_\alpha \cap A) - \eta(F_k \cap L_\alpha \cap A).$$

A routine computation and (4) show that ν_k is a countably additive, nonnegative, *finite-valued* measure on the algebra $\bigcup_{n=1}^{\infty} \mathscr{M}_n$, in the sense of (10.3). Let ν be the set function $\sum_{k=1}^{\infty} \nu_k$, also defined only on $\bigcup_{n=1}^{\infty} \mathscr{M}_n$. It is easy to see that ν is a nonnegative, countably additive, σ-finite measure on $\bigcup_{n=1}^{\infty} \mathscr{M}_n$. By (10.39.c), ν admits a unique [countably additive!] extension over the σ-algebra $\mathscr{M}_\omega$. All this implies that (4) holds not only for $A \in \bigcup_{n=1}^{\infty} \mathscr{M}_n$ but for all $A \in \mathscr{M}_\omega$. Thus condition (1) is established.

The proof of (2) is very like the proof of (1), and we leave it to the reader. □

The following special result is frequently useful.

(20.57) Corollary. *Suppose that $\eta_n \ll \mu_n$ for all $n \in N$ and that the equality*

(i) $\lim\limits_{n\to\infty} \int\limits_E f_n \, d\mu = \int\limits_E f_\omega \, d\mu$

holds for all E in the algebra $\bigcup\limits_{k=1}^{\infty} \mathcal{M}_k$. Then η_ω is absolutely continuous with respect to μ_ω.

Proof. For each fixed $k \in N$ and every $E \in \mathcal{M}_k$, we have

$$\eta(E) = \eta_k(E) = (\eta_k)_a(E) = \int\limits_E f_k \, d\mu_k = \int\limits_E f_k \, d\mu . \tag{1}$$

For $n > k$, E is also in $\mathcal{M}_n$, and so (1) can be extended to

$$\eta(E) = \eta_n(E) = (\eta_n)_a(E) = \int\limits_E f_n \, d\mu . \tag{2}$$

Take the limit as $n \to \infty$ in (2) and apply (i). This yields

$$\eta(E) = \lim_{n\to\infty} \int\limits_E f_n \, d\mu = \int\limits_E f_\omega \, d\mu . \tag{3}$$

Since f_ω is a derivative of η_ω with respect to μ_ω, we have

$$(\eta_\omega)_a(E) = \int\limits_E f_\omega \, d\mu ,$$

and from (3) we infer that

$$\eta(E) = \eta_\omega(E) = (\eta_\omega)_a(E) \tag{4}$$

for all $E \in \mathcal{M}_k$. Since k is arbitrary, we have proved that η_ω and $(\eta_\omega)_a$ agree on $\bigcup\limits_{k=1}^{\infty} \mathcal{M}_k$. By (10.39.c), η_ω is equal to $(\eta_\omega)_a$ on the entire σ-algebra $\mathcal{M}_\omega$. □

(20.58) Remark. Condition (20.57.i) is of course just the condition that $\int\limits_E \left(\lim\limits_{n\to\infty} f_n\right) d\mu = \lim\limits_{n\to\infty} \int\limits_E f_n \, d\mu$. If all of the functions $|f_n|$ are bounded by a fixed function in $\mathfrak{L}_1^+(X, \mathcal{M}, \mu)$, then Lebesgue's dominated convergence theorem (12.24) guarantees (20.57.i). If $\lim\limits_{m,n\to\infty} \|f_n - f_m\|_1 = 0$ and $\mu(X) < \infty$, then (13.39) implies that (20.57.i) holds. Other conditions under which (20.57.i) holds are set down in (13.39).

We have a second limit theorem, like (20.56), but dealing with descending instead of ascending sequences of σ-algebras.

(20.59) Theorem. *Suppose that*

(i) $\mathcal{M}_1 \supset \mathcal{M}_2 \supset \cdots \supset \mathcal{M}_n \supset \cdots$,

and write $\mathcal{M}_0$ for the σ-algebra $\bigcap\limits_{n=1}^{\infty} \mathcal{M}_n$. Let μ_0 and η_0 be μ and η, respectively, restricted to the σ-algebra $\mathcal{M}_0$. Suppose that μ_0 and $|\eta_0|$ are σ-finite. Let $\underline{f}$ and $\bar{f}$ be defined as in (20.56.ii). Then both $\underline{f}$ and $\bar{f}$ are derivatives of η_0 with respect to μ_0.

Proof. We note first that $\underline{f}$ and $\bar{f}$ are $\mathscr{M}_0$-measurable. To see this, write

$$h_n = \lim_{k\to\infty} \left[\min\{f_n, f_{n+1}, \ldots, f_{n+k}\}\right].$$

It is plain that h_n is $\mathscr{M}_n$-measurable, and also that $\lim_{n\to\infty} h_n = \underline{f}$ [cf. (6.83)]. By (i), all of the functions $h_m, h_{m+1}, \ldots$ are $\mathscr{M}_m$-measurable, and since $\underline{f} = \lim_{n\to\infty} h_{n+m}$, we infer that $\underline{f}$ is $\mathscr{M}_m$-measurable for all m, *i.e.*, $\underline{f}$ is $\mathscr{M}_0$-measurable. The proof for $\bar{f}$ is similar.

We borrow the notation L_α and G_α from the proof of (20.56). It is clear that L_α and G_α are in $\mathscr{M}_0$. As in the proof of (20.56), it is sufficient to prove (20.56.1) and (20.56.2) for all $A \in \mathscr{M}_0$. For each real number α, let

$$M_\alpha = \{x \in X : \inf\{f_1(x), f_2(x), \ldots\} < \alpha\}$$

and

$$H_\alpha = \{x \in X : \sup\{f_1(x), f_2(x), \ldots\} > \alpha\}.$$

The sets M_α and H_α are in $\mathscr{M}_1$, but need not be in $\mathscr{M}_0$. Suppose that the inequalities

$$\eta(M_\alpha \cap A) \leqq \alpha\mu(M_\alpha \cap A) \tag{1}$$

and

$$\eta(H_\alpha \cap A) \geqq \alpha\mu(H_\alpha \cap A) \tag{2}$$

obtain for all $A \in \mathscr{M}_0$. For all $\varepsilon > 0$, the inclusions $L_\alpha \subset M_{\alpha+\varepsilon}$ and $G_\alpha \subset H_{\alpha-\varepsilon}$ are evident, and so (1) implies that

$$\begin{aligned}\eta(L_\alpha \cap A) = \eta(M_{\alpha+\varepsilon} \cap L_\alpha \cap A) &\leqq (\alpha+\varepsilon)\,\mu(M_{\alpha+\varepsilon} \cap L_\alpha \cap A)\\ &= (\alpha+\varepsilon)\,\mu(L_\alpha \cap A).\end{aligned} \tag{3}$$

Since $|\eta|$ and μ are σ-finite on $\mathscr{M}_1$, a simple argument and (3) imply that

$$\eta(L_\alpha \cap A) \leqq \alpha\mu(L_\alpha \cap A),$$

i.e., (20.56.1) holds if (1) holds. Similarly we have

$$\eta(G_\alpha \cap A) \geqq (\alpha - \varepsilon)\,\mu(G_\alpha \cap A)$$

if (2) holds, and (20.56.2) follows from this. Therefore to prove the present theorem we need only to verify (1) and (2).

To prove (1), we write for each $n \in N$

$$J_n = \{x \in X : \min\{f_1(x), \ldots, f_n(x)\} < \alpha\},$$

$$J_{n,p} = \{x \in X : f_p(x) < \alpha, f_{p+1}(x) \geqq \alpha, \ldots, f_n(x) \geqq \alpha\}$$
$$(1 \leqq p < n),$$

and

$$J_{n,n} = \{x \in X : f_n(x) < \alpha\}.$$

Then $J_{n,p}$ is in $\mathscr{M}_p$, $\{J_{n,p}\}_{p=1}^n$ is a pairwise disjoint family, and $\bigcup_{p=1}^n J_{n,p} = J_n$.

From the definition of f_p and (20.54), we infer that

$$\begin{aligned}\eta(J_n \cap A) &= \sum_{p=1}^{n} \eta(J_{n,p} \cap A)\\ &= \sum_{p=1}^{n} \eta(\{x \in X : f_p(x) \leqq \alpha\} \cap J_{n,p} \cap A)\\ &\leqq \sum_{p=1}^{n} \alpha\mu(\{x \in X : f_p(x) \leqq \alpha\} \cap J_{n,p} \cap A)\\ &= \alpha\mu(J_n \cap A) . \end{aligned} \tag{4}$$

It is clear that $J_1 \subset J_2 \subset \cdots$ and that $\bigcup_{n=1}^{\infty} J_n = M_\alpha$. Formula (4) and countable additivity imply that

$$\eta(M_\alpha \cap A) \leqq \alpha\mu(M_\alpha \cap A),$$

i.e., (1) is verified. The inequality (2) is proved in like manner; we leave it to the reader. □

(20.60) Remarks. The limit theorems (20.56) and (20.59) are versions of what probabilists call *martingale* theorems. Our treatment is taken from a paper of SPARRE ANDERSEN and JESSEN [Danske Vid. Selsk. Mat.-Fys. Medd. 25 (1948), Nr. 5]. The interested reader may also consult J. L. DOOB's treatise *Stochastic Processes* [John Wiley and Sons, New York, 1953], Ch. VII. We have included these limit theorems primarily for their applications to infinite products of measure spaces [see § 22]. Several interesting and unexpected results follow immediately from them, however, and we shall now point out a few of these in the form of exercises with copious hints.

(20.61) Exercise: Differentiation on a net. Let $(X, \mathscr{A}, \mu)$ be a σ-finite measure space. Consider a sequence $(\mathscr{N}_n)_{n=1}^{\infty}$ of subfamilies of $\mathscr{A}$ having the following properties.

(i) The sets in each $\mathscr{N}_n$ are pairwise disjoint and $\bigcup \mathscr{N}_n = X$.

(ii) If $E \in \mathscr{N}_n$, then $0 < \mu(E) < \infty$.

(iii) For each $E \in \mathscr{N}_n$, $E = \bigcup\{F : F \in \mathscr{N}_{n+1}, F \subset E\}$.

Such a sequence $(\mathscr{N}_n)_{n=1}^{\infty}$ is called a *net*. Let $\mathscr{M}_n$ be the σ-algebra generated by $\mathscr{N}_n$, *i.e.*, the family of all unions of sets in $\mathscr{N}_n$.

(a) Let η be a σ-finite signed measure on $\mathscr{A}$. For each $n \in N$, let

$$f_n = \sum_{E \in \mathscr{N}_n} \frac{\eta(E)}{\mu(E)} \xi_E.$$

Prove that $\lim_{n\to\infty} f_n(x)$ exists for μ-almost all $x \in X$ and in fact that $\lim_{n\to\infty} f_n$ is a derivative of η with respect to μ for the smallest σ-algebra containing all $\mathscr{N}_n$. [From (ii) we see that each $\mathscr{N}_n$ is countable. Also if $E \in \mathscr{M}_n$ and $\mu(E) = 0$, then $E = \varnothing$. The relation $|\eta_n| \ll \mu_n$ is therefore trivially satisfied, and the function f_n is plainly the LEBESGUE-RADON-NIKODÝM derivative of η_n with respect to μ_n. Now apply (20.56) and the definition of derivative in (20.53).]

(b) Let φ be a Lebesgue measurable function on R such that $\int_{-a}^{a} |\varphi| \, d\lambda < \infty$ for all $a > 0$. For every $x \in R$, let $J(n, x)$ be the interval $[k \cdot 2^{-n}, (k+1) \cdot 2^{-n}[$ that contains x ($n \in N$, $k \in Z$). Prove that $\lim_{n \to \infty} 2^n \int_{J(n,x)} \varphi \, d\lambda = \varphi(x)$ for λ-almost all $x \in R$. [Hint. Apply part (a) with $X = R$, $\mathscr{A} = \mathscr{M}_\lambda$, $\mathscr{N}_n = \{[k2^{-n}, (k+1)2^{-n}[\}_{k=-\infty}^{\infty}$, $\eta(A) = \int_A \varphi \, d\lambda$.] Compare this result with (18.3).

(c) Let η be a σ-finite signed measure or a complex measure on $\mathscr{B}(R)$ such that $|\eta|$ and λ are mutually singular. Prove that $\lim_{n \to \infty} 2^n \eta(J(n, x)) = 0$ for λ-almost all $x \in R$. For η a signed measure, prove that $\lim_{n \to \infty} 2^n \eta(J(n, x)) = \pm\infty$ on a set B such that $|\eta| (B') = 0$. [Apply part (b), (20.56), and (20.53.iv).] Compare this with the behavior of α's described in (19.60).

(20.62) Exercise: Densities. Notation is as in (20.61). Let E be any set in $\mathscr{A}$. Let η be the measure on $\mathscr{A}$ such that

$$\mu(A) = \mu(E \cap A) \quad \text{for all} \quad A \in \mathscr{A} .$$

Plainly $\eta_n \ll \mu_n$; let w_n be a LEBESGUE-RADON-NIKODÝM derivative of η_n with respect to μ_n [here $n \in \{1, 2, \ldots, \omega\}$]. The function w_n is called a *density of E with respect to $\mathscr{M}_n$*.

(a) Prove that $\lim_{n \to \infty} w_n(x) = w_\omega(x)$ for μ-almost all $x \in X$. [This is a direct application of (20.56).]

(b) Suppose that each w_n is a constant real function μ-almost everywhere, in addition to being $\mathscr{M}_n$-measurable. Suppose also that there is a set $D \in \mathscr{M}_\omega$ such that $\mu(D \triangle E) = 0$. Then $\mu(E) = 0$ or $\mu(E') = 0$. [Plainly $\lim_{n \to \infty} w_n$ is a constant μ-almost everywhere. A simple argument shows that w_ω is equal to ξ_D μ-almost everywhere. Thus ξ_D is a constant μ-a.e.]

(c) Consider the space $X = [0, 1[$, and for each $n \in N$, let

$$\mathscr{N}_n = \left\{[k\,2^{-n}, (k+1)2^{-n}[\,:\, k \in \{0, 1, \ldots, 2^n - 1\}\right\} .$$

Let $\mathscr{M}_n$ be the [obviously finite] algebra of sets generated by $\mathscr{N}_n$. Let $\mathscr{M} = \mathscr{M}_\lambda([0, 1[)$. Let P be the Cantor-like set in $[0, 1[$ obtained by removing middle one-quarters at each step (6.62). Determine $\lambda(P)$. Let μ be λ and let $\eta(A) = \lambda(A \cap P)$. Compute w_n as defined above and find $\lim_{n \to \infty} w_n$.

(20.63) Exercise: Application of (20.59). Suppose in (20.59) that $\mu(X) = 1$ and that μ_0 assumes only the values 0 and 1 on the σ-algebra $\mathscr{M}_0$.

(a) Prove that $\lim_{n\to\infty} f_n$ is a constant μ-almost everywhere. [Hints. Consider the Lebesgue decomposition $\eta_0 = \eta_{0a} + \eta_{0s}$. Let $B_0 \in \mathscr{M}_0$ be a set such that $\mu_0(B_0) = 0$ and $|\eta_{0s}|(B_0') = 0$. Let g be a LEBESGUE-RADON-NIKODÝM derivative of η_{0a} with respect to μ_0. Then for every $E \in \mathscr{M}_0$, we have

$$\eta_{0a}(E) = \int_E g\, d\mu_0 = \int_{E\cap B_0'} g\, d\mu_0 .$$

Since μ_0 assumes only the values 0 and 1, it follows from (12.60) that there is a number α for which $\mu_0(\{x \in X : g(x) = \alpha\}) = 1$ and so $\lim_{n\to\infty} f_n = \alpha$ μ_0-almost everywhere on B_0' and so μ_0-almost everywhere on X.]

(b) Prove that the α of part (a) is equal to $\int_X f_1(x)\, d\mu_1(x) = \eta_{1a}(X)$, where η_{1a} is the μ_1-absolutely continuous part of η_1.

(20.64) Exercise. Let $X = [0, 1[$. For each $n \in N$, let $\mathscr{M}_n = \{A \subset X$: there is an $\mathscr{M}_\lambda$-measurable set $B \subset [0, 2^{-n}[$ such that $A = \bigcup_{k=0}^{2^n-1} (B + k2^{-n})\}$. Note that $\mathscr{M}_n$ is a σ-algebra and that $\mathscr{M}_1 \supset \mathscr{M}_2 \supset \cdots \supset \mathscr{M}_n \supset \cdots$. Prove that if $A \in \mathscr{M}_0 = \bigcap_{n=1}^{\infty} \mathscr{M}_n$, then $\lambda(A) = 0$ or $\lambda(A) = 1$. [Hints. Apply (20.62.b) to the set A, and use the increasing algebras of sets described in (20.62.c). It is evident that each w_n is a constant, and so $\lambda(A) = 0$ or $\lambda([0, 1[\cap A') = 0$.]

(20.65) Exercise [JESSEN]. Let f be any function in $\mathfrak{L}_1([0, 1[, \mathscr{M}_\lambda, \lambda)$ and extend f over R by the definition $f(x + k) = f(x)$ for all $k \in Z$ and $x \in [0, 1[$. Prove that

$$\text{(i)}\quad \lim_{n\to\infty} 2^{-n} \left[\sum_{k=0}^{2^n-1} f(x + k2^{-n})\right] = \int_0^1 f(t)\, dt$$

for λ-almost all $x \in R$. [Hints. Let $\mathscr{M}_n$ and $\mathscr{M}_0$ be as in (20.64). Let f_n be the function on the left side of (i). On the domain $[0, 1[$, f_n is plainly $\mathscr{M}_n$-measurable, and is the LEBESGUE-RADON-NIKODÝM derivative of the measure η_n, where $\eta(A) = \int_A f\, d\lambda$ for $A \in \mathscr{M}_1$. By (20.59), f_n converges λ-almost everywhere to the LEBESGUE-RADON-NIKODÝM derivative f_0 of η_0 with respect to λ_0. Since $\eta_0([0, 1[) = \int_0^1 f_0\, d\lambda_0 = \int_0^1 f\, d\lambda$, and since (20.64) and (20.63) hold, we have $\lim_{n\to\infty} f_n(x) = \int_0^1 f\, d\lambda$ λ-a.e. on $[0, 1[$; and by periodicity, λ-a.e. on R.]

(20.66) Exercise: A martingale theorem. Let $X, \mathscr{M}, \mathscr{M}_n, \mathscr{M}_\omega$ and μ be as in (20.56) and suppose that $\mu(X) < \infty$. Suppose that $(f_n)_{n=1}^{\infty}$ is a sequence of functions on X each with values in $[0, \infty[$ such that:

(i) f_n is $\mathscr{M}_n$-measurable for all $n \in N$;

(ii) $\int_A f_n \, d\mu = \int_A f_{n+1} \, d\mu$ for all $A \in \mathscr{M}_n$ and all $n \in N$;

(iii) $\int_X f_n \, d\mu = 1$ for all $n \in N$.

(a) Prove that there is a finitely additive measure η on $\bigcup_{n=1}^{\infty} \mathscr{M}_n$ such that

(iv) $\eta(A) = \int_A f_n \, d\mu$ for all $A \in \mathscr{M}_n$ and all $n \in N$.

(b) Prove by giving an example that n need not be countably additive on $\bigcup_{n=1}^{\infty} \mathscr{M}_n$.

(c) Prove that $\lim_{n \to \infty} f_n$ exists and is finite μ-almost everywhere on X. [Hints. Consider the set Ω of all finitely additive measures ω on $\bigcup_{n=1}^{\infty} \mathscr{M}_n$ that assume the values 0 and 1 and no other values and vanish for μ-null sets. Make Ω into a topological space by neighborhoods $\Delta_A = \{\omega \in \Omega : \omega(A) = 1\}$, for $A \in \bigcup_{n=1}^{\infty} \mathscr{M}_n$. Then Ω is a compact Hausdorff space, and μ and η can be transferred to Ω. Apply (20.56) appropriately and go back to X.]

CHAPTER SIX

Integration on Product Spaces

§ 21. The product of two measure spaces

(21.1) Remarks. Suppose that $(X, \mathscr{M}, \mu)$ and $(Y, \mathscr{N}, \nu)$ are two measure spaces. We wish to define a product measure space

$$(X \times Y, \mathscr{M} \times \mathscr{N}, \mu \times \nu),$$

where $\mathscr{M} \times \mathscr{N}$ is an appropriate σ-algebra of subsets of $X \times Y$ and $\mu \times \nu$ is a measure on $\mathscr{M} \times \mathscr{N}$ for which

$$\mu \times \nu(A \times B) = \mu(A) \cdot \nu(B)$$

whenever $A \in \mathscr{M}$ and $B \in \mathscr{N}$. That is, we wish to generalize the usual geometric notion of the area of a rectangle. We also wish it to be true that

$$\int_{X \times Y} f \, d\mu \times \nu = \int_X \int_Y f \, d\nu \, d\mu = \int_Y \int_X f \, d\mu \, d\nu , \tag{1}$$

for a reasonably large class of functions f on $X \times Y$. Thus we want a generalization of the classical formula

$$\int_{[a,b] \times [c,d]} f(x, y) \, dS = \int_a^b \int_c^d f(x, y) \, dy \, dx = \int_c^d \int_a^b f(x, y) \, dx \, dy ,$$

which, as we know from elementary analysis, is valid for all functions $f \in \mathfrak{C}([a, b] \times [c, d])$.

In the case that X and Y are locally compact Hausdorff spaces and μ and ν are measures constructed as in § 9 from nonnegative linear functionals I and J on $\mathfrak{C}_{00}(X)$ and $\mathfrak{C}_{00}(Y)$ respectively, this program can be carried out by first constructing a nonnegative linear functional $I \times J$ on $\mathfrak{C}_{00}(X \times Y)$ and then letting $\mu \times \nu$ be the outer measure induced on $X \times Y$ by $I \times J$ just as in § 9. A brief outline of this construction follows. For $f \in \mathfrak{C}_{00}(X \times Y)$ and $y \in Y$, the function $f^{[y]}$: $x \to f(x, y)$ is in $\mathfrak{C}_{00}(X)$. Thus we define a function on Y by

$$y \to I(f^{[y]}) .$$

To indicate that we "integrate" with respect to x, we denote this function by $I_x(f)$. Similarly we obtain a function $J_y(f)$ on X. Next choose open sets $U \subset X$ and $V \subset Y$ such that U^- and V^- are compact and $U \times V \supset \{(x, y) \in X \times Y : f(x, y) \neq 0\}^-$. Use (9.5) to find positive constants α

and β for which

$$|I(\varphi)| \leqq \alpha \|\varphi\|_u \tag{2}$$

for all $\varphi \in \mathfrak{C}_{00}(X)$ such that $\varphi(U') \subset \{0\}$ and

$$|J(\psi)| \leqq \beta \|\psi\|_u \tag{3}$$

for all $\psi \in \mathfrak{C}_{00}(Y)$ such that $\psi(V') \subset \{0\}$. Consider any $\varepsilon > 0$ and use the STONE-WEIERSTRASS theorem (7.30) to find a function g on $X \times Y$ of the form

$$g(x, y) = \sum_{j=1}^{n} \varphi_j(x) \psi_j(y),$$

where the φ's and ψ's satisfy the conditions of the preceding sentence, such that $\|f - g\|_u < \varepsilon$. Since $f^{[y]} - g^{[y]}$ is in $\mathfrak{C}_{00}(X)$ and vanishes on U' for every $y \in Y$, we deduce from (2) that

$$\begin{aligned} |I_x(f)(y) - \sum_{j=1}^{n} I(\varphi_j)\psi_j(y)| &= |I(f^{[y]}) - I(g^{[y]})| \\ &\leqq \alpha \cdot \sup\{|f(x, y) - g(x, y)| : x \in X\} \\ &\leqq \alpha \|f - g\|_u < \alpha \varepsilon \end{aligned} \tag{4}$$

for all $y \in Y$. Thus $I_x(f)$ is the uniform limit of a sequence of functions in $\mathfrak{C}_{00}(Y)$ all of which vanish on V'; hence $I_x(f) \in \mathfrak{C}_{00}(Y)$ and $I_x(f)$ vanishes on V'. Combining this fact with (4). we infer from (3) that

$$|J(I_x(f)) - \sum_{j=1}^{n} I(\varphi_j) J(\psi_j)| = |J(I_x(f) - \sum_{j=1}^{n} I(\varphi_j)\psi_j)| \leqq \beta \alpha \varepsilon.$$

Similarly we obtain

$$|I(J_y(f)) - \sum_{j=1}^{n} I(\varphi_j) J(\psi_j)| \leqq \beta \alpha \varepsilon.$$

We conclude that

$$J(I_x(f)) = I(J_y(f))$$

and we denote this common value by $I \times J(f)$. This defines a nonnegative linear functional $I \times J$ on $\mathfrak{C}_{00}(X \times Y)$. Let $\mu \times \nu$ denote the outer measure on $X \times Y$ constructed from $I \times J$ as in § 9. Using (12.35), it is easy to show that (1) holds for all $f \in \mathfrak{C}_{00}(X \times Y)$. It is also true that (1) holds for all $f \in \mathfrak{L}_1(X \times Y, \mathscr{M}_{\mu \times \nu}, \mu \times \nu)$. For a detailed treatment of this approach to our problem the reader should consult HEWITT and ROSS, *Abstract Harmonic Analysis I* [Springer-Verlag, Heidelberg, 1963], pp. 150–157.

While the above approach has aesthetic appeal, as well as more generality in that σ-finiteness of the measure spaces need not be supposed, it produces product measures only for measures constructed as in § 9 on locally compact Hausdorff spaces. We prefer a construction that can be carried out for any two σ-finite measure spaces. These measure spaces include most of those that arise in classical analysis. The subject exhibits several technicalities, which can be an annoyance or a source of fascination: depending upon one's point of view. We proceed to some exact definitions.

(21.2) Definitions. Let $(X, \mathscr{M})$ and $(Y, \mathscr{N})$ be any two measurable spaces. For $A \in \mathscr{M}$ and $B \in \mathscr{N}$, the set $A \times B \subset X \times Y$ is called a *measurable rectangle*. The smallest σ-algebra of subsets of $X \times Y$ containing all of the measurable rectangles is denoted by $\mathscr{M} \times \mathscr{N}$[1] and is called the *product σ-algebra*. For $E \subset X \times Y$ and $x \in X$, let

$$E_x = \{y \in Y : (x, y) \in E\};$$

similarly, for $y \in Y$ let

$$E^y = \{x \in X : (x, y) \in E\}.$$

These sets are called *X- and Y-sections of E*, respectively. For a function f on $X \times Y$ and a fixed $x \in X$, let $f_{[x]}$ be the function defined on Y by

$$f_{[x]}(y) = f(x, y);$$

similarly, for each $y \in Y$ define $f^{[y]}$ on X by

$$f^{[y]}(x) = f(x, y).$$

These functions are called *X- and Y-sections of f*, respectively.

(21.3) Theorem. *Let $(X, \mathscr{M})$ and $(Y, \mathscr{N})$ be measurable spaces. Then the family $\mathscr{A}$ of all finite, pairwise disjoint unions of measurable rectangles is an algebra of subsets of $X \times Y$.*

Proof. First we note that if $A \times B$ and $C \times D$ are two measurable rectangles, then

$$(A \times B) \cap (C \times D) = (A \cap C) \times (B \cap D),$$

as the reader can easily verify. Thus if $E = \bigcup_{i=1}^{m} (A_i \times B_i)$ and $F = \bigcup_{j=1}^{n} (C_j \times D_j)$ are in $\mathscr{A}$, where these are pairwise disjoint unions of measurable rectangles, then

$$E \cap F = \bigcup_{i=1}^{m} \bigcup_{j=1}^{n} [(A_i \cap C_j) \times (B_i \cap D_j)],$$

and so $E \cap F \in \mathscr{A}$. That is, $\mathscr{A}$ is closed under the formation of finite intersections.

If $A \times B$ is a measurable rectangle, one easily checks that

$$(A \times B)' = (A' \times B) \cup (X \times B'),$$

which is the union of two disjoint measurable rectangles. For

---

[1] Note that $\mathscr{M} \times \mathscr{N}$ is *not* the Cartesian product of $\mathscr{M}$ and $\mathscr{N}$. Convention demands that this notation be used; we trust it will cause no confusion.

$E = \bigcup_{i=1}^{m} (A_i \times B_i) \in \mathscr{A}$, we have $E' = \bigcap_{i=1}^{m} (A_i \times B_i)'$, which is a finite intersection of sets in $\mathscr{A}$, and so $E' \in \mathscr{A}$. □

(21.4) Theorem. *Let* $(X, \mathscr{M})$ and $(Y, \mathscr{N})$ *be measurable spaces and let* $E \in \mathscr{M} \times \mathscr{N}$. *Then*

(i) $E_x \in \mathscr{N}$ *for all* $x \in X$

and

(ii) $E^y \in \mathscr{M}$ *for all* $y \in Y$.

Proof. To prove (i), let $\mathscr{S} = \{E \in \mathscr{M} \times \mathscr{N} : E_x \in \mathscr{N} \text{ for all } x \in X\}$. For any measurable rectangle $A \times B$ and any $x \in X$, we have

$$(A \times B)_x = \begin{cases} B \text{ if } x \in A\,, \\ \varnothing \text{ if } x \notin A\,. \end{cases}$$

Thus $\mathscr{S}$ contains all measurable rectangles. To complete the proof of (i), we need only show that $\mathscr{S}$ is a σ-algebra. If $(E_n)_{n=1}^{\infty} \subset \mathscr{S}$, then it is easy to see that $\left(\bigcup_{n=1}^{\infty} E_n\right)_x = \bigcup_{n=1}^{\infty} (E_n)_x \in \mathscr{N}$ for all $x \in X$; hence $\mathscr{S}$ is closed under the formation of countable unions. For $E \in \mathscr{S}$, we have $(E')_x = (E_x)' \in \mathscr{N}$ for all $x \in X$, and so $\mathscr{S}$ is also closed under complementation. This proves (i). The proof of (ii) is similar. □

(21.5) Theorem. *Let* $(X, \mathscr{M})$ *and* $(Y, \mathscr{N})$ *be measurable spaces, and let* f *be an extended real- or complex-valued* $\mathscr{M} \times \mathscr{N}$*-measurable function on* $X \times Y$. *Then*

(i) $f_{[x]}$ *is* $\mathscr{N}$*-measurable for all* $x \in X$,

and

(ii) $f^{[y]}$ *is* $\mathscr{M}$*-measurable for all* $y \in Y$.

Proof. Suppose that $f = \xi_E$ for some $E \in \mathscr{M} \times \mathscr{N}$. Since the statements $f_{[x]}(y) = 1$; $(x, y) \in E$; $y \in E_x$; $\xi_{E_x}(y) = 1$ are mutually equivalent, it follows that $f_{[x]} = \xi_{E_x}$ for all $x \in X$. Thus (i) follows from (21.4.i) in the case that $f = \xi_E$. It is now plain that (i) holds if f is a simple function. The general case follows from the above by using (11.35), (11.14), and (11.18). A similar argument proves (ii). □

We will need the following purely set-theoretic fact.

(21.6) Theorem. *Let* T *be any set and let* $\mathscr{A}$ *be an algebra of subsets of* T. *Then the* σ*-algebra* $\mathscr{S}(\mathscr{A})$ *generated by* $\mathscr{A}$ *is the smallest family* $\mathscr{F}$ *of subsets of* T *that contains* $\mathscr{A}$ *and satisfies the following two conditions.*

(i) *If* $E_n \in \mathscr{F}$ *and* $E_n \subset E_{n+1}$ *for* $n = 1, 2, \ldots$, *then* $\bigcup_{n=1}^{\infty} E_n \in \mathscr{F}$.

(ii) *If* $F_n \in \mathscr{F}$ *and* $F_n \supset F_{n+1}$ *for* $n = 1, 2, \ldots$, *then* $\bigcap_{n=1}^{\infty} F_n \in \mathscr{F}$.

Thus, in particular, if the algebra $\mathscr{A}$ *satisfies* (i) *and* (ii), *then* $\mathscr{A}$ *is a* σ*-algebra*[1].

[1] Families that satisfy (i) and (ii) are called *monotone families*.

Proof. Note first that the family $\mathscr{F}$ exists, for $\mathscr{P}(T)$ is a monotone family and the intersection of all monotone families containing $\mathscr{A}$ is again a monotone family containing $\mathscr{A}$; this intersection is $\mathscr{F}$.

Since any σ-algebra is a monotone family and $\mathscr{S}(\mathscr{A}) \supset \mathscr{A}$, it follows that $\mathscr{S}(\mathscr{A}) \supset \mathscr{F} \supset \mathscr{A}$. To finish the proof it therefore suffices to prove that $\mathscr{F}$ is a σ-algebra. For a monotone sequence $(H_n)_{n=1}^{\infty}$, we write $\lim H_n$ for the set $\bigcup_{n=1}^{\infty} H_n$ or $\bigcap_{n=1}^{\infty} H_n$ according as (H_n) is an increasing or a decreasing sequence. If $E \in \mathscr{F}$, write

$$\mathscr{F}_E = \{F \in \mathscr{F} : F \cap E' \in \mathscr{F}, E \cap F' \in \mathscr{F}, E \cup F \in \mathscr{F}\}.$$

Note that $F \in \mathscr{F}_E$ if and only if $E \in \mathscr{F}_F$. It is also clear that if $(F_n)_{n=1}^{\infty}$ is a monotone sequence in $\mathscr{F}_E$, then

$$(\lim F_n) \cap E' = \lim (F_n \cap E') \in \mathscr{F},$$

$$E \cap (\lim F_n)' = E \cap (\lim F_n') = \lim (E \cap F_n') \in \mathscr{F},$$

and

$$E \cup (\lim F_n) = \lim (E \cup F_n) \in \mathscr{F},$$

since $\mathscr{F}$ is a monotone family. Thus $\mathscr{F}_E$ is a monotone family for every $E \in \mathscr{F}$.

If $E, F \in \mathscr{A}$, then, since $\mathscr{A}$ is an algebra, E belongs to $\mathscr{F}_F$ and F belongs to $\mathscr{F}_E$. It follows that $\mathscr{A} \subset \mathscr{F}_E$ for all $E \in \mathscr{A}$. Since $\mathscr{F}$ is the smallest monotone family containing $\mathscr{A}$, we therefore have $\mathscr{F} \subset \mathscr{F}_E$ for all $E \in \mathscr{A}$. Thus for any $F \in \mathscr{F}$ and $E \in \mathscr{A}$ we have $F \in \mathscr{F}_E$; hence $E \in \mathscr{F}_F$. This shows that $\mathscr{A} \subset \mathscr{F}_F$ for all $F \in \mathscr{F}$. Since each $\mathscr{F}_F$ is a monotone family, it follows that $\mathscr{F} \subset \mathscr{F}_F$ for all $F \in \mathscr{F}$. The definition of $\mathscr{F}_F$ now shows that $\mathscr{F}$ is an algebra.

To show that $\mathscr{F}$ is a σ-algebra, let $(F_n)_{n=1}^{\infty} \subset \mathscr{F}$. For each $n \in N$, write $G_n = F_1 \cup F_2 \cup \cdots \cup F_n$. Then $G_1 \subset G_2 \subset \cdots$ and, since $\mathscr{F}$ is an algebra, $(G_n)_{n=1}^{\infty} \subset \mathscr{F}$. Thus

$$\bigcup_{n=1}^{\infty} F_n = \bigcup_{n=1}^{\infty} G_n = \lim G_n \in \mathscr{F}. \quad \square$$

(21.7) Corollary. *Let $(X, \mathscr{M})$ and $(Y, \mathscr{N})$ be measurable spaces. Then $\mathscr{M} \times \mathscr{N}$ is the smallest monotone family of subsets of $X \times Y$ that contains all finite disjoint unions of measurable rectangles.*

Proof. This follows at once from (21.3) and (21.6). $\square$

(21.8) Theorem. *Let $(X, \mathscr{M}, \mu)$ and $(Y, \mathscr{N}, \nu)$ be σ-finite measure spaces and let $E \in \mathscr{M} \times \mathscr{N}$. Then the following assertions obtain:*

(i) *the function $x \to \nu(E_x)$ on X is $\mathscr{M}$-measurable;*

(ii) *the function $y \to \mu(E^y)$ on Y is $\mathscr{N}$-measurable;*

(iii) $\int_X \nu(E_x)\, d\mu(x) = \int_Y \mu(E^y)\, d\nu(y)$.

Proof. Let $\mathscr{F}$ be the family of all sets $E \in \mathscr{M} \times \mathscr{N}$ such that (i), (ii), and (iii) hold for E. We will show that $\mathscr{F}$ is a monotone family that contains all finite disjoint unions of measurable rectangles. Then (21.7) will show that $\mathscr{F} = \mathscr{M} \times \mathscr{N}$.

Suppose that $E = A \times B$ is a measurable rectangle. Since $E_x = B$ or $\varnothing$ according as $x \in A$ or $x \in A'$, and $E^y = A$ or $\varnothing$ according as $y \in B$ or $y \in B'$, we have $\nu(E_x) = \nu(B)\, \xi_A(x)$ and $\mu(E^y) = \mu(A)\, \xi_B(y)$; hence (i) and (ii) hold for this E. We may also write

$$\int_X \nu(E_x)\, d\mu(x) = \int_X \nu(B)\, \xi_A\, d\mu = \nu(B) \cdot \mu(A)$$
$$= \int_Y \mu(A)\, \xi_B\, d\nu = \int_Y \mu(E^y)\, d\nu(y)\,.$$

Thus (iii) holds for this E, and so $E \in \mathscr{F}$. Thus $\mathscr{F}$ contains all measurable rectangles.

Let $\{E_1, \ldots, E_p\}$ be a finite, pairwise disjoint subfamily of $\mathscr{F}$. Since $\left(\bigcup_{n=1}^{p} E_n\right)_x = \bigcup_{n=1}^{p} (E_n)_x$ and $\left(\bigcup_{n=1}^{p} E_n\right)^y = \bigcup_{n=1}^{p} (E_n)^y$ for all $x \in X$ and $y \in Y$, it is clear that $\left(\bigcup_{n=1}^{p} E_n\right) \in \mathscr{F}$. Thus, $\mathscr{F}$ contains all finite disjoint unions of measurable rectangles.

Now let $(E_n)_{n=1}^{\infty}$ be an increasing sequence in $\mathscr{F}$ and let $E = \bigcup_{n=1}^{\infty} E_n = \lim E_n$. For all $x \in X$, we have

$$\nu(E_x) = \lim_{n \to \infty} \nu\left((E_n)_x\right)$$

as (10.13) shows, and so (11.14) implies that (i) holds for E. In like manner, (ii) holds for E. Applying B. Levi's theorem (12.22), we have

$$\int_X \nu(E_x)\, d\mu(x) = \lim_{n \to \infty} \int_X \nu\left((E_n)_x\right) d\mu(x)$$
$$= \lim_{n \to \infty} \int_Y \mu((E_n)^y)\, d\nu(y)$$
$$= \int_Y \mu(E^y)\, d\nu(y)\,.$$

Therefore E is in $\mathscr{F}$, and so $\mathscr{F}$ is closed under the formation of unions of increasing sequences.

It remains only to show that $\mathscr{F}$ is closed under the formation of intersections of decreasing sequences. To do this, we must use our σ-finiteness hypothesis, since we shall call upon (10.15) and (12.24), each of which contains a finiteness hypothesis. Let $(F_n)_{n=1}^{\infty}$ be a decreasing sequence in $\mathscr{F}$ such that $F_1 \subset A \times B$ for some measurable rectangle $A \times B$ for which $\mu(A) < \infty$ and $\nu(B) < \infty$. Write $F = \bigcap_{n=1}^{\infty} F_n$. For each $x \in X$, we have $(F_1)_x \subset B$, and so $\nu\left((F_1)_x\right) < \infty$. It follows from (10.15) that

$$\nu(F_x) = \lim_{n\to\infty} \nu((F_n)_x)$$

for every $x \in X$. Thus the function $x \to \nu(F_x)$ is the pointwise limit of a sequence of $\mathscr{M}$-measurable functions, and so it is $\mathscr{M}$-measurable (11.14); hence (i) holds for F. Likewise (ii) holds for F. Since

$$\int_X \nu((F_1)_x)\, d\mu(x) \leqq \int_X \nu(B)\, \xi_A\, d\mu < \infty$$

and

$$\int_Y \mu((F_1)^y)\, d\nu(y) \leqq \int_Y \mu(A)\, \xi_B d\nu < \infty,$$

LEBESGUE's dominated convergence theorem (12.24) implies that

$$\begin{aligned}\int_X \nu(F_x)\, d\mu(x) &= \lim_{n\to\infty} \int_X \nu((F_n)_x)\, d\mu(x)\\ &= \lim_{n\to\infty} \int_Y \mu((F_n)^y)\, d\nu(y)\\ &= \int_Y \mu(F^y)\, d\nu(y),\end{aligned}$$

and therefore (iii) holds for F.

Use the σ-finiteness hypothesis to choose increasing sequences $(A_k)_{k=1}^{\infty} \subset \mathscr{M}$ and $(B_k)_{k=1}^{\infty} \subset \mathscr{N}$ such that $\mu(A_k) < \infty$ and $\nu(B_k) < \infty$ for all k, and such that $X = \bigcup_{k=1}^{\infty} A_k$ and $Y = \bigcup_{k=1}^{\infty} B_k$. Let $\mathscr{E} = \{E \in \mathscr{M} \times \mathscr{N} : E \cap (A_k \times B_k) \in \mathscr{F}$ for all $k \in N\}$. Since the family $\mathscr{A}$ of all finite disjoint unions of measurable rectangles is an algebra (21.3) and $\mathscr{A} \subset \mathscr{F}$, we have $\mathscr{A} \subset \mathscr{E}$. If $(E_n)_{n=1}^{\infty}$ is an increasing sequence in $\mathscr{E}$, then

$$\left(\bigcup_{n=1}^{\infty} E_n\right) \cap (A_k \times B_k) = \bigcup_{n=1}^{\infty} (E_n \cap (A_k \times B_k)) \in \mathscr{F}$$

because $\mathscr{F}$ is closed under the formation of unions of increasing sequences. Thus $\mathscr{E}$ is closed under limits of increasing sequences. If $(E_n)_{n=1}^{\infty}$ is a decreasing sequence in $\mathscr{E}$, then

$$\left(\bigcap_{n=1}^{\infty} E_n\right) \cap (A_k \times B_k) = \bigcap_{n=1}^{\infty} (E_n \cap (A_k \times B_k)) \in \mathscr{F},$$

as proved in the preceding paragraph. Thus $\mathscr{E}$ is also closed under the formation of intersections of decreasing sequences. By (21.7) we see that $\mathscr{E} = \mathscr{M} \times \mathscr{N}$. Now let $(F_n)_{n=1}^{\infty}$ be any decreasing sequence in $\mathscr{F}$ and let $F = \bigcap_{n=1}^{\infty} F_n$. Since $F \in \mathscr{E}$, we have $F \cap (A_k \times B_k) \in \mathscr{F}$ for all $k \in N$. From the fact that $\mathscr{F}$ is closed under the formation of unions of increasing sequences, it follows that

$$F = \bigcup_{k=1}^{\infty} (F \cap (A_k \times B_k)) \in \mathscr{F}.$$

This completes the proof that $\mathscr{F}$ is a monotone family. □

We now define a set function on $\mathcal{M} \times \mathcal{N}$ that turns out to be the desired product measure.

(21.9) Definition. Let $(X, \mathcal{M}, \mu)$ and $(Y, \mathcal{N}, \nu)$ be σ-finite measure spaces. For $E \in \mathcal{M} \times \mathcal{N}$, define

$$\mu \times \nu(E) = \int_X \nu(E_x)\, d\mu(x)\,.$$

According to (21.8.iii), we also have

$$\mu \times \nu(E) = \int_Y \mu(E^y)\, d\nu(y)\,.$$

(21.10) Theorem. *Notation is as in* (21.9). *The set function $\mu \times \nu$ is a [countably additive] σ-finite measure on $\mathcal{M} \times \mathcal{N}$.*

Proof. Let $\{E_n\}_{n=1}^{\infty}$ be a pairwise disjoint countable family of sets in $\mathcal{M} \times \mathcal{N}$. Then (12.21) implies that

$$\begin{aligned} \mu \times \nu\left(\bigcup_{n=1}^{\infty} E_n\right) &= \int_X \nu\left(\left[\bigcup_{n=1}^{\infty} E_n\right]_x\right) d\mu(x) \\ &= \int_X \sum_{n=1}^{\infty} \nu([E_n]_x)\, d\mu(x) \\ &= \sum_{n=1}^{\infty} \int_X \nu([E_n]_x)\, d\mu(x) \\ &= \sum_{n=1}^{\infty} \mu \times \nu(E_n)\,. \end{aligned}$$

Thus $\mu \times \nu$ is countably additive. Plainly $\mu \times \nu \geqq 0$ and $\mu \times \nu(\varnothing) = 0$; hence $\mu \times \nu$ is a measure on $\mathcal{M} \times \mathcal{N}$. To show that $\mu \times \nu$ is σ-finite, let $(A_k)_{k=1}^{\infty}$ and $(B_k)_{k=1}^{\infty}$ be as in the proof of (21.8). Then $X \times Y = \bigcup_{k=1}^{\infty} (A_k \times B_k)$ and $\mu \times \nu(A_k \times B_k) = \mu(A_k) \cdot \nu(B_k) < \infty$ for all $k \in N$. □

(21.11) Note. A simple computation made in the proof of (21.8) shows that $\mu \times \nu\,(A \times B) = \mu(A) \cdot \nu(B)$ for all measurable rectangles $A \times B$. [Recall that $0 \cdot \infty = 0$.] If ω is any measure on $\mathcal{M} \times \mathcal{N}$ such that $\omega(A \times B) = \mu(A) \cdot \nu(B)$ for all measurable rectangles $A \times B$, then $\mu \times \nu(E) = \omega(E)$ for all E in the algebra $\mathcal{A}$ of all finite disjoint unions of measurable rectangles. Since $\mu \times \nu$ is σ-finite and $\mathcal{S}(\mathcal{A}) = \mathcal{M} \times \mathcal{N}$, it follows from the uniqueness part of Hopf's extension theorem (10.39.c) that $\mu \times \nu(E) = \omega(E)$ for all $E \in \mathcal{M} \times \mathcal{N}$. Therefore the product measure $\mu \times \nu$ is uniquely determined by the requirements that it be a measure on $\mathcal{M} \times \mathcal{N}$ and that $\mu \times \nu(A \times B) = \mu(A) \cdot \nu(B)$.

We can now prove two versions of Fubini's theorem for integrals on product spaces.

(21.12) Theorem [Fubini]. *Let $(X, \mathcal{M}, \mu)$ and $(Y, \mathcal{N}, \nu)$ be σ-finite measure spaces and let $(X \times Y, \mathcal{M} \times \mathcal{N}, \mu \times \nu)$ be the product measure* *space constructed above. If f is a nonnegative, extended real-valued, $\mathscr{M}\times\mathscr{N}$-measurable function on $X\times Y$, then:*

(i) *the function $x \to f(x, y)$ is $\mathscr{M}$-measurable for each $y \in Y$;*

(ii) *the function $y \to f(x, y)$ is $\mathscr{N}$-measurable for each $x \in X$;*

(iii) *the function $y \to \int_X f(x, y)\, d\mu(x)$ is $\mathscr{N}$-measurable;*

(iv) *the function $x \to \int_Y f(x, y)\, d\nu(y)$ is $\mathscr{M}$-measurable;*

and

(v) *the equalities*

$$\int_{X\times Y} f(x, y)\, d\mu\times\nu(x, y) = \int_Y\int_X f(x, y)\, d\mu(x)\, d\nu(y)$$
$$= \int_X\int_Y f(x, y)\, d\nu(y)\, d\mu(x)\text{[1]}$$

hold.

Proof. Conclusions (i) and (ii) are just (21.5.i) and (21.5.ii): we have written them again only for the sake of completeness. To deal with (iii)–(v), suppose first that $f = \xi_E$ for some $E \in \mathscr{M}\times\mathscr{N}$. It is clear that

$$\int_X \xi_E(x, y)\, d\mu(x) = \int_X \xi_{E^y}(x)\, d\mu(x) = \mu(E^y)$$

for every $y \in Y$ and that

$$\int_Y \xi_E(x, y)\, d\nu(y) = \int_Y \xi_{E_x}(y)\, d\nu(y) = \nu(E_x)$$

for every $x \in X$. Thus (iii), (iv), and (v) for ξ_E follow at once from (21.8) and the definition (21.9) of $\mu\times\nu$. For a simple $\mathscr{M}\times\mathscr{N}$-measurable function f, assertions (iii)–(v) are clear from the linearity of all of our integrals.

Finally, let f be any nonnegative, extended real-valued $\mathscr{M}\times\mathscr{N}$-measurable function on $X\times Y$. Let $(\sigma_n)_{n=1}^{\infty}$ be an increasing sequence of nonnegative, real-valued, $\mathscr{M}\times\mathscr{N}$-measurable simple functions on $X\times Y$ such that $\sigma_n(x, y) \to f(x, y)$ for all $(x, y) \in X\times Y$ (11.35). For every $y \in Y$, (12.22) shows that

$$\int_X f(x, y)\, d\mu(x) = \lim_{n\to\infty} \int_X \sigma_n(x, y)\, d\mu(x)\,;$$

hence the function in (iii) is the pointwise limit of a sequence of $\mathscr{N}$-measurable functions, and so (iii) follows from (11.14). Likewise (iv)

---

[1] In interated integrals we treat the symbols $\int \cdots d$ like parentheses. For example, $\int_Y\int_X f(x, y)\, d\mu(x)\, d\nu(y) = \int_Y \left[\int_X f(x, y)\, d\mu(x)\right] d\nu(y)$.

holds. To prove (v), use (12.22) once more to write

$$\begin{aligned}\int\limits_{X\times Y} f(x,y)\,d\mu\times\nu(x,y) &= \lim_{n\to\infty}\int\limits_{X\times Y}\sigma_n(x,y)\,d\mu\times\nu(x,y)\\ &= \lim_{n\to\infty}\int\limits_Y\int\limits_X \sigma_n(x,y)\,d\mu(x)\,d\nu(y)\\ &= \int\limits_Y\Big[\lim_{n\to\infty}\int\limits_X\sigma_n(x,y)\,d\mu(x)\Big]d\nu(y)\\ &= \int\limits_Y\int\limits_X f(x,y)\,d\mu(x)\,d\nu(y)\,.\end{aligned}$$

A like computation proves the equality

$$\int\limits_{X\times Y} f(x,y)\,d\mu\times\nu(x,y) = \int\limits_X\int\limits_Y f(x,y)\,d\nu(y)\,d\mu(x)\,. \quad \square$$

The following version of Fubini's theorem is particularly useful.

(21.13) Fubini's Theorem. *Let $(X, \mathscr{M}, \mu)$ and $(Y, \mathscr{N}, \nu)$ be σ-finite measure spaces and let $(X\times Y, \mathscr{M}\times\mathscr{N}, \mu\times\nu)$ be the product measure space constructed above. Let f be a complex-valued $\mathscr{M}\times\mathscr{N}$-measurable function on $X\times Y$ and suppose that at least one of the three integrals*

$$\int\limits_{X\times Y} |f(x,y)|\,d\mu\times\nu(x,y)\,,$$

$$\int\limits_Y\int\limits_X |f(x,y)|\,d\mu(x)\,d\nu(y)\,,$$

and

$$\int\limits_X\int\limits_Y |f(x,y)|\,d\nu(y)\,d\mu(x)$$

is finite[1]. *Then:*

(i) *the function $x\to f(x,y)$ is in $\mathfrak{L}_1(X,\mathscr{M},\mu)$ for ν-almost all $y\in Y$;*

(ii) *the function $y\to f(x,y)$ is in $\mathfrak{L}_1(Y,\mathscr{N},\nu)$ for μ-almost all $x\in X$;*

(iii) *the function $y\to\int\limits_X f(x,y)\,d\mu(x)$ is in $\mathfrak{L}_1(Y,\mathscr{N},\nu)$*[2]*;*

(iv) *the function $x\to\int\limits_Y f(x,y)\,d\nu(y)$ is in $\mathfrak{L}_1(X,\mathscr{M},\mu)$;*

and

(v) *the equalities*

$$\begin{aligned}\int\limits_{X\times Y} f(x,y)\,d\mu\times\nu(x,y) &= \int\limits_Y\int\limits_X f(x,y)\,d\mu(x)\,d\nu(y)\\ &= \int\limits_X\int\limits_Y f(x,y)\,d\nu(y)\,d\mu(x)\end{aligned}$$

obtain.

Proof. Our hypothesis and (21.12) show that

$$\int\limits_{X\times Y}|f|\,d\mu\times\nu = \int\limits_Y\int\limits_X|f|\,d\mu\,d\nu = \int\limits_X\int\limits_Y|f|\,d\nu\,d\mu < \infty\,. \qquad (1)$$

---

[1] If one of these integrals is finite, then by (21.12.v), all are finite [and equal].

[2] It is to be understood that this function is defined only for those $y\in Y$ such that $x\to f(x,y)$ is in $\mathfrak{L}_1(X,\mathscr{M},\mu)$. A similar remark applies to assertion (iv).

Thus $f \in \mathfrak{L}_1(X \times Y, \mathscr{M} \times \mathscr{N}, \mu \times \nu)$. Write $f = f_1 - f_2 + i(f_3 - f_4)$, where $f_j \in \mathfrak{L}_1^+(X \times Y, \mathscr{M} \times \mathscr{N}, \mu \times \nu)$ and $f_j \leqq |f|$, for $j = 1, 2, 3, 4$. The functions described in (i) and (ii) are measurable by (21.5). From (1) we have

$$\int_X |f_j(x, y)|\, d\mu(x) \leqq \int_X |f(x, y)|\, d\mu(x) < \infty$$

for ν-almost all y, and so $f_j^{[y]} \in \mathfrak{L}_1(X, \mathscr{M}, \mu)$ for ν-almost all y $(j = 1, 2, 3, 4)$. Thus (i) holds. Assertion (ii) is proved in like manner. Because of (i), the function in (iii) is defined for ν-almost all $y \in Y$, and its $\mathscr{N}$-measurability follows upon applying (21.12.iii) to each f_j and taking linear combinations. Thus for ν-almost all y, we have

$$\left|\int_X f(x, y)\, d\mu(x)\right| \leqq \int_X |f(x, y)|\, d\mu(x)\ ,$$

and so (iii) follows upon applying (1). The proof of (iv) is similar. Finally, apply (12.12.v) to each f_j and then take linear combinations to obtain (v). This last step is legitimate since the integrals in question are linear on the various $\mathfrak{L}_1$-spaces and (i)—(iv) hold. □

(21.14) Remarks. (a) The product σ-algebra $\mathscr{M} \times \mathscr{N}$ may be "quite small" even when $\mathscr{M}$ and $\mathscr{N}$ are "very large". In fact, if X is a Hausdorff space such that $\overline{\overline{X}} > \mathfrak{c}$, then $\mathscr{P}(X) \times \mathscr{P}(X)$ does not contain all closed subsets of $X \times X$ [see (21.20.c)].

(b) The measure spaces $(X \times Y, \mathscr{M} \times \mathscr{N}, \mu \times \nu)$ are seldom complete (11.20) even when $(X, \mathscr{M}, \mu)$ and $(Y, \mathscr{N}, \nu)$ are both complete [see (21.21)].

As in (11.21), let $(X \times Y, \overline{\mathscr{M} \times \mathscr{N}}, \overline{\mu \times \nu})$ denote the completion of the measure space $(X \times Y, \mathscr{M} \times \mathscr{N}, \mu \times \nu)$. The following lemma allows us to extend FUBINI's theorem to this completed product space.

(21.15) Lemma. *Let $(X, \mathscr{M}, \mu)$ and $(Y, \mathscr{N}, \nu)$ be two complete, σ-finite measure spaces. Let $E \in \mathscr{M} \times \mathscr{N}$ be such that $\mu \times \nu(E) = 0$, and suppose that $F \subset E$. Then*

(i) *$\mu(F^y) = 0$ ν-a. e.*

and

(ii) *$\nu(F_x) = 0$ μ-a. e.*

Proof. We content ourselves with proving (ii). By (21.12), we have

$$0 = \mu \times \nu(E) = \int_X \int_Y \xi_E\, d\nu\, d\mu = \int_X \nu(E_x)\, d\mu(x)\ .$$

Since $\nu(E_x) \geqq 0$ for all $x \in X$, it follows that $\nu(E_x) = 0$ μ-a.e. Plainly $F_x \subset E_x$ for all x, and so $F_x \in \mathscr{N}$ and $\nu(F_x) = 0$ for μ-almost all x. □

(21.16) Theorem [FUBINI]. *Let $(X, \mathscr{M}, \mu)$ and $(Y, \mathscr{N}, \nu)$ be complete, σ-finite measure spaces and let f be a nonnegative, extended real-valued $\overline{\mathscr{M} \times \mathscr{N}}$-measurable function on $X \times Y$. Then:*

(i) *the function $x \to f(x, y)$ is $\mathscr{M}$-measurable for ν-almost all $y \in Y$;*

(ii) *the function $y \to f(x, y)$ is $\mathscr{N}$-measurable for μ-almost all $x \in X$;*

(iii) *the function* $y \to \int_X f(x, y)\, d\mu(x)$ *is* $\mathscr{N}$*-measurable;*

(iv) *the function* $x \to \int_Y f(x, y)\, d\nu(y)$ *is* $\mathscr{M}$*-measurable;*

and

(v) *the equalities*

$$\int_{X\times Y} f(x, y)\, d\overline{\mu \times \nu}(x, y) = \int_Y \int_X f(x, y)\, d\mu(x)\, d\nu(y)$$
$$= \int_X \int_Y f(x, y)\, d\nu(y)\, d\mu(x)$$

hold.

Proof. Let $H \in \overline{\mathscr{M}\times\mathscr{N}}$. According to (11.21), H has the form $G \cup F$, where $G \in \mathscr{M}\times\mathscr{N}$ and $F \subset E$ for some $E \in \mathscr{M}\times\mathscr{N}$ such that $\mu \times \nu(E) = 0$. For each $x \in X$, we have $H_x = G_x \cup F_x$, and so it follows from (21.15) and (21.4.i) that $H_x \in \mathscr{N}$ for μ-almost all $x \in X$. This proves (i) for the case that $f = \xi_H$ [(ii) is similar].

The preceding paragraph shows too that $\nu(H_x) = \nu(G_x)$ for μ-almost all $x \in X$. Since

$$\int_Y \xi_H(x, y)\, d\nu(y) = \nu(H_x),$$

(21.8.i) shows that the function

$$x \to \int_Y \xi_H(x, y)\, d\nu(y)$$

is equal μ-a.e. to the $\mathscr{M}$-measurable function $x \to \nu(G_x)$. This proves (iv) for $f = \xi_H$ [(iii) is similar].

To prove (v) for ξ_H, we note that

$$\int_{X\times Y} \xi_H d\overline{\mu \times \nu} = \overline{\mu \times \nu}(H) = \mu \times \nu(G) = \int_Y \int_X \xi_G d\mu d\nu = \int_Y \int_X \xi_H d\mu d\nu .$$

The second equality in (v) is similar. The remainder of the proof is like that of (21.12). □

(21.17) Note. Fubini's theorem (21.13) is also valid for $\overline{\mathscr{M}\times\mathscr{N}}$-measurable functions. This fact can be deduced from (21.16) in just the same way that (21.13) was deduced from (21.12). It seems unnecessary to repeat the details.

A great many arguments in the theory of integration depend upon regularity of the measure or measures under consideration. Measures ι as constructed in § 9 are automatically regular [cf. (9.24) and (10.30)]. Regularity is far from obvious, however, for products of such measures. In the next theorem, we prove a little more: namely, the *completion* of a product measure is regular if the factors are regular [and σ-finite]. The proof is rather long, in fact tedious, but it presents no technical difficulties.

(21.18) Theorem. *Let X and Y be locally compact Hausdorff spaces, and let $(X, \mathscr{M}_\mu, \mu)$ and $(Y, \mathscr{M}_\nu, \nu)$ be σ-finite measure spaces as in §§ 9 and 10. Then the completion $\overline{\mu \times \nu}$ of $\mu \times \nu$ is regular on $\overline{\mathscr{M}_\mu \times \mathscr{M}_\nu}$ in the sense that if $E \in \overline{\mathscr{M}_\mu \times \mathscr{M}_\nu}$, then*

(i) $\overline{\mu \times \nu}(E) = \inf\{\mu \times \nu(U): E \subset U, U \in \mathscr{M}_\mu \times \mathscr{M}_\nu, U \text{ is open}\}$

and

(ii) $\overline{\mu \times \nu}(E) = \sup\{\mu \times \nu(F): E \supset F, F \in \mathscr{M}_\mu \times \mathscr{M}_\nu, F \text{ is compact}\}$.

Proof. Let $\mathscr{R}$ be the family of all sets $E \in \mathscr{M}_\mu \times \mathscr{M}_\nu$ for which (i) and (ii) hold. We will prove that $\mathscr{R} = \mathscr{M}_\mu \times \mathscr{M}_\nu$. Suppose that $(E_n)_{n=1}^\infty \subset \mathscr{R}$ and write $E = \bigcup_{n=1}^\infty E_n$. If $\mu \times \nu(E_{n_0}) = \infty$ for some n_0, then (i) is trivial for E and, by (ii), E_{n_0} has compact subsets of arbitrarily large [finite!] measure; hence E is in $\mathscr{R}$. Thus suppose that $\mu \times \nu(E_n) < \infty$ for all $n \in N$. For arbitrary $\varepsilon > 0$ and for each $n \in N$, choose a compact set $F_n \in \mathscr{M}_\mu \times \mathscr{M}_\nu$ and an open set $U_n \in \mathscr{M}_\mu \times \mathscr{M}_\nu$ such that $F_n \subset E_n \subset U_n$ and $\mu \times \nu(U_n \cap F_n') < \frac{\varepsilon}{2^n}$. Let $U = \bigcup_{n=1}^\infty U_n$ and $F = \bigcup_{n=1}^\infty F_n$. Then we have: $F \subset E \subset U$; $U, F \in \mathscr{M}_\mu \times \mathscr{M}_\nu$; and also

$$\begin{aligned} \mu \times \nu(U \cap F') &\leqq \sum_{n=1}^\infty \mu \times \nu(U_n \cap F') \\ &\leqq \sum_{n=1}^\infty \mu \times \nu(U_n \cap F_n') < \sum_{n=1}^\infty \frac{\varepsilon}{2^n} = \varepsilon . \end{aligned}$$

It follows that

$$\mu \times \nu(U) = \mu \times \nu(E) + \mu \times \nu(U \cap E') \leqq \mu \times \nu(E) + \varepsilon \tag{1}$$

and

$$\mu \times \nu(E) = \mu \times \nu(F) + \mu \times \nu(E \cap F') \leqq \mu \times \nu(F) + \varepsilon . \tag{2}$$

Since U is open, (1) implies that (i) holds for E. Applying (2) and (10.13), we obtain

$$\mu \times \nu(E) \leqq \lim_{p \to \infty} \mu \times \nu(F_1 \cup \cdots \cup F_p) + \varepsilon ,$$

and since $F_1 \cup \cdots \cup F_p$ is compact for all p, (ii) holds for E. Hence $\mathscr{R}$ is closed under the formation of countable unions.

Next, let $A \times B$ be a measurable rectangle. Use our σ-finiteness hypothesis to select ascending sequences $(A_n)_{n=1}^\infty \subset \mathscr{M}_\mu$ and $(B_n)_{n=1}^\infty \subset \mathscr{M}_\nu$ such that $X = \bigcup_{n=1}^\infty A_n$, $Y = \bigcup_{n=1}^\infty B_n$, and $\mu(A_n) < \infty$ and $\nu(B_n) < \infty$ for all $n \in N$. Then

$$A \times B = \bigcup_{n=1}^\infty [(A \cap A_n) \times (B \cap B_n)] ,$$

and so if we can show that $\mathscr{R}$ contains each measurable rectangle having sides of finite measure, it will follow from the preceding paragraph that $A \times B \in \mathscr{R}$. Thus we suppose that $\mu(A) < \infty$ and $\nu(B) < \infty$. Use the regularity of μ and ν to select ascending sequences $(C_n)_{n=1}^{\infty}$ and $(D_n)_{n=1}^{\infty}$ of compact sets and descending sequences $(U_n)_{n=1}^{\infty}$ and $(V_n)_{n=1}^{\infty}$ of open sets such that:

$$C_n \subset A \subset U_n \subset X;$$

$$D_n \subset B \subset V_n \subset Y;$$

$$\mu(U_n) - \mu(C_n) < \frac{1}{n};$$

and

$$\nu(V_n) - \nu(D_n) < \frac{1}{n}.$$

Then for each $n \in N$, $U_n \times V_n$ is open, $C_n \times D_n$ is compact, and $C_n \times D_n \subset A \times B \subset U_n \times V_n$. Also we have

$$\begin{aligned} \mu \times \nu(A \times B) &= \mu(A) \cdot \nu(B) \\ &= \lim_{n\to\infty} [\mu(U_n) \cdot \nu(V_n)] = \lim_{n\to\infty} \mu \times \nu(U_n \times V_n) \end{aligned}$$

and

$$\begin{aligned} \mu \times \nu(A \times B) &= \mu(A) \cdot \nu(B) \\ &= \lim_{n\to\infty} [\mu(C_n) \cdot \nu(D_n)] = \lim_{n\to\infty} \mu \times \nu(C_n \times D_n). \end{aligned}$$

Thus $A \times B \in \mathscr{R}$, and so $\mathscr{R}$ contains all measurable rectangles.

To finish the proof that $\mathscr{R} = \mathscr{M}_\mu \times \mathscr{M}_\nu$, we need only to show that $\mathscr{R}$ is closed under complementation. Let $(A_n \times B_n)_{n=1}^{\infty}$ be as above. Define $\mathscr{R}_n = \{E \in \mathscr{R} : E \subset A_n \times B_n\}$. Clearly $\mathscr{R}_n$ is closed under the formation of countable unions. Let E be in $\mathscr{R}_n$ and let $\varepsilon > 0$ be given. Since $E \in \mathscr{R}$, there exist a compact set F and an open set U [both in $\mathscr{M}_\mu \times \mathscr{M}_\nu$] such that $F \subset E \subset U$ and $\mu \times \nu(U \cap F') < \varepsilon$. Since $A_n \times B_n \in \mathscr{R}$, there exist a compact set J and an open set W [both in $\mathscr{M}_\mu \times \mathscr{M}_\nu$] such that $J \subset A_n \times B_n \subset W$ and $\mu \times \nu(W \cap J') < \varepsilon$. Now $W \cap F'$ is open and $J \cap U'$ is compact. Furthermore it is clear that

$$J \cap U' \subset (A_n \times B_n) \cap E' \subset W \cap F'$$

and that

$$\begin{aligned} \mu \times \nu(W \cap F') - \mu \times \nu(J \cap U') &= \mu \times \nu((W \cap F') \cap (J \cap U')') \\ &\leq \mu \times \nu(W \cap F' \cap J') + \mu \times \nu(W \cap F' \cap U) \\ &< \varepsilon + \varepsilon = 2\varepsilon. \end{aligned}$$

This proves that $(A_n \times B_n) \cap E'$ is in $\mathscr{R}_n$. We conclude that $\mathscr{R}_n$ is a σ-algebra of subsets of $A_n \times B_n$.

Now let $\mathscr{E} = \{E \in \mathscr{M}_\mu \times \mathscr{M}_\nu : E \cap (A_n \times B_n) \in \mathscr{R}_n \text{ for every } n \in N\}$. It is easy to see from our previous results that $\mathscr{E}$ is a σ-algebra of subsets of $X \times Y$ that contains every measurable rectangle. Thus $\mathscr{E} = \mathscr{M}_\mu \times \mathscr{M}_\nu$. Then for any $E \in \mathscr{R}$ we have $E' \in \mathscr{E}$; hence $E' \cap (A_n \times B_n)$ is in $\mathscr{R}$ for all n, and so $E' = \bigcup_{n=1}^{\infty} [E' \cap (A_n \times B_n)] \in \mathscr{R}$. Therefore $\mathscr{R}$ is closed under complementation. Altogether we have proved that $\mathscr{R} = \overline{\mathscr{M}_\mu \times \mathscr{M}_\nu}$.

Finally, let H be any set in $\overline{\mathscr{M}_\mu \times \mathscr{M}_\nu}$. Then $H = G \cup F$ where $G \in \mathscr{M}_\mu \times \mathscr{M}_\nu$ and $F \subset E$ for some $E \in \mathscr{M}_\mu \times \mathscr{M}_\nu$ for which $\mu \times \nu(E) = 0$. If $\mu \times \nu(G) = \infty$, it is clear that (i) and (ii) hold for H; thus suppose that $\mu \times \nu(G) < \infty$. Given $\varepsilon > 0$, choose open sets U and V and a compact set C [all in $\mathscr{M}_\mu \times \mathscr{M}_\nu$] such that $C \subset G \subset U$, $E \subset V$, $\mu \times \nu(U \cap C') < \varepsilon$, and $\mu \times \nu(V) < \varepsilon$. Then we have

$$C \subset H \subset U \cup V$$

and

$$\mu \times \nu(U \cup V) - \mu \times \nu(C) \leqq \mu \times \nu(V) + \mu \times \nu(U \cap C') < 2\varepsilon .$$

It follows that

$$\mu \times \nu(U \cup V) < \overline{\mu \times \nu}(H) + 2\varepsilon$$

and

$$\overline{\mu \times \nu}(H) < \mu \times \nu(C) + 2\varepsilon ,$$

and so again (i) and (ii) hold for H. □

In the following exercises (21.19) and (21.20), the reader may get some idea of the size of σ-algebras $\mathscr{M} \times \mathscr{N}$: actually they are rather small.

(21.19) Exercise. Let X and Y be topological spaces, each having a countable base for its topology. Prove that $\mathscr{B}(X) \times \mathscr{B}(Y) = \mathscr{B}(X \times Y)$. [Use (6.41) and (10.42).]

(21.20) Exercise. (a) Let T be a set and let $\mathscr{E}$ be a family of subsets of T. Prove that the σ-algebra $\mathscr{S}(\mathscr{E})$ of subsets of T generated by $\mathscr{E}$ consists precisely of those sets $F \subset T$ such that $F \in \mathscr{S}(\mathscr{C})$ for some countable family $\mathscr{C} \subset \mathscr{E}$.

(b) Use (a) to prove that if X and Y are topological spaces and if $F \in \mathscr{B}(X) \times \mathscr{B}(Y)$, then there exists a countable family $\mathscr{C}$ of sets of the form $U \times V$, where U is open in X and V is open in Y, such that $F \in \mathscr{S}(\mathscr{C})$.

(c) Use (a) to prove that if X is a set and if $D = \{(x, x) : x \in X\}$ is the diagonal in $X \times X$, then $D \in \mathscr{P}(X) \times \mathscr{P}(X)$ if and only if $\overline{\overline{X}} \leqq \mathfrak{c}$. [If $\overline{\overline{X}} \leqq \mathfrak{c}$, suppose that $X \subset R$ and use (21.19). If $D \in \mathscr{P}(X) \times \mathscr{P}(X)$, choose a countable family $\mathscr{C} \subset \mathscr{P}(X)$ such that $D \in \mathscr{S}(\{A \times B : A, B \in \mathscr{C}\})$ and then show that the mapping $x \to \{C \in \mathscr{C} : x \in C\}$ from X into $\mathscr{P}(\mathscr{C})$ is one-to-one.]

(d) Combine (21.19) and parts (a), (b), and (c) to find all Hausdorff spaces X such that $\mathscr{B}(X) \times \mathscr{B}(X)$ contains all closed subsets of $X \times X$.

Product measures are as a rule incomplete. This is brought out in the following exercise.

(21.21) Exercise. (a) Let $(X, \mathscr{M}, \mu)$ and $(Y, \mathscr{N}, \nu)$ be σ-finite measure spaces. Suppose that there exists a set $A \subset X$ such that $A \notin \mathscr{M}$ and suppose that there exists a nonvoid set $B \in \mathscr{N}$ such that $\nu(B) = 0$. Prove that $(X \times Y, \mathscr{M} \times \mathscr{N}, \mu \times \nu)$ is incomplete.

(b) Prove that $(R^2, \mathscr{M}_\lambda \times \mathscr{M}_\lambda, \lambda \times \lambda)$ is an incomplete measure space.

(21.22) Exercise. For each of the functions f on $R \times R$ defined by the following formulas, compute $\int_0^1 \int_0^1 f(x, y)\, dx\, dy$, $\int_0^1 \int_0^1 f(x, y)\, dy\, dx$, $\int_0^1 \int_0^1 |f(x, y)|\, dx\, dy$, and $\int_0^1 \int_0^1 |f(x, y)|\, dy\, dx$:

(a) $f(x, y) = \dfrac{x^2 - y^2}{(x^2 + y^2)^2}$;

(b) $f(x, y) = \begin{cases} \dfrac{1}{\left(x - \frac{1}{2}\right)^3} & \text{for } 0 < y < \left|x - \dfrac{1}{2}\right|, \\ 0 & \text{otherwise;} \end{cases}$

(c) $f(x, y) = \dfrac{x - y}{(x^2 + y^2)^{3/2}}$;

(d) $f(x, y) = \dfrac{1}{(1 - xy)^p}$, where $p > 0$.

Compare your findings with (21.13).

(21.23) Exercise. Let $(X, \mathscr{M}, \mu)$ be a σ-finite measure space. Let f be a nonnegative, extended real-valued, $\mathscr{M}$-measurable function defined on X. Define

$$V^* f = \{(x, t) \in X \times R : 0 \leq t \leq f(x)\}$$

and

$$V_* f = \{(x, t) \in X \times R : 0 \leq t < f(x)\}.$$

Prove that $V^* f$ and $V_* f$ are in the σ-algebra $\mathscr{M} \times \mathscr{B}(R)$ and that

(i) $\mu \times \lambda(V^* f) = \mu \times \lambda(V_* f) = \int_X f\, d\mu$.

[The equality (i) asserts that the integral of f is the "area under the curve $y = f(x)$".]

(21.24) Exercise. Let μ be counting measure on $[0, 1]$ and let $D = \{(x, x) : x \in [0, 1]\}$. Write $X = Y = [0, 1]$. Prove that

$$\int_Y \int_X \xi_D\, d\mu\, d\lambda \neq \int_X \int_Y \xi_D\, d\lambda\, d\mu.$$

Why does this not contradict (21.12)?

(21.25) Exercise. Let $(X, \mathscr{M}_\mu, \mu)$ and $(Y, \mathscr{M}_\nu, \nu)$ be measure spaces as in (21.18). Prove the following.

(a) Every compact G_δ subset of $X \times Y$ is in $\mathscr{M}_\mu \times \mathscr{M}_\nu$.

(b) Every function in $\mathfrak{C}_{00}(X \times Y)$ is $\mathscr{M}_\mu \times \mathscr{M}_\nu$-measurable.

(21.26) Exercise. (a) Let $I = [0, 1]$. Suppose that $E \subset I \times I$ is such that $\lambda(E_x) = \lambda(I \cap (E^y)') = 0$ for all $x, y \in I$. Prove that E is not $\mathscr{M}_\lambda \times \mathscr{M}_\lambda$-measurable.

(b) Prove that sets E as described in part (a) exist. [Let Δ be the least ordinal number such that $\overline{\overline{P_\Delta}} = \mathfrak{c}$ [see (4.47) and (4.48)]. Let $\alpha \to x_\alpha$ be any one-to-one mapping of P_Δ onto I. Define $E = \{(x_\alpha, x_\beta): \beta < \alpha < \Delta\}$. Then $\overline{\overline{E}}_x < \mathfrak{c}$ and $\overline{\overline{I \cap (E^y)'}} < \mathfrak{c}$ for all $x, y \in I$. If we accept the continuum hypothesis (4.50), then all of these sets are countable. In any case, it follows from (10.30), (6.66), and (6.65) that if these sets are measurable, then they have measure zero.]

(21.27) Exercise. Prove that there exists a subset S of $I \times I$ $[I = [0, 1]]$ such that $\overline{\overline{S}}_x \leqq 1$ and $\overline{\overline{S}}^y \leqq 1$ for all $x, y \in I$, but $S \notin \overline{\mathscr{M}_\lambda \times \mathscr{M}_\lambda}$. [Supply the many missing details in the following outline. Let $\mathscr{F}$ be the family of all compact sets $F \subset I \times I$ such that $\lambda \times \lambda(F) > 0$. Let Δ be as in (21.26) and let $\alpha \to F_\alpha$ be a one-to-one mapping of P_Δ onto $\mathscr{F}$. Choose $(x_0, y_0) \in F_0$. If $\beta < \Delta$ and $S_\beta = \{(x_\alpha, y_\alpha) : \alpha < \beta\}$ have been chosen so that no vertical or horizontal section of S_β has more than one point, let $B_\beta = \{x \in I : \lambda((F_\beta)_x) > 0\}$. Since $\lambda(B_\beta) > 0$, there exists $x_\beta \in B_\beta \cap \{x_\alpha : \alpha < \beta\}'$. Now since $\lambda((F_\beta)_{x_\beta}) > 0$, there exists $y_\beta \in (F_\beta)_{x_\beta} \cap \{y_\alpha : \alpha < \beta\}'$. Let $S = \{(x_\alpha, y_\alpha) : \alpha < \Delta\}$. Clearly $S \cap F_\alpha \neq \varnothing$ for every α, and so, since any measurable set of positive measure contains some F_α, S cannot be measurable.]

(21.28) Exercise. Recall our discussion of ultrafilters and their corresponding finitely additive measures given in (20.37). Let $\mathscr{U}$ and $\mathscr{V}$ be ultrafilters of subsets of R such that $\mathscr{U} \supset \{[a, \infty[: a \in R\}$ and $\mathscr{V} \supset \{]-\infty, b] : b \in R\}$. For all $E \subset R$, define $\mu(E) = 1$ if $E \in \mathscr{U}$, $\mu(E) = 0$ if $E \notin \mathscr{U}$, $\nu(E) = 1$ if $E \in \mathscr{V}$, and $\nu(E) = 0$ if $E \notin \mathscr{V}$. Then μ and ν are finitely additive on $\mathscr{P}(R)$. Let f be any bounded real-valued function on R such that $\lim\limits_{x \to \infty} f(x) = \alpha$ and $\lim\limits_{x \to -\infty} f(x) = \beta$, where α and β are any given real numbers. Prove that:

(a) $\int\limits_R f(x+y)\, d\mu(x) = \alpha$ for every $y \in R$;

(b) $\int\limits_R f(x+y)\, d\nu(y) = \beta$ for every $x \in R$.

(c) $\int\limits_R \int\limits_R f(x+y)\, d\mu(x)\, d\nu(y) = \alpha$;

and

(d) $\int\limits_R \int\limits_R f(x+y)\, d\nu(y)\, d\mu(x) = \beta$.

Thus Fubini's theorem may fail completely for very simple functions in the absence of *countable* additivity.

The following theorem describes the behavior of absolute continuity and singularity under the formation of product measures.

(21.29) Theorem. *Let* $(X, \mathcal{M})$ *and* $(Y, \mathcal{N})$ *be measurable spaces. Let* μ *and* $\mu^\dagger$ *be* σ*-finite measures on* $(X, \mathcal{M})$ *and let* ν *and* $\nu^\dagger$ *be* σ*-finite measures on* $(Y, \mathcal{N})$. *If* $\mu^\dagger \ll \mu$ and $\nu^\dagger \ll \nu$, *then we have* $\mu^\dagger \times \nu^\dagger \ll \mu \times \nu$ *and*

(i) $$\frac{d(\mu^\dagger \times \nu^\dagger)}{d(\mu \times \nu)}(x, y) = \frac{d\mu^\dagger}{d\mu}(x) \cdot \frac{d\nu^\dagger}{d\nu}(y)$$

for all $(x, y) \in X \times Y$. *If* $\mu^\dagger \perp \mu$ *or* $\nu^\dagger \perp \nu$, *then we have* $\mu^\dagger \times \nu^\dagger \perp \mu \times \nu$. *Writing subscript a's and s's for the Lebesgue decomposition of* $\mu^\dagger$ *with respect to* μ, *etc., we thus have*

(ii) $(\mu^\dagger \times \nu^\dagger)_a = \mu_a^\dagger \times \nu_a^\dagger$

and

(iii) $(\mu^\dagger \times \nu^\dagger)_s = (\mu_a^\dagger \times \nu_s^\dagger) + (\mu_s^\dagger \times \nu_a^\dagger) + (\mu_s^\dagger \times \nu_s^\dagger)$.

Proof. Suppose that $\mu^\dagger \ll \mu$ and $\nu^\dagger \ll \nu$, and let $E \in \mathcal{M} \times \mathcal{N}$ be such that $\mu \times \nu(E) = 0$. By (21.9), we have

$$0 = \mu \times \nu(E) = \int_X \nu(E_x)\, d\mu(x) . \tag{1}$$

By (12.6), the set $A = \{x \in X : \nu(E_x) > 0\}$ has μ-measure 0. By hypothesis, we have $\mu^\dagger(A) = 0$ and also $\nu^\dagger(E_x) = 0$ if $x \in A'$. Therefore

$$\begin{aligned} \mu^\dagger \times \nu^\dagger(E) &= \int_X \nu^\dagger(E_x)\, d\mu^\dagger(x) = \int_A \nu^\dagger(E_x)\, d\mu^\dagger(x) + \int_{A'} \nu^\dagger(E_x)\, d\mu^\dagger(x) \\ &= 0 + \int_{A'} 0\, d\mu^\dagger(x) = 0 . \end{aligned}$$

This shows that $\mu^\dagger \times \nu^\dagger \ll \mu \times \nu$. To prove (i), write $\frac{d\mu^\dagger}{d\mu}$ as f_0 and $\frac{d\nu^\dagger}{d\nu}$ as g_0 [for brevity's sake]. Applying (21.13) to an arbitrary f belonging to $\mathfrak{L}_1(X \times Y, \mathcal{M} \times \mathcal{N}, \mu^\dagger \times \nu^\dagger)$, and taking note of (19.24), we obtain

$$\begin{aligned} \int_{X \times Y} f\, d\mu^\dagger \times \nu^\dagger &= \int_X \int_Y f(x, y)\, d\nu^\dagger(y)\, d\mu^\dagger(x) \\ &= \int_X \int_Y f(x, y) g_0(y)\, d\nu(y)\, d\mu^\dagger(x) \\ &= \int_X \int_Y f(x, y) g_0(y)\, d\nu(y) f_0(x)\, d\mu(x) \\ &= \int_X \int_Y f(x, y) f_0(x) g_0(y)\, d\nu(y)\, d\mu(x) . \end{aligned} \tag{1}$$

The function $(x, y) \to f(x, y) f_0(x) g_0(y)$ on $X \times Y$ is plainly $\mathcal{M} \times \mathcal{N}$-measurable, and so we can apply (21.13) to the last integral in (1), finding that

$$\begin{aligned} &\int_X \int_Y f(x, y) f_0(x) g_0(y)\, d\nu(y)\, d\mu(x) \\ &\qquad = \int_{X \times Y} f(x, y) f_0(x) g_0(y)\, d\mu \times \nu(x, y) . \end{aligned} \tag{2}$$

Now combine (1) and (2) and let $f = \xi_A$, where $A \in \mathscr{M} \times \mathscr{N}$ and $\mu^\dagger \times \nu^\dagger (A) < \infty$. This gives

$$\mu^\dagger \times \nu^\dagger(A) = \int\limits_{X \times Y} \xi_A(x, y) f_0(x) g_0(y) \, d\mu \times \nu(x, y) . \tag{3}$$

Since $\mu^\dagger \times \nu^\dagger$ is σ-finite, (3) holds for all $A \in \mathscr{M} \times \mathscr{N}$, and the uniqueness provision in (19.24) proves (i).

Now suppose that $\mu^\dagger \perp \mu$. Let B be a set in $\mathscr{M}$ such that $\mu(B) = 0$ and $\mu^\dagger(B') = 0$. Then, as noted in (21.11), we have

$$\mu \times \nu(B \times Y) = \mu(B) \nu(Y) = 0$$

and

$$\mu^\dagger \times \nu^\dagger((B \times Y)') = \mu^\dagger \times \nu^\dagger(B' \times Y) = \mu^\dagger(B') \nu^\dagger(Y) = 0 .$$

This implies that $\mu^\dagger \times \nu^\dagger \perp \mu \times \nu$. For arbitrary μ, $\mu^\dagger$, ν, and $\nu^\dagger$, we have

$$\begin{aligned} \mu^\dagger \times \nu^\dagger &= (\mu_a^\dagger + \mu_s^\dagger) \times (\nu_a^\dagger + \nu_s^\dagger) \\ &= (\mu_a^\dagger \times \nu_a^\dagger) + (\mu_a^\dagger \times \nu_s^\dagger) + (\mu_s^\dagger \times \nu_a^\dagger) + (\mu_s^\dagger \times \nu_s^\dagger) , \end{aligned}$$

from which (ii) and (iii) follow easily. □

FUBINI'S theorem and LEBESGUE'S dominated convergence theorem are cornerstones of analysis. The theory of Fourier transforms, as well as many other theories, depends in the last analysis on these two theorems. We devote the remainder of the present section to a number of applications of these theorems. Our first result is a simple lemma that is useful in establishing the $\mu \times \nu$-measurability of functions on spaces $X \times Y$.

(21.30) Lemma. *Let φ be a real-valued Borel measurable function defined on R^2 such that if $M \subset R$ and $\lambda(M) = 0$, then $\varphi^{-1}(M) \in \overline{\mathscr{M}_\lambda \times \mathscr{M}_\lambda}$ and $\overline{\lambda \times \lambda}(\varphi^{-1}(M)) = 0$. Then $f \circ \varphi$ is $\overline{\mathscr{M}_\lambda \times \mathscr{M}_\lambda}$-measurable for every complex-valued Lebesgue measurable function f defined λ-a. e. on R.*

Proof. First suppose that $f = \xi_A$, where $A \in \mathscr{M}_\lambda$. By (10.34), we have $A = B \cup M$ where $B \in \mathscr{B}(R)$ and $\lambda(M) = 0$. Then we may write $\xi_A \circ \varphi = \xi_{A_1}$, where

$$A_1 = \varphi^{-1}(A) = \varphi^{-1}(B) \cup \varphi^{-1}(M) .$$

Since

$$\varphi^{-1}(B) \in \mathscr{B}(R^2) \subset \overline{\mathscr{M}_\lambda \times \mathscr{M}_\lambda}$$

and

$$\varphi^{-1}(M) \in \overline{\mathscr{M}_\lambda \times \mathscr{M}_\lambda} ,$$

the lemma follows for $f = \xi_A$. Since $(f + g) \circ \varphi = f \circ \varphi + g \circ \varphi$, the lemma also holds if f is a simple function.

Finally let f be a complex-valued, Lebesgue measurable function defined on $R \cap F'$, where $F \subset R$ and $\lambda(F) = 0$. Use (11.35) to obtain a sequence (s_n) of complex-valued, Lebesgue measurable, simple functions on R such that $s_n(x) \to f(x)$ for all $x \in R \cap F'$. Then $s_n \circ \varphi$ is $\overline{\mathscr{M}_\lambda \times \mathscr{M}_\lambda}$-

measurable for all $n \in N$ and $s_n \circ \varphi \to f \circ \varphi$ except on $\varphi^{-1}(F)$. Since $\overline{\lambda \times \lambda}(\varphi^{-1}(F)) = 0$, it follows that $f \circ \varphi$ is $\overline{\mathcal{M}_\lambda \times \mathcal{M}_\lambda}$-measurable (11.24). □

(21.31) Theorem. *Let f and g be in $\mathfrak{L}_1(R)$. Then for almost all $x \in R$ the function $y \to f(x-y)\,g(y)$ is in $\mathfrak{L}_1(R)$. For all such x define*

(i) $f * g(x) = \int\limits_{-\infty}^{\infty} f(x-y)\,g(y)\,dy$.

Then $f * g \in \mathfrak{L}_1(R)$ and $\|f * g\|_1 \leq \|f\|_1 \cdot \|g\|_1$. [The function $f * g$ is called the convolution of f and g.]

Proof. Suppose for the moment that the function $(x,y) \to f(x-y)g(y)$ is $\overline{\mathcal{M}_\lambda \times \mathcal{M}_\lambda}$-measurable on R^2. We apply (21.12) to write

$$\int\limits_{-\infty}^{\infty}\int\limits_{-\infty}^{\infty} |f(x-y)\,g(y)|\,dx\,dy = \int\limits_{-\infty}^{\infty} |g(y)| \cdot \int\limits_{-\infty}^{\infty} |f(x-y)|\,dx\,dy$$
$$= \int\limits_{-\infty}^{\infty} |g(y)| \cdot \|f\|_1\,dy = \|f\|_1 \cdot \|g\|_1$$
$$< \infty .$$

Thus the hypothesis of FUBINI's theorem (21.13) is satisfied, and so $y \to f(x-y)\,g(y)$ is in $\mathfrak{L}_1(R)$ for almost all $x \in R$, $f * g \in \mathfrak{L}_1(R)$, and

$$\|f * g\|_1 = \int\limits_{-\infty}^{\infty} \left| \int\limits_{-\infty}^{\infty} f(x-y)\,g(y)\,dy \right| dx$$
$$\leq \int\limits_{-\infty}^{\infty}\int\limits_{-\infty}^{\infty} |f(x-y)\,g(y)|\,dy\,dx$$
$$= \int\limits_{-\infty}^{\infty}\int\limits_{-\infty}^{\infty} |f(x-y)\,g(y)|\,dx\,dy$$
$$= \|f\|_1 \cdot \|g\|_1 < \infty .$$

We now proceed to prove that the function $(x,y) \to f(x-y)\,g(y)$ is $\overline{\mathcal{M}_\lambda \times \mathcal{M}_\lambda}$-measurable. The function $(x,y) \to g(y)$ is $\overline{\mathcal{M}_\lambda \times \mathcal{M}_\lambda}$-measurable since for all $B \in \mathcal{B}(K)$, we have $\{(x,y) \in R^2 : g(y) \in B\} = R \times [g^{-1}(B)] \in \mathcal{M}_\lambda \times \mathcal{M}_\lambda$. Thus we have only to show that the function $(x,y) \to f(x-y)$ is $\overline{\mathcal{M}_\lambda \times \mathcal{M}_\lambda}$-measurable. Let $\varphi(x,y) = x - y$ for $(x,y) \in R^2$. Then φ, being continuous, is Borel measurable, and so the desired result will follow from (21.30) once we show that $\overline{\lambda \times \lambda}(\varphi^{-1}(M)) = 0$ whenever $\lambda(M) = 0$.

Let M be such a subset of R. Then

$$\varphi^{-1}(M) = \{(x,y) \in R^2 : (x-y) \in M\} = \bigcup_{n=1}^{\infty} P_n ,$$

where $P_n = \{(x,y) \in R^2 : (x-y) \in M,\ |y| \leq n\}$. We complete the proof by showing that $P_n \in \overline{\mathcal{M}_\lambda \times \mathcal{M}_\lambda}$ and $\overline{\lambda \times \lambda}(P_n) = 0$ for all $n \in N$. Fix $n \in N$ and choose a decreasing sequence $(U_k)_{k=1}^{\infty}$ of open subsets of R

such that $M \subset \bigcap_{k=1}^{\infty} U_k$ and $\lambda(U_k) \to 0$. Let $B_k = \{(x, y) \in R^2 : (x-y) \in U_k, |y| \leqq n\}$. We see at once that $(B_k) \subset \mathscr{B}(R^2)$ and $P_n \subset \bigcap_{k=1}^{\infty} B_k$. Use (10.15), (21.12), and (12.44) to write

$$\begin{aligned}\lambda \times \lambda\left(\bigcap_{k=1}^{\infty} B_k\right) &= \lim_{k\to\infty} \lambda \times \lambda(B_k) \\ &= \lim_{k\to\infty} \int_{-\infty}^{\infty}\int_{-\infty}^{\infty} \xi_{B_k}(x, y)\, dx\, dy \\ &= \lim_{k\to\infty} \int_{-n}^{n}\int_{-\infty}^{\infty} \xi_{U_k}(x-y)\, dx\, dy \\ &= \lim_{k\to\infty} \int_{-n}^{n}\int_{-\infty}^{\infty} \xi_{U_k}(x)\, dx\, dy \\ &= \lim_{k\to\infty} 2n\,\lambda(U_k) = 0\,.\end{aligned}$$

Since $P_n \subset \bigcap_{k=1}^{\infty} B_k$, we have proved that $\overline{\lambda \times \lambda}(P_n) = 0$. □

It is possible to convolve some pairs of functions not both of which are in $\mathfrak{L}_1(R)$. This is brought out in the following two theorems and in Exercise (21.56).

(21.32) Theorem. *Suppose that $1 < p < \infty$, that $f \in \mathfrak{L}_1(R)$, and that $g \in \mathfrak{L}_p(R)$. Then for almost every $x \in R$, the functions $y \to f(x-y)\,g(y)$ and $y \to f(y)\,g(x-y)$ are in $\mathfrak{L}_1(R)$. For all such x, we write*

$$f * g(x) = \int_R f(x-y)\, g(y)\, dy$$

and

$$g * f(x) = \int_R g(x-y)\, f(y)\, dy\,.$$

Then $f * g = g * f$ a. e., $f * g \in \mathfrak{L}_p(R)$, and $\|f * g\|_p \leqq \|f\|_1 \cdot \|g\|_p$.

Proof. As always, let $p' = \frac{p}{p-1}$, and let $h \in \mathfrak{L}_{p'}(R)$. As in (21.31), we see that the functions $(x, y) \to f(x-y)$ and $(x, y) \to g(x-y)$ are $\overline{\mathscr{M}_\lambda \times \mathscr{M}_\lambda}$-measurable. Applying (12.44), (21.12), and HÖLDER's inequality (13.4), we have

$$\begin{aligned}\int_R\int_R |f(x-y)\, g(y)\, h(x)|\, dy\, dx &= \int_R |h(x)| \int_R |f(x-y)\, g(y)|\, dy\, dx \\ &= \int_R |h(x)| \int_R |f(t)\, g(x-t)|\, dt\, dx \\ &= \int_R |f(t)| \int_R |g(x-t)\, h(x)|\, dx\, dt \\ &\leqq \int_R |f(t)| \cdot \|g_{-t}\|_p \cdot \|h\|_{p'}\, dt \\ &= \int_R |f(t)| \cdot \|g\|_p \cdot \|h\|_{p'}\, dt \\ &= \|f\|_1 \cdot \|g\|_p \cdot \|h\|_{p'} < \infty\,. \qquad (1)\end{aligned}$$

[As in (8.14) and (12.44), g_{-t} denotes the translate of g by $-t$.] Since h can be taken to never vanish, *e. g.* $h(x) = \exp(-x^2)$, (1) implies that the integrals $\int_R |f(x-y)\,g(y)|\,dy$ and $\int_R |f(t)\,g(x-t)|\,dt$ are both finite for almost all $x \in R$. This proves our first assertion.

It also follows from (1) and (21.13) that the mapping

$$h \to \int_R h(x)\, f * g(x)\, dx$$

is a bounded linear functional on $\mathfrak{L}_{p'}(R)$ with norm not exceeding $\|f\|_1 \cdot \|g\|_p$. Theorem (15.12) shows that there exists a function $\varphi \in \mathfrak{L}_p(R)$ such that

$$\int_R h(x)\,\varphi(x)\,dx = \int_R h(x)\, f * g(x)\,dx \tag{2}$$

for all $h \in \mathfrak{L}_{p'}(R)$ and such that

$$\|\varphi\|_p \leqq \|f\|_1 \cdot \|g\|_p.$$

From (2) and a standard argument, we infer that $\varphi = f * g$ a. e., and also that $f * g \in \mathfrak{L}_p(R)$ and $\|f * g\|_p \leqq \|f\|_1\,\|g\|_p$. Finally we have

$$f * g(x) = \int_R f(x-y)\,g(y)\,dy = \int_R f(t)\,g(x-t)\,dt = g * f(x)$$

for almost all x, *i. e.*, for all x such that these integrands are in $\mathfrak{L}_1(R)$. □

(21.33) Theorem. *Let* $1 \leqq p < \infty$, $f \in \mathfrak{L}_p(R)$, $g \in \mathfrak{L}_{p'}(R)$ *[where* $p' = \frac{p}{p-1}$ *if* $p > 1$ *and* $1' = \infty$*]. Define* $f * g$ *on* R *by*

$$f * g(x) = \int_R f(x-y)\,g(y)\,dy\,.$$

Then $f * g$ *is uniformly continuous on* R *and* $\|f * g\|_u \leqq \|f\|_p \cdot \|g\|_{p'}$. *If* $p > 1$, *then* $f * g \in \mathfrak{C}_0(R)$.

Proof. For $p > 1$ use (13.4), and for $p = 1$ use (20.16), to infer that $f * g(x)$ exists for all $x \in R$ and that

$$\|f * g\|_u \leqq \|f\|_p\,\|g\|_{p'}\,.$$

Now consider any $\varepsilon > 0$. By (13.24), there is a $\delta > 0$ such that $x, z \in R$ and $|x - z| < \delta$ imply $\|f_x - f_z\|_p\,\|g\|_{p'} < \varepsilon$. Then $|x - z| < \delta$ implies

$$\begin{aligned} |f * g(x) - f * g(z)| &\leqq \int_R |f(x-y) - f(z-y)| \cdot |g(y)|\,dy \\ &\leqq \|f_x - f_z\|_p \cdot \|g\|_{p'} < \varepsilon\,. \end{aligned}$$

Thus $f * g$ is uniformly continuous on R.

Now suppose that $p > 1$. Given $\varepsilon > 0$, choose a compact interval $F = [-a, a] \subset R$ such that

$$\int_{F'} |f|^p\,d\lambda < \varepsilon^p \quad \text{and} \int_{F'} |g|^{p'}\,d\lambda < \varepsilon^{p'}.$$

[The existence of F follows at once from (12.22) or (12.24).] Then if $x \in R$ and $|x| > 2a$, we have $[x - a, x + a] \subset F'$ and hence

$$\begin{aligned}
|f * g(x)| &\leqq \Big|\int_F f(x-y)\, g(y)\, dy\Big| + \Big|\int_{F'} f(x-y)\, g(y)\, dy\Big| \\
&\leqq \Big(\int_F |f(x-y)|^p\, dy\Big)^{\frac{1}{p}} \|g\|_{p'} + \|f\|_p \cdot \Big(\int_{F'} |g(y)|^{p'}\, dy\Big)^{\frac{1}{p'}} \\
&\leqq \left(\int_{x-a}^{x+a} |f(y)|^p\, dy\right)^{\frac{1}{p}} \cdot \|g\|_{p'} + \|f\|_p \cdot \varepsilon \\
&\leqq \Big(\int_{F'} |f(y)|^p dy\Big)^{\frac{1}{p}} \cdot \|g\|_{p'} + \|f\|_p \cdot \varepsilon \\
&\leqq (\|f\|_p + \|g\|_{p'})\, \varepsilon .
\end{aligned}$$

Thus $f * g$ vanishes at infinity, and so $f * g \,\varepsilon\, \mathfrak{C}_0(R)$. □

(21.34) Theorem. *With convolution as multiplication, $\mathfrak{L}_1(R)$ is a complex commutative Banach algebra.*

Proof. We leave it to the reader to make the necessary computations to show that $f * (g * h) = (f * g) * h$, $f * (g + h) = (f * g) + (f * h)$, and $\alpha (f * g) = (\alpha f) * g = f * (\alpha g)$ for all $f, g, h \in \mathfrak{L}_1(R)$ and $\alpha \in K$. We saw in (21.32) that convolution is commutative and we saw in (21.31) that $\|f * g\|_1 \leqq \|f\|_1 \cdot \|g\|_1$. Since $\mathfrak{L}_1(R)$ is a complex Banach space (13.11), all of the requirements of Definition (7.7) are fulfilled. □

The algebra $\mathfrak{L}_1(R)$ is often called the *group algebra of* R.

(21.35) Theorem. *The algebra $\mathfrak{L}_1(R)$ has no multiplicative unit.*

Proof. Assume that $\mathfrak{L}_1(R)$ has a multiplicative unit u, *i. e.*, $u \in \mathfrak{L}_1(R)$ and $u * f = f$ a. e. for all $f \in \mathfrak{L}_1(R)$. By (12.34), there exists a real number $\delta > 0$ such that $\int_{-2\delta}^{2\delta} |u(t)|\, dt < 1$. Let $f = \xi_{[-\delta,\delta]}$. Then $f \in \mathfrak{L}_1(R)$, and so for almost all $x \in R$ we have

$$\begin{aligned}
f(x) = u * f(x) &= \int_R u(x-y)\, f(y)\, dy \\
&= \int_{-\delta}^{\delta} u(x-y)\, dy = \int_{x-\delta}^{x+\delta} u(t)\, dt .
\end{aligned}$$

Since $\lambda([-\delta, \delta]) > 0$, there must be an $x \in [-\delta, \delta]$ such that

$$1 = f(x) = \int_{x-\delta}^{x+\delta} u(t)\, dt .$$

Since $[x - \delta, x + \delta] \subset [-2\delta, 2\delta]$, our choice of δ implies that

$$1 = \left|\int_{x-\delta}^{x+\delta} u(t)\, dt\right| \leqq \int_{x-\delta}^{x+\delta} |u(t)|\, dt \leqq \int_{-2\delta}^{2\delta} |u(t)|\, dt < 1 .$$

This contradiction proves the theorem. □

Even though $\mathfrak{L}_1(R)$ has no unit, it does have "approximate units", which for many purposes serve just as well. We give a precise definition.

(21.36) Definition. A sequence $(u_n)_{n=1}^{\infty} \subset \mathfrak{L}_1(R)$ is called an *approximate unit* [or a *positive kernel*] if:

(i) $u_n \geqq 0$ for all n;

(ii) $\|u_n\|_1 = 1$ for all n;

(iii) for each neighborhood V of 0 we have

$$\lim_{n\to\infty} \int_{V'} u_n(t)\, dt = 0 .$$

It is obvious that approximate units exist; *e.g.*, take $u_n = \frac{n}{2}\,\xi_{\left[-\frac{1}{n},\frac{1}{n}\right]}$. The next theorem justifies our terminology.

(21.37) Theorem. *Let* (u_n) *be an approximate unit in* $\mathfrak{L}_1(R)$. *If* $1 \leqq p < \infty$, *then* $\lim_{n\to\infty} \|f * u_n - f\|_p = 0$ *for all* $f \in \mathfrak{L}_p(R)$.

Proof. Let $f \in \mathfrak{L}_p(R)$ and let $\varepsilon > 0$ be given. Apply (13.24) to obtain a neighborhood V of 0 in R such that $2\|f_{-y} - f\|_p < \varepsilon$ whenever $y \in V$. Next use (21.36.iii) to choose an $n_0 \in N$ such that $4\|f\|_p \int_{V'} u_n(y)\, dy < \varepsilon$ for all $n \geqq n_0$. Now fix an arbitrary $n \geqq n_0$. Then (21.31) or (21.32) shows that $(f * u_n - f) \in \mathfrak{L}_p(R)$, and so for any $h \in \mathfrak{L}_{p'}(R)$ [recall that $1' = \infty$], FUBINI's theorem (21.13) and HÖLDER's inequality (13.4) show that

$$\begin{aligned}
\Big|\int_R (f * u_n(x) - f(x))\, h(x)\, dx\Big| &= \Big|\int_R\int_R (f(x-y)\, u_n(y) - f(x)\, u_n(y))\, dy\, h(x)\, dx\Big| \\
&\leqq \int_R |u_n(y)| \int_R |f(x-y) - f(x)|\, |h(x)|\, dx\, dy \\
&\leqq \int_R |u_n(y)|\, \|f_{-y} - f\|_p\, \|h\|_{p'}\, dy \\
&\leqq \int_V u_n(y)\, \frac{\varepsilon}{2}\, \|h\|_{p'}\, dy + \int_{V'} u_n(y)\, 2\,\|f\|_p \cdot \|h\|_{p'}\, dy \\
&< \varepsilon\, \|h\|_{p'} .
\end{aligned}$$

Thus the bounded linear functional $h \to \int_R (f * u_n - f)\, h\, d\lambda$ on $\mathfrak{L}_{p'}(R)$ has norm not exceeding ε, and so it follows from (15.1) that $\|f * u_n - f\|_p \leqq \varepsilon$ if $p > 1$. For the case $p = 1$, we simply take $h = 1$ in the above computation. □

(21.38) Remarks. (a) We now take up the Fourier transform for various classes of functions on R. This transform is of great importance in applications of analysis, and it is also very useful in describing the structure of the Banach algebra $\mathfrak{L}_1(R)$. There are close similarities, as well as some important differences, between the theories of Fourier series and Fourier transforms. We will point these out as they come up in our exposition.

(b) Recall our definition (16.36) of the Fourier transform $\hat{f}$ of a function $f \in \mathfrak{L}_1(R)$: for all $y \in R$,

(i) $\hat{f}(y) = (2\pi)^{-\frac{1}{2}} \int_R f(x) \exp(-ixy)\, dx$.

The factor $(2\pi)^{-\frac{1}{2}}$ in (i) is placed there as a matter of convenience. The reader will note the normalization used in the definition of Fourier coefficients $\hat{f}(n)$ (16.33): we divided all integrals by 2π. This was done to render the set $\{\exp(inx)\}_{n\in Z}$ orthonormal over $[-\pi, \pi]$, and had useful by-products in (16.37), (18.28), and (18.29). All of these theorems would be slightly more complicated to state had we used $\int_{-\pi}^{\pi} \cdots d\lambda$ instead of $(2\pi)^{-1} \int_{-\pi}^{\pi} \cdots d\lambda$. The situation is similar in the case of Fourier transforms. There are good reasons for "normalizing" our integrals with the factor $(2\pi)^{-\frac{1}{2}}$, and we will point them out at the appropriate places.

(c) It is in fact convenient to replace *all* integrals $\int_R \cdots d\lambda$ by $(2\pi)^{-\frac{1}{2}} \int_R \cdots d\lambda$. Let us agree to do this throughout (21.38)–(21.66). Let us also agree that in (21.38)–(21.66),

$$\|f\|_p = \left[(2\pi)^{-\frac{1}{2}} \int_R |f|^p d\lambda\right]^{\frac{1}{p}}$$

for $1 \leq p < \infty$. With this reinterpretation, we have

$$f * g(x) = (2\pi)^{-\frac{1}{2}} \int_R f(x-y)\, g(y)\, dy;$$

and the inequalities $\|f * g\|_1 \leq \|f\|_1 \|g\|_1$ from (21.31) and $\|f * g\|_p \leq \|f\|_1 \|g\|_p$ from (21.32) evidently remain valid.

Our first theorem is simple enough.

(21.39) Riemann-Lebesgue Lemma. *If f is in $\mathfrak{L}_1(R)$, then $\hat{f}$ is in $\mathfrak{C}_0(R)$*[1].

Proof. For nonzero $y \in R$, we have

$$\begin{aligned}
\hat{f}(y) &= (2\pi)^{-\frac{1}{2}} \int_R f(x) \exp(-ixy)\, dx \\
&= (-1) \exp(-\pi i)\ (2\pi)^{-\frac{1}{2}} \int_R f(x) \exp(-ixy)\, dx \\
&= (-1)\ (2\pi)^{-\frac{1}{2}} \int_R f(x) \exp\left(-i\left(x + \frac{\pi}{y}\right) y\right) dx \\
&= (-1)\ (2\pi)^{-\frac{1}{2}} \int_R f\left(x - \frac{\pi}{y}\right) \exp(-ixy)\, dx\,.
\end{aligned}$$

[1] Compare this fact with (16.35).

Thus

$$2|\hat{f}(y)| = \left|(2\pi)^{-\frac{1}{2}} \int_R f(x)\exp(-ixy)\,dx - (2\pi)^{-\frac{1}{2}} \int_R f\left(x - \frac{\pi}{y}\right)\exp(-ixy)\,dx\right|$$

$$\leq (2\pi)^{-\frac{1}{2}} \int_R \left|f(x) - f\left(x - \frac{\pi}{y}\right)\right| \cdot |\exp(-ixy)|\,dx$$

$$= (2\pi)^{-\frac{1}{2}} \int_R \left|f(x) - f\left(x - \frac{\pi}{y}\right)\right|\,dx\,.$$

A look at (13.24) shows that $|\hat{f}(y)|$ is arbitrarily small if $|y|$ is sufficiently large.

It remains to show that $\hat{f}$ is continuous. Given $\varepsilon > 0$, choose a compact interval $I = [-a, a]$ $(a > 0)$ such that

$$4 \int_{I'} |f(x)|\,dx < \varepsilon \tag{1}$$

and then choose $\delta > 0$ such that

$$2a\delta \int_{-a}^{a} |f(x)|\,dx < \varepsilon\,. \tag{2}$$

Since $|\sin(u)| \leq |u|$ for all $u \in R$, it follows from (1) and (2) that if $y, t \in R$ and $|t| < \delta$, then

$$(2\pi)^{\frac{1}{2}} |\hat{f}(y+t) - \hat{f}(y)| = \left|\int_R f(x)\exp(-iyx)\,(\exp(-itx) - 1)\,dx\right|$$

$$\leq \int_R |f(x)| \cdot |\exp(-itx) - 1|\,dx$$

$$= \int_R |f(x)| \cdot 2\left|\sin\left(\frac{tx}{2}\right)\right|\,dx$$

$$\leq 2\int_{I'} |f(x)|\,dx + \int_{-a}^{a} |f(x)| \cdot |tx|\,dx$$

$$< \frac{\varepsilon}{2} + a\delta \int_{-a}^{a} |f(x)|\,dx$$

$$< \varepsilon\,.$$

Thus $\hat{f}$ is [uniformly] continuous on R. □

(21.40) Remarks. It is plain that the Fourier transform $f \to \hat{f}$, which maps $\mathfrak{L}_1(R)$ into $\mathfrak{C}_0(R)$, is linear. It is also bounded, since

$$\|\hat{f}\|_u = \sup_{y \in R} \left|(2\pi)^{-\frac{1}{2}} \int_R f(x)\exp(-iyx)\,dx\right|$$

$$\leq (2\pi)^{-\frac{1}{2}} \int_R |f(x)|\,dx = \|f\|_1\,.$$

Our next theorem shows that this transformation also preserves products [convolution in $\mathfrak{L}_1$ turns into pointwise multiplication in $\mathfrak{C}_0$]. The Fourier transform is also one-to-one (21.47) and its range is uniformly dense in $\mathfrak{C}_0$ (21.62.b). There seems to be no simple way to describe intrinsically the functions in $\mathfrak{C}_0$ that have the form $\hat{f}$ for some $f \in \mathfrak{L}_1$.

(21.41) Theorem. *Let f, g be functions in $\mathfrak{L}_1(R)$. Then*

(i) $\widehat{f * g} = \hat{f} \cdot \hat{g}$.

Proof. For all $y \in R$ we have

$$\begin{aligned}
\widehat{f * g}(y) &= (2\pi)^{-\frac{1}{2}} \int_R f * g(x) \exp(-iyx)\, dx \\
&= (2\pi)^{-\frac{1}{2}} \int_R (2\pi)^{-\frac{1}{2}} \int_R f(x-t)\, g(t)\, dt \exp(-iyx)\, dx \\
&= \frac{1}{2\pi} \int_R g(t) \int_R f(x-t) \exp(-iyx)\, dx\, dt \\
&= \frac{1}{2\pi} \int_R g(t) \int_R f(u) \exp(-i(t+u)\, y)\, du\, dt \\
&= \frac{1}{2\pi} \int_R g(t) \exp(-iyt) \int_R f(u) \exp(-iyu)\, du\, dt \\
&= \hat{f}(y) \cdot \hat{g}(y);
\end{aligned}$$

we have made free use of FUBINI's theorem and of (12.44). □

We next take up the problem of reconstructing a function from its Fourier transform. The analogous problem for Fourier series was treated in (18.29) and (18.47) *supra.* The following lemma points up the close connection between this problem and that of approximating a function by convolving it with an approximate unit.

(21.42) Lemma. *Let f, k be functions in $\mathfrak{L}_1(R)$, write $u = \hat{k}$, and suppose that $u(t) = u(-t)$ for all $t \in R$. Then*

(i) $$(2\pi)^{-\frac{1}{2}} \int_R \hat{f}(y)\, k(y) \exp(ixy)\, dy = f * u(x) = (2\pi)^{-\frac{1}{2}} \int_R f(x-s)\, u(s)\, ds$$

for all $x \in R$.

Proof. Let $x \in R$. Since the function $(s, t) \to f(s)\, k(t)$ is in $\mathfrak{L}_1(R \times R)$, FUBINI's theorem can be applied in the following computation:

$$
\begin{aligned}
(2\pi)^{-\frac{1}{2}} \int_R f(y)\, k(y) \exp(ixy)\, dy
&= \frac{1}{2\pi} \int_R \int_R f(t) \exp(-iyt)\, dt\, k(y) \exp(ixy)\, dy \\
&= \frac{1}{2\pi} \int_R f(t) \int_R k(y) \exp(-i(t-x)y)\, dy\, dt \\
&= (2\pi)^{-\frac{1}{2}} \int_R f(t)\, \hat{k}(t-x)\, dt \\
&= (2\pi)^{-\frac{1}{2}} \int_R f(t)\, u(x-t)\, dt \\
&= u * f(x) = f * u(x). \quad \square
\end{aligned}
$$

(21.43) Pointwise summability Theorem. *Let $(k_n)_{n=1}^{\infty}$ be a sequence of functions in $\mathfrak{L}_1(R)$ and write $u_n = \hat{k}_n$. Suppose that for each $n \in N$ we have*

$$u_n \in \mathfrak{L}_1(R), \quad (2\pi)^{-\frac{1}{2}} \int_R u_n(t)\, dt = 1, \quad \textit{and} \quad u_n(-t) = u_n(t) \quad \textit{for all} \quad t \in R.$$

Furthermore, suppose that there is a function u such that:

(i) *$u \in \mathfrak{L}_1^+([0, \infty[)$;*

(ii) *u is nonincreasing and absolutely continuous on $[0, \infty[$;*

(iii) *$|u_n(t)| \leqq n u(nt)$ for all $t \geqq 0$ and all $n \in N$.*

Then if f is a function in $\mathfrak{L}_1(R)$ and x is a Lebesgue point for f, we have

(iv) $$\lim_{n \to \infty} (2\pi)^{-\frac{1}{2}} \int_R f(y)\, k_n(y) \exp(ixy)\, dy = f(x).$$

In particular, (iv) *holds if f is continuous at x.*

Proof. Let x be a fixed Lebesgue point of f. In view of Lemma (21.42), it suffices to prove that $\lim_{n \to \infty} f * u_n(x) = f(x)$. We have

$$
\begin{aligned}
f * u_n(x) - f(x) &= (2\pi)^{-\frac{1}{2}} \int_R f(x-t)\, u_n(t)\, dt - (2\pi)^{-\frac{1}{2}} \int_R f(x)\, u_n(t)\, dt \\
&= (2\pi)^{-\frac{1}{2}} \int_0^{\infty} [f(x-t) - f(x)]\, u_n(t)\, dt + (2\pi)^{-\frac{1}{2}} \int_{-\infty}^{0} [f(x-t) - f(x)]\, u_n(t)\, dt \\
&= (2\pi)^{-\frac{1}{2}} \int_0^{\infty} [f(x+t) + f(x-t) - 2f(x)]\, u_n(t)\, dt\,. \qquad (1)
\end{aligned}
$$

As in (18.29), we write $\varphi(t) = f(x+t) + f(x-t) - 2f(x)$. Since $|u_n(t)| \leqq nu(nt)$ for all $t \geqq 0$ and all $n \in N$, (1) yields

$$|f * u_n(x) - f(x)| \leqq (2\pi)^{-\frac{1}{2}} \int_0^\infty |\varphi(t)|\, nu(nt)\, dt\,. \tag{2}$$

Thus we need only to show that the right side of (2) has limit zero as $n \to \infty$. To this end, let $\varepsilon > 0$ be given and write

$$c = 6\varepsilon + 6u(0) + 3\int_0^\infty u(y)\, dy + 6\|f\|_1 + 6|f(x)| + 1.$$

For $h > 0$, set

$$\Phi(h) = \int_0^h |\varphi(t)|\, dt\,.$$

Since x is a Lebesgue point (18.6) for f, there exists a number $\alpha \in]0, 1]$ such that

$$\frac{1}{h}\Phi(h) < \frac{\varepsilon}{c} \quad \text{for} \quad h \in]0, \alpha]\,. \tag{3}$$

Since u is nonincreasing and in $\mathfrak{L}_1^+([0, \infty[)$, it follows at once from dominated convergence (12.24) that there exists an integer n_0 such that if $n > n_0$, then

$$\int_{n\alpha}^\infty u(y)\, dy < \frac{\varepsilon}{c}\,, \tag{4}$$

and also

$$nu(n\alpha) = \frac{2}{\alpha}\frac{n\alpha}{2} u(n\alpha) \leqq \frac{2}{\alpha}\int_{\frac{n\alpha}{2}}^{n\alpha} u(y)\, dy$$

$$\leqq \frac{2}{\alpha}\int_{\frac{n\alpha}{2}}^{\infty} u(y)\, dy < \frac{\varepsilon}{c}\,. \tag{5}$$

Now let n be any integer $> n_1 = \max\left\{\frac{1}{\alpha}, n_0\right\}$. We have

$$\int_0^\infty |\varphi(t)|\, nu(nt)\, dt = \int_0^{1/n} |\varphi(t)|\, nu(nt)\, dt + \int_{1/n}^{\alpha} |\varphi(t)|\, nu(nt)\, dt$$

$$+ \int_\alpha^\infty |\varphi(t)|\, nu(nt)\, dt$$

$$= I_1 + I_2 + I_3\,. \tag{6}$$

We apply (3) to I_1 and find that

$$I_1 \leqq \int_0^{1/n} |\varphi(t)|\, n u(0)\, dt = n\Phi\left(\frac{1}{n}\right) u(0) < u(0)\frac{\varepsilon}{c} < \frac{\varepsilon}{3}. \tag{7}$$

To majorize I_2, we integrate by parts (18.19), use (3) and (5), and integrate by parts again. This yields the following estimates:

$$\begin{aligned}
I_2 &= \int_{1/n}^{\alpha} |\varphi(t)|\, n u(nt)\, dt \\
&= \Phi(\alpha)\, n u(n\alpha) - \Phi\left(\frac{1}{n}\right) n u(1) - \int_{1/n}^{\alpha} \Phi(t)\, n^2 u'(nt)\, dt \\
&\leqq \frac{1}{\alpha}\Phi(\alpha)\, n u(n\alpha) + n\Phi\left(\frac{1}{n}\right) u(0) - \int_{1/n}^{\alpha} \Phi(t)\, n^2 u'(nt)\, dt \\
&< \frac{\varepsilon^2}{c^2} + \frac{\varepsilon}{c} u(0) - \int_{1/n}^{\alpha} t\frac{\varepsilon}{c} n^2 u'(nt)\, dt\,^{1} \\
&= \frac{\varepsilon^2}{c^2} + \frac{\varepsilon}{c} u(0) - \frac{\varepsilon}{c}\left[n\alpha\, u(n\alpha) - u(1)\right] + \frac{\varepsilon n}{c}\int_{1/n}^{\alpha} u(nt)\, dt \\
&< \frac{\varepsilon^2}{c} + \frac{\varepsilon}{c} u(0) + \frac{\varepsilon^2}{c} + \frac{\varepsilon}{c} u(0) + \frac{\varepsilon}{c}\int_1^{n\alpha} u(y)\, dy \\
&< \frac{\varepsilon}{c}\left(2\varepsilon + 2u(0) + \int_0^{\infty} u(y)\, dy\right) < \frac{\varepsilon}{3}.
\end{aligned} \tag{8}$$

Next we use obvious estimates and (4) to write

$$\begin{aligned}
I_3 &= \int_{\alpha}^{\infty} |\varphi(t)|\, n u(nt)\, dt \\
&\leqq \int_{\alpha}^{\infty} (|f(x+t)| + |f(x-t)|)\, n u(nt) dt + 2|f(x)|\, n\int_{\alpha}^{\infty} u(nt)\, dt \\
&\leqq n u(n\alpha)\cdot 2\|f\|_1 + 2|f(x)|\int_{n\alpha}^{\infty} u(y)\, dy \\
&\leqq \frac{\varepsilon}{c}\left(2\|f\|_1 + 2|f(x)|\right) < \frac{\varepsilon}{3}.
\end{aligned} \tag{9}$$

Thus, combining (2), (6), (7), (8), and (9), we infer that

$$|f * u_n(x) - f(x)| < \varepsilon$$

---

[1] Recall that $u'(nt) \leqq 0$.

if $n > n_1$. Since every point of continuity of f is a Lebesgue point of f, our last assertion also holds. □

(21.44) Notes. The reader may well have noticed the similarity between the proofs of (21.43) and (18.29) [(18.47) is different]. It would be no trouble to generalize (18.29) to a class of kernels $(u_n)_{n=1}^{\infty}$ of period 2π satisfying hypotheses like those imposed on $(u_n)_{n=1}^{\infty}$ in (21.43). The only essential difference in the arguments of (21.43) and (18.29) is that we need FUBINI's theorem to interchange the order of integration in the former, while only sums and integrals are involved in the latter. [Of course equalities $\sum \int = \int \sum$ are special cases of FUBINI's theorem.]

There are many sequences $(k_n)_{n=1}^{\infty}$ that satisfy the hypotheses of (21.43). We will now give three classical examples of such sequences. The reader should check in each case that the hypotheses of (21.43) hold.

(21.45) Examples. (a) Let $k_n(y) = \exp\left(-\frac{|y|}{n}\right)$. For each $\alpha > 0$, we have

$$\int_R \exp(-\alpha|y|)\exp(-ity)\,dy = \int_{-\infty}^{0} \exp((\alpha - it)y)\,dy + \int_0^{\infty} \exp((-\alpha - it)y)\,dy$$

$$= \lim_{A\to\infty}\left[\frac{1}{\alpha - it} - \frac{1}{\alpha - it}\exp(-(\alpha - it)A) + \frac{1}{\alpha+it} - \frac{1}{\alpha+it}\exp(-(\alpha+it)A)\right]$$

$$= \frac{2\alpha}{\alpha^2+t^2}.\text{[1]}$$

Hence

$$u_n(t) = \hat{k}_n(t) = \left(\frac{2}{\pi}\right)^{\frac{1}{2}} \frac{n}{1+n^2t^2}.$$

For the function u of (21.43), we may evidently take

$$u(t) = \left(\frac{2}{\pi}\right)^{\frac{1}{2}} \frac{1}{1+t^2}.$$

This sequence $(u_n)_{n=1}^{\infty}$ is known as *ABEL's kernel.*

(b) Let $k_n(y) = \left(1 - \frac{|y|}{n}\right)\xi_{[-n,n]}(y)$. For every $\alpha > 0$, integration by parts yields

$$\int_{-\alpha}^{\alpha}\left(1-\frac{|y|}{\alpha}\right)\exp(-ity)\,dy = 2\int_0^{\alpha}\left(1-\frac{y}{\alpha}\right)\cos(ty)\,dy$$

$$= 2\int_0^{\alpha}\frac{\sin(ty)}{\alpha t}\,dy$$

$$= \frac{2(1-\cos(\alpha t))}{\alpha t^2}$$

$$= \alpha\left[\frac{\sin(\frac{1}{2}\alpha t)}{\frac{1}{2}\alpha t}\right]^2.$$

---

[1] This is one of the few Fourier transforms that is computable by inspection.

Hence

$$u_n(t) = \hat{k}_n(t) = (2\pi)^{-\frac{1}{2}} n \left[\frac{\sin(\frac{1}{2}nt)}{\frac{1}{2}nt}\right]^2 .$$

In this case, we may take $u(t) = \frac{4}{1+t^2}$. This sequence $(u_n)_{n=1}^{\infty}$ is known as FEJÉR's *kernel*. See also (21.55) *infra*.

(c) Let $k_n(y) = \exp\left(-\frac{y^2}{2n^2}\right)$. It is shown in (21.60) *infra* that

$$u_n(t) = \hat{k}_n(t) = n \exp\left(-\frac{n^2 t^2}{2}\right).$$

Here we take $u = u_1$. This sequence is called GAUSS's *kernel*.

(21.46) Notes. (a) Theorem (21.43) and Examples (21.45) show why the factor $(2\pi)^{-\frac{1}{2}}$ is used in the integral defining $\hat{f}$. With this factor, we use the same integral for integrating Fourier transforms that we use for integrating the original functions. This is convenient, and it becomes useful in some later developments (21.53).

(b) All of the kernels listed in (21.45) can plainly be taken to depend upon an arbitrary positive real number α instead of a positive integer n. The equality (21.43.iv) holds as $\alpha \to \infty$ for all three kernels.

(21.47) Corollary [Uniqueness Theorem]. *If f and g are in $\mathfrak{L}_1(R)$ and $\hat{f} = \hat{g}$, then $f = g$ a.e. Thus the mapping $f \to \hat{f}$ is one-to-one.*

Proof. Let $h = f - g$. Then $\hat{h} = 0$, and so (21.43) and (21.45.a) imply that

$$h(x) = \lim_{n\to\infty} \int_R 0 \cdot k_n(y) \exp(ixy)\, dy = 0$$

for all Lebesgue points of h, *i.e.*, for almost all $x \in R$ (18.5). □

(21.48) Remarks. Our next theorem shows that sometimes a thoroughly simple-minded device will recapture a function $f \in \mathfrak{L}_1(R)$ from its Fourier transform $\hat{f}$. An analogue for Fourier transforms of the partial sum $s_n f$ of a Fourier *series* is evidently

(i) $$(2\pi)^{-\frac{1}{2}} \int_{-A}^{A} \hat{f}(y) \exp(ixy)\, dy ,$$

and the limit of this expression as $A \to \infty$, when it exists, is an analogue of the sum $\sum_{n=-\infty}^{\infty} \hat{f}(n) \exp(inx)$ of a Fourier series. Neither of these limits need exist, as is well known. In case $\hat{f}$ is in $\mathfrak{L}_1(R)$, however, the limit of (i) as $A \to \infty$ plainly exists, and remarkably it is $f(x)$ a.e., as we will now show.

(21.49) Fourier Inversion Theorem. *Let f be a function in $\mathfrak{L}_1(R)$. If $\hat{f}$ is also in $\mathfrak{L}_1(R)$, then*

$$\text{(i)} \qquad (2\pi)^{-\frac{1}{2}} \int_R \hat{f}(y) \exp(ixy)\, dy = f(x)$$

for every Lebesgue point x of f. Hence f is equal a.e. to a function in $\mathfrak{C}_0(R) \cap \mathfrak{L}_1(R)$. If f is continuous, then (i) *holds everywhere.*

Proof. Suppose that $\hat{f} \in \mathfrak{L}_1(R)$ and let x be a Lebesgue point of f. According to (21.43) and (21.45.a), we have

$$f(x) = \lim_{n\to\infty} (2\pi)^{-\frac{1}{2}} \int_R \hat{f}(y) \exp\left(-\frac{|y|}{n}\right) \exp(ixy)\, dy\,. \tag{1}$$

Moreover

$$\left|\hat{f}(y) \exp\left(-\frac{|y|}{n}\right) \exp(ixy)\right| \leqq |\hat{f}(y)| \tag{2}$$

for all $n \in N$ and all $y \in R$, and

$$\lim_{n\to\infty} \exp\left(-\frac{|y|}{n}\right) = 1 \tag{3}$$

for all $y \in R$. Since $|\hat{f}| \in \mathfrak{L}_1(R)$, it follows from (1), (2), (3), and LEBESGUE'S dominated convergence theorem (12.30) that (i) holds. Theorem (21.39) shows that the left side of (i) is a function in $\mathfrak{C}_0(R)$; and so the second assertion holds. Two continuous functions that are equal a.e. are the same function, and so the last statement holds. □

We propose now to define the Fourier transform for all functions in $\mathfrak{L}_2(R)$. We need two lemmas.

(21.50) Lemma. *Suppose that $f \in \mathfrak{L}_1(R) \cap \mathfrak{L}_\infty(R)$ and that $\hat{f}$ is real-valued and nonnegative. Then $\hat{f}$ is in $\mathfrak{L}_1(R)$ and so the conclusions of* (21.49) *hold for f.*

Proof. Consider ABEL'S kernel $u_n(t) = \left(\frac{2}{\pi}\right)^{\frac{1}{2}} \frac{n}{(1+n^2t^2)}$. We have

$$\begin{aligned}
|f * u_n(x)| &= (2\pi)^{-\frac{1}{2}} \left|\int_R f(x-t)\, u_n(t)\, dt\right| = \frac{1}{\pi} \left|\int_R f(x-t) \frac{n}{1+n^2t^2}\, dt\right| \\
&= \frac{1}{\pi} \left|\int_R f\left(x - \frac{s}{n}\right) \frac{1}{1+s^2}\, ds\right| \\
&\leqq \frac{\|f\|_\infty}{\pi} \int_R \frac{ds}{1+s^2} = \|f\|_\infty
\end{aligned}$$

for all $x \in R$ and $n \in N$. We set $x = 0$ in (21.42.i) and use (21.45.a) to infer that

$$(2\pi)^{-\frac{1}{2}} \int_R \hat{f}(y) \exp\left(-\frac{|y|}{n}\right) dy = f * u_n(0) \leqq \|f\|_\infty < \infty \tag{1}$$

for all $n \in N$. Since $\hat{f}$ is nonnegative, we may apply B. LEVI's theorem (12.22) to the left side of (1) to obtain

$$(2\pi)^{-\frac{1}{2}} \int_R \hat{f}(y)\, dy \leqq \|f\|_\infty < \infty .$$

Thus $\hat{f}$ is in $\mathfrak{L}_1(R)$. The rest follows from (21.49). □

(21.51) Lemma. *Let $f \in \mathfrak{L}_1(R)$. Define $\tilde{f}$ on R by $\tilde{f}(x) = \overline{f(-x)}$. Then $\hat{\tilde{f}}(y) = \overline{\hat{f}(y)}$ for all $y \in R$.*

Proof. We have

$$\begin{aligned} \hat{\tilde{f}}(y) &= (2\pi)^{-\frac{1}{2}} \int_R \tilde{f}(x) \exp(-iyx)\, dx \\ &= (2\pi)^{-\frac{1}{2}} \int_R \overline{f(-x)}\; \overline{\exp(iyx)}\, dx \\ &= (2\pi)^{-\frac{1}{2}} \overline{\int_R f(t) \exp(-iyt)\, dt} \\ &= \overline{\hat{f}(y)} . \quad \square \end{aligned}$$

A first step in defining $\hat{f}$ for $f \in \mathfrak{L}_2(R)$ follows.

(21.52) Theorem. *Let $f \in \mathfrak{L}_1(R) \cap \mathfrak{L}_2(R)$. Then $\hat{f} \in \mathfrak{L}_2(R)$ and*

(i) $$(2\pi)^{-\frac{1}{2}} \int_R |\hat{f}(y)|^2\, dy = (2\pi)^{-\frac{1}{2}} \int_R |f(x)|^2\, dx .^{1}$$

Proof. Let $\tilde{f}$ be as in (21.51) and let $g = f * \tilde{f}$. Since $f, \tilde{f} \in \mathfrak{L}_1(R)$, we have $g \in \mathfrak{L}_1(R)$ (21.31); and so (21.41) and (21.51) imply that

$$\hat{g} = \hat{f}\, \overline{\hat{f}} = |\hat{f}|^2 \geqq 0 .$$

Since $f, \tilde{f} \in \mathfrak{L}_2(R)$, (21.33) shows that $g \in \mathfrak{C}_0(R)$. Thus $g \in \mathfrak{L}_1(R) \cap \mathfrak{L}_\infty(R)$, and so (21.50) shows that $|\hat{f}|^2 = \hat{g} \in \mathfrak{L}_1(R)$ and that the inversion formula (21.49.i) holds for g everywhere, since g is continuous. Thus

$$\begin{aligned} (2\pi)^{-\frac{1}{2}} \int_R f(x+y)\, \overline{f(y)}\, dy &= (2\pi)^{-\frac{1}{2}} \int_R f(x-y)\, \overline{f(-y)}\, dy \\ &= f * \tilde{f}(x) = g(x) = (2\pi)^{-\frac{1}{2}} \int_R \hat{g}(y) \exp(ixy)\, dy \\ &= (2\pi)^{-\frac{1}{2}} \int_R |\hat{f}(y)|^2 \exp(ixy)\, dy \qquad (1) \end{aligned}$$

for all $x \in R$. Setting $x = 0$ in (1), we have (i). □

---

[1] The $(2\pi)^{-\frac{1}{2}}$ in (i) is of no consequence, but the equality would not hold as written without the factor $(2\pi)^{-\frac{1}{2}}$ in (21.38.i).

(21.53) Plancherel's Theorem. *There exists a unique bounded linear transformation T from $\mathfrak{L}_2(R)$ into $\mathfrak{L}_2(R)$ such that $Tf = \hat{f}$ for all f in $\mathfrak{L}_1(R) \cap \mathfrak{L}_2(R)$*[1]. *Moreover:*

(i) $\|Tf\|_2 = \|f\|_2$ *for all* $f \in \mathfrak{L}_2(R)$;

(ii) $\langle Tf, Tg\rangle = \langle f, g\rangle$ *for all* $f, g \in \mathfrak{L}_2(R)$;

(iii) T *carries* $\mathfrak{L}_2(R)$ onto $\mathfrak{L}_2(R)$.

Proof. Define T on $\mathfrak{L}_1 \cap \mathfrak{L}_2$ by

$$Tf = \hat{f}.$$

Since $\mathfrak{C}_{00} \subset \mathfrak{L}_1 \cap \mathfrak{L}_2$, it follows from (13.21) that $\mathfrak{L}_1 \cap \mathfrak{L}_2$ is dense in $\mathfrak{L}_2$. For $f \in \mathfrak{L}_2$, let (f_n) be any sequence in $\mathfrak{L}_1 \cap \mathfrak{L}_2$ such that $\|f - f_n\|_2 \to 0$. Then (f_n) is a Cauchy sequence in $\mathfrak{L}_2$, and so (21.52) implies that

$$\|Tf_n - Tf_m\|_2 = \|\hat{f}_n - \hat{f}_m\|_2 = \|f_n - f_m\|_2 \to 0 \quad \text{as } m, n \to \infty .$$

Thus (Tf_n) is a Cauchy sequence in $\mathfrak{L}_2$ and, since $\mathfrak{L}_2$ is complete (13.11), there is a unique function $Tf \in \mathfrak{L}_2$ such that

$$\|Tf - Tf_n\|_2 \to 0 \quad \text{as} \quad n \to \infty .$$

It is easy to see that Tf is independent of the particular sequence (f_n) that is used. Thus T is defined from $\mathfrak{L}_2$ into $\mathfrak{L}_2$. It is also easy to see that T is linear. Again (21.52) implies that

$$\|Tf\|_2 = \lim_{n\to\infty} \|Tf_n\|_2 = \lim_{n\to\infty} \|f_n\|_2 = \|f\|_2 ,$$

where (f_n) is as above, and so (i) holds. Since T preserves norms, it is bounded and one-to-one. Conclusion (ii) follows from (i) and the polar identity (16.5), which shows that the inner product is determined by the norm. The uniqueness statement follows from the fact that if two continuous mappings [into a Hausdorff space] agree on a dense subset of their common domain, then they agree throughout that domain.

It remains only to show that T carries $\mathfrak{L}_2$ onto all of $\mathfrak{L}_2$. Since T preserves norms and $\mathfrak{L}_2$ is complete, the range of T is closed in $\mathfrak{L}_2$. Thus it will suffice to show that the range of T is dense in $\mathfrak{L}_2$.

Let us compute the adjoint T^* of the operator T (16.40). For $f, \varphi \in \mathfrak{L}_1 \cap \mathfrak{L}_2$, write

$$g(x) = (2\pi)^{-\frac{1}{2}} \int_R \varphi(y) \exp(ixy)\, dy .\text{[2]} \qquad (1)$$

[1] The function Tf is called the *Fourier transform of* f. Some writers call it the *Plancherel transform*, but the term "Fourier transform" is more common, and we will retain it.

[2] That is, g is the inverse Fourier transform of φ, as defined in (21.49.i).

Then we have

$$\begin{aligned}\langle f, T^*\varphi\rangle = \langle Tf, \varphi\rangle &= (2\pi)^{-\frac{1}{2}} \int_R f(y)\,\overline{\varphi(y)}\,dy \\ &= \frac{1}{2\pi} \int_R \int_R f(x) \exp(-iyx)\,dx\,\overline{\varphi(y)}\,dy \\ &= \frac{1}{2\pi} \int_R f(x) \overline{\int_R \varphi(y) \exp(ixy)\,dy}\,dx \\ &= (2\pi)^{-\frac{1}{2}} \int_R f(x)\,\overline{g(x)}\,dx = \langle f, g\rangle\,, \end{aligned} \tag{2}$$

the change of order being justified by FUBINI's theorem [the integrand is in $\mathfrak{L}_1(R^2)$ because $f, \varphi \in \mathfrak{L}_1(R)$]. Since $\mathfrak{L}_1 \cap \mathfrak{L}_2$ is dense in $\mathfrak{L}_2$ and the mapping $f \to \langle f, h\rangle$ is continuous on $\mathfrak{L}_2$ for all $h \in \mathfrak{L}_2$, (1) and (2) imply that

$$[T^*\varphi](x) = (2\pi)^{-\frac{1}{2}} \int_R \varphi(y) \exp(ixy)\,dy = \hat{\varphi}(-x) \tag{3}$$

for all $\varphi \in \mathfrak{L}_1 \cap \mathfrak{L}_2$ and all $x \in R$. Let φ, ψ be in $\mathfrak{L}_1 \cap \mathfrak{L}_2$ and write $f = T^*\varphi$, $g = T^*\psi$. Then (21.41) and (3) imply that

$$\widehat{\varphi * \psi}(x) = \hat{\varphi}(x)\,\hat{\psi}(x) = f(-x)\,g(-x) \tag{4}$$

for all $x \in R$. Since f, g are in $\mathfrak{L}_2$, (13.4) shows that fg and so also $\widehat{\varphi * \psi}$ are in $\mathfrak{L}_1$. In addition $\varphi * \psi$ is continuous (21.33). Thus we can apply (21.49) to invert $\widehat{\varphi * \psi}$: for all $y \in R$, we have

$$\begin{aligned}\varphi * \psi(y) &= (2\pi)^{-\frac{1}{2}} \int_R \widehat{\varphi * \psi}(x) \exp(iyx)\,dx \\ &= (2\pi)^{-\frac{1}{2}} \int_R f(-x)\,g(-x) \exp(iyx)\,dx \\ &= (2\pi)^{-\frac{1}{2}} \int_R (fg)(x) \exp(-iyx)\,dx \\ &= \widehat{fg}(y)\,. \end{aligned} \tag{5}$$

Also, (4) and (21.39) show that $fg \in \mathfrak{C}_0$, and so

$$fg \in \mathfrak{L}_1 \cap \mathfrak{C}_0 \subset \mathfrak{L}_1 \cap \mathfrak{L}_2\,.$$

[We omit the easy verification of the above inclusion.] Hence (5) shows that $\varphi * \psi$ is in the range of T, for all $\varphi, \psi \in \mathfrak{L}_1 \cap \mathfrak{L}_2$.

Now let (ψ_n) be an approximate unit (21.36) such that $(\psi_n) \subset \mathfrak{L}_1 \cap \mathfrak{L}_2$ and let $\varphi \in \mathfrak{L}_1 \cap \mathfrak{L}_2$. The preceding paragraph shows that $\varphi * \psi_n \in \operatorname{rng} T$

for all $n \in N$, and (21.37) shows that

$$\lim_{n\to\infty} \|\varphi - \varphi * \psi_n\|_2 = 0;$$

hence, rng T being closed in $\mathfrak{L}_2$, we have $\varphi \in$ rng T. Thus $\mathfrak{L}_1 \cap \mathfrak{L}_2 \subset$ rng T, and so rng T is dense in $\mathfrak{L}_2$. □

(21.54) Remarks. (a) The proof of (21.53) shows that if $\varphi, \psi \in \mathfrak{L}_1 \cap \mathfrak{L}_2$, then there exists a function $h \in \mathfrak{L}_1 \cap \mathfrak{C}_0$ such that $\hat{h} = \varphi * \psi$. In fact, $h = fg$ where $f = T^* \varphi$ and $g = T^* \psi$.

(b) Theorem (21.53) is the analogue for the line of the RIESZ-FISCHER theorem (16.39). It is of course much harder to prove than (16.39). This is accounted for by the fact that $\mathfrak{L}_1([-\pi, \pi]) \supset \mathfrak{L}_2([-\pi, \pi])$, while $\mathfrak{L}_2(R)$ neither contains nor is contained in $\mathfrak{L}_1(R)$.

(21.55) Example. PLANCHEREL's theorem can be used to evaluate certain integrals. We illustrate by integrating FEJÉR's kernel (21.45.b). For $\alpha > 0$, it is obvious that

$$\|\xi_{[-\alpha,\alpha]}\|_2^2 = (2\pi)^{-\frac{1}{2}} \int_R \xi_{[-\alpha,\alpha]}(x)\, dx = \left(\frac{2}{\pi}\right)^{\frac{1}{2}} \alpha .$$

By (21.52.i), we have

$$\|\hat{\xi}_{[-\alpha,\alpha]}\|_2^2 = \|T \xi_{[-\alpha,\alpha]}\|_2^2 = \left(\frac{2}{\pi}\right)^{\frac{1}{2}} \alpha \tag{1}$$

as well. Plainly too

$$\begin{aligned} \hat{\xi}_{[-\alpha,\alpha]}(y) &= (2\pi)^{-\frac{1}{2}} \int_{-\alpha}^{\alpha} \exp(-iyx)\, dx \\ &= (2\pi)^{-\frac{1}{2}} \frac{\exp(-iy\alpha) - \exp(iy\alpha)}{-iy} \\ &= \left(\frac{2}{\pi}\right)^{\frac{1}{2}} \frac{\sin(\alpha y)}{y}. \end{aligned} \tag{2}$$

Combining (1) and (2), we obtain

$$\int_R \left[\frac{\sin(\alpha y)}{y}\right]^2 dy = \pi\alpha \tag{3}$$

for all positive real numbers α. For FEJÉR's kernel, then, we have

$$(2\pi)^{-\frac{1}{2}} \int_R (2\pi)^{-\frac{1}{2}} n \left[\frac{\sin(\frac{1}{2}nt)}{\frac{1}{2}nt}\right]^2 dt = 1, \tag{4}$$

for all $n \in N$.

Before presenting our next application of FUBINI's theorem, we give a number of exercises illustrating and extending the notions of convolution and Fourier transform.

(21.56) Exercise [W. H. YOUNG]. Let p, q, and r be real numbers such that $p > 1$, $q > 1$, and also $\frac{1}{p} + \frac{1}{q} - 1 = \frac{1}{r} > 0$. Suppose that $f \in \mathfrak{L}_p(R)$ and $g \in \mathfrak{L}_q(R)$. Prove that the convolution $f * g$ is in $\mathfrak{L}_r(R)$ and that $\|f * g\|_r \leqq \|f\|_p \cdot \|g\|_q$. [Hints. Let a, b, and c be real numbers such that $a = r$, $\frac{1}{p} = \frac{1}{a} + \frac{1}{b}$, and $\frac{1}{q} = \frac{1}{a} + \frac{1}{c}$. Note that $\frac{1}{a} + \frac{1}{b} + \frac{1}{c} = 1$, and use the generalized HÖLDER inequality (13.26) on the product

$$\left(|f(x-y)|^{\frac{p}{a}} |g(y)|^{\frac{q}{a}}\right)\left(|f(x-y)|^{p\left(\frac{1}{p}-\frac{1}{a}\right)}\right)\left(|g(y)|^{q\left(\frac{1}{q}-\frac{1}{a}\right)}\right).]$$

(21.57) Exercise. (a) For $f \in \mathfrak{L}_1(R)$ and $a \in R$, let $f_a(x) = f(x+a)$. Prove that $\widehat{(f_a)}(y) = \hat{f}(y) \exp(iay)$ for all $y \in R$.

(b) Prove that if $f \in \mathfrak{L}_1(R)$, $a \in R$, and $g(x) = \exp(-iax)$, then $\widehat{fg}(y) = (\hat{f})_a(y)$ for all $y \in R$.

(c) Let $f, g \in \mathfrak{L}_1(R)$. Prove that $\int_R f(x)\, \hat{g}(x)\, dx = \int_R \hat{f}(x)\, g(x)\, dx$.

(d) Let f be in $\mathfrak{L}_1(R)$. Find a necessary and sufficient condition on f for $\hat{f}$ to be real-valued; similarly for $\hat{f}$ to be even.

(21.58) Exercise. (a) Find two functions $f, g \in \mathfrak{L}_1(R)$, neither of which vanishes anywhere, such that $f * g = 0$. [Hint. Use (21.57.b), (21.45.b), (21.47), and (21.41).]

(b) Suppose that $f \in \mathfrak{L}_1(R)$ and that $f * f = f$ a. e. Prove that $f = 0$ a. e.

(c) Suppose that $f \in \mathfrak{L}_1(R)$ and that $f * f = 0$ a. e. Prove that $f = 0$ a. e.

(21.59) Exercise. (a) Let $f \in \mathfrak{L}_1(R)$, write $g(x) = -ixf(x)$ for all $x \in R$, and suppose that $g \in \mathfrak{L}_1(R)$. Prove that $\hat{f}$ has a finite derivative at every point of R and that $\hat{f}' = \hat{g}$. $\Big[$Hint. Prove that $\left|\frac{\exp(ixh)-1}{h}\right| \leqq |x|$ and use (12.30).$\Big]$

(b) Suppose that $f \in \mathfrak{L}_1(R)$, that f is absolutely continuous on R, and that $f' \in \mathfrak{L}_1(R)$. Prove that $\widehat{f'}(y) = iy\hat{f}(y)$ for all $y \in R$. $\Big[$Hints. Write $f(b) - f(a) = \int_a^b f'(x)\, dx$ and apply (12.30) to prove that $\lim_{b\to\infty} f(b) = \lim_{a\to-\infty} f(a) = 0$. Then write $\widehat{f'}(y) = \lim_{b\to\infty} (2\pi)^{-\frac{1}{2}} \int_{-b}^{b} f'(x) \exp(-iyx)\, dx$ and integrate by parts.$\Big]$

(21.60) Exercise. Define f on R by $f(x) = \exp\left(-\frac{x^2}{2}\right)$. Use (21.59.a) to prove that $\hat{f} = f$. $\Bigg[$Hints. Write $\hat{f} = \varphi$. By (21.59.a) we have

$$\varphi'(y) = i(2\pi)^{-\frac{1}{2}} \int\limits_R (-x) \exp\left(-\frac{x^2}{2}\right) \exp(-iyx)\, dx\,.$$

Integrate by parts to obtain $\varphi'(y) = -y\,\varphi(y)$ for all $y \in R$. Conclude that $\frac{d}{dy}\left[\varphi(y) \exp\left(\frac{y^2}{2}\right)\right] = 0$ for all $y \in R$. Show that $\varphi(0) = 1$ by noticing that

$$\begin{aligned}
2\pi\,\varphi(0)^2 &= \left(\int\limits_R \exp\left(-\frac{x^2}{2}\right) dx\right)\left(\int\limits_R \exp\left(-\frac{y^2}{2}\right) dy\right) \\
&= \int\limits_{R\times R} \exp\left(-\frac{(x^2+y^2)}{2}\right) d(x, y) \\
&= \int\limits_0^{2\pi} \int\limits_0^{\infty} \exp\left(-\frac{r^2}{2}\right) r\, dr\, d\theta = 2\pi. \Bigg]
\end{aligned}$$

(21.61) Exercise. (a) Let $\varphi \in \mathfrak{L}_1(R)$. Suppose that φ is twice differentiable on R, that φ', φ'' are in $\mathfrak{L}_1(R)$, and that φ and φ' are absolutely continuous on R. Prove that there exists a function $f \in \mathfrak{L}_1(R)$ such that $\hat{f} = \varphi$. [Use (21.59.b) to show that $\widehat{\varphi''}(y) = -y^2\,\hat{\varphi}(y)$ for all $y \in R$. Thus conclude that $\hat{\varphi} \in \mathfrak{L}_1(R)$, and then use the inversion theorem (21.49).]

(b) Prove that φ satisfies the hypothesis of (a) if φ, φ', and φ'' are all in $\mathfrak{L}_1(R) \cap \mathfrak{C}_0(R)$.

(21.62) Exercise. (a) Let F be a compact subset of R and let U be an open subset of R such that $F \subset U$. Prove that there exists a function $f \in \mathfrak{L}_1(R)$ such that $\hat{f}(y) = 1$ for all $y \in F$ and $\hat{f}(y) = 0$ for all $y \in R \cap U'$. [Use (21.61) or (21.54.a).]

(b) Let $\mathfrak{A}(R) = \{\hat{f} : f \in \mathfrak{L}_1(R)\}$. Prove that $\mathfrak{A}(R)$ is dense in $\mathfrak{C}_0(R)$ in the topology induced by the uniform norm. [Use part (a) and the STONE-WEIERSTRASS theorem.]

(21.63) Exercise. (a) For $a \geqq 0$, prove that

$$\int\limits_R \left[\frac{\sin(ay)}{y}\right]^4 dy = \frac{2a^3\pi}{3}\,.$$

$\Big[$Let $f(x) = \left(1 - \frac{|x|}{a}\right)\xi_{[-a,a]}$, compute $\hat{f}$, and apply PLANCHEREL'S theorem. Cf. (21.55).$\Big]$

(b) Compute

$$\int_R \left[\frac{\sin(ay)}{y}\right]^3 dy .$$

[Use (21.53.ii), (a), and (21.55).]

(c) Evaluate

$$\int_R \left[\frac{\sin(ay)}{y}\right]^n dy$$

for $a \in R$ and $n \in \{5, 6, 7, \ldots\}$.

(21.64) Exercise. Some rudimentary facts about analytic functions are needed in this exercise[1].

(a) Let f and g be functions in $\mathfrak{L}_1(R)$ such that $f(x) = g(x) = 0$ for all $x < 0$. Suppose that $f * g = 0$ a. e. Prove that $f = 0$ a. e. or $g = 0$ a. e. [Hints. Consider a complex number $z = s + it$ with $t \leqq 0$. The Fourier transform $\hat{f}$ can be extended to

$$\hat{f}(z) = (2\pi)^{-\frac{1}{2}} \int_0^\infty \exp(-izx) f(x)\, dx .$$

Show that $\hat{f}$ is an analytic function in $\{z : \operatorname{Im} z < 0\}$ and a continuous function in $\{z : \operatorname{Im} z \leqq 0\}$. Show that

$$\widehat{f * g}(z) = \hat{f}(z)\, \hat{g}(z) \quad \text{for} \quad \operatorname{Im} z \leqq 0 .$$

Thus the analytic function $\hat{f}\hat{g}$ vanishes identically in $\{z : \operatorname{Im} z < 0\}$, and this implies that $\hat{f} = 0$ or $\hat{g} = 0$, in $\{z : \operatorname{Im} z < 0\}$. If $\hat{f} = 0$, then $\hat{f}(s) = 0$ as well for all $s \in R$, and the uniqueness theorem (21.47) shows that $f = 0$ a. e.]

(b) Prove that the Hermite functions (16.25) are a complete orthonormal set in $\mathfrak{L}_2(R)$. [Hints. Let f be any element of $\mathfrak{L}_2(R)$. For all $z \in K$, let

$$F(z) = (2\pi)^{-\frac{1}{2}} \int_R \exp\left(-izx - \frac{x^2}{2}\right) \overline{f(x)}\, dx .$$

Show that $F(z)$ is defined for all $z \in K$, and that F is analytic in the entire z-plane. Show too that the nth derivative $F^{(n)}$ of F is given by

$$F^{(n)}(z) = (2\pi)^{-\frac{1}{2}} (-i)^n \int_R x^n \exp\left(-izx - \frac{x^2}{2}\right) \overline{f(x)}\, dx) .$$

[1] No knowledge of analytic functions is presupposed elsewhere in the text, although most readers will surely know the fundamentals of the subject. In any case, (21.64) is used nowhere else in the book.

If f is orthogonal to all of the Hermite functions, then we have

$$F^{(n)}(0) = (2\pi)^{-\frac{1}{2}}(-i)^n \int_R x^n \exp\left(-\frac{x^2}{2}\right) \overline{f(x)}\, dx = 0$$

for all $n \in \{0, 1, 2, \ldots\}$, and so F itself must vanish identically. For real z, this yields

$$0 = F(s + i\,0) = (2\pi)^{-\frac{1}{2}} \int_R \exp(-isx) \exp\left(-\frac{x^2}{2}\right) \overline{f(x)}\, dx$$

and so by (21.47), $\exp\left(-\frac{x^2}{2}\right) f(x) = 0$ a. e. Therefore $f = 0$ a. e.]

(21.65) Exercise: More on the structure of $\mathfrak{L}_1(R)$. In this exercise, we point out some algebraic properties of the Banach algebra $\mathfrak{L}_1(R)$ analogous to those obtained in (20.52) for $\mathfrak{C}_0(X)$.

(a) Let $\mathfrak{J}$ be a closed ideal in $\mathfrak{L}_1(R)$ [for the definition, see (20.52)]. Prove that $f \in \mathfrak{J}$ and $a \in R$ imply $f_a \in \mathfrak{J}$. [Hints. For $g, h \in \mathfrak{L}_1(R)$ and $a \in R$, check that

$$(g_a) * h = g * (h_a) = (g * h)_a .$$

Now if $(u^{(n)})$ is an approximate unit in $\mathfrak{L}_1(R)$ (21.36), the relations

$$u^{(n)} * (f_a) \to f_a, \quad (u_a^{(n)}) * f \in \mathfrak{J}$$

prove that $f_a \in \mathfrak{J}$.]

(b) Let M be a multiplicative linear functional on $\mathfrak{L}_1(R)$, *i. e.*, a nonzero linear functional on $\mathfrak{L}_1(R)$ for which $M(f * g) = M(f)\, M(g)$ for all $f, g \in \mathfrak{L}_1(R)$. Prove that $|M(f)| \leqq \|f\|_1$ for all $f \in \mathfrak{L}_1(R)$. [Repeat the proof sketched in (20.52) for $\mathfrak{C}_0(X)$.]

(c) Let M be as in part (b). Prove that $\{f \in \mathfrak{L}_1(R) : M(f) = 0\} = \mathfrak{Z}_M$ is a closed maximal ideal in $\mathfrak{L}_1(R)$.

(d) Let M be as in part (b), let x be any real number, and let f be in $\mathfrak{L}_1(R) \cap \mathfrak{Z}'_M$. Prove that the number $\chi(x) = M(f_x)/M(f)$ is independent of f, that the function $x \to \chi(x)$ is a continuous function on R such that $\chi(x + y) = \chi(x)\, \chi(y)$ for all $x, y \in R$, and that $|\chi| = 1$. [Hints. If $f, g \in \mathfrak{L}_1(R)$ and $x \in R$, we have $(f_x) * g = f * (g_x)$ and $M(f_x)\, M(g) = M(f)\, M(g_x)$. Thus

$$\frac{M(f_x)}{M(f)} = \frac{M(g_x)}{M(g)} .$$

Now consider $x, y \in R$. If $M(f) \neq 0$, then $M(f_x) \neq 0$, as parts (c) and (a) show. Hence:

$$\chi(x + y) = \frac{M(f_{x+y})}{M(f)} = \frac{M(f_{x+y})}{M(f_x)} \frac{M(f_x)}{M(f)} = \chi(y)\, \chi(x) .$$

Choose $f \in \mathfrak{L}_1(R)$ such that $M(f) = 1$. Then

$$\begin{aligned} |\chi(x + y) - \chi(x)| &= |M(f_{x+y}) - M(f_x)| \leqq \|f_{x+y} - f_x\|_1 \\ &= \|f_y - f\|_1 , \end{aligned}$$

and by (13.24) we have $\lim_{y\to 0} |\chi(x+y) - \chi(x)| = 0$. Thus χ is continuous. Also $|\chi(x)| = |M(f_x)| \leq \|f_x\|_1 = \|f\|_1$, so that χ is bounded. This implies that $|\chi| = 1$.]

(e) Let M and χ be as in parts (b) and (d). Then we have

$$M(f) = (2\pi)^{-\frac{1}{2}} \int_R \overline{\chi(x)}\, f(x)\, dx$$

for all $f \in \mathfrak{L}_1(R)$. [Hints. By (20.19), there is an $h \in \mathfrak{L}_\infty(R)$ such that $M(f) = (2\pi)^{-\frac{1}{2}} \int_R f h\, d\lambda$ for all $f \in \mathfrak{L}_1(R)$. Take $g \in \mathfrak{L}_1(R)$ such that $M(g) = 1$. Then for all $y \in R$,

$$\overline{\chi(y)} = M(g_{-y}) = (2\pi)^{-\frac{1}{2}} \int_R g(x-y)\, h(x)\, dx\,.$$

For an arbitrary $f \in \mathfrak{L}_1(R)$, we thus have

$$\begin{aligned} (2\pi)^{-\frac{1}{2}} \int_R f(y)\, \overline{\chi(y)}\, dy &= (2\pi)^{-1} \int_R \int_R g(x-y)\, h(x)\, dx\, f(y)\, dy \\ &= (2\pi)^{-1} \int_R \int_R g(x-y)\, f(y)\, dy\, h(x)\, dx \\ &= (2\pi)^{-\frac{1}{2}} \int_R g * f(x)\, h(x)\, dx \\ &= M(g * f) \\ &= M(f).] \end{aligned}$$

(f) Every multiplicative linear functional M on the Banach algebra $\mathfrak{L}_1(R)$ has the form

$$M(f) = (2\pi)^{-\frac{1}{2}} \int_R f(x) \exp(-iyx)\, dx$$

for some fixed $y \in R$. That is, the Fourier transform $\hat{f}$ of f describes the values at f of all multiplicative linear functionals. [Use part (e) and (18.46.a).]

(21.66) Note. The *closed* ideals of $\mathfrak{C}_0(X)$ are identified in (20.52.f). For $\mathfrak{L}_1(R)$ the closed ideals are far more complicated, and no complete description of them has as yet been found. An obvious class of such ideals is the following. For a closed set $F \subset R$, let $\mathfrak{I}_F = \{f \in \mathfrak{L}_1(R) : \hat{f}(y) = 0 \text{ for all } y \in F\}$. It is known that there are closed ideals in $\mathfrak{L}_1(R)$ not of this form, but little more can be said at present. For a discussion, see W. RUDIN, *Fourier Analysis on Groups* [Interscience Publishers, New York, 1962], Chapter 7.

We now turn to a quite different application of FUBINI's theorem, obtaining a formula for integration by parts under very general circumstances.

(21.67) Theorem [Integration by parts for Lebesgue-Stieltjes integrals]. *Let α and β be any two real-valued nondecreasing functions on R and let λ_α and λ_β be their corresponding Lebesgue-Stieltjes measures* (9.19). *Then $a < b$ in R implies*

(i) $\lambda_\alpha([a, b[) = \alpha(b-) - \alpha(a-)$;

(ii) $\lambda_\alpha(\{b\}) = \alpha(b+) - \alpha(b-)$;

(iii) $\lambda_\alpha([a, b]) = \alpha(b+) - \alpha(a-)$;

(iv) $$\int_{[a,b]} \beta(x+)\, d\lambda_\alpha(x) + \int_{[a,b]} \alpha(x-)\, d\lambda_\beta(x) = \alpha(b+)\,\beta(b+) - \alpha(a-)\,\beta(a-);$$

and

(v) $$\int_{[a,b]} \frac{\beta(x+) + \beta(x-)}{2}\, d\lambda_\alpha(x) + \int_{[a,b]} \frac{\alpha(x+) + \alpha(x-)}{2}\, d\lambda_\beta(x) = \alpha(b+)\,\beta(b+) - \alpha(a-)\,\beta(a-)\,.$$

Proof. For $x \in R$, define $\alpha_0(x) = \alpha(x-) - \alpha(0-)$. Then α_0 is a left continuous nondecreasing function on R such that $\alpha_0(0) = 0$; *i. e.*, α_0 is normalized in the sense of (19.46). Clearly $\alpha_0 = \alpha - \alpha(0-)$ except possibly at the [countably many] discontinuities of α. Thus (8.17) implies that the Riemann-Stieltjes integrals S_α and S_{α_0} agree on $\mathfrak{C}_{00}(R)$; hence (9.19) shows that $\lambda_\alpha = \lambda_{\alpha_0}$. It follows from (19.47) that $\lambda_\alpha([a, b[) = \lambda_{\alpha_0}([a, b[) = a_0(b) - \alpha_0(a) = \alpha(b-) - \alpha(a-)$; hence (i) holds. From (19.52) we have

$$\lambda_\alpha(\{b\}) = \lambda_{\alpha_0}(\{b\}) = \lim_{h \downarrow 0} \alpha_0(b + h) - \alpha_0(b) = \lim_{h \downarrow 0} \alpha((b + h) -) - \alpha(b-) = \alpha(b+) - \alpha(b-);$$

hence (ii) holds. We obtain (iii) by adding equalities (i) and (ii).

To prove (iv), let $E = \{(x, y) \in [a, b] \times [a, b] : y \leqq x\}$. Since E is compact, we have $E \in \mathscr{B}(R^2) = \mathscr{B}(R) \times \mathscr{B}(R)$. Applying (21.8.iii), we obtain

$$\int_{[a,b]} \lambda_\beta(E_x)\, d\lambda_\alpha(x) = \int_{[a,b]} \lambda_\alpha(E^y)\, d\lambda_\beta(y)\,. \tag{1}$$

Since $E_x = [a, x]$ and $E^y = [y, b]$ for all $x, y \in [a, b]$, (iii) applied to (1) [(iii) obviously holds for β as well as α] yields

$$\begin{aligned}\int_{[a,b]} \beta(x+)\, d\lambda_\alpha(x) - \beta(a-)[\alpha(b+) - \alpha(a-)] \\ &= \int_{[a,b]} [\beta(x+) - \beta(a-)]\, d\lambda_\alpha(x) \\ &= \int_{[a,b]} [\alpha(b+) - \alpha(y-)]\, d\lambda_\beta(y) \\ &= \alpha(b+)\,[\beta(b+) - \beta(a-)] - \int_{[a,b]} \alpha(y-)\, d\lambda_\beta(y)\,. \qquad (2)\end{aligned}$$

Now replace y by x in (2) and rearrange the terms to obtain (iv). To get (v), simply interchange the rôles of α and β in (iv), add this new equality to (iv), and then divide by 2. □

(21.68) Remarks. A theorem more general than (21.67) can be formulated by allowing α and β to be any functions that are of finite variation on each bounded interval of R and considering the corresponding signed or complex Lebesgue-Stieltjes measures. All that is involved is to reduce the more general case to that of (21.67) by invoking the Jordan decomposition theorem (17.16). We leave this to the reader. In case α and β are both absolutely continuous functions, *i. e.*, $\lambda_\alpha \ll \lambda$ and $\lambda_\beta \ll \lambda$, (21.67) reduces to (18.19).

(21.69) Theorem [First mean value theorem for integrals]. *Let μ be any finite [nonnegative, countably additive] Borel measure on $[a, b]$ and let $f \in \mathfrak{C}^r([a, b])$. Then there exists a real number ξ such that $a < \xi < b$ and*

$$\int_{[a,b]} f(x)\, d\mu(x) = f(\xi)\, \mu([a, b])\,.$$

Proof. This follows immediately from the fact that $f([a, b])$ is a closed interval [perhaps a single point] and from the obvious inequalities

$$\min\{f(x) : x \in [a, b]\} \leqq \frac{1}{\mu([a,b])} \int_{[a,b]} f(x)\, d\mu(x) \leqq \max\{f(x) : x \in [a, b]\}\,. \quad \square$$

(21.70) Theorem [Second mean value theorem for integrals]. *Let α and β be real-valued nondecreasing functions on $[a, b]$. Suppose that β is continuous and let λ_β be the Lebesgue-Stieltjes measure corresponding to β (9.19). Then there exists a $\xi \in\,]a, b[$ such that*

(i) $\displaystyle\int_{[a,b]} \alpha(x)\, d\lambda_\beta(x) = \alpha(a)[\beta(\xi) - \beta(a)] + \alpha(b)[\beta(b) - \beta(\xi)].$

Proof. Let $\alpha(x) = \alpha(a)$ and $\beta(x) = \beta(a)$ for all $x < a$, and let $\alpha(x) = \alpha(b)$ and $\beta(x) = \beta(b)$ for all $x > b$. Let λ_α and λ_β be the Lebesgue-Stieltjes measures corresponding to α and β. Then $\alpha(a-) = \alpha(a)$, $\alpha(b+) = \alpha(b)$, and $\beta(x+) = \beta(x-) = \beta(x)$ for all $x \in R$, and so (21.67.v) becomes

$$\int_{[a,b]} \beta(x)\, d\lambda_\alpha(x) + \int_{[a,b]} \frac{\alpha(x+) + \alpha(x-)}{2}\, d\lambda_\beta(x) = \alpha(b)\,\beta(b) - \alpha(a)\,\beta(a) \qquad (1)$$

By (21.69), there exists a $\xi \in \,]a, b[$ such that

$$\int_{[a,b]} \beta(x)\, d\lambda_\alpha(x) = \beta(\xi)\,\lambda_\alpha([a, b]) = \beta(\xi)[\alpha(b) - \alpha(a)]\,. \qquad (2)$$

Since β is continuous, we have $\lambda_\beta(\{x\}) = 0$ for all x (19.52), and so

$$\int_{[a,b]} \frac{\alpha(x+) + \alpha(x-)}{2}\, d\lambda_\beta(x) = \int_{[a,b]} \alpha(x)\, d\lambda_\beta(x) \qquad (3)$$

because the two integrands differ only on a countable set.

Applying (2) and (3) to (1), we obtain

$$\beta(\xi)[\alpha(b) - \alpha(a)] + \int_{[a,b]} \alpha(x)\, d\lambda_\beta(x) = \alpha(b)\,\beta(b) - \alpha(a)\,\beta(a)\,. \qquad (4)$$

Plainly (i) follows at once from (4). □

Our next application of FUBINI's theorem is of interest in its own right and also is needed for yet another application that we have in mind [Theorems (21.76) and (21.80) *infra*].

(21.71) Theorem. *Let $(X, \mathscr{M}, \mu)$ be a σ-finite measure space, let f be a nonnegative, real-valued, $\mathscr{M}$-measurable function on X, and let E be any set in $\mathscr{M}$. Let φ be a real-valued nondecreasing function with domain $[0, \infty[$ that is absolutely continuous on every interval $[0, a]$ for $a > 0$. Suppose also that $\varphi(0) = 0$. For $t \geqq 0$, let $G_t = \{x \in X : f(x) > t\}$. Then*

(i) $$\int_E \varphi \circ f(x)\, d\mu(x) = \int_0^\infty \mu(E \cap G_t)\, \varphi'(t)\, dt\,.$$

Proof. Using (18.16), we see that

$$\begin{aligned}\int_E \varphi \circ f(x)\, d\mu(x) &= \int_X \xi_E(x)\, (\varphi \circ f)(x)\, d\mu(x)\\ &= \int_X \xi_E(x) \int_0^{f(x)} \varphi'(t)\, dt\, d\mu(x)\\ &= \int_X \xi_E(x) \int_0^\infty \xi_{[0,f(x)[}(t)\, \varphi'(t)\, dt\, d\mu(x)\,. \qquad (1)\end{aligned}$$

The function $(x, t) \to \xi_{[0,f(x)[}(t)$ on $X \times [0, \infty[$ is the characteristic function of the set $\{(x, t) : f(x) > t\}$. This set is $V_* f$, defined as in (21.23). Hence it is $\mathscr{M} \times \mathscr{M}_\lambda$-measurable, and so the function

$$(x, t) \to \xi_E(x)\, \xi_{[0,f(x)[}(t)\, \varphi'(t)$$

on $X \times [0, \infty[$ is nonnegative and $\mathscr{M} \times \mathscr{M}_\lambda$-measurable, so that we may apply (21.12) to the last integral in (1). This yields

$$\int_X \xi_E(x) \int_0^\infty \xi_{[0,f(x)[}(t)\, \varphi'(t)\, dt\, d\mu(x) = \int_X \int_0^\infty \xi_E(x)\, \xi_{[0,f(x)[}(t)\, \varphi'(t)\, dt\, d\mu(x)$$
$$= \int_0^\infty \int_X \xi_E(x)\, \xi_{[0,f(x)[}(t)\, d\mu(x)\, \varphi'(t)\, dt$$
$$= \int_0^\infty \int_X \xi_E(x)\, \xi_{G_t}(x)\, d\mu(x)\, \varphi'(t)\, dt$$
$$= \int_0^\infty \mu(E \cap G_t)\, \varphi'(t)\, dt\,. \quad \square$$

(21.72) Corollary. *For* $(X, \mathscr{M}, \mu)$, E, f, *and* G_t *as in* (21.71) *and for* $p > 0$, *we have*

(i) $\int_E f^p\, d\mu = \int_0^\infty p t^{p-1} \mu(E \cap G_t)\, dt\,.$

Proof. Set $\varphi(t) = t^p$ in (21.71.i). $\square$

(21.73) Note. For $p = 1$, the equality (21.72.i) serves as a *definition* of $\int_E f\, d\mu$ if the Lebesgue integral on $[0, \infty[$ is known. This technique was used by J. Radon [*Theorie und Anwendungen der absolut additiven Mengenfunktionen*, Sitzungsberichte Akad. Wissenschaften Wien 122, 1295–1438 (1913)] to define Lebesgue-Stieltjes integrals. Note also that (21.72.i) holds for $f \in \mathfrak{L}_p(X, \mathscr{M}, \mu)$ even if X is not σ-finite, as the function f vanishes outside of a certain σ-finite $\mathscr{M}$-measurable set.

Our final application of Fubini's theorem is to the proof of a famous theorem due to the English mathematicians G. H. Hardy (1877–1947) and J. E. Littlewood (1885–).

(21.74) Notation and Definitions. We will adhere to the following notation and definitions throughout (21.74)–(21.83). First, f is a nonnegative, extended real-valued, Lebesgue measurable function on R such that $\int_F f\, d\lambda < \infty$ for all compact sets F. Define functions $f^{\Delta(r)}$, $f^{\Delta(l)}$, and f^{Δ} by the following rules:

$$f^{\Delta(r)}(x) = \sup\left\{\frac{1}{u-x}\int_x^u f\, d\lambda : u \in]x, \infty[\right\};$$

$$f^{\Delta(l)}(x) = \sup\left\{\frac{1}{x-u}\int_u^x f\, d\lambda : u \in]-\infty, x[\right\};$$

$$f^{\Delta}(x) = \max\{f^{\Delta(r)}(x), f^{\Delta(l)}(x)\}\,.$$

For each $t > 0$, let

$$G_t = \{x : f(x) > t\};$$

$$M_t^{(j)} = \{x : f^{\Delta(j)}(x) > t\} \quad (j = l, r);$$

and

$$M_t = \{x : f^{\Delta}(x) > t\}\,.$$

(21.75) Lemma. *The equality*

(i) $\lambda(M_t^{(j)}) = \frac{1}{t}\int\limits_{M_t^{(j)}} f\,d\lambda \quad (j = r, l)$

and the inequality

(ii) $\lambda(M_t) \leqq \frac{2}{t}\int\limits_{M_t} f\,d\lambda$

hold for every $t > 0$.

Proof. We prove (i) only for $j = r$, the case $j = l$ being almost the same. It is easy to see that the set $M_t^{(r)}$ is open, since the function $s \to \frac{1}{s-x}\int\limits_x^s f\,d\lambda$ is continuous on $]x, \infty[$. Let $\{]\beta_k, \gamma_k[\}_{k=1}^{\infty}$ be the unique family of pairwise disjoint intervals such that

$$M_t^{(r)} = \bigcup_{k=1}^{\infty}]\beta_k, \gamma_k[$$

(6.59). Consider an interval $]\beta_k, \gamma_k[$ [which may of course be unbounded]. For each $x \in]\beta_k, \gamma_k[$, the open set

$$N_x = \left\{s : \int\limits_x^s f\,d\lambda > t(s-x),\ s \in]x, \gamma_k] \cap R\right\}$$

is nonvoid. This is trivial if $\gamma_k = \infty$. If $\gamma_k < \infty$ and N_x is void for some $x \in]\beta_k, \gamma_k[$, there must be a $w > \gamma_k$ such that $\int\limits_x^w f\,d\lambda > t(w-x)$. We have

$$\int\limits_{\gamma_k}^{w} f\,d\lambda = \int\limits_x^w f\,d\lambda - \int\limits_x^{\gamma_k} f\,d\lambda > t(w-x) - t(\gamma_k - x) = t(w - \gamma_k)\,.$$

This inequality implies that $\gamma_k \in M_t^{(r)}$, a contradiction. Let $s_x = \sup N_x$. We will prove that $s_x = \gamma_k$. If $s_x < \gamma_k$, then the equality $\int\limits_x^{s_x} f\,d\lambda = t(s_x - x)$ holds; an obvious continuity argument proves this. The set N_{s_x} is nonvoid, so there is a real number $y \in]s_x, \gamma_k]$ such that $\int\limits_{s_x}^{y} f\,d\lambda > t(y - s_x)$. It follows that $\int\limits_x^y f\,d\lambda > t(y-x)$, a contradiction since $y > s_x$. Hence for all $x \in]\beta_k, \gamma_k[$, we have $s_x = \gamma_k$, and so the inequality

$$\int\limits_x^{\gamma_k} f\,d\lambda \geqq t(\gamma_k - x)$$

holds. Letting $x \to \beta_k$, we obtain

$$\int\limits_{\beta_k}^{\gamma_k} f\,d\lambda \geqq t(\gamma_k - \beta_k)\,.$$

If $\beta_k = -\infty$ or $\gamma_k = \infty$, the equality (i) follows. If $]\beta_k, \gamma_k[$ is bounded, we have

$$\int\limits_{\beta_k}^{\gamma_k} f\,d\lambda \leqq t(\gamma_k - \beta_k)\,,$$

since β_k is not in $M_t^{(r)}$. Hence in all cases we have

$$\int_{\beta_k}^{\gamma_k} f\,d\lambda = t(\gamma_k - \beta_k)\,.$$

The equality (i) follows.

To prove (ii), note that $M_t = M_t^{(r)} \cup M_t^{(l)}$. Hence we have

$$\begin{aligned}\lambda(M_t) &\leqq \lambda(M_t^{(r)}) + \lambda(M_t^{(l)}) \\ &= \frac{1}{t}\Big[\int_{M_t^{(r)}} f\,d\lambda + \int_{M_t^{(l)}} f\,d\lambda\Big] \\ &\leqq \frac{2}{t}\int_{M_t} f\,d\lambda.\ \square\end{aligned}$$

(21.76) Hardy-Littlewood Maximal Theorem for $\mathfrak{L}_p\,(p > 1)$. *Let p be a real number > 1. Notation is as in* (21.74). *Then*

(i) $$\Big[\int_R (f^{\Delta(j)})^p\,d\lambda\Big]^{\frac{1}{p}} \leqq \frac{p}{p-1}\Big[\int_R f^p\,d\lambda\Big]^{\frac{1}{p}}\ (j = r, l)$$

and

(ii) $$\Big[\int_R (f^{\Delta})^p\,d\lambda\Big]^{\frac{1}{p}} \leqq \frac{2p}{p-1}\Big[\int_R f^p\,d\lambda\Big]^{\frac{1}{p}}.$$

Proof. We use (21.72), (21.75.i), Fubini's theorem (21.12), and Hölder's inequality (13.4) [in that order] to calculate as follows:

$$\begin{aligned}\int_R (f^{\Delta(j)}(x))^p\,dx &= \int_0^\infty p\,t^{p-1}\,\lambda(M_t^{(j)})\,dt \\ &= p\int_0^\infty t^{p-2}\int_{M_t^{(j)}} f(x)\,dx\,dt \\ &= p\int_R\int_0^\infty \xi_{M_t^{(j)}}(x)\,f(x)\,t^{p-2}\,dt\,dx \\ &= p\int_R\int_0^\infty \xi_{]0,f^{\Delta(j)}(x)[}(t)\,t^{p-2}\,f(x)\,dt\,dx \\ &= p\int_R f(x)\,\frac{[f^{\Delta(j)}(x)]^{p-1}}{p-1}\,dx \\ &\leqq \frac{p}{p-1}\Big[\int_R f(x)^p\,dx\Big]^{\frac{1}{p}}\Big[\int_R (f^{\Delta(j)}(x))^{p'(p-1)}\,dx\Big]^{\frac{1}{p'}} \\ &= \frac{p}{p-1}\Big[\int_R f(x)^p\,dx\Big]^{\frac{1}{p}}\Big[\int_R (f^{\Delta(j)}(x))^p\,dx\Big]^{\frac{1}{p'}}.\end{aligned}$$

Since $1-\frac{1}{p'}=\frac{1}{p}$, the inequality (i) follows if $f^{\Delta(j)}\in\mathfrak{L}_p$. To check this, use (21.79.i) [which depends only upon (21.75)] to write

$$\int\limits_R (f^{\Delta(j)}(x))^p\,dx \leq \frac{p}{1-k}\int\limits_0^\infty t^{p-2}\int\limits_{G_{kt}} f(x)\,dx\,dt.$$

Then argue as above to obtain

$$\int\limits_R (f^{\Delta(j)}(x))^p\,dx \leq \frac{p\,k^{1-p}}{(p-1)\,(1-k)}\int\limits_R (f(x))^p\,dx.$$

The use of FUBINI's theorem is justified because the inverse image of the set $]a,\infty]$ under the map $(x,t)\to\xi_{M_t^{(j)}}(x)$ is $\varnothing$ if $a\geqq 1$, is $R\times[0,\infty[$ if $a<0$, and is $\{(x,t): f^{\Delta(j)}(x)>t\}$ if $0\leqq a<1$. Each of these sets is product measurable. A like calculation, based on (21.75.ii), proves (ii). □

The preceding theorem is ordinarily stated with R replaced by an interval and with smaller functions $f^{\Delta(r)}$, $f^{\Delta(l)}$, and f^{Δ}. This case is contained in the following corollary.

(21.77) Corollary. *Let f be as in* (21.74), *and suppose further that E is a Lebesgue measurable set such that $f(E')=\{0\}$. For $p>1$, we have*

(i) $\displaystyle\int\limits_E (f^{\Delta(j)}(x))^p\,dx \leq \left(\frac{p}{p-1}\right)^p\int\limits_E f(x)^p\,dx$

for $j=r,l$; and

(ii) $\displaystyle\int\limits_E f^{\Delta}(x)^p\,dx \leq \left(\frac{2p}{p-1}\right)^p\int\limits_E f(x)^p\,dx.$

Proof. Since $\int\limits_E g\,d\lambda \leq \int\limits_R g\,d\lambda$ for nonnegative g, the result is an immediate consequence of (21.76). □

(21.78) Remark. If the set E of (21.77) is contained in an interval $[\alpha,\beta]$, then it is clear that

$$f^{\Delta(r)}(x)=\sup\left\{\frac{1}{t-x}\int\limits_x^t f\,d\lambda : x<t\leqq\beta\right\}$$

if $\alpha<x\leqq\beta$. This is the customary definition of $f^{\Delta(r)}$ in the case of an interval; similarly with $f^{\Delta(l)}$ and f^{Δ}.

There is also a version of the HARDY-LITTLEWOOD maximal theorem for functions in $\mathfrak{L}_1$. As frequently happens [for some mysterious reason], less is true in the $\mathfrak{L}_1$ case than in the $\mathfrak{L}_p$ case for $p>1$. We need a lemma, as follows.

(21.79) Lemma. *Notation is as in* (21.74). *For all $k\in\,]0,1[$ and all $t>0$, we have*

(i) $\displaystyle\lambda(M_t^{(j)}) \leq \frac{1}{(1-k)\,t}\int\limits_{G_{kt}} f\,d\lambda \quad (j=r,l)$, *and*

(ii) $\lambda(M_t) \leqq \frac{2}{(1-k)\,t} \int\limits_{G_{kt}} f\,d\lambda$.

Proof. Define a function g on R by

$$g(x) = \begin{cases} f(x) & \text{if } f(x) > kt, \\ 0 & \text{otherwise}. \end{cases}$$

We have

$$f^{\Delta(r)}(x) = \sup\left\{\frac{1}{y-x}\int\limits_x^y g\,d\lambda + \frac{1}{y-x}\int\limits_{]x,y[\cap G_{kt}} f\,d\lambda : y > x\right\}$$
$$\leqq g^{\Delta(r)}(x) + kt.$$

Write $N_u = \{x : g^{\Delta(r)}(x) > u\}$ $(u > 0)$. It is plain that

$$M_t^{(r)} \subset N_{(1-k)t}.$$

Applying (21.75.i) to the function g, we obtain

$$\lambda(M_t^{(r)}) \leqq \frac{1}{(1-k)\,t}\int\limits_{N_{(1-k)t}} g\,d\lambda. \tag{1}$$

Since $g = 0$ on $\{x : f(x) \leqq kt\}$, the integral on the right side of (1) is

$$\int\limits_{N_{(1-k)t}\cap G_{kt}} f\,d\lambda;$$

it follows that

$$\lambda(M_t^{(r)}) \leqq \frac{1}{(1-k)\,t}\int\limits_{G_{kt}} f\,d\lambda.$$

Thus (i) is established for $j = r$; it is clear how to obtain (i) for $j = l$ and also how to prove (ii). □

(21.80) Hardy-Littlewood Maximal Theorem for $\mathfrak{L}_1$. *Let f be a function as in* (21.74) *and let E be any set in $\mathscr{M}_\lambda$. For each k such that $0 < k < 1$, we have*

(i) $\int\limits_E f^{\Delta(j)}\,d\lambda \leqq \frac{1}{k}\lambda(E) + \frac{1}{1-k}\int\limits_R f(x)\log^+ f(x)\,dx$ [1] $(j = r, l)$,

(ii) $\int\limits_E f^{\Delta}\,d\lambda \leqq \frac{1}{k}\lambda(E) + \frac{2}{1-k}\int\limits_R f(x)\log^+ f(x)\,dx$.

For $0 < p < 1$, we have

(iii) $\int\limits_E (f^{\Delta(j)})^p\,d\lambda \leqq \frac{\lambda(E)^{1-p}}{1-p}\left(\int\limits_R f\,d\lambda\right)^p$ $(j = r, l)$

and

(iv) $\int\limits_E (f^{\Delta})^p\,d\lambda \leqq 2^p\,\frac{\lambda(E)^{1-p}}{1-p}\left(\int\limits_R f\,d\lambda\right)^p$.

---

[1] Recall that $\log^+ t = \max\{\log t, 0\}$ $(0 < t < \infty)$: see (13.37).

Proof. To prove (i), we use (21.72) and (21.79) to compute as follows:

$$\int_E f^{\Delta(j)}\,d\lambda = \int_0^\infty \lambda(\{y : \xi_E f^{\Delta(j)}(y) > t\})\,dt$$

$$= \int_0^\infty \lambda(M_t^{(j)} \cap E)\,dt$$

$$= \int_0^{1/k} + \int_{1/k}^\infty \leq \frac{1}{k}\lambda(E) + \frac{1}{1-k}\int_{1/k}^\infty \frac{1}{t}\int_{G_{kt}} f(x)\,dx\,dt$$

$$= \frac{\lambda(E)}{k} + \frac{1}{1-k}\int_{1/k}^\infty \frac{1}{t}\int_R \xi_{G_{kt}}(x)\,f(x)\,dx\,dt$$

$$= \frac{\lambda(E)}{k} + \frac{1}{1-k}\int_R f(x)\left\{\int_{1/k}^\infty \xi_{G_{kt}}(x)\,\frac{1}{t}\,dt\right\}dx\,. \qquad (1)$$

The integral $\{\cdots\}$ in the last line of (1) is equal to

$$\int_{1/k}^{f(x)/k} \frac{1}{t}\,dt = \log f(x)$$

if $f(x) > 1$ and is 0 if $f(x) \leq 1$ [use elementary calculus or, if you like, (20.5)]. Thus the last line of (1) is equal to

$$\frac{\lambda(E)}{k} + \frac{1}{1-k}\int_R f(x)\log^+ f(x)\,dx\,. \qquad (2)$$

From (1) and (2), (i) is immediate. To prove (ii), use (21.79.ii) instead of (21.79.i) in (1).

To prove (iii), a slightly different argument is required. Let α be any positive real number. We may suppose that $\lambda(E) > 0$ and $\int_R f\,d\lambda < \infty$. Then, using (21.72), we write

$$\int_E (f^{\Delta(j)})^p\,d\lambda = p\int_0^\infty t^{p-1}\lambda(M_t^{(j)} \cap E)\,dt$$

$$= p\int_0^{\alpha/k} + p\int_{\alpha/k}^\infty$$

$$\leq \lambda(E)\,\frac{\alpha^p}{k^p} + \frac{p}{1-k}\int_{\alpha/k}^\infty t^{p-2}\int_{G_{kt}} f(x)\,dx\,dt$$

$$= \frac{\alpha^p}{k^p}\lambda(E) + \frac{p}{1-k}\int_R f(x)\left\{\int_{\alpha/k}^\infty t^{p-2}\,\xi_{G_{kt}}(x)\,dt\right\}dx\,. \qquad (3)$$

Corollary (20.5) shows that

$$\int_{\alpha/k}^{\infty} t^{p-2}\,\xi_{G_{kt}}(x)\,dt = \left(\frac{1}{k}\right)^{p-1}\int_{\alpha}^{\infty} s^{p-2}\,\xi_{G_s}(x)\,ds\,,$$

and it is easy to verify that

$$\int_{\alpha}^{\infty} s^{p-2}\,\xi_{G_s}(x)\,ds = \begin{cases}\frac{1}{p-1}\big((f(x))^{p-1}-\alpha^{p-1}\big) & \text{if} \quad f(x)>\alpha\,,\\ 0 & \text{if} \quad f(x)\leqq\alpha\,.\end{cases}$$

Since p is less than 1, the last line of (3) is therefore equal to

$$\frac{\alpha^p}{k^p}\,\lambda(E)+\frac{p}{1-p}\,\frac{1}{k^{p-1}(1-k)}\int_R f(x)\,\max\{0,\,\alpha^{p-1}-f(x)^{p-1}\}\,dx \tag{4}$$

and in turn (4) does not exceed

$$\frac{1}{k^p}\,\lambda(E)\,\alpha^p+\left(\frac{p}{(1-p)\,k^{p-1}(1-k)}\int_R f\,d\lambda\right)\alpha^{p-1}\,. \tag{5}$$

Regarding (5) as a function of α, we see that it has exactly one minimum value, attained at $\alpha=\frac{k}{1-k}\,(\lambda(E))^{-1}\int_R f\,d\lambda$. The value of (5) for this α is

$$\frac{1}{(1-k)^p}\,\frac{\lambda(E)^{1-p}}{1-p}\left(\int_R f\,d\lambda\right)^p,$$

so that for each k such that $0<k<1$ we have

$$\int_E (f^{\Delta(j)})^p\,d\lambda \leqq \frac{1}{(1-k)^p}\,\frac{\lambda(E)^{1-p}}{1-p}\left(\int_R f\,d\lambda\right)^p.$$

Letting $k\to 0$, we obtain (iii). Obvious modifications in the above proof, using (21.79.ii), yield (iv). □

(21.81) Exercise. Let

$$f(x)=\begin{cases}\frac{1}{x(\log x)^2} & \text{if} \quad x\in\left]0,\frac{1}{2}\right[\,,\\ 0 & \text{otherwise}\,.\end{cases}$$

Prove that $f\in\mathfrak{L}_1(R)$, but that $f^{\Delta(l)}\,\xi_{]0,\frac{1}{2}[}\notin\mathfrak{L}_1(R)$. [Hint. Show that $f^{\Delta(l)}(x)\geqq\frac{1}{x\,|\log x|}$ for $x\in\left]0,\frac{1}{2}\right[$.]

(21.82) Exercise [T. M. Flett]. For a function $f\in\mathfrak{L}_p^+(R)$, $p>1$, let $A^f(p)$ denote the number such that $\|f^{\Delta(l)}\|_p=A^f(p)\,\|f\|_p$. Define a sequence in $\mathfrak{L}_p(R)$ by

$$f_n(x)=\begin{cases}x^{(n^{-1}-1)\,p^{-1}} & \text{if} \quad x\in\,]0,1[\,,\\ 0 & \text{otherwise.}\end{cases}$$

Prove that $\lim\limits_{n\to\infty} A^{f_n}(p) = \frac{p}{p-1}$, thus proving that the constant $\frac{p}{p-1}$ in (21.76.i) is the best possible.

(21.83) Exercise [K. L. PHILLIPS]. Let f be as in (21.74). Prove that

$$f^{\Delta}(x) = \sup\left\{\frac{1}{u-t}\int\limits_t^u f\,d\lambda : -\infty < t \leqq x \leqq u < \infty,\ t \neq u\right\}.$$

§ 22. Products of infinitely many measure spaces

(22.1) Introductory Remarks. If one flips a coin n times, the number of possible outcomes is 2^n, the number of n-tuples having 0 and 1 as entries; or, it is the number of functions from $\{1, 2, 3, \ldots, n\}$ into $\{0, 1\}$. It is intuitively obvious that any two of these functions are equally likely to occur if the coin is unbiased. Thus it is just as likely that all n flips will yield heads as that they will alternate from heads to tails to heads .... It seems also intuitively clear that "in the long run" [as n goes to ∞] it is probable that the number of heads obtained is about one half the total number of tosses. We can interpret the expression "is probable that" to mean that for some appropriately chosen measure μ on $\{0, 1\}^N$ [= all sequences $\boldsymbol{t} = (t_1, t_2, \ldots, t_n, \ldots)$, where t_j is 0 or 1 for all j], we have

$$\mu\left(\left\{\boldsymbol{t} : \lim_{n\to\infty} \frac{1}{n}\left(\sum_{k=1}^{n} t_k\right) = \frac{1}{2}\right\}\right) = 1\,.$$

Perhaps the most convenient way to study probability measures such as the one indicated here is to consider *infinite* products of measure spaces, and this fact is a sufficient reason for including the present section. Furthermore, many important and useful constructions in measure theory and its applications throughout analysis depend upon measures on infinite products of measure spaces. The subject is too important to be ignored.

The ideas used in the study of infinite product measures are not at all recondite, but the notation is complicated and, it may be, a little forbidding. The reader should be sure to keep the notation of (22.2) firmly in mind throughout.

(22.2) Definitions and Notation. Throughout this section $(T_\gamma, \mathscr{M}_\gamma, \mu_\gamma)$ will denote a measure space for each γ contained in an index set Γ, where $\overline{\overline{\Gamma}} \geqq \aleph_0$ and $\mu_\gamma(T_\gamma) = 1$ for every $\gamma \in \Gamma$. Let $T = \mathop{\times}\limits_{\gamma\in\Gamma} T_\gamma$; we will often write $\boldsymbol{t} = (t_\gamma)$ for elements of T, where t_γ denotes the value of $\boldsymbol{t}$ at γ. The symbol Ω, with or without a subscript, will be reserved for *finite* subsets of Γ. The complement Ω' will always be with respect to

Γ: $\Omega' = \Gamma \cap \Omega'$. For any set Δ such that $\varnothing \neq \Delta \subset \Gamma$, we let $T_\Delta = \bigtimes_{\gamma \in \Delta} T_\gamma$ [thus in particular $T_\Gamma = T$]. Note that T_Δ is *not* a subset of T if $\Delta \subsetneqq \Gamma$. For $\Omega = \{\gamma_1, \gamma_2, \ldots, \gamma_m\} \subset \Gamma$ [the γ_j's are distinct], let $\mathscr{E}_\Omega$ be the family of all sets

$$A_{\gamma_1} \times A_{\gamma_2} \times \cdots \times A_{\gamma_m} \ (A_{\gamma_j} \in \mathscr{M}_{\gamma_j}) .$$

These are the analogues of measurable rectangles for the product of two spaces. For an arbitrary subset Δ of Γ, let $\mathscr{N}_\Delta$ be the smallest *algebra* [*not* σ-algebra] of subsets of T_Δ that contains all sets $A_\Omega \times T_{\Delta \cap \Omega'}$, where Ω runs through all finite subsets of Δ and A_Ω through all sets in $\mathscr{E}_\Omega$.[1] Let $\mathscr{M}_\Delta$ be the σ-algebra $\mathscr{S}(\mathscr{N}_\Delta)$. We write $\mathscr{N}$ for $\mathscr{N}_\Gamma$ and $\mathscr{M}$ for $\mathscr{M}_\Gamma$. Note that for finite Ω, the σ-algebra $\mathscr{M}_\Omega$ is just $\mathscr{S}(\mathscr{E}_\Omega)$.

(22.3) Discussion. Our aim is to construct a [countably additive] measure μ on $\mathscr{M}$ such that

(i) $\mu(T) = 1$

and

(ii) $\mu((A_{\gamma_1} \times \cdots \times A_{\gamma_n}) \times T_{\{\gamma_1, \gamma_2, \ldots, \gamma_n\}'})$
$= \mu_{\gamma_1}(A_{\gamma_1})\, \mu_{\gamma_2}(A_{\gamma_2}) \cdots \mu_{\gamma_n}(A_{\gamma_n})$

if $A_{\gamma_j} \in \mathscr{M}_{\gamma_j}$ $(j = 1, 2, \ldots, n)$. We will then prove two analogues of FUBINI's theorem for this *product measure*. We begin with a technical lemma.

(22.4) Lemma. *Let $\{\Delta_\varrho\}_{\varrho \in P}$ be a pairwise disjoint family of nonvoid subsets of Γ such that $\bigcup_{\varrho \in P} \Delta_\varrho = \Gamma$. Then the mapping*

(i) $(t_\gamma) \to ((t_\gamma)_{\gamma \in \Delta_\varrho})_{\varrho \in P} = \Phi(\boldsymbol{t})$

is a one-to-one mapping of T onto the product space

$$\bigtimes_{\varrho \in P} T_{\Delta_\varrho} = T^\dagger .$$

The mapping Φ carries $\mathscr{M}$ onto the smallest σ-algebra $\mathscr{M}^\dagger$ of subsets of $T^\dagger$ that contains all sets

$$\bigtimes_{\varrho \in P} A_{\Delta_\varrho} ,$$

where $A_{\Delta_\varrho} \in \mathscr{M}_{\Delta_\varrho}$ for all $\varrho \in P$ and only a finite number of the sets A_{Δ_ϱ} are different from T_{Δ_ϱ}.

Proof. The first assertion of the lemma is obvious on a little reflection: the mapping Φ merely "regroups" the terms t_γ in $(t_\gamma)_{\gamma \in \Gamma}$. To prove the second assertion, consider first a set

$$A = \bigtimes_{\gamma \in \Gamma} A_\gamma \subset T , \tag{1}$$

[1] We commit a slight inaccuracy in writing $A_\Omega \times T_{\Delta \cap \Omega'}$ as a subset of T_Δ, although the intended meaning is clear enough. See (22.4) for a justification.

where $A_\gamma \in \mathscr{M}_\gamma$ for all γ and the set $\Omega = \{\gamma \in \Gamma : A_\gamma \neq T_\gamma\}$ is finite. Plainly $\Phi(A)$ is the set

$$B = \Phi(A) = \underset{\varrho \in P}{\times}\Big(\underset{\gamma \in \Delta_\varrho}{\times} A_\gamma\Big). \tag{2}$$

Since Ω is finite, all but a finite number of the sets $\underset{\gamma \in \Delta_\varrho}{\times} A_\gamma$ are equal to T_{Δ_ϱ}, and it is plain that each $\underset{\gamma \in \Delta_\varrho}{\times} A_\gamma$ is in $\mathscr{N}_{\Delta_\varrho} \subset \mathscr{M}_{\Delta_\varrho}$. Therefore $\Phi(A)$ is in $\mathscr{M}^\dagger$. Obviously $\mathscr{M}$ is the smallest σ-algebra of subsets of T containing all sets A of the form (1). Since Φ is one-to-one, it preserves all Boolean operations, and therefore $\Phi(\mathscr{M}) \subset \mathscr{M}^\dagger$. It is easy to see that the family $\mathscr{M}^\dagger$ is the smallest σ-algebra of subsets of $T^\dagger$ containing all sets B of the form (2). Since $\Phi^{-1}(B)$ has the form A, we have $\Phi^{-1}(\mathscr{M}^\dagger) \subset \mathscr{M}$, and so $\mathscr{M}^\dagger \subset \Phi(\mathscr{M})$. □

Our first step in constructing the measure μ on T is to show that μ is uniquely determined for finite products by the requirements in (22.3).

(22.5) Lemma. *Let $\Omega = \{\gamma_1, \gamma_2, \ldots, \gamma_m\}$ be any finite nonvoid subset of Γ. There is a unique measure μ_Ω on $\mathscr{M}_\Omega$ such that*

(i) $\mu_\Omega\Big(\underset{j=1}{\overset{m}{\times}} A_{\gamma_j}\Big) = \prod_{j=1}^{m} \mu_{\gamma_j}(A_{\gamma_j})$

for all $\underset{j=1}{\overset{m}{\times}} A_{\gamma_j} \in \mathscr{E}_\Omega$.

Proof. (I) Suppose first that $\Omega = \{\gamma_1, \gamma_2\}$. Let μ_Ω be the product measure $\mu_{\gamma_1} \times \mu_{\gamma_2}$ on $\mathscr{M}_\Omega = \mathscr{M}_{\gamma_1} \times \mathscr{M}_{\gamma_2}$ defined in (21.9). The uniqueness of μ_Ω follows from (21.11).

(II) We complete the proof by induction on the number of elements in Ω. Suppose that the result holds for all subsets of Γ having n elements, and that $\Omega = \{\gamma_1, \gamma_2, \ldots, \gamma_n, \gamma_{n+1}\}$. Let μ' be the unique measure on $\mathscr{M}_{\{\gamma_1,\gamma_2,\ldots,\gamma_n\}}$ satisfying the inductive hypothesis, and let μ be the unique measure on $\mathscr{M}_{\{\gamma_1,\ldots,\gamma_n\}} \times \mathscr{M}_{\gamma_{n+1}}$ such that $\mu(B \times A_{\gamma_{n+1}}) = \mu'(B) \cdot \mu_{\gamma_{n+1}}(A_{\gamma_{n+1}})$ for all $B \in \mathscr{M}_{\{\gamma_1,\ldots,\gamma_n\}}$ and $A_{\gamma_{n+1}} \in \mathscr{M}_{\gamma_{n+1}}$. The existence and uniqueness of μ are guaranteed by part (I). Lemma (22.4) shows that $\mathscr{M}_{\{\gamma_1,\gamma_2,\ldots,\gamma_n\}} \times \mathscr{M}_{\gamma_{n+1}}$ can be identified with $\mathscr{M}_\Omega$. We can therefore define the measure μ on $\mathscr{M}_\Omega$. Doing this, we have

$$\begin{aligned} \mu(A_{\gamma_1} \times A_{\gamma_2} \times \cdots \times A_{\gamma_n} \times A_{\gamma_{n+1}}) & \\ &= \mu((A_{\gamma_1} \times A_{\gamma_2} \times \cdots \times A_{\gamma_n}) \times A_{\gamma_{n+1}}) \\ &= \mu'(A_{\gamma_1} \times \cdots \times A_{\gamma_n}) \cdot \mu_{\gamma_{n+1}}(A_{\gamma_{n+1}}) \\ &= \mu_{\gamma_1}(A_{\gamma_1}) \cdot \mu_{\gamma_2}(A_{\gamma_2}) \cdots \mu_{\gamma_n}(A_{\gamma_n}) \cdot \mu_{\gamma_{n+1}}(A_{\gamma_{n+1}}) \end{aligned}$$

for sets A_{γ_j} in $\mathscr{M}_{\gamma_j}$. Take the measure μ just constructed to be μ_Ω in the statement of the theorem. To prove that μ_Ω is unique, observe that any measure $\tilde{\mu}$ on $\mathscr{M}_\Omega$ satisfying (i) defines, in an obvious way, a measure on $\mathscr{M}_{\{\gamma_1,\ldots,\gamma_n\}}$. In view of the uniqueness of μ', this measure is equal to μ'. It follows that $\tilde{\mu} = \mu_\Omega$ by the uniqueness of μ_Ω, for the case $\Omega = \{\gamma_1, \gamma_2\}$. □

Our next result is also a useful technicality.

(22.6) Lemma. *If* $\Omega_1 \cap \Omega_2 = \varnothing$ *and* $B_{\Omega_j} \in \mathscr{M}_{\Omega_j} (j = 1, 2)$, *then*

(i) $\mu_{\Omega_1 \cup \Omega_2}(B_{\Omega_1} \times B_{\Omega_2}) = \mu_{\Omega_1}(B_{\Omega_1}) \cdot \mu_{\Omega_2}(B_{\Omega_2})$.

Proof. Use Lemma (22.4) to identify $\mathscr{M}_{\Omega_1 \cup \Omega_2}$ with $\mathscr{M}_{\Omega_1} \times \mathscr{M}_{\Omega_2}$. If B_{Ω_j} is in $\mathscr{E}_{\Omega_j}$ $(j = 1, 2)$, then $B_{\Omega_1} \times B_{\Omega_2}$ is in $\mathscr{E}_{\Omega_1 \cup \Omega_2}$, and it is clear that

$$\mu_{\Omega_1 \cup \Omega_2}(B_{\Omega_1} \times B_{\Omega_2}) = \mu_{\Omega_1}(B_{\Omega_1}) \cdot \mu_{\Omega_2}(B_{\Omega_2}):$$

both sides of this equality are the same products $\prod \mu_{\gamma_j}(A_{\gamma_j})$. Thus $\mu_{\Omega_1 \cup \Omega_2}$ and $\mu_{\Omega_1} \times \mu_{\Omega_2}$ are measures on $\mathscr{M}_{\Omega_1 \cup \Omega_2}$ that satisfy (22.5.i) for $\Omega = \Omega_1 \cup \Omega_2$. By (22.5), they are the same measure. □

We now consider the full infinite product $T = \bigtimes_{\gamma \in \Gamma} T_\gamma$, first proving a preliminary fact.

(22.7) Theorem. *There is a unique* finitely additive *measure* μ *on the algebra of sets* $\mathscr{N}$ *such that*

(i) $\mu(A_\Omega \times T_{\Omega'}) = \mu_\Omega(A_\Omega)$

for all Ω *and all* $A_\Omega \in \mathscr{M}_\Omega$.

Proof. Let the expression (i) define μ. We first show that this definition is unambiguous. Thus suppose that

$$A_{\Omega_1} \times T_{\Omega_1'} = A_{\Omega_2} \times T_{\Omega_2'} .\text{[1]} \tag{1}$$

Write $\Omega_3 = \Omega_1 \cup \Omega_2$. Then (1) can be rewritten as

$$A_{\Omega_1} \times T_{\Omega_3 \cap \Omega_1'} \times T_{\Omega_3'} = A_{\Omega_2} \times T_{\Omega_3 \cap \Omega_2'} \times T_{\Omega_3'} . \tag{2}$$

The sets $A_{\Omega_1} \times T_{\Omega_3 \cap \Omega_1'}$ and $A_{\Omega_2} \times T_{\Omega_3 \cap \Omega_2'}$ are in $\mathscr{M}_{\Omega_3}$, and (2) shows that they are equal. Lemma (22.6) shows that

$$\mu_{\Omega_3}(A_{\Omega_1} \times T_{\Omega_3 \cap \Omega_1'}) = \mu_{\Omega_1}(A_{\Omega_1}) \cdot 1 = \mu_{\Omega_1}(A_{\Omega_1})$$

and

$$\mu_{\Omega_3}(A_{\Omega_2} \times T_{\Omega_3 \cap \Omega_2'}) = \mu_{\Omega_2}(A_{\Omega_2}) \cdot 1 = \mu_{\Omega_2}(A_{\Omega_2}) .$$

Thus μ is well defined by (i).

For every set $A \in \mathscr{N}$, it is easy to see that there exists an Ω and a set $A_\Omega \in \mathscr{N}_\Omega$ such that

$$A = A_\Omega \times T_{\Omega'} ; \tag{3}$$

we omit the proof of this simple fact. If A_1 and A_2 are disjoint sets in $\mathscr{N}$, write

$$A_j = A_{\Omega_j} \times T_{\Omega_j'} \qquad (j = 1, 2)$$

---

[1] Again we commit a slight solecism in writing $A_{\Omega_1} \times T_{\Omega_1'} = A_{\Omega_2} \times T_{\Omega_2'}$, as the elements of these two sets are really different entities. Again Lemma (22.4) saves the day.

as in (3); next write $\Omega_3 = \Omega_1 \cup \Omega_2$; and then write

$$A_j = A^{(j)}_{\Omega_3} \times T_{\Omega'_3} \quad (j = 1, 2)\,; \tag{4}$$

it is clear that this can be done and that $A^{(j)}_{\Omega_3} \in \mathscr{N}_{\Omega_3}$. It is also clear that $A^{(1)}_{\Omega_3} \cap A^{(2)}_{\Omega_3} = \varnothing$, and so, using (i) and the additivity of μ_{Ω_3} (22.5), we have

$$\begin{aligned}\mu(A_1 \cup A_2) &= \mu_{\Omega_3}(A^{(1)}_{\Omega_3} \cup A^{(2)}_{\Omega_3}) \\ &= \mu_{\Omega_3}(A^{(1)}_{\Omega_3}) + \mu_{\Omega_3}(A^{(2)}_{\Omega_3}) \\ &= \mu(A_1) + \mu(A_2)\,.\end{aligned}$$

That is, μ is finitely additive on $\mathscr{N}$. □

It is tempting to use the technique of (22.7) to try to prove that μ is actually countably additive, for each μ_Ω is countably additive. This approach must fail, since we cannot necessarily obtain a finite subset Ω of Γ such that each of the countably many sets in question is a subset of T_Ω. Furthermore, one cannot apply (10.36), since there may exist pairwise disjoint families $\{C_n\}_{n=1}^{\infty} \subset \mathscr{N}$ such that $\bigcup\limits_{n=1}^{\infty} C_n \in \mathscr{N}$, but no $\mathscr{M}_\Omega$ contains all of the sets $A_{\Omega_n} \in \mathscr{M}_{\Omega_n}$ such that $C_n = A_{\Omega_n} \times T_{\Omega'_n}$. For example, let $\Gamma = N$, $T = [0, 1]^N$, and $B_n = \left\{\boldsymbol{t} \in T : 0 < t_k < \frac{1}{n} \text{ for } k = 1, 2, \ldots, n\right\}$. Then write

$$C_1 = B'_1,\ C_n = B'_n \cap (B_1 \cap \cdots \cap B_{n-1}) \quad \text{for } n > 1\,.$$

It is clear that $\bigcup\limits_{n=1}^{\infty} C_n = T$, that $C_n = A_n \times T_{\{1,\ldots,n\}'}$ for $A_n \in \mathscr{M}_{\{1,\ldots,n\}}$, and that C_n does not have the form $D_{n-1} \times T_{\{1,\ldots,n-1\}'}$ for any $D_{n-1} \in \mathscr{M}_{\{1,\ldots,n-1\}}$. We proceed to the theorem itself. The proof is perhaps complicated, but its basic idea is simple enough.

(22.8) Theorem. *The finitely additive measure μ on $\mathscr{N}$ admits a unique extension over $\mathscr{M}$ that is countably additive.*

Proof. The uniqueness of μ's extension over $\mathscr{M}$ [if it exists at all] is proved by an obvious application of (21.6). To prove that μ has *some* countably additive extension over $\mathscr{M}$, we need only show that

$$\lim_{n\to\infty} \mu(F_n) = 0 \tag{1}$$

for every sequence $(F_n)_{n=1}^{\infty}$ such that $F_n \in \mathscr{N}$, $F_1 \supset F_2 \supset \cdots \supset F_n \supset \ldots$, and $\bigcap\limits_{n=1}^{\infty} F_n = \varnothing$. This follows from (10.37).[1]

[1] The reader will recall that (10.37) is an exercise, for which he must supply the proof. We believe that all readers who have worked through the text to this point will now be able easily to prove (10.37), if they omitted to do so on first reading § 10.

In this paragraph, we make some reductions in order to simplify subsequent notation. Each F_n has the form $A_{\Omega_n} \times T_{\Omega_n'}$, where $A_{\Omega_n} \in \mathscr{N}_{\Omega_n}$. Let $\Delta = \bigcup_{n=1}^{\infty} \Omega_n$. By (22.7), there is a finitely additive measure μ_Δ on $\mathscr{N}_\Delta$ such that $\mu_\Delta(A_\Omega \times T_{\Delta \cap \Omega'}) = \mu_\Omega(A_\Omega)$ for all $\Omega \subset \Delta$ and $A_\Omega \in \mathscr{M}_\Omega$. For each n, let $F_n^\Delta = A_{\Omega_n} \times T_{\Delta \cap \Omega_n'} \subset T_\Delta$. It is then clear that:

each F_n^Δ belongs to $\mathscr{N}_\Delta$;

$\mu_\Delta(F_n^\Delta) = \mu(F_n)$;

$F_1^\Delta \supset F_2^\Delta \supset \cdots \supset F_n^\Delta \supset \cdots$;

and $\bigcap_{n=1}^{\infty} F_n^\Delta = \varnothing$.

It clearly suffices to prove that $\lim_{n\to\infty} \mu_\Delta(F_n^\Delta) = 0$. In other words, we lose no generality in supposing that Γ is countably infinite. It is now just a notational matter to suppose that $\Gamma = N = \{1, 2, \ldots\}$. Let $k_n = \max \Omega_n$. We may suppose with no loss of generality that $\Omega_n = \{1, 2, \ldots, k_n\}$ and that $k_1 < k_2 < k_3 < \cdots$. Define the sequence of sets $(E_m)_{m=1}^{\infty}$ by the following rule:

$$E_m = \begin{cases} T \text{ if } 1 \leqq m < k_1, \\ F_n \text{ if } k_n \leqq m < k_{n+1}. \end{cases}$$

Then we have $\bigcap_{m=1}^{\infty} E_m = \bigcap_{n=1}^{\infty} F_n = \varnothing$ and $\lim_{m\to\infty} \mu(E_m) = \lim_{n\to\infty} \mu(F_n)$. We must show that $\lim_{m\to\infty} \mu(E_m) = 0$. Let $\Theta_m = \{1, 2, \ldots, m\}$ for each m. Note that each E_m has the form $A_m \times T_{\Theta_m'}$, where $A_m \in \mathscr{N}_{\Theta_m}$.

As noted in the proof of (22.5), we have $\mu_{\Theta_{m+1}} = \mu_{\Theta_m} \times \mu_{m+1}$ for all m. It is also easy to see that a set in $\mathscr{N}_{\Theta_{m+1}}$ is $\mathscr{M}_{\Theta_m} \times \mathscr{M}_{m+1}$-measurable. Now apply (22.7) and (21.12.v) to write

$$\begin{aligned} \mu(E_m) = \mu_{\Theta_m}(A_m) &= \int_{T_{\Theta_m}} \xi_{A_m}(\boldsymbol{t})\, d\mu_{\Theta_m}(\boldsymbol{t}) \\ &= \int_{T_{\Theta_{m-1}}} \int_{T_m} \xi_{A_m}(\boldsymbol{t}^*, t_m)\, d\mu_m(t_m)\, d\mu_{\Theta_{m-1}}(\boldsymbol{t}^*), \end{aligned} \tag{2}$$

where $\boldsymbol{t}^*$ denotes a generic element of $T_{\Theta_{m-1}}$. By (21.12.iii), the inner integral in (2) is $\mathscr{M}_{\Theta_{m-1}}$-measurable, and so we can apply (22.5) and (21.12.v) again. Doing this $m-1$ times, we arrive at the equality

$$\mu(E_m) = \int_{T_1} \int_{T_2} \cdots \int_{T_m} \xi_{A_m}(t_1, t_2, \ldots, t_m)\, d\mu_m(t_m) \cdots d\mu_2(t_2)\, d\mu_1(t_1). \tag{3}$$

Assume that $\mu(E_m)$ does not go to zero. For $s_1 \in T_1$, let

$$f_{1,m}(s_1) = \int_{T_2} \cdots \int_{T_m} \xi_{A_m}(s_1, t_2, \ldots, t_m)\, d\mu_m(t_m) \cdots d\mu_2(t_2)$$

for $m = 1, 2, 3, \ldots$. That is, we carry out all but the outermost integration in (3), and leave the first variable unintegrated. It is plain that

$$\mu(E_m) = \int_{T_1} f_{1,m}(t_1)\, d\mu_1(t_1)$$

and that $f_{1,m}(T_1) \subset [0, 1]$. It is not the case that

$$\lim_{m\to\infty} f_{1,m}(t_1) = 0 \quad \text{everywhere on } T_1; \tag{4}$$

if (4) held, then LEBESGUE's dominated convergence theorem (12.24) would imply that $\lim_{m\to\infty} \mu(E_m) = 0$. Hence there is a point $a_1 \in T_1$ such that $f_{1,m}(a_1)$ does not go to zero as $m \to \infty$. Next define $f_{2,m}$ by

$$f_{2,m}(s_2) = \int_{T_3} \cdots \int_{T_m} \xi_{A_m}(a_1, s_2, t_3, \ldots, t_m)\, d\mu_m(t_m) \cdots d\mu_3(t_3)\,.$$

If it were the case that $\lim_{m\to\infty} f_{2,m}(s_2) = 0$ for all $s_2 \in T_2$, then by (12.24) we would also have $\lim_{m\to\infty} \int_{T_2} f_{2,m}(t_2)\, d\mu_2(t_2) = 0$; *i.e.*, we would have $\lim_{m\to\infty} f_{1,m}(a_1) = 0$. Hence there is a point $a_2 \in T_2$ such that $f_{2,m}(a_2)$ does not go to zero as $m \to \infty$.

In this way we construct a sequence of points $(a_1, a_2, \ldots, a_n, \ldots) = \boldsymbol{a}$ in T with the following property. For every $n \in N$, the sequence of numbers

$$\int_{T_{n+1}} \int_{T_{n+2}} \cdots \int_{T_m} \xi_{A_m}(a_1, \ldots, a_n, t_{n+1}, \ldots, t_m)\, d\mu_m(t_m) \cdots d\mu_{n+1}(t_{n+1}), \tag{5}$$

defined for $m > n$, does *not* have limit zero as $m \to \infty$. Now the integrand in (5) cannot vanish identically for all large m, and so

$$(a_1, a_2, \ldots, a_n, s_{n+1}, \ldots, s_m) \in A_m \tag{6}$$

for appropriately chosen $s_j \in T_j$, and for arbitrarily large m. Since $E_m = A_m \times T_{\Theta'_m}$, we can choose a point $\mathbf{s}^{(m,n)} \in T_{\Theta'_n}$ such that

$$(a_1, a_2, \ldots, a_n) \times \mathbf{s}^{(m,n)} \in E_m\,.$$

Since $E_m \subset E_n$, we also have

$$(a_1, a_2, \ldots, a_n) \times \mathbf{s}^{(m,n)} \in E_n\,. \tag{7}$$

Since $E_n = A_n \times T_{\Theta'_n}$, (7) implies that

$$(a_1, a_2, \ldots, a_n) \times T_{\Theta'_n} \subset E_n\,.$$

In particular, $\boldsymbol{a} \in E_n$. Since n is any positive integer, we infer that

$$\boldsymbol{a} \in \bigcap_{n=1}^{\infty} E_n\,.$$

This contradicts the equality $\bigcap_{n=1}^{\infty} E_n = \varnothing$ and completes the proof. □

Thus we have a countably additive measure μ on a σ-algebra of subsets of T that behaves like a product measure. A first and very simple application is to a classical problem in probability.

(22.9) Example. Let $\Gamma = \{1, 2, 3, \ldots\}$ and let $T_n = \{0, 1\}$ for each $n \in \Gamma$. Then $T = \{\boldsymbol{t} : \boldsymbol{t} = (t_n),\ t_n = 1 \text{ or } t_n = 0\}$. Define the measure μ_n on each T_n by setting $\mu_n(\{0\}) = \mu_n(\{1\}) = \frac{1}{2}$, and let $\mathscr{M}_n$ be all four subsets of $\{0, 1\}$. For a finite subset $\{k_1, \ldots, k_n\}$ of Γ [all k's distinct] and any sequence $(a_{k_1}, a_{k_2}, \ldots, a_{k_n})$ of 0's and 1's, write $E(a_{k_1}, a_{k_2}, \ldots, a_{k_n})$ for the set $\{\boldsymbol{t} \in T : t_{k_1} = a_{k_1},\ t_{k_2} = a_{k_2}, \ldots, t_{k_n} = a_{k_n}\}$. The definition of μ as given in (22.3.ii) and (22.7) shows at once that

$$\mu(E(a_{k_1}, a_{k_2}, \ldots, a_{k_n})) = 2^{-n}.$$

For $n \in N$, define f_n on T by $f_n(\boldsymbol{t}) = t_n - \frac{1}{2}$, and let $h_n = \frac{1}{n}(f_1 + \cdots + f_n)$. We will show that

$$\lim_{n\to\infty} \|h_n\|_2 = 0. \tag{1}$$

It is clear that $f_j = \xi_{E(1_j)} - \frac{1}{2}$, that

$$f_j f_k = \xi_{E(1_j, 1_k)} - \frac{1}{2}\xi_{E(1_j)} - \frac{1}{2}\xi_{E(1_k)} + \frac{1}{4} \quad (j \neq k),$$

and that $f_j^2 = \frac{1}{4}$.
Therefore

$$\|h_n\|_2^2 = \frac{1}{n^2}\sum_{j=1}^{n}\sum_{k=1}^{n}\int_T f_j f_k \, d\mu = \frac{1}{4n^2}\sum_{j=1}^{n}\sum_{k=1}^{n}\delta_{jk} = \frac{1}{4n}.$$

This proves (1). From (1) and (13.33), we infer that $h_n \to 0$ in measure. Thus for every $\varepsilon > 0$, the equality

$$\lim_{n\to\infty} \mu\left(\left\{\boldsymbol{t} \in T : \left|\frac{t_1 + t_2 + \cdots + t_n}{n} - \frac{1}{2}\right| > \varepsilon\right\}\right) = 0 \tag{2}$$

obtains.

If we let 0 and 1 correspond to obtaining heads or tails, respectively, upon flipping a coin, then $\frac{t_1 + \cdots + t_n}{n}$ is the proportion of tails obtained in n flips. The equality (2) then asserts that the probability that the proportion of tails is farther than ε from $\frac{1}{2}$ decreases to 0 as n [the number of flips] goes to infinity, for every $\varepsilon > 0$. If the coin is unbiased $\left[\mu_n(\{0\}) = \mu_n(\{1\}) = \frac{1}{2}\right]$ this is what we would expect.

The equality (2) is one form of the *weak law of large numbers.* [See also (22.32.b) *infra.*]

(22.10) Exercise. Consider a generalization of (22.9), as follows. Let Γ be arbitrary [but infinite of course]. For each γ, let A_γ be any set in $\mathscr{M}_\gamma$ and let f_γ be the function on T such that $f_\gamma(\mathbf{t}) = \xi_{A_\gamma}(t_\gamma) - \mu_\gamma(A_\gamma)$. For $\Omega = \{\gamma_1, \gamma_2, \ldots, \gamma_n\} \subset \Gamma$, let $w(\Omega)$ be a positive number and let

$$h_\Omega = w(\Omega) \sum_{\gamma \in \Omega} f_\gamma \,.$$

(a) Show that $\int\limits_T |h_\Omega|^2 \, d\mu = w(\Omega)^2 \sum\limits_{\gamma \in \Omega} \left[\mu_\gamma(A_\gamma) - (\mu_\gamma(A_\gamma))^2\right]$.

(b) Generalize the notion of convergence in measure: $h_\Omega \to 0$ in measure if for every $\delta > 0$ and every $\varepsilon > 0$ there is an $\Omega_0 \subset \Gamma$ such that

$$\mu(\{\mathbf{t} \in T : |h_\Omega(\mathbf{t})| \geqq \delta\}) < \varepsilon$$

for all $\Omega \supset \Omega_0$. Find reasonable conditions on $w(\Omega)$ for h_Ω to converge to 0 in measure. What simple form can you give $w(\Omega)$ if all $\mu_\gamma(A_\gamma)$ are equal? What happens if $\mu_\gamma(A_\gamma) = 0$? If $\mu_\gamma(A_\gamma) = 1$?

There are several quite distinct analogues of FUBINI'S theorem for infinite products, all of which coalesce trivially for finite products. These distinct versions arise because of the various different ways in which we can approximate $\int\limits_T f \, d\mu$ and f by integrals over finite numbers of coordinates. Our first Fubini-esque theorem deals with $\mathfrak{L}_p$-convergence and is quite general. We need three lemmas.

(22.11) Lemma. *Let* $(T_j, \mathscr{M}_j, \mu_j)$ $(j = 1, 2)$ *be measure spaces such that* $\mu_j(T_j) = 1$; *let* $(T, \mathscr{M}, \mu) = (T_1 \times T_2, \mathscr{M}_1 \times \mathscr{M}_2, \mu_1 \times \mu_2)$; *and let* p *be a real number* $\geqq 1$. *For* $f \in \mathfrak{L}_p(T, \mathscr{M}, \mu)$, *let* Sf *be the function on* T *such that*

$$Sf(s_1, s_2) = \int\limits_{T_2} f(s_1, t_2) \, d\mu_2(t_2)$$

for all $s_2 \in T_2$. *Then* Sf *is in* $\mathfrak{L}_p(T, \mathscr{M}, \mu)$ *and* $\|Sf\|_p \leqq \|f\|_p$, *so that* S *is a norm nonincreasing linear transformation of* $\mathfrak{L}_p(T, \mathscr{M}, \mu)$ *into itself.*

Proof. Since $\mu(T) = 1$, we have $\mathfrak{L}_p(T, \mathscr{M}, \mu) \subset \mathfrak{L}_1(T, \mathscr{M}, \mu)$. Thus f is in $\mathfrak{L}_1(T, \mathscr{M}, \mu)$, and it follows from (21.13.iv) that the function

$$s_1 \to \int\limits_{T_2} f(s_1, t_2) \, d\mu_2(t_2)$$

is in $\mathfrak{L}_1(T_1, \mathscr{M}_1, \mu_1)$. In particular this function is $\mathscr{M}_1$-measurable, and since the function

$$(s_1, s_2) \to \int\limits_{T_2} f(s_1, t_2) \, d\mu_2(t_2)$$

does not depend on s_2, it is plainly $\mathscr{M}$-measurable. Using (12.28.ii),

(13.17), and (21.12), we obtain

$$\begin{aligned}\|Sf\|_p^p &= \int_{T_1\times T_2} \left|\int_{T_2} f(s_1, t_2)\, d\mu_2(t_2)\right|^p d(\mu_1 \times \mu_2)(s_1, s_2)\\ &\leqq \int_{T_1\times T_2} \left[\int_{T_2} |f(s_1, t_2)|\, d\mu_2(t_2)\right]^p d(\mu_1 \times \mu_2)(s_1, s_2)\\ &\leqq \int_{T_1\times T_2} \int_{T_2} |f(s_1, t_2)|^p\, d\mu_2(t_2)\, d(\mu_1 \times \mu_2)(s_1, s_2)\\ &= \int_{T_2}\Big(\int_{T_1}\int_{T_2} |f(s_1, t_2)|^p\, d\mu_2(t_2)\, d\mu_1(s_1)\Big)\, d\mu_2(s_2)\\ &= \int_{T_2}\Big[\int_{T_1\times T_2} |f(s_1, t_2)|^p\, d(\mu_1 \times \mu_2)(s_1, t_2)\Big]\, d\mu_2(s_2)\\ &= \int_{T_2} \|f\|_p^p\, d\mu_2(s_2) = \|f\|_p^p\,.\end{aligned}$$

Hence Sf is in $\mathfrak{L}_p(T, \mathscr{M}, \mu)$ and $\|Sf\|_p \leqq \|f\|_p$. □

(22.12) Lemma. *Let Γ be an arbitrary infinite index set, and suppose that $\varnothing \subsetneqq \Delta \subsetneqq \Gamma$. Let μ_Δ and $\mu_{\Delta'}$ be the measures on the σ-algebras $\mathscr{M}_\Delta$ and $\mathscr{M}_{\Delta'}$ constructed as in* (22.7) *and* (22.8). *Identifying $\mathscr{M}$ and $\mathscr{M}_\Delta \times \mathscr{M}_{\Delta'}$ [the mapping Φ of* (22.4) *allows us to do this], we have $\mu = \mu_\Delta \times \mu_{\Delta'}$.*

Proof. The measures μ and $\mu_\Delta \times \mu_{\Delta'}$ agree on sets of the form $A_\Omega \times T_{\Omega'}$, where Ω is a finite subset of Γ, and so by the uniqueness of μ [(22.7) and (22.8)] they agree throughout $\mathscr{M}$. □

The next lemma is a necessary technicality.

(22.13) Lemma. *For every $B \in \mathscr{M}$ and every $\varepsilon > 0$, there is a set $A \in \mathscr{N}$ such that $\|\xi_A - \xi_B\|_1 = \mu(A \triangle B) < \varepsilon$.*

Proof. Define the family $\mathscr{P}$ as $\{B \in \mathscr{M}$: for all $\varepsilon > 0$ there exists $A \in \mathscr{N}$ such that $\mu(A \triangle B) < \varepsilon\}$. It is trivial that $\mathscr{P} \supset \mathscr{N}$; to prove that $\mathscr{P} = \mathscr{M}$, it suffices to prove that $\mathscr{P}$ is a σ-algebra. We do this by appealing to (21.6). Let (B_n) be a monotone sequence in $\mathscr{P}$ [either increasing or decreasing] and write $B = \lim B_n$. Given $\varepsilon > 0$, use (10.13) or (10.15) to select $m \in N$ such that $\mu(B \triangle B_m) < \frac{\varepsilon}{2}$. Since $B_m \in \mathscr{P}$, there exists a set $A_m \in \mathscr{N}$ such that $\mu(A_m \triangle B_m) < \frac{\varepsilon}{2}$. Then we have

$$\begin{aligned}A_m \triangle B &= (A_m \cap B') \cup (B \cap A_m')\\ &\subset (A_m \cap B_m') \cup (B \triangle B_m) \cup (B_m \cap A_m')\\ &= (A_m \triangle B_m) \cup (B \triangle B_m)\,,\end{aligned}$$

and so $\mu(A_m \triangle B) < \frac{\varepsilon}{2} + \frac{\varepsilon}{2} = \varepsilon$; hence $B \in \mathscr{P}$. Thus $\mathscr{P}$ is a monotone family. Since $\mathscr{P}$ contains the algebra $\mathscr{N}$, (21.6) implies that $\mathscr{P}$ contains

$\mathscr{S}(\mathscr{N})$, and so we have

$$\mathscr{M} \supset \mathscr{P} \supset \mathscr{S}(\mathscr{N}) = \mathscr{M}.^1 \quad \square$$

We can now state and prove a mean convergence version of FUBINI's theorem, due to B. JESSEN.

(22.14) Theorem. *Let Γ be an arbitrary infinite index set. For every finite subset Ω of Γ, regard $(T, \mathscr{M}, \mu)$ as $(T_\Omega \times T_{\Omega'}, \mathscr{M}_\Omega \times \mathscr{M}_{\Omega'}, \mu_\Omega \times \mu_{\Omega'})$ [making use of (22.12)]. For $1 \leq p < \infty$ and $f \in \mathfrak{L}_p(T, \mathscr{M}, \mu)$, define $f_{\Omega'}$ on T by*

$$f_{\Omega'}(\boldsymbol{t}_\Omega, \boldsymbol{t}_{\Omega'}) = \int_{T_\Omega} f(\boldsymbol{u}_\Omega, \boldsymbol{t}_{\Omega'})\, d\mu_\Omega(\boldsymbol{u}_\Omega).$$

That is, $f_{\Omega'}$ is a function of the form Sf as in (22.11). Then $f_{\Omega'}$ is in $\mathfrak{L}_p(T, \mathscr{M}, \mu)$ and

(i) $\lim_\Omega \| f_{\Omega'} - \int_T f\, d\mu \|_p = 0.$[2]

Also, let

$$f_\Omega(\boldsymbol{t}_\Omega, \boldsymbol{t}_{\Omega'}) = \int_{T_{\Omega'}} f(\boldsymbol{t}_\Omega, \boldsymbol{u}_{\Omega'})\, d\mu_{\Omega'}(\boldsymbol{u}_{\Omega'}).$$

Then f_Ω is in $\mathfrak{L}_p(T, \mathscr{M}, \mu)$ and

(ii) $\lim_\Omega \| f_\Omega - f \|_p = 0.$

Proof. (I) We first consider functions f of a very special kind: suppose that $f = \xi_{A_{\Omega_0} \times T_{\Omega_0'}}$, where $A_{\Omega_0} \in \mathscr{M}_{\Omega_0}$. For $\Omega \supset \Omega_0$, we have

$$\begin{aligned} f_{\Omega'}(\boldsymbol{t}_\Omega, \boldsymbol{t}_{\Omega'}) &= \int_{T_\Omega} \xi_{A_{\Omega_0} \times T_{\Omega_0'}}(\boldsymbol{u}_\Omega, \boldsymbol{t}_{\Omega'})\, d\mu_\Omega(\boldsymbol{u}_\Omega) \\ &= \int_{T_\Omega} \xi_{A_{\Omega_0} \times T_{\Omega \cap \Omega_0'} \times T_{\Omega'}}(\boldsymbol{u}_\Omega, \boldsymbol{t}_{\Omega'})\, d\mu_\Omega(\boldsymbol{u}_\Omega). \qquad (1) \end{aligned}$$

As a function on T_Ω, $\xi_{A_{\Omega_0} \times T_{\Omega \cap \Omega_0'} \times T_{\Omega'}}$ is merely the characteristic function of the set $A_{\Omega_0} \times T_{\Omega \cap \Omega_0'}$. Thus the integrals in (1) are equal to $\mu_\Omega(A_{\Omega_0} \times T_{\Omega \cap \Omega_0'}) = \mu_{\Omega_0}(A_{\Omega_0}) = \mu(A_{\Omega_0} \times T_{\Omega_0'}) = \int_T f\, d\mu$; *i. e.*,

$$f_{\Omega'}(\boldsymbol{t}_\Omega, \boldsymbol{t}_{\Omega'}) = \int_T f\, d\mu.$$

Thus (i) is established for our special function f. To establish (ii) for f, again let $\Omega \supset \Omega_0$ and observe that the integrand in

$$f_\Omega(\boldsymbol{t}_\Omega, \boldsymbol{t}_{\Omega'}) = \int_{T_{\Omega'}} \xi_{A_{\Omega_0} \times T_{\Omega \cap \Omega_0'} \times T_{\Omega'}}(\boldsymbol{t}_\Omega, \boldsymbol{u}_{\Omega'})\, d\mu_{\Omega'}(\boldsymbol{u}_{\Omega'})$$

---

[1] Observe that (22.13) holds for any finite measure space $(T, \mathscr{M}, \mu)$ and any algebra $\mathscr{N} \subset \mathscr{M}$ such that $\mathscr{S}(\mathscr{N}) = \mathscr{M}$.

[2] By this limit we mean that for every $\varepsilon > 0$, there is an $\Omega_0 \subset \Gamma$ such that if $\Omega \supset \Omega_0$, then $\| f_{\Omega'} - \int_T f\, d\mu \|_p < \varepsilon$. The limit in (ii) has a similar definition.

is equal to 1 for all $u_{\Omega'}$ if $t_\Omega|_{\Omega'_0} \in A_{\Omega_0}$; therefore in this case $f_\Omega = f = 1$. If $t_\Omega|_{\Omega_0} \notin A_{\Omega_0}$, the integrand vanishes and so $f_\Omega(t_\Omega, t_{\Omega'}) = 0$. Thus we have

$$f_\Omega(t_\Omega, t_{\Omega'}) = \xi_{A_{\Omega_0} \times T_{\Omega'_0}}(t_\Omega, t_{\Omega'}),$$

and (ii) is established.

(II) To establish (i) and (ii) for all $f \in \mathfrak{L}_p(T, \mathscr{M}, \mu)$, let $\mathfrak{Z}$ be the subset of $\mathfrak{L}_p$ for which (i) and (ii) are true. We prove first that $\mathfrak{Z}$ is a closed linear subspace of $\mathfrak{L}_p$. It is obvious that $\mathfrak{Z}$ is a linear subspace, and so we have to show that it is closed. Suppose that $\lim_{n\to\infty} \|f^{(n)} - f\|_p = 0$, where $f^{(n)} \in \mathfrak{Z}$ for $n = 1, 2, 3, \ldots$. For every $\varepsilon > 0$ there is a set Ω_n such that $\|f^{(n)}_{\Omega'_n} - \int_T f^{(n)}\, d\mu\|_p < \varepsilon$ $(n = 1, 2, 3, \ldots)$, and such that the same inequality obtains with Ω_n replaced by any larger finite set. By (22.11), the inequality $\|f^{(n)} - f\|_p < \delta$ implies the inequalities $\|f^{(n)}_{\Omega'_n} - f_{\Omega'}\|_p < \delta$ and $\|f^{(n)}_{\Omega'} - f_{\Omega'}\|_p < \delta$ $(n = 1, 2, 3, \ldots)$ for any $\delta > 0$ and any Ω. Choose n so large that $\|f^{(n)} - f\|_p < \varepsilon$, and let $\Omega \supset \Omega_n$. Then we have

$$\begin{aligned}
\|f_{\Omega'} - \int_T f\, d\mu\|_p &\leqq \|f_{\Omega'} - f^{(n)}_{\Omega'}\|_p + \|f^{(n)}_{\Omega'} - \int_T f^{(n)}\, d\mu\|_p \\
&\quad + \|\int_T f^{(n)}\, d\mu - \int_T f\, d\mu\|_p \\
&< \varepsilon + \varepsilon + |\int_T f^{(n)}\, d\mu - \int_T f\, d\mu| \\
&\leqq 2\varepsilon + \int_T |f^{(n)} - f|\, d\mu \\
&\leqq 2\varepsilon + \|f^{(n)} - f\|_p < 3\varepsilon\,.
\end{aligned}$$

Here we have used (13.17) and the fact that $\|C\|_p = |C|$ for any constant C. Thus the inclusion $\Omega \supset \Omega_n$ implies that $\|f_{\Omega'} - \int_T f\, d\mu\|_p < 3\varepsilon$, and as ε is arbitrary, (i) follows for the function f. The relation (ii) for f is proved in like manner, and so $\mathfrak{Z}$ is closed. By step (I), $\mathfrak{Z}$ contains all functions of the form ξ_A for $A \in \mathscr{N}$, and since $\mathfrak{Z}$ is closed, Lemma (22.13) and the trivial identity $\|\xi_E\|_p = \mu(E)^{\frac{1}{p}}$ prove that $\mathfrak{Z}$ contains all ξ_B for $B \in \mathscr{M}$. Thus $\mathfrak{Z}$ contains all $\mathscr{M}$-measurable simple functions, and as these are dense in $\mathfrak{L}_p$ (13.20), the proof is complete. □

(22.15) Note. Theorem (22.14), which is of course two theorems, tells us all we could hope for about mean convergence of integrals over partial products of $\mathop{\times}_{\gamma\in\Gamma} T_\gamma$, either to the integral (22.14.i) or to the integrand (22.14.ii). For finite products, (22.14.i) becomes trivial, and (22.14.ii) becomes meaningful [and immediately trivial] only if we agree that integration for a void set of coordinates does nothing at all.

For Γ countably infinite, the mean convergence of (22.14) can be replaced by pointwise convergence μ-almost everywhere. These results follow readily from (20.56) and (20.59), as we shall now show.

(22.16) Notation. Throughout (22.16)—(22.23), the following notation will be used. The set Γ will be $\{1, 2, 3, \ldots\}$, the set Ω_n will be $\{1, 2, 3, \ldots, n\}$ for $n \in N$, and f will be an arbitrary function in $\mathfrak{L}_1(T, \mathcal{M}, \mu)$. The function f_n will be the function f_{Ω_n} of (22.14), *i. e.*,

$$\text{(i)}\ f_n(\boldsymbol{t}) = f_n(t_1, \ldots, t_n, t_{n+1}, \ldots) = \int_{T_{\Omega_n'}} f(t_1, \ldots, t_n, u_{n+1}, \ldots)\, d\mu_{\Omega_n'}(u_{n+1}, \ldots)\,.$$

The function f_n' will be the function $f_{\Omega_n'}$ of (22.14), *i. e.*,

$$\text{(ii)}\ f_n'(\boldsymbol{t}) = f_n'(t_1, \ldots, t_n, t_{n+1}, \ldots) = \int_{T_{\Omega_n}} f(u_1, \ldots, u_n, t_{n+1}, \ldots)\, d\mu_{\Omega_n}(u_1, \ldots, u_n)\,.$$

(22.17) Theorem [JESSEN]. *The relation*

$$\text{(i)}\ \lim_{n\to\infty} f_n(\boldsymbol{t}) = f(\boldsymbol{t})$$

holds for μ-almost all $\boldsymbol{t} \in T$.

Proof. We wish to apply the limit theorem (20.56). To do this, consider the σ-algebra $\mathcal{M}_{\Omega_n}$ of subsets of $T_1 \times \cdots \times T_n$ as defined in (22.2), and let $\mathcal{M}^{(n)}$ be the family of all subsets of T having the form $A_{\Omega_n} \times T_{\Omega_n'}$, where $A_{\Omega_n} \in \mathcal{M}_{\Omega_n}$. It is evident that $\mathcal{M}^{(1)} \subset \mathcal{M}^{(2)} \subset \cdots \subset \mathcal{M}^{(n)} \subset \cdots$, that each $\mathcal{M}^{(n)}$ is a σ-algebra, and that $\mathcal{M}$ is the smallest σ-algebra containing $\bigcup_{n=1}^{\infty} \mathcal{M}^{(n)}$. Thus the hypotheses of (20.56) are satisfied, where the $\mathcal{M}_n$ of (20.56) is our present $\mathcal{M}^{(n)}$. The σ-algebra $\mathcal{M}_\omega$ is our present $\mathcal{M}$. As the measure μ of (20.56) we take our product measure μ, and we define the measure η of (20.56) by

$$\eta(A) = \int_A f\, d\mu$$

for all $A \in \mathcal{M}$. [To satisfy the hypothesis that η be a signed measure, we must consider first $f \in \mathfrak{L}_1^r(T, \mathcal{M}, \mu)$. The complex case obviously follows at once.] It is trivial that $|\eta| \ll \mu$.

Now look at the definition (22.16.i) of f_n, and use Lemma (22.12) and (21.12.iv). These assertions show that f_n is $\mathcal{M}^{(n)}$-measurable. We must show that f_n is a LEBESGUE-RADON-NIKODÝM derivative of $\eta^{(n)}$ with respect to $\mu^{(n)}$ [$\eta^{(n)}$ and $\mu^{(n)}$ are restrictions to $\mathcal{M}^{(n)}$, as in (20.55)], *i. e.*, we must show that

$$\eta(A) = \int_A f_n\, d\mu \tag{1}$$

for all $A \in \mathcal{M}^{(n)}$. Write $A = A_{\Omega_n} \times T_{\Omega_n'}$, where $A_{\Omega_n} \in \mathcal{M}_{\Omega_n}$. Applying

(22.12) and (21.13), we write

$$\begin{aligned}\eta(A) &= \int_T \xi_A(\boldsymbol{t})\, f(\boldsymbol{t})\, d\mu(\boldsymbol{t}) \\ &= \int_{T_{\Omega_n}\times T_{\Omega_n'}} \xi_{A_{\Omega_n}\times T_{\Omega_n'}}(\boldsymbol{t}_{\Omega_n}, \boldsymbol{t}_{\Omega_n'})\, f(\boldsymbol{t}_{\Omega_n}, \boldsymbol{t}_{\Omega_n'})\, d(\mu_{\Omega_n}\times\mu_{\Omega_n'})(\boldsymbol{t}_{\Omega_n}, \boldsymbol{t}_{\Omega_n'}) \\ &= \int_{T_{\Omega_n}}\int_{T_{\Omega_n'}} \xi_{A_{\Omega_n}\times T_{\Omega_n'}}(\boldsymbol{t}_{\Omega_n}, \boldsymbol{t}_{\Omega_n'})\, f(\boldsymbol{t}_{\Omega_n}, \boldsymbol{t}_{\Omega_n'})\, d\mu_{\Omega_n'}(\boldsymbol{t}_{\Omega_n'})\, d\mu_{\Omega_n}(\boldsymbol{t}_{\Omega_n}). \qquad (2)\end{aligned}$$

Since $\xi_{A_{\Omega_n}\times T_{\Omega_n'}}(\boldsymbol{t}_{\Omega_n}, \boldsymbol{t}_{\Omega_n'}) = \xi_{A_{\Omega_n}}(\boldsymbol{t}_{\Omega_n})$, the last integral in (2) is equal to

$$\begin{aligned}&\int_{T_{\Omega_n}} \xi_{A_{\Omega_n}}(\boldsymbol{t}_{\Omega_n}) \int_{T_{\Omega_n'}} f(\boldsymbol{t}_{\Omega_n}, \boldsymbol{t}_{\Omega_n'})\, d\mu_{\Omega_n'}(\boldsymbol{t}_{\Omega_n'})\, d\mu_{\Omega_n}(\boldsymbol{t}_{\Omega_n}) \\ &\qquad = \int_{T_{\Omega_n}} \xi_{A_{\Omega_n}}(\boldsymbol{t}_{\Omega_n})\, f_n(\boldsymbol{t}_{\Omega_n})\, d\mu_{\Omega_n}(\boldsymbol{t}_{\Omega_n}). \qquad (3)\end{aligned}$$

A similar but simpler computation using (22.12) and (21.13) shows that the right side of (3) is in fact

$$\int_T \xi_A(\boldsymbol{t})\, f_n(\boldsymbol{t})\, d\mu(\boldsymbol{t}).$$

Thus (1) holds. Since f is plainly the LEBESGUE-RADON-NIKODÝM derivative of η with respect to μ on the σ-algebra $\mathscr{M}$, we apply (20.56) to infer that $\lim_{n\to\infty} f_n(\boldsymbol{t}) = f(\boldsymbol{t})$ μ-a. e. on T. This is (i). □

(22.18) Exercise. Suppose that η is a σ-finite measure on T such that $\eta \ll \mu$. Express $\frac{d\eta}{d\mu}$ as an *iterated* limit of functions each of the form (22.16.i). [Hints. By (19.24), there is a nonnegative, real-valued, $\mathscr{M}$-measurable function f on T such that $\int_A f\, d\mu = \eta(A)$ for all $A \in \mathscr{M}$. Let $f^{(k)} = \min\{f, k\}$ for all $k \in N$. Define $f_n^{(k)}$ as in (22.16.i) for $f^{(k)}$. Then (22.17) shows that $\lim_{n\to\infty} f_n^{(k)} = f^{(k)}$ μ-a. e. and so

$$\lim_{k\to\infty}\left[\lim_{n\to\infty} f_n^{(k)}\right] = f \qquad \mu\text{-a. e.}]$$

(22.19) Exercise. Let T and μ_n be as in (22.9).

(a) Let L be an infinite subset of N, let $\boldsymbol{a}$ be a fixed element of T, and let $B = \{\boldsymbol{t} \in T : t_n = a_n \text{ for all } n \in L\}$. Prove that $\mu(B) = 0$.

(b) In the notation of (22.9), we have

$$E(0_1, 0_3, \ldots, 0_{2n-1}) = \{\boldsymbol{t} \in T : t_1 = t_3 = \cdots = t_{2n-1} = 0\}.$$

Write S_n for this set. Prove that

$$\lim_{n\to\infty} 2^n\, \xi_{S_n}(\boldsymbol{t}) = 0$$

for μ-almost all $\boldsymbol{t} \in T$. [Hints. Let η be the product measure on T defined from measures η_n on each $\{0, 1\}_n$ as follows. If n is even, then $\eta_n(\{0\})$

$= \eta_n(\{1\}) = \frac{1}{2}$. If n is odd, then $\eta_n(\{0\}) = 1$ and $\eta_n(\{1\}) = 0$. Show that η and μ are mutually singular. Let $\mathcal{M}^{(n)}$ be defined as in (22.16). Show that η on $\mathcal{M}^{(n)}$ is absolutely continuous with respect to μ on $\mathcal{M}^{(n)}$ and that $2^n \xi_{S_n}$ is its LEBESGUE-RADON-NIKODÝM derivative. Now apply (20.56).]

(c) Find a set D of η-measure 1 and μ-measure 0 such that $\lim\limits_{n\to\infty} 2^n \xi_{S_n}(\boldsymbol{t}) = \infty$ η-almost everywhere on D. Note that this is consistent with (20.53.iv).

We now present an important corollary of Theorem (22.17).

(22.20) Definition. Let $\boldsymbol{t} = (t_n)_{n=1}^{\infty}$ and $\boldsymbol{u} = (u_n)_{n=1}^{\infty}$ be points in T. We say that $\boldsymbol{t}$ und $\boldsymbol{u}$ are *ultimately equal* if there is some $n_0 \in N$ such that $t_n = u_n$ for all $n \geqq n_0$.

(22.21) Theorem: The Zero-One Law. *Let U be a set in $\mathcal{M}$ such that for all $\boldsymbol{t} \in T$, $\boldsymbol{t}$ is in U if and only if all points $\boldsymbol{u}$ ultimately equal to $\boldsymbol{t}$ are also in U. Then $\mu(U)$ is either* 0 *or* 1.

Proof. Consider the function $f = \xi_U$ and form the functions f_n as in (22.16.i). For all n and all $(t_1, \ldots, t_n)$ and $(t'_1, \ldots, t'_n)$ in T_{Ω_n}, the points $(t_1, \ldots, t_n, u_{n+1}, u_{n+2}, \ldots)$ and $(t'_1, \ldots, t'_n, u_{n+1}, u_{n+2}, \ldots)$ are ultimately equal, and so ξ_U has the same value at these two points. The definition (22.16.i) shows that $f_n(\boldsymbol{t})$ is actually a constant on all of T, and so $\xi_U(\boldsymbol{t})$, which by (22.17.i) is the limit μ-a.e. of $f_n(\boldsymbol{t})$, must be a constant μ-a.e. As ξ_U assumes only the values 0 and 1, we have $\mu(U) = 0$ or $\mu(U) = 1$. □

With Theorems (22.21) and (20.59), we can prove another pointwise limit theorem.

(22.22) Theorem [JESSEN]. *Notation is as in* (22.16). *The relation*

(i) $\lim\limits_{n\to\infty} f'_n(\boldsymbol{t}) = \int\limits_T f\,d\mu$

holds for μ-almost all $\boldsymbol{t} \in T$.

Proof. We wish to apply the limit theorem (20.59). To do this, consider the σ-algebra $\mathcal{M}_{\Omega'_n}$ of subsets of $T_{\Omega'_n}$, and now let $\mathcal{M}^{(n)}$ be the family of all subsets of T having the form $T_{\Omega_n} \times A_{\Omega'_n}$ for $A_{\Omega'_n} \in \mathcal{M}_{\Omega'_n}$. It is evident that $\mathcal{M} \supset \mathcal{M}^{(1)} \supset \mathcal{M}^{(2)} \supset \cdots \supset \mathcal{M}^{(n)} \supset \cdots$ and that each $\mathcal{M}^{(n)}$ is a σ-algebra of subsets of T. A set U in $\mathcal{M}_0 = \bigcap\limits_{n=1}^{\infty} \mathcal{M}^{(n)}$ clearly satisfies the hypotheses of (22.21) and so has μ-measure 0 or 1. Define the measure η on $\mathcal{M}$ by

$$\eta(A) = \int\limits_A f\,d\mu\,.$$

Modifying in an obvious way the computation used in proving (22.17), we see that f'_n is $\mathcal{M}^{(n)}$-measurable and that

$$\eta(A) = \int\limits_A f'_n(\boldsymbol{t})\,d\mu(\boldsymbol{t})$$

for all $A \in \mathcal{M}^{(n)}$.

Thus f'_n is a LEBESGUE-RADON-NIKODÝM derivative of η with respect to μ for $\mathscr{M}^{(n)}$, and (20.59) implies that $\lim\limits_{n\to\infty} f'_n(\boldsymbol{t}) = f_0(\boldsymbol{t})$ exists and is a LEBESGUE-RADON-NIKODÝM derivative of η with respect to μ on $\mathscr{M}_0$.

Since f_0 is $\mathscr{M}_0$-measurable and μ assumes only the values 0 and 1 on $\mathscr{M}_0$, it is easy to see that there is a number α such that $f_0(\boldsymbol{t}) = \alpha$ for all $\boldsymbol{t}$ in a set of μ-measure 1. Thus we have

$$\eta(T) = \int_T f\,d\mu = \int_T f_0\,d\mu = \alpha\,,$$

and so (i) is proved. □

(22.23) Exercise. Let f be an $\mathscr{M}$-measurable complex-valued function such that

(i) $f(u_1, u_2, \ldots, u_n, t_{n+1}, t_{n+2}, \ldots) = f(v_1, v_2, \ldots, v_n, t_{n+1}, t_{n+2}, \ldots)$

for all positive integers n and all choices of $u_1, u_2, \ldots, u_n$ and $v_1, v_2, \ldots, v_n$. Prove that f is a constant μ-a.e. [Hint. Use the argument of (22.21).]

(22.24) Remarks. Theorems (22.14), (22.17), and (22.22) assume a particularly simple form for functions that are products of functions depending on a single coordinate. Making no effort to be exhaustive, we list a few examples.

(a) Let Γ be an arbitrary infinite index set, let Ω be a nonvoid finite subset of Γ, and let f_γ be a function in $\mathfrak{L}_1(T_\gamma, \mathscr{M}_\gamma, \mu_\gamma)$ for each $\gamma \in \Omega$. Let g be the function $\boldsymbol{t} \to \prod\limits_{\gamma\in\Omega} f_\gamma(t_\gamma)$ on T. Then we have

(i) $\int\limits_T g\,d\mu = \prod\limits_{\gamma\in\Omega} \int\limits_{T_\gamma} f_\gamma\,d\mu_\gamma\,.$

This follows immediately from (22.14) and (21.13).

(b) Let $\Gamma = N = \{1, 2, 3, \ldots\}$, let f_n be a function in $\mathfrak{L}_1(T_n, \mathscr{M}_n, \mu_n)$ for each $n \in N$, and suppose that $\lim\limits_{p\to\infty} \prod\limits_{n=1}^{p} f_n(t_n) = \prod\limits_{n=1}^{\infty} f_n(t_n)$ exists and is finite for μ-almost all $(t_n) \in T$. The function g defined by $\boldsymbol{t} \to \prod\limits_{n=1}^{\infty} f_n(t_n)$ is certainly $\mathscr{M}$-measurable. Suppose that $g \in \mathfrak{L}_1(T, \mathscr{M}, \mu)$. Then we have

(ii) $\int\limits_T g\,d\mu = \lim\limits_{p\to\infty} \prod\limits_{n=1}^{p} \int\limits_{T_n} f_n\,d\mu_n\,.$

This too follows at once from (22.14). By applying (12.22), the reader can easily extend (ii) to the case in which $f_n \in \mathfrak{L}_1^+(T_n, \mathscr{M}_n, \mu_n)$ and $f_n \geqq 1$, with no assumption on g.

(c) A special case of (ii) is the equality

(iii) $\mu\Big(\bigtimes\limits_{n\in N} A_n\Big) = \prod\limits_{n=1}^{\infty} \mu_n(A_n)\,,$

which holds for all sequences $(A_n)_{n=1}^{\infty}$ of sets such that $A_n \in \mathscr{M}_n$ for all n.

(d) Nothing like (iii) holds for uncountable products even if we consider the completed measure space $(T, \overline{\mathscr{M}}, \bar{\mu})$. For example, suppose that $\overline{\overline{\Gamma}} > \aleph_0$ and $T_\gamma = [0, 1]$ for all $\gamma \in \Gamma$. For each γ, let E_γ be λ-measurable, $E_\gamma \subsetneqq [0, 1]$, and $\lambda(E_\gamma) = 1$. Then $\underset{\gamma\in\Gamma}{\times} E_\gamma$ is *not* measurable in the product space in which each coordinate has Lebesgue measure. For a proof of this rather delicate fact, see HEWITT and ROSS, *Abstract Harmonic Analysis I* [Springer-Verlag Heidelberg 1963], p. 228. Also, for general $(T_\gamma, \mathscr{M}_\gamma, \mu_\gamma)$, if $\overline{\overline{\Gamma}} > \aleph_0$, if $A_\gamma \in \mathscr{M}_\gamma$, and $\mu_\gamma(A_\gamma) < 1$ for uncountably many γ's, then $\bar{\mu}\big(\underset{\gamma\in\Gamma}{\times} A_\gamma\big) = 0$. This is very simple to show, and we omit the argument.

We continue with a fact (22.26) related to but not dependent on (22.21), for which an elementary lemma is needed. Lemma (22.26) is of independent interest and is also needed in the proof of (22.31).

(22.25) Lemma. *Suppose that* $\alpha_k \in [0, 1[$ *for all* $k \in N$. *Then if* $\sum_{k=1}^{\infty} \alpha_k = \infty$, *we have* $\prod_{k=1}^{\infty}(1 - \alpha_k) = 0$. *If* $\sum_{k=1}^{\infty} \alpha_k < \infty$, *then we have* $\prod_{k=1}^{\infty}(1 - \alpha_k) > 0$.

Proof. It is obvious that $\lim_{n\to\infty} \prod_{k=1}^{n}(1 - \alpha_k)$ exists. For $k = 1, 2, \ldots$, we have $(1 - \alpha_k)(1 + \alpha_k) = 1 - \alpha_k^2 \leqq 1$, so that $1 + \alpha_k \leqq \frac{1}{1 - \alpha_k}$. Thus

$$\sum_{k=1}^{n} \alpha_k \leqq \prod_{k=1}^{n}(1 + \alpha_k) \leqq \frac{1}{\prod_{k=1}^{n}(1 - \alpha_k)},$$

from which it follows that $\lim_{n\to\infty} \prod_{k=1}^{n}(1 - \alpha_k) = 0$ if $\sum_{k=1}^{\infty} \alpha_k = \infty$. Suppose conversely that $\sum_{k=1}^{\infty} \alpha_k < \infty$. Then for k such that $\alpha_k \leqq \frac{1}{2}$, we have

$$1 - \alpha_k \geqq \frac{1}{1 + 2\alpha_k}$$

and we also have $1 + 2\alpha_k \leqq \exp(2\alpha_k)$ for all k. Hence

$$\prod_{k=m}^{n}(1 - \alpha_k) \geqq \exp\left(-2\left(\sum_{k=m}^{n} \alpha_k\right)\right) \geqq \exp\left(-2\left(\sum_{k=1}^{\infty} \alpha_k\right)\right)$$

for a certain m and all $n \geqq m$. This obviously implies that $\prod_{k=1}^{\infty}(1 - \alpha_k) > 0$. □

(22.26) BOREL-CANTELLI Lemma. *Let* $\Gamma = N = \{1, 2, 3, \ldots\}$. *For each* $n \in N$, *let* $E^{(n)}$ *be a set in* $\mathscr{M}_n$, *and write* E_n *for the set* $E^{(n)} \times T_{\{n\}'}$. *Let*

F be the set $\bigcap_{n=1}^{\infty}\left(\bigcup_{k=n}^{\infty} E_k\right)$. Then we have

$$\mu(F) = \begin{cases} 0 \text{ if } \sum_{n=1}^{\infty} \mu_n(E^{(n)}) < \infty\,, \\ 1 \text{ if } \sum_{n=1}^{\infty} \mu_n(E^{(n)}) = \infty\,. \end{cases}$$

Proof. The set F plainly satisfies the hypotheses of (22.21), and so $\mu(F)$ must be 0 or 1. The present lemma tells us which. It is clear that

$$\mu(F) \leqq \mu\left(\bigcup_{k=n}^{\infty} E_k\right) \leqq \sum_{k=n}^{\infty} \mu(E_k)$$

for all n; therefore $\mu(F) = 0$ if $\sum_{n=1}^{\infty} \mu_n(E^{(n)}) < \infty$. $\Big[$A like result holds for $\bigcap_{n=1}^{\infty}\left(\bigcup_{k=n}^{\infty} A_k\right)$ where the A_k's are measurable sets in any measure space.$\Big]$ To prove the second statement, we need our special sets E_n. Observing that $F' = \bigcup_{n=1}^{\infty}\left(\bigcap_{k=n}^{\infty} E'_k\right)$ and that $E'_k = (E^{(k)}{}') \times T_{\{k\}'}$, we have

$$\begin{aligned} \mu(F') &= \lim_{n\to\infty} \mu\left(\bigcap_{k=n}^{\infty} E'_k\right) = \lim_{n\to\infty} \mu\left(\left(\mathop{\times}_{k=n}^{\infty} (E^{(k)})'\right) \times T_1 \times \cdots \times T_{n-1}\right) \\ &= \lim_{n\to\infty} \prod_{k=n}^{\infty} \mu_k((E^{(k)})') = \lim_{n\to\infty} \prod_{k=n}^{\infty} (1 - \mu_k(E^{(k)}))\,. \end{aligned}$$

If $\sum_{k=1}^{\infty} \mu_k(E^{(k)}) = \infty$, then it follows from (22.25) that $\prod_{k=n}^{\infty}(1 - \mu_k(E^{(k)})) = 0$ for $n = 1, 2, 3, \ldots$, and hence that $\mu(F') = 0$; thus $\mu(F) = 1$. □

Theorem (22.22) has many applications. As an example, we will use it to prove a famous limit theorem called *the strong law of large numbers.* We first present two elementary lemmas.

(22.27) Lemma. *Let $(\beta_n)_{n=1}^{\infty}$ be any sequence of complex numbers having a limit; say $\lim_{n\to\infty} \beta_n = \beta$. Then we also have*

$$\lim_{n\to\infty} \frac{1}{n}(\beta_1 + \beta_2 + \cdots + \beta_n) = \beta\,.$$

Proof. Given $\varepsilon > 0$, choose n_0 such that $k > n_0$ implies that $|\beta_k - \beta| < \frac{\varepsilon}{2}$. If $n > n_0$, then

$$\begin{aligned} \left|\frac{1}{n}\sum_{k=1}^{n} \beta_k - \beta\right| &= \left|\frac{1}{n}\sum_{k=1}^{n} \beta_k - \frac{1}{n}\sum_{k=1}^{n} \beta\right| \\ &\leqq \frac{1}{n}\sum_{k=1}^{n} |\beta_k - \beta| = \frac{1}{n}\sum_{k=1}^{n_0} |\beta_k - \beta| + \frac{1}{n}\sum_{k=n_0+1}^{n} |\beta_k - \beta| \\ &< \frac{1}{n}\sum_{k=1}^{n_0} |\beta_k - \beta| + \frac{\varepsilon}{2}\,. \end{aligned}$$

Thus for n so large that $\frac{1}{n}\sum_{k=1}^{n_0}|\beta_k - \beta| < \frac{\varepsilon}{2}$, we have

$$\left|\frac{1}{n}\sum_{k=1}^{n}\beta_k - \beta\right| < \varepsilon. \quad \square$$

(22.28) Lemma. *Let (α_k) be a sequence of complex numbers such that $\sum_{k=1}^{\infty}\frac{1}{k}\alpha_k$ converges. Then we have*

$$\lim_{n\to\infty}\frac{1}{n}(\alpha_1 + \alpha_2 + \cdots + \alpha_n) = 0\,.$$

Proof. Let $s_0 = 0$ and $s_n = \sum_{k=1}^{n}\frac{1}{k}\alpha_k$ for $n \geqq 1$, and let $t_n = \alpha_1 + \alpha_2 + \cdots + \alpha_n\,(n \geqq 1)$. With this notation we have

$$\alpha_k = k(s_k - s_{k-1})\,,$$

and therefore

$$t_{n+1} = \sum_{k=1}^{n+1} k(s_k - s_{k-1}) = (n+1)\,s_{n+1} + \sum_{k=1}^{n} k s_k - \sum_{k=2}^{n+1} k s_{k-1}$$

$$= (n+1)\,s_{n+1} + \sum_{k=1}^{n} k s_k - \sum_{k=1}^{n}(k+1)\,s_k = (n+1)\,s_{n+1} - \sum_{k=1}^{n} s_k\,;$$

i.e.,

$$t_{n+1} = (n+1)\,s_{n+1} - \sum_{k=1}^{n} s_k$$

for $n = 1, 2, 3, \ldots$. Thus we have

$$\frac{1}{n+1}t_{n+1} = s_{n+1} - \left(\frac{n}{n+1}\right)\frac{1}{n}\left(\sum_{k=1}^{n} s_k\right).$$

The partial sum s_{n+1} goes to $\gamma \in K$ as n goes to infinity, $\frac{1}{n}\left(\sum_{k=1}^{n} s_k\right)$ goes to γ by (22.27), and $\lim_{n\to\infty}\frac{n}{n+1} = 1$. Thus we have $\lim_{n\to\infty}\frac{1}{n+1}t_{n+1} = 0$, as we wished to show. $\square$

(22.29) Theorem: The Strong Law of Large Numbers. *Let $\Gamma = N$. For each $k \in N$, let g_k be a function in $\mathfrak{L}_2(T_k, \mathscr{M}_k, \mu_k)$ such that $\int_{T_k} g_k\,d\mu_k = 0$, and suppose that*

(i) $\sum_{k=1}^{\infty} k^{-2}\,\|g_k\|_2^2 < \infty$.

Let f_k be the function on T such that $f_k(\boldsymbol{t}) = g_k(t_k)$. Then we have

(ii) $\lim_{n\to\infty}\left[\frac{1}{n}\sum_{k=1}^{n} f_k(\boldsymbol{t})\right] = 0$ *for almost all $\boldsymbol{t} \in T$.*

Proof. We first prove that the series of functions $\sum_{k=1}^{\infty} \frac{f_k}{k}$ converges in the space $\mathfrak{L}_2(T, \mathscr{M}, \mu)$. If $m < n$, then we have

$$\left\| \sum_{k=1}^{n} \frac{f_k}{k} - \sum_{k=1}^{m} \frac{f_k}{k} \right\|_2^2 = \int_T \left| \sum_{k=m+1}^{n} \frac{f_k}{k} \right|^2 d\mu = \sum_{j,k=m+1}^{n} \frac{1}{jk} \int_T f_j \bar{f}_k \, d\mu$$

$$= \sum_{j \neq k} \frac{1}{jk} \int_{T_j} g_j \, d\mu_j \int_{T_k} \bar{g}_k \, d\mu_k + \sum_{k=m+1}^{n} \frac{1}{k^2} \int_{T_k} |g_k|^2 \, d\mu_k .$$

We have used (22.24.i) to write the last equality. The first term of the last expression above is zero by hypothesis, and the second term goes to zero as m and n go to ∞, by (i). Thus we have

$$\lim_{m,n \to \infty} \left\| \sum_{k=1}^{n} \frac{f_k}{k} - \sum_{k=1}^{m} \frac{f_k}{k} \right\|_2 = 0 ,$$

so that the partial sums of $\sum_{k=1}^{\infty} \frac{f_k}{k}$ form a Cauchy sequence in $\mathfrak{L}_2(T, \mathscr{M}, \mu)$. Let h be the $\mathfrak{L}_2$ limit of $\sum_{k=1}^{n} \frac{f_k}{k}$; then h is also in $\mathfrak{L}_1$. Theorem (13.17) shows that h is also the $\mathfrak{L}_1$ sum $\sum_{k=1}^{\infty} \frac{f_k}{k}$. We claim that $\int_T h \, d\mu = 0$. Write

$$\left| \int_T h \, d\mu \right| = \left| \sum_{k=1}^{n} \int_T \frac{1}{k} f_k \, d\mu + \int_T \left(h - \sum_{k=1}^{n} \frac{1}{k} f_k \right) d\mu \right|$$

$$\leq \left| \int_T \left(f_1 + \frac{1}{2} f_2 + \cdots + \frac{1}{n} f_n \right) d\mu \right| + \int_T \left| h - \sum_{k=1}^{n} \frac{1}{k} f_k \right| d\mu .$$

The first term on the right is zero and the second term goes to zero as n goes to ∞. It follows that $\int_T h \, d\mu = 0$. We now use Theorem (22.22) to write

$$0 = \int_T h \, d\mu = \lim_{n \to \infty} \int_{T_{\Omega_n}} h \, d\mu_{\Omega_n} \tag{1}$$

for almost all $\boldsymbol{t} \in T$. [Recall that $\Omega_n = \{1, 2, \ldots, n\}$ and note that the expression $\int_{T_{\Omega_n}} h \, d\mu_{\Omega_n}$ is a function of $\boldsymbol{t}$ which is independent of the first n coordinates.] We have

$$\int_{T_{\Omega_n}} h \, d\mu_{\Omega_n} = \int_{T_{\Omega_n}} \left(\sum_{k=1}^{n} \frac{1}{k} f_k \right) d\mu_{\Omega_n} + \int_{T_{\Omega_n}} \left(\sum_{k=n+1}^{\infty} \frac{1}{k} f_k \right) d\mu_{\Omega_n} . \tag{2}$$

The first integral on the right is zero, and the integrand in the second is independent of the first n coordinates. $\left[\text{Note that } \sum_{k=n+1}^{\infty} \frac{1}{k} f_k \text{ converges in}\right.$ the $\mathfrak{L}_1$ metric to a function in $\left.\mathfrak{L}_1.\right]$ The integral

$$\int\limits_{T_{\Omega_n}} \left(\sum_{k=n+1}^{\infty} \frac{1}{k} f_k \right) d\mu_{\Omega_n}$$

is plainly equal to the function $\sum_{k=n+1}^{\infty} \frac{1}{k} f_k$; thus (1) and (2) show that $\lim_{n\to\infty} \sum_{k=n+1}^{\infty} \frac{1}{k} f_k(\boldsymbol{t}) = 0$ for almost all $\boldsymbol{t} \in T$. That is, the series $\sum_{k=1}^{\infty} \frac{1}{k} f_k(\boldsymbol{t})$ converges a. e. in T. By (22.28), we have

$$\lim_{n\to\infty} \frac{1}{n} (f_1(\boldsymbol{t}) + \cdots + f_n(\boldsymbol{t})) = 0$$

a. e. in T. □

(22.30) Example. Let T_n, μ_n and f_n be just as in (22.9). We have $\|f_k\|_2^2 = \frac{1}{4}$, so that the hypotheses of (22.29) are satisfied. Hence we have

$$\lim_{n\to\infty} \frac{f_1 + \cdots + f_n}{n} = \lim_{n\to\infty} \frac{\left(t_1 - \frac{1}{2}\right) + \left(t_2 - \frac{1}{2}\right) + \cdots + \left(t_n - \frac{1}{2}\right)}{n} = 0$$

a. e.; *i. e.*,

$$\lim_{n\to\infty} \left(\frac{t_1 + t_2 + \cdots + t_n}{n} - \frac{1}{2} \right) = 0$$

a. e. Interpreting the occurrence of obtaining heads or tails upon flipping an unbiased coin as a 0 or 1, respectively, this result says that the portion of heads [or tails] obtained in n flips goes to $\frac{1}{2}$ as n goes to infinity for almost all sequences of flips. This result is far stronger than that obtained in (22.9).

The following is another version of the strong law of large numbers.

(22.31) Theorem. *Let $\Gamma = N$. For each $k \in N$, let g_k be a function in $\mathfrak{L}_1(T_k, \mathcal{M}_k, \mu_k)$. Write $g_k = \varphi_k + i\psi_k$, where φ_k and ψ_k are real-valued, and suppose that*

(i) *the numbers $\mu_k(\{t_k \in T_k : \varphi_k(t_k) > \alpha\})$ and $\mu_k(\{t_k \in T_k : \psi_k(t_k) > \alpha\})$ are independent of k for every real number α.*

Let f_k be the function on T such that $f_k(\boldsymbol{t}) = g_k(t_k)$. Then we have

(ii) $\lim_{n\to\infty} \left[\frac{1}{n} \sum_{k=1}^{n} f_k(\boldsymbol{t}) \right] = \int\limits_T f_1 \, d\mu$ *for almost all* $\boldsymbol{t} \in T$.

Proof. A moment's reflection shows that there is no harm in supposing that each g_k is real, and we do this. For each k, define

$$f_k'(\boldsymbol{t}) = \begin{cases} f_k(\boldsymbol{t}) \text{ if } & |f_k(\boldsymbol{t})| \leqq k, \\ 0 \text{ otherwise.} \end{cases}$$

We will show first that for almost all $\boldsymbol{t} \in T$, there is a positive integer m_0 [depending on $\boldsymbol{t}$] such that

$$f_m(\boldsymbol{t}) = f'_m(\boldsymbol{t}) \quad \text{if} \quad m \geqq m_0 . \tag{1}$$

Under the hypothesis (i), it is trivial to show that the numbers $\mu_k(\{t_k \in T_k : |g_k(t_k)| > \alpha\})$ are independent of k for every $\alpha \in R$. Using this fact, we have

$$\begin{aligned}\sum_{k=1}^{\infty} \mu(\{\boldsymbol{t} \in T : f_k(\boldsymbol{t}) \neq f'_k(\boldsymbol{t})\}) &= \sum_{k=1}^{\infty} \mu(\{\boldsymbol{t} \in T : |f_k(\boldsymbol{t})| > k\}) \\ &= \sum_{k=1}^{\infty} \mu_k(\{t_k \in T_k : |g_k(t_k)| > k\}) \\ &= \sum_{k=1}^{\infty} \left[\sum_{n=k}^{\infty} \mu_1(\{t_1 \in T_1 : n < |g_1(t_1)| \leqq n+1\})\right] \\ &= \sum_{k=1}^{\infty} k\mu_1(\{t_1 \in T_1 : k < |g_1(t_1)| \leqq k+1\}) \\ &\leqq \int_{T_1} |g_1| \, d\mu_1 < \infty . \end{aligned} \tag{2}$$

Next write $E^{(k)} = \{t_k \in T_k : |g_k(t_k)| > k\}$ and $E_k = E^{(k)} \times T_{\{k\}'}$. By (2), the series $\sum_{k=1}^{\infty} \mu_k(E^{(k)})$ converges, and so the BOREL-CANTELLI lemma (22.26) implies that $\mu\left(\bigcup_{n=1}^{\infty}\left(\bigcap_{k=n}^{\infty} E'_k\right)\right) = 1$. Thus almost all $\boldsymbol{t} \in T$ have the property that $\boldsymbol{t} \in \bigcap_{k=n}^{\infty} E'_k$ for some n. This is exactly the assertion (1).

From (1) it is immediate that

$$\lim_{n\to\infty} \left[\frac{1}{n}\sum_{k=1}^{n} f_k(\boldsymbol{t}) - \frac{1}{n}\sum_{k=1}^{n} f'_k(\boldsymbol{t})\right] = 0 \tag{3}$$

for almost all $\boldsymbol{t} \in T$.

We wish to apply (22.29) to the functions $f'_k - \int_T f'_k \, d\mu$. To establish (22.29.i), it suffices to show that

$$\sum_{k=1}^{\infty} k^{-2} \left\|f'_k - \int_T f'_k \, d\mu\right\|_2^2 < \infty . \tag{4}$$

We first write

$$\begin{aligned}\sum_{k=1}^{\infty} k^{-2} \left\|f'_k - \int_T f'_k \, d\mu\right\|_2^2 &= \sum_{k=1}^{\infty} k^{-2}\left[\int_T f'^2_k \, d\mu - \left(\int_T f'_k \, d\mu\right)^2\right] \\ &\leqq \sum_{k=1}^{\infty} k^{-2} \int_T f'^2_k \, d\mu . \end{aligned} \tag{5}$$

Now for each $k \in N$, define the function h_k on T_1 by

$$h_k(t_1) = \begin{cases} g_1(t_1) & \text{if} \quad |g_1(t_1)| \leqq k , \\ 0 & \text{otherwise.} \end{cases}$$

From (i), from the primeval definition (12.2) of the integral, and from (12.21), it is clear that

$$\sum_{k=1}^{\infty} k^{-2} \int_T f_k'^2 \, d\mu = \int_{T_1} \left(\sum_{k=1}^{\infty} k^{-2} h_k^2 \right) d\mu_1 . \tag{6}$$

We will show that the function $w = \sum_{k=1}^{\infty} k^{-2} h_k^2$ is in $\mathfrak{L}_1(T_1)$. Consider any point $t_1 \in T_1$ such that $|g_1(t_1)| > 0$. There is a [unique] positive integer p such that $p - 1 < |g_1(t_1)| \leqq p$. We have $h_1(t_1) = h_2(t_1) = \cdots = h_{p-1}(t_1) = 0$ and $h_p(t_1) = h_{p+1}(t_1) = \cdots = g_1(t_1)$, so that

$$\sum_{k=1}^{\infty} k^{-2} h_k^2(t_1) = \sum_{k=p}^{\infty} k^{-2} g_1^2(t_1) \leqq |g_1(t_1)| \sum_{k=p}^{\infty} k^{-2} p . \tag{7}$$

For every positive integer p, the relations

$$\sum_{k=p+1}^{\infty} k^{-2} < \int_0^{\infty} \frac{dx}{(x+p)^2} = \frac{1}{p}$$

are obvious, and so we have

$$\sum_{k=p}^{\infty} k^{-2} p < \frac{1}{p} + 1 \leqq 2 . \tag{8}$$

Combining (7) and (8), we see that

$$w \leqq 2\,|g_1| .$$

Since $g_1 \in \mathfrak{L}_1(T_1)$ by hypothesis, we can retrace our steps (6) and (5) to see that (4) does hold.

Thus the hypotheses of (22.29) are satisfied for the functions $f_k' - \int_T f_k' \, d\mu$; the conclusion (22.29.ii) assumes here the form

$$\lim_{n\to\infty} \left[\frac{1}{n} \sum_{k=1}^{n} f_k'(\mathbf{t}) - \frac{1}{n} \sum_{k=1}^{n} \int_T f_k' \, d\mu \right] = 0$$

for almost all $\mathbf{t} \in T$. In view of (3), the present proof will be completed by showing that

$$\lim_{n\to\infty} \frac{1}{n} \sum_{k=1}^{n} \int_T f_k' \, d\mu = \int_{T_1} f_1 \, d\mu_1 . \tag{9}$$

With h_k as defined above, we again have

$$\int_T f_k' \, d\mu = \int_{T_1} h_k \, d\mu_1 , \tag{10}$$

and (12.24) implies that

$$\lim_{k\to\infty} \int_{T_1} h_k \, d\mu_1 = \int_{T_1} g_1 \, d\mu_1 = \int_T f_1 \, d\mu . \tag{11}$$

The equality (9) follows from (10), (11), and (22.27). □

(22.32) Exercise. (a) Prove the following analogue of (22.29). Notation is as in (22.29). Replace the hypothesis (22.29.i) by

(i) $\sum_{k=1}^{\infty} \alpha_k \|g_k\|_2^2 < \infty$,

where $\alpha_k > 0$ and $\sum_{k=1}^{\infty} \alpha_k < \infty$. Then the infinite series

(ii) $\sum_{k=1}^{\infty} \alpha_k f_k(\boldsymbol{t})$

converges for almost all $\boldsymbol{t} \in T$.

(b) Prove the following analogue of (22.29), which is known as *the weak law of large numbers*. Again notation is as in (22.29). Replace (22.29.i) by the hypothesis

(iii) $\lim\limits_{n\to\infty} \left[\frac{1}{n^2} \sum_{k=1}^{n} \|f_k\|_2^2\right] = 0$.

Then for every $\varepsilon > 0$, the equality

$$\lim_{n\to\infty} \mu\left(\left\{\boldsymbol{t} \in T : \left|\frac{1}{n} \sum_{k=1}^{n} f_k(\boldsymbol{t})\right| > \varepsilon\right\}\right) = 0$$

holds. That is, the sequence of functions $\left(\frac{1}{n} \sum_{k=1}^{n} f_k\right)_{n=1}^{\infty}$ converges to zero in measure.

(22.33) Exercise. Let $T_n = \{0, 1, 2, \ldots, r-1\}$ and let $\mu_n(A) = \frac{1}{r} \bar{A}$ $(n = 1, 2, \ldots)$. For a fixed $l \in \{0, 1, \ldots, r-1\}$, define

$$g_n(\boldsymbol{t}) = \begin{cases} 1 \text{ if } t_n = l, \\ 0 \text{ if } t_n \neq l \end{cases}$$

for all $\boldsymbol{t}$ in the product space T.

Prove that

$$\lim_{n\to\infty} \mu\left(\left\{\boldsymbol{t} \in T : \left|\frac{1}{n} \sum_{k=1}^{n} g_k(\boldsymbol{t}) - \frac{1}{r}\right| > \varepsilon\right\}\right) = 0$$

for all $\varepsilon > 0$.

(22.34) Exercise. For $x \in]0,1[$, a fixed integer $r > 1$, and $l \in \{0,1, \ldots, r-1\}$, let $b_k(x)$ be the number of l's among the numbers $x_1, \ldots, x_k$, where

$$x = \sum_{n=1}^{\infty} r^{-n} x_n ,$$

$x_n \in \{0,1, \ldots, r-1\}$, and $x_n \neq 0$ for infinitely many n's. Prove that $\lim\limits_{k\to\infty} \frac{1}{k} b_k(x) = \frac{1}{r}$ for [Lebesgue] almost all $x \in]0,1[$. [This follows from (22.31).]

(22.35) We now present an application of the limit theorem (20.56) somewhat different from those given above. As in (22.16) and (22.17), let $\Gamma = \{1, 2, 3, \ldots\}$, let $\Omega_n = \{1, 2, , \ldots, n\}$, and let $\mathscr{M}^{(n)}$ be the σ-algebra of all sets $A_{\Omega_n} \times T_{\Omega'_n}$ for $A_{\Omega_n} \in \mathscr{M}_{\Omega_n}$. Now consider measures μ_n and η_n on $(T_n, \mathscr{M}_n)$ such that $\mu_n(T_n) = \eta_n(T_n) = 1$, and let μ und η be the product measures formed from the measures μ_n and η_n, respectively. Our first result deals with the case that $\eta_n \ll \mu_n$ for all n, and establishes a remarkable fact about η and μ.

(22.36) Theorem [S. Kakutani]. *Notation is as in* (22.35). *Suppose that* $\eta_n \ll \mu_n$ *for all* n. *Then we have either*

(i) $\eta \ll \mu$

or

(ii) $\eta \perp \mu$.

Let f_n *be a function in* $\mathfrak{L}_1^+ (T_n, \mathscr{M}_n, \mu_n)$ *such that*

(iii) $\int\limits_{E_n} f_n \, d\mu_n = \eta_n(E_n)$

for all $E_n \in \mathscr{M}_n$, i. e., *let* f_n *be a* Lebesgue-Radon-Nikodým *derivative* $\frac{d\eta_n}{d\mu_n}$ *in the sense of* (19.43). *Then* (i) *holds if and only if*

(iv) $\prod\limits_{k=1}^{\infty} \Big(\int\limits_{T_k} f_k^{\frac{1}{2}} \, d\mu_k \Big) > 0,$

and (ii) *holds if and only if*

(v) $\prod\limits_{k=1}^{\infty} \Big(\int\limits_{T_k} f_k^{\frac{1}{2}} \, d\mu_k \Big) = 0.$

Proof. We observe first of all that

$$0 < \int\limits_{T_k} f_k^{\frac{1}{2}} d\mu_k \leqq \Big[\int\limits_{T_k} f_k \, d\mu_k \Big]^{\frac{1}{2}} \Big[\int\limits_{T_k} 1^2 \, d\mu_k \Big]^{\frac{1}{2}} = \eta_k(T_k)^{\frac{1}{2}} \mu_k(T_k)^{\frac{1}{2}} = 1 ,$$

as (13.4) shows. Hence the infinite product in (iv) and (v) is a number in $[0,1]$. For each $n \in N$, consider the finite product T_{Ω_n} and the product measures $\mu_1 \times \cdots \times \mu_n$ and $\eta_1 \times \cdots \times \eta_n$ on $\mathscr{M}_{\Omega_n}$. It follows from (21.29) by induction on n that

$$\eta_1 \times \cdots \times \eta_n \ll \mu_1 \times \cdots \times \mu_n$$

and that the function

$$(t_1, t_2, \ldots, t_n) \to f_1(t_1) \, f_2(t_2) \cdots f_n(t_n) \tag{1}$$

is a Lebesgue-Radon-Nikodým derivative of $\eta_1 \times \cdots \times \eta_n$ with respect to $\mu_1 \times \cdots \times \mu_n$. Now consider the σ-algebra $\mathscr{M}^{(n)}$ of all sets $A_{\Omega_n} \times T_{\Omega'_n}$ where $A_{\Omega_n} \in \mathscr{M}_{\Omega_n}$. Let $f^{(n)}$ be the function on T such that

$$f^{(n)}(\mathbf{t}) = f_1(t_1) \, f_2(t_2) \cdots f_n(t_n) ,$$

and let $\mu^{(n)}$ and $\eta^{(n)}$ be the measures μ and η, respectively, restricted to the σ-algebra $\mathscr{M}^{(n)}$. It is clear from (1) and (22.24.i) that

$$\int_T f^{(n)}\,d\mu = \prod_{k=1}^{n} \int_{T_k} f_k\,d\mu_k = \prod_{k=1}^{n} \eta_k(T_k) = 1 \tag{2}$$

and that for $1 \leqq m < n$,

$$f^{(n)}(\boldsymbol{t})\,f^{(m)}(\boldsymbol{t}) = f_1^2(t_1)\cdots f_m^2(t_m)\,f_{m+1}(t_{m+1})\cdots f_n(t_n)\,. \tag{3}$$

It is evident from (21.29) that $\eta^{(n)} \ll \mu^{(n)}$ and that $f^{(n)}$ is a LEBESGUE-RADON-NIKODÝM derivative of $\eta^{(n)}$ with respect to $\mu^{(n)}$. We therefore cite (20.56) to assert that

$$\lim_{n\to\infty} f^{(n)}(\boldsymbol{t}) = f(\boldsymbol{t}) \tag{4}$$

exists for μ-almost all $\boldsymbol{t} \in T$ and is a derivative of η with respect to μ in the sense of (20.53).

Suppose that (v) holds. Then we apply (22.24.i), (12.23), and (4) to write

$$\begin{aligned} 0 &= \lim_{n\to\infty} \prod_{k=1}^{n}\Big(\int_{T_k} f_k^{\frac{1}{2}} d\mu_k\Big) = \lim_{n\to\infty} \int_T (f^{(n)})^{\frac{1}{2}} d\mu \\ &\geqq \int_T \lim_{n\to\infty} (f^{(n)})^{\frac{1}{2}} d\mu = \int_T f^{\frac{1}{2}} d\mu\,. \end{aligned} \tag{5}$$

From (5) it follows that $f = 0$ μ-a.e., and so (20.53) implies that $\eta \perp \mu$, since the μ-absolutely continuous part of η is obtained by integrating f.

Regardless of the value of $\prod_{k=1}^{\infty} \int_{T_k} f_k^{\frac{1}{2}}\,d\mu_k$, we can compute as follows. For $m < n$, we use (13.4) and (2) to write

$$\begin{aligned} \int_T |f^{(m)} - f^{(n)}|\,d\mu &= \int_T |(f^{(m)})^{\frac{1}{2}} + (f^{(n)})^{\frac{1}{2}}|\,|(f^{(m)})^{\frac{1}{2}} - (f^{(n)})^{\frac{1}{2}}|\,d\mu \\ &\leqq \Big[\int_T |(f^{(m)})^{\frac{1}{2}} + (f^{(n)})^{\frac{1}{2}}|^2\,d\mu\Big]^{\frac{1}{2}} \Big[\int_T |(f^{(m)})^{\frac{1}{2}} - (f^{(n)})^{\frac{1}{2}}|^2\,d\mu\Big]^{\frac{1}{2}} \\ &= 2\Big[1 + \int_T (f^{(m)})^{\frac{1}{2}}(f^{(n)})^{\frac{1}{2}}\,d\mu\Big]^{\frac{1}{2}} \Big[1 - \int_T (f^{(m)})^{\frac{1}{2}}(f^{(n)})^{\frac{1}{2}}\,d\mu\Big]^{\frac{1}{2}}. \end{aligned} \tag{6}$$

Now taking note of (3) and (22.24.i), we write

$$\begin{aligned} &\int_T (f^{(m)})^{\frac{1}{2}}(f^{(n)})^{\frac{1}{2}}\,d\mu \\ &\quad = \int_{T_1} f_1\,d\mu_1 \times\cdots\times \int_{T_m} f_m\,d\mu_m \times \int_{T_{m+1}} f_{m+1}^{\frac{1}{2}}\,d\mu_{m+1} \times\cdots\times \int_{T_n} f_n^{\frac{1}{2}}\,d\mu_n \\ &\quad = \prod_{k=m+1}^{n} \int_{T_k} f_k^{\frac{1}{2}}\,d\mu_k\,. \end{aligned} \tag{7}$$

Combining (6) and (7), we obtain

$$\int_T |f^{(n)} - f^{(m)}|\, d\mu \leqq 2\left[1 - \left(\prod_{k=m+1}^{n} \int_{T_k} f_k^{\frac{1}{2}}\, d\mu_k\right)^2\right]^{\frac{1}{2}}. \tag{8}$$

If (iv) holds, then it is clear that

$$\lim_{m,n\to\infty} \prod_{k=m+1}^{n} \left(\int_{T_k} f_k^{\frac{1}{2}}\, d\mu_k\right) = 1\,.$$

Hence (8) shows that $(f^{(n)})_{n=1}^{\infty}$ is a Cauchy sequence in $\mathfrak{L}_1(T, \mathscr{M}, \mu)$. We now appeal to (20.58) and (20.57) to infer that $\eta \ll \mu$. □

(22.37) Remarks. Notation is as in (22.35). If not all η_n are absolutely continuous with respect to μ_n, then η cannot be absolutely continuous with respect to μ, but it may still have a large absolutely continuous part. Suppose that for some $l \in N$, we have

(i) $\eta_l = \alpha_l \varrho_l + (1 - \alpha_l)\sigma_l$,

where ϱ_l and σ_l are measures on $(T_l, \mathscr{M}_l)$ such that $\varrho_l(T_l) = \sigma_l(T_l) = 1$, $\varrho_l \ll \mu_l$, $\sigma_l \perp \mu_l$, and $0 \leqq \alpha_l < 1$. If $\alpha_l = 0$, *i.e.*, if $\eta_l \perp \mu_l$, then (21.29) shows that $\eta \perp \mu$. Otherwise, let η' be the product of all η_k for $k \neq l$, on the space $T_{\{l\}'}$. It is easy to see that $\eta = \alpha_l(\varrho_l \times \eta') + (1 - \alpha_l)(\sigma_l \times \eta')$. As $(\sigma_l \times \eta') \perp \mu$, η cannot be absolutely continuous with respect to μ, but it is possible for $\varrho_l \times \eta'$ to be singular with respect to μ, absolutely continuous with respect to μ, or to be "mixed". A precise description of η in terms of the decomposition (i) for all $l \in N$ could be given: we leave the details to any interested readers.

(22.38) Exercise. Notation is as in (22.35). Let $T_n = \{0, 1\}$ for all $n \in N$, and let $\boldsymbol{\alpha}$ be a sequence $(\alpha_n)_{n=1}^{\infty}$ with values in $]0, 1[$. Let $\mu_{\boldsymbol{\alpha}}$ be the measure on $(T, \mathscr{M})$ that is the product of the measures μ_n on $\{0, 1\}$ such that $\mu_n(\{0\}) = \alpha_n$, $\mu_n(\{1\}) = 1 - \alpha_n$. Suppose that $\boldsymbol{\alpha}$ and $\boldsymbol{\beta}$ are any two such sequences.

(a) Prove that exactly one of the two following assertions holds:

(i) $\mu_{\boldsymbol{\alpha}} \ll \mu_{\boldsymbol{\beta}}$ and $\mu_{\boldsymbol{\beta}} \ll \mu_{\boldsymbol{\alpha}}$;

or

(ii) $\mu_{\boldsymbol{\alpha}} \perp \mu_{\boldsymbol{\beta}}$.

Prove also that (i) holds if and only if

(iii) $\sum_{n=1}^{\infty} \left(1 - \alpha_n^{\frac{1}{2}} \beta_n^{\frac{1}{2}} - (1 - \alpha_n)^{\frac{1}{2}}(1 - \beta_n)^{\frac{1}{2}}\right) < \infty$,

and that (ii) holds if and only if

(iv) the series in (iii) diverges.

[Hints. Apply (22.36) to the measures $\mu_{\boldsymbol{\alpha}}$ and $\mu_{\boldsymbol{\beta}}$. The factor measures are evidently absolutely continuous with respect to each other, and the integral over T_n of $f_n^{\frac{1}{2}}$ is $\alpha_n^{\frac{1}{2}} \beta_n^{\frac{1}{2}} + (1 - \alpha_n)^{\frac{1}{2}}(1 - \beta_n)^{\frac{1}{2}}$. Now apply (22.29). It makes no difference which of $\mu_{\boldsymbol{\alpha}}$ and $\mu_{\boldsymbol{\beta}}$ is taken as μ in (22.35).]

(b) Suppose that for some $\delta > 0$ the inequalities $\delta \leqq \alpha_n \leqq 1 - \delta$ and $\delta \leqq \beta_n \leqq 1 - \delta$ hold for all $n \in N$. Prove that (iii) holds if and only if

(v) $\sum_{n=1}^{\infty} (\alpha_n - \beta_n)^2 < \infty$.

[Hint. Use the identity

$$\text{(vi)}\quad 1 - \alpha^{\frac{1}{2}}\beta^{\frac{1}{2}} - (1-\alpha)^{\frac{1}{2}}(1-\beta)^{\frac{1}{2}} = \frac{1}{2}\left[(\alpha^{\frac{1}{2}} - \beta^{\frac{1}{2}})^2 + ((1-\alpha)^{\frac{1}{2}} - (1-\beta)^{\frac{1}{2}})^2\right]$$

and the mean value theorem of the differential calculus.]

(c) Suppose that β_n is constant: $\beta_n = \beta$ for some $\beta \in]0, 1[$. Show that (iii) holds if and only if (v) holds. [Hint. If $\lim_{n\to\infty} \alpha_n = \beta$, then part (b) can be applied. Otherwise (vi) shows that the terms of the series in (iii) do not have limit 0.]

(22.39) Exercise. Prove that there is a set $\boldsymbol{S}$ of measures on $(R, \mathscr{B}(R))$ such that: $\sigma(R) = 1$ for all $\sigma \in \boldsymbol{S}$; each $\sigma \in \boldsymbol{S}$ is regular; each $\sigma \in \boldsymbol{S}$ is continuous; each σ has support the interval $[0, 1]$; $\sigma \perp \sigma'$ for distinct σ and σ' in $\boldsymbol{S}$; and $\overline{\overline{\boldsymbol{S}}} = \mathfrak{c}$. [Hints. Let T be as in (22.38), and let $\varphi(\boldsymbol{t}) = \sum_{k=1}^{\infty} 2^{-k} t_k$ for $\boldsymbol{t} \in T$. The mapping φ carries T onto $[0, 1]$. For every number $\gamma \in]0, 1[$, let μ_γ be the measure on T constructed from the constant sequence $(\gamma, \gamma, \gamma, \ldots)$ as in (22.38). For $\gamma \neq \gamma'$, μ_γ and $\mu_{\gamma'}$ are obviously mutually singular. For $\gamma \in]0, 1[$ let σ_γ be the measure on $[0, 1]$ constructed as in (12.45) and (12.46) from the measure μ_γ on T and the continuous mapping φ. It is simple to verify that $\{\sigma_\gamma : \gamma \in]0, 1[\}$ can be taken for the set $\boldsymbol{S}$ of measures.]

(22.40) Exercise. A set $\boldsymbol{S}_1$ of measures on $\mathscr{B}(R)$ with all of the properties of $\boldsymbol{S}$ [see (22.39)] except that supports be $[0, 1]$ can be constructed without recourse to KAKUTANI's theorem (22.36). Fill in the details of the following argument. Let $T = \{0, 1\}^N$ and consider the measure $\mu_{\frac{1}{2}}$ on $(T, \mathscr{M})$ as in (22.39). For each $\boldsymbol{u} = (u_1, u_2, \ldots, u_n, \ldots) \in T$, Let $\varphi_{\boldsymbol{u}}$ be the mapping of T into $[0, 1]$ given by

$$\varphi_{\boldsymbol{u}}(\boldsymbol{t}) = 2\sum_{k=1}^{\infty} \frac{u_k}{3^{2k}} + 2\sum_{k=1}^{\infty} \frac{t_k}{3^{2k-1}}\,.$$

For $A \in \mathscr{B}(R)$, let $\sigma_{\boldsymbol{u}}(A) = \mu_{\frac{1}{2}}\left(\varphi_{\boldsymbol{u}}^{-1}(A \cap \varphi_{\boldsymbol{u}}(T))\right)$.

Then the set $\boldsymbol{S}_1 = \{\sigma_{\boldsymbol{u}} : \boldsymbol{u} \in T\}$ of measures has all of the asserted properties.

(22.41) Exercise. Consider the measure $\sigma_{\frac{1}{2}}$, constructed as in (22.39). Prove that $\sigma_{\frac{1}{2}}$ is Lebesgue measure on $[0, 1]$.

(22.42) Exercise. (a) Alter the construction of (22.39) in the following way. Let T and μ_γ be as in (22.38) and (22.39), but define the mapping φ of T into R by $\varphi(\mathbf{t}) = 2\sum_{k=1}^{\infty} 3^{-k} t_k$. For $\gamma \in]0, 1[$, let τ_γ be the image of μ_γ under φ as in (12.45) and (12.46). Prove that the support of each τ_γ is the Cantor ternary set. Prove too that $\tau_{\frac{1}{2}}$ is the Lebesgue-Stieltjes measure that corresponds to LEBESGUE's singular function (8.28).

(b) Prove that $\int_{[0,1]} x\, d\tau_\gamma(x) = 1 - \gamma$. [Hint. The following steps are easy to check:

$$\int_{[0,1]} x\, d\tau_\gamma(x) = \int_T \left(2\sum_{k=1}^{\infty} 3^{-k} t_k\right) d\mu_\gamma(\mathbf{t}) = 2\sum_{k=1}^{\infty} 3^{-k} \int_T t_k\, d\mu_\gamma(\mathbf{t})$$
$$= 2(1-\gamma)\sum_{k=1}^{\infty} 3^{-k} = 1 - \gamma.]$$

(c) Prove that

$$\int_{[0,1]} x^2\, d\tau_\gamma(x) = 2^{-1}(1-\gamma) + 2^{-1}(1-\gamma)^2.$$

[Hints. The computation can be made as follows:

$$\int_{[0,1]} x^2\, d\tau_\gamma(x) = 4\int_T \left(\sum_{k=1}^{\infty} 3^{-k} t_k\right)^2 d\mu_\gamma(\mathbf{t})$$
$$= 4\left[\sum_{k=1}^{\infty} 3^{-2k}\int_T t_k\, d\mu_\gamma(\mathbf{t}) + 2\sum_{l=1}^{\infty}\left(\sum_{k=l+1}^{\infty} 3^{-k-l}\int_T t_k t_l\, d\mu_\gamma(\mathbf{t})\right)\right]$$
$$= 4\left[(1-\gamma)\sum_{k=1}^{\infty} 3^{-2k} + 2(1-\gamma)^2\sum_{l=1}^{\infty} 3^{-2l-1}\left(\sum_{n=0}^{\infty} 3^{-n}\right)\right]$$
$$= \frac{1}{2}(1-\gamma) + \frac{1}{2}(1-\gamma)^2.]$$

(d) Prove that

$$\int_{[0,1]} x^3\, d\tau_\gamma(x) = \frac{8}{26}\left[(1-\gamma) + \frac{15}{8}(1-\gamma)^2 + \frac{3}{8}(1-\gamma)^3\right].$$

(22.43) Exercise. Notation is as in (22.42). (a) Let α be any complex number. Prove that

$$\int_{[0,1]} \exp(\alpha x)\, d\tau_\gamma(x) = \prod_{k=1}^{\infty}\left[\gamma + (1-\gamma)\exp(3^{-k} 2\alpha)\right].$$

(b) Prove that

(i) $\int_{[0,1]} \exp(2\pi i\, 3^p x)\, d\tau_\gamma(x) = \prod_{k=1}^{\infty}\left[\gamma + (1-\gamma)\exp(4\pi i\, 3^{-k})\right]$

for all $p \in N$.

(c) Prove that the value of (i) is a positive real number for $\gamma = \frac{1}{2}$. What can you say for other values of γ in $[0, 1]$?

(22.44) Exercise. (a) Generalize the result of (22.42) in the following way. Let a be any number in the interval $]-1, 1[$. Define a map φ of T into R by $\varphi(\boldsymbol{t}) = \sum_{k=1}^{\infty} a^k t_k$. Let ω_γ be the image of μ_γ under φ as in (12.45). Thus for every continuous f on R we have

$$\int_R f\,d\omega_\gamma = \int_T f \circ \varphi\,d\mu_\gamma\,.$$

Let $\Delta_n = \int_R x^n\,d\omega_\gamma(x)$ $(n = 0, 1, 2, \ldots)$. Prove that $\Delta_0 = 1$ and

(i) $\Delta_n = \dfrac{a^n(1-\gamma)}{1-a^n}\left(\sum_{j=1}^{n} \binom{n}{j} \Delta_{n-j}\right).$ [1]

[Hints. For $n \in N$, we have

$$(\varphi(\boldsymbol{t}))^n = a^n \left(t_1 + \sum_{k=2}^{\infty} a^{k-1} t_k\right)^n$$

$$= a^n \left[\left(\sum_{k=2}^{\infty} a^{k-1} t_k\right)^n + \sum_{j=1}^{n} \binom{n}{j} t_1 \left(\sum_{k=2}^{\infty} a^{k-1} t_k\right)^{n-j}\right],$$

since $t_1^2 = t_1$. Integrating over T, we find

$$\Delta_n = a^n \left[\int_T \left(\sum_{k=2}^{\infty} a^{k-1} t_k\right)^n d\mu_\gamma(\boldsymbol{t}) + \sum_{j=1}^{n} \binom{n}{j} \int_T t_1 \left(\sum_{k=2}^{\infty} a^{k-1} t_k\right)^{n-j} d\mu_\gamma(\boldsymbol{t})\right]$$

$$= a^n \left[\Delta_n + \sum_{j=1}^{n} \binom{n}{j} (1-\gamma)\,\Delta_{n-j}\right].$$

From this, (i) is immediate.]

(b) For the case $a = \gamma = \frac{1}{2}$, (22.41) shows that $\omega_{\frac{1}{2}} = \sigma_{\frac{1}{2}} = \lambda$ on $[0, 1]$. In this case, compute Δ_n directly and then verify formula (i).

(c) Let τ_γ be as in (22.42) and write $\Delta_n' = \int_{[0,1]} x^n\,d\tau_\gamma(x)$ $(n=0,1,2,\ldots)$. Prove that $\Delta_0' = 1$ and

(ii) $\Delta_n' = \dfrac{1-\gamma}{3^n-1} \sum_{j=1}^{n} \binom{n}{j} 2^j \Delta_{n-j}'$ $(n = 1, 2, \ldots)$.

(22.45) Exercise [H. S. Zuckerman]. Notation is as in (22.44). For brevity, write b_n for $\dfrac{a^n(1-\gamma)}{1-a^n}$.

(a) Prove that for $n \geq 1$,

(i) $\Delta_n = b_n \left\{1 + \sum_{k=1}^{n-1}{}' \dfrac{n!}{(n-i_1)!\,(i_1-i_2)!\cdots(i_{k-1}-i_k)!\,i_k!}\, b_{i_1} b_{i_2} \cdots b_{i_k}\right\},$

where the sum $\sum'$ is taken over all k-tuples $(i_1, i_2, \ldots, i_k)$ of positive integers such that $n > i_1 > i_2 > \cdots > i_k > 0$. [Hints. Rewrite (22.44.i)

[1] This recursion formula was kindly suggested to us by Professor R. M. Blumenthal.

in the form

$$\Delta_n = b_n \sum_{j=0}^{n-1} \binom{n}{j} \Delta_j ,$$

note that

$$\frac{n!}{(n - i_1)!\,(i_1 - i_2)!\,(i_2 - i_3)! \cdots (i_{k-1} - i_k)!\, i_k!} = \binom{n}{i_1}\binom{i_1}{i_2} \cdots \binom{i_{k-1}}{i_k},$$

and use induction on n.]

(b) Prove that for $n \geqq 1$,

(ii) $\Delta_n = \sum_{k=1}^{n} \sum'' \frac{n!}{j_1!\,j_2! \cdots j_k!}\, b_{j_1+j_2+\cdots+j_k}\, b_{j_2+j_3+\cdots+j_k}\, b_{j_3+j_4+\cdots+j_k} \cdots b_{j_k}$,

the sum $\sum''$ being taken over all k-tuples $(j_1, j_2, \ldots, j_k)$ of positive integers such that $j_1 + j_2 + \cdots + j_k = n$. [Hint. Rewrite (i).]

Index of Symbols

Index of authors and terms

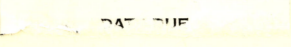